MATHEMATICAL PAPERS.

London: C. J. CLAY AND SONS,
CAMBRIDGE UNIVERSITY PRESS WAREHOUSE,
AVE MARIA LANE.
Glasgow: 263, ARGYLE STREET.

Cambridge: DEIGHTON, BELL AND CO.
Leipzig: F. A. BROCKHAUS.
New York: MACMILLAN AND CO.

THE COLLECTED

MATHEMATICAL PAPERS

OF

ARTHUR CAYLEY, Sc.D., F.R.S.,

SADLERIAN PROFESSOR OF PURE MATHEMATICS IN THE UNIVERSITY OF CAMBRIDGE.

VOL. VII.

CAMBRIDGE:
AT THE UNIVERSITY PRESS.
1894

CAMBRIDGE:
PRINTED BY C. J. CLAY, M.A. AND SONS,
AT THE UNIVERSITY PRESS.

ADVERTISEMENT.

THE present volume contains 69 papers numbered 417 to 485 published for the most part in the years 1866 to 1872; they include a series of astronomical papers published in the *Memoirs* and *Monthly Notices* of the Royal Astronomical Society.

The Portrait in Volume VI. is from the Painting in Oil by Mr Lowes Dickenson in the year 1874, presented by the Subscribers to Trinity College, Cambridge, and now in the Hall of the College: the portrait in the present volume is a photograph of a pencil sketch by Mr Lowes Dickenson in the year 1893.

The Table for the seven volumes is

Vol.					
Vol.	I.	Numbers	1	to	100.
,,	II.	,,	101	,,	158.
,,	III.	,,	159	,,	222.
,,	IV.	,,	223	,,	299.
,,	V.	,,	300	,,	383.
,,	VI.	,,	384	,,	416.
,,	VII.	,,	417	,,	485.

CONTENTS.

[An Asterisk denotes that the paper is not printed in full.]

CLASSIFICATION.

GEOMETRY :

ANALYSIS :

417.

ON THE LOCUS OF THE FOCI OF THE CONICS WHICH PASS THROUGH FOUR GIVEN POINTS.

[From the *Philosophical Magazine*, vol. XXXII. (1866), pp. 362—365.]

THE curve which is the locus of the foci of the conics which pass through four given points is, as appears from a general theorem of M. Chasles, a sextic curve having a double point at each of the circular points at infinity; and Professor Sylvester, in his "Supplemental Note on the Analogues in Space to the Cartesian Ovals *in plano*" (*Phil. Mag.*, May 1866), has further remarked that the lines (eight in all) joining the circular points at infinity with any one of the four points are all of them double tangents of the curve; whence each of these points is a focus (more accurately a quadruple focus) of the curve. It is to be added that, besides the circular points at infinity, the curve has 6 double points (3 of these are the centres of the quadrangles formed by the 4 points), in all 8 double points; the class is therefore $=14$. Hence also the number of tangents to the curve from a circular point at infinity is $=10$; viz. these are the 4 double tangents each reckoned twice, and 2 single tangents; and the theoretical number of foci is $=100$; viz. we have

16 quadruple foci, or intersections of a double tangent by a double tangent . .	$16\times4=64$
16 double foci, or intersections of a double tangent by a single tangent . .	$16\times2=32$
4 single foci, or intersections of a single tangent by a single tangent . . .	$4\times1=\ \ 4$
	100

To verify the foregoing results, consider any two given points I, J, and the series of conics which pass through four given points A, B, C, D; we have thus a curve the locus of the intersections of the tangents from I and the tangents from J to any conic of the series; which curve, if I, J are the circular points at infinity, is the required curve of foci. Taking $U+\lambda V=0$ for the equation of a conic of the series, the pair of tangents from I is given by an equation of the form

$$(\lambda,\ 1)^2\,(x,\ y,\ z)^2=0,$$

and the pair of tangents from J by an equation of the like form

$$(\lambda,\ 1)^2\,(x,\ y,\ z)^2=0\,;$$

and by eliminating λ from these equations, we obtain the equation of the required curve. This in the first instance presents itself as an equation of the eighth order; but it is to be observed that in the series of conics there are two conics each of them touching the line IJ, and that, considering the tangents drawn to either of these conics, the line IJ presents itself as part of the locus; that is, the line IJ twice repeated is part of the locus; and the residual curve is thus of the order $8-2$, $=6$; that is, the required curve is of the order 6. The consideration of the same two conics shows that each of the points I, J is a double point on the curve. Moreover, by taking for the conic any one of the line-pairs through the four points, it appears that each of the points $(AB\,.\,CD)$, $(AC\,.\,BD)$, $(AD\,.\,BC)$ is a double point on the curve: this establishes the existence of five double points. The two conics of the series which touch the line IA are a single conic taken twice, and the consideration of this conic shows that the line IA is a double tangent to the curve; similarly each of the eight lines $I\,(A,\ B,\ C,\ D)$ and $J\,(A,\ B,\ C,\ D)$ is a double tangent to the curve. Instead of seeking to establish directly the existence of the remaining three double points, the easier course is to show that, besides the four double tangents from I, the number of tangents from I to the curve is $=2$; for, this being so, the total number of tangents from I to the curve will be $(2\times 4+2=)\ 10$; that is, I being a double point, the class of the curve is $=14$; and assuming that the depression $(6\times 5-14=)\ 16$ in the class of the curve is caused by double points, the number of double points will be $=8$. But observing that in the series of conics there is one conic which passes through J, so that the tangents from J to this conic are the tangent at J twice repeated, then it is easy to see that the tangents from I to this conic, at the points where they meet the tangent at J, *touch* the required curve, and that these two tangents are in fact (besides the double tangents) the only tangents from I to the curve; that is, the number of tangents from I to the curve is $=2$.

Considering I, J as the circular points at infinity, and writing A, B, C, D to denote the squared distances of a point P from the four points A, B, C, D respectively, then, as remarked by Professor Sylvester, the equation

$$\lambda\sqrt{A}+\mu\sqrt{B}+\nu\sqrt{C}+\pi\sqrt{D}=0$$

(where λ, μ, ν, π are constants) is in general a curve of the order 8; but the ratios $\lambda:\mu:\nu:\pi$ may be so determined that the order of the curve in question shall be

$=6$; the resulting curve of the order 6 is (not one of a group of curves, but the very curve) the locus of the foci of the conics through the four points. And the determination of the ratios $\lambda : \mu : \nu : \pi$ is in fact quite simple; for writing

$$A = (x-a)^2 + (y-a_1)^2$$
$$= \rho^2 - 2(ax + a_1 y) + \&c.$$
$$(\text{if } \rho^2 = x^2 + y^2),$$

and therefore

$$\sqrt{A} = \rho - \frac{ax + a_1 y}{\rho} + \&c.,$$

with similar values for $\sqrt{B}$, $\sqrt{C}$, $\sqrt{D}$, it is easy to see that, considering λ, μ, ν, π as standing for $\pm\lambda$, $\pm\mu$, $\pm\nu$, $\pm\pi$ respectively, the conditions for the reduction to the order 6 are

$$\lambda + \mu + \nu + \pi = 0,$$
$$\lambda a + \mu b + \nu c + \pi d = 0,$$
$$\lambda a_1 + \mu b_1 + \nu c_1 + \pi d_1 = 0,$$

and hence that the required equation of the curve of foci is

$$\Sigma \left\{ \begin{vmatrix} 1, & 1, & 1 \\ b, & c, & d \\ b_1, & c_1, & d_1 \end{vmatrix} \sqrt{(x-a)^2 + (y-a_1)^2} \right\} = 0,$$

or, as this may also be written,

$$\Sigma \pm (B,\ C,\ D)\sqrt{A} = 0,$$

where (B, C, D), &c. are the areas of the triangles B, C, D, &c.

I remark, in conclusion, that the number of conditions to be satisfied in order that a curve may have for double points two given points I, J, may have besides six double points, and may have for double tangents eight given lines, is $(3+3+6+16=)\,28$; the number of constants contained in the general equation of the order 6 is $=27$. The conditions that a curve of the order 6 shall have for double points two given points I, J, shall besides have six double points, and shall have for double tangents four given lines through I and four given lines through J, are more than sufficient for the determination of the sextic curve; and the existence of a sextic curve satisfying these conditions is therefore a theorem.

In the case where the points I, J lie on a conic of the series, the consideration of this conic shows that the curve has a ninth double point, the pole of the line IJ in regard to the conic in question: in this case the sextic curve, as is known, breaks up into two cubic curves. [It need not do so, for a proper sextic curve may have nine (or indeed ten) double points.]

P.S. In general the curve $\lambda\sqrt{A}+\mu\sqrt{B}+\nu\sqrt{C}+\pi\sqrt{D}=0$ has (exclusively of multiple points at infinity) six double points; viz. these are situate at the intersections of the pairs of circles,

$$(\lambda\sqrt{A}+\mu\sqrt{B}=0,\quad \nu\sqrt{C}+\pi\sqrt{D}=0),$$

$$(\lambda\sqrt{A}+\nu\sqrt{C}=0,\quad \mu\sqrt{B}+\pi\sqrt{D}=0),$$

$$(\lambda\sqrt{A}+\pi\sqrt{D}=0,\quad \mu\sqrt{B}+\nu\sqrt{C}=0).$$

In the case of the curve of foci, the first, second, and third pairs of circles intersect respectively in the points $(AB.CD)$, $(AC.BD)$, $(AD.BC)$, which, as mentioned above, are double points on the curve; and they besides intersect in three other points, which are the other three double points mentioned above.

Professor Sylvester reminds me that he mentioned to me in conversation that he had himself obtained the foregoing equation $\Sigma \pm (B, C, D)\sqrt{A}=0$, for the locus of the foci of the conics which pass through the four points A, B, C, D.

Cambridge, October 10, 1866.

418.

A REMARK ON DIFFERENTIAL EQUATIONS.

[From the *Philosophical Magazine*, vol. XXXII. (1866), pp. 379—381.]

CONSIDER a differential equation $f(x, y, p)=0$, of the first order, but of the degree n, where f is a rational and integral function of (x, y, p) not rationally decomposable into factors: the integral equation contains an arbitrary constant c, and represents therefore a system of curves, for any one of which curves the differential equation is satisfied: the differential equation is assumed to be such that the curves are algebraical curves. The curves in question may be considered as undecomposable curves; in fact, if the curve $U^\alpha V^\beta W^\gamma \ldots = 0$ (composed of the undecomposable curves $U=0$, $V=0$, $W=0, \ldots$) satisfies the differential equation, then either the curves $U=0$, $V=0$, $W=0, \ldots$ each satisfy the differential equation, and instead of the curve $U^\alpha V^\beta W^\gamma \ldots = 0$ we have thus the undecomposable curves $U=0$, $V=0$, $W=0, \ldots$ each satisfying the differential equation; or if any of these curves, for instance $W=0$, &c., do not satisfy the differential equation, then W^γ, &c. are mere extraneous factors which may and ought to be rejected, and instead of the original curve $U^\alpha V^\beta W^\gamma \ldots = 0$, we have the undecomposable curves $U=0$, $V=0$ satisfying the differential equation. Assuming, as above, the existence of an algebraical solution, this may always be expressed in the form $\phi(x, y, c)=0$, where ϕ is a rational and integral function of (x, y, c), of the degree n as regards the arbitrary constant c: this appears by the consideration that for given values (x_0, y_0) of (x, y) the differential equation and the integral equation must each of them give the same number of values of p. It is to be observed that ϕ regarded as a function of (x, y, c) cannot be rationally decomposable into factors; for if the equation were $\phi=\Phi\Psi \ldots =0$, Φ, Ψ, &c. being each of them rational and integral functions of (x, y, c), then the differential equation would be satisfied by at least one of the equations $\Phi=0$, $\Psi=0, \ldots$ that is, by an equation of a degree less than n in the arbitrary constant c.

But the equation $\phi(x, y, c) = 0$ is not of necessity the equation of an undecomposable curve, and the undecomposable curve which constitutes the proper solution of the differential equation cannot always be represented by an equation of the form in question. For although ϕ regarded as a function of (x, y, c) is not rationally decomposable into factors, yet it may very well happen that ϕ regarded as a function of (x, y) is rationally decomposable into factors (geometrically the sections by the planes $z = c$ of the undecomposable surface $\phi(x, y, z) = 0$ may each of them be composed of two or more distinct curves); and assuming that the function ϕ is thus decomposed into its prime factors, then each factor equated to 0 gives an undecomposable curve satisfying the differential equation, and constituting the proper solution thereof.

It may be observed that, by the foregoing process of decomposition, we sometimes reduce the original equation $\phi(x, y, c) = 0$ into a like equation $\phi_1(x, y, c_1) = 0$ of a more simple form. Thus, for instance, if we have $\phi(x, y, c) = U^2 - c = 0$, U being a rational and integral function of (x, y), then instead of $\phi = U^2 - c = 0$ we have the equations $U + \sqrt{c} = 0$, $U - \sqrt{c} = 0$, each of which is an equation of the form $U - c_1 = 0$, or we pass from the original equation $\phi(x, y, c) = U^2 - c = 0$ to the simplified equation

$$\phi_1(x, y, c_1) = U - c_1 = 0.$$

Again, to take a somewhat more complicated instance, if the given integral equation be

$$\phi(x, y, c) = U^4 + c^2V^4 + (c+1)^2 W^2 - 2cU^2V^2 - 2(c+1)U^2W^2 - 2c(c+1)V^2W^2 = 0,$$

then the equation $U + V\sqrt{c} + W\sqrt{c+1} = 0$, writing therein $\sqrt{c} = \dfrac{2c_1}{c_1^2 - 1}$, and therefore $\sqrt{c+1} = \dfrac{c_1^2+1}{c_1^2-1}$, becomes

$$U(c_1^2 - 1) + V.2c_1 + W(c_1^2 + 1) = 0;$$

so that we pass from the original equation $\phi(x, y, c) = 0$ to the simplified equation

$$\phi_1(x, y, c_1) = U(c_1^2 - 1) + V.2c_1 + W(c_1^2 + 1) = 0.$$

But observe that the possibility of the rationalization depends on the form of the radicals $\sqrt{c}$ and $\sqrt{c+1}$; if we had had $\sqrt{c}$ and $\sqrt{c^2+1}$ (or c and $\sqrt{c^4+1}$), the rationalization could not have been effected.

Returning to the case of an integral equation $\phi(x, y, c) = 0$, where ϕ regarded as a function of (x, y) is decomposable into factors, then equating to zero any one of the prime factors of ϕ, we obtain an integral equation $\psi(x, y, c_1, c_2, \ldots c_k) = 0$, where $c_1, c_2 \ldots c_k$ are irrational functions (not of necessity representable by radicals, and without any superior limit to the number of these functions) of c: here ψ regarded as a function of (x, y) is of course undecomposable, and the equation $\psi(x, y, c_1, c_2, \ldots c_k) = 0$ belongs to the undecomposable curve which is the proper solution of the differential equation. The result may be stated under a quasi-geometrical form; viz. regarding $c_1, c_2, \ldots c_k$ as the coordinates of a point in k-dimensional space, then as these are functions of the single parameter c, the point to which they belong is an arbitrary point on a certain curve or $(k-1)$fold locus C in the k-dimensional space. And this curve must be such that to given values of (x, y) there shall correspond n points on the curve; that is, treating (x, y) as constants, the surface or onefold locus $\psi(x, y, c_1, c_2 \ldots c_k)=0$, and the curve or $(k-1)$fold locus C, shall meet in n points. The conclusion stated in the foregoing quasi-geometrical form is, that the solution of the differential equation may be exhibited in the form $\psi(x, y, c_1, c_2 \ldots c_k)=0$; viz. ψ is a rational and integral function of $(x, y, c_1, c_2 \ldots c_k)$, where $(c_1, c_2 \ldots c_k)$ are the coordinates of an arbitrary or variable point on a curve or $(k-1)$fold locus C in a k-dimensional space, which curve meets the surface or onefold locus $\psi(x, y, c_1, c_2 \ldots c_k)$ in n points, and where ψ regarded as a function of (x, y) is not rationally decomposable into factors.

Cambridge, October 13, 1866.

419.

A THEOREM ON DIFFERENTIAL OPERATORS.

[From a *paper by* PROF. SYLVESTER, "*Note on the Test Operators which occur in the Calculus of Invariants, &c.*," *Philosophical Magazine*, vol. XXXII. (1866), pp. 461—472, see p. 471.]

THE paper concludes with an Observation from Professor Cayley as follows:

"In the case of two variables, if

$$P_1 = (ax + by)\frac{d}{dx} + (cx + dy)\frac{d}{dy},$$

then in the notation of matrices,

$$P_1 = \begin{Bmatrix} a, & b \\ c, & d \end{Bmatrix} (x,\ y)\left(\frac{d}{dx},\ \frac{d}{dy}\right),$$

$$P_2 = \tfrac{1}{2}\begin{Bmatrix} a, & b \\ c, & d \end{Bmatrix}^2 (x,\ y)\left(\frac{d}{dx},\ \frac{d}{dy}\right),$$

$$P_3 = \tfrac{1}{6}\begin{Bmatrix} a, & b \\ c, & d \end{Bmatrix}^3 (x,\ y)\left(\frac{d}{dx},\ \frac{d}{dy}\right);$$

whence also

$$P * P_2 = P_2 * P_1 = \tfrac{1}{2}\begin{Bmatrix} a, & b \\ c, & d \end{Bmatrix}^3 (x,\ y)\left(\frac{d}{dx},\ \frac{d}{dy}\right) = 3P_3,$$

which accords with your theorem,

$$E_1 * E_2 * = E_2 * E_1 * = E_1 E_2 * + 3E_3 *."$$

I have taken the liberty of writing in the above $\frac{d}{dx}$, $\frac{d}{dy}$ for δ_x, δ_y, and P for δ in the original. It will be useful to bear in mind that in any operator such as E_1* or E_2*, *the asterisk forms an integral part* of the symbol. Thus $E_1 * E_2 *$, if we choose, may be written under the form of E_1* multiplied by E_2*, i.e. $(E_1*)\times(E_2*)$, where the cross is the sign of ordinary algebraical multiplication.

420.

ON RICCATI'S EQUATION.

[From the *Philosophical Magazine*, vol. XXXVI. (1868), pp. 348—351.]

THE following is, it appears to me, the proper form in which to present the solution of Riccati's equation.

The equation may be written

$$\frac{dy}{dx} + y^2 = x^{2q-2},$$

which is integrable by algebraic and exponential functions if $(2i+1)q = \pm 1$, i being zero, or a positive integer. To effect the integration, writing $y = \frac{1}{u}\frac{du}{dx}$, we have

$$\frac{d^2u}{dx^2} = x^{2q-2}u.$$

The peculiar advantage of this well-known transformation has not (so far as I am aware) been explicitly stated; it puts in evidence the form under which the sought-for function y contains the constant of integration. In fact if $u = P$, $u = Q$ be two particular solutions of the equation in u, then the general solution is $u = CP + DQ$; and denoting by P', Q' the derived functions, the value of y is

$$y = \frac{CP' + DQ'}{CP + DQ},$$

showing the form under which the constant of integration $C \div D$ is contained in y. To complete the solution, assume

$$u = ze^{\frac{1}{q}x^q};$$

we find

$$\frac{d^2z}{dx^2} + 2x^{q-1}\frac{dz}{dx} + (q-1)\,x^{q-2}\,z = 0:$$

considering first the particular integral of the form

$$z = A + Bx^q + Cx^{2q} + Dx^{3q} + \&c.,$$

we find that the equation will be satisfied if

$$\begin{aligned}
(\ q-1)\,A + \ q(\ q-1)\,B &= 0,\\
(3q-1)\,B + 2q\,(2q-1)\,C &= 0,\\
(5q-1)\,C + 3q\,(3q-1)\,D &= 0,\\
(7q-1)\,D + 4q\,(4q-1)\,E &= 0,\\
\&c.
\end{aligned}$$

If $A = 1$, this is

$$\begin{aligned}
A &= \ 1,\\
B &= -\frac{q-1}{q\,(q-1)},\\
C &= +\frac{(q-1)\,(3q-1)}{q\,(q-1)\,2q\,(2q-1)},\\
D &= -\frac{(q-1)\,(3q-1)\,(5q-1)}{q\,(q-1)\,2q\,(2q-1)\,3q\,(3q-1)},\\
&\&c.,
\end{aligned}$$

where it is to be noticed that the series may be considered to stop so soon as there is in the numerator a factor $=0$. For instance, if $5q-1=0$, then if the particular integral had been assumed to be $z = A + Bx^q + Cx^{2q}$, the only conditions to be satisfied by the coefficients are the first and second equations giving the foregoing values of A, B, C. It is immaterial that the analytical expressions of F and the subsequent coefficients contain in the denominators the evanescent factor $5q-1$; the coefficients after C do not ever come into consideration.

Thus if $(2i+1)\,q = +1$, the series terminates, and we have for u the finite particular solution

$$u = P = \left(1 - \frac{q-1}{q\,(q-1)}\,x^q + \frac{(q-1)\,(3q-1)}{q\,(q-1)\,2q\,(2q-1)}\,x^{2q} - \&c.\right)e^{\frac{1}{q}x^q}:$$

and it is easy to see that we may herein change the sign of x^q, thereby obtaining another finite particular solution,

$$u = Q = \left(1 + \frac{q-1}{q\,(q-1)}\,x^q + \frac{(q-1)\,(3q-1)}{q\,(q-1)\,2q\,(2q-1)}\,x^{2q} + \&c.\right)e^{-\frac{1}{q}x^q}.$$

Reverting to the equation in z, we have next a particular solution of the form

$$z = Ax + Bx^{q+1} + Cx^{2q+1} + Dx^{3q+1} + \&c.,$$

giving between the coefficients the relation

$$\begin{aligned}
(\ q+1)\,A + (\ q+1)\ \ q\,B &= 0,\\
(3q+1)\,B + (2q+1)\,2q\,C &= 0,\\
(5q+1)\,C + (3q+1)\,3q\,D &= 0,\\
(7q+1)\,D + (4q+1)\,4q\,E &= 0,\\
\&c.
\end{aligned}$$

If $A=1$, we have

$$\begin{aligned}
A &= \ 1,\\
B &= -\frac{(q+1)}{(q+1)\,q},\\
C &= +\frac{(q+1)(3q+1)}{(q+1)\,q\,(2q+1)\,2q},\\
D &= -\frac{(q+1)\,(3q+1)\,(5q+1)}{(q+1)\,q\,(2q+1)\,2q\,(3q+1)\,3q},\\
\&c.,
\end{aligned}$$

where, as in the former case, the series is considered to terminate as soon as there is an evanescent factor in the numerator, without any regard to the subsequent coefficients which contain in the denominators the same evanescent factor. [In particular, $q=-1$, we have the solution $z=x$.]

Hence if we have $(2i+1)\,q=-1$, the series terminates, and we have for u the finite particular solution,

$$u = P = x\left(1 - \frac{q+1}{(q+1)\,q}\,x^q + \frac{(q+1)(3q+1)}{(q+1)\,q\,(2q+1)\,2q}\,x^{2q} - \&c.\right)e^{\frac{1}{q}x^q},$$

from which, changing the sign of x^q, we deduce the other finite particular solution,

$$u = Q = x\left(1 + \frac{q+1}{(q+1)\,q}\,x^q + \frac{(q+1)(3q+1)}{(q+1)\,q\,(2q+1)\,2q}\,x^{2q} + \&c.\right)e^{-\frac{1}{q}x^q}.$$

Hence, in the equation

$$\frac{dy}{dx} + y^2 = x^{2q-2},$$

where $q\,(2i+1) = \pm 1$, we have (writing $D=1$)

$$y = \frac{CP'+Q'}{CP+Q},$$

where C is the constant of integration, P, Q are finite series as above, and P', Q' are the derived functions of P and Q. Writing successively $i=0$, $i=1$, $i=2$, &c., we may tabulate the solutions

$$\frac{dy}{dx}+y^2=1, \quad P=e^x, \quad Q=e^{-x},$$

$$\frac{dy}{dx}+y^2=x^{-4}, \quad P=xe^{-\frac{1}{x}}, \quad Q=xe^{\frac{1}{x}},$$

$$\frac{dy}{dx}+y^2=x^{-\frac{4}{3}}, \quad P=(1-3x^{\frac{1}{3}})\,e^{3x^{\frac{1}{3}}}, \quad Q=(1+3x^{\frac{1}{3}})\,e^{-3x^{\frac{1}{3}}},$$

$$\frac{dy}{dx}+y^2=x^{-\frac{8}{3}}, \quad P=x\,(1+3x^{-\frac{1}{3}})\,e^{-3x^{-\frac{1}{3}}}, \quad Q=x\,(1-3x^{-\frac{1}{3}})\,e^{3x^{-\frac{1}{3}}},$$

$$\frac{dy}{dx}+y^2=x^{-\frac{8}{5}}, \quad P=(1-5x^{\frac{1}{5}}+\tfrac{25}{3}\,x^{\frac{2}{5}})\,e^{5x^{\frac{1}{5}}}, \quad Q=(1+5x^{\frac{1}{5}}+\tfrac{25}{3}\,x^{\frac{2}{5}})\,e^{-5x^{\frac{1}{5}}},$$

&c.

It is hardly necessary to make the final step of calculating P' and Q' and substituting in y; but, as an example, take the above equation $\frac{dy}{dx}+y^2=x^{-\frac{4}{3}}$: we have

$$y=\frac{-3x^{-\frac{1}{3}}\,(Ce^{3x^{\frac{1}{3}}}+e^{-3x^{\frac{1}{3}}})}{C\,(1-3x^{\frac{1}{3}})\,e^{3x^{\frac{1}{3}}}+(1+3x^{\frac{1}{3}})\,e^{-3x^{\frac{1}{3}}}},$$

which is readily identified with the solution, p. 98 of Boole's *Differential Equations* (Cambridge, 1859).

Cambridge, September 29, 1868.

421.

NOTE ON THE SOLVIBILITY OF EQUATIONS BY MEANS OF RADICALS.

[From the *Philosophical Magazine*, vol. XXXVI. (1868), pp. 386, 387.]

IN regard to the theorem that the general quintic equation of the nth order is not solvible by radicals, I believe that the proofs which have been given depend, or at any rate that a proof may be given that shall depend, on the following two lemmas:

I. A one-valued (or symmetrical) function of n letters is a perfect kth power, only when the kth root is a one-valued function of the n letters.

There is an exception in the case $k=2$, whatever be the value of n: viz. the product of the squares of the differences is a one-valued function, a perfect square; but its square root, or the product of the simple differences, is a two-valued function. It is in virtue of this exception that a quadric equation is solvible by radicals; we have the one-valued function $(x_1-x_2)^2$, the square of a two-valued function x_1-x_2, and thence the two roots are each expressible in the form

$$\tfrac{1}{2}\{x_1+x_2+\sqrt{(x_1-x_2)^2}\}.$$

II. A two-valued function of n letters is a perfect kth power, only when the kth root is a two-valued function of the n letters.

There is an exception in the case $k=3$, when $n=3$ or 4: viz. for $n=3$ we have $(x_1+\omega x_2+\omega^2 x_3)^3$ (ω an imaginary cube root of unity) a two-valued function, and a perfect cube; whereas its cube root is the six-valued function $x_1+\omega x_2+\omega^2 x_3$. And similarly for $n=4$ we have, for instance,

$$\{x_1x_2+x_3x_4+\omega(x_1x_3+x_2x_4)+\omega^2(x_1x_4+x_2x_3)\}^3$$

a two-valued function, and a perfect cube, whereas its cube root is a six-valued function. And it is in virtue of this exception that a cubic or a quartic equation is solvible by radicals. But I assume that for $n > 4$ the lemma is true without exception.

The course of demonstration would be something as follows: Imagine, if possible, the root of an equation expressed, by means of radicals, in terms of the coefficients; the expression cannot contain any radical such as $\sqrt[p]{X}$, $p > 2$, where X is a one-valued (or rational) function of the coefficients, not a perfect pth power, for the reason that, expressing the coefficients in terms of the roots, such function $\sqrt[p]{X}$ is not a rational function of the roots; if it were so, by lemma I. it would be a one-valued (that is, a symmetrical) function of the roots; consequently a rational function of the coefficients, or X expressed in terms of the coefficients, would be a perfect pth power.

The expression may however contain a radical $\sqrt{X}$, X a one-valued (or rational) function of the coefficients, not a perfect square: viz. X may be any square function multiplied into that function of the coefficients which is equal to the product of the squared differences of the roots, or, say, multiplied into the discriminant; that is, we may have $X = Q^2\nabla$, or $\sqrt{X} = Q\sqrt{\nabla}$.

We have next to consider whether the expression can contain any radical $\sqrt[p]{X}$, where X, not being a rational function of the coefficients, is a function expressible by radicals. But the foregoing reasoning shows that if this be so, X cannot contain any radical other than the radical $\sqrt{Q^2\nabla}$ or $Q\sqrt{\nabla}$, as above; that is, X must be $= P + Q\sqrt{\nabla}$, where P and Q are rational functions of the coefficients, and where we may assume that $P + Q\sqrt{\nabla}$ is not a perfect pth power of a function of the like form $P' + Q'\sqrt{\nabla}$. But then, expressing the coefficients in terms of the roots, we have $P + Q\sqrt{\nabla}$, a (rational) two-valued function of the roots; and there is no radical $\sqrt[p]{P + Q\sqrt{\nabla}}$, which is a rational function of the roots; for by lemma II., if such radical existed we should have $\sqrt[p]{P + Q\sqrt{\nabla}}$ a (rational) two-valued function of the roots; that is, it would be $= P' + Q'\sqrt{\nabla}$, P' and Q' one-valued (symmetrical) functions of the roots, consequently rational functions of the coefficients; or $P + Q\sqrt{\nabla}$ would be a perfect pth power $(P' + Q'\sqrt[p]{\nabla})^p$.

The conclusion is that for $n > 4$ there is not (besides the function $P + Q\sqrt{\nabla}$) any function of the coefficients, expressible by means of radicals, which, when the coefficients are expressed in terms of the roots, will be a rational function of the roots, and consequently there is no possibility of expressing the roots in terms of the coefficients by means of radicals.

Cambridge, October 1, 1868.

422.

ON THE GEODESIC LINES ON AN OBLATE SPHEROID.

[From the *Philosophical Magazine,* vol. XL. (1870), pp. 329—340.]

THE theory of the geodesic lines on an oblate spheroid of any excentricity whatever was investigated by Legendre ([1]); and the general course of them is well known, viz. each geodesic line undulates between two parallels equidistant from the equator (being thus either a closed curve, or a curve of indefinite length, according to the distance between the two parallels): at a point of contact with the parallel the curve is, of course, at right angles to the meridian; say this is V, a vertex of the geodesic

FIG. 1.

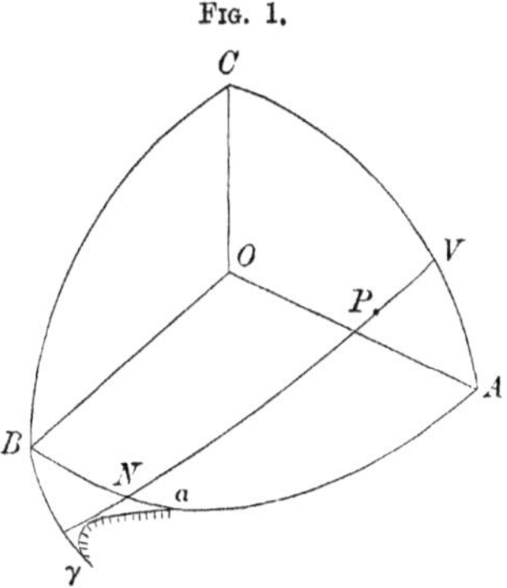

line, and let the meridian through V meet the equator in A; the geodesic line proceeds from V to meet the equator in a point N, the node, where AN is at most $=90°$; and the undulations are obtained by the repetition of this portion VN of the geodesic line alternately on each side of the equator and of the meridian.

[1] *Mém. de l'Inst.* 1806; see also the *Exer. de Calcul Intégral,* t. I. (1811), p. 178, and the *Traité des Fonctions Elliptiques,* t. I. (1825), p. 360.

I consider in the present paper the series of geodesic lines which cut at right angles a given meridian AC, or, say, a series of geodesic normals. It may be remarked that as V passes from the position A on the equator to the pole C, the angular distance AN increases from a certain determinate value (equal, as will appear, to $\frac{C}{A}90^\circ$, if C, A are the polar and equatorial axes respectively) up to the value 90°; and it thus appears that, attending only to their course after they first meet the equator, the geodesic normals have an envelope resembling in its general appearance the evolute of an ellipse (see fig. 1 and also fig. 2), the centre hereof being the point

FIG. 2.

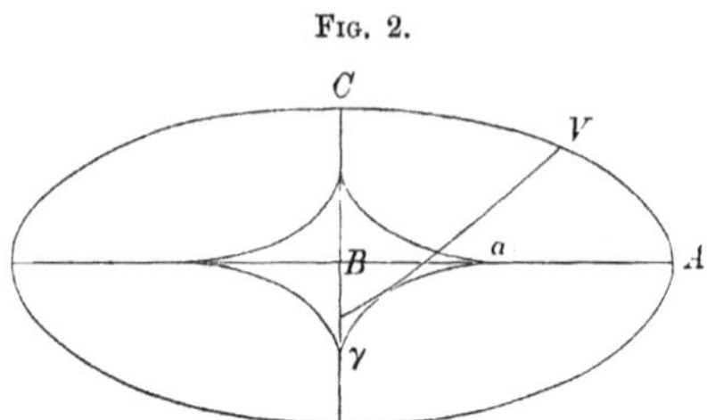

B at the distance $BA = 90^\circ$, and the axes coinciding in direction with the equator BA and meridian BC: this is in fact a real geodesic evolute of the meridian CA. The point α is, it is clear, the intersection of the equator by the geodesic line for which V is consecutive to the point A (so that $\angle BOA = \left(1 - \frac{C}{A}\right) 90^\circ$); and the point γ is the intersection of the meridian CB by the geodesic line for which V is consecutive to the point C; and its position will be in this way presently determined. I was anxious, with a view to the construction of a drawing and a model, to obtain some numerical results in relation to a spheroid of considerable excentricity, and I selected that for which $\frac{C}{A} = \frac{1}{2}$ (polar axis $= \frac{1}{2}$ equatorial).

Before proceeding further, I remark that Legendre's expression "reduced latitude" is used in what is not, I think, the ordinary sense; and I propose to substitute the

FIG. 3.

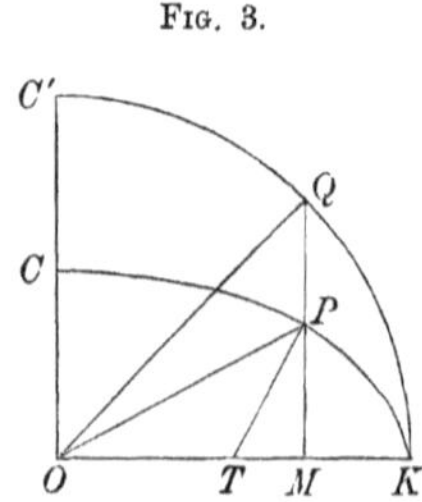

term "parametric latitude": viz., in fig. 3, referring the point P on the ellipse by means of the ordinate MPQ to a point Q on the circle, radius $OK (=OA$, fig. 1), and drawing the normal PT, then we have for the point P the three latitudes,

$$\lambda = \angle PTK, \text{ normal latitude,}$$

$$\lambda'' = \angle POK, \text{ central latitude,}$$

$$\lambda' = \angle QOK, \text{ parametric latitude;}$$

viz. λ' is the parameter most convenient for the expression of the values of the coordinates x, y ($x = A\cos\lambda'$, $y = C\sin\lambda'$) of a point P on the ellipse. The relations between the three latitudes are

$$\tan\lambda'' = \frac{C}{A}\tan\lambda' = \frac{C^2}{A^2}\tan\lambda,$$

so that λ'', λ', λ are in the order of increasing magnitude. I use in like manner l, l', l'' in regard to the vertex V. The course of a geodesic line is determined by the equation

$$\cos\lambda' \sin\alpha = \text{const.},$$

where λ' is the reduced latitude of any point P on the geodesic line, and α is at this point the azimuth of the geodesic line, or its inclination to the meridian. Hence, if l' be the parametric latitude of the vertex V, the equation is

$$\cos\lambda' \sin\alpha = \cos l'$$

(whence also, when $\lambda' = 0$, $\alpha = 90° - l'$; that is, the geodesic line cuts the equator at an angle $= l'$, the parametric latitude of the vertex). The equation in question, $\cos\lambda' \sin\alpha = \cos l'$, leads at once to Legendre's other equations: viz. taking, as above, A, C for the equatorial and polar semiaxes respectively, and δ for the excentricity, $\delta = \sqrt{1 - \frac{C^2}{A^2}}$; and to determine the position of P on the meridian, using (instead of the parametric latitude λ') the angle ϕ determined by the equation

$$\cos\phi = \frac{\sin\lambda'}{\sin l'},$$

and writing, moreover, s to denote the geodesic distance VP, and Λ to denote the longitude of P measured from the meridian CA which passes through the vertex V, these are

$$ds = d\phi\sqrt{C^2 + A^2\delta^2\sin^2 l' \cos^2\phi},$$

$$d\Lambda = \frac{\cos l'}{A} \cdot \frac{d\phi\sqrt{C^2 + A^2\delta^2\sin^2 l'\cos^2\phi}}{1 - \sin^2 l'\cos^2\phi};$$

which differential expressions are to be integrated from $\phi = 0$; and the equations then determine λ', s, and Λ, all in terms of the angle ϕ,—that is, virtually s and Λ, the length and longitude, in terms of the parametric latitude λ'.

Writing, with Legendre,

$$c^2 = \frac{A^2\delta^2 \sin^2 l'}{C^2 + A^2\delta^2 \sin^2 l'}, \quad = \quad \delta^2 \sin^2 l,$$

$$b^2 = 1 - c^2, \quad = \frac{C^2}{C^2 + A^2\delta^2 \sin^2 l'}, \quad = 1 - \delta^2 \sin^2 l;$$

also

$$n = \tan^2 l', \quad M = \frac{C}{bA \cos l'} = \frac{C}{A \cos l},$$

then the formulæ become

$$ds = \frac{C}{b} d\phi \sqrt{1 - c^2 \sin^2 \phi},$$

$$d\Lambda = M \frac{d\phi \sqrt{1 - c^2 \sin^2 \phi}}{1 + n \sin^2 \phi}.$$

Hence integrating from $\phi = 0$, and using the notations F, E, Π of elliptic functions, we have

$$s = \frac{C}{b} E(c, \phi),$$

$$\Lambda = \frac{M}{n} \{(n + c^2) \Pi(n, c, \phi) - c^2 F(c, \phi)\};$$

viz. these belong to any point P whatever on the geodesic line, parametric latitude of vertex $= l'$; and if we write herein $\phi = 90°$, then they will refer to the node N, or point of intersection with the equator.

The position of the point α is at once obtained by writing $l' = 0$: viz. this gives $c = 0$, $b = 1$, $M = \frac{C}{A}$, $n = 0$: the differential expressions are $ds = Cd\phi$, $d\Lambda = \frac{C}{A} d\phi$. Or integrating from $\phi = 0$ to $\phi = \frac{1}{2}\pi$, we have $s = A \cdot \frac{C}{A} \cdot \frac{1}{2}\pi$, $\Lambda = \frac{C}{A} \cdot \frac{1}{2}\pi$, agreeing with each other, and giving longitude of $\alpha = \frac{C}{A} \cdot \frac{1}{2}\pi$; or, what is the same thing, $\angle \alpha OB = \frac{1}{2}\pi\left(1 - \frac{C}{A}\right)$.

Writing in the formulæ $l' = 90°$, we have $c = \delta$, $b = \frac{C}{A}$, $\frac{M}{n} = 0$; whence $d\Lambda = 0$, or $\Lambda = \text{const.}$, $= \frac{1}{2}\pi$, since the geodesic line here coincides with the meridian CB; and moreover $s = AE(\delta, \phi)$; viz. this is merely the expression of the distance from C of a point P on the meridian CB. But we do not thus obtain the position of the point γ.

To find it we must consider a position of V consecutive to C, say, $l' = \frac{1}{2}\pi - \epsilon$, where ϵ is indefinitely small; n is thus indefinitely large, and the integral $\Pi(n, c, \phi)$ is not conveniently dealt with. But it may be replaced by an expression depending on

$\Pi\left(\frac{c^2}{n}, c, \phi\right)$, where $\frac{c^2}{n}$ is indefinitely small; viz. (Legendre, *Fonct. Ellipt.* vol. I. p. 69) we have

$$\Pi(n, c, \phi) = F(c, \phi) + \frac{1}{\sqrt{\alpha}} \tan^{-1} \frac{\sqrt{\alpha}\tan\phi}{\sqrt{1-c^2\sin^2\phi}} - \Pi\left(\frac{c^2}{n}, c, \phi\right),$$

where

$$\alpha = (1+n)\left(1+\frac{c^2}{n}\right).$$

We thus have

$$\Lambda = \frac{M}{n}\left\{nF(c, \phi) + \frac{\sqrt{n}\sqrt{c^2+n}}{\sqrt{1+n}} \tan^{-1} \frac{\sqrt{\alpha}\tan\phi}{\sqrt{1-c^2\sin^2\phi}} - (c^2+n)\,\Pi\left(\frac{c^2}{n}, c, \phi\right)\right\},$$

where, $\frac{c^2}{n}$ being small,

$$\Pi\left(\frac{c^2}{n}, c, \phi\right) = \int \frac{d\phi}{\left(1+\frac{c^2}{n}\sin^2\phi\right)\sqrt{1-c^2\sin^2\phi}},$$

$$= \int \frac{\left(1-\frac{c^2}{n}\sin^2\phi\right)d\phi}{\sqrt{1-c^2\sin^2\phi}}, \quad = \left(1-\frac{1}{n}\right)F(c, \phi) + \frac{1}{n}E(c, \phi).$$

And expanding also the $\tan^{-1}$ term, we thus have

$$\Lambda = \frac{M}{n}\left\{nF(c, \phi) + \frac{\sqrt{n}\sqrt{c^2+n}}{\sqrt{1+n}}\left[\tfrac{1}{2}\pi - \frac{\sqrt{1-c^2\sin^2\phi}}{\tan\phi}\,\frac{\sqrt{n}}{\sqrt{(1+n)(c^2+n)}}\right]\right.$$

$$\left. - (c^2+n)\left[\left(1-\frac{1}{n}\right)F(c, \phi) + \frac{1}{n}E(c, \phi)\right]\right\},$$

$$= \frac{M}{n}\left\{\left(b^2+\frac{c^2}{n}\right)F(c, \phi) - \left(1+\frac{c^2}{n}\right)E(c, \phi) + \frac{\sqrt{n}\sqrt{c^2+n}}{\sqrt{1+n}}\cdot\tfrac{1}{2}\pi - \frac{n}{n+1}\cot\phi\sqrt{1-c^2\sin^2\phi}\right\},$$

which, in the term in { } neglecting negative powers of n, becomes

$$\Lambda = \frac{M}{n}\left\{\sqrt{n}\,.\tfrac{1}{2}\pi + b^2F(c, \phi) - E(c, \phi) - \cot\phi\sqrt{1-c^2\sin^2\phi}\right\}.$$

We may moreover write $c = \delta$, $b = \frac{C}{A}$, $\phi = 90^\circ - \lambda'$, $n = \frac{1}{\epsilon^2}$, $M = \epsilon$, and therefore $\frac{M}{n} = \epsilon$, so that the formula is

$$\Lambda = \epsilon\left\{\frac{1}{\epsilon}\cdot\tfrac{1}{2}\pi + b^2F(c, 90^\circ-\lambda') - E(c, 90^\circ-\lambda') - \tan\lambda'\sqrt{1-c^2\cos^2\lambda'}\right\},$$

$$= \tfrac{1}{2}\pi - \epsilon\left\{\tan\lambda'\sqrt{1-c^2\cos^2\lambda'} + E(c, 90^\circ-\lambda') - b^2F(c, 90^\circ-\lambda')\right\},$$

where I retain c, b as standing for $\sqrt{1-\frac{C^2}{A^2}}$, $\frac{C}{A}$ respectively.

Writing herein $\lambda'=0$, we have

$$\Lambda = \tfrac{1}{2}\pi - \epsilon\,(E,c - b^2F,c),$$

where the coefficient $E,c-b^2F,c$ is

$$=\int_0^{\frac{1}{2}\pi} d\theta\left(\sqrt{1-c^2\sin^2\theta}-\frac{1-c^2}{\sqrt{1-c^2\sin^2\theta}}\right)=c^2\int_0^{\frac{1}{2}\pi}\frac{\cos^2\theta\,d\theta}{\sqrt{1-c^2\sin^2\theta}}$$

consequently positive; that is, Λ, the longitude of the node, is less than 90°, as it should be. Hence in order that Λ may be $=90°$, we must have λ' negative, say, $\lambda'=-\mu'$, where μ' is positive; and, observing that we may under the signs E, F write $90°-\mu'$ instead of $90°+\mu'$, we thus have

$$\tfrac{1}{2}\pi = \tfrac{1}{2}\pi + \epsilon\,\{\sqrt{1-c^2\cos^2\mu'}\tan\mu' - E\,(c,\ 90°-\mu') + b^2F\,(c,\ 90°-\mu')\};$$

that is, we must have

$$\tan\mu'\sqrt{1-c^2\cos^2\mu'} = E\,(c,\ 90°-\mu') - b^2F\,(c,\ 90°-\mu');$$

viz. μ' is here the parametric latitude (south) of the intersection of the meridian CB with the consecutive geodesic line—that is, of the point γ. As μ' increases from 0 to 90°, the left-hand side increases from 0 to ∞; and the right-hand side, beginning from a positive value and either attaining a maximum or not, ultimately decreases to 0; there is consequently a real root, which is easily found by trial.

Thus $\dfrac{C}{A}=\tfrac{1}{2}$, $C=\tfrac{1}{2}\sqrt{3}$ (angle of modulus $=60°$), $b=\tfrac{1}{2}$; or the equation is

$$\tan\mu'\sqrt{1-\tfrac{3}{4}\cos^2\mu'} = E\,(90°-\mu') - \tfrac{1}{4}F\,(90°-\mu').$$

Using Legendre's Table IX., we have

μ'.	$90°-\mu'$.	E.	F.	$E-\frac{1}{4}F$.	$\tan\mu'\sqrt{1-\frac{3}{4}\cos^2\mu'}$.
0°	90°	1·21105	2·15651	·6719	·0
10	80	1·12248	1·81252	·6693	
20	70	1·02663	1·49441	·6530	
30	60	·91839	1·21253	·6153	·3819
40	50	·79538	·96465	·5542	·6278

so that we see the required value is between 30° and 40°; and a rough interpolation gives the value $\mu'=37°\,40'$. But repeating the calculation with the values 37° and 38°, we have

μ'.	$90°-\mu'$.	E.	F.	$E-\frac{1}{4}F$.	$\tan\mu'\sqrt{1-\frac{3}{4}\cos^2\mu'}$.
37°	53°	·833879	1·035870	·57419	·54425
38	52	·821197	1·011849	·56823	·57108

whence, interpolating, $\mu'=37°\,55'$.

The semiaxes of the geodesic evolute, measured according to their longitude and parametric latitude respectively, are thus $B\alpha$, long. of $\alpha = 45°$; $B\gamma$, param. lat. $= 37° 55'$. But measuring them according to their geodesic distance, the equatorial radius A being taken $= 1$, we have

$$B\alpha = \tfrac{1}{4}\pi \qquad\qquad = \cdot 78540,$$

$$B\gamma = \left(\frac{C}{b} - 1\right)\{E_{,} - E(52° 5')\} = 1\cdot 21106 - \cdot 82225 = \cdot 38881.$$

Reverting to the general formulæ for s, Λ, but writing therein $A = 1$, and therefore $C = \sqrt{1-\delta^2}$; writing also $\phi = 90°$ (that is, making the formulæ to refer to the node N of the geodesic line), we have

$$s = \frac{\sqrt{1-\delta^2}}{\sqrt{1-\delta^2 \sin^2 l}} E_{,} c,$$

$$\Lambda = \frac{\sqrt{1-\delta^2}}{n \cos l} \{(n + c^2)\, \Pi_{,}(n,\ c) - c^2 F_{,} c\};$$

but for the calculation of the second of these formulæ by means of Legendre's Tables it is necessary to express $\Pi_{,}(n,\ c)$ in terms of the functions E, F.

The proper formula is given in *Fonct. Ellipt.* vol. I. p. 137; viz. this is

$$\frac{\Delta(b,\ \theta)}{\sin\theta\cos\theta}\Pi_{,}(n,\ c) = \tfrac{1}{2}\pi + \frac{\sin\theta}{\cos\theta}\Delta(b,\ \theta)\, F_{,}c + F_{,}c\, F(b,\ \theta) - F_{,}c\, E(b,\ \theta) - E_{,}c\, F(b,\ \theta),$$

where $\Delta(b,\ \theta) = \sqrt{1 - b^2\sin^2\theta}$. θ is an angle given by the equation $\cot\theta = \sqrt{n}$; we have $n = \tan^2 l'$; consequently $\theta = 90° - l'$. Substituting this value, except that for shortness I retain $E(b,\ \theta)$, $F(b,\ \theta)$ in place of $E(b,\ 90° - l')$, $F(b,\ 90° - l')$, we have

$$\Delta(b,\ \theta) = \sqrt{1 - b^2\cos^2 l'},$$

$$= \sqrt{1 - (1 - \delta^2\sin^2 l)\cos^2 l'}\ ,\ = \sin l;$$

and thence

$$\tan\theta\,\Delta(b,\ \theta) = \cot l \sin l = \frac{\cos l}{\sqrt{1-\delta^2}};$$

whence

$$\Pi_{,}(n,\ c) = \frac{\sin l' \cos l'}{\sin l}\left\{\tfrac{1}{2}\pi + F_{,}c\left[\frac{\cos l}{\sqrt{1-\delta^2}} + F(b,\ \theta) - E(b,\ \theta)\right] - E_{,}c\, F(b,\ \theta)\right\}.$$

But

$$n + c^2 = \tan^2 l' + \delta^2 \sin^2 l = \sin^2 l \sec^2 l.$$

Hence

$$(n + c^2)\,\Pi_{,}(n,\ c) - c^2 F_{,}c = \sin^2 l\,\{\sec^2 l'\,\Pi_{,}(n,\ c) - \delta^2 F_{,}c\};$$

and multiplying this by

$$\frac{\sqrt{1-\delta^2}}{n\cos l}, \quad = \frac{\sqrt{1-\delta^2}\cos l}{\tan^2 l' \cos^2 l},$$

the exterior factor is

$$\frac{\sqrt{1-\delta^2}\cos l \tan^2 l}{\tan^2 l'}, \quad = \frac{\cos l}{\sqrt{1-\delta^2}},$$

and we have

$$\Lambda = \frac{\cos l}{\sqrt{1-\delta^2}}\{\sec^2 l' \Pi_{,}(n,\ c) - \delta^2 F_{,}c\},$$

which is the formula I used in the calculations. It would, however, have been better to reduce a step further; viz. we have

$$\begin{aligned}\sec^2 l'\Pi_{,}(n,\ c) &= \frac{\tan l'}{\tan l \cos l}\{\ \},\\ &= \frac{\sqrt{1-\delta^2}}{\cos l}\left\{\tfrac{1}{2}\pi + F_{,}c\left[\frac{\cos l}{\sqrt{1-\delta^2}} + F(b,\ \theta) - E(b,\ \theta)\right] - E_{,}c\,F(b,\ \theta)\right\},\\ &= \frac{\sqrt{1-\delta^2}}{\cos l}\{\tfrac{1}{2}\pi + F_{,}c\,[F(b,\ \theta) - E(b,\ \theta)] - E_{,}c\,F(b,\ \theta)\} + F_{,}c,\end{aligned}$$

and thence

$$\sec^2 l'\Pi_{,}(n,\ c) - \delta^2 F_{,}c = \frac{\sqrt{1-\delta^2}}{\cos l}\{\tfrac{1}{2}\pi + F_{,}c\,[\sqrt{1-\delta^2}\cos l + F(b,\ \theta) - E(b,\ \theta)] - E_{,}cF(b,\ \theta)\};$$

or, finally,

$$\Lambda = \tfrac{1}{2}\pi + F_{,}c\,F(b,\ \theta) - F_{,}cE(b,\ \theta) - E_{,}cF(b,\ \theta) + \sqrt{1-\delta^2}\cos l F_{,}c.$$

It is easy with this expression of Λ to obtain the results already found for the extreme values $l' = 0°$, $l' = 90°$.

As Legendre's Tables have for argument, not the modulus c, but the angle of the modulus, say χ (that is, $\sin\chi = c = \delta\sin l$), it is convenient to replace $\sqrt{1-\delta^2\sin^2 l}$ by its value $\cos\chi$; and the formulæ thus are

$$s = \frac{\sqrt{1-\delta^2}}{\cos\chi}E_{,}c,$$

$$\Lambda = \tfrac{1}{2}\pi + F_{,}c\,[\sqrt{1-\delta^2}\cos l + F(b,\ \theta) - E(b,\ \theta)] - E_{,}c\,F(b,\ \theta),$$

where

$$C = \sin\chi = \delta\sin l, \quad \tan l' = \sqrt{1-\delta^2}\tan l, \quad \theta = 90° - l';$$

and in the case intended to be numerically discussed, $\delta = \frac{1}{2}\sqrt{3}$, $\sqrt{1-\delta^2} = \frac{1}{2}$. I take l' as the argument, giving it the values $0°$, $10°, \ldots 90°$, and perform the calculation as shown in the Table.

l'.	θ.	l.	$\chi = \angle c$.	$90^\circ - \chi = \angle b$.	$\frac{1}{2}\cos l$.	$F_{,}c$, log.	$E_{,}c$, log.	$F_{,}c$, nat.	$E_{,}c$, nat.	$F(b, \theta)$, nat.	$E(b, \theta)$, nat.
0°											
10	80°	19° 26′	16°·75	73°·25	·47151	20548	18686	1·6050	1·5376	2·08962	1·03762
20	70	36 3	30·63	59·37	·40425	22814	16532	1·6910	1·4633	1·48840	1·02962
30	60	49 6	40·90	49·10	·32737	25478	14167	1·7980	1·3857	1·16024	·95214
40	50	59 13	48·07	41·93	·25589	27626	12163	1·8891	1·3232	·92141	·82827
50	40	67 14	52·98	37·02	·19349	29935	10645	1·9923	1·2777	·71820	·67903
60	30	73 54	56·32	33·68	·13866	31479	09557	2·0644	1·2461	·53083	·51655
70	20	79 41	58·43	31·57	·08954	32540	08850	2·1154	1·2260	·35099	·34716
80	10	84 58	59·62	30·38	·04387	33169	08446	2·1463	1·2147	·17475	·17431
90			60·0				08316				

l'.	$\frac{1}{2}\cos l + F(b, \theta) - E(b, \theta)$, nat.	Do. log.	Add log $F_{,}c$.	(1) Nat.	Log $F(b, \theta)$.	Add log $E_{,}c$.	(2) Nat.	$\frac{1}{2}\pi + (1) - (2)$.	* Do. in (°).	Log $E_{,}c$ − log cos χ.	Nat.	* $\frac{1}{2}$ do., = length.
0°								·7854	45°			·7854
10	1·52308	·18272	·38820	2·4446	·32007	·50693	3·2132	·8022	45 58′	20569	1·6058	·8029
20	·86318	$\bar{1}$·93610	·16424	1·4596	·17272	·33804	2·1779	·8525	48 51	23075	1·7012	·8506
30	·53547	$\bar{1}$·72874	$\bar{1}$·98352	·9627	·06455	·20622	1·6078	·9257	53 2	26323	1·8333	·9166
40	·34903	$\bar{1}$·54286	$\bar{1}$·81912	·6594	$\bar{1}$·96445	·08608	1·2192	1·0110	57 56	29668	1·9801	·9900
50	·23266	$\bar{1}$·36672	$\bar{1}$·66607	·4635	$\bar{1}$·85625	$\bar{1}$·96270	·9177	1·1166	63 59	32682	2·1224	1·0612
60	·15292	$\bar{1}$·18446	$\bar{1}$·49925	·3157	$\bar{1}$·72496	$\bar{1}$·82053	·6615	1·2250	70 11	35178	2·2479	1·1239
70	·09327	$\bar{2}$·96974	$\bar{1}$·29514	·1973	$\bar{1}$·54529	$\bar{1}$·63379	·4303	1·3378	76 39	36939	2·3410	1·1705
80	·04431	$\bar{2}$·64650	$\bar{2}$·97819	·0951	$\bar{1}$·24242	$\bar{1}$·32688	·2123	1·4536	83 17	38454	2·4240	1·2120
90								1·5708	90			1·2111

χ, $90^\circ - \chi$ in degrees and decimals of a degree, to correspond with Legendre's Tables.

where the columns marked with an * show respectively the longitude of the node, and the length (or distance of node from vertex), for the geodesic lines belonging to the different values of the argument l'.

The remarks which follow have reference to the stereographic projection of the figure on the plane of the equator, the centre of projection being the pole (say the South Pole) of the spheroid. It is to be remarked that if a point P of the spheroid is projected as above, by means of an ordinate into the point Q of the sphere radius $OK (= OA)$, then projecting stereographically as to the spheroid and the sphere from the south poles thereof respectively, the points P and Q have the same projection. And it is hence easy to show that an azimuth α at a point of the meridian $\left(\text{parametric latitude } \lambda', \text{ normal latitude } \lambda, \text{ and therefore } \tan\lambda' = \frac{C}{A}\tan\lambda\right)$ is projected into an angle (α) such that

$$\tan(\alpha) = \frac{\sin\lambda'}{\sin\lambda}\tan\alpha.$$

In fact in fig. 3, if we take therein OK, OC for the axes of x, z respectively, and the axis of y at right angles to the plane of the paper, and if we have at P on the surface of the spheroid an element of length PR at the inclination α to the meridian PK, then if x, y, z are the coordinates of P, and $x + \delta x$, $y + \delta y$, $z + \delta z$ those of R, we have

$$\begin{aligned} \delta x &= \quad \rho\cos\alpha\sin\lambda, \\ \delta z &= -\rho\cos\alpha\cos\lambda, \\ \delta y &= \quad \rho\sin\alpha, \end{aligned}$$

and thence

$$\tan\alpha = \frac{\delta y}{\sqrt{\delta x^2 + \delta z^2}}.$$

Now, if the meridian and the points P, R are referred by lines parallel to Oz to the surface of the sphere radius OA, the only difference is that the ordinates z are increased in the ratio $C : A$; so that if the projected angle be (α), we have

$$\tan(\alpha) = \frac{\delta y}{\sqrt{\delta x^2 + \frac{A^2}{C^2}\delta z^2}};$$

and then projecting the sphere stereographically from its south pole, the angle in the projection is $= (\alpha)$. And according to the foregoing remark, the angle (α) thus obtained is also the projection of α from the south pole of the spheroid. We have thus

$$\frac{\tan(\alpha)}{\tan\alpha} = \frac{\sqrt{\delta x^2 + \delta z^2}}{\sqrt{\delta x^2 + \frac{A^2}{C^2}\delta z^2}}, \quad = \frac{\sqrt{\sin^2\lambda + \cos^2\lambda}}{\sqrt{\sin^2\lambda + \frac{A^2}{C^2}\cos^2\lambda}}, \quad = \sqrt{\frac{1 + \cot^2\lambda}{1 + \cot^2\lambda'}}, \quad = \frac{\sin\lambda'}{\sin\lambda},$$

which is the required relation.

The foregoing equations,

$$\cos\lambda'\sin\alpha = \cos l', \quad \tan\lambda' = \frac{C}{A}\tan\lambda,$$

$$\tan(\alpha) = \frac{\sin\lambda'}{\sin\lambda}\tan\alpha,$$

determine in the stereographic projection the inclination (α) to the radius, or projection of the meridian, of the geodesic line (parametric latitude of vertex $=l'$) at the point the parametric latitude of which is $=\lambda'$; viz. they enable the construction (in the projection) of the direction of the successive elements of the geodesic line. There would be no difficulty in performing the construction geometrically; but it would, I think, be more convenient to calculate (α) numerically for a given value of l' and for the successive values of λ'. Observe that for $\lambda'=0$ we have (as above) $90^\circ-\alpha=l'$, and then $\frac{\sin\lambda'}{\sin\lambda}=\frac{\tan\lambda'}{\tan\lambda}=\frac{C}{A}$, consequently $\tan(\alpha)=\frac{C}{A}\cot l'$: but we have also $\cot l'=\frac{A}{C}\cot l$, so that this equation becomes $\tan(\alpha)=\cot l$, or we have $90^\circ-(\alpha)=l$; viz. in the projection, the geodesic line cuts the equator at an angle $l=$the normal latitude of the vertex of the geodesic line.

The preceding formulæ and results have enabled me to construct a drawing, on a large scale, of the stereographic projection of the geodesic lines for the spheroid, polar axis $=\frac{1}{2}$ equatorial axis.

423.

ON THE PLANE REPRESENTATION OF A SOLID FIGURE.

[From the *Philosophical Magazine,* vol. XLI. (1871), pp. 286—290.]

WE represent *in plano* the position of a point P whose coordinates in space are (x, y, z) by drawing these coordinates, on the same scale or on different scales, and in given directions from a fixed origin in the plane; $OM = x$, $MP' = y$, $P'P'' = z$. But observe that the point P'' *alone* does not completely represent the point P; in fact P'' represents a whole series of points lying in a line; any one such point is the

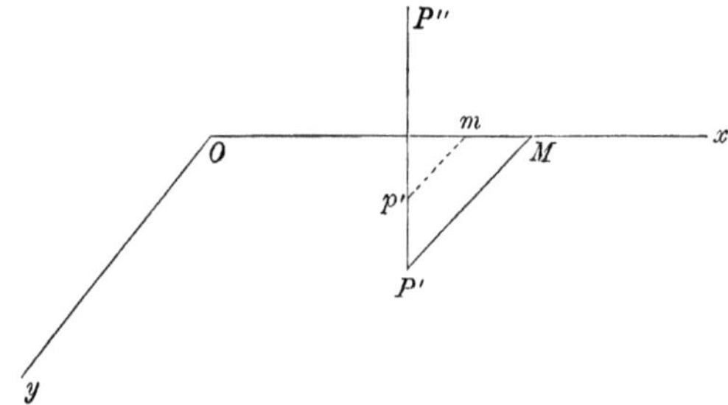

point whose coordinates are Om, mp', $p'P''$. For the complete representation of P we require the *two points* P', P'': these might be distinguished as the projection P'', and the foot-point P'. The two points P', P'' are obviously such that the line joining them is in a given direction.

The preceding is, of course, the ordinary method of orthogonal projection, or geometrical delineation of a solid figure: it may be used under various forms; for example, the coordinates x, y, z may be taken on the same scale and in directions inclined to each other at angles of 120° (isometrical projection); or the coordinates x, y may be drawn on the same scale and at their actual inclination, 90°, to each other;

and the coordinate z on the same or an altered scale in any given direction; the points P' then give a true ground-plan of the solid figure, and the lengths of the lines $P'P''$ give the altitudes of the several points P: this is also a method in ordinary use.

But it is to be observed that the points P', P'' are both of them *projections*, and that the general theory is as follows: we represent the position of the point P

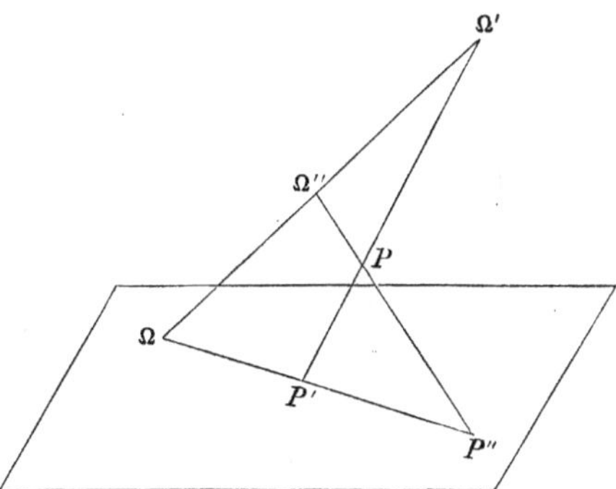

by means of its projections P', P'', from two fixed points Ω', Ω'' respectively; the line joining these points passes, it is clear, through a fixed point Ω which is the intersection of the plane of projection by the line which joins the two points Ω', Ω''.

Hence we say that a point P in space is represented *in plano* by any two points P', P'' which are such that the line joining them passes through a fixed point Ω. And we have thus a *system of constructive geometry* which is the more simple on account of the generality of its basis, and which is at once applicable to any of the special projections above referred to. I establish the fundamental notions of such a geometry, and by way of illustration apply it to the solution of the well-known problem of finding the lines which meet four given lines in space.

A point P (as already mentioned) is given by its projections P', P'', which are points such that the line joining them passes through the fixed point Ω.

A line L is given by its projections L', L'', which are any two lines in the plane. We speak of the point (P', P''), meaning the point P whose projections are P', P''; and similarly of the line (L', L''), meaning the line whose projections are L', L''.

If P', P'' coincide, then the point P is in the plane of projection; and so if L', L'' coincide, then the line L is in the plane of projection.

If through Ω we draw a line meeting L', L'' in the points P', P'' respectively, these are the projections of a point P on the line L. In particular the intersection of L', L'' (considered as two coincident points) represents the intersection of the line L with the plane of projection.

The line through the points (P', P'') and (Q', Q'') has for its projections the lines $P'Q'$ and $P''Q''$.

Two lines (L', L'') and (M', M'') intersect each other if only the intersections $L'M'$ and $L''M''$ are the projections of a point P—that is, if the line through the points $L'M'$ and $L''M''$ passes through Ω. And then clearly P is the intersection of the two lines.

A plane Π is conveniently given by means of its trace Θ on the plane of projection, and of the projections (P', P'') of a point on the plane; or, say, by means of the trace Θ, and of a point P on the plane.

Suppose, however, that a plane is given by means of a line L and a point P on the plane. The trace Θ passes through the point of intersection of the line L with the plane of projection—that is, through the point of intersection of the projections L', L''. To find another point on the trace, we have only to imagine on the line L a point Q, and, joining this with P, to suppose the line PQ produced to meet the plane of projection. The construction is obvious; but by way of illustration I give it in full. Through Ω draw a line meeting L', L'' in Q', Q'' respectively (then these are the projections of a point Q on the line L); the lines $P'Q'$ and $P''Q''$ are the projections of the line PQ, and the intersection of $P'Q'$ and $P''Q''$ is therefore the required point on the trace Θ.

The line of intersection of two planes passes through the point of intersection of their traces Θ_1, Θ_2; whence, if the planes have in common a point P, the line of intersection is the line joining P with the intersection of the traces Θ_1, Θ_2.

In what precedes we have the solution of the following problem:—"Given a point P, and two lines L_1, L_2, to find a line through P meeting the two lines L_1, L_2." The required line is in fact the line of intersection of the planes (P, L_1) and (P, L_2); we have seen how to construct the traces Θ_1 and Θ_2 of these planes respectively; and the required line is the line joining P with the intersection of Θ_1 and Θ_2.

I proceed now to the problem to find the two lines, each of them meeting four given lines, L_1, L_2, L_3, L_4 (these being, of course, given by means of their projections (L_1', L_1'') &c.). The question is in effect to find on the line L_1 a point P such that, drawing from it a line to meet L_2, L_3, and also a line to meet L_2, L_4, these shall be one and the same line.

Now, considering in the first instance P as an arbitrary point on the line L_1, the line from P to meet L_2, L_3 is any line whatever meeting the lines L_1, L_2, L_3: say it is a generating line of the hyperboloid whose directrices are L_1, L_2, L_3, or of the hyperboloid $L_1L_2L_3$. Hence projecting from any point Ω' whatever, the generating lines and directrices are projected into tangents of one and the same conic. We know the projections L_1', L_2', L_3' of the directrices; to find two other tangents of the conic, we take two arbitrary positions of P on the line L_1, and construct as above the projections M', N' of the lines from these to meet the lines L_2, L_3. The conic is then given as the conic touching the five lines L_1', L_2', L_3', M', N': say this is the conic Σ'. Similarly, instead of Ω', considering the point Ω'', we have the lines L_1'', L_2'', L_3'' and the lines M'', N'', which are the other projections of the lines through the two positions of P; and touching these five lines we have a conic Σ''. Each tangent T' of Σ', combined with the *corresponding* tangent T'' of Σ'', represents a line T meeting L_1, L_2, L_3; to establish the correspondence, observe that, inasmuch as the line T meets L_1, the intersections of T', L_1' and of T'', L_1'' must lie in a line with Ω; if T' be given, the point (T'', L_1'') is thus uniquely determined, and therefore also T'' (since L_1'' is a tangent of Σ''); and similarly if T'' be given, T' is uniquely determined; the correspondence T', T'' is thus, as it should be, a (1, 1) correspondence.

Considering in like manner the lines which meet L_1, L_2, L_4, we have touching L_1', L_2', L_4', $\overline{M}'$, $\overline{N}'$ a conic $\overline{\Sigma}'$; and touching L_1'', L_2'', L_4'', $\overline{M}''$, $\overline{N}''$ a conic $\overline{\Sigma}''$; each tangent T' of $\overline{\Sigma}'$, combined with the corresponding tangent $\overline{T}''$ of $\overline{\Sigma}''$, represents a line meeting L_1, L_2, L_4, the correspondence being a (1, 1) correspondence such as in the former case.

The conics Σ', $\overline{\Sigma}'$ both touch L_1', L_2'; hence they have in common two tangents. Say one of these is $T' = \overline{T}'$, the corresponding tangents T'' and $\overline{T}''$ will coincide with each other and be a common tangent of Σ'', $\overline{\Sigma}''$ (these conics both touch L_1'', L_2'', and have thus in common two tangents). We have thus $T' = \overline{T}'$, and $T'' = \overline{T}''$, as the projections of a line meeting L_1, L_2, L_3, L_4; and taking the other common tangents of Σ', $\overline{\Sigma}'$ and of Σ'', $\overline{\Sigma}''$, we have the projections of the other line meeting L_1, L_2, L_3, L_4.

The whole process is:—Construct M', M'' and N', N'' each of them the projections of a line through a point P of L_1, which meets L_2, L_3; and $\overline{M}'$, $\overline{M}''$ and N', $\overline{N}''$ each of them the projections of a line through a point P of L_1, which meets L_2, L_4; we have then the conics

Σ', Σ'' touching L_1', L_2', L_3', M', N', and L_1'', L_2'', L_3'', M'', N'' respectively,

$\overline{\Sigma}'$, $\overline{\Sigma}''$ „ L_1', L_2'. L_4', $\overline{M}'$, $\overline{N}'$, „ L_1'', L_2'', L_4'', $\overline{M}''$, $\overline{N}''$ „ ;

and then the projections of each of the required lines are $T' = \overline{T}'$, a common tangent of Σ', $\dot{\Sigma}'$, and $T'' = \overline{T}''$, the corresponding common tangent of Σ'', $\overline{\Sigma}''$.

It is material to remark how the construction is simplified when there is given one of the lines, say, M, which meets L_1, L_2, L_3, L_4. Here M is a common directrix of the two hyperboloids; we may for the hyperbolas Σ' and Σ'' consider, instead of L_1, L_2, L_3 and two new generating lines, the lines L_1, L_2, L_3, M, and a single new generating line N; and similarly for the hyperbolas $\overline{\Sigma}'$, $\overline{\Sigma}''$ the lines L_1, L_2, L_4, M and a single new generating line $\overline{N}$. Σ', $\overline{\Sigma}'$ have thus in common the three tangents L_1', L_2', M', and therefore only a single other common tangent, $T' = \overline{T}'$; and similarly Σ'', $\overline{\Sigma}''$ have in common the three tangents L_1'', L_2'', M'', and therefore only a single other common tangent, $T'' = \overline{T}''$; and we have thus the other line cutting the four given lines.

I take the opportunity of mentioning the following theorem:

"If in a given triangle we inscribe a variable triangle of given form, the envelope of each side of the variable triangle is a conic touching the two sides (of the given triangle) which contain the extremities of the variable side in question."

We have thence a solution of the problem (*Principia*, Book I. Sect. V. Lemma XXVII.), in a given quadrilateral to inscribe a quadrangle of given form. The question in effect is: in the triangle ABC to inscribe a triangle $\alpha\beta\gamma$ of given form; and in the triangle ADE a triangle $\alpha'\beta'\gamma'$ of given form, in such wise that the sides $\alpha\gamma$, $\alpha'\gamma'$

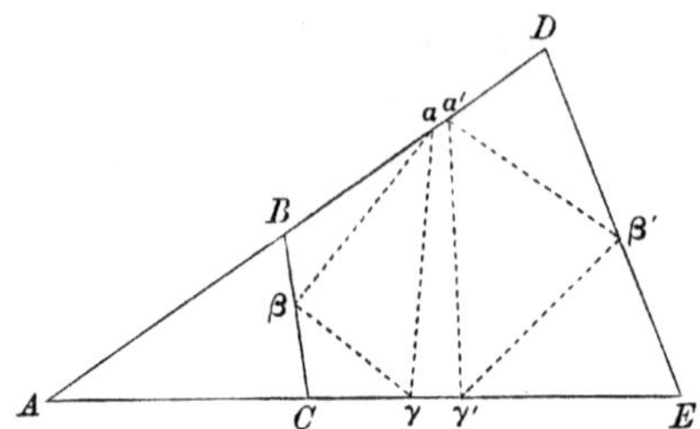

may be coincident. The envelope of $\alpha\gamma$ is a conic touching AD, AE, and the envelope of $\alpha'\gamma'$ a conic also touching AD, AE: there are thus two other common tangents, either of which may be taken for the position of the side $\alpha\gamma = \alpha'\gamma'$; and the problem admits accordingly of two solutions.

424.

ON THE ATTRACTION OF A TERMINATED STRAIGHT LINE.

[From the *Philosophical Magazine*, vol. XLI. (1871), pp. 358—360.]

WRITE for shortness $(a, b, c; \epsilon)$ to denote the shell included between the ellipsoids

$$\frac{x^2}{a^2}+\frac{y^2}{b^2}+\frac{z^2}{c^2}=1 \text{ and } \frac{x^2}{a^2}+\frac{y^2}{b^2}+\frac{z^2}{c^2}=(1+\epsilon)^2$$

(where ϵ is indefinitely small); then, if the ellipsoids

$$\frac{x^2}{a^2}+\frac{y^2}{b^2}+\frac{z^2}{c^2}=1 \text{ and } \frac{x^2}{a'^2}+\frac{y^2}{b'^2}+\frac{z^2}{c'^2}=1$$

are confocal, the attractions of the shells $(a, b, c; \epsilon)$ and $(a', b', c'; \epsilon)$ upon any exterior point P are proportional to their masses. Hence, considering a prolate spheroid of revolution, $c=b$, the attractions of the shell $(a, b, b; \epsilon)$ will be proportional to those of the shell $(\sqrt{a^2-h}, \sqrt{b^2-h}, \sqrt{b^2-h}; \epsilon)$; or if, as usual, $b^2=a^2(1-e^2)$, then, if h increases and becomes ultimately equal to b^2, to those of the shell $(ae, 0, 0; \epsilon)$; viz. this last is the portion of the axis of x included between the limits $x=-ae$, $x=+ae$; or say it is the terminated line $x=\pm ae$; and I say that the mass is distributed over this line *uniformly*.

To see that this is so, observe in general that, in the spheroid $\frac{x^2}{a'^2}+\frac{y^2+z^2}{b'^2}=1$, the volume included between the planes $x=\alpha$, $x=\alpha+d\alpha$, is $=(y^2+z^2)\,d\alpha$, $=\pi\left(b'^2-\frac{b'^2}{a'^2}\alpha^2\right)d\alpha$; and thence, writing $a'(1+\epsilon)$, $b'(1+\epsilon)$ for a', b', in the shell $(a', b', b'; \epsilon)$ the volume included between the planes $x=\alpha$, $x=\alpha+d\alpha$ is $=\pi b'^2 . 2\epsilon' d\alpha$; viz. this is independent of α, and simply proportional to $d\alpha$. Hence, writing $b'=0$, when the shell shrinks up into a line, the mass must be disturbed uniformly over the line. It follows that for a line of uniform density the equipotential surfaces are each of them a prolate spheroid of revolution having the extremities of the line for its foci, and that, if we have a shell bounded by any such surface and the consecutive *similar* surface, with its mass equal to that of the line, then such shell and the line will exert the same

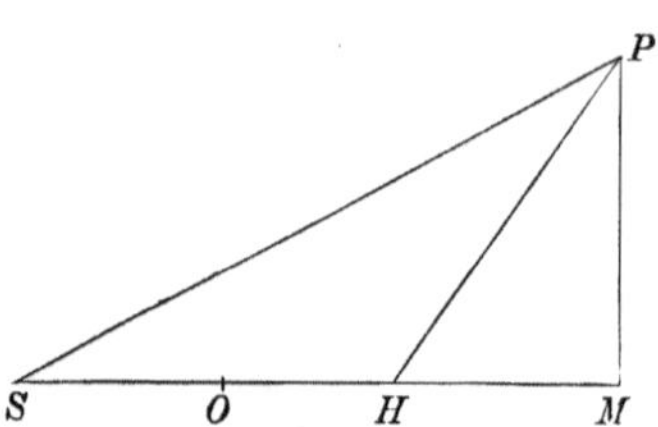

attractions upon any point P exterior to the shell. The attractions of the line are obtained most easily by means of its potential; viz. taking S, H for the extremities of the line, and, as above, the origin at the middle point, and the axis of x in the direction of the line, and writing $2ae$ for the length of the line, x, y, z for the coordinates of P, and r, s for the values of HP, SP (that is, $r = \sqrt{(x-ae)^2 + y^2 + z^2}$, $s = \sqrt{(x+ae)^2 + y^2 + z^2}$), then the potential is at once found to be

$$V = \log \frac{x + ae + s}{x - ae + r};$$

and we can hereby verify that the equipotential surface is in fact a spheroid of revolution having the foci S, H; for, taking the equation of such a spheroid to be

$$\frac{x^2}{a^2} + \frac{y^2 + z^2}{a^2(1-e^2)} = 1,$$

(a is an arbitrary parameter, since only the value of ae has been defined), we have

$$s = a + ex, \quad r = a - ex$$

and thence

$$x + ae + s = (1+e)(x+a),$$

$$x - ae + r = (1-e)(x+a),$$

and the quotient is $= \dfrac{1+e}{1-e}$, a constant value, as it should be. The equation $V = \text{const.}$ may in fact be written

$$\frac{1+e}{1-e} = \frac{x + ae + s}{x - ae + r};$$

viz. this equation, apparently of the fourth order, breaks up into the twofold plane $y^2 = 0$, and the spheroid $\dfrac{x^2}{a^2} + \dfrac{y^2 + z^2}{a^2(1-e^2)} = 1$.

The foregoing results in regard to the attraction of a line are not new. See Green's *Essay on Electricity*, 1828, and *Collected Works*, Cambridge, 1871, p. 68; also

Joachimsthal, "On the Attraction of a Straight Line," with Sir W. Thomson's Note, *Camb. and Dubl. Math. Journ.*, vol. III. (1848), p. 93; but it does not appear to have been noticed that they are, in fact, included in the theory of the attraction of ellipsoids.

The like considerations show that the attractions of the ellipsoidal shell $(a, b, c; \epsilon)$ upon an exterior point are equal to those of an elliptic disk $z=0$, $\frac{x^2}{a^2-c^2}+\frac{y^2}{b^2-c^2}=1$, the mass of which is equal to that of the shell, and which has the density at the point (x, y) proportional to $\left(1-\frac{x^2}{a^2-c^2}-\frac{y^2}{b^2-c^2}\right)^{-\frac{1}{2}}$.

Sir W. Thomson informs me that the foregoing results have long been familiar to him.

425.

NOTE ON THE GEODESIC LINES ON AN ELLIPSOID.

[From the *Philosophical Magazine*, vol. XLI. (1871), pp. 534, 535.]

THE general configuration of the geodesic lines on an ellipsoid is established by means of the known theorem (an immediate consequence of Jacobi's fundamental formulæ, but which was first given by Mr Michael Roberts, *Comptes Rendus*, vol. XXI. p. 1470, Dec. 1845) that every geodesic line touches a curve of curvature; that is, attending to the two opposite ovals which constitute the curve of curvature, the geodesic line is in general an infinite curve undulating between these opposite ovals, and so touching each of them an infinite number of times (but possibly in particular cases it is a reentrant curve touching each oval a finite number of times). The geodesic lines thus divide themselves into two kinds, accordingly as they touch a curve of curvature of the one or the other kind; and there is besides a third limiting kind, the lines which pass through an umbilicus: any such geodesic line passes through the opposite umbilicus, and is in general an infinite curve passing an infinite number of times alternately through the two umbilici; but possibly it is in particular cases a reentrant curve passing a finite number of times through the two umbilici. I annex a figure giving a general idea of the configuration of the geodesic lines drawn in different directions from a given point P on the surface of the ellipsoid: this is drawn (as it were) on the plane of the greatest and least axes; but it is not a perspective or geometrical representation of any kind, but a mere diagram for the purpose in question. We have A, A, B, C, C the extremities of the axes; U_1, U_2, U_3, U_4 the umbilici; P the point on the surface; $1P2$ and $1P4$ the curves of curvature through P, viz. these are ovals containing the umbilici U_1, U_2 and U_1, U_4 respectively. Then U_1PU_3 and U_2PU_4 are the limiting geodesics passing through the umbilici; the line TPT' represents a

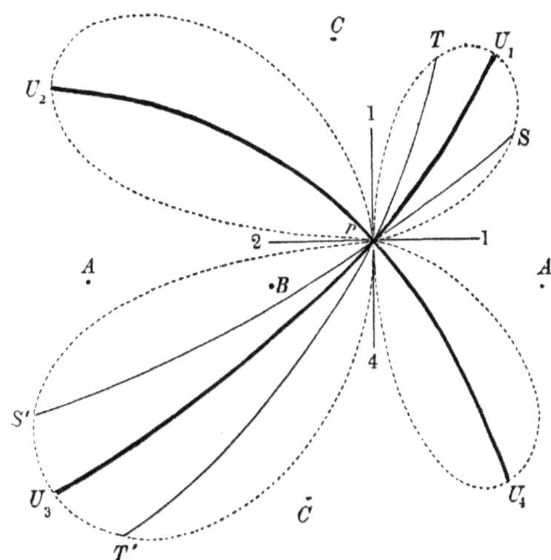

geodesic line of the one kind, viz. this at T touches an oval (curve of curvature) U_1U_4, and at T' the conjugate oval U_2U_3. Similarly SPS' is a geodesic line of the other kind, viz. this at S touches an oval (curve of curvature) U_1U_2, and at S' the conjugate oval U_3U_4; the dotted figure-of-eight curves are the loci of the points of contact T, T', S, S'.

426.

ON A SUPPOSED NEW INTEGRATION OF DIFFERENTIAL EQUATIONS OF THE SECOND ORDER.

[From the *Philosophical Magazine*, vol. XLII. (1871), pp. 197—199.]

THIS refers to a paper, Challis, "On the Application of a new Integration of Differential Equations of the Second Order to some unsolved Problems in the Calculus of Variations," *Phil. Mag.* same volume, pp. 28—40.

427.

ON GAUSS'S PENTAGRAMMA MIRIFICUM.

[From the *Philosophical Magazine*, vol. XLII. (1871), pp. 311, 312.]

TAKE on a sphere (in the northern hemisphere) two points, A, B, whose longitudes differ by 90°, and refer them to the equator by the meridians AE and BC respectively; join A, B by an arc of great circle, and take in the southern hemisphere the pole D of this circle; and join D with E and C respectively by arcs of great circle. We have a spherical pentagon $ABCDE$, which is in fact the "Pentagramma mirificum," considered by Gauss, as appearing vol. III. pp. 481—490 of the *Collected Works*. Among its properties we have

$$\left.\begin{array}{l}\text{the distance of any two non-adjacent summits}\\ \text{the inclination of any two non-adjacent sides}\end{array}\right\}=90^\circ;$$

so that each summit is the pole of the opposite side, or the pentagon is its own reciprocal.

Each angle is the supplement of the opposite side.

If the squared tangents of the sides (or angles) taken in order are α, β, γ, δ, ϵ, then

$$1+\alpha=\gamma\delta,\quad 1+\beta=\delta\epsilon,\quad 1+\gamma=\epsilon\alpha,\quad 1+\delta=\alpha\beta,\quad 1+\epsilon=\beta\gamma,$$

equivalent to three independent equations, so that any three of the quantities may be expressed in terms of the remaining two. (This agrees with the foregoing construction, where the arbitrary quantities are the latitudes of A, B respectively.)

Projecting from the centre of the sphere upon any plane, we have a plane pentagon which is such that the perpendiculars let fall from the summits upon the opposite sides respectively meet in a point. This (as easily seen) implies that the two portions into which each perpendicular is divided by the point in question have the same product.

Conversely, starting from the plane pentagon, and erecting from the point of intersection a perpendicular to the plane, the length of this perpendicular being equal to the square root of the product in question, we have the centre of a sphere such that the projection upon it of the plane polygon is the pentagramma mirificum.

I remark as to the analytical theory, that, taking the origin at the intersection of the perpendiculars, and for the coordinates of the summits $(\alpha_1, \beta_1), \dots (\alpha_5, \beta_5)$ respectively, then we have

$$\alpha_1\alpha_4 + \beta_1\beta_4 = \alpha_2\alpha_5 + \beta_2\beta_5 = \alpha_3\alpha_1 + \beta_3\beta_1 = \alpha_4\alpha_2 + \beta_4\beta_2 = \alpha_5\alpha_3 + \beta_5\beta_3, \quad = -\gamma^2,$$

where γ^2 is the above-mentioned product, or γ is the radius of the sphere.

Cambridge, September 14, 1871.

428.

NOTE SUR LA CORRESPONDANCE DE DEUX POINTS SUR UNE COURBE.

[From the *Comptes Rendus de l'Académie des Sciences de Paris*, tom. LXII. (*Janvier—Juin*, 1866), pp. 586—590.]

THIS is to the same effect with the paper 385, "On the Correspondence of two Points on a Curve"; four examples of the theory are given, the first, second, and third of them the same as in this paper—the fourth example is as follows:

4°. *Recherche du nombre des points* sextactiques, c'est-à-dire des points qui sont tels que par chacun passe une conique qui a dans ce point un contact du cinquième ordre avec la courbe. Il faut prendre pour les points P les intersections avec la courbe de la conique qui a au point P' un contact du quatrième ordre: les points unis seront ceux dont il s'agit. La courbe $\Theta = 0$ est la conique qui a au point P' un contact du quatrième ordre. On a ainsi parmi les intersections le point P' 5 fois; donc $k = 5$. A chaque point P' correspondent $2m - 5$ points P; à chaque point P $10m^2 - 20m - 5 - 20\delta$ points P' (j'emprunte le terme -20δ d'une formule que vient de donner M. Zeuthen); donc la formule donne pour le nombre des points unis

$$10\,m^2 - 18\,m - 10 - 20\,\delta + 10\,D,$$

c'est-à-dire

$$15\,m^2 - 33\,m - 30\,\delta.$$

Mais cette expression comprend le nombre $3m(m-2) - 6\,\delta$ des inflexions; en effet pour un point d'inflexion la conique avec contact du quatrième ordre se réduit à la tangente prise deux fois, ce qui est une conique avec contact du cinquième ordre. Donc enfin le nombre des points sextactiques sera

$$m(12\,m - 27) - 24\,\delta,$$

ou, pour une courbe sans points doubles

$$m(12\,m - 27),$$

ce qui s'accorde avec la valeur que j'ai trouvée par d'autres moyens.

429.

SUR LES CONIQUES DÉTERMINÉES PAR CINQ CONDITIONS DE CONTACT AVEC UNE COURBE DONNÉE.

[From the *Comptes Rendus de l'Académie des Sciences à Paris,* tom. LXIII. (*Juillet—Décembre,* 1866), pp. 9—12.]

This paper (dated Cambridge, 26 June 1866), contains the expressions for the numbers (5), (4, 1), (3, 2) (3, 1, 1), (2, 2, 1), (2, 1, 1, 1) and (1, 1, 1, 1, 1), of the conics which satisfy five conditions of contact with a given curve, as obtained in the paper 406 "On the Curves which satisfy given conditions," see p. 214, and which expressions were found by the same process, viz. by consideration of functional equations obtained by supposing the given curve to break up into two curves of the orders m and m' respectively; there was in the expression for (1, 1, 1, 1, 1) a numerical error as mentioned in the footnote of the same page. The paper contains also the formula $\mu'' - \frac{3}{2}\nu'' + \frac{3}{2}\rho'' - \sigma'' = 0$, and the expression for $(2X, 3Z)$ given, pp. 203, 204.

430.

NOTE SUR QUELQUES FORMULES DE M. E. DE JONQUIÈRES, RELATIVES AUX COURBES QUI SATISFONT À DES CONDITIONS DONNÉES.

[From the *Comptes Rendus de l'Académie des Sciences de Paris*, tom. LXIII. (*Juillet—Décembre*, 1866), pp. 666—670.]

LES formules dont il s'agit sont publiées dans les *Comptes Rendus*, séances du 3 et du 17 septembre 1866. En faisant une simple transformation algébrique pour y introduire la classe $M (= m^2 - m)$ de la courbe donnée U^m, et en changeant un peu la forme, les théorèmes de M. de Jonquières peuvent s'énoncer comme il suit:

1°. Le nombre des contacts des courbes C^r qui ont un contact de l'ordre n avec une courbe fixe U^m, et qui passent en outre par $\frac{1}{2}r(r+3) - n$ points donnés, est

$$= \tfrac{1}{2}(n+1)\,[nM + (2r - 2n)\,m].$$

OBSERVATION. Énoncé de cette manière, le théorème s'applique même au cas $n = 0$. En effet, pour $n = 0$, le nombre donné par le théorème est $= mr$, qui est le nombre des contacts de l'ordre 0 (intersections simples) de la courbe donnée U^m avec une courbe déterminée de l'ordre r.

2°. Le nombre des contacts de l'ordre $n' (=$ ou $< n)$ des courbes C^r qui ont deux contacts des ordres n et n' respectivement avec une courbe fixe U^m, et qui passent en outre par $\frac{1}{2}r(r+3) - n - n'$ points donnés est

$$\begin{aligned} = \tfrac{1}{4}(n+1)(n'+1)\{&[nM + (2r-2n)\,m]\,[n'M + (2r-2n')\,m] \\ &- 2\,(n^2 + nn' + n'^2 + n + n')\,M \\ &+ \ [-4r\,(n+n'+1) + 4\,(n^2 + nn' + n'^2 + n + n')]\,m\}. \end{aligned}$$

OBSERVATION. Énoncé de cette manière, le théorème s'applique même aux cas $n' = 0$, et $n' = n$. En effet, pour $n' = 0$, le nombre donné par le théorème est $= (rm - n - 1) . \frac{1}{2}(n+1)[nM + (2r-2n)m]$, ce qui est égal au nombre des courbes C^r qui ont avec la courbe donnée U^m un contact de l'ordre n, multiplié par $rm - n - 1$, nombre des contacts de l'ordre 0 (intersections simples) de chacune de ces courbes avec la courbe U^m. Et pour $n' = n$, le nombre des contacts est le double du nombre des courbes C^r.

Je remarque que les deux théorèmes peuvent se démontrer de la manière dont je me suis servi en cherchant le nombre des coniques qui satisfont à cinq conditions données; car, en remplaçant la courbe m par l'ensemble de deux courbes m et m', on trouve que pour le théorème 1° le nombre cherché est

$$= \alpha M + \beta m,$$

où les coefficients (α, β) ne dépendent que de (r, n); et puis, en supposant que ce théorème soit connu, on trouve que pour le théorème 2° le nombre cherché est

$$= \tfrac{1}{4}(n+1)(n'+1)[nM + (2r-2n)m][n'M + (2r-2n')m] + \alpha M + \beta m,$$

où de même les coefficients (α, β) ne dépendent que de (r, n).

Or voici comment on peut déterminer les coefficients dans les deux théorèmes:

Pour le théorème 1°, on démontre que pour U^m une droite, le nombre cherché est $= (n+1)(r-n)$; et que pour U^m une conique, le nombre cherché se déduit de là en écrivant $2r$ au lieu de r; c'est-à-dire, que pour la conique, le nombre est $= (n+1)(2r-n)$. On a donc

$$\beta = (n+1)(r-n), \quad = \tfrac{1}{2}(n+1)(2r-2n),$$

$$2\alpha + 2\beta = (n+1)(2r-n),$$

et de là

$$\alpha = \tfrac{1}{2}(n+1)\,n\,;$$

ce qui achève la démonstration.

Pour le théorème 2°, on démontre que pour U^m une droite, le nombre cherché est $= (n+1)(n'+1)(r-n-n')(r-n-n'-1)$, et que pour U^m une conique, le nombre cherché se déduit de là en écrivant $2r$ au lieu de r; c'est-à-dire, pour la conique, le nombre est

$$= (n+1)(n'+1)(2r-n-n')(2r-n-n'-1).$$

On a donc

$$(n+1)(n'+1)(\;r-n-n')(\;r-n-n'-1) = (n+1)(n'+1)(\;r-n)(\;r-n') + \beta,$$

$$(n+1)(n'+1)(2r-n-n')(2r-n-n'-1) = (n+1)(n'+1)(2r-n)(2r-n') + 2\alpha + 2\beta\,;$$

cela donne pour α et β les valeurs

$$\alpha = \tfrac{1}{4}(n+1)(n'+1)[-2\;(n^2 + nn' + n'^2 + n + n')],$$

$$\beta = \tfrac{1}{4}(n+1)(n'+1)[-4r(n+n'+1) + 4(n^2 + nn' + n'^2 + n + n')]\,;$$

et la démonstration est ainsi achevée.

Je remarque que sous les formes ici données les deux théorèmes s'appliquent à une courbe U^m avec des points doubles, mais sans point de rebroussement.

Le théorème dont je me suis servi pour la détermination des coefficients peut s'énoncer sous la forme plus générale que voici, savoir:

En dénotant par $\phi(r, n, n', \ldots)$ le nombre des courbes C^r qui ont avec une droite donnée des contacts des ordres $n, n', \ldots$, et qui passent en outre par $\frac{1}{2}r(r+3)-n-n'\ldots$ points donnés, alors si, au lieu de la droite donnée, on a une conique donnée, le nombre des courbes C^r sera $=\phi(2r, n, n', \ldots)$.

En effet, l'équation de la courbe cherchée C^r contient des coefficients indéterminés, lesquels, par les conditions de passer par les points donnés, se réduisent linéairement à $n+n'\ldots+1$ coefficients; en dénotant par $(A, B, \ldots)$ ces coefficients, l'équation de la courbe contiendra linéairement $(A, B, \ldots)$ et sera ainsi de la forme $(A, B, \ldots \between x, y, z)^r=0$. L'équation de la droite donnée est satisfaite en prenant pour (x, y, z) des fonctions linéaires déterminées d'un paramètre variable θ; donc, en coupant la courbe C^r par la droite donnée, on obtient une équation $(A, B, \ldots \between \theta, 1)^r=0$, et en exprimant que cette équation ait n racines égales, n' racines égales, etc., on obtient entre $(A, B, C, \ldots)$ des équations, lesquelles, en éliminant tous les coefficients, excepté deux quelconques (A, B), conduisent à une équation finale $(A, B)^p=0$, et le degré p de cette équation est ce qu'il s'agissait de trouver, le nombre des courbes C^r. Si au lieu d'une droite donnée on a une conique donnée, il n'y a rien à changer, sinon que les coordonnées (x, y, z) doivent être remplacées par des fonctions quadratiques de θ; on a ainsi une équation $(A, B, \ldots \between \theta, 1)^{2r}=0$, qui conduit à une équation finale $(A, B)^{p'}=0$, où p' est la même fonction de $(2r, n, n', \ldots)$ qu'est p de $(r, n, n', \ldots)$; et le nombre des courbes C^r est $=p'$. Le théorème est donc démontré. Et, précisément de la même manière, on démontre le théorème encore plus général:

En dénotant par $\phi(r, n, n', \ldots)$ le nombre des courbes C^r qui ont avec une droite donnée des contacts des ordres $n, n', \ldots$, et qui passent en outre par $\frac{1}{2}r(r+3)-n-n'\ldots$ points donnés, alors si, au lieu de la droite donnée, on a une courbe *unicursale* donnée de l'ordre m, le nombre des courbes C^r est $=\phi(mr, n, n', \ldots)$.

On aurait pu se servir directement de cela pour démontrer les théorèmes 1° et 2°. Par exemple, pour le théorème 1°, la considération de la courbe unicursale U^m donne

$$\alpha M+\beta m=\alpha(2m-2)+\beta m=(n+1)(mr-n);$$

c'est-à-dire

$$\alpha=\tfrac{1}{2}(n+1)n, \quad \beta=\tfrac{1}{2}(n+1)(2r-2n),$$

comme auparavant.

431.

SUR LA TRANSFORMATION CUBIQUE D'UNE FONCTION ELLIPTIQUE.

[From the *Comptes Rendus de l'Académie des Sciences de Paris*, tom. LXIV. *Janvier—Juin*, 1867, pp. 560—563.]

SOIT $U = (a, b, c, d, e \between x, 1)^4$ une fonction quartique quelconque de x: I, J les deux invariants:

$$(I = ae - 4bd + 3c^2, \quad J = ace - ad^2 - b^2e + 2bcd - c^3),$$

et prenons $\Omega = \dfrac{I^3 - 27J^2}{I^3}$ pour l'invariant absolu de U. Soient de même $U' = (a', b', \ldots \between x, 1)^4$ et $\Omega' = \dfrac{I'^3 - 27J'^2}{I'^3}$ l'invariant absolu de U'. En supposant que $\sqrt{U}$, $\sqrt{U'}$ soient les radicaux des deux fonctions elliptiques liées par la transformation du troisième ordre ou *cubique*, on peut se proposer la question quelle est la relation entre les deux invariants absolus Ω, Ω'? J'ai trouvé cette relation d'abord par des considérations géométriques qui me furent suggérées par une lettre de M. Sylvestre; puis je l'ai déduite des formules pour la transformation cubique données par M. Hermite, *Crelle*, t. LX., 1862, p. 304), et enfin, à l'aide d'une considération tirée de ces formules, j'ai réussi à l'obtenir à moyen des formules des *Fundamenta Nova*. Je vais donner ici cette dernière investigation de la relation dont il s'agit.

En supposant que les fonctions U, U' soient transformées linéairement en $(1 - x^2)(1 - k^2x^2)$, $(1 - y^2)(1 - \lambda^2y^2)$ respectivement, pour exprimer la liaison entre les modules k^2, λ^2, au lieu de l'équation explicite entre $\sqrt[4]{k}$, $\sqrt[4]{\lambda}$ (*Fund. Nova*, p. 23), je me sers des formules, p. 25, lesquelles en y écrivant

$$-\beta = \frac{\alpha + 2}{2\alpha + 1},$$

c'est-à-dire

$$2\alpha\beta + \alpha + \beta = 2,$$

deviennent

$$k^2 = -\alpha^3\beta, \quad \lambda^2 = -\alpha\beta^3.$$

Les transformations linéaires donnent sans peine

$$\Omega = \frac{108k^2(k^2-1)^4}{(k^4+14k^2+1)^3}, \quad \Omega' = \frac{108\lambda^2(\lambda^2-1)^4}{(\lambda^4+14\lambda^2+1)^3},$$

et il s'agit entre ces équations d'éliminer α, β, k, λ de manière à obtenir une équation entre Ω, Ω'.

J'écris

$$\alpha' = \frac{\frac{1}{2}(2\alpha+1)(\alpha+2)(\alpha-1)^4}{(\alpha^2+4\alpha+1)^3}, \quad \beta' = \frac{\frac{1}{2}(2\beta+1)(\beta+2)(\beta-1)^4}{(\beta^2+4\beta+1)^3}.$$

L'équation entre α, β donne

$$2\beta+1 = \frac{-3}{2\alpha+1}, \quad \beta+2 = \frac{3\alpha}{3\alpha+1}, \quad \beta-1 = \frac{-3(\alpha+1)}{3\alpha+1}, \quad \beta^2+4\beta+1 = -\frac{3(\alpha^2+4\alpha+1)}{(2\alpha+1)^2},$$

et on a de là

$$\beta' = \frac{\frac{27}{2}\alpha(\alpha+1)^4}{(\alpha^2+4\alpha+1)^3};$$

puis, en faisant attention à l'identité

$$(2\alpha+1)(\alpha+2)(\alpha-1)^4 + 27\alpha(\alpha+1)^4 = 2(\alpha^2+4\alpha+1)^3,$$

on obtient entre α', β', la relation très simple $\alpha'+\beta'=1$.

L'expression de k^2 donne

$$k^2 = \frac{\alpha^3(\alpha+2)}{2\alpha+1},$$

$$k^2 - 1 = \frac{(\alpha-1)(\alpha+1)^3}{2\alpha+1},$$

$$k^4+14k^2+1 = \frac{1}{(2\alpha+1)^2}\{\alpha^6(\alpha+2)^2 + 14\alpha^3(\alpha+2)(2\alpha+1) + (2\alpha+1)^2\},$$

$$= \frac{1}{(2\alpha+1)^2}\{(\alpha^2+4\alpha+1)(\alpha^6+3\alpha^4+16\alpha^3+3\alpha^2+1)\},$$

et on a de là

$$\Omega = \frac{108\alpha^3(2\alpha+1)(\alpha+2)(\alpha-1)^4(\alpha+1)^{12}}{(\alpha^2+4\alpha+1)^3 . (\alpha^6+3\alpha^4+16\alpha^3+3\alpha^2+1)^3}.$$

Or on a

$$\alpha' - 1 = \frac{\frac{1}{2}(2\alpha+1)(\alpha+2)(\alpha-1)^4 - (\alpha^2+4\alpha+1)^3}{(\alpha^2+4\alpha+1)^3}, \quad = \frac{-\frac{27}{2}\alpha(\alpha+1)^4}{(\alpha^2+4\alpha+1)^3}$$

$$8\alpha' + 1 = \frac{4(2\alpha+1)(\alpha+2)(\alpha-1)^4 + (\alpha^2+4\alpha+1)^3}{(\alpha^2+4\alpha+1)^3}, \quad = \frac{9(\alpha^6+3\alpha^4+16\alpha^3+3\alpha^2+1)}{(\alpha^2+4\alpha+1)^3},$$

et de là, en formant l'expression de la fonction $\frac{-64\alpha'(\alpha'-1)^4}{(8\alpha'+1)^3}$, on la trouve égale à la valeur qui vient d'être donnée pour Ω en termes de α: on a donc

$$\Omega = -\frac{64\alpha'(\alpha'-1)^3}{(8\alpha'+1)^3}$$

et de même

$$\Omega' = -\frac{64\beta'(\beta'-1)^3}{(8\beta'+1)^3}.$$

Avec la relation $\alpha' + \beta' = 1$, l'élimination de α', β' entre ces équations ne présente pas de difficulté.

432.

THÉORÈME RELATIF À LA THÉORIE DES SUBSTITUTIONS.

Extrait d'une lettre adressée à M. J. A. SERRET.

[From the *Comptes Rendus de l'Académie des Sciences de Paris,* tom. LXVII. (*Juillet—Décembre,* 1868), pp. 784, 785.]

ON peut énoncer par rapport aux substitutions un théorème qui comprend les trois théorèmes III. IV. V., t. II. pp. 260—263 de votre *Cours d'Algèbre Supérieure.*

Pour un nombre quelconque μ on peut former avec les $\phi(\mu)$ nombres inférieurs et premiers à μ plusieurs systèmes de nombres lesquels sont chacun un système conjugué (mod. μ); c'est-à-dire que le produit de deux nombres quelconques d'un tel système est congru suivant le module μ, à un nombre du système. Comme cas extrêmes, l'unité est un tel système, et les $\phi(\mu)$ nombres forment aussi un système conjugué.

Pour μ premier, en dénotant par α une racine primitive de μ et par f un diviseur quelconque de $f-1$, les nombres $\equiv a^{fa}$ (mod. μ), a étant un entier quelconque, forment un système conjugué. Et généralement pour un nombre μ quelconque ce nombre a des racines quasi-primitives α, β, $\gamma, \ldots$, aux exposants A, B, $C, \ldots$, tels que $\alpha^A \equiv 1$ (mod. μ), $\beta^B \equiv 1$ (mod. μ), ... et $ABC \ldots = \phi(\mu)$. En choisissant une combinaison quelconque, par exemple α, β de ces racines, soient f, g des diviseurs quelconques de A, B respectivement, les nombres $\equiv a^{fa} b^{g\beta}$ (mod. μ) forment un système conjugué, l'ordre du système ou nombre des termes étant $=\dfrac{AB}{fg}$.

Cela étant, on a ce théorème :

Une substitution T quelconque de l'ordre μ étant formée avec n lettres, l'on forme toutes les substitutions S telles que le produit STS^{-1} se réduise à une puissance de T dont l'exposant soit un nombre quelconque appartenant à un système conjugué (mod. μ), les substitutions S constitueront un système conjugué de l'ordre θM, où θ dénote l'ordre du système conjugué (mod. μ) et M le nombre des substitutions échangeables avec T.

La démonstration est tout à fait la même que celle que vous donnez p. 62 pour votre théorème IV, en y ajoutant seulement que les nombres i, j qui appartiennent au système conjugué (mod. μ) auront leur produit ij congru à un nombre de ce même système conjugué.

433.

SUR LES SURFACES TÉTRAÉDRALES.

[*Notes to the work* De la Gournerie, "*Recherches sur les surfaces réglées tétraédrales symétriques*," 8vo. Paris, 1867.]

Premier Mémoire. *Notes* pp. 190—193.

ÉQUATIONS DE CERTAINES

1°. L'équation

$$
\begin{array}{ll}
0 = + b^2c^2f^2x^8 & + c^2a^2g^2y^8 \\
\\
\quad \ldots & - 2c^2bf(af - bg)\,x^6y^2 \\
+ 2c^2ag\;(af - bg)\,y^6\,x^2 & \quad \ldots \\
- 2b^2ah\,(ch - af)\,z^6\,x^2 & + 2a^2bh\,(bg - ch)\,z^6y^2 \\
+ 2f^2gh\,(bg - ch)\,w^6x^2 & + 2g^2hf\,(ch - af)\,w^6y^2
\end{array}
$$

$$
\begin{array}{ll}
+ a^2(b^2g^2 + c^2h^2 - 4bgch)\,y^4z^4 & + b^2(c^2h^2 + a^2f^2 \\
+ f^2(b^2g^2 + c^2h^2 - 4bgch)\,w^4x^4 & + g^2(c^2h^2 + a^2f^2
\end{array}
$$

$$
\begin{array}{ll}
\quad \ldots & + 2bf(afbg + c^2h^2 - 2ch\chi)\,x^4z^2w^2 \\
- 2ag(afbg + c^2h^2 - 2ch\chi)\,y^4w^2z^2 & \quad \ldots
\end{array}
$$

$$+ 2ah(chaf + b^2g^2 - 2bg\chi)\,z^4y^2w^2 - 2bh(bgch + a^2f^2 - 2af\chi)\,z^4w^2x^2$$

$$- 2gh(bcgh + a^2f^2 - 2af\chi)\,w^4y^2z^2 - 2hf(cahf + b^2g^2 - 2bg\chi)\,w^4z^2x^2$$

$$+ 2\Omega x^2y^2z^2w^2$$

SURFACES DU HUITIÈME ORDRE.

$$+\,a^2b^2h^2z^8 \qquad\qquad +\,f^2g^2h^2w^8$$

$$\begin{array}{ll}
+\,2b^2cf\,(ch-af)\,x^6z^2 & -\,2bcf^2\,(bg-ch)\,x^6w^2\\
-\,2a^2cg\,(bg-ch)\,y^6z^2 & -\,2cag^2\,(ch-af)\,y^6w^2\\
\quad\ldots & -\,2abh^2\,(af-bg)\,z^6w^2\\
+\,2h^2fg\,(af-bg)\,w^6z^2 & \quad\ldots
\end{array}$$

$$\begin{array}{ll}
-\,4chaf)\,z^4x^4 & +\,c^2\,(a^2f^2+b^2g^2-4afbg)\,x^4y^4\\
-\,4chaf)\,w^4y^4 & +\,h^2\,(a^2f^2+b^2g^2-4afbg)\,w^4z^4
\end{array}$$

$$\begin{array}{ll}
-\,2cf\,(chaf+b^2g^2-2bg\chi)\,x^4w^2y^2 & +\,2bc\,(bcgh+a^2f^2-2af\chi)\,x^4y^2z^2\\
+\,2cg\,(bgch+a^2f^2-2af\chi)\,y^4x^2w^2 & +\,2ca\,(cahf+b^2g^2-2bg\chi)\,y^4z^2x^2\\
\quad\ldots & +\,2ab\,(abfg+c^2h^2-2ch\chi)\,z^4x^2y^2\\
-\,2fg\,(abfg+c^2h^2-2ch\chi)\,w^4x^2y^2 & \quad\ldots
\end{array}$$

(où, pour abréger, on a écrit $\chi = af + bg + ch$, et Ω est une quantité quelconque) est celle d'une surface du huitième ordre qui a pour courbes doubles les quatre coniques

$$\begin{array}{ll} x = 0, & + cy^2 - bz^2 + fw^2 = 0\,; \\ y = 0, & - cx^2 \quad + az^2 + gw^2 = 0\,; \\ z = 0, & + bx^2 - ay^2 \quad + hw^2 = 0\,; \\ w = 0, & - fx^2 - gy^2 - hz^2 \quad = 0. \end{array}$$

En déterminant Ω, savoir, en écrivant $\lambda + \mu + \nu = 0$, $af\lambda^2 + bg\mu^2 + ch\nu^2 = 0$ (ce qui donne deux systèmes de valeurs de $\lambda : \mu : \nu$), et puis

$$6\Omega = 4\Sigma \frac{\mu - \lambda}{\nu} (af + bg)^2 ch - 2\Sigma \frac{\mu^2 bg - \lambda^2 af}{\nu^2} (af + bg)^2 - 4 (af - bg)(bg - ch)(ch - af),$$

la surface devient une surface réglée, savoir, la *quadrispinale* de M. de la Gournerie ; et, en particulier, en supposant $\frac{1}{af} + \frac{1}{bg} + \frac{1}{ch} = 0$, on obtient

$$\lambda : \mu : \nu = \frac{1}{af} : \frac{1}{bg} : \frac{1}{ch},$$

et de là

$$6\Omega = (-2 - 4 =) - 6 (af - bg)(bg - ch)(ch - af),$$

et la surface sera développable.

2°. L'équation

$$\begin{array}{llll} 0 = + a^2y^4z^4 & + b^2z^4x^4 & + c^2x^4y^4 + f^2x^4w^4 + g^2y^4w^4 + h^2z^4w^4 & \\ \quad \ldots & + 2bfx^4z^2w^2 & - 2cfx^4y^2w^2 & + 2bcx^4y^2z^2 \\ - 2agy^4z^2w^2 & \quad \ldots & + 2cgy^4x^2w^2 & + 2cax^2y^4z^2 \\ + 2ahz^4y^2w^2 & - 2bhz^4x^2w^2 & \quad \ldots & + 2abx^2y^2z^4 \\ - 2ghw^4y^2z^2 & - 2hfw^4z^2x^2 & - 2fgw^4x^2y^2 & \quad \ldots \\ + 2\Omega x^2y^2z^2w^2 & & & \end{array}$$

(où Ω est une quantité quelconque) est celle d'une surface du huitième ordre ayant pour courbes doubles les quatre courbes du quatrième ordre

$$\begin{array}{ll} x = 0, & hz^2w^2 - gw^2y^2 + ay^2z^2 = 0\,; \\ y = 0, & - hz^2w^2 \quad + fw^2x^2 + bz^2x^2 = 0\,; \\ z = 0, & + gy^2w^2 - fw^2x^2 \quad + cx^2y^2 = 0\,; \\ w = 0, & - ay^2z^2 - bz^2x^2 - cx^2y^2 \quad = 0. \end{array}$$

En écrivant

$$\lambda + \mu + \nu = 0, \quad \frac{af}{\lambda^2} + \frac{bg}{\mu^2} + \frac{ch}{\nu^2} = 0$$

(ce qui donne quatre systèmes de valeurs de $\lambda : \mu : \nu$), et puis

$$\Omega = af\frac{\nu-\mu}{\lambda} + bg\,\frac{\lambda-\nu}{\mu} + ch\,\frac{\mu-\lambda}{\nu}:$$

la surface devient une surface réglée, savoir, la *quadricuspidale* de M. de la Gournerie; et, en supposant

$$(af)^{\frac{1}{3}} + (bg)^{\frac{1}{3}} + (ch)^{\frac{1}{3}} = 0,$$

et

$$\lambda : \mu : \nu = (af)^{\frac{1}{3}} : (bg)^{\frac{1}{3}} : (ch)^{\frac{1}{3}},$$

ce qui donne

$$\Omega = \left[(af)^{\frac{1}{3}} - (bg)^{\frac{1}{3}}\right]\left[(bg)^{\frac{1}{3}} - (ch)^{\frac{1}{3}}\right]\left[(ch)^{\frac{1}{3}} - (af)^{\frac{1}{3}}\right]:$$

la surface devient développable.

Cambridge, 18 *Octobre* 1866.

Deuxième Mémoire. Notes pp. 279—283.

Note I. Sur la décomposition du lieu des génératrices en surfaces tétraédrales distinctes.

Il me semble qu'une de vos conclusions a besoin d'être modifiée. Ainsi la surface tétraédrale dérivée de deux courbes triangulaires à exposant $\frac{1}{m}$ (m étant un entier positif), laquelle, selon un de vos théorèmes, serait de l'ordre $2m^2$, paraît se décomposer en m surfaces chacune de l'ordre $2m$. Il y a pour cela une raison *à priori*; en effet,

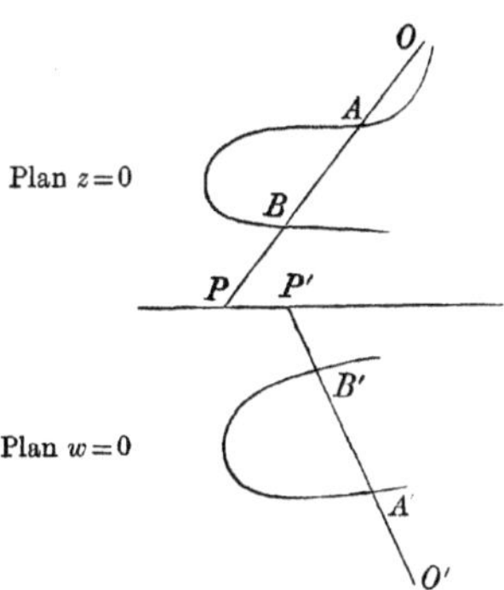

pour deux triangulaires de la forme en question, en employant votre construction, on peut établir une correspondance non-seulement entre les deux systèmes de points

$A, B, \ldots$, et $A', B', \ldots$, mais aussi entre chaque point A et un seul point correspondant A', car l'équation de la première courbe étant de la forme

$$fx^{\frac{1}{m}} + gy^{\frac{1}{m}} + hz^{\frac{1}{m}} = 0,$$

on satisfait à cette équation en écrivant

$$x : y : z = a(\theta + \alpha)^m : b(\theta + \beta)^m : c(\theta + \gamma)^m,$$

où θ est un paramètre variable. De même, l'équation de la seconde courbe étant

$$f'x^{\frac{1}{m}} + g'y^{\frac{1}{m}} + k'w^{\frac{1}{m}} = 0,$$

on satisfait à cette équation en écrivant

$$x : y : w = a'(\theta' + \alpha')^m : b'(\theta' + \beta')^m : d'(\theta' + \delta')^m,$$

où θ' est aussi un paramètre variable.

Pour la droite OP, on a

$$\frac{x}{y} = \frac{a(\theta + \alpha)^m}{b(\theta + \beta)^m},$$

et pour la droite $O'P'$

$$\frac{x}{y} = \frac{a'(\theta' + \alpha')^m}{b'(\theta' + \beta')^m};$$

donc, la condition pour la correspondance des droites est

$$\frac{a(\theta + \alpha)^m}{b(\theta + \beta)^m} = \lambda \frac{a'(\theta' + \alpha')^m}{b'(\theta' + \beta')^m},$$

ce qui donne m valeurs différentes pour θ' en termes de θ. Mais chacune de ces valeurs est de la forme

$$\theta' = \frac{A\theta + B}{C\theta + D},$$

et, en ne faisant attention qu'à une seule valeur de θ', on a le point

$$x : y : z = a(\theta + \alpha)^m : b(\theta + \beta)^m : c(\theta + \gamma)^m,$$

qui correspond à un point unique

$$x : y : w = a'(\theta' + \alpha')^m : b'(\theta' + \beta')^m : d'(\theta' + \delta')^m.$$

Pour le cas de l'exposant $\frac{1}{3}$, on a, de cette manière, une surface de l'ordre 6. J'ai vérifié cela dans le cas particulier de la surface développable. Il est très-singulier (c'est M. Salmon qui m'a fait cette remarque) qu'en écrivant dans cette équation (x^2, y^2, z^2, w^2) au lieu de (x, y, z, w), on obtient l'équation d'une surface du douzième ordre, lieu des centres de courbure d'un ellipsoïde.

Cambridge, 15 *Février* 1866.

NOTE II. À L'OCCASION DE L'ORDRE DES SURFACES TÉTRAÉDRALES.

Je crois que j'ai négligé de vous faire connaître un théorème assez général au sujet de l'ordre de ces surfaces. En considérant dans l'espace deux courbes (planes ou à double courbure) des ordres m et m' respectivement, et en supposant qu'il y ait entre les points de ces deux courbes une correspondance (α, α'), (c'est-à-dire qu'à un point donné de la courbe m correspondent α' points sur la courbe m', et à un point donné de la courbe m' correspondent α points sur la courbe m), alors la surface réglée que l'on obtient en unissant par des droites les points correspondants des courbes m et m' sera de l'ordre $m\alpha' + m'\alpha$.

Cambridge, 18 *Octobre*, 1866.

NOTE III. SUR LA SURFACE COMPLÉMENTAIRE.

Je puis reconnaître, par mes propres formules, que, des pq^2 surfaces de l'ordre $2p^2q$, il n'y en a que pq qui passent par la troisième directrice. En effet, le rapport anharmonique k est donné en termes des paramètres de la troisième directrice, au moyen d'une équation qui contient la quantité irrationnelle $k^{\frac{p}{q}}$. En rationalisant cette équation, on obtient pour k une équation de l'ordre pq; à chaque racine k_1 correspondent q surfaces, savoir celles qui appartiennent aux q valeurs de $k_1^{\frac{p}{q}}$; mais l'équation irrationnelle n'est satisfaite que par une seule valeur de $k_1^{\frac{p}{q}}$, à savoir la valeur de $k_1^{\frac{p}{q}}$ donnée par l'équation irrationnelle, en y substituant pour k, en tant que k y entre rationnellement, la valeur $k = k_1$. Donc, à chaque racine k_1 correspond une seule surface qui passe par la troisième directrice. La question à laquelle donne lieu cette circonstance paraît très-intéressante. La surface déterminée par les trois directrices est composée de pq surfaces chacune de l'ordre $2p^2q$, et d'une surface résiduale de l'ordre $2p^3(q^3 - q^2)$. Quelles sont la nature et les propriétés de cette surface résiduale? Je serais bien aise de savoir si vous avez fait des recherches à ce sujet.

Cambridge, 29 *Mars* 1866.

434.

ON CERTAIN SKEW SURFACES, OTHERWISE SCROLLS.

[From the *Transactions of the Cambridge Philosophical Society*, vol. XI. Part II. (1869), pp. 277—289. Read Nov. 11, 1867.]

THE investigations contained in the present Memoir were suggested to me by the Memoirs of M. De la Gournerie, presented by him to the Academy of Sciences in 1865 and 1866, published in extract in the *Comptes Rendus*, and reproduced in his work "*Recherches sur les surfaces réglées tétraédrales symétriques*, par Jules De la Gournerie, avec des notes par Arthur Cayley," 8vo. Paris, 1867. Although the results or the greater part of them, agree with those in the work just referred to, the mode of treatment is different, and more general, the orders, &c. of the different scrolls being obtained by considerations founded on the theory of Correspondence, and I have thought it not improper to submit to geometers in this altered form the theory of the very interesting class of Scrolls for which they are indebted to M. De la Gournerie's researches.

Article Nos. 1 to 10. *Geometrical Construction of a Class of Scrolls.*

1. Consider any two curves (plane or of double curvature) U, U', of the orders m, m' respectively, and let the points of U have with those of U' an (α, α') correspondence; viz. let the points of the two curves be so related that to each point of U correspond α' points of U', and to each point of U' correspond α points of U: then the lines joining the corresponding points of U, U' form a scroll the order of which is $= m\alpha' + m'\alpha$.

2. In particular let U, U' be plane curves in the planes Π, Π' respectively; and let the correspondence between the points of the two curves be established as follows; viz. consider in the plane Π a curve Ω of the class μ, and in the plane Π' a curve Ω' of the class μ'; and (to avoid useless generality) let the tangents of these two curves Ω, Ω' have to each other a (1, 1) correspondence; that is, to each tangent of

Ω there corresponds a single tangent of Ω', and to each tangent of Ω' a single tangent of Ω (this assumes that the curves Ω, Ω' are rational transformations one of the other, and that they have consequently the same Deficiency). This being so, let the points of U which lie on any tangent of Ω and the points of U' which lie on the corresponding tangent of Ω' be taken to be corresponding points of U, U'. The correspondence is then $(\mu m', \mu' m)$: in fact through a given point of U there pass μ tangents of Ω, and the corresponding μ tangents of Ω' meet U' in $\mu m'$ points, that is, to a given point of U correspond $\mu m'$ points of U'; and similarly to a given point of U' correspond $\mu' m$ points of U. And hence the order of the scroll formed by the lines joining the corresponding points of U, U' is $=(\mu+\mu')\,mm'$.

3. This conclusion may be otherwise established as follows; let K, K' be any two corresponding points of U, U', so that the scroll we are concerned with is that generated by the series of lines KK'; and let I denote the line of intersection of the planes Π, Π'. The line I meets the curve U in m points, and taking one of these points for a point K we may from this point draw μ tangents to the curve Ω, that is, the point in question is a point K in respect of μ different tangents of the curve Ω; to each of these tangents there corresponds a single tangent of Ω', and such tangent of Ω' meets the curve U' in m' points, that is, to the point K in question there correspond $\mu m'$ points K' and consequently $\mu m'$ lines KK' in the plane Π'; hence to each of the m points K on the line I there correspond $\mu m'$ lines KK' in the plane Π'; and we have thus $\mu mm'$ generating lines in the plane Π'; there are in like manner $\mu' mm'$ generating lines in the plane Π.

Take K an arbitrary point on the curve U; there are $\mu m'$ corresponding points K', and consequently $\mu m'$ generating lines through K, that is, through each point of the curve U; or the curve U (which is of the order m) is a $\mu m'$-tuple line on the scroll; similarly the curve U' (which is of the order m') is a $\mu' m$-tuple line on the scroll.

The complete section of the scroll by the plane Π consists of the curve U taken $\mu m'$ times (order $\mu mm'$) and of the $\mu' mm'$ generating lines in the plane Π; that is, the order of the section is $=(\mu+\mu')\,mm'$; and we thus see that the order of the scroll is $=(\mu+\mu')\,mm'$. Of course in like manner the complete section of the scroll by the plane Π' consists of the curve U' taken $\mu' m$ times (order $\mu' mm'$) and of the $\mu mm'$ generating lines in the plane Π', the order of the section being thus $=(\mu+\mu')mm'$.

4. There are on the scroll certain singular tangent planes; viz. if we have two corresponding tangents of Ω, Ω' meeting the line I in the same point, then we have m points K and m' points K' all lying in the plane of the two tangents; and of course the mm' lines KK' will all lie in the plane of the two tangents; that is, the intersection of the scroll by the plane in question will be made up of the mm' lines, and of a curve of the order $(\mu+\mu'-1)\,mm'$; and the plane in question is thus a singular tangent plane.

5. The number of these singular tangent planes is $=\mu+\mu'$; in fact considering as corresponding points on the line I, the intersection of this line by any tangent of Ω and the intersection by the corresponding tangent of Ω', the correspondence is

obviously (μ, μ'); viz. through a given point P considered as belonging to the first system there pass μ tangents of Ω, and corresponding thereto we have μ tangents of Ω' each intersecting I in a single point P'; that is, to a given point P correspond μ points P'; and similarly to a given point P' correspond μ' points P'. And this being so, the number of united points, that is, points of I through which pass corresponding tangents of Ω, Ω', is $=\mu+\mu'$.

6. In particular the curves Ω, Ω' may reduce themselves each to a point: the tangents to the two curves are here the lines passing through the points Ω, Ω' respectively: and the condition for the (1, 1) correspondence of the two tangents is that the pencils of lines shall be homographically related; or, what is the same thing, that these two pencils shall determine on the line I two ranges which are homographically related; the entire construction is then as follows:

Given in the plane Π a curve U and a point Ω, and in the plane Π' a curve U' and a point Ω'; and taking in the plane Π a pencil of lines through Ω, and in the plane Π' a pencil of lines through Ω', in such wise that the two pencils correspond homographically; then if a line of the first pencil meets the curve U in the m points K, and the corresponding line of the second pencil meets the curve U' in the m' points K', the scroll in question is that generated by the mm' lines KK'.

7. By what precedes, the scroll is of the order $2mm'$; the curve U is a m'-tuple line, and the complete section by the plane Π is made up of this curve taken m' times and of mm' generating lines; similarly the curve U' is a m-tuple line, and the complete section by the plane Π' is made up of this curve taken m times and of mm' generating lines; there are two singular tangent planes such that the section by each of them is made up of mm' generating lines and of a curve of the order mm'; the planes in question are obviously those through the lines $\Omega\Omega'$ and the coincident points of the two ranges on the line I, say the points A, B respectively.

8. The foregoing results will be modified in special cases. Suppose, for instance, that the curve U passes ω times, α times, and β times through the points Ω, A, B, respectively, and that the curve U' passes ω' times, α' times, and β' times through the points Ω', A, B respectively. Then to each point on the curve U there correspond the $m'-\omega'$ intersections (other than the point Ω') on a line through Ω', so that U' is a $(m'-\omega')$tuple line on the surface. The curve U' meets the line I in m' points and corresponding to each of them we have a line through Ω meeting the curve U in $(m-\omega)$ points, exclusive of the point Ω; this would give $m'(m-\omega)$ generating lines in the plane Π; but among the m' points are included the point $A\alpha'$ times, and the point $B\beta'$ times; the $(m-\omega)$ points corresponding to A include the point $A\alpha$ times, and we have thus the point A corresponding to itself $\alpha\alpha'$ times, and giving a reduction $=\alpha\alpha'$ in the number $m'(m-\omega)$ of generating lines: similarly the $m-\omega$ points corresponding to B include the point $B\beta$ times, and we have thus the point B corresponding to itself $\beta\beta'$ times and giving a reduction $=\beta\beta'$ in the number $m'(m-\omega)$ of generating lines; the number of generating lines in the plane Π is thus $=m'(m-\omega)-\alpha\alpha'-\beta\beta'$. The complete section by the plane Π is made up of the

curve U $(m-\omega')$ times (order $m(m'-\omega')$) and of the $m'(m-\omega)-\alpha\alpha'-\beta\beta'$ generating lines; the order of the section, and consequently also the order of the scroll, is thus $=2mm'-m\omega'-m'\omega-\alpha\alpha'-\beta\beta'$. It is clear that in like manner the curve U' is a $(m-\omega)$tuple line on the surface, and that the complete section by the plane Π' is made up of this curve taken $(m-\omega)$ times, order $m'(m-\omega)$, and of $m(m'-\omega')-\alpha\alpha'-\beta\beta'$ generating lines.

9. The section by the tangent plane through A is made up of $(m-\omega)(m'-\omega')-\alpha\alpha'$ generating lines (viz. these are, the line ΩA $\alpha'(m-\omega-\alpha)$ times, the line $\Omega' A$ $\alpha(m'-\omega'-\alpha')$ times, and $(m-\omega-\alpha)(m'-\omega'-\alpha')$ other generating lines) and of a curve of the order $mm'-\omega\omega'-\beta\beta'$: similarly the section by the tangent plane through B is made up of $(m-\omega)(m'-\omega')-\beta\beta'$ generating lines (viz. these are, the line ΩB $\beta'(m-\omega-\beta)$ times, the line $\Omega' B$ $\beta(m'-\omega'-\beta')$ times, and $(m-\omega-\beta)(m'-\omega'-\beta')$ other generating lines), and of a curve of the order $mm'-\omega\omega'-\alpha\alpha'$.

10. A very interesting case is when (m, m' being each even) we have

$$\omega=\alpha=\beta=\frac{1}{2m}, \quad \omega'=\alpha'=\beta'=\tfrac{1}{2}m'.$$

Here the curve U is a $\frac{1}{2}m'$-tuple line on the scroll, and the complete section by the plane Π is this curve taken $\frac{1}{2}m'$ times; the order of the section, and therefore of the scroll is thus $=\frac{1}{2}mm'$; of course in like manner the curve U' is a $\frac{1}{2}m$-tuple line on the scroll, and the complete section by this plane is the curve U' taken $\frac{1}{2}m$ times: the section by each of the planes $\Omega\Omega' A$, $\Omega\Omega' B$ is a curve of the order $\frac{1}{2}mm'$, the planes in question being in the present case no longer singular tangent planes, or even tangent planes at all, of the scroll.

Article Nos. 11 to 14. *Analytical Theory.*

11. It will be convenient to denote by D, C respectively the points heretofore called Ω, Ω' respectively: this being so, we have a tetrahedron $ABCD$, of which the faces ABD, ABC are the planes heretofore called Π, Π' respectively, and the other two faces CDA, CDB are the singular tangent planes $\Omega\Omega' A$, $\Omega\Omega' B$ respectively. And then, taking

$$x=0, \quad y=0, \quad z=0, \quad w=0$$

for the equations of the faces BCD, CDA, DAB, ABC of the tetrahedron, we may write for the equations of the curve U, $z=0$, $f_3(x, y, w)=0$, for those of the curve U', $w=0$, $f_4(x, y, z)=0$; and take the homographic ranges on the line I $(z=0,\ w=0)$ to be given as the intersections of this line with the pencils of planes $x-\theta y$, $x-k\theta y=0$ respectively (θ a variable parameter, k a constant). The points K are therefore given by

$$x-\theta y=0, \quad z=0, \quad f_3(x, y, w)=0,$$

the points K' by

$$x-k\theta y=0, \quad w=0, \quad f_4(x, y, z)=0;$$

and then the lines KK' belonging to the different values of the parameter θ generate the scroll.

12. Or, what is the same thing, taking (X, Y, Z, W) as the coordinates of K, (X', Y', Z', W') as the coordinates of K', we have

$$X - \theta Y = 0, \quad Z = 0, \quad f_3(X, Y, W) = 0,$$
$$X' - k\theta Y = 0, \quad W' = 0, \quad f_4(X', Y', Z') = 0,$$

and then the equations of the line KK' are

$$\left\| \begin{matrix} x, & y, & z, & w \\ X, & Y, & 0, & W \\ X', & Y', & Z', & 0 \end{matrix} \right\| = 0:$$

or, as these may be written,

$$\begin{array}{rrrr} . & -WZ'y + & WY'z + & YZ'w = 0, \\ WZ'x + & . & -WX'z - & XZ'w = 0, \\ -W'Yx + & WX'y & . & +(XY' - X'Y)w = 0, \\ -YZ'x + & XZ'y & -(XY' - X'Y)z & . \quad = 0, \end{array}$$

equivalent of course to two equations. The elimination of $X, Y, W, X', Y', Z', \theta$ from all the equations gives the equation of the scroll.

13. Substituting the values $X = \theta Y$, $X' = k\theta Y'$, we have

$$f_3(\theta Y, Y, W) = 0, \quad f_4(k\theta Y', Y', Z') = 0,$$

$$\begin{array}{rrrr} . & -WZ'y + & WY'z + & YZ'w = 0, \\ WZ'x & . & -k\theta WY'z - & \theta YZ'w = 0, \\ -WY'x + & k\theta WY'y & . & +\theta(1-k)YY'w = 0, \\ -YZ'x + & \theta YZ'y & -\theta(1-k)YY'z & . \quad = 0; \end{array}$$

or, what is the same thing, writing $\frac{W}{Y} = \omega$, and $\frac{Z'}{Y'} = \zeta$, we have

$$f_3(\theta, 1, \omega) = 0, \quad f_4(k\theta, 1, \zeta) = 0,$$

$$\begin{array}{rrrr} . & -\omega\zeta y + & \omega z + & \zeta w = 0, \\ \omega\zeta x & . & -k\theta\omega z - & \theta\zeta w = 0, \\ -\omega x + & k\theta\omega y & . & +\theta(1-k)w = 0, \\ -\zeta x + & \theta\zeta y - & \theta(1-k)z & . \quad = 0. \end{array}$$

Recollecting that the last four equations are equivalent to two equations only, and substituting for ω, ζ their values in terms of θ, we have in effect two equations, which by the elimination of θ lead to a relation in (x, y, z, w), the equation of the scroll.

14. We may find the sections of the scroll by the planes $x = 0$, $y = 0$ respectively. Writing first $x = 0$, we have

$$\omega y = \frac{k-1}{k} w, \quad \zeta y = -(k-1) z.$$

Hence taking the other two equations in the form

$$f_3(\theta y,\ y,\ \omega y)=0,\quad f_4\left(\theta y,\ \frac{y}{k},\ \frac{\zeta y}{k}\right)=0,$$

and putting $\theta y=u$, we have

$$f_3\left(u,\ y,\ \frac{(k-1)\,w}{k}\right)=0,\quad f_4\left(u,\ \frac{y}{k},\ \frac{-(k-1)\,z}{k}\right)=0,$$

from which eliminating u we obtain an equation $f_1(y,\ z,\ w)=0$, the equation of the section by the plane $x=0$.

Similarly, writing $y=0$, we have

$$\frac{x\omega}{\theta}=-(k-1)\,w,\quad \frac{x\zeta}{\theta}=(k-1)\,z,$$

whence taking the equations in the form

$$f_3\left(x,\ \frac{x}{\theta},\ \frac{x\omega}{\theta}\right)=0,\quad f_4\left(kx,\ \frac{x}{\theta},\ \frac{x\zeta}{\theta}\right)=0,$$

and writing $\frac{x}{\theta}=v$, we have

$$f_3\left(x,\ v,\ -(k-1)\,w\right)=0,\quad f_4\left(kx,\ v,\ (k-1)\,z\right)=0,$$

from which, eliminating v, we obtain an equation $f_2(x,\ z,\ w)=0$, the equation of the section by the plane $y=0$.

Article Nos. 15 to 29. *The Curves U, U' are henceforward "triangular" curves.*

15. Let $r=\pm\frac{p}{q}$, where p, q are positive integers prime to each other, and let the given sections be

$$\begin{aligned} z&=0,\quad A\,x^r+B\,y^r\quad .\quad+Dw^r=0,\\ w&=0,\quad A'x^r+B'y^r+C'z^r\quad .\quad=0,\end{aligned}$$

where it is to be observed that r being $=+\frac{p}{q}$, the two given sections are of the order pq, the order of the scroll is $=2p^2q^2$, each of the given sections is a pq-tuple line on the scroll, and the plane thereof meets the scroll in the section taken pq times, and in the pq generating lines: but r being $=-\frac{p}{q}$, the two given sections are each of the order $2pq$, with three pq-tuple points ($\omega=\alpha=\beta=pq$, $\omega'=\alpha'=\beta'=pq$), and thence the order of the scroll is $\frac{1}{2}(2pq)^2$, $=2p^2q^2$; each of the sections is a pq-tuple line on the scroll, and the plane meets the scroll only in the section taken pq times. But in either case, if q be >1, that is, if r be fractional, it will presently appear that the scroll of the order $2p^2q^2$ breaks up into q scrolls each of the order $2p^2q$.

16. To find the section by the plane $x=0$, we have

$$A\,u^r + B\quad y^r \qquad . \qquad + D\left(\frac{k-1}{k}\,w\right)^r = 0,$$

$$A'u^r + B'\quad \left(\frac{y}{k}\right)^r + C'\quad \left(-\frac{k-1}{k}\,z\right)^r \qquad . \qquad = 0,$$

and eliminating u we obtain

$$\left(AB'\frac{1}{k^r} - A'B\right)y^r + AC'\left(-\frac{k-1}{k}\right)^r z^r - A'D\left(\frac{k-1}{k}\right)^r w^r = 0:$$

writing $\left(\frac{k-1}{k}\right)^r = (-)^r\left(\frac{1-k}{k}\right)^r$, this is

$$\left(AB'\frac{1}{k^r} - A'B\right)y^r + AC'\quad\left(\frac{1-k}{k}\right)^r z^r - (-)^r\left(\frac{1-k}{k}\right)^r w^r = 0,$$

or, what is the same thing, it is

$$-(-)^r\,\frac{AB' - A'Bk^r}{(1-k)^r}\,y^r - (-)^{-r}AC'\qquad z^r + \qquad A'Dw^r = 0.$$

And in regard to this and the other equations which contain $(-)^{-r}$, it is to be observed that r being integral we have $(-)^{-r} = (-)^r$, and that r being fractional, every value of $(-)^{-r}$ is also a value of $(-)^r$; so that we may in every case write $(-)^r$ in place of $(-)^{-r}$.

Similarly for the section by the plane $y=0$, we have

$$Ax^r + Bv^r \qquad . \qquad + D\left(-(k-1)\ w\right)^r = 0,$$

$$A'(kx)^r + B'v^r + C\left((k-1)\,z\right)^r \qquad . \qquad = 0,$$

and eliminating v, we have

$$(AB' - A'Bk^r)\,x^r \qquad - BC'(k-1)^r z^r + B'D\left(-(k-1)\right)^r w^r = 0\,;$$

or, what is the same thing,

$$\frac{AB' - A'Bk^r}{(1-k)^r}\,x^r \qquad - (-)^r\,BC'z^r + B'D \qquad w^r = 0.$$

17. The four sections thus are

$$x=0,\qquad . \qquad -(-)^r\,\frac{AB' - A'Bk^r}{(1-k)^r}\,y^r - (-)^r\,AC'z^r + A'Dw^r = 0,$$

$$y=0,\quad \frac{AB' - A'Bk^r}{(1-k)^r}\,x^r \qquad . \qquad - (-)^r\,BC'z^r + B'Dw^r = 0,$$

$$z=0,\qquad Ax^r \qquad + By^r \qquad . \qquad + Dw^r = 0,$$

$$w=0,\qquad A'x^r \qquad + B'y^r + \qquad C'z^r \qquad . \qquad = 0.$$

It will be convenient to speak of these four curves as *directrices* of the scroll.

18. Suppose for a moment that r is integral; as either of the given equations may be multiplied by a constant, we may assume that $D = -(-)^r C'$; substituting this value and dividing the first and second equations each by C', we have

$$\begin{array}{lllll} x=0, & . & -(-)^r \dfrac{AB'-A'Bk^r}{(1-k)^r C'} y^r & -(-)^r Az^r & -(-)^r A'w^r = 0, \\ y=0, & \dfrac{AB'-A'Bk^r}{(1-k)^r C'} x^r & . & -(-)^r Bz^r & -(-)^r B'w^r = 0, \\ z=0, & A x^r + & By^r & . & -(-)^r C'w^r = 0, \\ w=0, & A'x^r + & B'y^r + & C'z^r & . \quad = 0, \end{array}$$

so that the diagonally opposite coefficients differ only by the factor $-(-)^r$; viz. the matrix is symmetrical or skew symmetrical according as r is odd or even.

19. If r be fractional, it is to be observed that, although the three symbols $(-)^r$ and the two symbols $(1-k)^r$ which enter into the first and second equations of No. 17, do not in the first instance represent of necessity the same values of $(-)^r$ and $(1-k)^r$ respectively, yet there is no loss of generality in assuming that they do so—the irrational equations are mere symbols for the rational equations to which they respectively give rise—and the irrationalities $(-)^r$ and $(1-k)^r$ will on the rationalisation of the equations disappear along with the irrationalities x^r, y^r, z^r, to which they are attached. But the case is otherwise with the irrationality k^r involved in the expression $AB'-A'Bk^r$; writing as before $r = \pm\dfrac{p}{q}$ (p and q positive integers prime to each other), the symbol k^r has q different values; and there is not in the first instance any relation between the k^r of the first equation and the k^r of the second equation: for each of these equations the rationalised equation (that is, the equation rationalised in regard to the coordinates) will contain the irrationality k^r, and will thus for each of the q values of k^r represent a distinct curve. The given equations (viz. the first and second equations) represent each of them a single curve of the order pq or $2pq$, according as r is positive or negative; the first and second equations represent each of them q such curves.

20. Hence, starting from the two given curves in the planes $z=0$ and $w=0$, respectively, and with a given value of k, the section of the scroll by the plane $y=0$ is made up of q curves, viz. the curves obtained from the second equation of No. 17, by assigning to the radical k^r each of its q different values; the scroll consequently breaks up into q different scrolls, viz. the lines passing through the two given curves, and any one of the q curves in the plane $y=0$, constitute a distinct scroll. The lines in question meet the plane $x=0$, not indifferently in any one of the q curves in that plane, but in a certain one of these curves, viz. in that curve for which the radical k^r has the same value as for the curve in question in the plane $y=0$. Hence we may in the first and second equations regard the radicals k^r as having the same meaning, and the system of four equations in effect breaks up into q systems, viz. the systems obtained by giving to the radical k^r its q different values; each of these systems gives a scroll, and the scroll derived from the two given curves with a given

value of k is made up of these q scrolls. And hence, attaching a unique value to each of the symbols $(-)^r$, $(1-k)^r$, and k^r, we may, as before, write $D=(-)^r C'$, and so reduce the original equations as in the case r integral, to the form No. 18, in which the diagonally opposite coefficients differ only by the factor $-(-)^r$.

21. Let the two given equations be taken to be

$$\begin{array}{llll} z=0, & bx^r-(-)^r ay^r & . & +hw^r=0, \\ w=0, & -(-)^r fx^r-(-)^r gy^r-(-)^r hz^r & . & =0; \end{array}$$

we have then

$$\frac{AB'-A'Bk^r}{(1-k)^r C'}=\frac{(-)^r bg+afk^r}{(-)^r h\,(1-k)^r},$$

or, putting this $=-(-)^r c$, that is, writing

$$afk^r+bg\,(-1)^r+ch\,(1-k)^r=0,$$

the four equations become

$$\begin{array}{lllll} x=0, & . & cy^r-(-)^r bz^r+fw^r=0, \\ y=0, & -(-)^r cx^r & . \quad + \quad az^r+gw^r=0, \\ z=0, & bx^r-(-)^r ay^r & . \quad +hw^r=0, \\ w=0, & -(-)^r fx^r-(-)^r gy^r-(-)^r hz^r & . \quad =0; \end{array}$$

where c being considered as given, k is determined as mentioned above, or, what is the same thing, $k:-1:1-k=\lambda:\mu:\nu$, we have $\lambda:\mu:\nu$, and thence k, determined by the equations

$$\lambda+\ \mu+\ \nu=0,$$
$$af\lambda^r+bg\mu^r+ch\nu^r=0.$$

22. Consider for a moment λ, μ, ν, as the coordinates of a point in a plane, then $\left(r=\pm\frac{p}{q}\text{ as before}\right)$, the equation $af\lambda^r+bg\mu^r+ch\nu^r=0$, is that of a curve of the order pq or $2pq$, according as r is positive or negative: and this curve is met by the line $\lambda+\mu+\nu=0$, in pq or $2pq$ points, that is, k has this number pq or $2pq$, of values: but to each of these values of k there corresponds (not q values but) only a single value of k^r, viz. that value for which $afk^r+bg\,(-1)^r+ch\,(1-k)^r=0$; that is, starting from the two directrices in the planes $z=0$, $w=0$, respectively, and a given third directrix in the plane $y=0$ (or in the plane $x=0$), we may by means of each of the pq or $2pq$ values of k construct a scroll passing through the three directrices, and which will also pass through the fourth directrix in the plane $x=0$ (or in the plane $y=0$), but such scroll is only one (not each) of the q scrolls which can be constructed from the two given sections in the planes $z=0$, $w=0$, respectively, and from the assumed value of k. It has been mentioned that whether r is $=+\frac{p}{q}$, or $=-\frac{p}{q}$, the total scroll constructed from the two given directrices in the planes $z=0$, $w=0$, and from a given value of k is of the order

$2p^2q^2$, and that such scroll breaks up into q distinct scrolls, hence the order of each of the distinct scrolls is $=2p^2q$. Whence, starting with the given directrices in the planes $z=0$, $w=0$, and a given third directrix in the plane $y=0$ (or in the plane $x=0$), we have pq or $2pq$ scrolls each of the order $2p^2q$, and passing through these three directrices, and through the given fourth directrix in the plane $x=0$ (or in the plane $y=0$).

23. It is to be observed that when q is >1, then considering the three directrices as given, the pq or $2pq$ scrolls each of the order $2p^2q$, do not make up the total scroll generated by the lines which pass through the three given directrices. I call to mind that for three given directrices the orders of which are m, n, p, respectively, and which meet, the second and third, the third and first, and the first and second, in α points, β points, and γ points respectively, the order of the scroll generated by the lines which meet the three directrices is $=2mnp-\alpha m-\beta n-\gamma p$. Suppose first, that $r=+\frac{p}{q}$, then the directrices are each of the order pq, and they do not any two of them meet; the order of the scroll is $=2p^3q^3$. Suppose secondly, $r=-\frac{p}{q}$, then the directrices are each of the order $2pq$, but each two of them have in common two pq-tuple points counting as $2p^2q^2$ intersections; the order of the scroll is thus $(16-3\,.\,4)\,p^3q^3=4p^3q^3$. In the first case the lines which meet the three directrices generate a residuary scroll of the order $2p^3(q^3-q^2)$, and the pq scrolls each of the order $2p^2q$; in the second case they generate a residuary scroll of the order $4p^3(q^3-q^2)$, and the $2pq$ scrolls each of the order $2pq$.

24. In the case $r=+\frac{p}{q}$, by way of illustration of the origin of the pq scrolls each of the order $2p^2q$, I consider the particular case $p=1$, that is, $r=\frac{1}{q}$, the reciprocal of a positive integer q, and where it is to be shown that we have q scrolls each of the order $2q$. The given directrices are here

$$z=0,\quad Ax^{\frac{1}{q}}+By^{\frac{1}{q}}\quad .\ +Dw^{\frac{1}{q}}=0,$$

$$w=0,\quad A'x^{\frac{1}{q}}+B'y^{\frac{1}{q}}+C'z^{\frac{1}{q}}\quad .\ =0,$$

each of them a unicursal curve; we may in fact satisfy the two equations respectively, by writing in the first of them

$$x:y:w=a(\phi+\alpha)^q : b(\phi+\beta)^q : d(\phi+\delta)^q;$$

and in the second

$$x:y:z=a'(\phi'+\alpha')^q : b'(\phi'+\beta')^q : c'(\phi'+\gamma')^q,$$

where a, b, d, α, β, δ, a', b', c', α', β', γ' are properly determined constants, ϕ, ϕ' are variable parameters. It follows that, considering the points K, K' which are the intersections of the first curve by the line $x-\theta y=0$, and of the second curve by the corresponding line $x-k\theta y=0$, we have not only a correspondence of q points K with

k points K', but we may establish, and that in q different manners, a correspondence between single points K and K'. For, substituting the foregoing values of $x : y$ in the equations $x - \theta y = 0$ and $x - k\theta y = 0$ respectively, we have

$$\theta = \frac{a}{b}\frac{(\phi+\alpha)^q}{(\phi+\beta)^q},\quad k\theta = \frac{a'}{b'}\frac{(\phi'+\alpha')^q}{(\phi'+\beta')^q},$$

and thence

$$\frac{a}{b}\frac{(\phi+\alpha)^q}{(\phi+\beta)^q} = \frac{1}{k}\frac{a'}{b'}\frac{(\phi'+\alpha')^q}{(\phi'+\beta')^q};$$

so that, extracting the qth root of each side, we have, in q different ways corresponding to the q values of the radical $\left(\frac{1}{k}\frac{ba'}{b'a}\right)^{\frac{1}{q}}$, a relation of the form $\phi' = \frac{l\phi + m}{n\phi + r}$; and considering ϕ' as having this value, the points K, K' as given by the equations

$$z = 0,\quad x : y : w = a(\phi+\alpha)^q : b(\phi+\beta)^q : d(\phi+\delta)^q,$$

and

$$w = 0,\quad x : y : z = a'(\phi'+\alpha')^q : b'(\phi'+\beta')^q : c'(\phi'+\gamma')^q,$$

respectively, correspond as single points to each other. We have thus in q different ways a series of corresponding points K, K', and consequently q series of lines KK' each of them generating a scroll which (as the order of the scroll generated by all the q series is $= 2q^2$), must be each of them of the order $2q$; and the decomposition in question is thus explained.

25. In the scroll of the order $2q^2$, each directrix is a q-tuple line, and the complete section by the plane of the directrix is made up of the directrix q times (order q^2), and of q^2 generating lines, in fact, of q q-fold generating lines: to show that this is so, consider the directrix in the plane $z = 0$, viz. the equation of this is $Ax^{\frac{1}{q}} + By^{\frac{1}{q}} + Dw^{\frac{1}{q}} = 0$. Writing herein $w = 0$, we have $Ax^{\frac{1}{q}} + By^{\frac{1}{q}} = 0$, that is, $A^q x - (-)^q B^q y = 0$; it is clear that the rationalised equation must reduce itself to $\{A^q x - (-)^q B^q y\}^q = 0$, and that the line $w = 0$, is thus a tangent of q-pointic intersection at the point $w = 0$, $A^q x - (-)^q B^q y = 0$. Taking K at this point we have, in each of the scrolls of the order $2q$, q coincident positions of K', that is, a q-fold line KK' in the plane $w = 0$; and the like for the plane $z = 0$, so that the total section by the plane $z = 0$ is made up of the directrix q times and of q q-fold generating lines; and it follows that for each of the scrolls of the order $2q$, the section by the plane $z = 0$ is made up of the directrix once, and of a q-fold generating line.

26. It is easy to see that in the general case $r = +\frac{p}{q}$, the like conclusion holds; for the scroll of the order $2p^2q^2$, the section by the plane of the directrix consists of the directrix pq times (order p^2q^2), and of p^2q q-fold generating lines; whence for each of the q component scrolls of the order $2p^2q$, the section is made up of the directrix p times (that is, the directrix is a p-tuple line on the scroll) and of p^2 q-fold generating lines.

27. In the case $r = -\frac{p}{q}$, where the order of the directrix is $= 2pq$, then in the scroll of the order $2p^2q^2$, the directrix is a pq-tuple line on the scroll, and taken pq times it constitutes the complete section by the plane of the directrix; whence in each of the q component scrolls of the order $2p^2q$, the directrix is a p-tuple line; and taken p times it constitutes the complete section by the plane of the directrix.

28. It is convenient to exhibit the foregoing results in a tabular form as follows:

$r = +\frac{p}{q}$.	$r = -\frac{p}{q}$.
Each directrix is of order pq.	Each directrix is of order $2pq$, with three pq-tuple points.

Scroll belonging to two directrices, and a given value of k, is of the order

$2p^2q^2$,	$2p^2q^2$,
breaking up into q scrolls each of order $2p^2q$, each which scroll of the order $2p^2q$ has each directrix for a p-tuple line and has besides p^2 q-fold generating lines in the plane of the directrix.	breaking up into q scrolls each of the order $2p^2q$, each which scroll of the order $2p^2q$ has each directrix for a p-tuple line, and consequently no generating line in the plane of the directrix.

Considering two directrices and a given third directrix,

k has pq values.	k has $2pq$ values.

Total scroll for the three directrices is made up of

pq scrolls each of order $2p^2q$ (viz. one for each value of k), and residuary scroll of order $2p^3(q^3 - q^2)$.	$2pq$ scrolls each of order $2p^2q$ (viz. one for each value of k), and residuary scroll of order $4p^3(q^3 - q^2)$.

29. The following are noticeable cases; $r = 1$ gives the hyperboloid as derived from three directrix lines; $r = -1$ the hyperboloid as derived from three plane sections thereof; $r = 2$, an octic surface, M. De la Gournerie's Quadrispinal; $r = -2$, an octic surface, his Quadricuspidal; $r = \frac{1}{3}$, a sextic surface which (as remarked by Dr Salmon), on writing therein (x^2, y^2, z^2, w^2), in place of (x, y, z, w), is converted into a surface of the twelfth order, locus of the centres of curvature of an ellipsoid.

435.

ON THE SIX COORDINATES OF A LINE.

[From the *Transactions of the Cambridge Philosophical Society*, vol. XI. Part II. (1869), pp. 290—323. Read November 11, 1867.]

THE notion of the six coordinates of a line was, so far as I am aware, first established in my paper "On a new analytical representation of Curves in Space," *Quart. Math. Jour.* t. III. (1860), pp. 225—236, [284]; see p. 226, where writing p, q, r, s, t, u for the six determinants of the matrix $\begin{Bmatrix} x, & y, & z, & w \\ \alpha, & \beta, & \gamma, & \delta \end{Bmatrix}$, I remark that these values give identically $ps+qt+ru=0$; and I consider a cone as represented by a homogeneous equation $V=0$ between the six coordinates (p, q, r, s, t, u); and many of the investigations of the present memoir, in which these coordinates are employed, have been in my possession for some years past. But these coordinates presented themselves independently to Prof. Plücker, and the theory of them is set forth in his most interesting and valuable memoir, "On a new Geometry of Space," *Phil. Trans.* t. CLV. (1865), pp. 725—791; the course of development there given to the theory is however altogether different from that in the present memoir. They have also more recently been made use of in a paper by Herr Lüroth, "Zur Theorie der windschiefen Flächen," *Crelle*, t. LXII. (1867), pp. 130—152.

I have in the present memoir applied these coordinates to the question of the Involution of six lines; the notion of this relation of six lines is due to Prof. Sylvester, to whom it presented itself in the year 1861, in connexion with a theorem in the *Lehrbuch der Statik*, by Möbius (Leipzig, 1837), that if four forces acting on a solid body are in equilibrium the lines along which the forces act are the generating lines of a hyperboloid. Prof. Sylvester was thereby led to consider six lines such that (regarding them as lines in a solid body) there exist along them forces which are in equilibrium; and he thence obtained, by the statical considerations reproduced in the present memoir, the construction (when five of the lines are given) of a sixth line to pass through a given point or to be situate in a given plane.

Article, Nos. 1 to 8. *The Six Coordinates of a Line; definition and general notions.*

1. Using any quadriplanar coordinates (x, y, z, w) whatever, consider a line; on the line two points the coordinates of which are $(\alpha, \beta, \gamma, \delta)$ and $(\alpha', \beta', \gamma', \delta')$ respectively; and through the line two planes, the equations whereof are $(A, B, C, D \between x, y, z, w) = 0$, and $(A', B', C', D' \between x, y, z, w) = 0$ respectively; we have

$$\begin{aligned}(A, B, C, D \between \alpha, \beta, \gamma, \delta) &= 0,\\ (A, B, C, D \between \alpha', \beta', \gamma', \delta') &= 0,\\ (A', B', C', D' \between \alpha, \beta, \gamma, \delta) &= 0,\\ (A', B', C', D' \between \alpha', \beta', \gamma', \delta') &= 0.\end{aligned}$$

2. From the first and second equations, eliminating successively A, B, C, D, we find

$$\left|\begin{array}{cccc} 0, & \alpha\beta' - \alpha'\beta, & -(\gamma\alpha' - \gamma'\alpha), & \alpha\delta' - \alpha'\delta \\ -(\alpha\beta' - \alpha'\beta), & 0, & \beta\gamma' - \beta'\gamma, & \beta\delta' - \beta'\delta \\ \gamma\alpha' - \gamma'\alpha, & -(\beta\gamma' - \beta'\gamma), & 0, & \gamma\delta' - \gamma'\delta \\ -(\alpha\delta' - \alpha'\delta), & -(\beta\delta' - \beta'\delta), & -(\gamma\delta' - \gamma'\delta), & 0 \end{array}\right| (A, B, C, D) = 0,$$

and from the third and fourth equations we find the like system with (A', B', C', D') in place of (A, B, C, D). Comparing the corresponding equations of the two systems, we find an equality of ratios, as will presently be mentioned.

3. From the first and third equations, eliminating successively $\alpha, \beta, \gamma, \delta$, we find

$$\left|\begin{array}{cccc} 0, & AB' - A'B, & -(CA' - C'A), & AD' - A'D \\ -(AB' - A'B), & 0, & BC' - B'C, & BD' - B'D \\ CA' - C'A, & -(BC' - B'C), & 0, & CD' - C'D \\ -(AD' - A'D), & -(BD' - B'D), & -(CD' - C'D), & 0 \end{array}\right| (\alpha, \beta, \gamma, \delta) = 0,$$

and from the second and fourth equations we find the like system with $(\alpha', \beta', \gamma', \delta')$ in place of $(\alpha, \beta, \gamma, \delta)$: comparing the corresponding equations of the two systems, we find the same equality of ratios as before, viz.

4. This is

$$\begin{aligned}&\beta\gamma' - \beta'\gamma : \gamma\alpha' - \gamma'\alpha : \alpha\beta' - \alpha'\beta : \alpha\delta' - \alpha'\delta : \beta\delta' - \beta'\delta : \gamma\delta' - \gamma'\delta\\ =\ &AD' - A'D : BD' - B'D : CD' - C'D : BC' - B'C : CA' - C'A : AB' - A'B,\end{aligned}$$

and putting each of these two equal sets of ratios

$$= a : b : c : f : g : h,$$

then the quantities (a, b, c, f, g, h), which it is easy to see satisfy the condition

$$af + bg + ch = 0,$$

are said to be the 'six coordinates' of the line: as only the ratios of the six quantities are material, and as the last-mentioned equation establishes a single relation between these ratios, the system of the six coordinates contain four arbitrary ratios or parameters, for the determination of the particular line.

5. A line is thus determined by its six coordinates (a, b, c, f, g, h), which are such that $af + bg + ch = 0$; and conversely any six quantities (a, b, c, f, g, h) satisfying this relation may be taken to be the six coordinates of a line.

6. It is proper to show that the ratios $a : b : c : f : g : h$ are independent of the particular two points on the line, or two planes through the line, used for their determination. In fact, if instead of the points

$$\begin{array}{cccc} \alpha, & \beta, & \gamma, & \delta, \\ \alpha', & \beta', & \gamma', & \delta', \end{array}$$

we have any other two points on the line, say the points

$$\begin{array}{cccc} \lambda\alpha + \mu\alpha', & \lambda\beta + \mu\beta', & \lambda\gamma + \mu\gamma', & \lambda\delta + \mu\delta', \\ \nu\alpha + \rho\alpha', & \nu\beta + \rho\beta', & \nu\gamma + \rho\gamma', & \nu\delta + \rho\delta', \end{array}$$

then the six determinants have their original values each multiplied by $\lambda\rho - \mu\nu$; and the ratios are unaltered.

And the like is the case, if instead of the planes

$$\begin{array}{cccc} A, & B, & C, & D, \\ A', & B', & C', & D', \end{array}$$

we have any other two planes through the line, say the planes

$$\begin{array}{cccc} \lambda A + \mu A', & \lambda B + \mu B', & \lambda C + \mu C', & \lambda D + \mu D', \\ \nu A + \rho A', & \nu B + \rho B', & \nu C + \rho C', & \nu D + \rho D', \end{array}$$

the determinants have their original values each multiplied by $\lambda\rho - \mu\nu$; and the ratios are unaltered.

7. It may be remarked, that the theory of the six coordinates considered as derived from the two points $(\alpha, \beta, \gamma, \delta)$, $(\alpha', \beta', \gamma', \delta')$, and as derived from the two planes (A, B, C, D), (A', B', C', D'), is precisely the same in each case; and we may confine ourselves to the first point of view, regarding therefore the six coordinates as derived from the two points $(\alpha, \beta, \gamma, \delta)$, $(\alpha', \beta', \gamma', \delta')$. I further remark, that I do not at present in anywise fix the absolute magnitudes of the coordinates $(a, b\ c, f, g, h)$: it is only the ratios that we are concerned with.

8. The values of the ratios $a : b : c : f : g : h$ of the six coordinates do however depend on the particular coordinate planes $x = 0$, $y = 0$, $z = 0$, $w = 0$, made use of for their determination; and in the sequel it will be necessary to investigate the

formulæ of transformation to a new set of coordinate planes $x_0=0$, $y_0=0$, $z_0=0$, $w_0=0$. And I shall also show in what manner the absolute magnitudes of the coordinates may be fixed. But deferring the consideration of these questions, I consider the planes $x=0$, $y=0$, $z=0$, $w=0$ as given planes, and take the six coordinates (a, b, c, f, g, h) of a line to be determined as above in reference to these given planes, the absolute values of these coordinates remaining indeterminate, and their ratios only being attended to. And I proceed to consider the various questions which present themselves in the geometry of the line, considered as thus determined by means of its six coordinates (a, b, c, f, g, h).

Article, Nos. 9 to 18. (*Various Sub-headings.*) *Elementary Theorems.*

Condition that a line may be in a given plane.

9. Taking the line to be (a, b, c, f, g, h), the equation of the given plane to be

$$(A, B, C, D \between x, y, z, w)=0;$$

then if $(\alpha, \beta, \gamma, \delta)$, $(\alpha', \beta', \gamma', \delta')$ are the coordinates of any two points on the line, we have the system of equations *ante*, No. 2, and substituting therein for $\beta\gamma'-\beta'\gamma$, &c. the values (a, b, c, f, g, h), we find

$$\left|\begin{matrix} 0, & c, & -b, & f \\ -c, & 0, & a, & g \\ b, & -a, & 0, & h \\ -f, & -g, & -h, & 0 \end{matrix}\right|(A, B, C, D)=0;$$

which equations, equivalent to a twofold relation, are the required condition. It may be remarked that, treating (A, B, C, D) as current plane coordinates, each equation of the system is that of a point lying in the line.

Condition that a line may pass through a given point.

10. The coordinates of the given point are taken to be $(\alpha, \beta, \gamma, \delta)$. If

$$(A, B, C, D \between x, y, z, w)=0, \quad (A', B', C', D' \between x, y, z, w)=0,$$

are the equations of any two planes through the line, then we have the system of equations *ante* No. 3, and substituting therein for $AB'-A'B$, &c. their values in terms of the coordinates (a, b, c, f, g, h) of the line, we have

$$\left|\begin{matrix} 0, & h, & -g, & a \\ -h, & 0, & f, & b \\ g, & -f, & 0, & c \\ -a, & -b, & -c, & 0 \end{matrix}\right|(\alpha, \beta, \gamma, \delta)=0;$$

which equations, equivalent to a twofold relation, are the required condition. It is obvious that, treating $(\alpha, \beta, \gamma, \delta)$ as current point coordinates, each equation of the system is the equation of a plane through the given line.

Condition for the intersection of two lines.

11. The coordinates of the lines are taken to be (a, b, c, f, g, h), and $(a_{,}, b_{,}, c_{,}, f_{,}, g_{,}, h_{,})$, respectively. If $(\alpha, \beta, \gamma, \delta)$, $(\alpha', \beta', \gamma', \delta')$, are the coordinates of any two points in the first line, and $(\alpha_{,}, \beta_{,}, \gamma_{,}, \delta_{,})$, $(\alpha_{,}', \beta_{,}', \gamma_{,}', \delta_{,}')$, are the coordinates of any two points on the second line, then the four points are in a plane, that is, we have

$$\begin{vmatrix} \alpha\ , & \beta\ , & \gamma\ , & \delta \\ \alpha'\ , & \beta'\ , & \gamma'\ , & \delta' \\ \alpha_{,}\ , & \beta_{,}\ , & \gamma_{,}\ , & \delta_{,} \\ \alpha_{,}'\ , & \beta_{,}'\ , & \gamma_{,}'\ , & \delta_{,}' \end{vmatrix} = 0,$$

that is, expanding the determinant and substituting for $\beta\gamma' - \beta'\gamma$, &c. and $\beta_{,}\gamma_{,}' - \beta_{,}'\gamma_{,}$, &c. their values in terms of the coordinates of the two lines respectively, we have

$$af_{,} + bg_{,} + ch_{,} + fa_{,} + gb_{,} + hc_{,} = 0,$$

or, as this may also be written,

$$(f_{,}, g_{,}, h_{,}, a_{,}, b_{,}, c_{,} \between a, b, c, f, g, h) = 0,$$

for the condition that the two lines may intersect.

12. The same result will be obtained if we take

$$(A, B, C, D \between x, y, z, w) = 0, \quad (A', B', C', D' \between x, y, z, w) = 0,$$

for the equations of any two planes through the first line, and

$$(A_{,}, B_{,}, C_{,}, D_{,} \between x, y, z, w) = 0, \quad (A_{,}', B_{,}', C_{,}', D_{,}' \between x, y, z, w) = 0,$$

for the equations of any two planes through the second line. The four planes will meet in a point, that is, we have

$$\begin{vmatrix} A\ , & B\ , & C\ , & D \\ A'\ , & B'\ , & C'\ , & D' \\ A_{,}\ , & B_{,}\ , & C_{,}\ , & D_{,} \\ A_{,}'\ , & B_{,}'\ , & C_{,}'\ , & D_{,}' \end{vmatrix} = 0,$$

or, expanding and substituting, we have the same condition as before.

13. In the case of any two lines (a, b, c, f, g, h), and $(a_{,}, b_{,}, c_{,}, f_{,}, g_{,}, h_{,})$, we may define the 'moment' of the two lines to be the function

$$af_{,} + bg_{,} + ch_{,} + fa_{,} + gb_{,} + hc_{,},$$

it being understood that we have not as yet any complete quantitative definition of the moment; this being so, we have, in what precedes, the theorem that the moment of two intersecting lines is $= 0$.

Plane through two intersecting lines.

14. Let $(A, B, C, D \between x, y, z, w) = 0$ be the equation of the plane through the two intersecting lines (a, b, c, f, g, h) and $(a_,, b_,, c_,, f_,, g_,, h_,)$. We have two systems of equations, as in No. 9, and comparing the corresponding equations of the two systems, we find in the first instance

$$\begin{aligned} A : B : C : D &= \lambda : bf_, - b_,f : cf_, - c_,f : -(bc_, - b_,c) \\ &= ag_, - a_,g : \mu : cg_, - c_,g : -(ca_, - c_,a) \\ &= ah_, - a_,h : bh_, - b_,h : \nu : -(ab_, - a_,b) \\ &= gh_, - g_,h : hf_, - h_,f : fg_, - f_,g : \rho \quad , \end{aligned}$$

where λ, μ, ν, ρ, are in the first instance unknown; the different sets of ratios are of course identical in virtue of the relation

$$(f_,, g_,, h_,, a_,, b_,, c_, \between a, b, c, f, g, h) = 0,$$

and comparing them we have equations which lead to the values of λ, μ, ν, ρ; and we thus obtain more completely,

$$\begin{aligned} A : B : C : D &= f_,a + b_,g + c_,h : bf_, + b_,f : cf_, - c_,f : -(bc_, - b_,c) \\ &= ag_, - a_,g : a_,f + g_,b + c_,h : cg_, - c_,g : -(ca_, - c_,a) \\ &= ah_, - a_,h : bh_, - b_,h : a_,f + b_,g + h_,c : -(ab_, - a_,b) \\ &= gh_, - g_,h : hf_, - h_,f : fg_, - f_,g : af_, + bg_, + ch_, . \end{aligned}$$

15. It is in these equations easy to verify the identity of the different sets of values: we ought, for instance, to have

$$\frac{a_,f + b_,g + h_,c}{fg_, - f_,g} = - \frac{ab_, - a_,b}{af_, + bg_, + ch_,};$$

that is,

$$(h_,c + a_,f + b_,g)(h_,c + af_, + bg_,) + (ab_, - a_,b)(fg_, - f_,g) = 0,$$

and, observing that

$$\begin{aligned} &(a_,f + b_,g)(af_, + bg_,) + (ab_, - a_,b)(fg_, - f_,g) \\ &\quad = (af + bg)(a_,f_, + b_,g_,), \; = ch \cdot c_,h_,, \end{aligned}$$

the left-hand side is

$$\begin{aligned} &= ch_,(ch_, + af_, + bg_, + a_,f + b_,g + c_,h), \\ &= ch_,(af_, + bg_, + ch_, + fa_, + gb_, + hc_,), \; = 0. \end{aligned}$$

Point on two intersecting lines.

16. Let $(\alpha, \beta, \gamma, \delta)$ be the coordinates of the point of intersection of the two intersecting lines (a, b, c, f, g, h) and $(a_,, b_,, c_,, f_,, g_,, h_,)$. We have two systems of

equations such as in No. 10, and comparing the corresponding equations of the two systems, we find

$$\begin{array}{rlllll}\alpha : \beta : \gamma : \delta = & L & : \; ag_{,} - a_{,}g & : \; ah_{,} - a_{,}h & : \; gh_{,} - g_{,}h & \\ = & bf_{,} - b_{,}f & : \; M & : \; bh_{,} - b_{,}h & : \; hf_{,} - h_{,}f & \\ = & cf_{,} - c_{,}f & : \; cg_{,} - c_{,}g & : \; N & : \; fg_{,} - f_{,}g & \\ = & -(bc_{,} - b_{,}c) & : \; -(ca_{,} - c_{,}a) & : \; -(ab_{,} - a_{,}b) & : \; P & ,\end{array}$$

where L, M, N, P, are in the first instance unknown; but, comparing the different sets of values, we have equations for finding the values of these quantities, and we thus obtain the more complete system

$$\begin{array}{rllll}\alpha : \beta : \gamma : \delta = & f_{,}a + b_{,}g + c_{,}h & : \; ag_{,} - a_{,}g & : \; ah_{,} - a_{,}h & : \; gh_{,} - g_{,}h \\ = & bf_{,} - b_{,}f & : \; a_{,}f + g_{,}b + c_{,}h & : \; bh_{,} - b_{,}h & : \; hf_{,} - h_{,}f \\ = & cf_{,} - c_{,}f & : \; cg_{,} - c_{,}g & : \; a_{,}f + b_{,}g + h_{,}c & : \; fg_{,} - f_{,}g \\ = & -(bc_{,} - b_{,}c) & : \; -(ca_{,} - c_{,}a) & : \; -(ab_{,} - a_{,}b) & : \; f_{,}a + g_{,}b + h_{,}c,\end{array}$$

where it is to be observed that the right-hand side considered as a matrix is the transposed matrix of that which occurs in No. 13, in the formula for $A : B : C : D$. The verification of the identity of the different sets of values can of course be effected as in No. 15.

Expression for an arbitrary plane through a line.

17. The condition in order that the plane $(A, B, C, D \between x, y, z, w) = 0$, may pass through the line (a, b, c, f, g, h), is the twofold relation given, No. 9; it is satisfied by any one of the four systems

$$\begin{array}{rl} A : B : C : D = & 0 : h : g : a, \\ \text{or} = & -h : 0 : f : b, \\ \text{or} = & g : -f : 0 : c, \\ \text{or} = & -a : -b : -c : 0;\end{array}$$

and consequently also by

$$\begin{array}{rl} A : B : C : D = & (\;0,\; -h,\; g,\; -a \between \xi, \eta, \zeta, \omega) \\ & :(\;h,\; 0,\; -f,\; -b \between \xi, \eta, \zeta, \omega) \\ & :(-g,\; f,\; 0,\; -c \between \xi, \eta, \zeta, \omega) \\ & :(\;a,\; b,\; c,\; 0 \between \xi, \eta, \zeta, \omega);\end{array}$$

or, what is the same thing, by

$$\begin{array}{rl} A : B : C : D = & (\;0,\; h,\; -g,\; a \between \xi, \eta, \zeta, \omega) \\ & :(-h,\; 0,\; f,\; b \between \xi, \eta, \zeta, \omega) \\ & :(\;g,\; -f,\; 0,\; c \between \xi, \eta, \zeta, \omega) \\ & :(-a,\; -b,\; -c,\; 0 \between \xi, \eta, \zeta, \omega)\end{array}$$

where $(\xi, \eta, \zeta, \omega)$ are arbitrary: there is, however, no loss of generality in putting any two of these quantities $= 0$.

Expression for an arbitrary point in a line.

18. The condition in order that the point $(\alpha, \beta, \gamma, \delta)$, may lie in the line (a, b, c, f, g, h), is the twofold relation given, No. 10; it is satisfied by any one of the four systems

$$\begin{array}{rrrrr} \alpha : \beta : \gamma : \delta = & 0 : & c : & -b : & f, \\ \text{or} = & -c : & 0 : & a : & g, \\ \text{or} = & b : & -a : & 0 : & h, \\ \text{or} = & -f : & -g : & -h : & 0; \end{array}$$

and consequently, also by

$$\begin{array}{rl} \alpha : \beta : \gamma : \delta = & (\ \ 0,\ \ -c,\ \ \ b,\ \ -f \between x, y, z, w) \\ : & (\ \ c,\ \ \ \ 0,\ \ -a,\ \ -g \between x, y, z, w) \\ : & (-b,\ \ \ \ a,\ \ \ \ 0,\ \ -h \between x, y, z, w) \\ : & (\ \ f,\ \ \ \ g,\ \ \ \ h,\ \ \ \ \ 0 \between x, y, z, w); \end{array}$$

or, what is the same thing, by

$$\begin{array}{rl} \alpha : \beta : \gamma : \delta = & (\ \ 0,\ \ \ \ c,\ \ -b,\ \ \ f \between x, y, z, w) \\ : & (-c,\ \ \ \ 0,\ \ \ \ a,\ \ \ g \between x, y, z, w) \\ : & (\ \ b,\ \ -a,\ \ \ \ 0,\ \ \ h \between x, y, z, w) \\ : & (-f,\ \ -g,\ \ -h,\ \ \ 0 \between x, y, z, w) \end{array}$$

where (x, y, z, w) are arbitrary: there is, however, no loss of generality in putting two of these quantities $=0$.

Article Nos. 19 to 25. *Geometrical considerations in regard to three, four, five, and six lines.*

Before proceeding further, I will establish certain geometrical notions in regard to three, four, five, and six lines. I use the term 'tractor' to denote a line which meets any given lines.

19. Three given lines have an infinity of tractors; viz. these are the generating lines of a hyperboloid having the three given lines for directrices.

20. Four given lines may be directrices (generating lines) of the same hyperboloid, viz. every tractor of any three of the four lines is then a tractor of all the four lines. But in general, four given lines have a pair of tractors; viz. considering the tractors of any three of the four lines, these form a hyperboloid having the three lines for directrices; the fourth line meets this hyperboloid in two points, and the generating line through either of these points is a line meeting each of the four given lines, that is, it is a tractor of the four given lines.

21. The fourth line may however touch the hyperboloid; and in this case, instead of a pair of tractors, the four lines have a twofold tractor. The relation of the four lines to each other is a symmetrical one; and we have thence the theorem, that if any one of four given lines touch the hyperboloid through the other three lines, then will each of the four given lines touch the hyperboloid through the other three lines. But the relation to each other of four lines having a twofold tractor may be otherwise expressed as follows; viz. considering a tractor of the four given lines, each line determines with the tractor a point, the intersection of the line and tractor; and it also determines a plane, viz. the plane containing the line and tractor; we have therefore a range of four points on the tractor, and a pencil of four planes through the tractor; and if the tractor be a two-fold tractor, the range and pencil will be homographic; and conversely, if the range and pencil are homographic, the tractor will be a twofold tractor. This is easily obtained as a limiting case from the general one where the four lines have a pair of tractors; each line determines with the one tractor a point and a plane as above, and this plane intersects the second tractor in a point; we have thus through the first tractor a pencil of planes, and on the second tractor a range of points, and these two are homographic. But, in the case of a twofold tractor, the range on the second tractor coincides with that on the first tractor; that is, the range of points on the tractor is homographic with the pencil of planes through the tractor.

22. Given any four lines, and a point O, then either in the general case where the four lines have a pair of tractors, or in the special case where they have a twofold tractor, there exists and can be found through the point O a single fifth line such that the five lines have (as the case may be) a pair of tractors, or a twofold tractor. And similarly, given the four lines and a plane Ω, there exists and can be found in the plane Ω a single fifth line such that the five lines have (as the case may be) a pair of tractors, or a twofold tractor.

23. Five given lines have not in general any tractor; the five lines may be directrices (generating lines) of the same hyperboloid, and they have then an infinity of tractors; or they may have a pair of tractors, viz. the fifth line may be a line meeting the tractors of the other four lines; or (as a particular case of the last relation) the five lines may have a twofold tractor; or the five lines may have a single tractor.

24. Given any five lines and a point O; then, selecting any four of the given lines, we may through O draw a line having with the four lines a pair of tractors. Treating in this manner each of the five sets of four lines, we obtain through the point O five lines constructed as above; we have the theorem which will be proved in the sequel, that these five lines lie in a plane Ω. And similarly, given the five lines, and a plane Ω, then selecting any four of the five lines, we may in the plane Ω draw a line having with the four lines a pair of tractors; treating in this manner each of the five sets of four lines, we obtain in the plane Ω five lines; and we have then the theorem that these five lines meet in a point O.

25. In the case of six given lines, we may have between the lines the like relations to those for the case of five given lines; or we may have the more general relation of the involution of six lines, depending on the last-mentioned theorems, viz. given any five lines, and the point O or the plane Ω, then determining in the one case the plane Ω and in the other case the point O, and taking as a sixth line any line whatever through the point O and in the plane Ω, the six lines are said to be in involution, or to form an involution of six lines. I now revert to the analytical theory of the line.

Article Nos. 26 to 51. (*Various sub-headings.*) *Cases of a linear relation or linear relations between the six Coordinates.*

26. If the coordinates (a, b, c, f, g, h) of a line are regarded as variable quantities connected by a single equation or by two or three equations, we have a system of lines with three or two arbitrary parameters or with a single arbitrary parameter; and so if there are four equations the system consists of a determinate number of lines. For a linear relation, the coefficients may be either (F, G, H, A, B, C), not the coordinates of a line, that is, not satisfying the relation $AF + BG + CH = 0$, or they may be the coordinates of a line, satisfying the relation in question. I consider the several cases in order as follows:

Linear relation $(F, G, H, A, B, C \between a, b, c, f, g, h) = 0$, *where* (A, B, C, F, G, H) *are not the coordinates of a line.*

27. Considering any six lines which satisfy the relation in question, we may eliminate the coefficients F, G, H, A, B, C, and thus obtain an equation $\nabla = 0$, where ∇ is the determinant formed with the coordinates of the six lines; this equation, regarding therein the coordinates of five of the six lines as given, is in regard to the coordinates of the remaining line, say the original line (a, b, c, f, g, h), a linear relation equivalent to the original linear relation $(F, G, H, A, B, C \between a, b, c, f, g, h) = 0$. The equation in its new form, viz. the equation $\nabla = 0$, establishes between the six lines a relation which is in fact the relation of involution already referred to; viz. it will be shown in the sequel that, starting from the equation $\nabla = 0$ as the definition of the relation of involution, we are led to a construction for a line in involution with five given lines the same as the construction explained *ante* No. 25.

Linear relation $(F, G, H, A, B, C \between a, b, c, f, g, h) = 0$, *where* (A, B, C, F, G, H) *are the coordinates of a line.*

28. The linear relation expresses that the two lines $(a, b, c, f, g, h \between A, B, C, F, G, H)$ intersect, or what is the same thing, that the line (a, b, c, f, g, h) is any line whatever meeting the line (A, B, C, F, G, H).

Two linear relations $(F, G, H, A, B, C \between a, b, c, f, g, h) = 0$,

$$(F_1, G_1, H_1, A_1, B_1, C_1 \between a, b, c, f, g, h) = 0,$$

where the two sets of coefficients respectively are or are not the coordinates of a line.

29. If the two sets of coefficients are each of them the coordinates of a line, then the two equations express that the line (a, b, c, f, g, h) is any line whatever cutting each of the two given lines. And the general case is in fact reducible to this particular one; for suppose that neither set of coefficients belongs to a line, then we may from the two given linear relations form the relation

$$(\lambda F+\lambda_1 F_1,\ \lambda G+\lambda_1 G_1,\ \lambda H+\lambda_1 H_1,\ \lambda A+\lambda_1 A_1,\ \lambda B+\lambda_1 B_1,\ \lambda C+\lambda_1 C_1 \between a,\ b,\ c,\ f,\ g,\ h)=0,$$

and if the ratio $\lambda : \lambda_1$ be properly determined, then $(\lambda A+\lambda_1 A_1, \ldots)$ will be the coordinates of a line. This will in fact be the case if

$$(\lambda A+\lambda_1 A)(\lambda F+\lambda_1 F_1)+(\lambda B+\lambda_1 B_1)(\lambda G+\lambda_1 G_1)+(\lambda C+\lambda_1 C_1)(\lambda H+\lambda_1 H_1)=0,$$

that is, if

$$(AF+BG+CH,\ AF_1+BG_1+CH_1+FA_1+GB_1+CH_1,\ A_1F_1+B_1G_1+C_1H_1 \between \lambda,\ \lambda_1)^2=0,$$

a quadric equation giving two values of the ratio $\lambda : \lambda_1$, that is, two linear relations in each of which the coefficients are the coordinates of a line: we have thus two derived lines, and the line (a, b, c, f, g, h) meets each of these derived lines.

There is no real difference if one or the other of the given systems of coefficients, say the system (A, B, C, F, G, H), are the coordinates of a line. We have then $AF+BG+CH=0$; the quadric equation in $\lambda : \lambda_1$ has a root $\lambda_1 : \lambda=0$, and rejecting it, the other root is determined by a simple equation: this only means that the line (A, B, C, F, G, H) is itself one of the two derived lines.

But there is a real difference in the case where the equation in $\lambda : \lambda_1$ has equal roots; to explain this special case, observe that if in the general case we consider the two derived lines as a pair of tractors of any four lines, then the linear relations express that the line (a, b, c, f, g, h) has with these four lines a pair of tractors; and in the special case under consideration the linear relations express that the line (a, b, c, f, g, h) has with the four lines, or (what is the same thing) with any three of them, that is with some three lines, a twofold tractor. According to what precedes (No. 21), the construction of the line (a, b, c, f, g, h) is in fact as follows, viz. if on the twofold tractor considered as given, we take a series of points p, and through the tractor, homographic with the range, a pencil of planes P, then the sought-for line will be any line through a point p, in the corresponding plane P. But it is proper to give an analytical proof of the construction.

30. I observe that we may without loss of generality assume $A_1F_1+B_1G_1+C_1H_1=0$, and this being so, the condition for the equality of the roots of the quadric equation is

$$AF_1+BG_1+CH_1+FA_1+BG_1+CH_1=0,$$

that is, writing $(a_1, b_1, c_1, f_1, g_1, h_1)$ in place of $(A_1, B_1, C_1, F_1, G_1, H_1)$, the case in question may be taken to be that of

Two linear relations

$$(f_1,\ g_1,\ h_1,\ a_1,\ b_1,\ c_1 \between a,\ b,\ c,\ f,\ g,\ h)=0,$$
$$(F,\ G,\ H,\ A,\ B,\ C \between a,\ b,\ c,\ f,\ g,\ h)=0,$$

where $(a_1, b_1, c_1, f_1, g_1, h_1)$ *are,* (A, B, C, F, G, H) *are not, the coordinates of a point, and where*

$$(f_1, g_1, h_1, a_1, b_1, c_1 \between A, B, C, F, G, H) = 0;$$

that is, where the twofold derived line is in fact the original line

$$(a_1, b_1, c_1, f_1, g_1, h_1).$$

31. To simplify, we may take $x=0$, $y=0$ for the equations of the line; the coordinates of the line then are $(a_1, b_1, c_1, f_1, g_1, h_1) = (0, 0, 0, 0, 0, 1)$. Taking moreover $x=0$, $y=0$, $\frac{z}{\gamma} = \frac{w}{\delta}$ for the coordinates of the point p, and $\frac{x}{\alpha} = \frac{y}{\beta}$ for the equation of the plane P, the homographic relation of the point and plane is given by an equation of the form

$$-F\beta\gamma + G\alpha\gamma - A\alpha\delta - B\beta\delta = 0,$$

or, as this may be written,

$$(F, G, H, A, B, 0 \between -\beta\gamma, \alpha\gamma, 0, -\alpha\delta, -\beta\delta, \omega) = 0,$$

where H and ω, being each multiplied by 0, do not really enter into the equation.

The equations of any line whatever through the point p and in the plane P may be written $\beta x - \alpha y = 0$, $A'x + B'y + \delta z - \gamma\omega = 0$, where A', B' are arbitrary: hence arranging the coefficients in the order

$$\begin{matrix} \beta, & -\alpha, & 0, & 0, \\ A', & B', & \delta, & -\gamma, \end{matrix}$$

the coordinates (a, b, c, f, g, h) of the line in question are

$$(-\beta\gamma, \alpha\gamma, 0, -\alpha\delta, -\beta\delta, A'\alpha + B'\beta);$$

so that we have

$$\begin{aligned} &(f_1, g_1, h_1, a_1, b_1, c_1 \between a, b, c, f, g, h) \\ &= (0, 0, 1, 0, 0, 0 \between a, b, c, f, g, h), = c, = 0; \end{aligned}$$

and morever the homographic relation, replacing therein the arbitrary quantity ω by $A'\alpha + B'\beta$, becomes

$$(F, G, H, A, B, 0 \between a, b, c, f, g, h) = 0.$$

Hence the linear relations satisfied by the coordinates (a, b, c, f, g, h) of the line in question are

$$\begin{aligned} (f_1, g_1, h_1, a_1, b_1, c_1 \between a, b, c, f, g, h) &= 0, \\ (F, G, H, A, B, C \between a, b, c, f, g, h) &= 0, \end{aligned}$$

with the coefficients

$$\begin{aligned} (f_1, g_1, h_1, a_1, b_1, c_1) &= (0, 0, 1, 0, 0, 0), \\ (A, B, C, F, G, H) &= (A, B, 0, F, G, H), \end{aligned}$$

values which satisfy the condition

$$(f_1, g_1, h_1, a_1, b_1, c_1 \between A, B, C, F, G, H) = 0.$$

Hence the line (a, b, c, f, g, h) through the point p and in the plane P is a line the coordinates of which satisfy two linear relations as mentioned in the heading; and the theorem is thus proved. The demonstration would be simplified by taking, as is allowable, the homographic relation to be $\frac{\alpha}{\beta} = k\frac{\gamma}{\delta}$.

32. It appears from the foregoing examination of the case of two linear relations that in the following cases of three or more linear relations there is no real loss of generality in assuming that the coefficients of each set are the coordinates of a line; for if originally this be not so, we have only to replace the given relations by linear functions of these relations, and to assign such values to the multipliers $\lambda, \lambda_1, \lambda_2 \ldots$ as in each case to make the new coefficients to be the coordinates of a line; and as there are two or more arbitrary ratios $\lambda : \lambda_1 : \lambda_2 \ldots$ to be assigned at pleasure and only a single condition to be satisfied, no cases of failure can arise. The remaining cases may consequently be stated in a more simple form.

Three linear relations, the coefficients of each set being the coordinates of a line.

33. The three relations express that the line (a, b, c, f, g, h) meets each of the three given lines; that is, that the line is any generating line of a hyperboloid having the three given lines for directrices.

Four linear relations, the coefficients of each set being the coordinates of a line.

34. The four relations express that the line (a, b, c, f, g, h) meets each of four given lines; or what is the same thing, that the line is a tractor of four given lines. It is to be noticed that the four linear relations serve to express the ratios $a : b : c : f : g : h$ linearly in terms of any one of these ratios, or what is the same thing, to express the several ratios in terms of an arbitrary ratio $u : v$. Substituting the resulting values in the equation

$$af + bg + ch = 0,$$

we have a quadric equation for the determination of the remaining ratio, or of the ratio $u : v$; and then each of the ratios of the coordinates can be expressed rationally in terms of either root of the quadric equation; we thus obtain the coordinates of each of the two tractors of the four given lines; or we have a complete analytical solution of the problem, to find the tractors of four given lines. The quadric equation may have equal roots; that is, the four given lines may have a twofold tractor, which is then determined linearly.

35. The theory of the linear relations of the coordinates (a, b, c, f, g, h) of a line may be considered in a different manner. It will be convenient to take the different cases in a reverse order, beginning with the extreme case (not before mentioned) of a fivefold relation and ascending to the case of a onefold or single relation.

Case of the fivefold relation.

36. The fivefold relation

$$\left\| \begin{matrix} a, & b, & c, & f, & g, & h \\ a_1, & b_1, & c_1, & f_1, & g_1, & h_1 \end{matrix} \right\| = 0,$$

expresses that the quantities (a, b, c, f, g, h) are proportional to $(a_1, b_1, c_1, f_1, g_1, h_1)$. As the former set are by hypothesis the coordinates of a line, the given set $(a_1, b_1, c_1, f_1, g_1, h_1)$ must, it is clear, also be the coordinates of a line, and the relation then expresses that the line (a, b, c, f, g, h) coincides with the given line.

Case of the fourfold relation.

37. The fourfold relation is

$$\left\| \begin{matrix} a, & b, & c, & f, & g, & h \\ a_1, & b_1, & c_1, & f_1, & g_1, & h_1 \\ a_2, & b_2, & c_2, & f_2, & g_2, & h_2 \end{matrix} \right\| = 0,$$

or what is the same thing, we have the six equations $\lambda a + \lambda_1 a_1 + \lambda_2 a_2 = 0$, &c., involving the indeterminate quantities $\lambda, \lambda_1, \lambda_2$. If the coefficients

$$(a_1, b_1, c_1, f_1, g_1, h_1), \quad (a_2, b_2, c_2, f_2, g_2, h_2)$$

are not either set the coordinates of a line; then substituting the foregoing values $-\lambda a = \lambda_1 a_1 + \lambda_2 a_2$, &c. in the equation $af + bg + ch = 0$, we have a quadric equation in $(\lambda_1 : \lambda_2)$: and for each root of this equation, the coefficients $\lambda_1 a_1 + \lambda_2 a_2$, &c. will be the coordinates of a line. There are thus in general two derived lines; and the fourfold relation expresses that the line (a, b, c, f, g, h) coincides with one or other of these derived lines. There is no real difference if one or the other of the two sets $(a_1, b_1, c_1, f_1, g_1, h_1)$, $(a_2, b_2, c_2, f_2, g_2, h_2)$, or if each set, are the coordinates of a line; one of the derived lines or both of them will in these cases coincide with one or both of the given lines. And if the quadric equation has equal roots, then instead of two derived lines there is a twofold derived line, and the line (a, b, c, f, g, h) must coincide with this twofold line.

38. A case presenting peculiarity is however that in which the coefficients of the quadric equation vanish identically; this is only so when the coefficients $(a_1, b_1, c_1, f_1, g_1, h_1)$ and $(a_2, b_2, c_2, f_2, g_2, h_2)$ are the coordinates of two intersecting lines. The equations $-\lambda a = \lambda_1 a_1 + \lambda_2 a_2$, &c. here show that every line whatever which meets each of the two lines $(a_1, b_1, c_1, f_1, g_1, h_1)$ and $(a_2, b_2, c_2, f_2, g_2, h_2)$ meets also the line (a, b, c, f, g, h); that is, the line (a, b, c, f, g, h) is any line whatever in the plane and through the point of intersection of the two intersecting lines. We see moreover that not only

$$a_1 f_2 + b_1 g_2 + c_1 h_2 + a_2 f_1 + b_2 g_1 + c_2 h_1 = 0,$$

but also that $af_1+bg_1+ch_1+fa_1+gb_1+ch_1=0$ and $af_2+bg_2+ch_2+fa_2+gb_2+ch_2=0$; that is, the moment of each pair of lines is $=0$. It may be remarked that the ratios $\lambda : \lambda_1 : \lambda_2$ may be determined from any two of the six equations

$$\lambda a+\lambda_1 a_1+\lambda_2 a_2=0, \ldots \lambda h+\lambda_1 h_1+\lambda_2 h_2=0\,;$$

but that in consequence of the moments being each $=0$, there is not for the determination of these ratios any such set of equations as occur in the cases subsequently considered of a threefold relation, &c.

39. In what follows we have three or more sets $(a_1, b_1, c_1, f_1, g_1, h_1)$, &c.; and we may without loss of generality assume that each of these are the coordinates of a line: for replacing the several coefficients $a_1, \ldots$ by linear functions $\mu_1 a_1+\mu_2 a_2+\mu_3 a_3+$ &c., &c., the multipliers may be determined so that these are the coordinates of a point: and since for each set there is only a single condition to be satisfied by the two or more ratios $\mu_1 : \mu_2 : \mu_3 \ldots$, it is easy to see that no cases of failure will arise.

Case of the threefold relation.

40. The threefold relation is

$$\left\| \begin{array}{cccccc} a, & b, & c, & f, & g, & h \\ a_1, & b_1, & c_1, & f_1, & g_1, & h_1 \\ a_2, & b_2, & c_2, & f_2, & g_2, & h_2 \\ a_3, & b_3, & c_3, & f_3, & g_3, & h_3 \end{array} \right\| = 0,$$

where $(a_1, \ldots)$, $(a_2, \ldots)$ $(a_3, \ldots)$ are each the coordinates of a line. Here writing

$$\lambda a+\lambda_1 a_1+\lambda_2 a_2+\lambda_3 a_3=0 \ldots,$$

it is clear that every line which meets each of the lines $(a_1, \ldots)$, $(a_2, \ldots)$, $(a_3, \ldots)$ will also meet the line (a, b, c, f, g, h); the lines which meet the first-mentioned three lines are the generating lines of a hyperboloid having these three lines for directrices, and it hence appears that the line (a, b, c, f, g, h) is any directrix line whatever of the hyperboloid in question.

41. Using the notations 01, 02, 12, &c. to denote the moments of the several pairs of lines, viz.

$$\begin{aligned} 01 &= af_1+bg_1+ch_1+fa_1+gb_1+hc_1, \\ 12 &= a_1f_2+b_1g_2+c_1h_2+f_1a_2+g_1b_2+h_1c_2, \\ &\&\text{c.}, \end{aligned}$$

then from the equations $\lambda a+\lambda_1 a_1+\lambda_2 a_2+\lambda_3 a_3=0$, &c., we deduce

$$\begin{array}{ccccccc} \cdot & & \lambda_1 01 & + & \lambda_2 02 & + & \lambda_3 03=0, \\ \lambda\, 10 & + & \cdot & + & \lambda_2 12 & + & \lambda_3 13=0, \\ \lambda\, 20 & + & \lambda_1 21 & & \cdot & + & \lambda_3 23=0, \\ \lambda\, 30 & + & \lambda_1 31 & + & \lambda_2 32 & & \cdot \quad =0, \end{array}$$

and hence eliminating $\lambda, \lambda_1, \lambda_2, \lambda_3$, we find

$$\left|\begin{array}{cccc} . & 01, & 02, & 03 \\ 10, & . & 12, & 13 \\ 20, & 21, & . & 23 \\ 30, & 31, & 32, & . \end{array}\right| = 0,$$

a relation between the moments satisfied in virtue of the given threefold relation; but which as a mere onefold relation is of course not equivalent to the threefold relation. It will subsequently appear that the equation expresses that any one of the four lines, say the line (a, b, c, f, g, h) touches the hyperboloid having the other three lines for generatrices; this condition is satisfied in virtue of the threefold relation which, as we have seen, expresses that the line (a, b, c, f, g, h) lies wholly in the hyperboloid in question.

42. The last mentioned determinant is the Norm of

$$\sqrt{01 \,.\, 23} + \sqrt{02 \,.\, 31} + \sqrt{03 \,.\, 12};$$

so that the equation may be written

$$\sqrt{01 \,.\, 23} + \sqrt{02 \,.\, 31} + \sqrt{03 \,.\, 12} = 0,$$

or, what is the same thing,

$$\sqrt{01}\sqrt{23} + \sqrt{02}\sqrt{31} + \sqrt{03}\sqrt{12} = 0,$$

it being of course understood that the signs of the radicals must be determined in accordance with this equation; we then find

$$\lambda : \lambda_1 : \lambda_2 : \lambda_3 = \sqrt{23 \,.\, 31 \,.\, 12} : \sqrt{02 \,.\, 03 \,.\, 23} : \sqrt{03 \,.\, 01 \,.\, 31} : \sqrt{01 \,.\, 02 \,.\, 12},$$

or say

$$= \sqrt{23}\sqrt{31}\sqrt{12} : \sqrt{02}\sqrt{03}\sqrt{23} : \sqrt{03}\sqrt{01}\sqrt{31} : \sqrt{01}\sqrt{02}\sqrt{12};$$

in fact, substituting these last values in the linear equations for $\lambda, \lambda_1, \lambda_2, \lambda_3$, we find that the equations are all satisfied in virtue of the single equation

$$\sqrt{01}\sqrt{23} + \sqrt{02}\sqrt{31} + \sqrt{03}\sqrt{12} = 0.$$

Case of the twofold relation.

43. We have here

$$\left\|\begin{array}{cccccc} a, & b, & c, & f, & g, & h \\ a_1, & b_1, & c_1, & f_1, & g_1, & h_1 \\ a_2, & b_2, & c_2, & f_2, & g_2, & h_2 \\ a_3, & b_3, & c_3, & f_3, & g_3, & h_3 \\ a_4, & b_4, & c_4, & f_4, & g_4, & h_4 \end{array}\right\| = 0,$$

where $(a_1, \ldots)(a_2, \ldots)(a_3, \ldots)(a_4, \ldots)$, are each the coordinates of a line. Here, writing

$$\lambda a + \lambda_1 a_1 + \lambda_2 a_2 + \lambda_3 a_3 + \lambda_4 a_4 = 0;$$

it is clear that every line which meets each of the four given lines, will also meet the line (a, b, c, f, g, h); but the only lines meeting the four given lines are two determinate lines, the tractors of the four given lines; and the conclusion is, that the line (a, b, c, f, g, h) is any line whatever which meets the two tractors.

44. If, however, the four given lines have a twofold tractor, then the line (a, b, c, f, g, h) is still a line having two conditions imposed upon it; it is in fact a line determined as in No. 21, viz. if on the tractor we take a series of points p, and through the tractor a series of planes P, corresponding homographically to the points, then the line (a, b, c, f, g, h) is any line through a point p, in the corresponding plane P.

45. Using as before 01, 02, ... 12, &c. to denote the moments of the several pairs of lines, we have

$$\begin{aligned}
. \quad \lambda_1 01 + \lambda_2 02 + \lambda_3 03 + \lambda_4 04 &= 0,\\
\lambda 10 \quad . \quad + \lambda_2 12 + \lambda_3 13 + \lambda_4 14 &= 0,\\
\lambda 20 + \lambda_1 21 \quad . \quad + \lambda_3 23 + \lambda_4 24 &= 0,\\
\lambda 30 + \lambda_1 31 + \lambda_2 32 \quad . \quad + \lambda_4 34 &= 0,\\
\lambda 40 + \lambda_1 41 + \lambda_2 42 + \lambda_3 43 \quad . \quad &= 0,
\end{aligned}$$

and thence also

$$\begin{vmatrix} . & 01, & 02, & 03, & 04 \\ 10, & . & 12, & 13, & 14 \\ 20, & 21, & . & 23, & 24 \\ 30, & 31, & 32, & . & 34 \\ 40, & 41, & 42, & 43, & . \end{vmatrix} = 0,$$

a relation between the moments satisfied in virtue of the original twofold relation; but which, as a single equation, is of course not equivalent to the twofold relation. It is in fact easy to see that this equation expresses that the five lines have a common tractor; this is true, since in virtue of the twofold relation there are really two common tractors.

I have not obtained from the linear equations any symmetrical expressions for the ratios $\lambda : \lambda_1 : \lambda_2 : \lambda_3$.

Case of a onefold relation.

46. The onefold relation is

$$\begin{vmatrix} a, & b, & c, & f, & g, & h \\ a_1, & b_1, & c_1, & f_1, & g_1, & h_1 \\ a_2, & b_2, & c_2, & f_2, & g_2, & h_2 \\ a_3, & b_3, & c_3, & f_3, & g_3, & h_3 \\ a_4, & b_4, & c_4, & f_4, & g_4, & h_4 \\ a_5, & b_5, & c_5, & f_5, & g_5, & h_5 \end{vmatrix} = 0,$$

where $(a_1, \ldots)$, $(a_2, \ldots)$, $(a_3, \ldots)$, $(a_4, \ldots)$, $(a_5, \ldots)$, are each the coordinates of points in a line. The preceding mode of dealing with the question is inapplicable, since there is not in general any line which meets the five given lines; in the particular case, however, where the five given lines are met by a single line, say when they have a common tractor, then the line (a, b, c, f, g, h) is any line meeting this common tractor. The general case is that of the involution of six lines, mentioned No. 25, and the consideration of which was deferred.

47. The onefold relation implies that we can find multipliers λ, μ, ν, ρ, σ, τ, such that

$$\begin{aligned}
\lambda a + \mu b + \nu c + \rho f + \sigma g + \tau h &= 0,\\
\lambda a_1 + \mu b_1 + \nu c_1 + \rho f_1 + \sigma g_1 + \tau h_1 &= 0,\\
&\vdots\\
\lambda a_5 + \mu b_5 + \nu c_5 + \rho f_5 + \sigma g_5 + \tau h_5 &= 0,
\end{aligned}$$

we may by means of the last five equations determine the ratios of λ, μ, ν, ρ, σ, τ, viz. these quantities will be proportional to the determinants formed out of the matrix

$$\left|\begin{matrix}
a_1, & b_1, & c_1, & f_1, & g_1, & h_1\\
a_2, & b_2, & c_2, & f_2, & g_2, & h_2\\
a_3, & b_3, & c_3, & f_3, & g_3, & h_3\\
a_4, & b_4, & c_4, & f_4, & g_4, & h_4\\
a_5, & b_5, & c_5, & f_5, & g_5, & h_5
\end{matrix}\right|$$

and the first equation is then a linear relation in (a, b, c, f, g, h), expressing the relation that exists between these coordinates.

48. Consider an arbitrary point O on the line (a, b, c, f, g, h); taking this point as origin, the coordinates of O are $0, 0, 0, 1$; and if x, y, z, w, are the coordinates of any other point on the line, then writing

$$\begin{matrix} x, & y, & z, & w,\\ 0, & 0, & 0, & 1,\end{matrix}$$

we find

$$a : b : c : f : g : h = 0 : 0 : 0 : x : y : z;$$

and the equation

$$\lambda a + \mu b + \nu c + \rho f + \sigma g + \tau h = 0$$

becomes simply $\rho x + \sigma y + \tau z = 0$; viz. this equation expresses that the line (a, b, c, f, g, h), assumed to pass through a given point O, lies in a determinate plane Ω through this point.

49. To construct this plane Ω, I consider any four of the five given lines, say the lines 2, 3, 4, 5, and I endeavour to find the line OQ_1 through O, which has with these lines a pair of tractors; *quà* line through O, the coordinates of the line in question may be taken to be $0, 0, 0, F_1, G_1, H_1$ (where F_1, G_1, H_1, are in fact the

11—2

coordinates x, y, z, of any point on the line OQ_1); and then the condition for the pair of tractors may be written

$$\begin{aligned}
p_2a_2 + p_3a_3 + p_4a_4 + p_5a_5 &= 0,\\
p_2b_2 + p_3b_3 + p_4b_4 + p_5b_5 &= 0,\\
p_2c_2 + p_3c_3 + p_4c_4 + p_5c_5 &= 0,\\
p_2f_2 + p_3f_3 + p_4f_4 + p_5f_5 &= F_1,\\
p_2g_2 + p_3g_3 + p_4g_4 + p_5g_5 &= G_1,\\
p_2h_2 + p_3h_3 + p_4h_4 + p_5h_5 &= H_1,
\end{aligned}$$

where p_2, p_3 ... are arbitrary coefficients; and we hence deduce

$$\rho F_1 + \sigma G_1 + \tau H_1 = 0;$$

but in precisely the same way, if the line OQ_2 have with the lines 1, 3, 4, 5, a pair of tractors, and if F_2, G_2, H_2, be the coordinates of a point on the line OQ_2, and similarly for the lines OQ_3, OQ_4, OQ_5, and the coordinates (F_3, G_3, H_3), (F_4, G_4, H_4) (F_5, G_5, H_5), we have

$$\begin{aligned}
\rho F_2 + \sigma G_2 + \tau H_2 &= 0,\\
\rho F_3 + \sigma G_3 + \tau H_3 &= 0,\\
\rho F_4 + \sigma G_4 + \tau H_4 &= 0,\\
\rho F_5 + \sigma G_5 + \tau H_5 &= 0,
\end{aligned}$$

and these equations show that the five lines OQ_1, OQ_2, OQ_3, OQ_4, OQ_5, lie in the plane

$$\rho x + \sigma y + \tau z = 0;$$

so that this plane is given as the plane through the lines OQ_1, OQ_2, OQ_3, OQ_4, OQ_5; and we have thus (given the lines 1, 2, 3, 4, 5, and the arbitrary point O) the construction of the line (a, b, c, f, g, h) through O in involution with the given lines.

50. The original onefold relation may be replaced by the six equations

$$\begin{aligned}
&\lambda a + \lambda_1 a_1 + \lambda_2 a_2 + \lambda_3 a_3 + \lambda_4 a_4 + \lambda_5 a_5 = 0,\\
&\lambda b + \lambda_1 b_1 + \lambda_2 b_2 + \lambda_3 b_3 + \lambda_4 b_4 + \lambda_5 b_5 = 0,\\
&\vdots\\
&\lambda h + \lambda_1 h_1 + \lambda_2 h_2 + \lambda_3 h_3 + \lambda_4 h_4 + \lambda_5 h_5 = 0,
\end{aligned}$$

and hence denoting as before the moments by 01, 02, 12, &c. we have

$$\begin{array}{llllll}
. & \lambda_1 01 + \lambda_2 02 + \lambda_3 03 + \lambda_4 04 + \lambda_5 05 = 0,\\
\lambda 10 & . \quad + \lambda_2 12 + \lambda_3 13 + \lambda_4 14 + \lambda_5 15 = 0,\\
\lambda 20 + \lambda_1 21 & . \quad + \lambda_3 23 + \lambda_4 24 + \lambda_5 25 = 0,\\
\lambda 30 + \lambda_1 31 + \lambda_2 32 & . \quad + \lambda_4 34 + \lambda_5 35 = 0,\\
\lambda 40 + \lambda_1 41 + \lambda_2 42 + \lambda_3 43 & . \quad + \lambda_5 45 = 0,\\
\lambda 50 + \lambda_1 51 + \lambda_2 52 + \lambda_3 53 + \lambda_4 54 & . \quad = 0,
\end{array}$$

which lead to

$$\begin{vmatrix} . & 01, & 02, & 03, & 04, & 05 \\ 10, & . & 12, & 13, & 14, & 15 \\ 20, & 21, & . & 23, & 24, & 25 \\ 30, & 31, & 32, & . & 34, & 35 \\ 40, & 41, & 42, & 43, & . & 45 \\ 50, & 51, & 52, & 53, & 54, & . \end{vmatrix} = 0,$$

a relation between the moments equivalent to the original onefold relation, and consequently expressing that the six lines are in involution. I have not obtained a symmetrical system of values for the ratios $\lambda : \lambda_1 : \lambda_2 : \lambda_3 : \lambda_4 : \lambda_5$.

51. Reverting to the relation which exists between the point O and the plane Ω, it is proper to remark that, since to any given point O there corresponds a single plane Ω, and to any given plane Ω a single point O, it follows that the point O and plane Ω are reciprocal figures; viz. they are reciprocals of the particular kind treated of by Möbius, wherein the reciprocal of a point is a plane through the point, and the reciprocal of a plane a point in the plane; and of which the analytical character is that the reciprocal of the point $(\alpha, \beta, \gamma, \delta)$ is the plane

$$\begin{aligned} &(\quad . \quad h\beta - g\gamma + l\delta)\,x \\ +&(-h\alpha \quad . \quad + f\gamma + m\delta)\,y \\ +&(\;\; g\alpha - f\beta \quad . \quad + n\delta)\,z \\ +&(-l\alpha - m\beta - n\gamma \quad . \quad)\,w = 0. \end{aligned}$$

Article No. 52. *A geometrical property of an involution of six lines.*

52. The figure of six lines in involution is connected in various ways with the theory of cubic curves in space, for instance, considering a point A of the curve, this determines with any given line l a plane meeting the curve in two other points, and the line λ which joins these two points may be called the projection of the line l. This being so, if in any osculating plane of the cubic we have six lines, $l, l_1, l_2, l_3, l_4, l_5$, tangents of a conic in that plane, the six projections $\lambda, \lambda_1, \lambda_2, \lambda_3, \lambda_4, \lambda_5$ of these tangents will be a set of lines in involution. I do not stop to prove this theorem or to develope any of its consequences.

Article No. 53. *To find the condition that four given lines may have a twofold tractor.*

53. Taking the coordinates of the given lines to be

(a, b, c, f, g, h), $(a_1, b_1, c_1, f_1, g_1, h_1)$, $(a_2, b_2, c_2, f_2, g_2, h_2)$, $(a_3, b_3, c_3, f_3, g_3, h_3)$

then if (A, B, C, F, G, H) be the coordinates of a tractor of these lines, we have

$$(F, G, H, A, B, C \between a, b, c, f, g, h) = 0,$$
$$(F, G, H, A, B, C \between a_1, b_1, c_1, f_1, g_1, h_1) = 0,$$
$$(F, G, H, A, B, C \between a_2, b_2, c_2, f_2, g_2, h_2) = 0,$$
$$(F, G, H, A, B, C \between a_3, b_3, c_3, f_3, g_3, h_3) = 0.$$

In virtue of these relations the ratios $A : B : C : F : G : H$ are given linear functions of any one of these ratios or of an arbitrary ratio $u : v$; and we then have $AF + BG + CH = 0$, a quadric equation for determining the unknown ratio. In the case of a twofold tractor, this equation must have equal roots; whence employing as usual the method of indeterminate multipliers, we find

$$A + \lambda a + \lambda_1 a_1 + \lambda_2 a_2 + \lambda_3 a_3 = 0,$$
$$B + \lambda b + \lambda_1 b_1 + \lambda_2 b_2 + \lambda_3 b_3 = 0,$$
$$C + \lambda c + \lambda_1 c_1 + \lambda_2 c_2 + \lambda_3 c_3 = 0,$$
$$F + \lambda f + \lambda_1 f_1 + \lambda_2 f_2 + \lambda_3 f_3 = 0,$$
$$G + \lambda g + \lambda_1 g_1 + \lambda_2 g_2 + \lambda_3 g_3 = 0,$$
$$H + \lambda h + \lambda_1 h_1 + \lambda_2 h_2 + \lambda_3 h_3 = 0.$$

Hence representing as before the moments of the pairs of lines by 01, 02, &c., we deduce

$$\begin{array}{l} \quad . \qquad\ \ \lambda_1 01 + \lambda_2 02 + \lambda_3 03 = 0, \\ \lambda 10 + \quad . \quad\ + \lambda_2 12 + \lambda_3 13 = 0, \\ \lambda 20 + \lambda_1 21 \quad . \quad + \lambda_3 23 = 0, \\ \lambda 30 + \lambda_1 31 + \lambda_2 32 \quad . \quad = 0, \end{array}$$

so that, as already mentioned, we have

$$\begin{vmatrix} . & 01, & 02, & 03 \\ 10, & . & 12, & 13 \\ 20, & 21, & . & 23 \\ 30, & 31, & 32, & . \end{vmatrix} = 0,$$

as the condition that the four given lines may have a twofold tractor.

Article Nos. 54 to 56. *Hyperboloid passing through three given lines.*

54. The direct investigation is somewhat tedious; but I write down, and will afterwards verify, the equation of the hyperboloid passing through the three given lines

$$(a_1, b_1, c_1, f_1, g_1, h_1), (a_2, b_2, c_2, f_2, g_2, h_2), (a_3, b_3, c_3, f_3, g_3, h_3).$$

Writing for shortness (agh), &c. to denote the determinants

$$\begin{vmatrix} a_1, & g_1, & h_1 \\ a_2, & g_2, & h_2 \\ a_3, & g_3, & h_3 \end{vmatrix}, \text{ \&c.}$$

the equation of the hyperboloid is

$$\begin{aligned}
&(agh)\,x^2 + (bhf)\,y^2 + (cfg)\,z^2 + (abc)\,w^2\\
&+ [(abg) - (cah)]\,xw + [(bfg) + (chf)]\,yz\\
&+ [(bch) - (abf)]\,yw + [(cgh) + (afg)]\,zx\\
&+ [(caf) - (bcg)]\,zw + [(ahf) + (bgh)]\,xy = 0.
\end{aligned}$$

In fact, we have

$$\begin{aligned}
(agh) &= a_1(g_2h_3 - g_3h_2) + g_1(h_2a_3 - h_3a_2) + h_1(a_2g_3 - a_3g_2)\\
&= a\,.\,gh + g\,.\,ha + h\,.\,ag,
\end{aligned}$$

where a, &c. stand for a_1, &c. and gh, &c. for $g_2h_3 - g_3h_2$, &c. Hence the foregoing equation may be written

$$\begin{aligned}
&x^2\,(a\,.\,gh + g\,.\,ha + h\,.\,ag)\\
&+ y^2\,(b\,.\,hf + h\,.\,fb + f\,.\,bh)\\
&+ z^2\,(c\,.\,fg + f\,.\,gc + g\,.\,cf)\\
&+ w^2\,(a\,.\,bc + b\,.\,ca + c\,.\,ab)\\
&+ xw\begin{pmatrix} a\,.\,bg + b\,.\,ga + g\,.\,ab\\ - c\,.\,ah - a\,.\,hc - h\,.\,ca\end{pmatrix} + yz\begin{pmatrix} b\,.\,fg + f\,.\,gb + g\,.\,bf\\ + c\,.\,hf + h\,.\,fc + f\,.\,ch\end{pmatrix}\\
&+ yw\begin{pmatrix} b\,.\,ch + c\,.\,hb - h\,.\,bc\\ - a\,.\,bf - b\,.\,fa - f\,.\,ab\end{pmatrix} + zx\begin{pmatrix} c\,.\,gh + g\,.\,hc + h\,.\,cg\\ + a\,.\,fg + f\,.\,ga + g\,.\,af\end{pmatrix}\\
&+ zw\begin{pmatrix} c\,.\,af + a\,.\,fc + f\,.\,ca\\ - b\,.\,cg - c\,.\,gb - g\,.\,bc\end{pmatrix} + xy\begin{pmatrix} a\,.\,hf + h\,.\,fa + f\,.\,ch\\ + b\,.\,gh + g\,.\,hb + h\,.\,bg\end{pmatrix} = 0.
\end{aligned}$$

55. This is

$$\begin{aligned}
&bc\,.\,w\,(hy - gz + aw)\\
&+ ca\,.\,w\,(-hx + fz + bw)\\
&+ ab\,.\,w\,(gx - fy + cw)\\
&+ gh\,.\,x\,(ax + by + cz)\\
&+ hf\,.\,y\,(ax + by + cz)\\
&+ fg\,.\,z\,(ax + by + cz)\\
&+ af\,\{w\,(ax + by + cz) - x\,(hy - gz + aw)\}\\
&+ bg\,\{w\,(ax + by + cz) - y\,(-hx + fz + bw)\}\\
&+ ch\,\{w\,(ax + by + cz) - z\,(gx - fy + cw)\}\\
&- bf\,.\,y\,(hy - gz + aw)\\
&- cf\,.\,z\,(hy - gz + aw)\\
&- cg\,.\,z\,(-hx + fz + bw)\\
&- ag\,.\,x\,(-hx + fz + bw)\\
&- ah\,.\,x\,(gx - fy + cw)\\
&- bh\,.\,y\,(gx - fy + cw) = 0.
\end{aligned}$$

56. Hence writing

$$(X, Y, Z, W) = \left| \begin{matrix} \cdot\,, & h, & -g, & a, \\ -h, & \cdot\,, & f, & b, \\ g, & -f, & \cdot\,, & c, \\ -a, & -b, & -c, & \cdot \end{matrix} \right| (x, y, z, w),$$

the foregoing equation is

$$\begin{aligned} & bc\,.\,wX + ca\,.\,wY + ab\,.\,wZ - gh\,.\,xW - hf\,.\,yW - fg\,.\,zW \\ & - af(wW + xX) \quad - bg(wW + yY) \quad - ch(wW + zZ) \\ & - bf\,.\,yX - cg\,.\,zY - ah\,.\,xZ - cf\,.\,zX - ag\,.\,xY - bh\,.\,yZ = 0\,; \end{aligned}$$

or, collecting and arranging, this is

$$\begin{aligned} & X\,\{-af\,.\,x - bf\,.\,y - cf\,.\,z + bc\,.\,w\} \\ + & Y\,\{-ag\,.\,x - bg\,.\,y - cg\,.\,z + ca\,.\,w\} \\ + & Z\,\{-ah\,.\,x - bh\,.\,y - ch\,.\,z + ab\,.\,w\} \\ + & W\,\{-gh\,.\,x - hf\,.\,y - fg\,.\,z + (af\,. + bg\,. + ch\,.)\,w\} = 0, \end{aligned}$$

which is satisfied by $X = 0$, $Y = 0$, $Z = 0$, $W = 0$; that is, since (a, b, c, f, g, h) have been written in place of $(a_1, b_1, c_1, f_1, g_1, h_1)$, by $X_1 = 0$, $Y_1 = 0$, $Z_1 = 0$, $W_1 = 0$ (if we thus denote the corresponding functions of $(a_1, b_1, c_1, f_1, g_1, h_1)$), that is, the hyperboloid passes through the line $(a_1, b_1, c_1, f_1, g_1, h_1)$; and similarly it passes through the other two lines.

Article Nos. 57 and 58. *The six coordinates defined as to their absolute magnitudes.*

57. In all that precedes, the absolute magnitudes of the coordinates have been left indeterminate, only the ratios being attended to. But the magnitudes of the six coordinates may be fixed in a very simple manner as follows; viz. using ordinary rectangular coordinates, then for any line, if x_0, y_0, z_0 are the coordinates of a particular point on this line, and α, β, γ the inclinations of the line to the axes, the coordinates of another point on the line are

$$x_0 + r\cos\alpha,\ y_0 + r\cos\beta,\ z_0 + r\cos\gamma\,;$$

and hence writing

$$\begin{matrix} x_0 + r\cos\alpha, & y_0 + r\cos\beta, & z_0 + r\cos\gamma, & 1, \\ x_0 & , y_0 & , z_0 & , 1, \end{matrix}$$

we have

$$a : b : c : f : g : h = z_0\cos\beta - y_0\cos\gamma : x_0\cos\gamma - z_0\cos\alpha : y_0\cos\beta - x_0\cos\alpha : \cos\alpha : \cos\beta : \cos\gamma.$$

Or we may take

$$\begin{aligned} a &= z_0\cos\beta - y_0\cos\gamma, & f &= \cos\alpha, \\ b &= x_0\cos\gamma - z_0\cos\alpha, & g &= \cos\beta, \\ c &= y_0\cos\alpha - x_0\cos\beta, & h &= \cos\gamma, \end{aligned}$$

values which of course satisfy, as they should do, the relation $af+bg+ch=0$. It is hardly necessary to remark, that the values of a, b, c are not altered on substituting for x_0, y_0, z_0 the coordinates $x_0+s\cos\alpha$, $y_0+s\cos\beta$, $z_0+s\cos\gamma$ of any other point on the line.

58. Considering any two lines (a, b, c, f, g, h), $(a_1, b_1, c_1, f_1, g_1, h_1)$, if we define the moment of the two lines to be the product of the perpendicular distance into the sine of the inclination of the two lines, then we have, Moment

$$=af_1+bg_1+ch_1+fa_1+gb_1+hc_1,$$

viz. we have now a quantitative definition of the function of the coordinates previously called the moment of the two lines.

For the demonstration of this formula it is to be remarked, that taking on the first line a segment of the length r, the coordinates of its extremities being (x_0, y_0, z_0) and $(x_0+r\cos\alpha,\ y_0+r\cos\beta,\ z_0+r\cos\gamma)$,

and on the second line a segment of the length r_1 the coordinates of its extremities being (x_0', y_0', z_0') and $(x_0'+r_1\cos\alpha_1, y_0'+r_1\cos\beta_1, z_0'+r_1\cos\gamma_1)$ and joining the extremities of these segments so as to form a tetrahedron, the volume of the tetrahedron is

$$=\tfrac{1}{6}rr_1(af_1+bg_1+ch_1+fa_1+gb_1+hc_1).$$

But the volume of the tetrahedron is also equal to $\frac{1}{6}$ of the product of the opposite edges into their perpendicular distance into the sine of the inclination of the two edges [1]; that is, it is $=\frac{1}{6}rr_1$ into the moment of the two lines, and we have thus the formula in question.

Article Nos. 59 to 75. *Statical and Kinematical Applications.*

The coordinates (a, b, c, f, g, h), as last defined, are peculiarly convenient in kinematical and mechanical questions, as will appear from the following investigations.

59. Using the term rotation to denote an infinitesimal rotation, I say first that a rotation λ round the line (a, b, c, f, g, h) produces in the point (x, y, z) rigidly connected with this line the displacements

$$\begin{aligned}\delta x&=\lambda(\ \ .\ \ -hy+gz-a),\\ \delta y&=\lambda(\ hx\ \ .\ \ -fz-b),\\ \delta z&=\lambda(-gx+fy\ \ .\ \ -c).\end{aligned}$$

[1] I take the opportunity of mentioning a very simple demonstration of this formula: taking the opposite edges to be r, r_1, their inclination $=\theta$, and perpendicular distance $=h$; the section of the tetrahedron by a plane parallel to the two edges at the distances z, $h-z$ from the two edges respectively is a parallelogram, the sides of which are $\frac{r(h-z)}{h}$ and $\frac{r_1 z}{h}$ respectively, and their inclination is $=\theta$; the area of the section is therefore $\frac{rr_1}{h^2}\sin\theta\,.\,z(h-z)$ and the volume of the tetrahedron is $=\frac{rr_1}{h^2}\sin\theta\int_0^h z(h-z)\,dz$, $=\frac{1}{6}rr_1 h\sin\theta$. The same result is however obtained still more simply by drawing a plane through one of the two edges perpendicular to the other edge; the volume is then equal to the sum or the difference of the volumes of two tetrahedra standing on a common triangular base; and the required result at once follows.

In fact assuming for a moment that the axis of rotation passes through the origin, then for the point P coordinates x, y, z, the square of the perpendicular distance from the axis is

$$\begin{array}{l} (\quad . \qquad - y\cos\gamma + z\cos\beta)^2 \\ + (\ x\cos\gamma \quad . \qquad - z\cos\alpha\ )^2 \\ + (-x\cos\beta + y\cos\alpha \quad . \qquad)^2, \end{array}$$

and the expressions which enter into this formula denote as follows; viz. if through the point P at right angles to the plane through P and the axis of rotation we draw a line PQ, = perpendicular distance of P from the axis of rotation, then the coordinates of Q referred to P as origin are

$$\begin{array}{l} . \qquad - y\cos\gamma + z\cos\beta, \\ x\cos\gamma \quad . \qquad - z\cos\alpha, \\ -x\cos\beta + y\cos\alpha \quad . \quad , \end{array}$$

respectively. Hence the foregoing quantities each multiplied by λ are the displacements of the point P in the directions of the axes, produced by the rotation λ.

60. Suppose that the axis of rotation (instead of passing through the origin) pass through the point (x_0, y_0, z_0); the only difference is that we must in the formula write $(x - x_0, y - y_0, z - z_0)$ in place of (x, y, z): and attending to the significations of the six coordinates, it thus appears that the displacements produced by the rotation are equal to λ into the expressions

$$\begin{array}{l} . \quad - hy + gz - a, \\ hx \quad . \quad -fz - b, \\ -gx + fy \quad . \quad -c, \end{array}$$

respectively; which is the theorem in question.

61. I say secondly that considering in a solid body the point (x, y, z) situate in the line (a, b, c, f, g, h), and writing

$$a, b, c, f, g, h = z\cos\beta - y\cos\gamma,\ x\cos\gamma - z\cos\alpha,\ y\cos\alpha - x\cos\beta,\ \cos\alpha,\ \cos\beta,\ \cos\gamma,$$

then for any infinitesimal motion of the solid body the displacement of the point in the direction of the line is

$$= ap + bq + cr + fl + gm + hn,$$

where p, q, r, l, m, n are constants depending on the infinitesimal motion.

In fact for any infinitesimal motion of a solid body the displacements of the point (x, y, z) are

$$\begin{array}{l} \delta x = l \quad . \quad + ry - qz, \\ \delta y = m - rx \quad . \quad + pz, \\ \delta z = n + qx - py \quad . \ , \end{array}$$

and hence the displacement in the direction of the line is

$$= \cos\alpha\, \delta x + \cos\beta\, \delta y + \cos\gamma\, \delta z,$$

which attending to the significations of (a, b, c, f, g, h) is

$$= ap + bq + cr + fl + gm + hn,$$

and we have thus the theorem in question.

62. It thus appears that for a system of rotations

$$\begin{array}{lll} \lambda_1 \text{ about the line} & (a_1, b_1, c_1, f_1, g_1, h_1), \\ \lambda_2 \quad " & (a_2, b_2, c_2, f_2, g_2, h_2), \\ \&\text{c.} \quad " & \&\text{c.} \end{array}$$

the displacements of the point (x, y, z) rigidly connected with the several lines are

$$\begin{aligned} \delta x &= \quad . \qquad - y\Sigma h\lambda + z\Sigma g\lambda - \Sigma a\lambda, \\ \delta y &= \quad x\Sigma h\lambda \quad . \quad - z\Sigma f\lambda - \Sigma b\Sigma, \\ \delta z &= - x\Sigma g\lambda + y\Sigma f\lambda \quad . \quad - \Sigma c\lambda, \end{aligned}$$

and when the rotations are in equilibrium then the displacements $(\delta x, \delta y, \delta z)$ of any point (x, y, z) whatever must each of them vanish; that is, we must have

$$\Sigma\lambda a = 0,\ \ \Sigma\lambda b = 0,\ \ \Sigma\lambda c = 0,\ \ \Sigma\lambda f = 0,\ \ \Sigma\lambda g = 0,\ \ \Sigma\lambda h = 0,$$

which are therefore the conditions for the equilibrium of the system of rotations λ_1, λ_2, &c.

63. And it further appears that for a system of forces acting on a rigid body,

$$\begin{array}{ll} \lambda_1 \text{ along the line} & (a_1, b_1, c_1, f_1, g_1, h_1), \\ \lambda_2 \quad " & (a_2, b_2, c_2, f_2, g_2, h_2), \\ \&\text{c.} & \end{array}$$

the conditions of equilibrium as given by the Principle of Virtual Velocities is

$$\Sigma\lambda\, (ap + bq + cr + fl + gm + hn) = 0,$$

or what is the same thing, that we have

$$\Sigma\lambda a = 0,\ \ \Sigma\lambda b = 0,\ \ \Sigma\lambda c = 0,\ \ \Sigma\lambda f = 0,\ \ \Sigma\lambda g = 0,\ \ \Sigma\lambda h = 0,$$

for the conditions of equilibrium of the system of forces λ_1, λ_2, &c. The conditions of equilibrium are thus precisely the same in the case of a system of rotations (infinitesimal rotations) and in that of a system of forces.

64. It now appears that the greater portion of the investigations in the first part of the present paper are applicable, and may be considered as relating, to the equilibrium of forces (or of rotations; but as the two theories are identical, it is sufficient to attend to one of them), and that we have in effect solved the following

question, "Given any system of two, three, four, five or six lines considered as belonging to a solid body, to determine the relations between these lines in order that there may exist along them forces which are in equilibrium;" but for greater clearness I will consider the several cases in order; it is hardly necessary to remark that when the forces exist the equilibrium will depend on the ratios only, and that the absolute magnitude of any one of the forces may be assumed at pleasure.

65. The condition in the case of two lines is of course that these shall coincide together, or form one and the same line; and the forces are then equal and opposite forces.

66. In the case of three lines, these must meet in a point and lie in a plane; and the force along each line must then be as the sine of the angle between the other two lines.

67. Supposing that the forces are λ along the line (a, b, c, f, g, h), λ_1 along the line $(a_1, b_1, c_1, f_1, g_1, h_1)$, and λ_2 along the line $(a_2, b_2, c_2, f_2, g_2, h_2)$, the conditions of equilibrium are $\lambda a + \lambda_1 a_1 + \lambda_2 a_2 = 0, \ldots\ldots \lambda h + \lambda_1 h_1 + \lambda_2 h_2 = 0$, any two of which determine the ratios $\lambda : \lambda_1 : \lambda_2$; these ratios were not worked out *ante* No. 38 for the reason that with the coordinates there made use of, a symmetrical solution was not obtainable; but in the present case, selecting the last three equations, these are

$$\begin{aligned}
\lambda \cos \alpha + \lambda_1 \cos \alpha_1 + \lambda_2 \cos \alpha_2 &= 0, \\
\lambda \cos \beta + \lambda_1 \cos \beta_1 + \lambda_2 \cos \beta_2 &= 0, \\
\lambda \cos \gamma + \lambda_1 \cos \gamma_1 + \lambda_2 \cos \gamma_2 &= 0,
\end{aligned}$$

giving in the first instance an equation which expresses that the three lines (assumed to meet in a point) lie in the same plane: and then if 01, 02, 12 be the angles between the pairs of lines respectively, giving by an easy transformation

$$\begin{aligned}
\lambda \qquad\quad + \lambda_1 \cos 01 + \lambda_2 \cos 02 &= 0, \\
\lambda \cos 10 + \lambda_1 \qquad\quad + \lambda_2 \cos 12 &= 0, \\
\lambda \cos 20 + \lambda_1 \cos 21 + \lambda_2 \qquad\quad &= 0.
\end{aligned}$$

68. Putting for shortness A, B, C in the place of 12, 20, 01 respectively, we thence find

$$\begin{vmatrix} 1 , & \cos C , & \cos B \\ \cos C , & 1 , & \cos A \\ \cos B , & \cos A , & 1 \end{vmatrix} = 0,$$

that is

$$1 - \cos^2 A - \cos^2 B - \cos^2 C + 2 \cos A \cos B \cos C = 0,$$

equivalent to $A + B + C = 2\pi$; and then from the first and second equations

$$\begin{aligned}
\lambda : \lambda_1 : \lambda_2 &= \cos A \cos C - \cos B : \cos B \cos C - \cos A : 1 - \cos^2 C, \\
&= \sin A \sin C : \sin B \sin C : \sin^2 C, \\
&= \sin A : \sin B : \sin C,
\end{aligned}$$

which is the required formula.

69. In the case of four given lines the condition (as noticed by Möbius) is that the four lines shall be generating lines of the same hyperboloid. In fact every line which meets three of the four lines must also meet the fourth line; for otherwise the moment of the system about such line would not be $=0$. Calling the lines 0, 1, 2, 3 and writing as before 01, 02, &c. for the moments of the several pairs of lines, then taking the moments of the system about the four lines respectively, we obtain directly the before-mentioned system of equations

$$\begin{array}{llll} \cdot & \lambda_1 01 + \lambda_2 02 + \lambda_3 03 = 0, \\ \lambda 10 \quad \cdot & + \lambda_2 12 + \lambda_3 13 = 0, \\ \lambda 20 + \lambda_1 21 \quad \cdot & + \lambda_3 23 = 0, \\ \lambda 30 + \lambda_1 31 + \lambda_1 41 \quad \cdot & = 0, \end{array}$$

leading as before to the relation

$$\sqrt{01}\sqrt{23} + \sqrt{02}\sqrt{31} + \sqrt{03}\sqrt{12} = 0,$$

and to the values

$$\lambda : \lambda_1 : \lambda_2 : \lambda_3 = \sqrt{12}\sqrt{23}\sqrt{31} : \sqrt{23}\sqrt{30}\sqrt{02} : \sqrt{30}\sqrt{01}\sqrt{13} : \sqrt{01}\sqrt{12}\sqrt{20}$$

for the proportional magnitudes of the forces. These last equations give

$$\lambda\lambda_1\, 01 = \lambda_2\lambda_3\, 23,$$

which, representing each force by a segment on the line along which the force acts, denotes that the tetrahedron of any two of the forces is equal to the tetrahedron of the other two forces; this is in fact equivalent to the theorem of M. Chasles, that if a system of forces be in any manner whatever reduced to two forces, the tetrahedron formed by these two forces has a constant volume.

70. In the case of five given lines, the lines must have a pair of tractors. Any four of the lines have in fact two tractors; and each of these tractors must meet the fifth line, for otherwise the moment of the system about the tractor would not be $=0$. In the case where the four lines have a twofold tractor, the foregoing consideration shows only that the fifth line meets the twofold tractor, but it fails to show that the twofold tractor is a twofold tractor in regard to the fifth line.

71. I stop to consider this particular case under the present statical point of view. Taking the twofold tractor for the axis of z; let the line 0 meet this line in the point $(0, 0, c)$, the coordinates (a, b, c, f, g, h) of this line being consequently

$$(c\cos\beta, \quad -c\cos\alpha, \quad 0, \quad \cos\alpha, \quad \cos\beta, \quad \cos\gamma)$$

and the like for the other four lines 1, 2, 3, 4. Using the sign Σ to refer to the last-mentioned four lines the equations of equilibrium become

$$\begin{aligned} \lambda c\cos\beta + \Sigma\lambda_1 c_1 \cos\beta_1 &= 0, \\ \lambda c\cos\alpha + \Sigma\lambda_1 c_1 \cos\alpha_1 &= 0, \\ \lambda \cos\alpha + \Sigma\lambda_1 \cos\alpha_1 &= 0, \\ \lambda \cos\beta + \Sigma\lambda_1 \cos\beta_1 &= 0, \\ \lambda \cos\gamma + \Sigma\lambda_1 \cos\gamma_1 &= 0. \end{aligned}$$

These equations give

$$\frac{\Sigma\lambda_1 c_1 \cos\beta_1}{\Sigma\lambda_1 \cos\alpha_1} = \frac{c\cos\beta}{\cos\alpha};$$

we may without loss of generality take the homographic conditions which express that the axis of z is a twofold tractor of the four lines to be

$$\frac{c_1\cos\beta_1}{\cos\alpha_1} = \frac{c_2\cos\beta_2}{\cos\alpha_2} = \frac{c_3\cos\beta_3}{\cos\alpha_3} = \frac{c_4\cos\beta_4}{\cos\alpha_4} = k,$$

and this being so, the last-mentioned equation becomes

$$\frac{c\cos\beta}{\cos\alpha} = k;$$

and it thus appears that the axis of z is a twofold tractor in regard also to the line 0.

72. In the case of six lines such that there exist along them forces which are in equilibrium, taking this as a definition of the involution of six lines, we may very readily obtain from statical considerations the before-mentioned construction of the sixth line; viz. it may be shown that given any five of the lines, say the lines 1, 2, 3, 4, 5 and a point O, we can through the point O determine a plane Ω, such that any line whatever through the point O and in the plane Ω is in involution with the five given lines. Consider the tractors of any four of the lines, say the lines 2, 3, 4, 5; we may through the point O draw a line OA meeting the two tractors; that is, the lines 2, 3, 4, 5 and the line OA will have a pair of common tractors. There consequently exist along these lines forces which are in equilibrium; and since only the ratios are material, the absolute magnitude of the force along the line OA may be anything whatever. Similarly, considering the tractors of the lines 1, 3, 4, 5, and through O a line OB meeting these tractors, then there exist along the lines 1, 3, 4, 5 and the line OB forces which are in equilibrium, and the absolute magnitude of the force along the line OB may be anything whatever. Hence, combining the two sets of forces, we have, along a line through O in the plane OA, OB, but otherwise indeterminate in its direction, a force in equilibrium with forces along the lines 1, 2, 3, 4, 5; that is, the line found as above is a line in involution with the lines 1, 2, 3, 4, 5.

73. It is to be added, that through O we cannot, out of the plane OA, OB, draw a line in involution with the lines 1, 2, 3, 4, 5; for if any such line OK existed, then we should have along each of the lines OA, OB, OK forces in equilibrium with forces along the lines 1, 2, 3, 4, 5; and the magnitudes of the three forces being each of them anything whatever, it would follow that along any line whatever through the point O there would exist a force in equilibrium with forces along the lines 1, 2, 3, 4, 5; that is, any line whatever through the point O would be a line in involution with these lines.

74. It hence appears, that drawing OA to meet the tractors of 2, 3, 4, 5; OB to meet those of 3, 4, 5, 1; OC to meet those of 4, 5, 1, 2; OD to meet those of 5, 1, 2, 3; and OE to meet those of 1, 2, 3, 4; the lines OA, OB, OC, OD, OE will be in one plane, say the plane Ω: and that any line through O in the plane Ω will be a line in involution with the lines 1, 2, 3, 4, 5.

75. There is another statical representation of the involution of six lines. If a system of forces act on a solid body, then taking six lines at random, the system will be in equilibrium if the sum of the moments be $=0$ in regard to each of the six lines. But if the six lines be in involution; then, for the very reason that a rotation about one of these lines is resolvable into rotations about the other five lines, if the sum of the moments be $=0$ for each of the five lines, it will also be $=0$ for the sixth line: that is, it is *not* sufficient for the equilibrium of the forces that the sum of the moments shall be $=0$ for each of the six lines. And we thus see that six lines in involution are lines such that the equilibrium of a system of forces about each of the six lines as axes does *not* insure the equilibrium of the system.

Article Nos. 76 and 77. *Transformation of Coordinates.*

76. Reverting to the general definition of the six coordinates (a, b, c, f, g, h) of a line by means of the points $(\alpha, \beta, \gamma, \delta)$ and $(\alpha', \beta', \gamma', \delta')$ on the line; suppose that instead of the original coordinate planes $x=0$, $y=0$, $z=0$, $w=0$ (forming a tetrahedron $ABCD$) we have new coordinate planes $x_0=0$, $y_0=0$, $z_0=0$, $w_0=0$ (forming a tetrahedron $A_0B_0C_0D_0$); and that the relations between the two sets of current coordinates are given by the equations

$$\begin{aligned} x : y : z : w = &\ (\lambda_1, \mu_1, \nu_1, \rho_1 \between x_0, y_0, z_0, w_0) \\ &: (\lambda_2, \mu_2, \nu_2, \rho_2 \between x_0, y_0, z_0, w_0) \\ &: (\lambda_3, \mu_3, \nu_3, \rho_3 \between x_0, y_0, z_0, w_0) \\ &: (\lambda_4, \mu_4, \nu_4, \rho_4 \between x_0, y_0, z_0, w_0), \end{aligned}$$

with, of course, the like relations between the original coordinates $(\alpha, \beta, \gamma, \delta)$ and new coordinates $(\alpha_0, \beta_0, \gamma_0, \delta_0)$, and between the original coordinates $(\alpha', \beta', \gamma', \delta')$ and the new coordinates $(\alpha_0', \beta_0', \gamma_0', \delta_0')$, of the two points on the line (a, b, c, f, g, h); then taking $(a_0, b_0, c_0, f_0, g_0, h_0)$ as the new values of the six coordinates of the line, viz. writing

$$\begin{aligned} &a_0 : b_0 : c_0 : f_0 : g_0 : h_0 \\ &= \beta_0\gamma_0' - \beta_0'\gamma_0 : \gamma_0\alpha_0' - \gamma_0'\alpha_0 : \alpha_0\beta_0' - \alpha_0'\beta_0 : \alpha_0\delta_0' - \alpha_0'\delta_0 : \beta_0\delta_0' - \beta_0'\delta_0 : \gamma_0\delta_0' - \gamma_0'\delta_0, \end{aligned}$$

we obtain a system of formulæ which may be conveniently written as follows:

$$\begin{aligned} &a : b : c : f : g : h \\ &= \frac{\mu\nu}{23} a_0 + \frac{\nu\lambda}{23} b_0 + \frac{\lambda\mu}{23} c_0 + \frac{\lambda\rho}{23} f_0 + \frac{\mu\rho}{23} g_0 + \frac{\nu\rho}{23} h_0 \\ &: 31 \\ &: 12 \\ &: 14 \\ &: 24 \\ &: 34 \end{aligned}$$

viz. the top line stands for $(\mu_2\nu_3 - \mu_3\nu_2) a_0 + (\nu_2\lambda_3 - \nu_3\lambda_2) b_0 + \&c.$, and the other lines are obtained from this by mere alterations of the suffixes.

77. As to the interpretation of these formulæ, taking

$A\,B\,C\,D$ as the fundamental tetrahedron for $(x\,,\ y\,,\ z\,,\ w\,)$,

$A_0B_0C_0D_0$ " " $(x_0,\ y_0,\ z_0,\ w_0)$,

then

$(\lambda_1,\ \mu_1,\ \nu_1,\ \rho_1 \between x_0,\ y_0,\ z_0,\ w_0) = 0$ is the equation of plane BCD,

$(\lambda_2,\ \mu_2,\ \nu_2,\ \rho_2 \between$ " $) = 0$ " CDA,

$(\lambda_3,\ \mu_3,\ \nu_3,\ \rho_3 \between$ " $) = 0$ " DAB,

$(\lambda_4,\ \mu_4,\ \nu_4,\ \rho_4 \between$ " $) = 0$ " ABC,

whence, observing that the second and third equations belong to two planes each passing through the line DA, it appears that the coefficients

$$\frac{\mu\nu}{23},\ \frac{\nu\lambda}{23},\ \frac{\lambda\mu}{23},\ \frac{\lambda\rho}{23},\ \frac{\mu\rho}{23},\ \frac{\nu\rho}{23},$$

are the six coordinates of the line DA, expressed in regard to the tetrahedron $(A_0B_0C_0D_0)$; and similarly that the coefficients in the six expressions of the transformation formula are the six coordinates of the lines AD, BD, CD, BC, CA, AB respectively in regard to the tetrahedron $(A_0B_0C_0D_0)$.

In the preceding formulæ for the transformation of coordinates the ratios only have been attended to, no determinate absolute magnitudes have been assigned to the coordinates $(a,\ b,\ c,\ f,\ g,\ h)$. But I will nevertheless show how we may attribute absolute magnitudes to these coordinates.

Article Nos. 78 to 80. *New definition of the six coordinates as to their absolute magnitudes.*

78. I assume $(x,\ y,\ z,\ w)$ to be "volume" coordinates; viz. taking as before $ABCD$ for the fundamental tetrahedron, and denoting the point $(x,\ y,\ z,\ w)$ by P, I assume that we have

$$x : y : z : w : 1 = PBCD : APCD : ABPD : ABCP : ABCD,$$

where $PBCD$, &c. denote the volumes of the several tetrahedra $PBCD$, &c. It is to be noticed that the volume is in every case taken with a determinate sign: analytically the sign may be fixed by taking $(x_a,\ y_a,\ z_a)$, &c., as the Cartesian coordinates of the points A, &c. and writing

$$PBCD = \begin{vmatrix} x_p, & x_b, & x_c, & x_d \\ y_p, & y_b, & y_c, & y_d \\ z_p, & z_b, & z_c, & z_d \\ 1\ , & 1\ , & 1\ , & 1 \end{vmatrix},\ \text{\&c.}$$

(whence of course $PBCD = PCDA = -PCBD$, &c. according to the rule of signs): or we may in an equivalent manner, but less easily, determine the sign, by considering the sense of the rotation about CD (considered as an axis drawn from C to D) which would be produced by a force along PB (from P to B).

79. It is to be observed that the foregoing values give identically $x+y+z+w=1$, so that the equation of the plane infinity is $x+y+z+w=0$. The values of the coordinates (x, y, z, w) may be written

$$x : y : z : w : 1 = PBCD : PCAD : PABD : PCBA : ABCD;$$

or in the original form

$$x : y : z : w : 1 = PBCD : APCD : ABPD : ABCP : ABCD,$$

as may be most convenient.

80. Denoting the points $(\alpha, \beta, \gamma, \delta)$ and $(\alpha', \beta', \gamma', \delta')$ by Q, Q' respectively, we have

$$\alpha : \beta : \gamma : \delta : 1 = QBCD : AQCD : ABQD : ABCQ : ABCD$$

and

$$\alpha' : \beta' : \gamma' : \delta' : 1 = Q'BCD : AQ'CD : ABQ'D : ABCQ' : ABCD,$$

and writing

$$(a, b, c, f, g, h) = (\beta\gamma' - \beta'\gamma,\ \gamma\alpha' - \gamma'\alpha,\ \alpha\beta' - \alpha'\beta,\ \alpha\delta' - \alpha'\delta,\ \beta\delta' - \beta'\delta,\ \gamma\delta' - \gamma'\delta),$$

viz. the two sets being taken to be *equal*, $a = \beta\gamma' - \beta'\gamma$, &c. instead of merely proportional, then it is easily seen that we obtain

$$\begin{aligned} &a : b : c : f : g : h : 1 \\ &= AQQ'D : Q'BQD : QQ'CD : QBCQ' : AQCQ' : ABQQ' : ABCD, \end{aligned}$$

that is, in order to form the first six combinations we successively replace

$$(B, C), (C, A), (A, B), (A, D), (B, D), (C, D)$$

in $ABCD$ by (Q, Q').

Article No. 81. *Resulting formulæ of Transformation.*

81. For the transformation of coordinates if we assume

$$\begin{aligned} x &= (\lambda_1, \mu_1, \nu_1, \rho_1 \between x_0, y_0, z_0, w_0), \\ y &= (\lambda_2, \mu_2, \nu_2, \rho_2 \between \quad " \quad), \\ z &= (\lambda_3, \mu_3, \nu_3, \rho_3 \between \quad " \quad), \\ w &= (\lambda_4, \mu_4, \nu_4, \rho_4 \between \quad " \quad), \end{aligned}$$

and take also (a, b, c, f, g, h), $(a_0, b_0, c_0, f_0, g_0, h_0)$ respectively *equal*, instead of merely proportional, to the foregoing values, then, observing that for the point A_0 we have $(x_0, y_0, z_0, w_0) = (1, 0, 0, 0)$ we see that $\lambda_1, \lambda_2, \lambda_3, \lambda_4$ are the $ABCD$-coordinates of A_0; and the like as to the other sets of coefficients; viz. we have

$$\begin{aligned} \lambda_1 : \lambda_2 : \lambda_3 : \lambda_4 : 1 &= A_0BCD : AA_0CD : ABA_0D : ABCA_0 : ABCD \\ \mu_1 : \mu_2 : \mu_3 : \mu_4 : 1 &= B_0 \quad " \quad : \ " \ B_0 \ " \ : \ " \ B_0 \ " \ : \ " \ B_0 : \ " \\ \nu_1 : \nu_2 : \nu_3 : \nu_4 : 1 &= C_0 \quad " \quad : \ " \ C_0 \ " \ : \ " \ C_0 \ " \ : \ " \ C_0 : \ " \\ \rho_1 : \rho_2 : \rho_3 : \rho_4 : 1 &= D_0 \quad " \quad : \ " \ D_0 \ " \ : \ " \ D_0 \ " \ : \ " \ D_0 : \ " \end{aligned}$$

and we hence find

$$\frac{\mu\nu}{23} : \frac{\nu\lambda}{23} : \frac{\lambda\mu}{23} : \frac{\lambda\rho}{23} : \frac{\mu\rho}{23} : \frac{\nu\rho}{23} : 1$$

$$= AB_0C_0D : AC_0A_0D : AA_0B_0D : AA_0D_0D : AB_0D_0D : AC_0D_0D : ABCD$$

viz. multiplying the last-mentioned set of terms by $A_0B_0C_0D_0 \div ABCD$, in order to make the last term equal to unity, we see that the coefficients $\frac{\mu\nu}{23}, \frac{\nu\lambda}{23}$ &c. are equal to $\frac{A_0B_0C_0D_0}{ABCD}$ into the six $(ABCD)_0$ – coordinates respectively of the line AD by means of the points A, D thereof. And similarly in the six expressions which enter into the formula of transformation, the coefficients are $= \frac{A_0B_0C_0D_0}{ABCD}$ into the six $(ABCD)_0$ – coordinates of the

line	AD	in regard to points	A, D thereof
,,	BD	,,	B, D ,,
,,	CD	,,	C, D ,,
,,	BC	,,	B, C ,,
,,	CA	,,	C, A ,,
,,	AB	,,	A, B ,,

The foregoing theory of the transformation of coordinates seemed to me interesting for its own sake, and I have developed it in preference to the more simple theory which might easily be established of the case in which the coordinates are quantitatively defined as being equal to

$$(z_0 \cos\beta - y_0 \cos\gamma,\ x_0 \cos\gamma - z_0 \cos\alpha,\ y_0 \cos\beta - x_0 \cos\alpha,\ \cos\alpha,\ \cos\beta,\ \cos\gamma)$$

respectively.

436.

ON A CERTAIN SEXTIC TORSE.

[From the *Transactions of the Cambridge Philosophical Society*, vol. XI. Part III. (1871), pp. 507—523. Read Nov. 8, 1869.]

THE torse (developable surface) intended to be considered is that which has for its edge of regression an excubo-quartic curve, or say a *unicursal* quartic curve. I call to mind that (excluding the plane quartic) a quartic curve is either a quadriquadric, viz. it is the complete intersection of two quadric surfaces; or else it is an excubo-quartic, viz. there is through the curve only one quadric surface, and the curve is the partial intersection of this quadric surface with a cubic surface through two generating lines (of the same kind) of the quadric surface. Returning to the quadriquadric curve, this may be general, nodal, or cuspidal; viz. if the two quadric surfaces have an ordinary contact, the curve of intersection is a nodal quadriquadric; if they have a stationary contact, the curve is a cuspidal quadriquadric.

The unicursal quartic is a curve such that the coordinates (x, y, z, w) of any point thereof are proportional to rational and integral quartic functions $(*\between\theta, 1)^4$ of a variable parameter θ; and the general unicursal quartic is in fact the excubo-quartic; but included as particular cases of the unicursal curve (although not as cases of the excubo-quartic as above defined) we have the nodal quadriquadric and the cuspidal quadriquadric. The torse having for its edge of regression a unicursal curve is a sextic torse; and this is in fact the order of the torse derived from the excubo-quartic, and from the nodal quadriquadric; but for the cuspidal quadriquadric, there is a depression of *one*, and the torse becomes a quintic torse. The equations have been obtained of (1) the sextic torse derived from the nodal quadriquadric, (2) the quintic torse derived from the cuspidal quadriquadric, (3) the sextic torse derived from a certain special excubo-quartic; but the equation of the torse derived from the general unicursal quartic has not yet been found. To show at the outset what the analytical problem is, I anticipate the remark that the coordinates (x, y, z, w) of a point on the curve may by an obvious reduction be rendered proportional to the fourth powers $(\theta+\alpha)^4$, $(\theta+\beta)^4$, $(\theta+\gamma)^4$, $(\theta+\delta)^4$ in the parameter θ; this leads to an equation

$$\frac{x}{(\theta+\alpha)^2}+\frac{y}{(\theta+\beta)^2}+\frac{z}{(\theta+\gamma)^2}+\frac{w}{(\theta+\delta)^2}=0,$$

for the osculating plane at the point (x, y, z, w); or observing that this equation, when integralised, is of the form $(x, y, z, w \between \theta, 1)^6=0$, we see that the equation is obtained by equating to zero the discriminant of a certain sextic function in θ; the discriminant is of the order 10 in the coordinates (x, y, z, w), but it obviously contains the factor $xyzw$, or throwing this out we have an equation of the order 6, so that the torse is (as above stated) a sextic torse.

Theorem relating to Four Binary Quartics.

1. Consider the four quartics:

$$(a_1, b_1, c_1, d_1, e_1 \between x, y)^4,$$
$$(a_2, b_2, c_2, d_2, e_2 \between x, y)^4,$$
$$(a_3, b_3, c_3, d_3, e_3 \between x, y)^4,$$
$$(a_4, b_4, c_4, d_4, e_4 \between x, y)^4,$$

then if $\lambda_1, \lambda_2, \lambda_3, \lambda_4$ are any four quantities, these may be determined, and that in four different ways, so that

$$\lambda_1(a_1, \ldots \between x, y)^4+\lambda_2(a_2, \ldots \between x, y)^4+\lambda_3(a_3, \ldots \between x, y)^4+\lambda_4(a_4, \ldots \between x, y)^4=(\beta x+\alpha y)^4,$$

a perfect fourth power; in fact, equating the coefficients of the different powers of $(x, y)^4$, we have five equations, which determine the ratios of the unknown quantities $\lambda_1, \lambda_2, \lambda_3, \lambda_4$; α, β: eliminating $\lambda_1, \lambda_2, \lambda_3, \lambda_4$, we find the equation

$$\begin{vmatrix} \beta^4, & \beta^3\alpha, & \beta^2\alpha^2, & \beta\alpha^3, & \alpha^4 \\ a_1, & b_1, & c_1, & d_1, & e_1 \\ a_2, & b_2, & c_2, & d_2, & e_2 \\ a_3, & b_3, & c_3, & d_3, & e_3 \\ a_4, & b_4, & c_4, & d_4, & e_4 \end{vmatrix}=0,$$

giving four different values of the ratio $\alpha:\beta$; or, assigning at pleasure a value to α or β (say $\beta=1$), then to each of the four sets of values of (α, β) there correspond a determinate set of values of $(\lambda_1, \lambda_2, \lambda_3, \lambda_4)$; that is, we have as stated four sets of values of $\lambda_1, \lambda_2, \lambda_3, \lambda_4$; α, β.

Standard Equation of the Unicursal Quartic.

2. The coordinates (x, y, z, w) being originally taken to be proportional to any four given quartic functions $(*\between\theta, 1)^4$ of the parameter θ, then forming a linear function of the coordinates, we have four sets of values of the multipliers, each reducing the function of θ to a perfect fourth power; that is, writing (X, Y, Z, W) for the linear functions of the original coordinates, and taking (X, Y, Z, W) as coordinates, it appears that the unicursal quartic may be represented by the equations

$$X : Y : Z : W = (\theta+\alpha)^4 : (\theta+\beta)^4 : (\theta+\gamma)^4 : (\theta+\delta)^4.$$

Tangent Line, and Osculating Plane of the Unicursal Quartic.

3. The equations of the tangent line at the point (θ) (that is, the point the coordinates whereof are as $(\theta+\alpha)^4 : (\theta+\beta)^4 : (\theta+\gamma)^4 : (\theta+\delta)^4$) are at once seen to be

$$\left\|\begin{matrix} X, & Y, & Z, & W \\ (\theta+\alpha)^4, & (\theta+\beta)^4, & (\theta+\gamma)^4, & (\theta+\delta)^4 \\ (\theta+\alpha)^3, & (\theta+\beta)^3, & (\theta+\gamma)^3, & (\theta+\delta)^3 \end{matrix}\right\| = 0,$$

and that of the osculating plane to be

$$\left|\begin{matrix} X, & Y, & Z, & W \\ (\theta+\alpha)^4, & (\theta+\beta)^4, & (\theta+\gamma)^4, & (\theta+\delta)^4 \\ (\theta+\alpha)^3, & (\theta+\beta)^3, & (\theta+\gamma)^3, & (\theta+\delta)^3 \\ (\theta+\alpha)^2, & (\theta+\beta)^2, & (\theta+\gamma)^2, & (\theta+\delta)^2 \end{matrix}\right| = 0.$$

Writing as in the sequel

$$\begin{aligned} a &= \beta-\gamma, & f &= \alpha-\delta, \\ b &= \gamma-\alpha, & g &= \beta-\delta, \\ c &= \alpha-\beta, & h &= \gamma-\delta, \end{aligned}$$

the equations of the tangent line become

$$\begin{array}{rrrrl} \cdot & \dfrac{hY}{(\theta+\beta)^3} & -\dfrac{gZ}{(\theta+\gamma)^3} & +\dfrac{aZ}{(\theta+\delta)^3} & =0, \\ -\dfrac{hX}{(\theta+\alpha)^3} & \cdot & +\dfrac{fZ}{(\theta+\gamma)^3} & +\dfrac{bZ}{(\theta+\delta)^3} & =0, \\ \dfrac{gX}{(\theta+\alpha)^3} & -\dfrac{fY}{(\theta+\beta)^3} & \cdot & +\dfrac{cZ}{(\theta+\delta)^3} & =0, \\ -\dfrac{aX}{(\theta+\alpha)^3} & -\dfrac{bY}{(\theta+\beta)^3} & -\dfrac{cZ}{(\theta+\gamma)^3} & \cdot & =0, \end{array}$$

(equivalent of course to two equations), and the equation of the osculating plane becomes

$$\frac{ahgX}{(\theta+\alpha)^2}+\frac{hbfY}{(\theta+\beta)^2}+\frac{gfcZ}{(\theta+\gamma)^2}+\frac{abcW}{(\theta+\delta)^2}=0.$$

Modification of the foregoing notation, and final form for the Unicursal Quartic.

4. If instead of the coordinates (X, Y, Z, W) we introduce the coordinates (x, y, z, w) connected therewith by the relations

$$ahgX : hbfY : gfcZ : abcW = x : y : z : w,$$

or, what is the same thing,

$$X : Y : Z : W = bcfx : cagy : abhz : fghw,$$

then the curve is given by the equations

$$x : y : z : w = ahg(\theta+\alpha)^4 : hbf(\theta+\beta)^4 : cfg(\theta+\gamma)^4 : abc(\theta+\delta)^4.$$

The equations of the tangent line are

$$\begin{array}{rrrrl} \cdot & \dfrac{cy}{(\theta+\beta)^3} & -\dfrac{bz}{(\theta+\gamma)^3} & +\dfrac{fw}{(\theta+\delta)^3} & =0, \\ -\dfrac{cx}{(\theta+\alpha)^3} & \cdot & +\dfrac{az}{(\theta+\gamma)^3} & +\dfrac{gw}{(\theta+\delta)^3} & =0, \\ \dfrac{bx}{(\theta+\alpha)^3} & -\dfrac{ay}{(\theta+\beta)^3} & \cdot & +\dfrac{hw}{(\theta+\delta)^3} & =0, \\ -\dfrac{fx}{(\theta+\alpha)^3} & -\dfrac{gy}{(\theta+\beta)^3} & -\dfrac{hz}{(\theta+\gamma)^3} & \cdot & =0, \end{array}$$

and the equation of the osculating plane is

$$\frac{x}{(\theta+\alpha)^2}+\frac{y}{(\theta+\beta)^2}+\frac{z}{(\theta+\gamma)^2}+\frac{w}{(\theta+\delta)^2}=0.$$

Determination of the Sextic Torse.

5. Starting from the equation of the osculating plane written under the form

$$\begin{aligned} & x(\theta+\beta)^2(\theta+\gamma)^2(\theta+\delta)^2 \\ & + y(\theta+\gamma)^2(\theta+\delta)^2(\theta+\alpha)^2 \\ & + z(\theta+\delta)^2(\theta+\alpha)^2(\theta+\beta)^2 \\ & + w(\theta+\alpha)^2(\theta+\beta)^2(\theta+\gamma)^2=0, \end{aligned}$$

the equation of the torse is obtained by equating to zero the discriminant of the sextic function. Writing as before

$$\begin{aligned} a &= \beta-\gamma, & f &= \alpha-\delta, \\ b &= \gamma-\alpha, & g &= \beta-\delta, \\ c &= \alpha-\beta, & h &= \gamma-\delta, \end{aligned}$$

equations which give

$$\begin{array}{rrrrl} . & h & -g & +a & =0, \\ -h & . & +f & +b & =0, \\ g & -f & . & +c & =0, \\ -a & -b & -c & . & =0, \end{array}$$

$$\begin{array}{rrrrl} . & h\beta & -g\gamma & +a\delta & =0, \\ -h\alpha & . & +f\gamma & +b\delta & =0, \\ g\alpha & -f\beta & . & +c\delta & =0, \\ -a\alpha & -b\beta & -c\gamma & . & =0, \end{array}$$

and also

$$af+bg+ch=0,$$

then the discriminant is a function of (x, y, z, w), (a, b, c, f, g, h) of the degree 10 in (x, y, z, w) and the degree 30 in (a, b, c, f, g, h). But the equation in θ has two equal roots, or the discriminant vanishes, if any one of the quantities (x, y, z, w) is $=0$; and again, if any one of the differences $\alpha-\beta$, &c. (that is any one of the quantities a, b, c, f, g, h) is $=0$: the discriminant thus contains the factors $xyzw$ and $(abcfgh)^2$, and throwing these out, we have an equation of the form

$$\Delta=(a, b, c, f, g, h)^{18}(x, y, z, w)^6=0,$$

which is the equation of the sextic torse.

Principal Sections of the Torse.

6. Consider for instance the section by the plane $w=0$. Writing $w=0$, the equation of the osculating plane is

$$(\theta+\delta)^2[x(\theta+\beta)^2(\theta+\gamma)^2+y(\theta+\gamma)^2(\theta+\alpha)^2+z(\theta+\alpha)^2(\theta+\beta)^2]=0.$$

The discriminant of the sextic function vanishes identically in virtue of the double factor $(\theta+\delta)^2$. But omitting this factor, the equation becomes

$$x(\theta+\beta)^2(\theta+\gamma)^2+y(\theta+\gamma)^2(\theta+\alpha)^2+z(\theta+\alpha)^2(\theta+\beta)^2=0.$$

The discriminant of this quartic function of θ is a function of x, y, z, a, b, c of the degree 6 in (x, y, z) and 12 in (a, b, c); it contains however the factors xyz, $a^2b^2c^2$, and the remaining factor is of the degree 3 in (x, y, z) and 6 in (a, b, c); this remaining factor is as will presently be seen

$$=(a^2x+b^2y+c^2z)^3-27a^2b^2c^2\,xyz.$$

The last mentioned sextic equation in θ will have a triple root $\theta=-\delta$, if only the value $\theta=-\delta$ makes to vanish the factor in [ ], that is if we have

$$0=g^2h^2x+h^2f^2y+f^2g^2z.$$

The foregoing results lead to the conclusion that for $w=0$, we have

$$\Delta = (g^2h^2x + h^2f^2y + f^2g^2z)^3 \left[(a^2x + b^2y + c^2z)^3 - 27a^2b^2c^2\, xyz\right];$$

but this will appear more distinctly as follows.

7. First, as to the factor $(a^2x + b^2y + c^2z)^3 - 27a^2b^2c^2\, xyz$: writing in the equation of the osculating plane $w=0$, the equation becomes

$$\frac{x}{(\theta+\alpha)^2} + \frac{y}{(\theta+\beta)^2} + \frac{z}{(\theta+\gamma)^2} = 0,$$

which equation is therefore that of the trace of the osculating plane on the plane $w=0$; the envelope of the trace in question is *a part* of the section of the torse by the plane $w=0$. To find the equation of this envelope we must eliminate θ from the foregoing, and its derived equation

$$\frac{x}{(\theta+\alpha)^3} + \frac{y}{(\theta+\beta)^3} + \frac{z}{(\theta+\gamma)^3} = 0;$$

the two equations give

$$x : y : z = a(\theta+\alpha)^3 : b(\theta+\beta)^3 : c(\theta+\gamma)^3,$$

and thence

$$(a^2x)^{\frac{1}{3}} + (b^2y)^{\frac{1}{3}} + (c^2z)^{\frac{1}{3}} = a(\theta+\alpha) + b(\theta+\beta) + c(\theta+\gamma) = 0,$$

that is, we have

$$(a^2x)^{\frac{1}{3}} + (b^2y)^{\frac{1}{3}} + (c^2z)^{\frac{1}{3}} = 0,$$

or, what is the same thing,

$$(a^2x + b^2y + c^2z)^3 - 27a^2b^2c^2\, xyz = 0$$

for a part of the section in question.

8. I have said that the foregoing cubic is *a part* of the section; the equations

$$x : y : z : w = ahg(\theta+\alpha)^4 : bhf(\theta+\beta)^4 : cfg(\theta+\gamma)^4 : abc(\theta+\delta)^4,$$

which for $w=0$ give $\theta = -\delta$, and thence $x : y : z = af^3 : bg^3 : ch^3$, show that the last mentioned point is a four-pointic intersection of the curve with the plane $w=0$. But the curve, having four consecutive points, will have three consecutive tangents in the plane $w=0$; that is, the tangent at the point in question will present itself as a threefold factor in the equation of the torse. Writing in the equations of the tangent $w=0$, $\theta=-\delta$, we find for the equation of the tangent in question

$$\frac{x}{f^2} + \frac{y}{g^2} + \frac{z}{h^2} = 0,$$

or, what is the same thing,

$$g^2h^2x + h^2f^2y + f^2g^2z = 0.$$

Hence the section by the plane $w=0$ is made up of this line taken three times, and of the last mentioned cubic curve.

9. By symmetry, we conclude that the sections by the principal planes $x=0$, $y=0$, $z=0$, $w=0$, are each made up of a line taken three times, and of a cubic curve: viz. these are

$$\begin{array}{lllll|llll}
x=0, & . & b^2f^2y & +c^2f^2z & +b^2c^2w=0, & . & (h^2y)^{\frac{1}{3}} & +(g^2z)^{\frac{1}{3}} & +(a^2w)^{\frac{1}{3}}=0,\\
y=0, & a^2g^2x & . & c^2g^2z & +c^2a^2w=0, & (h^2x)^{\frac{1}{3}} & . & +(f^2z)^{\frac{1}{3}} & +(b^2w)^{\frac{1}{3}}=0,\\
z=0, & a^2h^2x & +b^2h^2y & . & +a^2b^2w=0, & (g^2x)^{\frac{1}{3}} & +(f^2y)^{\frac{1}{3}} & . & +(c^2w)^{\frac{1}{3}}=0,\\
w=0, & g^2h^2x & +h^2f^2y & +f^2g^2z & . \quad =0, & (a^2x)^{\frac{1}{3}} & +(b^2y)^{\frac{1}{3}} & +(c^2z)^{\frac{1}{3}} & =0,
\end{array}$$

where for shortness I have written the equations of the four cubics in their irrational forms respectively.

Partial Determination of the Equation.

10. As the value of Δ is known when any one of the coordinates x, y, z, w is put $=0$, we in fact know all the terms of Δ, except those which contain the factor $xyzw$, which unknown terms, as Δ is of the degree 6, are of the form $(*\between x, y, z, w)^2$.

I remark that if $(xyzw)$ is any homogeneous function $(*\between x, y, z, w)^p$, and (xyz), (xy), (x) are what $(xyzw)$ become on putting therein $(w=0)$, $(z=0,\ w=0)$, $(y=0,\ z=0,\ w=0)$ respectively, and the like for the other similar symbols, then that

$$\begin{aligned}
(xyzw) = {} & (x)+(y)+(z)+(w)\\
& -(xy)-(xz)-(xw)-(yz)-(yw)-(zw)\\
& +(xyz)+(xyw)+(xzw)+(yzw)\\
& +\text{terms multiplied by } xyzw\,;
\end{aligned}$$

in fact, omitting the last line, this equation on writing therein $x=0$ or $y=0$ or $z=0$ or $w=0$, becomes an identity, that is, the difference of the two sides vanishes when any one of these equations is satisfied, and such difference contains therefore the factor $xyzw$; which proves the theorem. It hence appears that the equation $\Delta=0$ of the torse is

$$\begin{aligned}
\Delta = {} & a^6g^6h^6x^6+b^6h^6f^6y^6+c^6f^6g^6z^6+a^6b^6c^6w^6\\
& -(g^2h^2x\ +h^2f^2y)^3\,(a^2x+b^2y)^3\\
& -(h^2f^2y+f^2g^2z)^3\,(b^2y\ +c^2z)^3\\
& -(g^2h^2x\ +f^2g^2z)^3\,(a^2x\ +c^2z)^3\\
& -(a^2h^2x\ +a^2b^2w)^3\,(g^2x\ +c^2w)^3\\
& -(b^2h^2y\ +a^2b^2w)^3\,(f^2y+c^2w)^3\\
& -(c^2g^2z\ +c^2a^2w)^3\,(f^2z+b^2w)^3\\
& +(b^2f^2y+c^2f^2z+b^2c^2w)^3\,[(h^2y+g^2z+a^2w)^3-27a^2g^2h^2yzw]\\
& +(a^2g^2x+c^2g^2z\ +c^2a^2w)^3\,[(h^2x+f^2z+b^2w)^3-27b^2h^2f^2zxw]\\
& +(a^2h^2x+b^2h^2y\ +a^2b^2w)^3\,[(g^2x+f^2y+c^2w)^3-27c^2f^2g^2xyw]\\
& +(g^2h^2x+h^2f^2y+f^2g^2z)^3\,[(a^2x+b^2y+c^2z\,)^3-27a^2b^2c^2xyz\,]\\
& +xyzw)\,(*\between x, y, z, w)^2,
\end{aligned}$$

where the ten coefficients of $(*\between x, y, z, w)^2$ remain to be found.

Process for the Determination of the Unknown Coefficients.

11. At a point of the cubic curve in the plane $w=0$, we have

$$x : y : z = a(\theta+\alpha)^3 : b(\theta+\beta)^3 : c(\theta+\gamma)^3;$$

and the tangent plane at this point is the osculating plane of the curve; that is, it is the plane

$$\frac{x'}{(\theta+\alpha)^2}+\frac{y'}{(\theta+\beta)^2}+\frac{z'}{(\theta+\gamma)^2}+\frac{w'}{(\theta+\delta)^2}=0$$

if for a moment (x', y', z', w') are the current coordinates of a point in the tangent plane. But the equation of the tangent plane as deduced from the equation $\Delta=0$ is

$$x'\frac{d\Delta}{dx}+y'\frac{d\Delta}{dy}+z'\frac{d\Delta}{dz}+w'\frac{d\Delta}{dw}=0,$$

where in the differential coefficients of Δ, the coordinates (x, y, z, w) are considered as having the values

$$x : y : z : w = a(\theta+\alpha)^3 : b(\theta+\beta)^3 : c(\theta+\gamma)^3 : 0.$$

Hence, with these values of (x, y, z, w), we have

$$\frac{d\Delta}{dx}:\frac{d\Delta}{dy}:\frac{d\Delta}{dz}:\frac{d\Delta}{dw}=\frac{1}{(\theta+\alpha)^2}:\frac{1}{(\theta+\beta)^2}:\frac{1}{(\theta+\gamma)^2}:\frac{1}{(\theta+\delta)^2};$$

conditions which determine the values of certain of the coefficients of $(*\between x, y, z, w)^2$, viz. the six coefficients of the terms independent of w; and when these are known the values of the remaining four coefficients are at once obtained by symmetry.

12. To develope this process, disregarding the higher powers of w, we may write

$$\Delta=\Theta+3w\Phi+xyzw(*\between x, y, z)^2,$$

where Θ denotes the terms independent of w, $3w\Phi$ the known terms which contain the factor w, and $xyzw\,(*\between x, y, z)^2$ the unknown terms which contain this same factor; the value of $(*\between x, y, z)^2$ being clearly $=(*\between x, y, z, 0)^2$.

We have, moreover,

$$\Theta = (g^2h^2x+h^2f^2y+f^2g^2z)^3[(a^2x+b^2y+c^2z)^3-27a^2b^2c^2xyz],$$

and

$$\begin{aligned}\Phi = \;& f^4(b^2y+c^2z)^2[a^2f^2(h^4y^2-7h^2g^2yz+g^4z^2)(b^2y+c^2z)+b^2c^2(h^2y+g^2z)^3]\\ &+g^4(a^2x+c^2z)^2[b^2g^2(h^4x^2-7h^2f^2xz+f^4z^2)(a^2x+c^2z)+c^2a^2(h^2x+f^2z)^3]\\ &+h^4(a^2x+b^2y)^2[c^2h^2(g^4x^2-7g^2h^2xy+f^4y^2)(a^2x+b^2y)+a^2b^2(g^2x+f^2y)^3]\\ &-a^6h^4g^4(b^2g^2+c^2h^2)x^5\\ &-b^6h^4f^4(c^2h^2+a^2f^2)y^5\\ &-c^6f^4g^4(a^2f^2+b^2g^2)z^5.\end{aligned}$$

13. The equations, putting after the differentiations $w=0$, and writing for shortness $(*)$ in place of $(*\between x, y, z)^2$, become

$$\frac{d\Theta}{dx} : \frac{d\Theta}{dy} : \frac{d\Theta}{dz} : 3\Phi + xyz\,(*) = \frac{1}{(\theta+\alpha)^2} : \frac{1}{(\theta+\beta)^2} : \frac{1}{(\theta+\gamma)^2} : \frac{1}{(\theta+\delta)^2}.$$

Now, observing that the second factor of Θ vanishes for the values

$$a(\theta+\alpha)^3,\quad b(\theta+\beta)^3,\quad c(\theta+\gamma)^3 \text{ of } (x, y, z),$$

we have simply

$$\frac{d\Theta}{dx} = (g^2h^2x + h^2f^2y + f^2g^2z)^3 \,.\, 3a^2\,[(a^2x + b^2y + c^2z)^2 - 9b^2c^2yz].$$

But

$$\begin{aligned} a^2x + b^2y + c^2z &= a^3(\theta+\alpha)^3 + b^3(\theta+\beta)^3 + c^3(\theta+\gamma)^3, \\ &= 3abc\,(\theta+\alpha)(\theta+\beta)(\theta+\gamma), \end{aligned}$$

in virtue of the relation $a(\theta+\alpha)+b(\theta+\beta)+c(\theta+\gamma)=0$ and hence

$$\begin{aligned} [(a^2x + b^2y + c^2z)^2 - 9b^2c^2yz] &= 9b^2c^2(\theta+\beta)^2(\theta+\gamma)^2 \,.\, [a^2(\theta+\alpha)^2 - bc\,(\theta+\beta)(\theta+\gamma)], \\ &= 9b^2c^2(\theta+\beta)^2(\theta+\gamma)^2\,Q, \end{aligned}$$

where

$$\begin{aligned} Q &= a^2(\theta+\alpha)^2 - bc\,(\theta+\beta)(\theta+\gamma), \\ &= b^2(\theta+\beta)^2 - ca\,(\theta+\gamma)(\theta+\alpha), \\ &= c^2(\theta+\gamma)^2 - ab\,(\theta+\alpha)(\theta+\beta). \end{aligned}$$

Hence

$$\frac{d\Theta}{dx} = 27a^2b^2c^2\,(g^2h^2x + h^2f^2y + f^2g^2z)^3\,(\theta+\beta)^2(\theta+\gamma)^2\,Q,$$

and similarly

$$\frac{d\Theta}{dy} = 27a^2b^2c^2\,(g^2h^2x + h^2f^2y + f^2g^2z)^3\,(\theta+\gamma)^2(\theta+\alpha)^2\,Q,$$

$$\frac{d\Theta}{dz} = 27a^2b^2c^2\,(g^2h^2x + h^2f^2y + f^2g^2z)^3\,(\theta+\alpha)^2(\theta+\beta)^2\,Q;$$

whence the above-mentioned conditions reduce themselves to the single condition

$$(\theta+\delta)^2\,\{3\Phi + xyz\,(*)\} = 27a^2b^2c^2\,(g^2h^2x + h^2f^2y + f^2g^2z)^3\,(\theta+\alpha)^2(\theta+\beta)^2(\theta+\gamma)^2\,Q.$$

14. But we have

$$\begin{aligned} &g^2h^2x + h^2f^2y + f^2g^2z \\ &\quad = g^2h^2a\,(\theta+\delta+f)^3 + h^2f^2b\,(\theta+\delta+g)^3 + f^2g^2c\,(\theta+\delta+h)^3, \\ &\quad = (\theta+\delta)^2\,[(g^2h^2a + h^2f^2b + f^2g^2c)(\theta+\delta) + 3\,(gha + hfb + fgc)fgh], \\ &\quad = -\,abc\,(\theta+\delta)^2\,[(gh + hf + fg)(\theta+\delta) + 3fgh], \\ &\quad = -\,abc\,(\theta+\delta)^2\,[gh\,(\theta+\alpha) + hf\,(\theta+\beta) + fg\,(\theta+\gamma)], \\ &\quad = -\,abc\,(\theta+\delta)^2\,P, \end{aligned}$$

if for shortness

$$P = gh(\theta+\alpha) + hf(\theta+\beta) + fg(\theta+\gamma).$$

Hence, substituting,

$$3\Phi + abc(\theta+\alpha)^3(\theta+\beta)^3(\theta+\gamma)^3(*) = -27(abc)^5(\theta+\delta)^4(\theta+\alpha)^2(\theta+\beta)^2(\theta+\gamma)^2 P^3 Q;$$

which when the values $a(\theta+\alpha)^3$, $b(\theta+\beta)^3$, $c(\theta+\gamma)^3$ for (x, y, z) are substituted in the functions Φ and $(*)$, will be an identical equation in θ.

15. It is right to remark that what we require is the expression of $(*)$, $=(*\between x, y, z)^2$; the foregoing equation leads to the value of $(*)$ expressed in terms of θ; and it is necessary to show that this leads back to the expression for $(*)$ as a function of (x, y, z); in fact, that the function of θ is transformable in a *definite* manner into a function of (x, y, z). Suppose that the function of θ could be expressed in two different manners as a function of (x, y, z); then we should have two different functions $(x, y, z)^2$ each equivalent to the same function of θ; and the difference of these functions would be identically $=0$; that is, we should have a function $(x, y, z)^2$ vanishing identically by the substitution

$$x : y : z = a(\theta+\alpha)^3 : b(\theta+\beta)^3 : c(\theta+\gamma)^3;$$

but these relations are equivalent to the single relation

$$(a^2x + b^2y + c^2z)^3 - 27a^2b^2c^2\,xyz = 0,$$

which, *quà* cubic equation, is not equivalent to any equation whatever of the form

$$(x, y, z)^2 = 0;$$

that is, the function of θ is equivalent to a definite functi n $(x, y, z)^2$.

16. To proceed with the reduction, I remark that we have

$$\Phi = (a^2x + b^2y + c^2z)^2 \left\{\begin{array}{l} f^4\,[a^2f^2\,(h^4y^2 - 7h^2g^2yz + g^4z^2)\,(b^2y + c^2z) + b^2c^2\,(h^2y + g^2z)^3] \\ + g^4\,[b^2g^2\,(h^4x^2 - 7h^2f^2xz + f^4z^2)\,(a^2x + c^2z) + c^2a^2\,(h^2x + f^2z)^3] \\ + h^4\,[c^2h^2\,(g^4x^2 - 7g^2f^2xy + f^4y^2)\,(a^2x + b^2y) + a^2b^2\,(g^2x + f^2y)^3] \\ - a^2h^4g^4\,(b^2g^2 + c^2h^2)\,x^3 \\ - b^2h^4f^4\,(c^2h^2 + a^2f^2)\,y^3 \\ - c^2f^4g^4\,(a^2f^2 + b^2g^2)\,z^3 \end{array}\right\}$$

$$+ xyz\Omega$$

where

$$\begin{aligned} -\Omega = {} & 2\,(b^2c^2Ax^2 + c^2a^2By^2 + a^2b^2Cz^2) \\ & + (My + Nz)\,(a^4x + 2a^2b^2y + 2a^2c^2z) \\ & + (Oz + Px)\,(2a^2b^2x + b^4y + 2b^2c^2z) \\ & + (Qx + Ry)\,(2a^2c^2x + 2b^2c^2y + cz), \end{aligned}$$

if for shortness

$$A = a^2g^4h^4\,(b^2g^2 + c^2h^2),$$
$$B = b^2h^4f^4\,(c^2h^2 + a^2f^2),$$
$$C = c^2f^4g^4\,(a^2f^2 + b^2g^2),$$

$$M = f^4h^2\,(a^2f^2c^2h^2 - 7a^2f^2b^2g^2 + 3b^2g^2c^2h^2), \qquad N = f^4g^2\,(-7a^2f^2c^2h^2 + a^2f^2b^2g^2 + 3b^2g^2c^2h^2),$$
$$O = g^4f^2\,(b^2g^2a^2f^2 - 7b^2g^2c^2h^2 + 3c^2h^2a^2f^2), \qquad P = g^4h^2\,(-7b^2g^2a^2f^2 + b^2g^2c^2h^2 + 3c^2h^2a^2f^2),$$
$$Q = h^4g^2\,(c^2h^2b^2g^2 - 7c^2h^2a^2b^2 + 3a^2f^2b^2g^2), \qquad R = h^4f^2\,(-7c^2h^2b^2g^2 + c^2h^2a^2f^2 + 3a^2f^2b^2g^2);$$

and I represent the foregoing equation by

$$\Phi = (a^2x + b^2y + c^2z)^2\,U + xyz\Omega.$$

Hence, writing for x, y, z the foregoing values, we have

$$\Phi = 9a^2b^2c^2\,(\theta+\alpha)^2\,(\theta+\beta)^2\,(\theta+\gamma)^2\,U + abc\,(\theta+\alpha)^3\,(\theta+\beta)^3\,(\theta+\gamma)^3\,\Omega;$$

and thence

$$27U + \frac{1}{abc}\,(\theta+\alpha)\,(\theta+\beta)\,(\theta+\gamma)\,\big(3\Omega + (*)\big) = -27\,(abc)^3\,(\theta+\delta)^4\,P^3Q;$$

that is

$$27\,[U + (abc)^3\,(\theta+\delta)^4\,P^3Q] + \frac{1}{abc}\,(\theta+\alpha)\,(\theta+\beta)\,(\theta+\gamma)\,\big(3\Omega + (*)\big) = 0.$$

In order that this may be the case, it is clear that we must have

$$U + (abc)^3\,(\theta+\delta)^4\,P^3Q = (\theta+\alpha)\,(\theta+\beta)\,(\theta+\gamma)\,M,$$

viz. the left-hand side expressed as a function of θ must be divisible by the product $(\theta+\alpha)\,(\theta+\beta)\,(\theta+\gamma)$. Assuming for a moment that this is so, the quotient M will be a function $(\theta, 1)^6$ expressible in a unique manner in the form $(x, y, z)^2$, and assuming it to be so expressed, we have

$$27Mabc + 3\Omega + (*) = 0;$$

which equation, without any further substitution of the θ-values of (x, y, z), gives $(*)$ in its proper form as a function of (x, y, z).

Reduction of the Equation, $U + (abc)^3\,(\theta+\delta)^4\,P^3\,Q = (\theta+\alpha)\,(\theta+\beta)\,(\theta+\gamma)\,M$.

17. We have by an easy transformation

$$\begin{aligned} U = \;& (a^2x + b^2y + c^2z)\left\{\begin{array}{l} a^2f^6\,(h^4y^2 - 7h^2g^2yz + g^4z^2) \\ + b^2g^6\,\;(f^4z^2 - 7f^2h^2zx + h^4x^2) \\ + c^2h^6\,\;(g^4x^2 - 7g^2f^2xy + f^4y^2) \end{array}\right\} \\ & + 7\,(a^4f^4 + b^4g^4 + c^4h^4)\,f^2g^2h^2\,xyz \\ & + U', \end{aligned}$$

if for shortness

$$\begin{aligned} U' = \;& y^2z\,.\,c^2h^4f^4\,(3b^2g^2 - c^2h^2) \\ &+ yz^2\,.\,b^2g^4f^4\,(3c^2h^2 - b^2g^2) \\ &+ z^2x\,.\,a^2f^4g^4\,(3c^2h^2 - a^2f^2) \\ &+ zx^2\,.\,c^2h^4g^4\,(3a^2f^2 - c^2h^2) \\ &+ x^2y\,.\,b^2g^4h^4\,(3a^2f^2 - b^2g^2) \\ &+ xy^2\,.\,a^2f^4h^4\,(3b^2g^2 - a^2f^2). \end{aligned}$$

Substituting the θ-values, the terms of U, other than U', are at once seen to contain the factor $(\theta+\alpha)(\theta+\beta)(\theta+\gamma)$, and we have

$$\begin{aligned} M = \;& 3abc\left\{\begin{array}{l} a^2f^6\,(h^4y^2 - 7h^2g^2yz + g^4z^2) \\ + b^2g^6\,(f^4z^2 - 7f^2h^2zx + h^4x^2) \\ + c^2h^6\,(g^4x^2 - 7g^2f^2xy + f^4y^2) \end{array}\right\} \\ &+ 7\,(a^4f^4 + b^4g^4 + c^4h^4)\,f^2g^2h^2\,abc\,(\theta+\alpha)^2(\theta+\beta)^2(\theta+\gamma)^2 \\ &+ M', \end{aligned}$$

where

$$U' + (abc)^3(\theta+\delta)^4P^3Q = (\theta+\alpha)(\theta+\beta)(\theta+\gamma)\,M'.$$

18. Write for shortness $p, q, r = (af, bg, ch)$; after a complicated reduction, I obtain

$$\begin{aligned} 3abc\,M' = \;& a^2g^2h^2\,(r-p)(p-q)(-2p^4 + 5p^2qr - 6q^2r^2)\,x^2 \\ &+ b^2h^2f^2\,(p-q)(q-r)(-2q^4 + 5q^2rp - 6q^2p^2)\,y^2 \\ &+ c^2f^2g^2\,(q-r)(r-p)(-2r^4 + 5r^2pq - 6p^2q^2)\,z^2 \\ &+ 2f^2g^2h^2b^2c^2\,(7p^4 - 20p^2qr + 4q^2r^2)\,yz \\ &+ 2f^2g^2h^2c^2a^2\,(7q^4 - 20pq^2r + 4r^2p^4)\,zx \\ &+ 2f^2g^2h^2a^2b^2\,(7r^4 - 20pqr^2 + 4p^2q^2)\,xy \\ &- 2f^2g^2h^2\,(p^4+q^4+r^4)\,(a^2x + b^2y + c^2z)^2. \end{aligned}$$

We then have

$$9abc\,M = \text{terms } (x, y, z)^2 + 9abc\,M', \quad \Omega = \text{terms } (x, y, z)$$

as above; and

$$27M\,abc + 3\Omega + (*) = 0,$$

which gives $(*)$.

19. After all reductions we find:

$$\begin{aligned} -\tfrac{1}{3}(*) = \;& a^2g^2h^2\,(28p^6 - 84p^4qr + 62p^2q^2r^2 - 28q^3r^3)\,x^2 \\ &+ b^2h^2f^2\,(28q^6 - 84pq^4r + 62p^2q^2r^2 - 28r^3p^3)\,y^2 \\ &+ c^2f^2g^2\,(28r^6 - 84pqr + 62p^2q^2r - 28p^3q^3)\,z^2 \\ &+ f^2\,(-3p^8 + 14p^6qr - 130p^4q^2r^2 - 136p^2q^3r^3 + 42q^4r^4)\,yz \\ &+ g^2\,(-3q^8 + 14pq^6r - 130p^2q^4r^2 - 136p^3q^2r^3 + 42r^4p^4)\,zx \\ &+ h^2\,(-3r^8 + 14pqr^6 - 130p^2q^2r^4 - 136p^3q^3r^2 + 42p^4q^4)\,xy\,; \end{aligned}$$

or observing that the coefficients of $a^2g^2h^2x^2$, $b^2h^2f^2y^2$ and $c^2f^2g^2z^2$ are equal to each other and to

$$62p^2q^2r^2 - 28(q^3r^3 + r^3p^3 + p^3q^3),$$

the equation becomes

$$\begin{aligned}(*) = \quad & -3(62p^2q^2r^2 - 28(q^3r^3 + r^3p^3 + p^3q^3)(a^2g^2h^2x^2 + b^2h^2f^2y^2 + c^2f^2g^2z^2)\\ & + 3(3p^8 - 14p^6qr + 130p^4q^2r^2 + 136p^2q^3r^3 - 42q^4r^4)f^2yz\\ & + 3(3q^8 - 14q^6pr + 130q^4p^2r^2 + 136q^2p^3r^3 - 42r^4p^4)g^2zx\\ & + 3(3r^8 - 14r^6pq + 130r^4p^2q^2 + 136r^2p^3q^3 - 42p^4q^4)h^2xy;\end{aligned}$$

and we thence obtain by symmetry the complete value of $(*\between x, y, z, w)^2$, viz. we have only to complete the literal parts of the foregoing expression into the forms

$$\begin{aligned}& a^2g^2h^2x^2 + b^2h^2f^2y^2 + c^2f^2g^2z^2 + a^2b^2c^2w^2,\\ & f^2yz + a^2xw,\\ & g^2zx + b^2yw,\\ & c^2xy + a^2zw,\end{aligned}$$

respectively.

20. The equation of the torse thus is

$$\begin{aligned}\Delta = & \; a^6g^6h^6x^6 + b^6h^6f^6y^6 + c^6f^6g^6z^6 + a^6b^6c^6w^6\\ & - f^6(h^2y + g^2z)^3(b^2y + c^2z)^3\\ & - g^6(f^2z + h^2x)^3(c^2z + a^2x)^3\\ & - h^6(g^2x + f^2y)^3(a^2x + b^2y)^3\\ & - a^6(g^2x + c^2w)^3(h^2x + b^2w)^3\\ & - b^6(h^2y + a^2w)^3(f^2y + c^2w)^3\\ & - c^6(f^2z + b^2w)^3(g^2z + a^2w)^3\end{aligned}$$

$$\begin{aligned}& + (b^2f^2y + c^2f^2z + b^2c^2w)^3[(h^2y + g^2z + a^2w)^3 - 27a^2g^2h^2\,yzw]\\ & + (a^2g^2x + c^2g^2z + c^2a^2w)^3[(f^2z + h^2x + b^2w)^3 - 27b^2h^2f^2\,zxw]\\ & + (a^2h^2x + b^2h^2y + a^2b^2w)^3[(g^2x + f^2y + c^2w)^3 - 27c^2f^2g^2\,xyw]\\ & + (g^2h^2x + h^2f^2y + f^2g^2z)^3[(a^2x + b^2y + c^2z)^3 - 27a^2b^2c^2\,xyz]\\ & + xyzw(*\between x, y, z, w)^2 = 0.\end{aligned}$$

I recall that

$$\begin{aligned}& a = \beta - \gamma, \quad f = \alpha - \delta, \qquad p = af = (\alpha - \delta)(\beta - \gamma),\\ & b = \gamma - \alpha, \quad g = \beta - \delta, \qquad q = bg = (\beta - \delta)(\gamma - \alpha),\\ & c = \alpha - \beta, \quad h = \gamma - \delta, \qquad r = ch = (\gamma - \delta)(\alpha - \beta).\end{aligned}$$

Developing, we have finally the equation of the torse in the form following.

Equation of the Sextic Torse.

21. The equation is

$$\begin{aligned}
0 =\ & (a^6g^6h^6,\ b^6h^6f^6,\ c^6f^6g^6,\ a^6b^6c^6 \between x^6,\ y^6,\ z^6,\ w^6) \\
&+ 3(q^2+r^2)\ (b^4h^4f^6,\ c^4g^4f^6,\ g^4h^4a^6,\ b^4c^4a^6 \between y^5z,\ yz^5,\ x^5w,\ xw^5) \\
&+ 3(r^2+p^2)\ (c^4f^4g^6,\ a^4h^4g^6,\ h^4f^4b^6,\ c^4a^4b^6 \between z^5x,\ zx^5,\ y^5w,\ yw^5) \\
&+ 3(p^2+q^2)\ (a^4g^4h^6,\ b^4f^4h^6,\ f^4g^4c^6,\ a^4b^4c^6 \between x^5y,\ xy^5,\ z^5w,\ zx^5) \\
&+ 3(q^4+3q^2r^2+r^4)(b^2h^2f^6,\ c^2g^2f^6,\ g^2h^2a^6,\ b^2c^2a^6 \between y^4z^2,\ y^2z^4,\ x^4w^2,\ x^2w^4) \\
&+ 3(r^4+3r^2p^2+p^4)(c^2f^2g^6,\ a^2h^2g^6,\ h^2f^2b^6,\ c^2a^2b^6 \between z^4x^2,\ x^2z^4,\ y^4w^2,\ y^2w^4) \\
&+ 3(p^4+3p^2q^2+q^4)(a^2g^2h^6,\ b^2f^2h^6,\ f^2g^2c^6,\ a^2b^2c^6 \between x^4y^2,\ x^2y^4,\ z^4w^2,\ z^2w^4) \\
&+ (6p^4+9p^2q^2+9p^2r^2-21q^2r^2)(a^2g^4h^4,\ a^2b^4c^4,\ f^2h^4b^4,\ f^2g^4c^4 \between x^4yz,\ w^4yz,\ y^4wx,\ z^4wy) \\
&+ (6q^4+9q^2r^2+9q^2p^2-21r^2p^2)(b^2h^4f^4,\ b^2c^4a^4,\ g^2f^4c^4,\ g^2h^4a^4 \between y^4zx,\ w^4zx,\ z^4wy,\ x^4wz) \\
&+ (6r^4+9r^2p^2+9r^2q^2-21p^2q^2)(c^2f^4g^4,\ c^2a^4b^4,\ h^2g^4a^4,\ h^2f^4b^4 \between z^4xy,\ w^4xy,\ x^4wz,\ y^4wx) \\
&+ (q^6+9q^4r^2+9q^2r^4+r^6)\ (f^6,\ a^6 \between y^3z^3,\ x^3w^3) \\
&+ (r^6+9r^4p^2+9r^2p^4+p^6)\ (g^6,\ b^6 \between z^3x^3,\ y^3w^3) \\
&+ (p^6+9p^4q^2+9p^2q^4+q^6)\ (h^6,\ c^6 \between x^3y^3,\ z^3w^3) \\
&+ 9(q^4r^2+q^2r^4+r^4p^2+r^2p^4+p^4q^2+p^2q^4-14p^2q^2r^2) \\
&\qquad \times\ (f^2g^2h^2,\ f^2b^2c^2,\ g^2c^2a^2,\ h^2a^2b^2 \between x^2y^2z^2,\ y^2z^2w^2,\ z^2x^2w^2,\ x^2y^2w^2) \\
&+ 3\{p^6+3p^4(2q^2+r^2)+3p^2(q^4-7q^2r^2)+q^4r^2\} \\
&\qquad \times\ (g^2h^4,\ h^4b^2,\ g^2c^4,\ c^4b^2 \between x^3y^2z,\ y^3wx^2,\ z^3w^2x,\ w^3yz^2) \\
&+ 3\{q^6+3q^4(2r^2+p^2)+3q^2(r^4-7r^2p^2)+r^4p^2\} \\
&\qquad \times\ (h^2f^4,\ f^4c^2,\ h^2a^4,\ a^4c^2 \between xy^3z^2,\ z^3wy^2,\ x^3w^2y,\ w^3zx^2) \\
&+ 3\{r^6+3r^4(2p^2+q^2)+3r^2(p^4-7p^2q^2)+p^4q^2\} \\
&\qquad \times\ (f^2g^4,\ g^4a^2,\ f^2b,\ b^4a^2 \between x^2yz^3,\ x^3wz^2,\ y^3w^2z,\ w^3xy^2) \\
&+ 3\{p^6+3p^4(2r^2+q^2)+3p^2(r^4-7q^2r^2)+q^2r^4\} \\
&\qquad \times\ (g^4h^2,\ h^2b^4,\ g^4c^2,\ c^2b^4 \between x^3yz^2,\ y^3w^2x,\ z^3wx^2,\ w^3y^2z) \\
&+ 3\{q^6+3q^4(2p^2+r^2)+3q^2(p^4-7r^2p^2)+r^2p^4\} \\
&\qquad \times\ (h^4f^2,\ f^2c^4,\ h^4a^2,\ a^2c^4 \between x^2y^3z,\ z^3wy^2,\ x^3wy^2,\ w^3z^2x) \\
&+ 3\{r^6+3r^4(2q^3+p^2)+3r^2(q^4-7p^2q^2)+p^2q^4\} \\
&\qquad \times\ (f^4g^2,\ g^2a^4,\ f^4b^2,\ b^2a^4 \between xy^2z^3,\ x^3w^2z,\ y^3wz^2,\ w^3x^2y) \\
&+ xyzw \left\{ \begin{aligned}
&-3\{62p^2q^2r^2-28(q^3r^3+r^3p^3+p^3q^3)\}(a^2g^2h^2,\ b^2h^2f^2,\ c^2f^2g^2,\ a^2b^2c^2 \between x^2,\ y^2,\ z^2,\ w^2) \\
&+3(3p^8-14p^6qr+130p^4q^2r^2+136p^2q^3r^3-42q^4r^4)\ (f^2,\ a^2 \between yz,\ xw) \\
&+3(3q^8-14q^6rp+130q^4r^2p^2+136q^2r^3p^3-42r^4p^4)\ (g^2,\ b^2 \between zx,\ yw) \\
&+3(3r^8-14r^6pq+130r^4p^2q^2+136r^2p^3q^3-42p^4q^4)\ (h^2,\ c^2 \between xy,\ zw)
\end{aligned} \right\}.
\end{aligned}$$

Comparison with the Equation of the Centro-surface of an Ellipsoid.

22. In the Equation

$$\frac{x}{(\theta+\alpha)^2}+\frac{y}{(\theta+\beta)^2}+\frac{z}{(\theta+\gamma)^2}+\frac{w}{(\theta+\delta)^2}=0,$$

for x, y, z, w write ξ^2, η^2, ζ^2, ω^2, and then $\delta=\infty$, the equation is converted into

$$\frac{\xi^2}{(\theta+\alpha)^2}+\frac{\eta^2}{(\theta+\beta)^2}+\frac{\zeta^2}{(\theta+\gamma)^2}+\omega^2=0\,;$$

or writing a^2, b^2, c^2 for α, β, γ, and understanding ξ^2, η^2, ζ^2, ω^2 to mean a^2x^2, b^2y^2, c^2z^2, -1, this is

$$\frac{a^2x^2}{(\theta+a^2)^2}+\frac{b^2y^2}{(\theta+b^2)^2}+\frac{c^2z^2}{(\theta+c^2)^2}-1=0.$$

This is an equation, the envelope of which in regard to the variable parameter θ, gives the surface which is the locus of the centres of curvature of the ellipsoid $\frac{x^2}{a^2}+\frac{y^2}{b^2}+\frac{z^2}{c^2}=1$, or say the Centro-surface of the Ellipsoid. (Salmon's *Solid Geometry*, Ed. 2, p. 400, [Ed. 4, p. 465].)

Making the same substitution in the foregoing equation $(*\between x, y, z, w)^6=0$, the quantities f, g, h become equal to $-\delta$, and p, q, r to $-a\delta$, $-b\delta$, $-c\delta$ respectively, and the whole equation divides by δ^{12}; throwing out this factor, we have a result which is obtained more simply by changing

$$x,\ y,\ z,\ w,\quad a,\ b,\ c,\ f,\ g,\ h,\quad p,\ q,\ r,$$

into

$$\xi^2,\ \eta^2,\ \zeta^2,\ \omega^2,\quad \alpha,\ \beta,\ \gamma,\ 1,\ 1,\ 1,\quad \alpha,\ \beta,\ \gamma,$$

where α, β, γ now signify b^2-c^2, c^2-a^2, a^2-b^2 respectively, and ξ^2, η^2, ζ^2, ω^2 are retained as standing for a^2x^2, b^2y^2, c^2z^2, -1 respectively; viz. the equation of the centro-surface is found to be

$$\begin{array}{rl}
0= & (\alpha^6,\ \beta^6,\ \gamma^6,\ \alpha^6\beta^6\gamma^6\between \xi^{12},\ \eta^{12},\ \zeta^{12},\ \omega^{12})\\
+3\,(\beta^2+\gamma^2) & (\beta^4,\ \gamma^2,\ \alpha^6,\ \alpha^6\beta^4\gamma^4\between \eta^{10}\zeta^2,\ \eta^2\zeta^{10},\ \xi^{10}\omega^2,\ \xi^2\omega^{10})\\
+3\,(\gamma^2+\alpha^2) & (\gamma^4,\ \alpha^2,\ \beta^6,\ \beta^6\gamma^4\alpha^4\between \zeta^{10}\xi^2,\ \zeta^2\xi^{10},\ \eta^{10}\omega^2,\ \eta^2\omega^{10})\\
+3\,(\alpha^2+\beta^2) & (\alpha^4,\ \beta^2,\ \gamma^6,\ \gamma^6\alpha^4\beta^4\between \xi^{10}\eta^2,\ \xi^2\eta^{10},\ \zeta^{10}\omega^2,\ \zeta^2\omega^{10})\\
+3\,(\beta^4+3\beta^2\gamma^2+\gamma^4) & (\beta^2,\ \gamma^2,\ \alpha^6,\ \alpha^6\beta^2\gamma^2\between \eta^8\zeta^4,\ \eta^4\zeta^8,\ \xi^8\omega^4,\ \xi^4\omega^8)\\
+3\,(\gamma^4+3\gamma^2\alpha^2+\alpha^4) & (\gamma^2,\ \alpha^2,\ \beta^6,\ \beta^6\gamma^2\alpha^2\between \zeta^8\xi^4,\ \zeta^4\xi^8,\ \eta^8\omega^4,\ \eta^4\omega^8)\\
+3\,(\alpha^4+3\alpha^2\beta+\beta^4) & (\alpha^2,\ \beta^2,\ \gamma^6,\ \gamma^6\alpha^2\beta^2\between \xi^8\eta^4,\ \xi^4\eta^8,\ \zeta^8\omega^4,\ \zeta^4\omega^8)\\
+3\,(2\alpha^4+3\alpha^2\beta^2+3\alpha^2\gamma^2-7\beta^2\gamma^2) & (\alpha^2,\ \alpha^2\beta^4\gamma^4,\ \beta^4,\ \gamma^4\between \xi^8\eta^2\zeta^2,\ \omega^8\eta^2\zeta^2,\ \eta^8\omega^2\xi^2,\ \zeta^8\omega^2\eta^2)\\
+3\,(2\beta^4+3\beta^2\gamma^2+3\beta^2\alpha^2-7\gamma^2\alpha^2) & (\beta^2,\ \beta^2\gamma^4\alpha^4,\ \gamma^4,\ \alpha^4\between \eta^8\zeta^2\xi^2,\ \omega^8\zeta^2\xi^2,\ \zeta^8\omega^2\eta^2,\ \xi^8\omega^2\zeta^2)\\
+3\,(2\gamma^4+3\gamma^2\alpha^2+3\gamma^2\beta^2-7\alpha^2\beta^2) & (\gamma^2,\ \gamma^2\alpha^4\beta^4,\ \alpha^4,\ \beta^4\between \zeta^8\xi^2\eta^2,\ \omega^8\xi^2\eta^2,\ \xi^8\omega^2\zeta^2,\ \eta^8\omega^2\xi^2)
\end{array}$$

$$
\begin{aligned}
&+ (\beta^6 + 9\beta^4\gamma^2 + 9\beta^2\gamma^4 + \gamma^6) \quad (1,\ \alpha^6 \between \eta^3\zeta^3,\ \xi^3\omega^3)\\
&+ (\gamma^6 + 9\gamma^4\alpha^2 + 9\gamma^2\alpha^4 + \alpha^6) \quad (1,\ \beta^6 \between \zeta^3\xi^3,\ \eta^3\omega^3)\\
&+ (\alpha^6 + 9\alpha^4\beta^2 + 9\alpha^2\beta^4 + \beta^6) \quad (1,\ \gamma^6 \between \xi^3\eta^3,\ \zeta^3\omega^3)
\end{aligned}
$$

$$
\begin{aligned}
&+ 3\,\{\alpha^6 + 3\alpha^4(2\beta^2 + \gamma^2) + 3\alpha^2(\beta^4 - 7\beta^2\gamma^2) + \beta^4\gamma^2\}\,(1,\ \beta^2,\ \gamma^4,\ \beta^2\gamma^4 \between \xi^6\eta^4\zeta^2,\ \eta^6\omega^2\xi^4,\ \zeta^6\omega^4\xi^2,\ \omega^6\eta^2\zeta^4)\\
&+ 3\,\{\beta^6 + 3\beta^4(2\gamma^2 + \alpha^2) + 3\beta^2(\gamma^4 - 7\gamma^2\alpha^2) + \gamma^4\alpha^2\}\,(1,\ \gamma^2,\ \alpha^4,\ \gamma^2\alpha^4 \between \eta^6\zeta^4\xi^2,\ \zeta^6\omega^2\eta^4,\ \xi^6\omega^4\eta^2,\ \omega^6\zeta^2\xi^4)\\
&+ 3\,\{\gamma^6 + 3\gamma^4(2\alpha^2 + \beta^2) + 3\gamma^2(\alpha^4 - 7\alpha^2\beta^2) + \alpha^4\beta^2\}\,(1,\ \alpha^2,\ \beta^4,\ \alpha^2\beta^4 \between \zeta^6\xi^4\eta^2,\ \xi^6\omega^2\zeta^4,\ \eta^6\omega^4\zeta^2,\ \omega^6\xi^2\eta^4)\\
&+ 3\,\{\alpha^6 + 3\alpha^4(2\gamma^2 + \beta^2) + 3\alpha^2(\gamma^4 - 7\beta^2\gamma^2) + \beta^2\gamma^4\}\,(1,\ \gamma^2,\ \beta^4,\ \beta^4\gamma^2 \between \xi^6\eta^2\zeta^4,\ \zeta^6\omega^2\xi^4,\ \eta^6\omega^4\xi^2,\ \omega^6\eta^4\zeta^2)\\
&+ 3\,\{\beta^6 + 3\beta^4(2\alpha^2 + \gamma^2) + 3\beta^2(\alpha^4 - 7\gamma^2\alpha^2) + \gamma^2\alpha^4\}\,(1,\ \alpha^2,\ \gamma^4,\ \gamma^4\alpha^2 \between \eta^6\zeta^2\xi^4,\ \xi^6\omega^2\eta^4,\ \zeta^6\omega^4\eta^2,\ \omega^6\zeta^4\xi^2)\\
&+ 3\,\{\gamma^6 + 3\gamma^4(2\beta^2 + \alpha^2) + 3\gamma^2(\beta^4 - 7\alpha^2\beta^2) + \alpha^2\beta^4\}\,(1,\ \beta^2,\ \alpha^4,\ \alpha^4\beta^2 \between \zeta^6\xi^2\eta^4,\ \eta^6\omega^2\zeta^4,\ \xi^6\omega^4\zeta^2,\ \omega^6\xi^4\eta^2)\\
&+ 9\,(\beta^4\gamma^2 + \beta^2\gamma^4 + \gamma^4\alpha^2 + \gamma^2\alpha^4 + \alpha^4\beta^2 + \alpha^2\beta^4 - 14\alpha^2\beta^2\gamma^2)\\
&\qquad\qquad (1,\ \beta^2\gamma^2,\ \gamma^2\alpha^2,\ \alpha^2\beta^2 \between \xi^4\eta^4\zeta^4,\ \eta^4\zeta^4\omega^4,\ \zeta^4\xi^4\omega^4,\ \xi^4\eta^4\omega^4)
\end{aligned}
$$

$$
+\,\xi^2\eta^2\zeta^2\omega^2 \left\{
\begin{aligned}
&- 3\,\{62\alpha^2\beta^2\gamma^2 - 28(\beta^3\gamma^3 + \gamma^3\alpha^3 + \alpha^3\beta^3)\}\,(\alpha^2,\ \beta^2,\ \gamma^2,\ \alpha^2\beta^2\gamma^2 \between \xi^4,\ \eta^4,\ \zeta^4,\ \omega^4)\\
&+ 3\,(3\alpha^8 - 14\alpha^6\beta\gamma + 130\alpha^4\beta^2\gamma^2 + 136\alpha^2\beta^3\gamma^3 - 42\beta^4\gamma^4)\ (1,\ \alpha^2 \between \eta^2\zeta^2,\ \xi^2\omega^2)\\
&+ 3\,(3\beta^8 - 14\beta^6\gamma\alpha + 130\beta^4\gamma^2\alpha^2 + 136\beta^2\gamma^3\alpha^3 - 42\gamma^4\alpha^4)\ (1,\ \beta^2 \between \zeta^2\xi^2,\ \eta^2\omega^2)\\
&+ 3\,(3\gamma^8 - 14\gamma^6\alpha\beta + 130\gamma^4\alpha^2\beta^2 + 136\gamma^2\alpha^3\beta^3 - 42\alpha^4\beta^4)\ (1,\ \gamma^2 \between \xi^2\eta^2,\ \zeta^2\omega^2)
\end{aligned}
\right\}.
$$

This agrees with the result given in Salmon's *Solid Geometry*, Ed. 2, p. 151, [Ed. 4, p. 178], and *Quarterly Mathematical Journal*, vol. II. p. 220 (1858); in the latter place, however, the term

$$\beta^2\eta^8\zeta^4 + \gamma^2\eta^4\zeta^8 + \alpha^6\xi^8\omega^4 + \alpha^6\beta^2\gamma^2\xi^4\omega^8$$

is by mistake written

$$\beta^2\eta^8\zeta^4 + \gamma^2\eta^4\zeta^8 + \alpha^6\xi^8\omega^4 + \quad \beta^2\gamma^2\xi^4\omega^8;$$

viz. a factor α^6 is omitted in one of the coefficients.

Some of the coefficients are presented under slightly different forms; viz. instead of

$$62\alpha^2\beta^2\gamma^2 - 28(\beta^3\gamma^3 + \gamma^3\alpha^3 + \alpha^3\beta^3)$$

Salmon has

$$14(\beta^4\gamma^2 + \beta^2\gamma^4 + \gamma^4\alpha^2 + \gamma^2\alpha^4 + \alpha^4\beta^2 + \alpha^2\beta^4) + 20\alpha^2\beta^2\gamma^2;$$

and instead of

$$3\alpha^8 - 14\alpha^6\beta\gamma + 130\alpha^4\beta^2\gamma^2 + 136\alpha^2\beta^3\gamma^3 - 42\beta^4\gamma^4,$$

he has

$$-4\alpha^8 + 7\alpha^6(\beta^2 + \gamma^2) + 196\alpha^4\beta^2\gamma^2 - 68\alpha^2\beta^2\gamma^2(\beta^2 + \gamma^2) - 42\beta^4\gamma^4,$$

but these different forms are respectively equivalent in virtue of the relation

$$\alpha + \beta + \gamma = 0.$$

437.

DÉMONSTRATION NOUVELLE DU THÉORÈME DE M. CASEY PAR RAPPORT AUX CERCLES QUI TOUCHENT À TROIS CERCLES DONNÉS.

[From the *Annali di Matematica pura ed applicata*, tom. I. (1867), pp. 132—134.]

THIS is in fact the investigation contained in the paper 414, "On Polyzomal Curves otherwise the curves $\sqrt{U}+\sqrt{V}+\&c.=0$," Annex II. pp. 568—573, "On Casey's theorem for the circle which touches three given circles," viz. it is based on the identity of the two problems 1° to find a circle touching three given circles, 2° to find a cone-sphere (sphere of radius zero) passing through three given points in space.

438.

NOTE SUR QUELQUES TORSES SEXTIQUES.

[From the *Annali di Matematica pura ed applicata*, tom. II. (1868), pp. 99, 100.]

JE désire d'appeler attention aux surfaces développables, ou torses, données par l'équation

$$(ae - 4bd + 3c^2)^3 - 27(ace - ad^2 - b^2e + 2bcd - c^3)^2 = 0.$$

Dans cette équation (a, b, c, d, e) sont des fonctions linéaires quelconques des quatre coordonnées (x, y, z, t); ces quantités sont donc liées par une équation linéaire

$$Aa + 4Bb + 6Cc + 4Dd + Ee = 0,$$

et je remarque que la classification des torses comprises sous l'équation mentionnée dépend des propriétés invariantives de la fonction $(A, B, C, D, E \between \tau, 1)^4$.

En effet la torse a une courbe cuspidale, ou arête de rebroussement, donnée par les équations

$$ae - 4bd + 3c^2 = 0, \quad ace - ad^2 - b^2e + 2bcd - c^3 = 0,$$

et une courbe nodale donnée par les équations

$$\frac{ac - b^2}{a} = \frac{ad - bc}{2b} = \frac{ae + 2bd - 3c^2}{6c} = \frac{be - cd}{2d} = \frac{ce - d^2}{e}:$$

ces deux courbes se rencontrent dans les points donnés par les équations

$$\frac{a}{b} = \frac{b}{c} = \frac{c}{d} = \frac{d}{e},$$

lesquels sont des points stationnaires de la courbe cuspidale. Pour trouver ces points, en écrivant $a : b : c : d : e = \tau^4 : \tau^3 : \tau^2 : \tau : 1$, on obtient pour le paramètre τ l'équation

$$(A, B, C, D, E \between \tau, 1)^4 = 0$$

et l'on voit ainsi qu'il y a un rapport entre la théorie de la surface et cette équation; à toute particularité invariantive de l'équation, il y correspondra quelque particularité de la torse.

Les cas à considérer sont:

1°. Racines inégales, sans aucune relation invariantive. C'est le cas général; je l'ai considéré dans le Mémoire, "On a certain Sextic Developable," *Quart. Math. Journ.*, t. IX. (1868), pp. 129—142, [398].

2°. Deux racines égales. Ce cas n'a pas été considéré; je remarque que la courbe cuspidale est du cinquième ordre. En effet on peut supposer que les racines égales soient $=0$, ce qui revient à prendre $D=0$, $E=0$. On a donc $Aa+4Bb+6Cc=0$, c'est-à-dire les équations $a=0$, $b=0$ impliquent l'équation $c=0$; et on voit de là que les surfaces $ae-4bd+3c^2=0$, $ace-ad^2-b^2e+2bcd-c^3=0$ se coupent selon la droite $a=0$, $b=0$; il reste ainsi une courbe du cinquième ordre pour la courbe cuspidale.

3°. Trois racines égales. On peut supposer que ces racines sont $=0$, ce qui donne $C=0$, $D=0$, $E=0$; et l'on a ainsi $Aa+4Bb=0$, c'est-à-dire les plans $a=0$, $b=0$ sont ici un seul plan. L'équation de la torse contient le facteur a, et en l'écartant elle se réduit au cinquième ordre; on obtient ainsi la torse générale du cinquième ordre.

4°. Deux paires de racines égales. On peut supposer que ces racines sont $=\infty$, 0; cela donne $A=0$, $B=0$, $D=0$, $E=0$, et l'on a ainsi identiquement $c=0$. L'équation de la torse est $(ae-4bd)^3-27(-ad^2-b^2e)^2=0$. J'ai considéré ce cas dans le Mémoire, "On a Special Sextic Developable," *Quart. Math. Journ.*, t. VII. (1866), pp. 105—113, [373]; la courbe cuspidale est du quatrième ordre, une courbe excubo-quartique d'une forme particulière.

5°. Quatre racines égales; en prenant ces racines $=0$, on a $B=C=D=E=0$, donc identiquement $a=0$; l'équation de la torse contient le facteur b^2, et en l'écartant elle se réduit au quatrième ordre: on a dans ce cas la torse générale du quatrième ordre. Il y a encore deux cas à considérer.

6°. L'invariant I de la fonction $(A, B, C, D, E \between \tau, 1)^4$ est $=0$;

7°. L'invariant J de cette fonction est $=0$;

Mais je n'ai pas encore examiné ce que cela veut dire (1). Il n'y a pas le cas à considérer où l'on a à la fois $I=0$, $J=0$; car cela revient au cas 3° de trois racines égales.

Cambridge, le 19 *mai* 1868.

1 La courbe cuspidale étant du genre 0 (unicursale), on peut considérer la série des points de la courbe comme correspondant anharmoniquement aux points d'une droite. Si le système des quatre points stationnaires est harmonique on a $J=0$; si ce système est équi-anharmonique, on a $I=0$.

439.

ADDITION À LA NOTE SUR QUELQUES TORSES SEXTIQUES.

[From the *Annali di Matematica pura ed applicata*, tom. II. (1868), pp. 219—221.]

JE viens de trouver ce que signifie la condition $J=0$. Considérons deux surfaces quadriques qui se touchent (d'un contact ordinaire). Les équations tangentielles peuvent s'écrire sous la forme

$$a\,\xi^2+b\,\eta^2+c\,\zeta^2+2n\,\zeta\omega=0,$$

$$a'\xi^2+b'\eta^2+c'\zeta^2+2n'\zeta\omega=0,$$

et l'on satisfait à ces équations par des valeurs de $\xi,\ \eta,\ \zeta,\ \omega$ qui contiennent un paramètre arbitraire θ, en écrivant

$$\xi : \eta : \zeta : \omega = \alpha\left(\theta+\frac{1}{\theta}\right) : \beta\left(\theta-\frac{1}{\theta}\right) : \epsilon : \gamma\left(\theta^2+\frac{1}{\theta^2}\right)+\delta\,;$$

en effet cela donne

$$a\,\alpha^2+b\,\beta^2+2n\,\gamma\epsilon=0,\qquad 2a\,\alpha^2-2b\,\beta^2+c\,\epsilon^2+2n\,\delta\epsilon=0,$$

$$a'\alpha^2+b'\beta^2+2n'\gamma\epsilon=0,\qquad 2a'\alpha^2-2b'\beta^2+c'\epsilon^2+2n'\delta\epsilon=0,$$

ce qui détermine les valeurs de $\alpha : \beta : \gamma : \delta : \epsilon$. L'équation du plan tangent commun sera donc

$$x\alpha\left(\theta+\frac{1}{\theta}\right)+y\beta\left(\theta-\frac{1}{\theta}\right)+z\epsilon+w\left[\gamma\left(\theta^2+\frac{1}{\theta^2}\right)+\delta\right]=0,$$

ou, en multipliant par $12\theta^2$, cette équation sera

$$(a,\ b,\ c,\ d,\ e\between\theta,\ 1)^4=0,$$

les valeurs des coefficients étant

$$a = 12\ \gamma w,$$
$$b = 3\,(\alpha x + \beta y),$$
$$c = 2\,(\delta w + \epsilon z),$$
$$d = 3\,(\alpha x - \beta y),$$
$$e = 12\ \gamma w.$$

En représentant par $Aa + 4Bb + 6Cc + 4Dd + Ee = 0$ l'équation qui lie les fonctions linéaires a, b, c, d, e, cette équation sera $a - e = 0$; on a donc $A = -E = 1$, $B = C = D = 0$; l'invariant J de la fonction $(A, B, C, D, E \between \tau, 1)^4$ est donc $= 0$.

Nous arrivons ainsi à la conclusion que la torse sextique

$$(ae - 4bd + 3c^2)^3 - 27\,(ace - ad^2 - b^2e - c^3 + 2bcd)^2 = 0,$$

où les fonctions linéaires a, b, c, d, e sont liées par une équation

$$Aa + 4Bb + 6Cc + 4Dd + Ee = 0,$$

telle que l'invariant

$$J = ACE - AD^2 - EB^2 - C^3 - 2BCD$$

de la fonction $(A, B, C, D, E \between \tau, 1)^4$ est $= 0$ (cas 7° de la Note), est la torse enveloppée par le plan tangent commun de deux surfaces quadriques qui se touchent d'un contact ordinaire. J'ai trouvé l'équation de cette torse dans le Mémoire, "On the Developable Surfaces which arise from two Surfaces of the Second Order," *Camb. and Dubl. Math. Jour.*, t. v. (1850), pp. 46—57, voir p. 56, [85]: $a, b, c, n, a', b', c', n'$ y dénotent les mêmes coefficients comme à présent, et en écrivant

$$bc' - b'c = f,\quad an' - a'n = p,$$
$$ca' - c'a = g,\quad bn' - b'n = q,$$
$$ab' - a'b = h,\quad cn' - c'n = r,$$

(et de là $pf + qg + rh = 0$), l'équation trouvée est

$$f^2g^2h^2w^6 \ldots + 4pqr\,(qx^2 + py^2)^3 = 0.$$

Je vais vérifier ces termes. Partant de l'équation $(a, b, c, d, e \between \theta, 1)^4 = 1$, l'équation de la torse, en y introduisant pour commodité le facteur $-\frac{1}{216}pqr$, sera

$$-\tfrac{1}{216}pqr\,\{4\,[(\delta w + \epsilon z)^2 + 12\gamma^2w^2 - 3\,(\alpha^2x^2 - \beta^2y^2)]^3$$
$$-\,[2\,(\delta w + \epsilon z)^3 - 9\,(\delta w + \epsilon z)(\alpha^2x^2 - \beta^2y^2) - 27\,(\delta w + \epsilon z)\,\gamma^2w^2 + 54\,(\alpha^2x^2 + \beta^2y^2)\,\gamma w]^2\} = 0.$$

En prenant $\gamma\epsilon = ab' - a'b = h$, on obtient pour α, β, γ, δ, ϵ les valeurs

$$\alpha^2 = 2q,\quad \beta^2 = -2p,\quad \gamma\epsilon = h,\quad \epsilon^2 = -\frac{8pq}{r},\quad \delta\epsilon = \frac{2(fp-gq)}{r},$$

et de là

$$\gamma^2 = -\frac{h^2 r}{8pq},\quad \delta^2 = -\frac{(fp-gq)^2}{2pqr} = -\frac{h^2r^2 + 4fgpq}{2pqr} = 4\gamma^2 + \frac{2fg}{r},\quad \delta^2 - 4\gamma^2 = \frac{2fg}{r}.$$

Les termes en w^6 et $(x^2, y^2)^3$ sont

$$-\tfrac{1}{216}\, pqr\,\{[4(\delta^2 + 12\gamma^2)^3 - (2\delta^3 - 72\gamma^2\delta)^2]\, w^6 - 108\,(\alpha^2x^2 - \beta^2y^2)^3\},$$
$$= -\tfrac{1}{216}\, pqr\,\{432\gamma^2(\delta^2 - 4\gamma^2)^2\, w^6 - 108\,(\alpha^2x^2 - \beta^2y^2)^3\};$$

ces termes sont donc

$$= -\tfrac{1}{216}\, pqr\left\{-432\,\frac{h^2r}{8pq}\cdot\frac{4f^2g^2}{r^2}\, w^2 + 864\,(qx^2 + pq^2)^2\right\}$$
$$= f^2g^2h^2w^6 + 4pqr\,(qx^2 + py^2)^3,$$

comme cela doit être.

Cambridge, le 22 *septembre* 1868.

440.

NOTE SUR UNE TRANSFORMATION GÉOMÉTRIQUE.

[From the *Journal für die reine und angewandte Mathematik* (Crelle), tom. LXVII. (1867), pp. 95, 96.]

LA lecture de la Note de M. Hesse, "Ein Uebertragungsprincip" (t. LXVI. p. 15 de ce Journal) m'a suggéré les remarques suivantes:

Soient (a_1, b_1, c_1, d_1), (a_2, b_2, c_2, d_2), (a_3, b_3, c_3, d_3) des constantes données, on peut supposer que les coordonnées (x, y) d'un point quelconque dans un plan soient exprimées en fonctions des paramètres variables (u, v) par les équations

$$x = \frac{a_1 + b_1u + c_1v + d_1uv}{a_3 + b_3u + c_3v + d_3uv}, \qquad y = \frac{a_2 + b_2u + c_2v + d_2uv}{a_3 + b_3u + c_3v + d_3uv}.$$

En introduisant une nouvelle indéterminée s, ces équations peuvent être écrites dans la forme

$$sx = a_1 + b_1u + c_1v + d_1uv,$$

$$sy = a_2 + b_2u + c_2v + d_2uv,$$

$$s = a_3 + b_3u + c_3v + d_3uv;$$

pour des valeurs données des coordonnées (x, y) la quantité s est en général déterminée par une équation quadratique, et les paramètres u et v sont des fonctions linéaires données de s; il y a cependant deux cas particuliers qu'il convient de distinguer.

1°. L'équation quadratique en s peut avoir la racine $s=0$ et, débarrassée de ce facteur, se réduire par conséquent à une équation linéaire; ce cas particulier a lieu si la condition $(abc)(bcd) = (abd)(acd)$ est remplie, où la notation (abc) désigne le déterminant

$$\begin{vmatrix} a_1, & b_1, & c_1 \\ a_2, & b_2, & c_2 \\ a_3, & b_3, & c_3 \end{vmatrix}.$$

Dans ce cas u et v sont des fonctions rationnelles de (x, y) et la transformation a la signification géométrique suivante :

En considérant deux droites quelconques L, M dans l'espace et en menant par le point donné (x, y) la droite unique G qui rencontre ces deux droites, on peut supposer que u et v soient des paramètres qui déterminent les positions des points de rencontre sur les deux droites respectivement ; c. à. d. que u soit la distance d'un point fixe sur la droite L au point de rencontre avec la droite G, et de même que v soit la distance d'un point fixe sur la droite M au point de rencontre avec la droite G.

2°. Supposons $b_1 : c_1 = b_2 : c_2 = b_3 : c_3$, ou ce qui est au fond la même chose $b_1 - c_1 = 0$, $b_2 - c_2 = 0$, $b_3 - c_3 = 0$; alors s est déterminée par une équation simple, mais u et v ne sont plus des fonctions rationnelles de s ; on voit que dans ce cas $u + v$ et uv sont des fonctions rationnelles de (x, y), et que par conséquent u et v sont les racines d'une équation quadratique qui contient (x, y) linéairement. On peut supposer que u et v soient les paramètres de deux points sur une droite donnée, c. à. d. que u et v soient les distances de ces deux points respectivement à un point fixe situé sur la droite donnée ; on a ainsi la transformation de M. Hesse.

Je n'ai pas cherché la signification géométrique des formules générales.

Cambridge, 10 *octobre* 1866.

441.

NOTE SUR L'ALGORITHME DES TANGENTES DOUBLES D'UNE COURBE DU QUATRIÈME ORDRE.

[From the *Journal für die reine und angewandte Mathematik* (Crelle), tom. LXVIII. (1868), pp. 176—179.]

ON n'a pas, je crois, assez fait attention à l'algorithme (tiré de la considération d'une figure dans l'espace) qu'a trouvé M. Hesse (dans le mémoire "Ueber die Doppeltangenten der Curven vierter Ordnung," t. XLIX de ce Journal, 1855) pour dénoter les tangentes doubles (ou bitangentes) d'une courbe du quatrième ordre. Voici en quoi cet algorithme consiste. En employant les huit symboles 1, 2, 3, ... 8, les 28 bitangentes sont représentées par les combinaisons binaires 12, 13, 14, ... 78. Cela posé, considérons une expression quelconque 12 . 13 . 14, ou 12 . 34, ... ou disons un "terme" qui représente un système d'une seule ou de plusieurs des bitangentes. On peut opérer sur ce terme avec deux espèces de substitutions; la substitution ordinaire qui consiste à changer l'arrangement 12345678 des huit symboles en un autre arrangement quelconque; et la substitution "bifide" représentée par un symbole tel que 1234.5678, lequel dénote qu'il faut entrechanger les combinaisons 12 et 34, 13 et 24, 14 et 23, 56 et 78, 57 et 68, 58 et 67, en ne changeant pas les autres combinaisons. Par exemple en opérant avec 1234.5678 sur 34.45.56.17 on obtient 12.45.78.17. Le nombre de ces substitutions bifides est 35, ou en comptant la substitution, unité, qui ne change aucune des combinaisons, ce nombre est 36.

Appelons "homotypiques" deux termes qui se dérivent l'un de l'autre par une substitution ordinaire; "syntypiques" qui se dérivent l'un de l'autre par une substitution ordinaire ou bifide; "sous-groupe" le système entier des termes homotypiques à un terme donné: "groupe" le système entier des termes syntypiques à un terme donné. Un groupe peut contenir un seul sous-groupe, ou plusieurs sous-groupes; mais il importe de remarquer que la notion du sous-groupe n'a pas de signification géométrique, et ne sert que comme moyen de former les termes du groupe. Cela étant, le théorème géométrique est celui-ci; "les systèmes de bitangentes représentées par des termes syntypiques (ou autrement dit, par des termes qui appartiennent au même groupe) ont les mêmes propriétés géométriques."

Par exemple, en considérant les bitangentes deux à deux, on a deux sous-groupes, l'un composé de termes homotypiques à 12.13; l'autre, de termes homotypiques à 12.34—ou disons, le sous-groupe 12.13 de 168 termes et le sous-groupe 12.34 de 210 termes; mais ces deux sous-groupes ne forment qu'un seul groupe: pour montrer cela il suffit d'opérer sur 12.13, par exemple avec la substitution 1245.3678, ce qui donne 45.13, terme homotypique à 12.34. Cela veut dire qu'il n'y a pas de combinaison de deux bitangentes qui se distingue d'une manière quelconque de toute autre combinaison de deux bitangentes.

Mais en combinant les bitangentes trois à trois, on a les deux sous-groupes 12.34.56 (420 termes) et 12.23.34 (840 termes) qui forment un groupe de 1260 termes; les trois bitangentes représentées par un quelconque des 1260 termes ont leurs six points de contact sur une même conique. Les trois autres sous-groupes 12.23.31 (56 termes), 12.23.45 (1680 termes) et 12.13.14 (280 termes) forment un groupe de 2016 termes, et pour trois bitangentes représentées par un terme quelconque de ce groupe, les six points de contact ne sont pas situés sur une même conique.

Comme un autre exemple j'explique la constitution des 63 "groupes" de Steiner (voir le mémoire de Steiner, "Eigenschaften der Curven vierten Grades rücksichtlich ihrer Doppeltangenten," t. XLIX. de ce journal, 1855) ou (pour éviter l'emploi de ce mot *groupe* dans une nouvelle signification) disons les 63 termes G de Steiner, chaque terme composé de 6 paires de bitangentes. On a ici un sous-groupe de 35 termes G_1 de la forme

12.34; 13.42; 14.23; 56.78; 57.86; 58.67

(pour abréger on peut dénoter ce terme par 1234.5678), et un sous-groupe de 28 termes G_2 de la forme

13.32; 14.42; 15.52; 16.62; 17.72; 18.82

(pour abréger on peut de même dénoter ce terme par 12.345678), les deux sous-groupes forment le groupe des 63 termes G.

Steiner a de plus considéré les "systèmes" ou disons les termes S_1, S_2, composés chacun de trois termes G; savoir 315 termes S_1 et 336 termes S_2. Les 315 termes S_1 sont ici un groupe composé d'un sous-groupe de 105 termes $3G_1$ de la forme

1234.5678; 1256.3478; 1278.3456

et un sous-groupe de 210 termes $2G_2+G_1$ de la forme

12.345678; 34.125678 et 1234.5678.

Et de même les 336 termes S_2 sont un groupe composé d'un sous-groupe de 280 termes $2G_1+G_2$ de la forme

1234.5678; 5234.1678 et 15.234678

et un sous-groupe de 56 termes $3G_2$ de la forme

$$12.345678;\quad 13.245678;\quad 31.245678.$$

Il va sans dire que je me suis servi de l'abréviation 1234.5678 pour dénoter le terme 12.34; 13.42; 14.23; 56.78; 57.86; 58.67; et de même pour les autres termes G_1 ou G_2.

M. Aronhold (dans le mémoire "Ueber den gegenseitigen Zusammenhang der 28 Doppeltangenten einer allgemeinen Curve vierten Grades," *Berl. Monatsber.* Juli 1864), partant de 7 bitangentes données, a trouvé une construction pour les autres 21 bitangentes. Les bitangentes données doivent être indépendantes; savoir pour trois quelconques de ces 7 bitangentes, les six points de contact ne sont pas situés sur une même conique. Les bitangentes représentées par les termes 12, 13, 14, 15, 16, 17, 18 sont un tel système de bitangentes indépendantes; et en dénotant de cette manière les 7 bitangentes données, la bitangente construite par le moyen de la conique qui touche cinq de ces droites, par exemple les droites 38, 48, 58, 68, 78, (ou conique 34567) peut être dénotée par 12, et de même pour les autres bitangentes cherchées; on a ainsi le système entier des bitangentes dénotées comme auparavant par 12, 13, 14,... 78.

J'ajoute que le groupe qui contient 18, 28, 38, 48, 58, 68, 78 est composé d'un sous-groupe 18, 28, 38, 48, 58, 68, 78 de 8 termes, et d'un sous-groupe 12, 23, 31, 48, 58, 68, 78 de 280 termes; le groupe contient donc 288 termes; savoir il y a ce nombre 288 de systèmes de sept bitangentes indépendantes qui peuvent chacun servir à trouver par la construction d'Aronhold les autres 21 bitangentes.

P.S. J'ai trouvé à propos de la méthode de M. Aronhold une forme commode pour l'équation de la conique qui touche cinq droites données; en supposant que l'on ait identiquement $x+y+z+w=0$, et que les droites données soient $x=0$, $y=0$, $z=0$, $w=0$, et $ax+by+cz+dw=0$, l'équation de la conique est

$$(a-d)^2(b-c)^2(xw+yz)+(b-d)^2(c-a)^2(yw+zx)+(c-d)^2(a-b)^2(zw+xy)=0.$$

J'ajoute qu'en écrivant pour abréger

$$\alpha : \beta : \gamma = (a-d)(b-c) : (b-d)(c-a) : (c-d)(a-b)$$

(d'où $\alpha+\beta+\gamma=0$) les coordonnées (x, y, z, w) des points de contact avec les droites

$x=0$, $y=0$, $z=0$, $w=0$ sont $(0, \gamma, \beta, \alpha)$, $(\gamma, 0, \alpha, \beta)$, $(\beta, \alpha, 0, \gamma)$, $(\alpha, \beta, \gamma, 0)$

respectivement; et que les coordonnées du point de contact avec la droite $ax+by+cz+dw=0$ sont

$$x : y : z : w = (bcd) : -(cda) : (dab) : -(abc)$$

où, pour abréger, (bcd) dénote $(b-c)(c-d)(d-b)$, et de même pour (cda), (dab), (abc).

Cambridge, le 23 *septembre* 1867.

442.

NOTE SUR LA SURFACE DU QUATRIÈME ORDRE DOUÉE DE SEIZE POINTS SINGULIERS ET DE SEIZE PLANS SINGULIERS.

[From the *Journal für die reine und angewandte Mathematik* (Crelle), tom. LXXIII. (1871), pp. 292—293.]

L'ÉQUATION de M. Kummer se transforme sans difficulté en celle-ci

$$\sqrt{\alpha x\left(\gamma'\gamma''y-\beta'\beta''z-\frac{w}{\alpha}\right)}+\sqrt{\beta y\left(\alpha'\alpha''z-\gamma'\gamma''x-\frac{w}{\beta}\right)}+\sqrt{\gamma z\left(\beta'\beta''x-\alpha'\alpha''y-\frac{w}{\gamma}\right)}=0,$$

où

$$\alpha+\beta+\gamma=0,\quad \alpha'+\beta'+\gamma'=0,\quad \alpha''+\beta''+\gamma''=0.$$

Or cette équation rendue rationnelle prend, après toutes les réductions nécessaires, la forme suivante:

$$\begin{aligned}&w^2(x^2+y^2+z^2-2yz-2zx-2xy)\\&+2w\,(\alpha\alpha'\alpha''(y^2z-yz^2)+\beta\beta'\beta''(z^2x-zx^2)+\gamma\gamma'\gamma''(x^2y-xy^2)+\theta xyz)\\&+\quad(\alpha\alpha'\alpha''yz+\beta\beta'\beta''zx+\gamma\gamma'\gamma''xy)^2=0,\end{aligned}$$

où, pour abréger, l'on a écrit

$$\begin{aligned}\theta&=(\beta-\gamma)\,\alpha'\alpha''+(\gamma-\alpha)\,\beta'\beta''+(\alpha-\beta)\,\gamma'\gamma'',\\&=(\beta'-\gamma')\,\alpha''\alpha+(\gamma'-\alpha')\,\beta''\beta+(\alpha'-\beta')\,\gamma''\gamma,\\&=(\beta''-\gamma'')\,\alpha\alpha'+(\gamma''-\alpha'')\,\beta\beta'+(\alpha''-\beta'')\,\gamma\gamma',\\&=-\tfrac{1}{3}\{(\beta-\gamma)(\beta'-\gamma')(\beta''-\gamma'')+(\gamma-\alpha)(\gamma'-\alpha')(\gamma''-\alpha'')+(\alpha-\beta)(\alpha'-\beta')(\alpha''-\beta'')\},\end{aligned}$$

l'identité de ces différentes valeurs de θ étant facile à vérifier.

En représentant par $Aw^2 + 2Bw + C = 0$ la forme rationnelle de l'équation de la surface, on trouve pour le discriminant $AC - B^2$ de cette équation du second degré en w la valeur

$$AC - B^2 = 4\alpha\alpha'\alpha''\beta\beta'\beta''\gamma\gamma'\gamma''\, xyz \left(\frac{x}{\alpha} + \frac{y}{\beta} + \frac{z}{\gamma}\right)\left(\frac{x}{\alpha'} + \frac{y}{\beta'} + \frac{z}{\gamma'}\right)\left(\frac{x}{\alpha''} + \frac{y}{\beta''} + \frac{z}{\gamma''}\right).$$

L'équation de la surface rendue rationnelle est symétrique par rapport aux trois systèmes de quantités (α, β, γ), $(\alpha', \beta', \gamma')$, $(\alpha'', \beta'', \gamma'')$; la forme irrationnelle de la même équation peut donc être présentée de trois manières différentes, savoir:

$$\sqrt{\alpha x\left(\gamma'\gamma''y - \beta'\beta''z - \frac{w}{\alpha}\right)} + \sqrt{\beta y\left(\alpha'\alpha''z - \gamma'\gamma''x - \frac{w}{\beta}\right)} + \sqrt{\gamma z\left(\beta'\beta''x - \alpha'\alpha''y - \frac{w}{\gamma}\right)} = 0,$$

$$\sqrt{\alpha' x\left(\gamma''\gamma y - \beta''\beta z - \frac{w}{\alpha'}\right)} + \sqrt{\beta' y\left(\alpha''\alpha z - \gamma''\gamma x - \frac{w}{\beta'}\right)} + \sqrt{\gamma' z\left(\beta''\beta x - \alpha''\alpha y - \frac{w}{\gamma'}\right)} = 0,$$

$$\sqrt{\alpha'' x\left(\gamma\gamma' y - \beta\beta' z - \frac{w}{\alpha''}\right)} + \sqrt{\beta'' y\left(\alpha\alpha' z - \gamma\gamma' x - \frac{w}{\beta''}\right)} + \sqrt{\gamma'' z\left(\beta\beta' x - \alpha\alpha' y - \frac{w}{\gamma''}\right)} = 0$$

et l'on voit de plus que les équations des seize plans singuliers sont

$$x = 0, \quad y = 0, \quad z = 0, \quad w = 0,$$

$$\frac{x}{\alpha} + \frac{y}{\beta} + \frac{z}{\gamma} = 0, \qquad \frac{x}{\alpha'} + \frac{y}{\beta'} + \frac{z}{\gamma'} = 0, \qquad \frac{x}{\alpha''} + \frac{y}{\beta''} + \frac{z}{\gamma''} = 0,$$

$$\gamma'\gamma''y - \beta'\beta''z - \frac{w}{\alpha} = 0, \quad \alpha'\alpha''z - \gamma'\gamma''x - \frac{w}{\beta} = 0, \quad \beta'\beta''x - \alpha'\alpha''y - \frac{w}{\gamma} = 0,$$

$$\gamma''\gamma y - \beta''\beta z - \frac{w}{\alpha'} = 0, \quad \alpha''\alpha z - \gamma''\gamma x - \frac{w}{\beta'} = 0, \quad \beta''\beta x - \alpha''\alpha y - \frac{w}{\gamma'} = 0,$$

$$\gamma\gamma' y - \beta\,\beta' z - \frac{w}{\alpha''} = 0, \quad \alpha\alpha' z - \gamma\gamma' x - \frac{w}{\beta''} = 0, \quad \beta\beta' x - \alpha\alpha' y - \frac{w}{\gamma''} = 0,$$

les quantités α, β, γ, etc. étant liées entre elles par les trois équations

$$\alpha + \beta + \gamma = 0, \quad \alpha' + \beta' + \gamma' = 0, \quad \alpha'' + \beta'' + \gamma'' = 0.$$

Voilà ce me semble la forme la plus simple pour l'équation de cette surface.

Cambridge, le 23 *février* 1871.

443.

NOTE ON THE SOLUTION OF THE QUARTIC EQUATION $\alpha U + 6\beta H = 0$.

[From the *Mathematische Annalen*, vol. I. (1869), pp. 54, 55.]

IF U denote the quartic function $(a, b, c, d, e \between x, y)^4$, H its Hessian

$$= (ac - b^2,\ 2(ad - bc),\ ae + 2bd - 3c^2,\ 2(be - cd),\ ce - d^2 \between x, y)^4,$$

α and β constants, then we may find the linear factors of the function $\alpha U + 6\beta H$ (or what is the same thing solve the equation $\alpha U + 6\beta H = 0$) by a formula almost identical with that given by me (Fifth Memoir on Quantics, *Phil. Trans.* vol. CXLVIII. (1858), see p. 446, [156]) in regard to the original quartic function U.

In fact (reproducing the investigation) if I, J are the two invariants, $M = \dfrac{I^3}{4J^2}$, Φ the cubicovariant

$$= (-a^2d + 3abc - 2b^3,\ \&c \between x, y)^6,$$

then the identical equation $JU^3 - IU^2H + 4H^3 = -\Phi^2$, may be written $(1, 0, -M, M \between IH, JU)^3 = -\tfrac{1}{4}I^3\Phi^2$, whence if ω_1, ω_2, ω_3 are the roots of the equation $(1, 0, -M, M \between \omega, 1)^3 = 0$, or what is the same thing $\omega^3 - M(\omega - 1) = 0$; then the functions

$$IH - \omega_1 JU,\ IH - \omega_2 JU,\ IH - \omega_3 JU$$

are each of them a square: writing

$$(\omega_2 - \omega_3)(IH - \omega_1 JU) = X^2,$$

$$(\omega_3 - \omega_1)(IH - \omega_2 JU) = Y^2,$$

$$(\omega_1 - \omega_2)(IH - \omega_3 JU) = Z^2,$$

so that identically $X^2+Y^2+Z^2=0$, the expression $\alpha X+\beta Y+\gamma Z$ will be a square if only $\alpha^2+\beta^2+\gamma^2=0$. (To see this observe that in virtue of the equation $X^2+Y^2+Z^2=0$, we have $X+iY$, $X-iY$ each of them a square, and thence

$$\alpha X+\beta Y+\gamma Z, \ =\tfrac{1}{2}(\alpha+i\beta)(X-iY)+\tfrac{1}{2}(\alpha-i\beta)(X-iY)-\gamma i\sqrt{X^2+Y^2},$$

is a square if the condition in question be satisfied.)

Hence in particular writing

$$\sqrt{\omega_2-\omega_3}\sqrt{\alpha I+6\beta\omega_1 J}, \ldots, \quad \sqrt{\omega_1-\omega_2}\sqrt{\alpha I+6\beta\omega_3 J},$$

for α, β, γ, we have

$$(\omega_2-\omega_3)\sqrt{\alpha I+6\beta\omega_1 J}\sqrt{IH+\omega_1 JU}+\ldots+(\omega_1-\omega_2)\sqrt{\alpha I+6\beta\omega_3 J}\sqrt{IH+\omega_3 JU}$$

a perfect square, and since the product of the four different values is a multiple of $(\alpha U+6\beta H)^2$ (this is most readily seen by observing that for $\alpha U+6\beta H=0$, the irrational expression omitting a factor is $(\omega_2-\omega_3)(\alpha I+6\beta\omega_1 J)+\ldots+(\omega_1-\omega_2)(\alpha I+6\beta\omega_3 J)$, which vanishes identically) it follows that the expression in question is the square of a linear factor of $\alpha U+6\beta H$.

It thus appears that the radicals (other than those arising from the solution of $U=0$) contained in the solution of the equation $\alpha U+6\beta H=0$ are the three roots

$$\sqrt{\alpha I+6\beta\omega_1 J}, \quad \sqrt{\alpha I+6\beta\omega_2 J}, \quad \sqrt{\alpha I+6\beta\omega_3 J}.$$

Cambridge, September 2, 1868.

444.

ON THE CENTRO-SURFACE OF AN ELLIPSOID.

[From the *Proceedings of the London Mathematical Society*, vol. III. (1869—1871), pp. 16—18.]

THE President [Prof. Cayley] gave an account of his investigations on the centro-surface of an ellipsoid (locus of the centres of curvature of the ellipsoid). The surface has been studied by Dr Salmon, and also by Prof. Clebsch, but in particular the theory of the nodal curve on the surface admits of further development. The position of a point on the ellipsoid is determined by means of the parameters, or elliptic coordinates, h, k; viz., if as usual a, b, c are the semi-axes, and if X, Y, Z are the coordinates of the point in question, then

$$\frac{X^2}{a^2+h}+\frac{Y^2}{b^2+h}+\frac{Z^2}{c^2+h}=1,$$

$$\frac{X^2}{a^2+k}+\frac{Y^2}{b^2+k}+\frac{Z^2}{c^2+k}=1;$$

and hence

$$-\beta\gamma X^2=a^2(a^2+h)(a^2+k),$$

$$-\gamma\alpha Y^2=b^2(b^2+h)(b^2+k),$$

$$-\alpha\beta Z^2=c^2(c^2+h)(c^2+k),$$

if for shortness

$$\alpha=b^2-c^2,\quad \beta=c^2-a^2,\quad \gamma=a^2-b^2,\quad (\alpha+\beta+\gamma=0).$$

This being so, the coordinates of the point of intersection of the normal at (X, Y, Z) by the normal at the consecutive point of the curve of curvature

$$\frac{X^2}{a^2+k}+\frac{Y^2}{b^2+k}+\frac{Z^2}{c^2+k}=1$$

are given by the formulæ

$$-\beta\gamma a^2x^2 = (a^2+h)^3(a^2+k),$$
$$-\gamma\alpha b^2y^2 = (b^2+h)^3(b^2+k),$$
$$-\alpha\beta c^2z^2 = (c^2+h)^3(c^2+k);$$

viz., these equations, considering therein (h, k) as arbitrary parameters, determine the coordinates (x, y, z) of a point on the centro-surface. The principal sections (as is known) consist each of them of an ellipse counting three times, and of an evolute of an ellipse; the evolute and ellipse have four contacts (two-fold intersections) and four simple intersections, but the contacts and intersections respectively are in the different sections real and imaginary; and if (as we may without loss of generality assume) $a^2+c^2>2b^2$, then the form of the principal sections is as shown in the figure (which

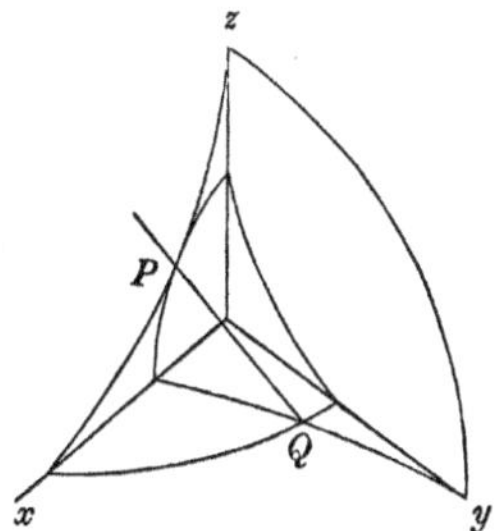

represents only an octant of the surface); viz., there is a real contact at P in the plane of xz, and a real intersection at Q in the plane of xy. The surface has thus an exterior and an interior sheet, but (instead of meeting in a conical point, as in the wave surface) these intersect in a nodal curve QP. The curve has a cusp at Q, and a node at P; viz., the curve extends beyond P, but from that point is acnodal, or without any real sheet of the surface passing through it. For the nodal curve there must be two values (h, k), (h_1, k_1), giving the same values of (x, y, z); viz., there must exist the relations

$$(a^2+h)^3(a^2+k) = (a^2+h_1)^3(a^2+k_1),$$
$$(b^2+h)^3(b^2+k) = (b^2+h_1)^3(b^2+k_1),$$
$$(c^2+h)^3(c^2+k) = (c^2+h_1)^3(c^2+k_1);$$

from which equations eliminating h_1 and k_1, we should have between h, k a relation which, combined with the expressions of x, y, z in terms of (h, k), determines the nodal curve. But the better course is to eliminate k, k_1, thus obtaining a relation between h and h_1, in virtue whereof h_1 may be regarded as a known function of h; k and k_1 can then be readily expressed in terms of h, h_1; that is, we have k as a function of h, h_1, or in effect as a function of h. The relation between h, h_1 (after a reduction of some complexity) assumes ultimately a form which is very simple and remarkable; viz., writing

$$P = a^2 + b^2 + c^2, \quad Q = b^2c^2 + c^2a^2 + a^2b^2, \quad R = a^2b^2c^2,$$

the relation is

$$\begin{aligned} &(6R + 3Qh + Ph^2) \\ &+ h_1\,(3Q + 4Ph + 3h^2\,) \\ &+ h_1{}^2(P \ + 3h \qquad\quad) = 0\,; \end{aligned}$$

this is a (2, 2) correspondence between the two parameters h, h_1; the united values $h_1 = h$, are given by the equation $6\,(R + Qh + Ph^2 + h^3) = 0$, that is

$$(a^2 + h)\,(b^2 + h)\,(c^2 + h) = 0\,;$$

viz., the two points on the ellipsoid which have their common centre of curvature on the nodal curve are only situate on the same curve of curvature when this curve is a principal section of the ellipsoid.

{Since the date of the foregoing communication, Prof. Cayley has found that the squared coordinates x^2, y^2, z^2 of a point on the nodal curve can be expressed as rational functions of a single variable parameter σ.}

445.

A MEMOIR ON QUARTIC SURFACES.

[From the *Proceedings of the London Mathematical Society*, vol. III. (1869—1871), pp. 19—69. Read February 10, 1870.]

THE present Memoir is intended as a commencement of the theory of the quartic surfaces which have nodes (conical points). A quartic surface may be without nodes, or it may have any number of nodes up to 16. I show that this is so, and I consider how many of the nodes may be given points. Although it would at first sight appear that the number is 8, it is in fact 7; viz., we can, with 7 given points as nodes (but not in a proper sense with 8 or more given points), find a quartic surface; such surface contains in its equation 6 constants, which may be such that the surface has an additional node or nodes. Suppose that the surface has an 8th node:—there are two distinct cases; viz., (1) the 8 nodes are the points of intersection of 3 quadric surfaces, or say they are an octad, and the surface is said to be octadic; (2) the 8th node is any point whatever on a certain sextic surface determined by means of the 7 given nodes, and called the dianodal surface of these 7 points; the quartic surface is said to be a dianome. The two cases are in general exclusive of each other; viz., the 7 given points being any points whatever, the dianodal surface does not pass through the 8th point of the octad; and thus the quartic surface with the 8 nodes is either octadic or else a dianome. Assuming it to be a dianome, the constants may be further determined so that there shall be a 9th node; it is necessary to examine whether this forms with 7 of the 8 nodes an octad. Supposing that it does not (viz., that there are not any 8 nodes in regard to which the surface is octadic), the 9th node is then any point whatever on a certain curve of the order 18, determined by means of the 8 nodes, and called the dianodal curve of these 8 points. And, finally, the constants may be further determined so that there shall be a 10th node; supposing, as before, that this does not form an octad with any 7 of the 9 nodes (viz., that there are not any 8 nodes in regard to which the surface is octadic), the 10th node is then any one of a system of 22 [should be 13] points determined by means of the 9 nodes, and called the dianodal system of these 9 points. But the quartic surface is now completely determined; viz., starting with any 7 given points as nodes, we have a dianome with 8 nodes, 9 nodes, or 10 nodes, say, an octodianome, enneadianome, or decadianome, but not with any greater number of nodes; these can only present themselves when particular conditions are satisfied in regard to the 7 given nodes, and to the 8th and 9th node; and the consideration of the quartic surfaces with more than 10 nodes would thus form a separate branch of the subject.

The case of the decadianome (or quartic surface with 10 nodes formed as above with 7 given points as nodes) is peculiarly interesting. I identify this with the surface which I call a symmetroid; viz., the surface represented by an equation $\Delta = 0$, where Δ is a *symmetrical* determinant of the 4th order the several terms whereof are linear functions of the coordinates (x, y, z, w); this surface is related to the Jacobian surface of 4 quadric surfaces (itself a very remarkable surface), and this theory of the symmetroid and the Jacobian, and of questions connected therewith, forms a large portion of the present Memoir.

The theory of the Jacobian is connected also with the researches in regard to nodal quartic surfaces in general; and, for greater clearness, it has seemed to me proper to commence the Memoir with certain definitions, &c., in regard to this theory. It will be seen in what manner I extend the notion of the Jacobian.

I remark that the present researches on Quartic Surfaces were suggested to me by Professor Kummer's most interesting Memoir "Ueber die algebraischen Strahlensysteme u.s.w.," *Berl. Abh.* 1866, in which, without entering upon the general theory, he is led to consider the quartic surfaces, or certain quartic surfaces, with 16, 15, 14, 13, 12, or 11 nodes; the last of these, or surface with 11 nodes, being in fact a particular case of the symmetroid.

Considerations in regard to the Jacobian of four, or more or less than four, Surfaces.

1. In the case of any four surfaces, $P = 0$, $Q = 0$, $R = 0$, $S = 0$, the differential coefficients of P, Q, R, S in regard to the coordinates (x, y, z, w) may be arranged as a square matrix in either of the ways

$$\begin{array}{c|c} & P,\ Q,\ R,\ S \\ \hline \delta_x & \\ \delta_y & \\ \delta_z & \\ \delta_w & \end{array} \quad ; \quad \begin{array}{c|c} \delta_x,\ \delta_y,\ \delta_z,\ \delta_w & \\ \hline & P \\ & Q \\ & R \\ & S \end{array}$$

and with either arrangement we may form one and the same determinant, the Jacobian determinant $J(P, Q, R, S)$, or, equating it to zero, the Jacobian surface $J(P, Q, R, S) = 0$, of the four surfaces.

2. In the case of more than four surfaces, adopting the arrangement

$$\begin{array}{c|l} & P,\ Q,\ R,\ S,\ T,\ldots \\ \hline \delta_x & \\ \delta_y & \\ \delta_z & \\ \delta_w & \end{array}$$

and considering the several determinants which can be formed with any four columns of the matrix, these equated to zero establish a more than one-fold relation between the coordinates; viz., in the case of five surfaces, we have $J(P, Q, R, S, T) \equiv 0$, a twofold relation representing a curve; and in the case of six surfaces, $J(P, Q, R, S, T, U) \equiv 0$, a threefold relation representing a point-system; and (since with four coordinates a relation is at most threefold) these are the only cases to be considered.

3. In the case of fewer than four surfaces, adopting the arrangement

$$\begin{array}{l|c} \delta_x,\ \delta_y,\ \delta_z,\ \delta_w & \\ \hline & P \\ & Q \\ & \vdots \end{array}$$

and considering the several determinants which can be formed with any 3 or 2 columns of the matrix, and equating these to zero, we have in like manner a more than one-fold relation between the coordinates; viz., in the case of three surfaces, we have $J(P, Q, R) \equiv 0$, a twofold relation representing a curve; and in the case of two surfaces $J(P, Q) \equiv 0$, a threefold equation representing a point-system, (viz., this denotes the points $\delta_x P : \delta_y P : \delta_z P : \delta_w P = \delta_x Q : \delta_y Q : \delta_z Q : \delta_w Q$); for a single surface we should have a fourfold relation, and the case is not considered. But observe that if the notation were used, $J(P) \equiv 0$ would denote $\delta_x P = 0$, $\delta_y P = 0$, $\delta_z P = 0$, $\delta_w P = 0$, equations which are satisfied simultaneously by the coordinates (x, y, z, w) of any node of the surface $P = 0$. Although in what precedes I have used the sign $\equiv$, there is no objection to using, and I shall in the sequel use, the ordinary sign $=$, it being understood that while $J(P, Q, R, S) = 0$ denotes a single equation or onefold relation, $J(P, Q, R, S, T) = 0$ or $J(P, Q, R) = 0$ will each denote a twofold relation, and $J(P, Q, R, S, T, U) = 0$ or $J(P, Q) = 0$ each of them a threefold relation.

4. It is not asserted that $\ldots J(P, Q, R) = 0$, $J(P, Q, R, S) = 0$, $J(P, Q, R, S, T) = 0, \ldots$ form a continuous series of analogous relations; and there might even be a propriety in using, in regard to four or more surfaces, J, and in regard to four or fewer surfaces an inverted J (viz., in regard to four surfaces, either symbol indifferently); but there is no ambiguity in, and I have preferred to adopt, the use of the single symbol J.

5. Suppose that the orders of the surfaces $P=0$, $Q=0, \ldots$ are $a+1$, $b+1, \ldots$ (so that the orders of the differential coefficients of P, $Q \ldots$ are a, $b, \ldots$), then we have for the orders of the several loci,

$$\begin{array}{lll} J(P,\ Q)=0, & \text{point-system, order} & a^3+a^2b+ab^2+b^3; \\ J(P,\ Q,\ R)=0, & \text{curve,} \quad\quad ,, & a^2+b^2+c^2+bc+ca+ab; \\ J(P,\ Q,\ R,\ S)=0, & \text{surface,} \quad\quad ,, & a+b+c+d; \\ J(P,\ Q,\ R,\ S,\ T)=0, & \text{curve,} \quad\quad ,, & ab+ac \ldots +de; \\ J(P,\ Q,\ R,\ S,\ T,\ U)=0, & \text{point-system,} \quad ,, & abc+abd \ldots +def; \end{array}$$

see, as to this, Salmon's *Solid Geometry*, Ed. 2, (1865), Appendix IV., "On the Order of Systems of Equations" [not reproduced in the later editions]. In particular, if $a=b=c \ldots =1$, then the orders are 4, 6, 4, 10, 20.

As to the Surface obtained by equating to zero a Symmetrical Determinant.

6. It is also shown (Salmon, Ed. 2, p. 495) that the surface obtained by equating to zero any symmetrical determinant has a determinate number of nodes; viz., if the orders of the terms in the diagonal be a, b, c, &c., then the number of nodes is $=\tfrac{1}{2}(\Sigma a \,.\, \Sigma ab - \Sigma abc)$, or, as this may also be written, $\tfrac{1}{2}(\Sigma a^2 b + 2\Sigma abc)$. In particular, the formula applies to the case of the surface

$$\begin{vmatrix} A, & H, & G, & L \\ H, & B, & F, & M \\ G, & F, & C, & N \\ L, & M, & N, & D \end{vmatrix} = 0,$$

(a, b, c, d) being here the orders of A, B, C, D respectively, and the orders of F, G, &c., being $\tfrac{1}{2}(b+c)$, $\tfrac{1}{2}(a+c)$, &c. If the terms are all of them linear functions of the coordinates, or $a=b=c=d=1$, then the number of nodes is $=10$.

7. That the surface has nodes is, in fact, clear from the consideration that any point for which the minors of the determinant all vanish will be a node; and that (for the symmetrical determinant), by making the minors all of them vanish, we establish only a threefold relation between the coordinates. The expression for the number of the nodes is, I think, obtained most readily as follows:

The nodes will be points of intersection of the curve and surface

$$\left\| \begin{array}{cccc} A, & H, & G, & L \\ H, & B, & F, & M \\ G, & F, & C, & N \end{array} \right\| = 0, \qquad \begin{vmatrix} B, & F, & M \\ F, & C, & N \\ M, & N, & D \end{vmatrix} = 0,$$

these, however, contain in common the points

$$\left\| \begin{array}{cccc} H, & B, & F, & M \\ G, & F, & C, & N \end{array} \right\| = 0;$$

and not only so, but they touch at the points in question; so that, multiplying together the orders of the curve and surface, and subtracting *twice* the order of the point-system, we obtain the expression for the number of nodes. In the particular case where the functions are all linear, we have a sextic curve and cubic surface intersecting in 18 points; but the curve and surface touch in 4 points, and the number of nodes is $(18-2.4)=10$. And in the same way the formula may be established for the general case.

8. The subsidiary theorem of the contact of the curve and surface requires, however, to be proved. Seeking for the equation of the tangent plane of the surface at any one of the points in question, we have first

$$\left|\begin{array}{ccc}\delta B, & \delta F, & \delta M \\ F, & C, & N \\ M, & N, & D\end{array}\right| + \left|\begin{array}{ccc}B, & F, & M \\ \delta F, & \delta C, & \delta N \\ M, & N, & D\end{array}\right| + \left|\begin{array}{ccc}B, & F, & M \\ F, & C, & N \\ \delta M, & \delta N, & \delta D\end{array}\right| = 0,$$

where, in virtue of the equations

$$\left\|\begin{array}{cccc}H, & B, & F, & M \\ G, & F, & C, & N\end{array}\right\| = 0,$$

the last term vanishes. Expanding the other two terms, the equation becomes

$$D(C\delta B + B\delta C - 2F\delta F) - (N^2\delta B - 2MN\delta F + M^2\delta C) + \delta M(FN - CM) + \delta N(BN - MF) = 0;$$

but, in virtue of the same equations, the coefficients of δM and δN each of them vanish, and we have also

$$N^2\delta B + M^2\delta C - 2MN\delta F = \frac{N^2}{C}(C\delta B + B\delta C - 2F\delta F);$$

so that the equation becomes finally $C\delta B + B\delta C - 2F\delta F = 0$. Investigating by a like process the equation of the tangent of the curve

$$\left\|\begin{array}{cccc}A, & H, & G, & L \\ H, & B, & F, & M \\ G, & F, & C, & N\end{array}\right\| = 0,$$

we find between the differentials δA, δB, &c., a twofold linear relation, expressible by means of the foregoing equation $C\delta B + B\delta C - 2F\delta F = 0$, and one other equation; that is, at each of the points in question the tangent of the curve lies in the tangent plane of the surface, or, what is the same thing, the curve and surface touch at these points.

Surfaces represented by an equation $F(P, Q) = 0$, &c.

9. In the remarks which follow as to the surfaces $F(P, Q) = 0$, $F(P, Q, R) = 0$, &c., the function F is a rational and integral function of (P, Q), (P, Q, R), &c., not in general homogeneous in regard to $P, Q, R, \ldots.$ but of such degrees in regard to these functions respectively as to be homogeneous in regard to the coordinates (x, y, z, w).

The surface $F(P, Q) = 0$ has in general a nodal curve $\delta_P F = 0$, $\delta_Q F = 0$; and if it has besides any nodes, these are points of the point-system $J(P, Q) = 0$.

The surface $F(P, Q, R) = 0$ has in general nodes $\delta_P F = 0$, $\delta_Q F = 0$, $\delta_R F = 0$; and if it has besides any nodes, these are points on the curve $J(P, Q, R) = 0$.

The surface $F(P, Q, R, S) = 0$ has not in general, but it may have, nodes $\delta_P F = 0$, $\delta_Q F = 0$, $\delta_R F = 0$, $\delta_S F = 0$; if it has any other nodes, these are points on the surface $J(P, Q, R, S) = 0$.

Nodes of a Quartic Surface; Circumscribed Cone having its vertex at a Node.

10. A quartic surface may be without nodes; or it may have any number of nodes up to 16. Consider a quartic surface having a node or nodes; and take the single node, or (if more nodes than one) any one of the nodes, as the vertex of a circumscribed cone; then, considering any plane through the vertex, the section will be a quartic curve having a node at the vertex, and the generating lines in the plane will be the tangents from the node to the quartic curve; the number of them is therefore 6, and the order of the circumscribed cone is thus $= 6$. Each tangent intersects the quartic curve in the node counting as two intersections, and in the point of contact counting as two intersections; there are consequently no singular tangents; and therefore in the circumscribed cone no singular lines arising from a singular tangency of the generating line. Hence, in the case of a single node on the surface, the circumscribed cone is a cone of the order 6 without nodal or stationary lines; and the class is $= 30$. But in the case of more than one node, say k nodes, the circumscribed cone passes through the remaining $k-1$ nodes, and the generating line through each of these nodes is a nodal line of the cone; that is, the cone has $k-1$ nodal lines, and its class is $= 30 - 2k + 2$. The cone is not of necessity a proper cone; the maximum number of nodal lines is when it breaks up into 6 planes, and we have then $k - 1 = 15$; that is, the number of nodes of the surface is at most $= 16$.

11. It is easy to form a table of the different *primâ facie* possible forms of the sextic cone, according to the number of nodes of the surface; viz., writing 6 for a proper sextic cone without nodal lines, 6_1, $6_2 \ldots 6_{10}$ for the proper sextic cone with 1, 2, ... or 10 nodal lines; and so 5, $5_1 \ldots 5_6$ for the proper quintic cones, 4, 4_1, 4_2, 4_3, 3, 3_1, 2 for the quartic, cubic, and quadric cones, and 1 for the plane, the table is

Circumscribed Sextic Cone.

Nodes of Surface.								
1	6							
2	6_1							
3	6_2							
4	6_3							
5	6_4							
6	6_5 ;	5 , 1						
7	6_6 ;	5_1, 1						
8	6_7 ;	5_2, 1						
9	6_8 ;	5_3, 1;	4 , 2					
10	6_9 ;	5_4, 1;	4_1, 2;	4 , 1, 1;	3 , 3			
11	6_{10};	5_5, 1;	4_2, 2;	4_1, 1, 1;	3_1, 3			
12	...	5_6, 1;	4_3, 2;	4_2, 1, 1;	3_1, 3_1;	3 , 2, 1		
13	...	...	...	4_3, 1, 1;	...	3_1, 2, 1;	3 , 1, 1, 1;	2 , 2, 2
14	...	...	...	...	...	...	3_1, 1, 1, 1;	2 , 2, 1, 1
15	...	...	...	...	...	...	...	2_1, 1, 1, 1, 1
16	...	...	...	...	...	...	...	1 , 1, 1, 1, 1, 1;

and moreover, in the cases where there are two or more forms of the sextic cone, then the k sextic cones may be of the different forms in various combinations. The total number of cases *primâ facie* possible is thus very great; but only a comparatively small number of them actually exist.

12. In the case where there is a plane 1, the sextic cone breaks up into this plane, and into a (proper or improper) quintic cone intersecting the plane in 5 lines; that is, there will be in the plane 6 nodes; the plane is, in fact, a singular tangent plane meeting the surface in a conic twice repeated; and the 6 nodes lie on this conic. Taking any one of these nodes as vertex, the corresponding sextic cone breaks up into the plane, and into a (proper or improper) quintic cone.

13. In the cases $k=1, 2, 3, 4, 5$, and $k=15, 16$, there is only one form of sextic cone; so that each node (at least so far as appears) stands in the same relation to the surface. Considering the last mentioned two cases; $k=16$,—each of the 16 nodes gives 6 singular tangent planes, but each of these passes through 6 nodes; therefore the number of planes is $=16$: similarly, $k=15$, the number of singular tangent planes is $15 \times 4 \div 6, =10$.

For $k=14$, the cones are 3_1, 1, 1, 1, or 2, 2, 1, 1: it is easy to see that we have only the three cases

Cones 3_1, 1, 1, 1 : 2, 2, 1, 1				Singular tangent planes
No. may be	14 ,	0	gives	$(14.3+0.2)\div 6, =7$
"	8 ,	6	"	$(8.3+6.2)\div 6, =6$
"	2 ,	12	"	$(2.3+12.2)\div 6, =5$

and we may in the like manner limit the number of possible cases, for other values of k. But I do not at present further pursue the inquiry.

As to the Number of Constants contained in a Surface.

14. We say that a surface $P=0$ contains or depends upon a certain number of constants; viz., this is the number of constants contained in the equation $P=0$ of the surface, taking the coefficient of any one term to be equal to unity; thus the general quadric surface contains 9 constants; the surface can in fact be determined so as to satisfy 9 conditions; or, as we might express it, the *Postulation* of the surface is $=9$. [I have elsewhere said *Postulandum* and *Capacity*: I prefer this last expression.] And if, in the general equation so containing 9 constants, k of these are given, or, what is the same thing, if the quadric surface be made to satisfy any k conditions, then the number of constants, or postulation of the surface, is $=9-k$.

15. But a different form of expression is sometimes convenient; the conditions to be satisfied are frequently such that, being satisfied by the surfaces $P=0$, $Q=0, \ldots$, they will be satisfied by the surface $\alpha P+\beta Q+\ldots=0$, where α, β, ... are any constant multipliers whatever. When this is so, there will be a certain number of solutions $P=0$, $Q=0, \ldots$ not connected by any such relation, or say of asyzygetic solutions, such that the general surface satisfying the conditions in question is $\alpha P+\beta Q+\ldots=0$; and hence, taking one of these coefficients as unity, the number of constants, or postulation of the surface, is equal to the number of the remaining coefficients, or, what is the same thing, it is less by unity than the number of the asyzygetic solutions $P=0$, $Q=0 \ldots$. Instead of considering the number of constants, or postulation, we may consider the number of solutions (that is, asyzygetic solutions) or surfaces $P=0$, $Q=0, \ldots$ which satisfy the conditions in question.

16. Thus, for the quadric not subjected to any conditions, there are 10 surfaces (for example, these may be taken to be the surfaces $x^2=0$, $y^2=0$, $z^2=0$, $w^2=0$, $yz=0$, $zx=0$, $xy=0$, $xw=0$, $yw=0$, $zw=0$); and the general quadric surface is by means of these expressed linearly in the form $(a, \ldots \between x, y, z, w)^2=0$. So for the quadric surfaces through k given points, the number of these is $=10-k$; thus for the surfaces through 4 given points, say the points (1, 0, 0, 0), (0, 1, 0, 0), (0, 0, 1, 0), (0, 0, 0, 1), the 6 given surfaces may be taken to be $yz=0$, $zx=0$, $xy=0$, $xw=0$, $yw=0$, $zw=0$, and every other quadric surface through the 4 points is by means of these expressed linearly in the form $(a, \ldots \between yz, zx, xy, xw, yw, zw)=0$; for the quadric surfaces through 8 points there are two surfaces $P=0$, $Q=0$; and every quadric surface through the

8 points is by means of these expressed linearly in the form $\alpha P+\beta Q=0$; and (as the extreme case) if the quadric surface passes through 9 given points, then there is the single quadric surface $P=0$.

17. In the questions in which such quadric surfaces present themselves, it is in general quite immaterial what particular surfaces are selected as the surfaces $P=0$, $Q=0, \ldots$; the selection may be made at pleasure and, being so made, the surfaces are to be regarded as completely determinate; viz., there would be no gain of generality if these were replaced by any other surfaces $\alpha P+\beta Q \ldots=0$. For instance, in the theory of the quartic surfaces with 6 given points as nodes, we have through the 6 given points the 4 quartic surfaces $P=0$, $Q=0$, $R=0$, $S=0$, and we consider the quartic functions $(a, \ldots \between P, Q, R, S)^2$ and $J(P, Q, R, S)$: each of these is unaltered as to its form when P, Q, R, S are replaced each of them by any linear function of these quantities; viz., $(a, \ldots \between P, Q, R, S)^2$ is changed into a new quadric function $(a', \ldots \between P, Q, R, S)^2$, and $J(P, Q, R, S)$ into a mere constant multiple of its original value. We have herein a justification of the expressions in question, through 6 given points there are 4 quadric surfaces, &c.

General theory of the Quartic Surface with a given Node or Nodes.

18. A quartic surface contains 34 constants; and the number of conditions to be satisfied in order that a given point may be a node is $=4$. Hence, if the surface has k given points as nodes, the number of constants is $=34-4k$; and it would at first sight appear that k might be $=8$, and that with the 8 given points as nodes we should have a quartic surface containing 2 constants. But this is not so in a proper sense; for through the 8 given points we have 2 quadric surfaces $P=0$, $Q=0$; and we can by means of these form a quartic surface $(a, b, c \between P, Q)^2=0$, containing 2 constants, and having in a sense the 8 points as nodes. This, however, is no proper quartic surface, but is a system of 2 quadric surfaces, each of them passing through the 8 points, and the two quadric surfaces therefore intersecting in a quadri-quadric curve through the 8 points; which curve is therefore a nodal curve on the compound surface; and it is only as points on this nodal curve, and not in a proper sense, that the 8 given points are nodes of the quartic surface. The greatest value of k is thus $k=7$.

19. Of course, if $k=0$, we have the general quartic surface $U=0$, containing 34 constants. The cases $k=1$, $k=2$, $k=3$ (viz., a single given node, 2 given nodes, 3 given nodes), may be at once disposed of; taking for instance the 1st node to be the point (1, 0, 0, 0), the 2nd node the point (0, 1, 0, 0), the 3rd node the point (0, 0, 1, 0), we find at once an equation $U=0$, with 30, 26, or 22 constants, having the given node or nodes.

Four given Nodes.

20. The case of 4 given nodes is just as easy; but in reference to what follows, it is proper to consider it more in detail. The equation should contain 18 constants; we have through the 4 given points 6 quadric surfaces, $P=0$, $Q=0$, $R=0$, $S=0$, $T=0$,

$U=0$, and we can by means of them form a quartic equation $(a, \ldots \between P, Q, R, S, T, U)^2=0$, having the 4 given points as nodes; this contains, however, $(21-1=)$ 20 constants; the reduction to the right number 18 occurs by reason that the functions (P, Q, R, S, T, U), although linearly independent, are connected by two quadric equations

$$(*\between P, Q, R, S, T, U)^2=0, \quad (*'\between P, Q, R, S, T, U)^2=0;$$

hence writing the equation of the quartic surface in the form

$$(a, \ldots \between _{,,})^2 - \lambda (*\between _{,,})^2 - \mu (*\between _{,,})^2 = 0,$$

the coefficients λ, μ may be so determined as to reduce to zero the coefficients of any two terms of the equation, and the number of constants really is $20-2=18$, as it should be.

21. In proof, observe that, taking the 4 given nodes to be the points (1, 0, 0, 0), (0, 1, 0, 0), (0, 0, 1, 0), (1, 0, 0, 0), the quadric surfaces may be taken to be $yz=0$, $zx=0$, $xy=0$, $xw=0$, $yw=0$, $zw=0$; the equation of the quartic surface will thus be

$$(a, \ldots \between yz, zx, xy, xw, yw, zw)^2=0;$$

but we have between the functions xy, &c., the two identical relations

$$xy \,.\, zw - xz \,.\, yw = 0, \quad xy \,.\, zw - xw \,.\, yz = 0;$$

and the number of constants is thus $=18$.

Five given Nodes.

22. In the case of 5 given nodes, the number of constants should be $=14$. We have through the 5 given points, 5 quadric surfaces $P=0$, $Q=0$, $R=0$, $S=0$, $T=0$, and we form herewith the quartic equation $(a, \ldots \between P, Q, R, S, T)^2=0$, containing the right number 14 of arbitrary constants. The functions P, Q, &c. are in this case not connected by any quadric relation, and the equation just written down is in fact the general equation of the quartic surface with the 5 given nodes.

23. In verification, take the first 4 nodes to be as above, and the 5th node to be the point (1, 1, 1, 1); we may write

$$(P, Q, R, S, T) = \{x(y-z),\ x(y-w),\ y(x-z),\ y(x-w),\ xy-zw\};$$

and if from the 5 equations $P=x(y-z)$, &c., we eliminate (x, y, z, w), we obtain one, and only one, relation between the functions P, Q, R, S, T; this is found to be

$$PS(Q+R-T) - \quad QR(P+S-T)=0,$$

or, what is the same thing,

$$R(P-Q)(S-T) - P(R-S)(Q-T)=0;$$

viz., it is a cubic relation, and there is consequently no quadric relation between the 5 functions.

Six given Nodes.

24. In the case of 6 given nodes, the quartic surface should contain 10 constants. We have through the 6 given points 4 quadric surfaces $P=0$, $Q=0$, $R=0$, $S=0$; but if we form herewith the quartic surface $(a, \dots \between P, Q, R, S)^2=0$, this contains only 9 constants. It is to be shown that the Jacobian surface $J(P, Q, R, S)=0$ of the 4 quadric surfaces (or say of the 6 points) is a quartic surface having the 6 given points as nodes, and not included in the foregoing form $(a, \dots \between P, Q, R, S)^2=0$; this being so, we have the quartic surface

$$(a, \dots \between P, Q, R, S)^2+\theta J(P, Q, R, S)=0,$$

having the 6 given points as nodes, and containing the complete number of constants, viz., 10.

25. The 6 given nodes being any points whatever, their coordinates may be taken to be (1, 0, 0, 0), (0, 1, 0, 0), (0, 0, 1, 0), (0, 0, 0, 1), (1, 1, 1, 1), and $(\alpha, \beta, \gamma, \delta)$. I proceed to find the Jacobian of these 6 points. For this purpose, let (a, b, c, f, g, h) be the 6 coordinates of the line through the points (1, 1, 1, 1) and $(\alpha, \beta, \gamma, \delta)$, viz.,

$$\begin{aligned} a&=\beta-\gamma, & f&=\alpha-\delta, \\ b&=\gamma-\alpha, & g&=\beta-\delta, \\ c&=\alpha-\beta, & h&=\gamma-\delta, \end{aligned}$$

whence $af+bg+ch=0$, and also

$$\begin{array}{rrrrl} . & h & -g & +a & =0, \\ -h & . & +f & +b & =0, \\ g & -f & . & +c & =0, \\ -a & -b & -c & . & =0, \end{array}$$

we have through the 6 points the plane pairs

$$\begin{aligned} x\,(\quad . \quad -hx-gz+aw)&=0, \\ y\,(-hx \quad . \quad +fz+bw)&=0, \\ z\,(\ gx-fy \quad . \quad +cw)&=0, \\ w\,(-ax-by-cz \quad . \quad)&=0, \end{aligned}$$

where, adding the four equations, we have identically $0=0$. For this reason, we cannot take these to be the equations of the 4 quadric surfaces, but we may take the first 3 of them for the surfaces $P=0$, $Q=0$, $R=0$; and for the 4th surface $S=0$, I take the quadric cone having its vertex at the point (0, 0, 0, 1); viz., the equation is

$$a\alpha yz + b\beta zx + c\gamma xy = 0;$$

that is, I write

$$(P, Q, R, S)=\{x(hy-gz+aw),\ y(-hx+fz+bw),\ z(gx-fy+cw),\ (a\alpha yz+b\beta zx+c\gamma xy)\}.$$

26. The Jacobian is then easily found to be

$$(b\beta zx + c\gamma xy)(-agh,\ bhf,\ cfg,\ abc,\ -af^2,\ -gB,\ hC,\ aA,\ b^2g,\ -c^2h \between x,\ y,\ z,\ w)^2$$
$$+ (c\gamma xy + a\alpha yz)(agh,\ -bhf,\ cfg,\ abc,\ fA,\ -bg^2,\ -hC,\ -a^2f,\ bB,\ c^2h \between x,\ y,\ z,\ w)^2$$
$$+ (a\alpha yz + b\beta zx)(agh,\ bhf,\ -cfg,\ abc,\ -fA,\ gB,\ -ch^2,\ a^2f,\ -b^2g,\ cC \between x,\ y,\ z,\ w)^2 = 0\,;$$

where for the moment A, B, C denote $bg-ch$, $ch-af$, $af-bg$ respectively. Collecting and reducing, the whole divides by $2abc$; and if finally we replace a, b, c, f, g, h by their values, the result is

$$J = \left\{\begin{array}{l} \ \ \ (\beta-\gamma)\,yz\,(\alpha w^2 - \delta x^2) + (\alpha-\delta)\,xw\,(\beta z^2 - \gamma y^2) \\ + (\gamma-\alpha)\,zx\,(\beta w^2 - \delta y^2) + (\beta-\delta)\,yw\,(\gamma x^2 - \alpha z^2) \\ + (\alpha-\beta)\,xy\,(\gamma w^2 - \delta z^2) + (\gamma-\delta)\,zw\,(\alpha y^2 - \beta x^2) \end{array}\right\} = 0.$$

27. It may be shown *à posteriori* that J is not a quadric function of P, Q, R, S. For, attempting to express it in this form, J does not contain the terms x^2w^2, y^2w^2, z^2w^2, and it thence at once appears that the coefficients of P^2, Q^2, R^2 each of them vanish. Hence, introducing for convenience the factor 2, I assume $(0, 0, 0, D, F, G, H, L, M, N \between P, Q, R, S)^2 = 2J$. Comparing the terms in $w^2(yz,\ zx,\ xy)$, we obtain

$$bcF = a\alpha,\quad caG = b\beta,\quad abH = c\gamma\,;$$

and comparing the coefficients of $w\,(y^2z,\ z^2x,\ x^2y,\ yz^2,\ zx^2,\ xy^2)$, we obtain

$$-Ff + a\alpha M = \frac{h\alpha}{b},\qquad Ff + a\alpha N = -\frac{g\alpha}{c},$$

$$-Gg + b\beta N = \frac{f\beta}{c},\qquad Gg + b\beta L = -\frac{h\beta}{a},$$

$$-Hh + c\gamma L = \frac{g\gamma}{a},\qquad Hh + c\gamma M = -\frac{f\gamma}{b}\,;$$

substituting for F, G, H their values, we obtain from the first 3 equations L, M, $N = \frac{-f}{bc},\ \frac{-g}{ca},\ \frac{-h}{ab}$, and from the second 3 equations, L, M, $N = \frac{f}{bc},\ \frac{g}{ca},\ \frac{h}{ab}$; that is, the equations are inconsistent, and the function J is not expressible in the form in question.

Jacobian Surface of Six given Points.

28. The equation $J=0$ is the locus of the vertices of the quadric cones which pass through the given 6 points; calling these 1, 2, 3, 4, 5, 6, we see at once that the surface passes through the 15 lines 12, 13, ... 56, and also through the ten lines 123.456 (viz., line of intersection of the planes through 1, 2, 3, and through 4, 5, 6), &c. In fact, taking the vertex at any point O in the line 1, 2, the lines drawn to the six points are $O1 = O2$, $O3$, $O4$, $O5$, $O6$; viz., there are only five lines, so that these lie in a quadric cone. And taking the vertex at any point in the line 123.456,

the lines to the 6 points lie in these planes 123 and 456 respectively, and the quadric cone is in fact this plane-pair. Moreover, the surface containing the lines 12, 13, 14, 15, 16, must have the point 1 for a node; and similarly, the points 2, 3, 4, 5, 6 are each of them a node on the surface. It is to be added that the surface contains the skew cubic through the 6 points, or say the skew cubic 123456. See, as to this, *post* No. 108.

29. The surface in question (the Jacobian of the 6 points) is a particular case of the Jacobian of any 4 quadric surfaces. This more general surface will be considered in the sequel; I only remark here that it contains 10 lines, corresponding to the 10 lines 123.456, &c., but it has not any other lines, or any nodes.

Jacobian Curve of Seven given Points, or of an Octad of Points.

30. In connexion with what precedes, we may here consider a curve which presents itself in the sequel; viz., the curve which is the locus of the vertices of the quadric cones which pass through seven given points. The general case is when no one of the points is the vertex of a quadric cone through the other 6 points. We have through the 7 points the three quadric surfaces $P=0$, $Q=0$, $R=0$; hence, forming the equation $\alpha P+\beta Q+\gamma R=0$ of the general quadric surface through the 7 points, and making this a cone, we find as the locus of the vertex $J(P, Q, R)=0$; the analytical form shows that this is a sextic curve. It appears, moreover, that the curve is symmetrically related to all the 8 points $P=0$, $Q=0$, $R=0$; and instead of calling it the Jacobian of the 7 points, we may call it the Jacobian of the octad. But in further explanation, take the points to be 1, 2, 3, 4, 5, 6, 7; the vertex will lie on each of the Jacobian surfaces 123456 and 123457; and it is at present assumed that 7 is not a point on the first surface, nor 6 a point on the second surface. The two surfaces have in common the lines 12, 13, ... 45, and they consequently besides intersect in a curve of the 6th order, or sextic curve, which is the locus in question. At the point 1 there is on the first surface a tangent cone through the lines 12, 13, 14, 15, 16, and on the second surface a tangent cone through the lines 12, 13, 14, 15, 17; these two cones have for their complete intersection the lines 12, 13, 14, 15, which lines belong to the complete intersection of the two surfaces, but not to the sextic curve. It thus appears, *à posteriori*, that the sextic curve does not pass through the point 1; and similarly, that it does not pass through any of the points 2, 3, 4, or 5. As to the points 6 and 7, each of these is on only one of the quartic surfaces, and therefore the curve of intersection does not pass through either of these points.

31. Suppose, however, that one of the seven points is the vertex of a cone through the other six; it is of course the same thing whether we take this to be one of the points 1, 2, 3, 4, 5, or one of the points 6 and 7, but the result comes out more easily in the latter case; viz., in the former case, taking 1 to be the point in question, the two tangent cones at 1 are one and the same cone, and all that appears is that there is nothing to hinder a branch or branches of the sextic curve from passing through the point 1. But in the latter case, taking 7 for the point in question, then 7 lies on the surface 123456, being a simple point on this surface, but a node on the surface 123457; and it thus appears that there are through 7 two

branches of the sextic curve; so that any one of the seven points, being the vertex of a cone through the other six, is an actual double point on the sextic curve.

32. In the case where two of the points are each of them the vertex of a cone through the other six points, then the seven points lie on a skew cubic; and the sextic curve of the general case becomes this skew cubic twice repeated.

Seven given Nodes.

33. In the case of 7 given nodes, the number of constants should be $=6$; the 7 given points determine 3 quadric surfaces $P=0$, $Q=0$, $R=0$; and we have hence the quartic surface $(a, \ldots \between P, Q, R)^2=0$, containing 5 constants only. That this is not the general quartic surface with the 7 given nodes, is also clear from the consideration that the surface in question has 8 nodes; viz., the 8 points of intersection of the three quadric surfaces. Suppose that a particular quartic surface, having the 7 given nodes, but not of the last mentioned form, is $\Delta=0$; then a quartic surface having the 7 given nodes is

$$(a, \ldots \between P, Q, R)^2 + \theta\Delta = 0;$$

and this, as containing 6 constants, will be the general quartic surface with the 7 given nodes.

34. It follows that, if $\Delta'=0$ be another quartic surface having the 7 given nodes, we must have identically $\Delta' - p\Delta = (* \between P, Q, R)^2$, where p is a determinate constant and $(* \between P, Q, R)^2$ a determinate quadric function of (P, Q, R). The formula extends to the case where $\Delta'=0$ has the 8 nodes $(P=0, Q=0, R=0)$, but we have then $p=0$, and the meaning is simply that the general quartic surface having the 8 nodes is $(* \between P, Q, R)^2=0$.

35. A particular quartic surface having (in an improper sense) the 7 given nodes, but not having the 8th node, is $M\Omega=0$, where $M=0$ in the plane through any 3 of the 7 points and $\Omega=0$ is the cubic surface through these same 3 points, and having the remaining 4 points as nodes. The equation of the cubic surface, if the 4 points are taken to be (1, 0, 0, 0), (0, 1, 0, 0), (0, 0, 1, 0), (0, 0, 0, 1), is obviously of the form

$$\frac{a}{x}+\frac{b}{y}+\frac{c}{z}+\frac{d}{w}=0, \text{ (that is, } ayzw+bzxw+cxyw+dxyz=0),$$

and by making the surface pass through the 3 points we determine linearly the coefficients (a, b, c, d), that is, their ratios. The equation of the quartic surface thus is

$$(a, \ldots \between P, Q, R)^2 + \theta M\Omega = 0,$$

the 7 given points being here proper nodes; and the formula being precisely equivalent to the preceding one containing Δ.

36. We can with the 7 given points form 35 such combinations $M\Omega=0$ of a plane and a cubic surface, and so present the equation of the quartic surface under 35 different forms; these are of course equivalent in virtue of the before mentioned formula for $\Delta' - p\Delta$; viz., we must have identically $M\Omega - pM'\Omega' = (* \between P, Q, R)^2$: a theorem of some interest, which it might be difficult to verify *à posteriori*.

Investigation of the cases of 8 *Nodes.*

37. It has already been shown that a quartic surface cannot in a proper sense have 8 given nodes. In regard to the quartic surfaces with 8 nodes, we start from the surface with 7 given nodes; viz.,

$$(a, \ldots \between P, Q, R)^2 + \theta\nabla = 0,$$

or, what is the same thing,

$$(a, \ldots \between P, Q, R)^2 + \theta M\Omega = 0;$$

and we inquire in what cases this surface has an 8th node. Obviously if $\theta = 0$. that is, if the surface is $(a, \ldots \between P, Q, R)^2 = 0$, the surface will have an 8th node, the remaining intersection of the quadric surfaces $P=0$, $Q=0$, $R=0$ (observe that this is a point in no wise depending on the particular quadric surfaces, but uniquely determined by means of the 7 given points); and we have thus one kind, say the "octadic" surface, of the quartic surfaces with 8 nodes; viz., the nodes are the 8 points of intersection of any 3 quadric surfaces, or they are an octad of points. By what precedes, 7 of the nodes may be given points, and the remaining node is then a uniquely determinate point, the 8th point of the octad.

38. But if θ be not $=0$, there may still be an 8th node; viz., this must then be a point on the Jacobian surface $J(P, Q, R, \nabla) = 0$, which is of the order 6. It is clear *à priori* that this must be a surface depending only on the 7 points, but independent of the particular surfaces $P=0$, $Q=0$, $R=0$, $\nabla=0$; to verify this, observe that, substituting for ∇ the function ∇', $=p\nabla + (*\between P, Q, R)^2$, we in fact leave the Jacobian unaltered; I call it the dianodal surface of the 7 points.

39. I say that the 8th node may be any point whatever on the dianodal surface; in fact, regarding for a moment the coordinates of the node as given, and expressing that the point is a node on the quartic surface, we have 4 equations containing

$$aP_0 + hQ_0 + gR_0, \quad hP_0 + bQ_0 + fR_0, \quad gP_0 + fQ_0 + cR_0,$$

(P_0, Q_0, R_0 the values of P, Q, R at the node,) but which, if only the point be on the dianodal surface, reduce themselves to three equations; viz., we have between the coefficients (a, b, c, f, g, h) and θ three equations which being satisfied, the point in question will be a node. And it thus appears that, taking the 8th node to be a given point on the dianodal surface, the equation $(a, \ldots \between P, Q, R)^2 + \theta\nabla = 0$ of the quartic surface will contain 3 constants. Observe that we may through the 8 nodes draw 2 quadric surfaces $P=0$, $Q=0$; and this being so if $\Delta = 0$ be a particular quartic surface with the 8 nodes, then the general quartic surface will be

$$(a, b, c\between P, Q)^2 + \theta\Delta = 0,$$

containing the right number 3 of constants. But there is not here any simple form of the surface $\Delta = 0$, such as the form $M\Omega = 0$ for the surface through 7 given points.

40. It is clear *à priori* that the relation between the 8 nodes is a symmetrical one; so that the 8th point being situate anywhere on the dianodal surface of the 7 points, each of the points will be situate on the dianodal surface of the remaining 7 points. This is a remarkable property of the dianodal surface, which will have to be again considered.

41. In what precedes, we have the second kind of quartic surfaces with 8 nodes, say the "dianome"; viz., each node is a point on the dianodal surface of the remaining 7 nodes; any 7 of the nodes may be taken to be given points, and the remaining node to be any point whatever on the dianodal surface of the 7 points.

The Dianodal Surface.

42. Consider the seven points 1, 2, 3, 4, 5, 6, 7. As already mentioned, through three of these, say 1, 2, 3, we may draw a plane $M=0$; and through the same three points, with the remaining points 4, 5, 6, 7 as nodes $(3+4.4=19$ conditions), a cubic surface $\Omega=0$; this surface passing through the six lines, 45, 46, ... 67. Hence we have $\Delta, =M\Omega, =0$, a quartic surface with the seven points as nodes. And using this form of Δ, it may be shown that the dianodal $J(P, Q, R, \Delta)=0$ passes through the 21 lines 12, 13, ... 67, and through 35 plane cubics such as $M=0$, $\Omega=0$; viz., this is a cubic in the plane 123 passing through the points 1, 2, 3, and through the intersections of the plane with each of the six lines 45, 46, ... 67 (nine points determining the cubic); the complete intersection by the plane 123 being therefore composed of this cubic and of the three lines 12, 13, 23. For the passage through the cubic, we have only to observe that

$$J(P,\ Q,\ R,\ M\Omega)=J(P,\ Q,\ R,\ \Omega)\,M+J(P,\ Q,\ R,\ M)\,\Omega=0$$

is satisfied by $M=0$, $\Omega=0$; and for the passage through the lines, taking $x=0$, $y=0$, $z=0$, $w=0$ for the equations of the planes 567, 674, 745, and 456 respectively, each of the functions P, Q, R is of the form $ayz+bzx+cxy+fxw+gyw+hzw$, and the function Ω is of the form $Ayzw+Bzwx+Cwxy+Dxyz$. Hence, writing in the derived functions for instance $z=0$, $w=0$, the first and second lines of the determinant $J(P, Q, R, \Omega)$ will be of the form

$$\begin{vmatrix} cy, & c'y, & c''y, & 0 \\ cx, & c'x, & c''x, & 0 \end{vmatrix},$$

or the determinant vanishes for $z=0$, $w=0$; that is, for any point of the line 45 we have $\Omega=0$ and also $J(P, Q, R, \Omega)=0$; consequently $J(P, Q, R, M\Omega)=0$, and the like for the other lines. The theorem is thus proved.

43. I say that the dianodal surface passes through each of the 7 skew cubics, such as 123456. To prove this, it is only necessary to show that the skew cubic

123456 lies on the dianodal surface. For this purpose it will be enough to show that the skew cubic meets the plane 712 in a point of the surface; for then it will, in like manner, meet each of the 15 planes 712, 713, ... 756 in a point of the surface; that is, we shall have 15 intersections of the curve and surface, and there are, besides, the intersections 1, 2, 3, 4, 5, 6, in all 21 intersections; that is, the skew cubic must lie on the surface.

44. The plane 712 meets the surface in three lines and in a plane cubic determined by the points 7, 1, 2 and the six intersections of the plane with the lines 34, 35, ... 56. We have therefore to show that this plane cubic meets the skew cubic 123456. Consider for a moment the points 1, 2, 3, 4, 5, 6 and another point 7′. As seen above, we have in general, through the points 1, 2, 7′ and with the points 3, 4, 5, 6 as nodes, a determinate cubic surface, which surface passes through the lines 34, 35, ... 56. But the cubic surface becomes indeterminate if the points 1, 2, 7′, 3, 4, 5, 6 are on the same skew cubic; that is, if 7′ is any point whatever on the skew cubic 123456 (the proof presently). Taking, then, 7′ as the intersection of the skew cubic by the plane 712, we have in this plane the points 7′, 1, 2, and the intersections of the plane by the lines 34, 35, ... 56, nine points through which there pass an infinity of plane cubics; that is, the plane cubic determined by the points 7, 1, 2 and the six intersections will pass through the point 7′; viz., it meets the skew cubic 123456.

45. For the subsidiary theorem, taking X, Y, Z, W as current coordinates, viz., $X=0$, $Y=0$, $Z=0$, $W=0$ as the equations of the planes 456, 563, 634, 345 respectively, (x_1, y_1, z_1, w_1) and (x_2, y_2, z_2, w_2) as the coordinates of the points 1 and 2 respectively, and (x, y, z, w) for those of 7′; the equation of the cubic surface passing through 7′, 1, 2, and having the nodes 3, 4, 5, 6, is

$$\begin{vmatrix} \frac{1}{X}, & \frac{1}{Y}, & \frac{1}{Z}, & \frac{1}{W} \\ \frac{1}{x}, & \frac{1}{y}, & \frac{1}{z}, & \frac{1}{w} \\ \frac{1}{x_1}, & \frac{1}{y_1}, & \frac{1}{z_1}, & \frac{1}{w_1} \\ \frac{1}{x_2}, & \frac{1}{y_2}, & \frac{1}{z_2}, & \frac{1}{w_2} \end{vmatrix} = 0;$$

and this ceases to be a determinate function if only

$$\begin{Vmatrix} \frac{1}{x}, & \frac{1}{y}, & \frac{1}{z}, & \frac{1}{w} \\ \frac{1}{x_1}, & \frac{1}{y_1}, & \frac{1}{z_1}, & \frac{1}{w_1} \\ \frac{1}{x_2}, & \frac{1}{y_2}, & \frac{1}{z_2}, & \frac{1}{w_2} \end{Vmatrix} = 0;$$

viz., considering (x_1, y_1, z_1, w_1), (x_2, y_2, z_2, w_2) as given, this is a twofold relation between the coordinates (x, y, z, w) of the point 7′. The relation may be represented by the four equations $(yzw)=0$, $(zwx)=0$, $(wxy)=0$, $(xyz)=0$, if for shortness

$$(yzw)=\begin{vmatrix} yz, & zw, & wy \\ y_1z_1, & z_1w_1, & w_1y_1 \\ y_2z_2, & z_2w_2, & w_2y_2 \end{vmatrix}$$

and the like as to the other symbols. The four equations represent quadric surfaces, each two intersecting in a line [e.g., $(yzw)=0$, $(zwx)=0$ in the line $z=0$, $w=0$], and the four surfaces besides intersecting in a skew cubic, which is the required locus of the point 7′, and which, as is seen at once, passes through the points 1, 2, 3, 4, 5, 6.

46. By what precedes, we have on the dianodal surface through the point 1 the lines 12, 13, 14, 15, 16, 17, and the skew cubics 123456, &c. The six lines are not on the same quadric cone, and it thus appears that the point 1 must be a cubic-node (point where, instead of the tangent plane, we have a cubic cone) on the surface. It is to be remarked that the lines 12, 13, 14, 15, 16, and the tangent at 1 to the skew cubic 123456, lie in a quadric cone; viz., this tangent is given as the sixth intersection of the cubic cone with the quadric cone through the lines 12, 13, 14, 15, 16.

47. I revert to the equation of the dianodal surface as given in the form $J=J(P, Q, R, M\Omega)=0$, where $M=0$ is the plane through the points 1, 2, 3, and $\Omega=0$ the cubic surface through these points, and having the points 4, 5, 6, 7, as nodes. We can find the orders of the several functions P, Q, R, M, Ω in the coordinates (x_1, y_1, z_1, w_1), &c., of the several points; viz., writing for shortness x_1^2 to denote the order 2 in regard to (x_1, y_1, z_1, w_1), and so in other cases, we have

$$\begin{aligned} P=Q=R&=x^2(x_5, x_6, x_7)^2(x_1, x_2, x_3, x_4)^2, \\ M&=x\,(x_5, x_6, x_7), \\ \Omega&=x^3(x_5, x_6, x_7)^3(x_1, x_2, x_3, x_4)^9; \end{aligned}$$

{where, of course, the x^2, x, x^3 show in like manner the orders in regard to the current coordinates (x, y, z, w); the proof in regard to Ω is easily supplied.} The order of J is equal that of $PQRM\Omega$, less 4 as regards the current coordinates, by reason of the differentiations; that is, we have $J=x^6(x_1x_2x_3)^{10}(x_4x_5x_6x_7)^{15}$; and we thus see that the equation of the dianodal surface as above obtained is encumbered with a constant factor of the form $(x_1x_2x_3)^4(x_4x_5x_6x_7)^9$. In fact, the relation between the 7 points and the current point (x, y, z, w), or say the point 8, as expressing that the 8 points are the nodes of a dianome, should be a symmetrical one in regard to the coordinates of the several points; and being of the order 6 in regard to the coordinates (x, y, z, w), it should be of the same order in regard to the other coordinates; that is, the true form would be $J=(xx_1x_2x_3x_4x_5x_6x_7)^6=0$.

48. It is possible that taking the 4 points, say 1, 2, 3, 4, to be (1, 0, 0, 0), (0, 1, 0, 0), (0, 0, 1, 0), (0, 0, 0, 1), and the 3 points, say 5, 6, 7, to be (1, 1, 1, 1), $(\alpha, \beta, \gamma, \delta)$, $(\alpha', \beta', \gamma', \delta')$, the extraneous factor might exhibit itself, and that the

equation divested of this factor might be of a tolerably simple form. I have not, however, worked this out, but I have, by an independent process, obtained in regard to the dianodal surface of the 7 points a result which may be interesting.

49. The dianodal surface, *quà* surface having the first-mentioned 4 points for cubic nodes, has its equation of the form

$$yzw(y, z, w)^3 + zxw(z, x, w)^3 + xyw(x, y, w)^3 + xyz(x, y, z)^3 + xyzw(x, y, z, w)^2 = 0;$$

where in the cubic functions the terms x^3, y^3, z^3, w^3 none of them appear. If for instance $w = 0$, the equation becomes $(x, y, z)^3 = 0$, which, by what precedes, is a known cubic curve, viz., the curve through the points 1, 2, 3 and the intersections of the plane 123 by the lines 45, 46, 47, 56, 57, 67; and we can by this consideration find the cubic function $(x, y, z)^3$, and thence by symmetry the other cubic functions. I take

$$\left.\begin{matrix}(a, b, c, f, g, h)\\(a', b', c', f', g', h')\\(\mathrm{a}, \mathrm{b}, \mathrm{c}, \mathrm{f}, \mathrm{g}, \mathrm{h})\end{matrix}\right\}\text{ for coordinates of line through }\left\{\begin{matrix}(1, 1, 1, 1), & (\alpha, \beta, \gamma, \delta)\\(1, 1, 1, 1), & (\alpha', \beta', \gamma', \delta')\\(\alpha, \beta, \gamma, \delta), & (\alpha', \beta', \gamma', \delta')\end{matrix}\right.$$

respectively; viz., I write

$$\begin{array}{ll|ll|ll} a=\beta-\gamma, & f=\alpha-\delta & a'=\beta'-\gamma', & f'=\alpha'-\delta' & \mathrm{a}=\beta\gamma'-\beta'\gamma, & \mathrm{f}=\alpha\delta'-\alpha'\delta \\ b=\gamma-\alpha, & g=\beta-\delta & b'=\gamma'-\alpha', & g'=\beta'-\delta' & \mathrm{b}=\gamma\alpha'-\gamma'\alpha, & \mathrm{g}=\beta\delta'-\beta'\delta \\ c=\alpha-\beta, & h=\gamma-\delta & c'=\alpha'-\beta', & h'=\gamma'-\delta' & \mathrm{c}=\alpha\beta'-\alpha'\beta, & \mathrm{h}=\gamma\delta'-\gamma'\delta \end{array}$$

and I write moreover

$$\begin{aligned} \lambda &= \ \ .\ \ \ \mathrm{h}-\mathrm{g}+\mathrm{a},\\ \mu &= -\mathrm{h}\ \ .\ +\mathrm{f}+\mathrm{b},\\ \nu &= \ \ \mathrm{g}-\mathrm{f}\ \ .\ +\mathrm{c},\\ \varpi &= -\mathrm{a}-\mathrm{b}-\mathrm{c}\ \ . \end{aligned}$$

50. This being so, the cubic curve through the last-mentioned six points has its equation of the form

$$\frac{A}{ax+by+cz}+\frac{B}{a'x+b'y+c'z}+\frac{C}{\mathrm{a}x+\mathrm{h}y+\mathrm{g}z}+\frac{D}{\lambda x+\mu y+\nu z}=0;$$

and to make this pass through the points 1, 2, 3, we write therein successively $(y=0, z=0)$, $(z=0, x=0)$, $(x=0, y=0)$; viz., we have for the ratios $A : B : C : D$ the three equations

$$\frac{A}{a}+\frac{B}{a'}+\frac{C}{\mathrm{a}}+\frac{D}{\lambda}=0,$$

$$\frac{A}{b}+\frac{B}{b'}+\frac{C}{\mathrm{h}}+\frac{D}{\mu}=0,$$

$$\frac{A}{c}+\frac{B}{c'}+\frac{C}{\mathrm{g}}+\frac{D}{\nu}=0.$$

In eliminating, for instance, B for the first and second equations, the resulting equation divides by $ab' - a'b$, $= \mathrm{a} + \mathrm{b} + \mathrm{c}$, and we thus obtain, between A, C, D, the three equations (equivalent to two)

$$\frac{A}{bc} + \frac{C\alpha'}{\mathrm{bc}} + \frac{Df'}{\mu\nu} = 0,$$

$$\frac{A}{ca} + \frac{C\beta'}{\mathrm{ca}} + \frac{Dg'}{\nu\lambda} = 0,$$

$$\frac{A}{ab} + \frac{C\gamma'}{\mathrm{ab}} + \frac{Dh'}{\lambda\mu} = 0,$$

from which the ratios $A : C : D$ may be obtained by actual calculation. After all reductions, we have

$$\begin{aligned} A &= \quad a\,b\,c\ \{(\alpha'\delta' + \beta'\gamma')\,\mathrm{af} + (\beta'\delta' + \gamma'\alpha')\,\mathrm{bg} + (\gamma'\delta' + \alpha'\beta')\,\mathrm{ch}\},\\ B &= -\,a'b'c'\ \{(\alpha\,\delta + \beta\,\gamma)\,\mathrm{af} + (\beta\,\delta + \gamma\,\alpha)\,\mathrm{bg} + (\gamma\,\delta + \alpha\,\beta)\,\mathrm{ch}\},\\ C &= \quad \mathrm{a\,b\,c}\ \{(\alpha\alpha'\lambda + \beta\beta'\mu + \gamma\gamma'\nu + \delta\delta'\varpi\},\\ D &= -\,\lambda\,\mu\,\nu\ \{(\alpha\alpha'\mathrm{a} + \beta\beta'\mathrm{b} + \gamma\gamma'\mathrm{c}\}\,; \end{aligned}$$

viz., A, B, C, D are proportional to these values respectively. Multiplying by the product of the denominators, I find without much difficulty that the resulting cubic function is divisible by $\mathrm{a} + \mathrm{b} + \mathrm{c}$; hence, introducing the factor xyz, and an indeterminate multiplier l, I write

$$xyz\,(x,\ y,\ z)^3 = \frac{l}{\mathrm{a} + \mathrm{b} + \mathrm{c}}\, xyz\,(ax + by + cz)\,(a'x + b'y + c'z)\,(\mathrm{a}x + \mathrm{b}y + \mathrm{c}z)\,(\lambda x + \mu y + \nu z)$$

$$\times \left\{\frac{A}{ax + by + cz} + \frac{B}{a'x + b'y + c'z} + \frac{C}{\mathrm{a}x + \mathrm{b}y + \mathrm{c}z} + \frac{D}{\lambda x + \mu y + \nu z}\right\},$$

where A, B, C, D have the values above written down.

51. Considering the orders in regard to $(\alpha, \beta, \gamma, \delta)$, $(\alpha', \beta', \gamma', \delta')$, and observing that a, b, c and a', b', c' are linear functions of the two sets respectively, but that a, $\mathrm{b} \ldots \mathrm{h}$, $\lambda \ldots \varpi$, are linear in the two sets conjointly, or say

$$a, \ldots = \alpha,\ a', \ldots = \alpha'\,;\ \mathrm{a}, \ldots = \alpha\alpha'\,;$$

we have

$$A\,a'\mathrm{a}\lambda = \alpha^5\alpha'^4\,.\,\alpha^2\alpha'^3 = \alpha^7\alpha'^7,$$

so that after the division by $\mathrm{a} + \mathrm{b} + \mathrm{c}$, $= \alpha\alpha'$, the order will be $\alpha^6\alpha'^6$. Hence l will be a mere numerical factor, and the last-mentioned equation gives, without any extraneous factor, the terms $xyz\,(x, y, z)^3$ in the equation of the dianodal surface of the seven points.

Octadic Surfaces with 9 *or* 10 *Nodes.*

52. In regard to the surfaces with 9 and 10 nodes, I consider first the octadic surfaces. Starting as before with the given points 1, 2, 3, 4, 5, 6, 7, we have a determinate point 8 completing the octad, and the surface with the 8 nodes is

$$(a, \ldots \between P,\ Q,\ R)^2 = 0,$$

(5 constants). Suppose that there is another node 9; this must be a point on the Jacobian curve $J(P, Q, R)=0$, which (as was seen) is a sextic curve not passing through any of the 8 points; the node 9 may be *any* point on this curve, viz., taking its coordinates as given, the condition of its being a node gives 4 equations, and these for the very reason that the point is on the Jacobian curve, reduce themselves to 2 equations, which can be satisfied by means of the constants $(a, \ldots)$; the resulting equation should therefore contain 3 constants.

53. In order to find it, taking as above 9 a given point on the Jacobian curve, this will be the vertex of a quadric cone, say $P=0$, through the 8 points; we may draw through the 9 points another quadric surface $Q=0$, and through the 8 points a quadric surface $R=0$; this being so, we have the quartic surface $(a, b, 0, 0, g, h \between P, Q, R)^2=0$, having the 9 nodes, and containing, as it should do, 3 constants; this may be written

$$(aP+2hQ+2gR)\,P+bQ^2=0\,;$$

viz., if $bR'=aP+2hQ+2gR$, that is, if $R'=0$ be the general quadric surface through the 8 points, then the equation is $Q^2-PR'=0$, where observe that R' is considered as containing implicitly 3 constants.

54. If there is a 10th node, say 10, this is also a point on the Jacobian curve $J(P, Q, R)=0$, and it may be any point whatever on the curve; taking it as a given point on the curve, the resulting equation should contain 1 constant. We may take $P=0$ to be the quadric cone, vertex 9, through the 8 points, $R=0$ the quadric cone, vertex 10, through the 8 points, $Q=0$ the quadric surface through the 8 points and the points 9 and 10 (viz., the surface through 9, 10 and any 7 of the 8 points will pass through the remaining 8th point). The equation of the quartic surface then is

$$(0, b, 0, 0, g, 0 \between P, Q, R)^2=0\,;$$

that is, $bQ^2+2gPR=0$, containing 1 constant; we may reduce this to $Q^2-PR=0$, the constant being considered as contained implicitly in one of the functions. It is clear that the constant cannot be so determined as to give rise to an 11th node, nor indeed to any other singularity in the surface.

55. In the case of the surface with 9 nodes, it is clear that this is octadic in one way only; the node 9 cannot form an octad with any 7 of the remaining nodes. But in the case of the surface with 10 nodes, the question arises whether the nodes 9 and 10 may not be such as to form an octad with some six, say with the nodes 1, 2, 3, 4, 5, 6 of the remaining 8 nodes; that is, whether we can have 1, 2, 3, 4, 5, 6, 7, 8 forming an octad, and also 1, 2, 3, 4, 5, 6, 9, 10 forming an octad. I will show that this is impossible if only the points 1, 2, 3, 4, 5, 6 are given points, that is, points assumed at pleasure and not specially related to each other. For this purpose, assuming that the points form 2 octads as above, take through 1, 2, 3, 4, 5, 6, 7, 9 the quadric surfaces $P=0$, $Q=0$, then each of these passes through 8, 10; take $R=0$ any other quadric surface through 1, 2, 3, 4, 5, 6, 7, 8, and $S=0$ any other quadric surface through 1, 2, 3, 4, 5, 6, 9, 10. Then $P=0$, $Q=0$, $R=0$ intersect in the 1st octad,

and $P=0$, $Q=0$, $S=0$ intersect in the 2nd octad; the quartic surface (if it exists) must be simultaneously of the forms $(*\between P, Q, R)^2=0$, $(*\between P, Q, S)^2=0$; and this implies an identical equation $(*\between P, Q, R, S)^2=0$. The quadric surfaces are surfaces through the points 1, 2, 3, 4, 5, 6, and taking through these six points any other quadric surfaces $A=0$, $C=0$, $E=0$, $H=0$, we have P, Q, R, S each of them a linear function of A, C, E, H; and the relation between P, Q, R, S gives a like relation $(*\between A, C, E, H)^2=0$ between A, C, E, H. I assume $A=123.456$, $E=134.256$, $H=145.236$, $C=152.346$; viz., $A=0$ is the plane-pair formed by the planes through 1, 2, 3 and 4, 5, 6 respectively; and so for the others: we have to show that there is not any such identical relation $(*\between A, C, E, H)^2=0$.

56. We may through 3 draw the lines LM, QT to meet 14, 26 and 12, 46 respectively; and through 5 the lines RS, NP to meet 14, 26 and 12, 46 respectively. Observe that the points O in the figure are apparent intersections only; viz., NP does

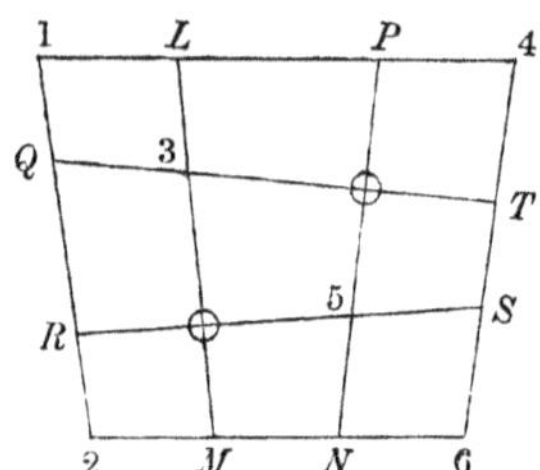

not meet QT, *nor* LM meet RS. In fact, if NP met QT it would be a line in the series of lines meeting 14, QT, 26; or 5 would be situate in a hyperboloid, determined by means of the points 1, 2, 4, 6, 3; viz., 5 would not be an arbitrary point: and so LM does not meet RS. Now the quadrics E, H meet in the lines 14, 26, LM, NP, and the quadrics A, C in the lines 12, 46, QT, RS. Suppose that we had identically $(*\between A, C, E, H)^2=0$; putting therein $E=0$, $H=0$, we should have $(*\between A, C)^2=0$, viz., $(A+\lambda C)(A+\mu C)=0$; or there would exist quadrics of the forms $A+\lambda C=0$ containing the lines 14, 26, LM, NP. Now there is no quadric surface $A+\lambda C=0$ containing the line NP; for $A+\lambda C=0$ is a quadric containing the sides of the quadrilateral $QRST$; the generating lines of the one kind meet each of the lines RS, QT; those of the other kind neither. Hence NP, which meets RS but not QT, cannot be a generating line of either kind; and we have no identical relation $(A, C, E, H)^2=0$.

57. In the octadic surface with 9 nodes; starting with any 7 nodes of the octad, 9 is not the 8th point of the octad, and hence (by the theory of the dianome) it must lie in the dianodal surface of the 7 points; that is, the dianodal surface of the 7 points must pass through 9, viz., through any point whatever of the Jacobian curve of the 7 points, that is, of the octad; or (what is the same thing) the dianodal surface of the 7 points passes through the Jacobian curve of the octad. This is an obvious property of the dianodal surface, the surface $J(P, Q, R, \nabla)=0$ contains the Jacobian curve $J(P, Q, R)=0$. But it further appears that, starting with any 6 points of the octad and with the point 9 (that is, any point whatever of the Jacobian curve), the

dianodal surface of these 7 points must contain the remaining 2 points of the octad. And in the octadic surface with 10 nodes, starting with any 5 points of the octad and with the points 9 and 10 (that is, any two points on the Jacobian curve) the dianodal surface of these 7 points must contain the remaining three points of the octad. I have not attempted to verify these last properties of the dianodal surface.

Dianomes with 9 *or* 10 *Nodes.*

58. I now consider the dianomes with 9 and 10 nodes. Starting from the general form

$$(a,\ b,\ c\between P,\ Q)^2 + \theta\Delta = 0,$$

where $\Delta = 0$ is a particular quartic surface having the 8 nodes, it at once appears that if there is a 9th node, say 9, this must be a point on the Jacobian curve $J(P, Q, \Delta) = 0$, or say on the dianodal curve of the 8 points, viz. ($a = b = 1$, $c = 3$, in the formula No. 5), this is a curve of the order 18; the node may be any point whatever on this curve, and taking it to be a given point on the curve, the number of constants in the resulting equation should be 1. Hence if $P = 0$ be the quadric surface through the 9 points, and $\Delta = 0$ a particular quartic surface having the 9 points as nodes, the general equation is $aP^2 + \theta\Delta = 0$.

59. But we may consider the question somewhat differently. Starting with the 7 given points 1, 2, 3, 4, 5, 6, 7 and with 8 a given point on the dianodal surface of the 7 points; it is clear that 9 must be on the dianodal surface 1234567, and also on the dianodal surface 1234568; the complete intersection is of the order 36, and we have to consider how this breaks up so as to contain as part of itself the dianodal curve of the order 18.

Dianodal Curve of 8 *Points.*

60. Consider first any 8 points whatever 1, 2, 3, 4, 5, 6, 7, 8; where 8 is not on the dianodal surface 1234567, nor 7 on the dianodal surface 1234568. The two surfaces have in common the 15 lines 12, 13,... 56 and the skew cubic 123456, they therefore besides intersect in a curve of the order 18. At the point 1 the tangent cubic cones of the two surfaces intersect in the lines 12, 13, 14, 15, 16 and the tangent to the skew cubic 123456, 6 lines lying in a quadric cone; they therefore besides intersect in 3 lines lying in a plane; that is, the point 1 is on the curve of the order 18 an actual triple point, the 3 tangents lying *in plano;* and the like of course in regard to each of the points 2, 3, 4, 5, 6. But as 7, 8 lie each of them on only one of the two surfaces, the curve of the order 18 does not pass through 7 or 8.

61. If, however, 8 lies on the dianodal surface 1234567, then each of the 8 points will lie on the dianodal surface of the other 7; and in particular 7 will lie on the dianodal surface 1234568. The surfaces intersect as before in a residual curve of the order 18; the only difference is that 7 and 8 are now points on each surface; viz., each of them is on one of the surfaces an ordinary point, and on the other a cubic node; the points 7 and 8 are thus each of them an actual triple point on the curve; and at each of them the 3 tangents are *in plano.* We thus see that the dianodal

curve 12345678 is a curve of the order 18, such that each of the 8 points is a triple point on the curve, the tangents at each of them being *in plano*.

Ten Nodes.

62. Suppose there is a 10th node, say 10; starting from the equation $aP^2 + \theta\Delta = 0$ ($P = 0$ the quadric surface through the 9 points, $\Delta = 0$ a particular quartic surface having the 9 points as nodes), it at once appears that the node must be one of the points $J(P, \Delta) = 0$; hence, taking it to be one of these points, we have 4 equations, which, in virtue of the node being one of the points in question, reduce themselves to a single equation determining the ratio $a : \theta$; we have thus a completely determinate surface, say $\square = 0$ having the 10 points as nodes. The number of points $J(P, \Delta)$, writing in the formula No. 5, $a = 1$, $b = 3$, is obtained as $1 + 3 + 9 + 27 = 40$, but it is to be observed that the surface $P = 0$ passes through each of the 9 nodes of the surface $\Delta = 0$; these count twice among the points $J(P, \Delta) = 0$, and the number of residual points (or say the dianodal centres of the 9 points) is $40 - 18 = 22$; viz., this is the number of positions of the node 10. [The nine points count each *three* times and the number of residual points, or positions of the node 10, is thus not $40 - 18 = 22$, but $40 - 27, = 13$.]

Dianodal Centres of 9 *Points.*

63. In further explanation, observe that 9 is any point on the dianodal curve 12345678; the node 10 must lie on this same curve, and also on the dianodal surface 1234569. Take $P = 0$ the quadric through all the 9 points, $Q = 0$ a quadric through all but the point 9, $R = 0$ through all but the point 8, $S = 0$ through all but the point 7. The dianodal curve 12345678 is $J(P, Q, \nabla) = 0$, and the dianodal surface 1234569 is $J(P, R, S, \nabla) = 0$; the total number of intersections is $6 \times 18 = 108$; these include the $4 \times 18 = 72$ points of intersection of the dianodal curve $J(P, Q, \Delta) = 0$ with the Jacobian surface $J(P, Q, R, S) = 0$, except the four points $J(P, Q) = 0$, which are the vertices of the 4 quadric cones through 1, 2, 3, 4, 5, 6, 7, 8 (which 4 points are not situate on the curve $J(P, R, S) = 0$), and there are besides 40 points $\{108 = (72 - 4) + 40\}$ which are the before mentioned points $J(P, \Delta) = 0$; viz., these are the 9 points each twice [three times], and the residual 22 [13] points which are the dianodal centres of the 9 points.

General result as to the Dianomes.

64. We have thus established the theory of the dianome quartic surfaces; viz., we have

The octodianome, 8 nodes, 7 of them arbitrary, and the 8th an arbitrary point on the dianodal surface (order 6) of the 7 points.

The enneadianome, 9 nodes, the 9th an arbitrary point on the dianodal curve (order 18) of the 8 points.

The decadianome, 10 nodes, the 10th any one of the 22 [13] dianodal centres of the 9 points.

And as already mentioned, so long as the first 7 nodes are arbitrary, there cannot be more than 10 nodes in all.

THE SYMMETROID.

The Lineolinear Correspondence of Quartic Surfaces.

65. I consider four equations $S=0$, $T=0$, $U=0$, $V=0$, lineolinear in regard to the two sets of coordinates (x, y, z, w) and $(\alpha, \beta, \gamma, \delta)$; viz., each of these equations is of the form

$$(*\between x, y, z, w\between \alpha, \beta, \gamma, \delta)=0.$$

This implies that the point (x, y, z, w) lies on a certain quartic surface $\Theta=0$, and the point $(\alpha, \beta, \gamma, \delta)$ on a certain quartic surface $\Delta=0$, and that the two surfaces correspond point to point to each other. In fact, writing the four equations in the form

$$\begin{aligned}
L\alpha+M\beta+N\gamma+P\delta&=0,\\
L'\alpha+M'\beta+N'\gamma+P'\delta&=0,\\
L''\alpha+M''\beta+N''\gamma+P''\delta&=0,\\
L'''\alpha+M'''\beta+N'''\gamma+P'''\delta&=0,
\end{aligned}$$

where L, &c., are linear functions of (x, y, z, w), then eliminating $(\alpha, \beta, \gamma, \delta)$, we obtain the equation

$$\Theta=\begin{vmatrix} L, & M, & N, & P\\ L', & M', & N', & P'\\ L'', & M'', & N'', & P''\\ L''', & M''', & N''', & P''' \end{vmatrix}=0;$$

and similarly, writing the four equations in the form

$$\begin{aligned}
Ax+By+Cz+Dw&=0,\\
A'x+B'y+C'z+D'w&=0,\\
A''x+B''y+C''z+D''w&=0,\\
A'''x+B'''y+C'''z+D'''w&=0,
\end{aligned}$$

where A, &c., are linear functions of $(\alpha, \beta, \gamma, \delta)$, then eliminating (x, y, z, w), we obtain the equation

$$\Delta=\begin{vmatrix} A, & B, & C, & D\\ A', & B', & C', & D'\\ A'', & B'', & C'', & D''\\ A''', & B''', & C''', & D''' \end{vmatrix}=0.$$

Moreover, Θ being $=0$, the four linear equations in $(\alpha, \beta, \gamma, \delta)$ are equivalent to three equations, and give for instance $(\alpha, \beta, \gamma, \delta)$ proportional to the determinants formed with the matrix

$$\begin{vmatrix} L', & M', & N', & P'\\ L'', & M'', & N'', & P''\\ L''', & M''', & N''', & P''' \end{vmatrix};$$

and similarly, Δ being $=0$, the four linear equations in (x, y, z, w) are equivalent to three equations, and give for instance (x, y, z, w) proportional to the determinants formed with the matrix

$$\left|\begin{array}{cccc} A', & B', & C', & D' \\ A'', & B'', & C'', & D'' \\ A''', & B''', & C''', & D''' \end{array}\right|;$$

which establishes the point-to-point correspondence of the two surfaces.

66. It would at first sight appear that any quartic surface $(*\between\alpha, \beta, \gamma, \delta)^4=0$ whatever might have its equation expressed in the foregoing determinant form $\Delta=0$. This equation seems, in fact, to contain homogeneously as many as 64 constants. But if we multiply the determinant line into line by a constant determinant

$$\left|\begin{array}{cccc} a, & b, & c, & d \\ a', & b', & c', & d' \\ a'', & b'', & c'', & d'' \\ a''', & b''', & c''', & d''' \end{array}\right|$$

and then column into column by another constant determinant, the coefficients, all but one of them, of these constant determinants may be used to specialize the form of the resulting equation, [say they are apoclastic constants]; this equation will really contain $64-(2.16-1)=33$ constants; and in order that the quartic surface $(*\between\alpha, \beta, \gamma, \delta)^2=0$ may have its equation expressible in the form $\Delta=0$, a single relation must hold good among the coefficients: but this in passing([1]).

67. Returning to the quartic surface

$$\Delta=\left|\begin{array}{cccc} A, & B, & C, & D \\ A', & B', & C', & D' \\ A'', & B'', & C'', & D'' \\ A''', & B''', & C''', & D''' \end{array}\right|=0,$$

we may connect this not only with the foregoing surface $\Theta=0$, but in a similar manner with another quartic surface $\Phi=0$; viz., taking the current coordinates $(\xi, \eta, \zeta, \omega)$, we may form the lineolinear equations

$$\begin{aligned} A\xi+A'\eta+A''\zeta+A'''\omega&=0,\\ B\xi+B'\eta+B''\zeta+B'''\omega&=0,\\ C\xi+C'\eta+C''\zeta+C'''\omega&=0,\\ D\xi+D'\eta+D''\zeta+D'''\omega&=0,\end{aligned}$$

[1] Applying the same reasoning to a cubic determinant $\Delta=0$, the number of constants is $36-(2.9-1)=19$; so that a cubic surface is expressible in the form in question. And so for the quadric determinant $\Delta=0$, the number of constants is $16-(2.4-1)=9$; so that a quadric surface is expressible in the form in question, as is otherwise obvious.

which, by the elimination of $(\xi, \eta, \zeta, \omega)$, give $\Delta = 0$, and by the elimination of $(\alpha, \beta, \gamma, \delta)$ a determinant quartic equation $\Phi = 0$ between the coordinates $(\xi, \eta, \zeta, \omega)$; and of course the two surfaces $\Delta = 0$, $\Phi = 0$ have a point-to-point correspondence such as exists between the surfaces $\Theta = 0$, $\Delta = 0$. The relation of the point $(\alpha, \beta, \gamma, \delta)$ on the surface $\Delta = 0$ to the point (x, y, z, w) on the surface $\Theta = 0$, and to the point $(\xi, \eta, \zeta, \omega)$ on the surface $\Phi = 0$, may be conveniently indicated by means of the diagram

$$\begin{array}{cccc|c}
\multicolumn{4}{c}{\overbrace{x\ ,\quad y\ ,\quad z\ ,\quad w}^{\Theta}} & \\
\hline
A\ , & B\ , & C\ , & D & \xi \\
A'\ , & B'\ , & C'\ , & D' & \eta \\
A''\ , & B''\ , & C''\ , & D'' & \zeta \\
A'''\ , & B'''\ , & C'''\ , & D''' & \omega
\end{array}\left.\vphantom{\begin{array}{c}A\\A\\A\\A\end{array}}\right\}\Phi.$$

68. It is to be observed that, writing for A, B, ... their values as linear functions of $(\alpha, \beta, \gamma, \delta)$, we have in all 64 constant coefficients, which we may conceive arranged in the form of a cube, thus:

$$\begin{array}{lll}
a & b & \text{———} \\
\mid\ a' & b' & \\
\mid\quad \diagdown & & \\
a_1 & b_1 & \text{———} \\
\mid\ a_1' & b_1' & \\
\mid\quad \diagdown & &
\end{array}$$

and taking these in fours height-wise, (a, a_1, a_2, a_3), &c., we compose with them the linear functions $a\alpha + a_1\beta + a_2\gamma + a_3\delta$, &c., which enter into the equation $\Delta = 0$; taking them in fours length-wise, (a, b, c, d), &c., we compose the linear functions $ax+by+cz+dw$, &c., which enter into the equation $\Theta = 0$; and taking them in fours breadth-wise (a, a', a'', a'''), &c., we compose the linear functions $a\xi + a'\eta + a''\zeta + a'''\omega$, &c., which enter into the equation $\Phi = 0$.

69. The process may be indefinitely repeated; we obtain always the same three surfaces over and over again, but on them an indefinite series of corresponding points; viz., we may write

$$\begin{array}{l}
\ldots\Theta\ ,\ \Delta\ ,\ \Phi\ ,\quad \Theta,\ \Delta,\ \Phi,\quad \Theta,\ \Delta,\ \Phi\ldots \\
\ldots P_1,\ Q_1,\ R_1,\quad P,\ Q,\ R,\quad P',\ Q',\ R'\ldots
\end{array}$$

viz., a point Q on Δ corresponds to a point P on Θ and to a point R on Φ; R corresponds to Q on Δ and to a new point P' on Θ; P' to R on Φ and to a new point Q' on Δ, and so on. And in the opposite direction P corresponds to Q on Δ, and to a new point R_1 on Φ; R_1 to P on Θ and to a new point Q_1 on Δ; and so on. And of course the correspondence of any two points of the series, whether belonging to the same surface or to different surfaces, is a one-to-one correspondence.

The Symmetrical Case; Symmetroid and Jacobian.

70. I have established the foregoing general theory; but it is only a particular case of it which connects itself with the theory of nodal quartics; viz., the cube of coefficients is a symmetrically arranged cube

$$\begin{array}{lllllll} a & h & g & l & & & \\ \cdot & h & b & f & m & & \\ \vdots & & g & f & c & n & \\ \vdots & & & l & m & n & d \\ & & & & & & \\ a_1 & h_1 \ldots & & & & & \\ \vdots & h_1 & b_1 & & & & \\ \vdots & & \ddots & & & & \end{array}$$

or say its upper face is the symmetrical square matrix

$$\left|\begin{array}{cccc} a, & h, & g, & l \\ h, & b, & f, & m \\ g, & f, & c, & n \\ l, & m, & n, & d \end{array}\right|$$

and the other horizontal planes, the like squares with the several terms affected by suffixes.

The surface $\nabla = 0$ is here a surface of the form

$$\nabla = \left|\begin{array}{cccc} A, & H, & G, & L \\ H, & B, & F, & M \\ G, & F, & C, & N \\ L, & M, & N, & P \end{array}\right| = 0$$

$\{A, B,$ &c. linear functions of $(\alpha, \beta, \gamma, \delta)\}$ viz., ∇ is a symmetrical determinant; I call this a symmetroid; the surfaces $\nabla = 0$, $\Phi = 0$ are one and the same surface, the Jacobian of 4 quadric surfaces; moreover the points P and R are one and the same point, and the correspondence R to P' is a reciprocal one; so that, instead of the indefinite series of points, we have only 2 points Q, Q' on the surface ∇, and 2 points P, P' on the surface Θ $(= \Phi)$; viz., the diagram is

$$\ldots \Delta, \Theta, \Theta, \Delta, \Theta, \Theta, \Delta \ldots$$
$$\ldots Q', P', P, Q, P, P', Q' \ldots$$

moreover the symmetroid surface $\nabla = 0$ is a surface with 10 nodes, which is clearly not octadic, and which is therefore the decadianome.

71. Consider the quadric surfaces

$$\begin{aligned} S &= (a, b, c, d, f, g, h, l, m, n \between x, y, z, w)^2 = 0, \\ T &= (a_1, \ldots \quad \between \quad " \quad)^2 = 0, \\ U &= (a_2, \ldots \quad \between \quad " \quad)^2 = 0, \\ V &= (a_3, \ldots \quad \between \quad " \quad)^2 = 0, \end{aligned}$$

and a point $(\alpha, \beta, \gamma, \delta)$ in the same or in a different space, such that the surface $\alpha S + \beta T + \gamma U + \delta V = 0$ is a cone, or say for shortness,

$$\alpha S + \beta T + \gamma U + \delta V = \text{cone};$$

$(\alpha, \beta, \gamma, \delta)$ is said to be the determining point, or determinator of the cone. And if we establish the equations

$$\begin{aligned} \delta_x (\alpha S + \beta T + \gamma U + \delta V) &= 0, \\ \delta_y (\quad " \quad) &= 0, \\ \delta_z (\quad " \quad) &= 0, \\ \delta_w (\quad " \quad) &= 0, \end{aligned}$$

which express that the surface is a cone, then the point (x, y, z, w) is the vertex of the cone. We have thus 4 equations lineolinear in (x, y, z, w) and also in $(\alpha, \beta, \gamma, \delta)$, so that the relation between the 2 points is of the nature of that above considered. The relation between (x, y, z, w) is given by the equation

$$J(S, T, U, V) = 0;$$

viz., the locus is the Jacobian of the 4 quadric surfaces. The relation between $(\alpha, \beta, \gamma, \delta)$ is given by the equation

$$\nabla = \begin{vmatrix} a\alpha + a_1\beta + a_2\gamma + a_3\delta, & h\alpha + \ldots, & g\alpha + \ldots, & l\alpha + \ldots \\ h\alpha + \ldots \quad , & b\alpha + \ldots, & f\alpha + \ldots, & m\alpha + \ldots \\ g\alpha + \ldots \quad , & f\alpha + \ldots, & c\alpha + \ldots, & n\alpha + \ldots \\ l\alpha + \ldots \quad , & m\alpha + \ldots, & n\alpha + \ldots, & d\alpha + \ldots \end{vmatrix} = 0,$$

so that the locus is (by the foregoing definition) the symmetroid. And the determinator point on the symmetroid thus corresponds to the cone-vertex on the Jacobian.

72. But the Jacobian may be obtained in a different manner; viz., if we establish the equations

$$\begin{aligned} (\xi\delta_x + \eta\delta_y + \zeta\delta_z + \omega\delta_w)\, S &= 0, \\ (\quad " \quad)\, T &= 0, \\ (\quad " \quad)\, U &= 0, \\ (\quad " \quad)\, V &= 0, \end{aligned}$$

then the elimination of $(\xi, \eta, \zeta, \omega)$ leads to the equation $J(S, T, U, V) = 0$ of the Jacobian surface. And since each of the equations is symmetrical in regard to (x, y, z, w)

and $(\xi, \eta, \zeta, \omega)$, it appears that the point $(\xi, \eta, \zeta, \omega)$ is also a point on the Jacobian surface. We have on the symmetroid a point related to $(\xi, \eta, \zeta, \omega)$ in the same way that $(\alpha, \beta, \gamma, \delta)$ on the symmetroid is related to the point (x, y, z, w); and this completes the system of the 4 points, Q on the symmetroid, P and P' on the Jacobian, Q' on the symmetroid; but in what follows I make no use of this last point Q'.

73. The points (x, y, z, w), $(\xi, \eta, \zeta, \omega)$ on the Jacobian correspond in such wise that, taking the polar planes of either of them in regard to the quadrics $S=0$, $T=0$, $U=0$, $V=0$, these intersect in a single point, viz., in the other of the two corresponding points. Or, what is the same thing, the line joining the two points cuts each of the four quadrics harmonically, whence also it cuts harmonically any quadric surface whatever of the series $\alpha S+\beta T+\gamma U+\delta V=0$, ($\alpha$, β, γ, δ being here arbitrary multipliers); viz., this property is an immediate interpretation of the equation

$$(\xi\delta_x+\eta\delta_y+\zeta\delta_z+\omega\delta_w)(\alpha S+\beta T+\gamma U+\delta V)=0,$$

or, as this is more conveniently written,

$$(a, \dots)(\xi, \eta, \zeta, \omega)(x, y, z, w)=0,$$

if for a moment $(a, \dots)$ denote the coefficients of the quadric function $\alpha S+\beta T+\gamma U+\delta V$.

74. Consider any 6 pairs of points (x_1, y_1, z_1, w_1), $(\xi_1, \eta_1, \zeta_1, \omega_1)$, &c., related as above; the quartic surfaces $S=0$, $T=0$, $U=0$, $V=0$ are surfaces cutting harmonically the lines joining the two pairs of points respectively; or say they are quadrics cutting harmonically 6 given segments; and the general quadric surface which cuts harmonically the 6 given segments is $\alpha S+\beta T+\gamma U+\delta V=0$. We thus see that the Jacobian surface $J(S, T, U, V)=0$ is in fact the locus of the vertices of the quadric cones which cut harmonically 6 given segments. The surface so defined by M. Chasles (*Comptes Rendus*, tom. LII., 1861, pp. 1157—62), and shown by him to be a quartic surface, is thus identified with the Jacobian of any 4 quartic surfaces; and included herein we have the particular case, also considered by him, of the locus of the vertices of the quadric cones which pass through 6 given points, or Jacobian of the 6 given points.

75. It is to be shown that there are 10 systems of values $(\alpha, \beta, \gamma, \delta)$, or, what is the same thing, 10 points on the symmetroid, for each of which the quartic surface $\alpha S+\beta T+\gamma U+\delta V=0$ is a plane-pair. For any such system of values the plane-pair may be regarded as a cone, having its vertex at any point whatever on the line which is the axis of the plane-pair; that is, each point of this line is the vertex of a cone of the system of surfaces $\alpha S+\beta T+\gamma U+\delta V=0$; or, what is the same thing, the axis of the plane-pair lies on the Jacobian surface; viz., there will be on the Jacobian surface 10 lines. Moreover, to the point $(\alpha, \beta, \gamma, \delta)$ on the symmetroid there corresponds indifferently any point whatever on the axis of the plane-pair. The analytical expressions for (x, y, z, w) in terms of $(\alpha, \beta, \gamma, \delta)$ must therefore, for the values in question of $(\alpha, \beta, \gamma, \delta)$, become indeterminate; and this can only happen if for the values in question the first minors of the determinant ∇ all of them vanish. But a point $(\alpha, \beta, \gamma, \delta)$, for which the minors of ∇ all of them vanish, is obviously a node on the symmetroid; and it thus appears that there are on the symmetroid 10 nodes,

each corresponding to a line on the Jacobian, and that the condition for determining these is

$$\alpha S + \beta T + \gamma U + \delta V = \text{plane-pair};$$

viz., the values of $(\alpha, \beta, \gamma, \delta)$, which satisfy this condition, belong to a node of the symmetroid, and the line on the Jacobian is the axis of the plane-pair.

76. Reverting to the equation $\nabla = 0$ of the symmetroid, where ∇ is a symmetrical determinant the terms of which are linear functions of the coordinates $(\alpha, \beta, \gamma, \delta)$, it has already been shown, *ante* No. 7, that this is a surface with 10 nodes; but this may be also proved as follows. Writing as before

$$\alpha S + \beta T + \gamma U + \delta V = (A, B, C, D, F, G, H, L, M, N \between x, y, z, w)^2 = 0,$$

the condition that this shall be a plane-pair implies a threefold relation between the coefficients A, B, &c., and the required number of nodes is equal to the order of this threefold relation. Establishing between the coefficients A, B, &c., any 6 linear relations whatever, we should have a ninefold relation to determine the ratios of the 10 quantities; and the number of solutions would be equal to the order of the threefold relations. But taking the 6 linear relations to be of the form $(A, \ldots \between x_1, y_1, z_1, w_1)^2 = 0$, the question is in fact to find the number of the plane-pairs which pass through 6 given points; and this is clearly $= 10$.

77. Applying the conclusion to the system of quadric surfaces $\alpha S + \beta T + \gamma U + \delta V = 0$, we see that there are in the system 10 plane-pairs; and that the lines of intersection, or axes of the plane-pairs, are lines upon the Jacobian surface.

78. The equation $\nabla = 0$ of the symmetroid seems to contain homogeneously 40 constants. But starting with any given symmetrical determinant, we may multiply it line into line by a constant determinant, and then column into column by the same constant determinant, in such wise that the resulting product is still a symmetrical determinant; and the coefficients of the constant determinant may then be used to specialise the form of the equation. The equation $\nabla = 0$ of the symmetroid thus really contains $40 - 16 = 24$ constants; this is as it should be, for the symmetroid, *quà* quartic surface with 10 nodes, contains $34 - 10 = 24$ constants.

Symmetroid with given Nodes.

79. A symmetroid can be formed with 7 given points as nodes; but there is no proper symmetroid with 8 given points as nodes. If we endeavour to form such a symmetroid, we obtain a system of 2 quadric cones, each of them passing through the 8 points; viz., these are any 2 out of the 4 quadric cones which pass through the 8 points. This will be shown in a moment; for the complete *à posteriori* identification with the decadianome, it would be necessary to show that a symmetroid could be found having for nodes 7 given points, an 8th point anywhere on the dianodal surface, and a 9th point anywhere on the dianodal curve; but this I have not succeeded in effecting.

80. We have for any node $(\alpha, \beta, \gamma, \delta)$ of the symmetroid,

$$\alpha S + \beta T + \gamma U + \delta V = \text{plane-pair}.$$

If, then, 4 of the given nodes are (1, 0, 0, 0), (0, 1, 0, 0), (0, 0, 1, 0), (0, 0, 0, 1), we must have S, T, U, V each of them a plane-pair. We may without loss of generality assume $S = x^2 + y^2$, $T = z^2 + w^2$; this, however, does not determine the signification of the coordinates (x, y, z, w), for S will remain unaltered if we write therein

$$x \cos\theta + y \sin\theta, \quad x \sin\theta - y \cos\theta \text{ for } x, y;$$

and similarly T will remain unaltered if we write therein

$$z \cos\theta_1 + w \sin\theta_1, \quad z \sin\theta_1 - w \cos\theta_1 \text{ for } z, w.$$

Hence, if we go on to assume

$$U = k\,(x + m\,y + n\,z + p\,w)(x + m'\,y + n'\,z + p'\,w),$$
$$V = k_1\,(x + m_1 y + n_1 z + p_1 w)(x + m_1' y + n_1' z + p_1' w),$$

we may imagine the θ, θ_1 so determined that, for instance,

$$m + m' = 0, \quad p_1 + p_1' = 0;$$

we have thus

$$\begin{aligned}
S &= x^2 + y^2,\\
T &= z^2 + w^2,\\
U &= k\,(x + m\,y + n\,z + p\,w)(x - m\,y + n'\,z + p'w),\\
V &= k_1(x + m_1 y + n_1 z + p_1 w)(x + m_1' y + n_1' z - p_1 w);
\end{aligned}$$

formulæ which contain the 12 constants

$$(k, m, n, p, n', p', k_1, m_1, n_1, p_1, m_1', n_1').$$

This is right, for the symmetroid containing 24 constants, the symmetroid with 4 given nodes should contain $(24 - 4 . 3 =)$ 12 constants. And each additional given node will determine 3 constants: hence for 4 new given nodes the expressions become determinate (not of necessity uniquely so).

81. But for any 4 new nodes, the equations may be satisfied by writing therein $n = n'$, $p = -p'$, $m = -m_1'$, $n_1 = n_1'$; viz., they then assume the form

$$\begin{aligned}
S &= x^2 + y^2,\\
T &= z^2 + w^2,\\
U &= (a\,x + c\,z)^2 + (b\,y + d\,w)^2,\\
V &= (a'x + c'z)^2 + (b'y + d'w)^2,
\end{aligned}$$

containing 8 constants, which may be determined so that the nodes shall be the 4 given points. If now with the last mentioned values we form the value of $\alpha S + \beta T + \gamma U + \delta V$, this will consist of two terms $(*\between x, z)^2$ and $(*\between y, w)^2$, the first of which will be a square if

$$(\alpha + \gamma a^2 + \delta a'^2)(\beta + \gamma c^2 + \delta c'^2) - (\gamma ac + \delta a'c')^2 = 0, \text{ say this is } \Lambda = 0,$$

and the second will be a square if

$$(\alpha+\gamma b^2+\delta b'^2)(\beta+\gamma d^2+\delta d'^2)-(\gamma bd+\delta b'd')^2=0, \text{ say this is } \Lambda'=0;$$

so that the condition

$$\alpha S+\beta T+\gamma U+\delta V=\text{cone}$$

will be satisfied if $\Lambda=0$, or if $\Lambda'=0$; that is, the equation of the symmetroid will be $\Lambda\Lambda'=0$, or the symmetroid breaks up into the 2 quadric surfaces $\Lambda=0$, $\Lambda'=0$, each of which is a cone.

82. It is to be further observed that, considering the first mentioned 4 points (1, 0, 0, 0), &c., and any other 4 given points whatever, the equation of any one of the 4 quadric cones through these 8 points will be of the form

$$(*\between\beta\gamma,\ \gamma\alpha,\ \alpha\beta,\ \alpha\delta,\ \beta\delta,\ \gamma\delta)=0;$$

viz., any equation of this form, being a cone, will admit of being expressed, and that in one way only, in the form $\Lambda=0$. Consider then any one of the 4 cones through the 8 points, and let its equation be thus expressed; we have the values of the coefficients a, c, a', c', which enter into the expressions of S, T, U, V; and similarly, considering any other of the 4 cones, and expressing its equation in the like form, we have the values of the coefficients b, d, b', d' which enter into the expressions of S, T, U, V.

83. If instead of taking 2 different cones through the 8 points, we take in each case the same cone, the expressions for S, T, U, V would be

$$\begin{aligned}
S&=x^2+y^2,\\
T&=z^2+w^2,\\
U&=(a\,x+c\,z)^2+(a\,y+c\,w)\\
V&=(a'x+c'z)^2+(a'y+c'w)^2;
\end{aligned}$$

and we have identically

$$(ac'-a'c)(aa'S-cc'T)-a'c'U+acV=0.$$

This solution may be disregarded.

84. Instead of the assumption $S=x^2+y^2$, $T=z^2+w^2$, we may take $x=0$, $y=0$, $z=0$, $w=0$ to be planes of the plane-pairs S, T, U, V respectively; it is then easy to fix the remaining constants so that the 5th and 6th nodes of the symmetroid shall be given points. Suppose that the coordinates of the 5th node are (1, 1, 1, 1); to obtain the result in the most simple manner, I take for the moment Ω an arbitrary quadric function $(x, y, z, w)^2$, and I write

$$\begin{aligned}
S&=x\,(\delta_x\Omega\qquad\ \ +hy-gz+aw),\\
T&=y\,(\delta_y\Omega-hx\qquad\ \ +fz+bw),\\
U&=z\,(\delta_z\Omega+gx-fy\qquad\ \ +cw),\\
V&=w\,(\delta_w\Omega-ax-by-cz\qquad\ \ ),
\end{aligned}$$

where the coefficients are arbitrary. We have identically $S+T+U+V=2\Omega$; wherefore the given point (1, 1, 1, 1) will be a node of the symmetroid if only $\Omega=0$ be a plane-pair; and it is easy to see that we may without loss of generality take one factor to be $x+y+z+w$, and write

$$\Omega=(x+y+z+w)(lx+my+nz+pw);$$

viz., Ω having this value, the symmetroid, $\alpha S+\beta T+\gamma U+\delta V=$ cone, will have the 5 given nodes; the equation contains, as it should do, 9 constants.

85. In order that the symmetroid may have a 6th given node $(\alpha_1, \beta_1, \gamma_1, \delta_1)$, I observe that the constants may be determined so that $\alpha_1 S+\beta_1 T+\gamma_1 U+\delta_1 V$ shall be equal to an arbitrary quadric function, say

$$\alpha_1 S+\beta_1 T+\gamma_1 U+\delta_1 V=(\mathrm{a},\ \mathrm{b},\ \mathrm{c},\ \mathrm{d},\ \mathrm{f},\ \mathrm{g},\ \mathrm{h},\ \mathrm{l},\ \mathrm{m},\ \mathrm{n} \between x,\ y,\ z,\ w)^2:$$

this in fact gives

$$(\mathrm{l},\ \mathrm{m},\ \mathrm{n},\ \mathrm{p})=\left(\frac{\mathrm{a}}{\alpha_1},\ \frac{\mathrm{b}}{\beta_1},\ \frac{\mathrm{c}}{\gamma_1},\ \frac{\mathrm{d}}{\delta_1}\right):$$

and then, completing the comparison,

$$\begin{aligned}
S=x\Bigg\{&\left[\frac{\mathrm{a}}{\alpha_1}\right]x+\left[\frac{2\mathrm{h}}{\alpha_1-\beta_1}-\frac{\beta_1}{\alpha_1-\beta_1}\left(\frac{\mathrm{a}}{\alpha_1}+\frac{\mathrm{b}}{\gamma_1}\right)\right]y\\
&+\left[\frac{2\mathrm{g}}{\alpha_1-\gamma_1}-\frac{\gamma_1}{\alpha_1-\gamma_1}\left(\frac{\mathrm{a}}{\alpha_1}+\frac{\mathrm{c}}{\gamma_1}\right)\right]z+\left[\frac{2\mathrm{l}}{\alpha_1-\delta_1}-\frac{\delta_1}{\alpha_1-\delta_1}\left(\frac{\mathrm{a}}{\alpha_1}+\frac{\mathrm{d}}{\delta_1}\right)\right]w\Bigg\},\\
T=y\Bigg\{&\left[\frac{2\mathrm{h}}{\beta_1-\alpha_1}-\frac{\alpha_1}{\beta_1-\alpha_1}\left(\frac{\mathrm{b}}{\beta_1}+\frac{\mathrm{a}}{\alpha_1}\right)\right]x+\left[\frac{\mathrm{b}}{\beta_1}\right]y\\
&+\left[\frac{2\mathrm{f}}{\beta_1-\gamma_1}-\frac{\gamma_1}{\beta_1-\gamma_1}\left(\frac{\mathrm{b}}{\beta_1}+\frac{\mathrm{c}}{\gamma_1}\right)\right]z+\left[\frac{2\mathrm{m}}{\beta_1-\delta_1}-\frac{\delta_1}{\beta_1-\delta_1}\left(\frac{\mathrm{b}}{\beta_1}+\frac{\mathrm{d}}{\delta_1}\right)\right]w\Bigg\},\\
U=z\Bigg\{&\left[\frac{2\mathrm{g}}{\gamma_1-\alpha_1}-\frac{\alpha_1}{\gamma_1-\alpha_1}\left(\frac{\mathrm{c}}{\gamma_1}+\frac{\mathrm{a}}{\alpha_1}\right)\right]x+\left[\frac{2\mathrm{f}}{\gamma_1-\beta_1}-\frac{\beta_1}{\gamma_1-\beta_1}\left(\frac{\mathrm{c}}{\gamma_1}+\frac{\mathrm{b}}{\beta_1}\right)\right]y\\
&+\left[\frac{\mathrm{c}}{\gamma_1}\right]z+\left[\frac{2\mathrm{n}}{\gamma_1-\delta_1}-\frac{\delta_1}{\gamma_1-\delta_1}\left(\frac{\mathrm{c}}{\gamma_1}+\frac{\mathrm{d}}{\delta_1}\right)\right]w\Bigg\},\\
V=w\Bigg\{&\left[\frac{2\mathrm{l}}{\delta_1-\alpha_1}-\frac{\alpha_1}{\delta_1-\alpha_1}\left(\frac{\mathrm{d}}{\delta_1}+\frac{\mathrm{a}}{\alpha_1}\right)\right]x+\left[\frac{2\mathrm{m}}{\delta_1-\beta_1}-\frac{\beta_1}{\delta_1-\beta_1}\left(\frac{\mathrm{d}}{\delta_1}+\frac{\mathrm{b}}{\beta_1}\right)\right]y\\
&+\left[\frac{2n}{\delta_1-\gamma_1}-\frac{\gamma_1}{\delta_1-\gamma_1}\left(\frac{\mathrm{d}}{\delta_1}+\frac{\mathrm{c}}{\gamma_1}\right)\right]z+\left[\frac{d}{\delta_1}\right]w\Bigg\};
\end{aligned}$$

viz., these values give

$$S+\quad T+\quad U+\quad V=(x+y+z+w)\left(\frac{\mathrm{a}}{\alpha_1}x+\frac{\mathrm{b}}{\beta_1}y+\frac{\mathrm{c}}{\gamma_1}z+\frac{\mathrm{d}}{\delta_1}w\right),$$

$$\alpha_1 S+\beta_1 T+\gamma_1 U+\delta_1 V=(\mathrm{a},\ \mathrm{b},\ \mathrm{c},\ \mathrm{d},\ \mathrm{f},\ \mathrm{g},\ \mathrm{h},\ \mathrm{l},\ \mathrm{m},\ \mathrm{n} \between x,\ y,\ z,\ w);$$

hence, taking the function $(\mathrm{a}, \ldots \between x, y, z, w)^2$ to be a plane-pair equal to $(x+iy+jz+kw)(x+i_1y+j_1z+k_1w)$ suppose, or considering the coefficients (a, ...) as given functions of (i, j, k, i_1, j_1, k_1), we have the symmetroid having the 6 given nodes and containing the last mentioned 6 constants.

The Jacobian with given Lines.

86. The Jacobian contains 24 constants; obviously it is uniquely determined if 4 of the plane-pairs thereof are given; and it is also determined, but not uniquely, if 6 of the lines thereof are given. We may enquire how many given nodes of the symmetroid may be considered as corresponding to given plane-pairs, or lines of the Jacobian. Take as given any 4 nodes of the symmetroid; the corresponding 4 plane-pairs may be taken to be given plane-pairs; and we may besides take as given a 5th node of the symmetroid. For let the first 4 nodes of the symmetroid be (1, 0, 0, 0), (0, 1, 0, 0), (0, 0, 1, 0), (0, 0, 0, 1); the given plane-pairs $P_1Q_1=0$, $P_2Q_2=0$, $P_3Q_3=0$, $P_4Q_4=0$; (l_1, l_2, l_3, l_4) any system of values such that we have

$$l_1P_1Q_1+l_2P_2Q_2+l_3P_3Q_3+l_4P_4Q_4=\text{plane-pair};$$

and (1, 1, 1, 1) the 5th node of the symmetroid; we have only to assume

$$(S,\ T,\ U,\ V)=(l_1P_1Q_1,\ l_2P_2Q_2,\ l_3P_3Q_3,\ l_4P_4Q_4).$$

87. Suppose, however, that on the Jacobian we have given, not the 4 plane-pairs, but only the 4 axes of the plane-pairs; the plane-pairs may be taken to be

$$(1,\ b_1,\ c_1 \between P_1,\ Q_1)^2=0, \ldots\ldots (1,\ b_4,\ c_4 \between P_4,\ Q_4)^2=0,$$

where the 8 constants $(b_1, b_2, b_3, b_4, c_1, c_2, c_3, c_4)$ are in the first instance undetermined. If we attempt to find l_1, l_2, l_3, l_4, so that

$$l_1(1,\ b_1,\ c_1 \between P_1,\ Q_1)^2 \ldots\ldots + l_4(1,\ b_4,\ c_4 \between P_4,\ Q_4)^2=\text{plane-pair of given axis},$$

we have between the coefficients (b, c) 4 equations; and similarly, if we attempt to find m_1, m_2, m_3, m_4 such that

$$m_1(1,\ b_1,\ c_1 \between P_1,\ Q_1)^2 \ldots\ldots + m_4(1,\ b_4,\ c_4 \between P_4,\ Q_4)^2=\text{plane-pair of another given axis},$$

we have 4 more equations between the coefficients (b, c); viz., these will be determined by the 8 equations (this is in fact the before mentioned property that 6 lines of the Jacobian may be taken to be given lines). But considering only the first system of equations; in order that to the given axis may correspond a given node on the symmetroid, say the node (1, 1, 1, 1), we have only to write

$$S=l_1(1,\ b_1,\ c_1 \between P_1,\ Q_1)^2, \ldots\ldots V=l_4(1,\ b_4,\ c_4 \between P_4,\ Q_4)^2;$$

that is, we may take as given 5 nodes of the symmetroid, and the corresponding 5 lines of the Jacobian; the formulæ will contain 4 constants; we may by means of them make the Jacobian have a 6th given line, thus determining the constants; or we may make the symmetroid have a 6th given node, leaving in this case one constant arbitrary.

Correspondence on the Jacobian: Lines and Skew Cubics.

88. I consider the correspondence of two points on the Jacobian; it is to be shown that when one of the points is on a line of the Jacobian, the corresponding point will be on a skew cubic; that is, that corresponding to each line of the Jacobian we have (on the Jacobian) a skew cubic. Call the plane-pairs of the system of quadric surfaces 1, 2, 3, ... 10; selecting any 4 of these, say 1, 2, 3, 4, the polar planes of any point of the Jacobian in regard to these 4 plane-pairs will meet in a point which will be the required corresponding point. And observe that, in regard to any one of the plane-pairs, say 1, the polar plane of a point P is the plane through the axis harmonic to the plane through the axis and the point P. Hence, for a point on the axis of 1, the polar plane in regard to 1 is indeterminate; the polar planes in regard to the plane-pairs 2, 3, 4 respectively meet in a point which is the required corresponding point. We may for any point whatever take the polar planes in regard to the plane-pairs 2, 3, 4 respectively, and call the intersection of these planes the corresponding point; this being so, if the first mentioned point moves along a line, the corresponding point moves along a curve, which is easily shown to be a skew cubic cutting the axis of each plane-pair twice; that is, in regard to the plane-pairs 2, 3, 4, the locus corresponding to any line whatever is a skew cubic cutting the axis of each plane-pair twice. In particular, the corresponding curve of the axis of 1, is a skew cubic cutting the axis of the plane-pairs 2, 3, 4 each twice; but the axis of 1 does not stand in any special relation to the plane-pairs 2, 3, 4, as distinguished from the remaining plane-pairs 5, 6...10; we have therefore the more complete theorem, that the skew cubic cuts the axes of the plane-pairs 2, 3,...10 each twice; or, instead of the plane-pairs, speaking of the line 1, 2, 3,...10, we may say that corresponding to any one of the lines we have a skew cubic meeting the other 9 lines each of them twice.

89. I stop for a moment to prove the subsidiary theorem assumed in the foregoing demonstration. Let the 3 plane-pairs be $PQ=0$, $RS=0$, $TU=0$, and let the line be that joining the points (x_0, y_0, z_0, w_0) and (x_1, y_1, z_1, w_1); the coordinates of any point in the line may be taken to be $\lambda x_0+\mu x_1$, $\lambda y_0+\mu y_1$, $\lambda z_0+\mu z_1$, $\lambda w_0+\mu w_1$; and hence for the polar plane in regard to the plane-pair $PQ=0$ we have

$$\{(\lambda x_0+\mu x_1)\,\delta_x \ldots + (\lambda w_0+\mu w_1)\,\delta_w\}\, PQ=0\,;$$

viz., this equation may be written

$$\lambda\,(PQ_0+P_0Q)+\mu\,(PQ_1+P_1Q)=0\,;$$

forming the like equations in regard to the other 2 plane-pairs respectively, and eliminating λ, μ, we obtain for the required locus

$$\left\|\begin{matrix} PQ_0+P_0Q, & RS_0+R_0S, & TU_0+T_0U \\ PQ_1+P_1Q, & RS_1+R_1S, & TU_1+T_1U \end{matrix}\right\|=0,$$

a skew cubic; and on writing herein $P=0$, $Q=0$, the equations become

$$\begin{vmatrix} RS_0+R_0S, & TU_0+T_0U \\ RS_1+R_1S, & TU_1+T_1U \end{vmatrix}=0;$$

viz., the line ($P=0$, $Q=0$) meets the skew cubic in the points where the line meets the quadric surface determined by this last equation, that is in 2 points.

90. We have thus on the Jacobian the 10 lines 1, 2, ... 9, 10, and corresponding thereto respectively the 10 skew cubics 1′, 2′, ... 9′, 10′, where each line meets twice each of the skew cubics except that denoted by the same number; a relation similar to that which exists between the lines 1, 2, 3, 4, 5, 6 and 1′, 2′, 3′, 4′, 5′, 6′, which compose a double-sixer on a cubic surface.

Suppose that there are given on the Jacobian the lines 1, 2, 3, 4, 5, 6; meeting each of these twice, we have the skew cubics 7′, 8′, 9′, 10′; and then

	7		8′, 9′, 10′
the lines	8	meet twice each of the cubics	9′, 10′, 7′
	9		10′, 7′, 8′
	10		7′, 8′, 9′

so that the determination of the remaining 4 lines depends upon that of the skew cubics 7′, 8′, 9′, 10′, which meet each of the given lines twice.

91. To determine a skew cubic cutting twice each of 6 given lines, I proceed as follows. Let the lines be 1, 2, 3, 4, 5, 6; take $U=0$ the general quadric surface through the lines 1 and 2, $V=0$ the general quadric surface through the lines 1, 3 (the equations contain each of them homogeneously 4 constants). The 2 surfaces intersect in the line 1, and in a skew cubic cutting twice each of the lines 1, 2, 3; we have therefore to determine the constants so that the 2 surfaces may meet the line 4 in the same 2 points, the line 5 in the same 2 points, the line 6 in the same two points. Imagine for a moment the equations of any one of the lines 4, 5, 6 to be $z=0$, $w=0$; the equations of the 2 surfaces, substituting therein these values, would assume the forms

$$(a, b, c \between x, y)^2=0, \quad (a', b', c' \between x, y)^2=0;$$

and the conditions for the intersection in the same 2 points would be $\frac{a}{a'}=\frac{b}{b'}=\frac{c}{c'},=p$ suppose. This is in fact the form of the conditions, understanding a, b, c to be linear functions of the coefficients of U, and a', b', c' to be linear functions of the coefficients of V. We have in this manner 3 sets of equations involving respectively the indeterminate quantities p, q, r; viz., these may be represented by

$$a=pa',\ b=pb',\ c=pc';\ d=qd',\ e=qe',\ f=qf';\ g=rg',\ h=rh',\ i=ri';$$

where the unaccented letters $a, b, \ldots i$ are linear functions of the coefficients of U, and the accented letters $a', b', \ldots i'$ linear functions of the coefficients of V. Eliminating the coefficients of U, V, we have between p, q, r a twofold relation, which may be represented as follows:

$$\left\| \begin{array}{ccccccccc} 1 & 1 & 1 & 1 & 1 & 1 & 1 & 1 & 1 \\ 1 & 1 & 1 & 1 & 1 & 1 & 1 & 1 & 1 \\ 1 & 1 & 1 & 1 & 1 & 1 & 1 & 1 & 1 \\ 1 & 1 & 1 & 1 & 1 & 1 & 1 & 1 & 1 \\ p & p & p & q & q & q & r & r & r \\ p & p & p & q & q & q & r & r & r \\ p & p & p & q & q & q & r & r & r \\ p & p & p & q & q & q & r & r & r \end{array} \right\| = 0,$$

it being understood that the 1's represent constants, and the p's, q's, and r's linear functions of these variables respectively. The several equations of the system, regarding therein p, q, r as coordinates, represent each of them a quartic curve; any 2 of these intersect in 16 points; but the number of points common to all the curves is $=10$. But each of the curves passes through the 3 points $(1, 0, 0)$, $(0, 1, 0)$, $(0, 0, 1)$; these are consequently included among the 10 points, but they do not give a proper solution of the question; and the number of solutions is thus reduced to $10-3=7$. There is yet another solution to be rejected; viz., $U=0$ being a quadric surface through the lines 1, 2, and $V=0$ the quadric surface through the lines 1, 3, it is possible to determine the coefficients of U, V so that each of these surfaces shall be the quadric surface through the lines 1, 2, 3; and if we then have identically $U=\theta V$, it is clear that corresponding values of p, q, r are $p=q=r\,(=\theta)$. We have thus the point $p=q=r$ common to all the curves of the system; this solution counts, I believe, once only, and the number of relevant solutions is $7-1=6$.

92. It may be observed, in regard to the foregoing solution, that if we take $123=0$ as the equation of the quadric surface through the lines 1, 2, 3, and so in other cases, then the equation of the surfaces $U=0$ and $V=0$ may be taken to be

$$\lambda\,.\,123+\mu\,.\,124+\nu\,.\,125+\rho\,.\,126=0,$$

$$\lambda'\,.\,132+\mu'\,.\,134+\nu'\,.\,135+\rho'\,.\,136=0,$$

respectively, the coefficients of the two surfaces being here put in evidence. And it is clear that for $\mu=\nu=\rho=0$, $\mu'=\nu'=\rho'=0$, the surfaces become each of them the surface through the lines 1, 2, 3.

93. The conclusion is, that touching twice each of the six lines 1, 2, 3, 4, 5, 6, we have six skew cubics; it would appear that any four of these may be taken for the skew cubics 7′, 8′, 9′, 10′ (so that there are 15 such tetrads of cubics). I am not, however, able to verify that we then have the remaining 4 lines each cutting twice 3 of the 4 skew cubics; assuming that for each system of 4 skew cubics there is one and only one, such system of lines, then of course to the given system of lines 1, 2, 3, 4, 5, 6, there will belong 15 systems of lines 7, 8, 9, 10, and therefore also 15 Jacobian surfaces.

Further Investigations as to the Jacobian, &c.

94. Taking $(\xi, \eta, \zeta, \omega)$ as plane-coordinates, two quadric surfaces

$$(a, b, c, d, f, g, h, l, m, n \between \xi, \eta, \zeta, \omega)^2 = 0$$

and

$$(A, B, C, D, F, G, H, L, M, N \between x, y, z, w)^2 = 0$$

are said to be interverts (or interverse) one of the other, when we have between the coefficients the relation

$$(a, b, c, d, f, g, h, l, m, n \between A, B, C, D, F, G, H, L, M, N) = 0,$$

that is

$$aA + \ldots + 2fF + \ldots = 0.$$

The condition that the two surfaces may be interverts of each other is linear in regard to the coefficients of each surface separately; hence, using a before explained locution, we may say—interverse to a given quadric surface we have 9 quadrics; interverse to two given quadrics 8 quadrics; or generally, that interverse to k given quadrics we have $10-k$ quadrics. And, moreover, if the quadrics of the two systems be $L=0$, $M=0$, &c., and $S=0$, $T=0$, $U=0$, &c., then every quadric $\lambda L + \mu M + \ldots = 0$ is interverse to each of the quadrics $\alpha S + \beta T + \gamma U + \ldots = 0$.

If the quadric $(a, \ldots \between \xi, \eta, \zeta, \omega)^2 = 0$ be an intervert of the plane-pair

$$(lx + my + nz + pw \between l'x + m'y + n'z + p'w) = 0,$$

the condition is

$$(a, \ldots \between l, m, n, p \between l', m', n', p') = 0;$$

viz., this expresses that the two planes are harmonics in regard to the pair of planes drawn through the axis of the plane-pair to touch the quadric surface; or say, that the plane-pair is harmonic in regard to the quadric.

95. To apply this to the Jacobian surface, I recall that, starting with the given quadric surfaces $S=0$, $T=0$, $U=0$, $V=0$, and taking $(\alpha, \beta, \gamma, \delta)$ to be such that

$$\alpha S + \beta T + \gamma U + \delta V = \text{plane-pair},$$

there are 10 such plane-pairs, and that the axes of these are the lines of the Jacobian. If instead of the given quadric surfaces, we consider the six interverse surfaces $(a_1, \ldots \between \xi, \eta, \zeta, \omega)^2 = 0$, $\ldots (a_6, \ldots \between \xi, \eta, \zeta, \omega)^2 = 0$, then the condition is that the plane-pair shall be harmonic in regard to each of these surfaces. Let the quadric surfaces be called 1, 2, 3, 4, 5, 6; then, attending to any three of these, say 1, 2, 3, the plane-pair is harmonic in regard to these three surfaces. Through the axis of the plane-pair draw tangent planes to 1, 2, and 3 respectively; each of these pairs of planes is harmonic in regard to the planes of the plane-pair; that is, the three pairs of tangent planes are in involution; or, as we may also express it, the axis is (*quoad* its planes) in involution in regard to the three quadric surfaces. Conversely, when the axis is thus in involution in regard to the surfaces 1, 2, and 3, we may by means of the surfaces 1 and 2 determine the two planes of the plane-pair, and then these will be harmonics in regard to the surface 3. It thus appears that the axis is given as a line which is (*quoad* its planes) in involution in regard to the surfaces 1, 2, 3, to the surfaces 1, 2, 4, the surfaces 1, 2, 5, and the surfaces 1, 2, 6, respectively; or, as we may express it, as a line which is (*quoad* its planes) in involution in regard to the surfaces 1, 2, 3, 4, 5, 6.

96. It is substantially the same thing, but it is rather easier, to consider the whole question under the reciprocal form; viz., instead of a plane-pair and a quadric surface represented by an equation in plane-coordinates, to take a point-pair and a quadric surface represented by an equation in point-coordinates; we have thus a line which is (*quoad* its points) in involution in regard to three given quadric surfaces, or as we may more simply express it, which cuts in involution the three given surfaces; and we thus arrive at the problem of finding a line which cuts in involution six given quadric surfaces; viz., this is equivalent to the above problem where the line has to satisfy (*quoad* its planes) the like condition; and in each problem the number of solutions should be $= 10$.

97. Consider a line which cuts in involution the three given surfaces $(a_1, \ldots \between x, y, z, w)=0$, $(a_2, \ldots \between x, y, z, w)^2 = 0$, $(a_3, \ldots \between x, y, z, w)^2 = 0$. I will presently show that this implies a cubic relation $(* \between \mathrm{a, b, c, f, g, h})^3$ between the six coordinates of the line. But assuming it for the moment, suppose that the line cuts in involution the three surfaces and a fourth quadric surface $(a_4, \ldots \between x, y, z, w)^2 = 0$. Considering the line as cutting in involution the surfaces 1, 2, 4, we have between the six coordinates a second cubic relation; there is, however, a reduction, and the order of the resulting twofold relation between the coordinates is $3 \,.\, 3 - 4 = 5$. To explain this, observe that every line which cuts in the same two points the surfaces 1 and 2 respectively (that is, which cuts the curve of intersection twice) will in an improper sense cut in involution the surfaces 1, 2, 3, and also the surfaces 1, 2, 4. There is thus a reduction equal to the order in the six coordinates of the twofold relation which expresses that the line cuts twice the curve of intersection of the surfaces 1 and 2. Join hereto the relations that the line meets each of two given lines; the coordinates of the line are determined by the twofold relation (say its order is $= \lambda$) two linear equations, and the universal equation $\mathrm{af} + \mathrm{bg} + \mathrm{ch} = 0$; the number of solutions is $= 2\lambda$. But the number of solutions is equal to that of the lines which meet the quadri-quadric curve of intersection twice, and meet also each of two given lines; or what is the same thing, it is equal to the order of the scroll generated by the lines which meet the curve twice, and also a given line. We have for the curve of intersection (m the order, h the number of apparent double points) $m = 4$, $h = 2$; whence order of the scroll is $2 + \frac{1}{2} \,.\, 4 \,.\, 3 = 8$; that is, $2\lambda = 8$, or $\lambda = 4$, which is the required reduction.

98. If the line cut in involution 5 given quadric surfaces {say the 5th surface is $(a_5, \ldots \between x, y, z, w)^2 = 0$}; then we have between the 6 coordinates a threefold relation, the order of which is $3.5 -$ reduction. This should be $= 10$, and consequently the reduction $= 5$; for admitting the value to be 10, the order (in the ordinary sense) of the scroll generated by the lines which cut in involution the 5 given quadrics should be $= 20$;

and conversely. But the value 20 may be verified without difficulty. For the question may be transformed as follows:—If a point-pair be harmonic in regard to each of 5 given quadrics, how many of the axes (or lines through the 2 points of a point-pair) cut a given line. Take (x, y, z, w), (x', y', z', w') as the coordinates of the 2 points of a point-pair; the harmonic condition in regard to a quadric surface $U=0$ is $x'\delta_x U + y'\delta_y U + z'\delta_z U + w'\delta_w U = 0$ {where U is regarded as a function of the (x, y, z, w) belonging to a point of the point-pair}; the condition for the intersection with a given line is a lineolinear equation in the coordinates (x, y, z, w) and (x', y', z', w'), or say it is $Lx' + My' + Nz' + Pw' = 0$, where L, M, N, P are linear functions of the coordinates; we have thence for (x, y, z, w) the threefold relation

$$\left\| \begin{array}{cccccc} L, & \delta_x U_1, & \delta_x U_2, & \delta_x U_3, & \delta_x U_4, & \delta_x U_5 \\ M, & \delta_y U_1 & . & . & & \\ N, & \delta_z U_1 & . & & & \\ P, & \delta_w U_1 & & & & \end{array} \right\| = 0,$$

which denotes a system of $\frac{1}{6}.6.5.4 = 20$ points.

It would seem that if the line cuts in involution 6 given quadrics, there should be between the 6 coordinates a fourfold relation of the order $\frac{1}{2}.10 = 5$; this would imply a reduction 25, viz. we should have $5 = 3.10 - 25$. I do not understand this, and I drop the question.

99. I return to the question to find the relation between the coordinates (a, b, c, f, g, h) of a line which cuts in involution the 3 quadric surfaces

$$(a_1, b_1, c_1, d_1, f_1, g_1, h_1, l_1, m_1, n_1 \between x, y, z, w)^2 = 0, \quad (a_2, \ldots \between x, y, z, w)^2 = 0, \quad (a_3, \ldots \between x, y, z, w)^2 = 0.$$

Writing down any two of the equations of the line, for instance

$$\begin{aligned} \mathrm{h}y - \mathrm{g}z + \mathrm{a}w &= 0, \\ -\mathrm{h}x \quad + \mathrm{f}z + \mathrm{b}w &= 0, \end{aligned}$$

if we substitute the values of (x, y) in the equation of the first surface, it becomes

$$(a_1, \ldots \between \mathrm{f}z + \mathrm{b}w,\ \mathrm{g}z - \mathrm{a}w,\ \mathrm{h}z,\ \mathrm{h}w)^2 = 0;$$

or if we write for shortness

$$\Pi = (\mathrm{f, g, h, o}), \quad \Pi' = (\mathrm{b, -a, o, h}),$$

then the equation is

$$(a_1, \ldots \between \Pi)^2 . z^2 + 2(a_1, \ldots \between \Pi \between \Pi') . zw + (a_1, \ldots \between \Pi')^2 . w^2 = 0,$$

and forming the like equations for the other two surfaces, the condition of involution is at once found to be

$$\left| \begin{array}{ccc} (a_1, \ldots \between \Pi)^2, & (a_1, \ldots \between \Pi \between \Pi'), & (a_1, \ldots \between \Pi')^2 \\ (a_2, \ldots \between \Pi)^2, & (a_2, \ldots \between \Pi \between \Pi'), & (a_2, \ldots \between \Pi')^2 \\ (a_3, \ldots \between \Pi)^2, & (a_3, \ldots \between \Pi \between \Pi'), & (a_3, \ldots \between \Pi')^2 \end{array} \right| = 0.$$

100. It is convenient, in working this out, to consider Π, Π' as standing, in the first instance, for (x, y, z, w), (x', y', z', w'), these symbols being ultimately replaced by the above-mentioned values. Writing also, for shortness, (abc) to denote the determinant $a_1(b_2c_3 - b_3c_2) + \&c.$, and so in other cases, it is at once seen that the function on the right-hand side is a sum of such determinants each into a proper factor, containing the coordinates (a, b, c, f, g, h), originally of the order 6, but where each term contains the factor h^3, which may be omitted; or finally the result is of the order 3 in the coordinates. Thus we have a term

$$(abc)\begin{vmatrix} x^2, & xx', & x'^2 \\ y^2, & yy', & y'^2 \\ z^2, & zz', & z'^2 \end{vmatrix}$$

where the second factor is

$$x^2y'z'(yz' - y'z) + y^2z'x'(zx' - z'x) + z^2x'y'(xy' - x'y), \quad = z^2x'y',(xy' - x'y),$$
$$= \text{h}^2(-\text{ab})(-\text{af} - \text{bg}), \quad = -\text{abch}^3,$$

or, omitting the factor $-\text{h}^3$, the term is $(abc)\,\text{abc}$.

101. There are in all 120 terms, but 16 of these are found to vanish (viz., these are the terms in *agh, bhf, cfg*; *ahl, bfm, cgn*; *agl, bhm, cfn*; *dmn, dnl, dlm*; *fgn, ghl, hfm*). The final result contains therefore 104 terms; viz., as a further abbreviation writing *abc* &c., instead of (abc) &c., to denote the above-mentioned determinants, the equation is

$$\begin{aligned}
& abc\,.\,\text{abc} - bcd\,.\,\text{agh} - cad\,.\,\text{bhf} - abd\,.\,\text{cfg} \\
& + bcf\,.\,\text{a}^3 + cag\,.\,\text{b}^3 + abh\,.\,\text{c}^3 + adl\,.\,\text{f}^3 + bdm\,.\,\text{g}^3 + cdn\,.\,\text{h}^3 \\
& + abn\,.\,\text{c}(\text{bg} - \text{af}) + adf\,.\,\text{f}(\text{ch} - \text{bg}) \\
& + bcl\,.\,\text{a}(\text{ch} - \text{bg}) + bdg\,.\,\text{g}(\text{af} - \text{ch}) \\
& + cam\,.\,\text{b}(\text{af} - \text{ch}) + cdh\,.\,\text{h}(\text{bg} - \text{af}) \\
& - bcg\,.\,\text{a}^2\text{b} - bch\,.\,\text{a}^2\text{c} + bcm\,.\,\text{a}^2\text{g} - bcn\,.\,\text{a}^2\text{h} \\
& - cah\,.\,\text{b}^2\text{c} - caf\,.\,\text{b}^2\text{a} + can\,.\,\text{b}^2\text{h} - cal\,.\,\text{b}^2\text{f} \\
& - abf\,.\,\text{c}^2\text{a} - abg\,.\,\text{c}^2\text{b} + abl\,.\,\text{c}^2\text{f} - abm\,.\,\text{c}^2\text{g} \\
& - adg\,.\,\text{bf}^2 + adh\,.\,\text{cf}^2 + adm\,.\,\text{f}^2g + adn\,.\,\text{f}^2\text{h} \\
& - bdh\,.\,\text{cg}^2 + bdf\,.\,\text{ag}^2 + bdn\,.\,\text{g}^2\text{h} + bdl\,.\,\text{g}^2\text{f} \\
& - cdf\,.\,\text{ah}^2 + cdg\,.\,\text{bh}^2 + cdl\,.\,\text{h}^2\text{f} + cdm\,.\,\text{h}^2\text{g} \\
& + 2\left\{\begin{array}{l} \, afg\,.\,\text{b}^2\text{c} - afh\,.\,\text{bc}^2 + afl\,.\,\text{bcf} - afm\,.\,\text{c}^2\text{h} - afn\,.\,\text{b}^2\text{g} \\ + bgh\,.\,\text{c}^2\text{a} - bgf\,.\,\text{ca}^2 + bgm\,.\,\text{cag} - bgl\,.\,\text{a}^2\text{f} - bgl\,.\,\text{c}^2\text{h} \\ + chf\,.\,\text{a}^2\text{b} - chg\,.\,\text{ab}^2 + chn\,.\,\text{abh} - chm\,.\,\text{b}^2\text{g} - chm\,.\,\text{a}^2\text{f} \end{array}\right\}
\end{aligned}$$

$$\begin{aligned}
&+2\left\{\begin{array}{l} agm\,.\,\mathrm{bcf}-agn\,.\,\mathrm{b^2f}-ahm\,.\,\mathrm{c^2f}+ahn\,.\,\mathrm{bcf} \\ +bhn\,.\,\mathrm{cag}-bfl\,.\,\mathrm{c^2g}-bfn\,.\,\mathrm{a^2g}+bfl\,.\,\mathrm{cag} \\ +cfl\,.\,\mathrm{abh}-chm\,.\,\mathrm{a^2h}-cgl\,.\,\mathrm{b^2h}+cgm\,.\,\mathrm{abh} \end{array}\right\} \\
&+2\left\{\begin{array}{l} -amn\,.\,\mathrm{af^2}-anl\,.\,\mathrm{bf^2}-alm\,.\,\mathrm{cf^2}+dfg\,.\,\mathrm{ch^2} \\ -bnl\,.\,\mathrm{bg^2}-blm\,.\,\mathrm{cg^2}-bmn\,.\,\mathrm{ag^2}+dgh\,.\,\mathrm{af^2} \\ -clm\,.\,\mathrm{ch^2}-cmn\,.\,\mathrm{ah^2}-cnl\,.\,\mathrm{bh^2}+dhf\,.\,\mathrm{bg^2} \end{array}\right\} \\
&+2\left\{\begin{array}{l} -dfl\,.\,\mathrm{fgh}-dfm\,.\,\mathrm{g^2h}-dfn\,.\,\mathrm{gh^2} \\ -dgm\,.\,\mathrm{fgh}-dgn\,.\,\mathrm{h^2f}-dgl\,.\,\mathrm{hf^2} \\ -dhn\,.\,\mathrm{fgh}-dhl\,.\,\mathrm{f^2g}-dhm\,.\,\mathrm{fg^2} \end{array}\right\} \\
&+4\left\{\begin{array}{l} fgh\,.\,\mathrm{bch}-fgm\,.\,\mathrm{ach}-fmn\,.\,\mathrm{agh}-fnl\,.\,\mathrm{bgh}-flm\,.\,\mathrm{cgh} \\ +ghm\,.\,\mathrm{caf}-ghn\,.\,\mathrm{baf}-gnl\,.\,\mathrm{bhf}-glm\,.\,\mathrm{chf}-gmn\,.\,\mathrm{ahf} \\ +hfn\,.\,\mathrm{abg}-hfl\,.\,\mathrm{cbg}-hlm\,.\,\mathrm{cfg}-hmn\,.\,\mathrm{afg}-hnl\,.\,\mathrm{bfg} \end{array}\right\} \\
&-4fgh\,.\,\mathrm{abc} \qquad\qquad = 0.
\end{aligned}$$

And observe, by what precedes, this triple system of lines contains each of the following double systems: viz., the lines which meet the quadriquadric curve (2, 3) twice, those which meet the curve (3, 1) twice, those which meet the curve (1, 2) twice.

Persymmetrical Case: the Hessian of a Cubic.

102. Reverting to the general equation

$$\alpha S+\beta T+\gamma U+\delta V=\text{cone},$$

which connects the symmetroid and Jacobian, it is evident that if S, T, U, V are the derivatives, in regard to the coordinates, of a single cubic function U, $=(*\between x, y, z, w)^3$, then the symmetroid and the Jacobian become one and the same surface; viz., this is the Hessian surface $H=0$ derived from the given cubic surface. The two corresponding points on the symmetroid and the Jacobian respectively, and the two corresponding points on the Jacobian, become one and the same pair of corresponding points on the Hessian; viz., either of these points is such that its first polar surface in regard to the cubic is a quadric cone having for its vertex the other corresponding point. And the Hessian surface unites the properties of the Jacobian and the symmetroid, viz., it has 10 nodes and 10 lines. It is, in fact, known that there are five planes such that the intersection of every two of them is a line on the Hessian surface, and the intersection of every three of them a node on the surface; viz., if the equations of the five planes are $x=0$, $y=0$, $z=0$, $w=0$, $u=0$, then the equation of the Hessian surface is

$$xyzwu\left(\frac{a}{x}+\frac{b}{y}+\frac{c}{z}+\frac{d}{w}+\frac{e}{u}\right)=0,$$

a form which puts in evidence the properties just referred to.

Quartics with 11 *or more Nodes.*

103. I mention two results which, although they relate to quadric surfaces with more than 10 nodes, present themselves in such immediate connexion with the present Memoir, that it is natural to speak of them. If, in the equation

$$\begin{vmatrix} A, & H, & G, & L \\ H, & B, & F, & M \\ G, & F, & C, & N \\ L, & M, & N, & D \end{vmatrix} = 0,$$

of the symmetroid (A, B,... linear functions of the coordinates), we have identically $A = 0$, then the surface has evidently a node $H = 0$, $G = 0$, $L = 0$; viz., this is a node in addition to the usual 10 nodes, or the surface has in all 11 nodes. And so also if (identically in every case) B is $= 0$, there are 12 nodes; if C is $= 0$, there are 13 nodes; and if D is $= 0$, there are 14 nodes. These are, in fact, quartic surfaces with 11, 12, 13, and 14 nodes respectively, mentioned in Kummer's Memoir.

104. We may consider the symmetroid derived from the quadric surfaces which pass through 6 given points; viz., taking as before (see No. 25) the coordinates of the 6 points to be (1, 0, 0, 0), (0, 1, 0, 0), (0, 0, 1, 0), (0, 0, 0, 1), (1, 1, 1, 1), (α, β, γ, δ), and (a, b, c, f, g, h) as the coordinates of the line joining the last-mentioned two points; and, to avoid confusion, taking for the present purpose (X, Y, Z, W) instead of (α, β, γ, δ) for the coordinates of a point on the symmetroid, the equation is obtained by arranging in the form of a determinant the coefficients of the quadric form

$$\begin{aligned} &Xx\,(\qquad\qquad hy - gz + aw) \\ &+ Yy\,(-hx \qquad + fz + bw) \\ &+ Zz\,(\; gx - fy \qquad + cw) \\ &+ W\,(a\alpha\, yz + b\beta\, zx + c\gamma\, xy \qquad); \end{aligned}$$

viz., the equation in question is

$$\begin{vmatrix} \cdot & , h(X-Y)+c\gamma W, & g(Z-X)+b\beta W, & aX \\ h(X-Y)+c\gamma W, & \cdot & , f(Y-Z)+a\alpha W, & bY \\ g(Z-X)+b\beta W, & f(Y-Z)+a\alpha W, & \cdot & , cZ \\ aX \quad , & bY \quad , & cZ \quad , & \cdot \end{vmatrix} = 0;$$

or, as it may be more simply written,

$$\sqrt{aX\,\{f(Y-Z)+a\alpha W\}} + \sqrt{bY\,\{g(Z-X)+b\beta W\}} + \sqrt{cZ\,\{h(X-Y)+c\gamma W\}} = 0.$$

This is, in fact, a surface with 16 nodes. It would appear that additional nodes correspond to the six common intersections of the quadric surfaces, or nodes of the Jacobian; and it would seem that for four quadric surfaces having in common 1, 2, 3, 4, 5, or 6 points, the corresponding symmetroid would have 11, 12, 13, 14, 15, or 16 nodes. But I reserve this for future consideration.

I take the opportunity of mentioning some results which have a connexion, although not an immediate one, with the subject of the present Memoir.

Quadric Surface through three given Lines.

105. To find the equation to the quadric surface through the three lines $(a_1, b_1, c_1, f_1, g_1, h_1)$, $(a_2, b_2, c_2, f_2, g_2, h_2)$, $(a_3, b_3, c_3, f_3, g_3, h_3)$. Take on one of the lines the points $(\alpha, \beta, \gamma, \delta)$ and $(\alpha', \beta', \gamma', \delta')$; then the equation of a quadric surface through this line will be of the form

$$\left|\begin{array}{cccccccccc} x^2 & y^2 & z^2 & w^2 & yz & zx & xy & xw & yw & zw \\ \alpha^2 & \beta^2 & \gamma^2 & \delta^2 & \beta\gamma & \gamma\alpha & \alpha\beta & \alpha\delta & \beta\delta & \gamma\delta \\ 2\alpha\alpha' & 2\beta\beta' & 2\gamma\gamma' & 2\delta\delta' & \beta\gamma'+\beta'\gamma & \gamma\alpha'+\gamma'\alpha & \alpha\beta'+\alpha'\beta & \alpha\delta'+\alpha'\delta & \beta\delta'+\beta'\delta & \gamma\delta'+\gamma'\delta \\ \alpha'^2 & \beta'^2 & \gamma'^2 & \delta'^2 & \beta'\gamma' & \gamma'\alpha' & \alpha'\beta' & \alpha'\delta' & \beta'\delta' & \gamma'\delta' \\ \vdots & & & & & & & & & \end{array}\right| = 0;$$

and if we form thus a determinant with three of its lines relating to the line 1, three of them to the line 2, and three to the line 3, we have the equation of the quadric surface through the three lines. But considering in the determinant the three lines which refer to the line 1, it is clear that the determinant is a function of the order 3 of the coordinates $(a_1, b_1, c_1, f_1, g_1, h_1)$ of the line in question; and the like as regards the other two lines respectively. Now observe that if two of the lines intersect, the problem becomes indeterminate (in fact, the plane of the intersecting lines, and any plane whatever through the third line, constitute a solution); the condition for the intersection of the lines 1 and 2 is $a_1f_2+a_2f_1+b_1g_2+b_2g_1+c_2h_1+c_1h_2=0$; hence, if this condition be satisfied, the determinant must vanish; it therefore divides by the factor a_1f_2+ &c.; but, similarly, it divides by the factors a_2f_3+ &c. and a_3f_1+ &c.; and throwing out the three factors, the result should be of the order 1, that is linear, in regard to the three sets of coordinates respectively. I have obtained this reduced result in my "Memoir on the Six Coordinates of a Line" (*Camb. Phil. Trans.*, t. XI., 1869, p. 311 [435]); viz., writing (abc) to denote the determinant $a_1(b_2c_3-b_3c_2)+$ &c., and so for the other like determinants, the result is

$$\begin{aligned} &(agh)\,x^2+(bhf)\,y^2+(cfg)\,z^2+(abc)\,w^2 \\ &+[(abg)-(cah)]\,xw+[(bfg)+(chf)]\,yz \\ &+[(bch)-(abf)]\,yw+[(cgh)+(afg)]\,zx \\ &+[(caf)-(bcg)]\,zw+[(ahf)+(bgh)]\,xy=0. \end{aligned}$$

Condition that five given lines may lie in a Cubic Surface.

106. Taking the lines to be $(a_1, b_1, c_1, f_1, g_1, h_1), \ldots (a_5, b_5, c_5, f_5, g_5, h_5)$, and $(\alpha, \beta, \gamma, \delta)$, $(\alpha', \beta', \gamma', \delta')$ the coordinates of any two points on one of the lines, the equation of a cubic surface through this line would be

$$\left|\begin{array}{llll} x^3, \ldots\ldots & x^2y, \quad\ldots\ldots & xyz, & \ldots\ldots \\ \alpha^3, & \alpha^2\beta, & \alpha\beta\gamma, & \\ 3\alpha^2\alpha', & 2\alpha\alpha'\beta+\alpha^2\beta', & \alpha'\beta\gamma+\alpha\beta'\gamma+\alpha\beta\gamma', & \\ 3\alpha\alpha'^2, & 2\alpha\alpha'\beta'+\alpha'^2\beta, & \alpha\beta'\gamma'+\alpha'\beta\gamma'+\alpha\beta'\gamma', & \\ \alpha'^3, & \alpha'^2\beta', & \alpha'\beta'\gamma', & \end{array}\right| = 0\,;$$

and hence it at once appears that, forming a determinant of 20 lines, wherein four lines relate to the line 1, four to the line 2,......, four to the line 5, and equating this to zero, we have the required condition. But the condition so obtained is of the order $(\tfrac{1}{2}4\,.\,3=)\,6$ in regard to the coordinates of each line; and, as for the quadric, it is satisfied identically if we have any such equation as $a_1f_2+\&c.=0$; it consequently contains the several factors $a_1f_2+\&c.$, which can be formed with the coordinates of any two of the five lines; and throwing out these factors, the condition should be of the order 2 in regard to the coordinates of each line. We in fact know that the required relation between the five lines is that they shall all of them be cut by a sixth line; and moreover that, writing $a_1f_2+a_2f_1+b_1g_2+b_2g_1+c_1h_2+c_2h_1=12$, &c., then that the condition for this is

$$\left|\begin{array}{ccccc} ., & 12, & 13, & 14, & 15 \\ 21, & ., & 23, & 24, & 25 \\ 31, & 32, & ., & 34, & 35 \\ 41, & 42, & 43, & ., & 45 \\ 51, & 52, & 53, & 54, & . \end{array}\right| = 0,$$

being, as it should be, of the order 2 in regard to the coordinates of each line.

Condition that 7 given lines shall lie on a Quartic Surface.

107. Taking the lines to be $(a_1, b_1, c_1, f_1, g_1, h_1), \ldots (a_7, b_7, c_7, f_7, g_7, h_7)$, then in precisely the same way we form a determinant of the order $(\tfrac{1}{2}5\,.\,4=)$ 10 in regard to the coordinates of each line; this determinant however divides out by the several factors $a_1f_2+\&c.$, which can be formed with the seven lines; or throwing these out and equating the quotient to zero, we have an equation of the order 4 in regard to the coordinates of each line. It would not be practicable to obtain the reduced equation in this manner, and I do not know how to obtain it otherwise, but the material conclusion is that the order is $=4$.

The Jacobian of 6 points.

108. Any 6 points whatever may be regarded as points on a skew cubic; and the coordinates (x, y, z, w) may be taken so that the equations of the skew cubic shall be $\left\|\begin{array}{ccc} x, & y, & z \\ y, & z, & w \end{array}\right\|=0$. This being so, the coordinates of the 6 given points may be taken to be $(1, t_1, t_1^2, t_1^3), \ldots (1, t_6, t_6^2, t_6^3)$; and the equation of the Jacobian surface of

the 6 points can then be expressed in a very simple form, putting in evidence the passage of the surface through the skew cubic; viz. writing

$$p_1 = \Sigma t_1,\quad p_2 = \Sigma t_1 t_2,\quad p_3 = \Sigma t_1 t_2 t_3,\quad p_4 = \Sigma t_1 t_2 t_3 t_4,\quad p_5 = \Sigma t_1 t_2 t_3 t_4 t_5,\quad p_6 = t_1 t_2 t_3 t_4 t_5 t_6\,;$$

moreover,

$$\square = \tfrac{1}{2}(6xyzw - 4xz^3 - 4y^3w + 3y^2z^2 - x^2w^2),$$

and therefore

$$\begin{aligned}
\delta_x \square &= -\ \ xw^2 - 2z^3 \ + 3yzw,\\
\delta_y \square &= \ \ 3yz^2 - 6y^2w + 3xzw,\\
\delta_z \square &= \ \ 3y^2z - 6xz^2 + 3xyz,\\
\delta_w \square &= -\ \ x^2w - 2y^3 \ + 3xyz\,;
\end{aligned}$$

then the equation of the Jacobian surface is

$$\begin{aligned}
&3\,(xp_3 + zp_1 - 2w)\,\delta_x\square\\
&+ (2zp_2 - wp_1)\,\delta_y\square\\
&+ (xp_5 - 2yp_4)\,\delta_z\square\\
&+ 3\,(2xp_6 - yp_5 - wp_3)\,\delta_w\square = 0.
\end{aligned}$$

There is not much difficulty in the direct investigation; but a simple verification may be obtained by showing that the surface contains upon it the 15 lines 12, 13,... 56. Write in the equation

$$(x,\ y,\ z,\ w) = (\lambda+\mu,\ \lambda s + \mu t,\ \lambda s^2 + \mu t^2,\ \lambda s^3 + \mu t^3),$$

the values $\delta_x\square$ &c. are found to contain the factor $\lambda\mu(s-t)^3$, and omitting this common factor the values are as

$$\tfrac{1}{3}(\lambda s^3 - \mu t^3),\quad -(\lambda s^2 - \mu t^2),\quad (\lambda s - \mu t),\quad -\tfrac{1}{3}(\lambda - \mu):$$

the equation thus becomes

$$\begin{aligned}
&\{\lambda(-2s^3 + s^2p_1 + p_3) + \mu(-2t^3 + t^2p_1 + p_3)\}\,(\lambda s^3 - \mu t^3)\\
&-\{\lambda(-s^3p_1 + 2s^2p_2) + \mu(-t^3p_1 + 2t^2p_2)\}\,(\lambda s^2 - \mu t^2)\\
&+\{\lambda(-2sp_4 + p_5) + \mu(-2tp_4 + p_5)\}\,(\lambda s - \mu t)\\
&-\{\lambda(-s^3p_3 - sp_5 + 2p_6) + \mu(-t^3p_3 - tp_5 + 2p_6)\}\,(\lambda - \mu) = 0,
\end{aligned}$$

viz., collecting the terms, the coefficient of $\lambda\mu$ vanishes, and the whole is

$$\begin{aligned}
&-2\lambda^2(1,\ p_1,\ p_2,\ p_3,\ p_4,\ p_5,\ p_6 \between s,\ -1)^6\\
&+2\mu^2(1,\ p_1,\ p_2,\ p_3,\ p_4,\ p_5,\ p_6 \between t,\ -1)^6 = 0\,;
\end{aligned}$$

viz., this equation is satisfied if s denote any one of the quantities $(t_1,\ t_2,\ t_3,\ t_4,\ t_5,\ t_6)$, and t any one of the same 6 quantities; that is, the equation of the surface is satisfied when $(x,\ y,\ z,\ w)$ are the coordinates of a point on the line joining any 2 of the 6 points.

Locus of the vertex of a Quadric Cone which touches each of Six given Lines.

109. Representing as before each line by means of its six coordinates, let (x, y, z, w) be the coordinates of the vertex, and (X, Y, Z, W) current coordinates. Suppose that (a, b, c, f, g, h) are the coordinates of any one of the lines, the equation of the plane through this line and the vertex is

$$\begin{aligned}&a(xW-wX)+b(yW-wY)+c(zW-wZ)\\&+f(yZ-zY)+g(zX-xZ)+h(xY-yX)=0;\end{aligned}$$

or, what is the same thing, writing for shortness

$$\begin{aligned}P&= \quad . \quad hy-gz+aw,\\Q&=-hx \quad . \quad +fz+bw,\\R&=\quad gx-fy \quad . \quad +cw,\\S&=-ax-by-cz \quad .\end{aligned}$$

the equation is

$$PX+QY+RZ+SW=0.$$

The plane in question is a tangent plane to the cone touched by the 6 lines. Now when 6 planes touch a quadric cone, their traces on any plane whatever touch a conic the intersection of the cone by that plane. Hence taking the plane $W=0$, the equation of the trace is

$$PX+QY+RZ=0,$$

and forming in like manner the equations belonging to each of the given lines, the condition that the 6 traces may touch a conic is

$$(P^2,\ Q^2,\ R^2,\ QR,\ RP,\ PQ)=0,$$

where the left-hand side represents a determinant of 6 lines, the several lines being respectively P_1^2, Q_1^2, R_1^2, Q_1R_1, R_1P_1, P_1Q_1, P_2^2, &c.... Or more simply we may denote the equation by

$$[(P,\ Q,\ R)^2]=0.$$

To ascertain the form of this, write for a moment $y=0$, $z=0$; the equation is

$$[(aw,\ -hx+bw,\ gx+cw)^2]=0,$$

or attending only to the highest and lowest powers of w, this is

$$w^{12}[(a,\ b,\ c)^2]\ldots+w^4x^8[(a,\ -h,\ g)^2]=0;$$

and it is thence easy to infer that the whole equation divides by w^4; so that, omitting this factor, the form of the equation is

$$((a,\ b,\ c,\ f,\ g,\ h)^2 \between x,\ y,\ z,\ w)^8=0;$$

viz., the equation is of the order 8 in the coordinates (x, y, z, w), and of the degree 2 in the coordinates (a, b, c, f, g, h) of each of the lines. It would not be very difficult to actually develope the equation; in fact, starting from the term $w^8[(a, b, c)^2]$ the other terms are obtained therefrom by changing a, b, c into $a+\frac{1}{w}(hy-gz)$, $b+\frac{1}{w}(-hx+fz)$, $c+\frac{1}{w}(gx-fy)$ respectively; the equation may therefore be written in the symbolic form

$$w^8 \,.\, \text{exp.}\frac{1}{w}\{(hy-gz)\,\delta_a+(-hx+fz)\,\delta_b+(gx-fy)\,\delta_c\}\,.\,[(a,\ b,\ c)^2]=0.$$

or, what is the same thing,

$$w^8 \,.\, \text{exp.}\frac{1}{w}\{x\,(g\delta_c-h\delta_b)+y\,(h\delta_a-f\delta_c)+z\,(f\delta_b-g\delta_a)\}\,.\,[(a,\ b,\ c)^2]=0,$$

where exp. θ (read exponential) denotes e^θ, and $[(a,\ b,\ c)^2]$ represents a determinant as above explained. The equation contains, it is clear, the four terms

$$x^8[(a,\ -h,\ g)^2]+y^8[(-h,\ b,\ -f)^2]+z^8[(-g,\ f,\ c)^2]+w^8[(a,\ b,\ c)^2].$$

I am not sure whether this surface of the eighth order has been anywhere considered.

446.

ON THE MECHANICAL DESCRIPTION OF A NODAL BICIRCULAR QUARTIC.

[From the *Proceedings of the London Mathematical Society*, vol. III. (1869—1871), pp. 100—106.]

THE ingenious method, devised by Mr S. Roberts (*Proceedings*, vol. II. p. 133) for the description of a nodal bicircular quartic suggests a further investigation. We have a quadrilateral $OAA'O'$, in which the adjacent sides OA, AA' are equal to each other, and the other two adjacent sides OO', $O'A'$ are also equal to each other; O, O' are fixed points; and we have thus a link AA', the extremities of which are connected

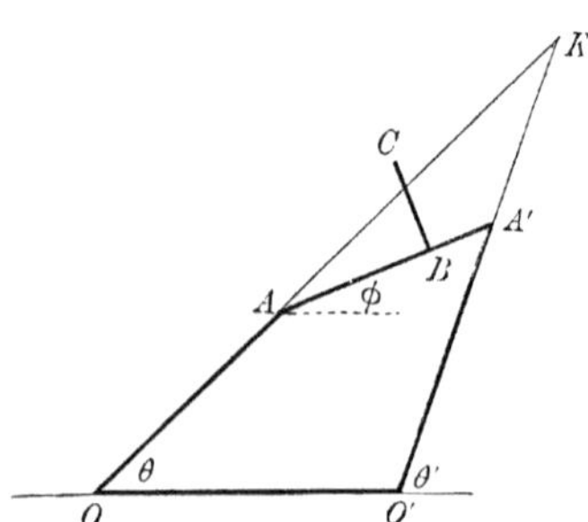

with the radii OA, $O'A'$ respectively, and consequently describe circles about the centres O, O' respectively, the radius OA of the one circle being equal to the length AA' of the link, and the radius $O'A'$ of the other circle being equal to the distance OO' of the centres. The theorem is, that any point C, rigidly connected with the link AA', describes a nodal bicircular quartic, that is, a quartic curve with three nodes (or unicursal quartic), two of the nodes being the circular points at infinity. Any such curve is the inverse of a conic, and it is also the antipode of a conic; viz., if at each point of the curve we draw a line at right angles to the radius vector from the node, these lines envelope a conic having for its pedal the curve in question. It is worth noticing at the outset that to a given position of A' there correspond two positions of A, viz., the broken line OAA' may occupy two positions situate symmetrically on the opposite sides of the line OA'. But to a given position of A, there corresponds only one position of A'; viz., the broken line $AA'O'$ is situate symmetrically with AOO' on the opposite side of the axis of symmetry $O'A$; the only other position would be A' coinciding with O, that is, AA' with AO, and the locus of C would then be a circle. If the equalities $OA = AA'$, $O'A' = OO'$ did not subsist, then to a given position of A' there would correspond two positions of A, and to a given position of A two positions of A', and the locus of C would be of a higher order than in the actual problem.

I have called AA' the link; OO' may be called the bar. OA is then the link-radius, $O'A'$ the bar-radius; moreover $AA'C$ may be called the constant triangle; and, producing OA, $O'A'$ to meet in K, then $AA'K$ may be called the variable triangle. Since at any instant the motion of A is normal to KA, and the motion of A' normal to KA', it is clear that the motion at that instant of the constant triangle is a motion of rotation about the point K.

Imagine any two positions of the link; say these are A_1A_1', and A_2A_2'. Join A_1A_2, and at its mid-point draw a perpendicular thereto; join in like manner $A_1'A_2'$, and at its mid-point draw a perpendicular thereto; and let these two perpendiculars meet in Γ; we have the two equal triangles $A_1A_1'\Gamma$, $A_2A_2'\Gamma$ (viz., $\Gamma A_1 = \Gamma A_2$, $\Gamma A_1' = \Gamma A_2'$, $A_1A_1' = A_2A_2$) with the common vertex Γ, and which may be brought to coincide with each other by a finite rotation about this point Γ. Considering any particular given position of Γ, if we take the constant triangle $AA'C$ equal to $A_1A_1'\Gamma$ or $A_2A_2'\Gamma$ (viz., $AC = A_1\Gamma$, $A'C = A_1'\Gamma$), then the constant triangle $AA'C$ will, in the course of its motion, come at two different times to coincide with the triangles $A_1A_1'\Gamma$ and $A_2A_2'\Gamma$ respectively; that is, Γ will be a node on the locus described by the point C; and moreover, if K_1 and K_2 be the corresponding positions of K, then by what precedes, the directions of the motion (or tangents at the node) will be normal to $K_1\Gamma$ and $K_2\Gamma$ respectively.

It is to be observed that the point Γ is determined by means of two arbitrary positions A_1A_1', A_2A_2' of the link; that is, the position of Γ depends upon two arbitrary parameters, and therefore Γ may be any point whatever in the plane; if, for an assumed position of Γ, the two positions A_1A_1', A_2A_2' of the link are real, then Γ is a crunode on the locus; but if imaginary, then Γ is an acnode on the locus. The transition case is when the two positions A_1A_1', A_2A_2', coincide with each other, Γ being in this case a cusp on the locus. But from the foregoing general construction for Γ, it appears that when A_1A_1' and A_2A_2' coincide, Γ is in fact the point K, the vertex of the variable triangle. I find that the locus of K is a nodal bicircular quartic, symmetrical in regard to the axis OO', and having the point O for a node; viz., when, as in the figure, AA' is $< OO'$, then the point O is an acnode, but when AA' is $> OO'$, then the point O is a crunode. The curve in question—say the "cuspidal locus"—

is a curve such that any point whatever thereof is a cusp on the curve described by some point C; it separates those points Γ, such that each of them is a crunode on the curve described by some point C, from the points Γ which are such that each of them is an acnode of the curve, described by some point C. If (as in the figure) AA' is $< OO'$, then the cuspidal curve is a closed curve (the inverse of an ellipse), the interior region being crunodal, and the exterior region acnodal. If AA' is $> OO'$, then the cuspidal curve is a figure of eight (inverse of a hyperbola), the two interior regions being crunodal, and the exterior region acnodal.

Passing now to the analytical investigation, I take the origin at O, the axis of x being in the direction from O to O', and the axis of y, at right angles thereto, upwards from O. The inclinations of OA, AA', $O'A'$ to the axis Ox, are taken to be θ, ϕ, θ' respectively. I write also $OA = AA' = a$, and $OO' = O'A' = a'$; and

$$m = \frac{a'-a}{a'+a},$$

or, what is the same thing,

$$m : 1 : 1+m : 1-m = a'-a : a'+a : 2a' : 2a;$$

and finally $AB = b$, $BC = c$.

Observing that the angle $AA'O'$ is $=\theta$, we have $\theta' = \theta + \phi$; and then, in the quadrilateral $OAA'O'$, the angles A, O' are $= \pi - \theta + \phi$, $\pi - \theta - \phi$ respectively; whence, projecting on the diagonal OA', we have

$$a \cos \tfrac{1}{2}(\theta - \phi) = a' \cos \tfrac{1}{2}(\theta + \phi),$$

which, attending to the value of m, is

$$\tan \tfrac{1}{2}\theta \tan \tfrac{1}{2}\phi = m;$$

whence, writing

$$\tan \tfrac{1}{2}\theta = u,$$

we have

$$\tan \tfrac{1}{2}\phi = \frac{m}{u},$$

and the sines and cosines of the angles θ, ϕ, θ' can be all of them expressed in terms of the single parameter u.

For the locus of C we have

$$x = a \cos\theta + b\cos\phi - c\sin\phi,$$
$$y = a\sin\theta + b\sin\phi + c\cos\phi,$$

or, instead of θ, ϕ introducing u, we have

$$\left.\begin{aligned} x &= -a\frac{u^2-1}{u^2+1} + b\frac{u^2-m^2}{u^2+m^2} - c\frac{2mu}{u^2+m^2} \\ y &= a\frac{2u}{u^2+1} + b\frac{2mu}{u^2+m^2} + c\frac{u^2-m^2}{u^2+m^2} \end{aligned}\right\},$$

which, in fact, show that the locus is a bicircular quartic. To put in evidence the third node, I assume that the values belonging thereto are $u=u_1$, $u=u_2$, and that the coordinates of the node are α, β; we have thus

$$\alpha = -a\frac{u_1^2-1}{u_1^2+1}+b\frac{u_1^2-m^2}{u_1^2+m^2}-c\frac{2mu_1}{u_1^2+m^2},\ = -a\frac{u_2^2-1}{u_2^2+1}+b\frac{u_2^2-m^2}{u_2^2+m^2}-c\frac{2mu_2}{u_2^2+m^2},$$

$$\beta = \ a\frac{2u_1}{u_1^2+1}+b\frac{2mu_1}{u_1^2+m^2}+c\frac{u_1^2-m^2}{u_1^2+m^2},\ = -a\frac{2u_2}{u_2^2+1}+b\frac{2mu_2}{u_2^2+m^2}+c\frac{u_2^2+m^2}{u_2^2+m^2}.$$

These give b, c, α, β in terms of a, m, u_1, u_2; and we may then express the values of $x-\alpha$, $y-\beta$ in terms of a, m, u_1, u_2, u. I find

$$b=\frac{a}{m}\left\{-1+\frac{m+1}{(u_1^2+1)(u_2^2+1)}[(u_1+u_2)^2+(1-m)(1-u_1u_2)]\right\},$$

$$c=\frac{a}{m}\left\{\qquad\frac{m+1}{(u_1^2+1)(u_2^2+1)}[\quad-(u_1+u_2)(m-u_1u_2)\quad]\right\};$$

and then

$$x=-a\frac{u^2-1}{u^2+1}+\frac{a}{m}\left\{-1+\frac{m+1}{(u_1^2+1)(u_2^2+1)}[(u_1+u_2)^2+(1-m)(1-u_1u_2)]\right\}\frac{u^2-m^2}{u^2+m^2}$$
$$+\frac{a}{m}\left\{\qquad\frac{m+1}{(u_1^2+1)(u_2^2+1)}[\quad(u_1+u_2)(m-u_1u_2)\quad]\right\}\frac{2mu}{u^2+m^2},$$

$$y=\ a\frac{2u}{u^2+1}+\frac{a}{m}\left\{-1+\frac{m+1}{(u_1^2+1)(n_2^2+1)}[(u_1+u_2)^2+(1-m)(1-u_1u_2)]\right\}\frac{2mu}{u^2+m^2}$$
$$+\frac{a}{m}\left\{\qquad\frac{m+1}{(u_1^2+1)(u_2^2+1)}[\quad-(u_1+u_2)(m-u_1u_2)\quad]\right\}\frac{u^2-m^2}{u^2+m^2},$$

$$\alpha=\frac{a}{m}\frac{m+1}{(u_1^2+1)(u_2^2+1)}(1-u_1u_2)(m+u_1u_2),$$

$$\beta=\frac{a}{m}\frac{m+1}{(u_1^2+1)(u_2^2+1)}(u_1+u_2)(m+u_1u_2);$$

and then

$$(x-\alpha)=-\frac{2(m+1)a}{(u_1^2+1)(u_2^2+1)}\frac{(u-u_1)(u-u_2)}{(u^2+1)(u^2+m^2)}\times[(1-u_1u_2)(u^2+m)+(1-m)(u_1+u_2)u],$$

$$(y-\beta)=-\frac{2(m+1)a}{(u_1^2+1)(u_2^2+1)}\frac{(u-u_1)(u-u_2)}{(u^2+1)(u^2+m^2)}\times[(u_1+u_2)(u^2+m)-(1-m)(1-u_1u_2)u],$$

where, of course, the factors $(u-u_1)$, $(u-u_2)$ indicate the node (α, β). We have moreover

$$(x-\alpha)^2+(y-\beta)^2=\frac{4(m+1)^2a^2}{(u_1^2+1)(u_2^2+1)}\frac{(u-u_1)^2(u-u_2)^2}{(u^2+1)(u^2+m^2)},$$

so that, writing

$$\frac{x-\alpha}{(x-\alpha)^2+(y-\beta)^2}=-\frac{1}{2(m+1)a}\frac{[(1-u_1u_2)(u^2+m)+(1-m)(u_1+u_2)u]}{(u-u_1)(u-u_2)},$$

$$\frac{y-\beta}{(x-\alpha)^2+(y-\beta)^2}=-\frac{1}{2(m+1)a}\frac{[(u_1+u_2)(u^2+m)-(1-m)(1-u_1u_2)u]}{(u-u_1)(u-u_2)},$$

we have the locus as the inverse of a conic. To exhibit it as the antipode of a conic, taking X, Y as current coordinates *measured from the node as origin*, the equation of the line through a point of the locus, at right angles to the radius vector from the node, is

$$X(x-\alpha)+Y(y-\beta)-(x-\alpha)^2-(y-\beta)^2=0;$$

or, substituting for $(x-\alpha)$, $(y-\beta)$ their values, this is

$$\begin{aligned}&X\,[(1-u_1u_2)(u^2+m)+(1-m)(u_1+u_2)\,u]\\&+Y\,[(u_1+u_2)(u^2+m)-(1-m)(1-u_1u_2)\,u]+2(m+1)\,a\,(u-u_1)(u-u_2)=0;\end{aligned}$$

and the antipodal conic is thus the envelope of the line represented by this equation. Putting for shortness

$$P=X(1-u_1u_2)+Y(u_1+u_2),\quad Q=X(u_1+u_2)-Y(1-u_1u_2),$$

the equation is

$$u^2\{P+2(m+1)\,a\}+u\,\{(1-m)\,Q-2(m+1)\,a\,(u_1+u_2)\}+mP+2(m+1)\,a\,u_1u_2=0,$$

and the equation of the conic therefore is

$$4\,\{P+2(m+1)\,a\}\,\{mP+2(m+1)\,a\,u_1u_2\}-\{(1-m)\,Q-2(m+1)\,a\,(u_1+u_2)\}^2=0,$$

so that the conic touches each of the lines $P+2(m+1)\,a=0$, $mP+2(m+1)\,a\,u_1u_2=0$ at its intersection with the line $(1-m)\,Q-2(m+1)\,au_1=0$. If these lines were constructed, one other condition would suffice for the construction of the conic.

The before-mentioned equations

$$\alpha=\frac{a}{m}\,\frac{m+1}{(u_1^2+1)(u_2^2+1)}(1-u_1u_2)(m+u_1u_2),$$

$$\beta=\frac{a}{m}\,\frac{m+1}{(u_1^2+1)(u_2^2+1)}(u_1+u_2)(m+u_1u_2),$$

give

$$\alpha^2+\beta^2=\frac{a^2}{m^2}\,\frac{(m+1)^2}{(u^2+1)(u_2^2+1)}(m+u_1u_2)^2;$$

and thence

$$\frac{\alpha}{\alpha^2+\beta^2}=\frac{m}{(m+1)\,a}\,\frac{1-u_1u_2}{m+u_1u_2},$$

$$\frac{\beta}{\alpha^2+\beta^2}=\frac{m}{(m+1)\,a}\,\frac{u_1+u_2}{m+u_1u_2},$$

which determine u_1+u_2 and u_1u_2 rationally in terms of α, β. For the cuspidal curve, writing $u_1=u_2=v$, we have

$$\frac{\alpha}{\alpha^2+\beta^2}=\frac{m}{(m+1)\,a}\,\frac{1-v^2}{m+v^2},$$

$$\frac{\beta}{\alpha^2+\beta^2}=\frac{m}{(m+1)\,a}\,\frac{2v}{m+v^2},$$

which show that the cuspidal curve is the inverse of a conic (viz., of an ellipse, if, as in the figure, m is positive). The result in the very same form would be obtained by considering the curve as the locus of the vertex K of the variable triangle.

If we imagine a plane rigidly connected with the link AA', and carried along with it, then (b, c) are the coordinates of the point C in this moveable plane; and if, as above, (α, β) are the coordinates of the node, then (b, c) and also (α, β), are given functions of (u_1, u_2). We have thus (b, c) functions of (α, β), and reciprocally (α, β) functions of (b, c); that is, we have a correspondence between the points of the fixed plane and those of the variable plane. It is worth while to investigate the nature of this correspondence, although the result does not appear to be one of any elegance.

Writing

$$A = \frac{(m+1)a}{m}\,\frac{\alpha}{\alpha^2+\beta^2},$$

$$B = \frac{(m+1)a}{m}\,\frac{\beta}{\alpha^2+\beta^2},$$

we may, in place of (α, β), consider the point in the fixed plane as given by means of the inverse coordinates (A, B). And then, if $p = u_1 + u_2$, $q = 1 - u_1u_2$, we have

$$A = \frac{q}{m+1-q}, \qquad B = \frac{q}{m+1-q};$$

whence

$$p = \frac{(m+1)B}{1+A}, \qquad q = \frac{(m+1)A}{1+A},$$

$$p^2 + q^2 = (1+u_1^2)(1+u_2^2) = \frac{(m+1)^2(A^2+B^2)}{(1+A)^2}.$$

Hence

$$\frac{m}{a}\left(b + \frac{a}{m}\right) = \frac{m+1}{p^2+q^2}[p^2 - (m-1)q],$$

$$-\frac{m}{a}c = \frac{m+1}{p^2+q^2}p(m-1+q),$$

which determine (b, c) in terms of (p, q); that is, of (A, B) or of (α, β).

In reference to some other constructions given in Mr Roberts' paper, it may be remarked that if we have a moveable plane Π_1 always coincident with a fixed plane Π, and if a condition of the motion is that a circle C_1, fixed in the plane Π_1 and carried along with it, always touches a fixed circle C in the plane Π, then this same condition may be expressed indifferently in either of the forms—(1) a circle C_1 in the plane Π_1 always passes through a fixed point of Π; (2) a point in the plane Π_1 is always situate on a fixed circle C in the plane Π. But if either of the circles C, C_1 reduce itself to a line, then we have two distinct forms of condition; viz., *first*, if a fixed line L_1 in the plane Π_1 always touches a fixed circle C in the plane Π, this is equivalent to the condition that a fixed line L_1 in the plane Π_1 always passes through a fixed point of the plane Π. And secondly, if a fixed circle C_1 in the plane Π_1 always touches a fixed line L in the plane Π, this is equivalent to the condition that a fixed point in the plane Π_1 is always situate in a fixed line L_1 in the plane Π_1. The different forms of condition therefore are:

(α) A fixed circle C_1 in the plane Π_1 always touches a fixed circle C in the plane Π (where, as above, either circle indifferently may be reduced to a point).

(β) A fixed line L_1 in the plane Π_1 always passes through a fixed point C in the plane Π.

(γ) A fixed point C_1 in the plane Π_1 is always situate in a fixed line L of the plane Π.

Hence, if the motion of the plane Π_1 satisfy any two such conditions (of the same form or of different forms, viz., the conditions may be each α, or they may be α and β, &c.), then the motion of the plane Π_1 will depend on a single variable parameter, and the question arises as to the locus described by a given point, or enveloped by a given line, of the plane Π; and again of the locus traced out, or enveloped, on the moving plane Π_1 by a given point of the plane Π. The case considered in the present paper is of course a particular case of the two conditions being each of them of the form α.

It may be remarked, that if the two conditions be each of them β, then there will be in the plane Π_1 a fixed point C_1 which describes a circle; and similarly, if the two conditions be each of them γ, then there will be in the plane Π_1 a fixed point C_1 which describes a circle([1]); that is, the combination $\beta\beta$ is a particular case of $\alpha\beta$, and the combination $\gamma\gamma$ a particular case of $\alpha\gamma$.

[1] The theorem is, that if an isosceles triangle, on the base AA' and with angle $=2\omega$ at the vertex C, slide between two lines OA, OA' inclined to each other at an angle ω, in such manner that C is the centre of the circle circumscribed about OAA', then the locus of C is a circle having O for its centre.

447.

ON THE RATIONAL TRANSFORMATION BETWEEN TWO SPACES.

[From the *Proceedings of the London Mathematical Society*, vol. III. (1869—1871), pp. 127—180. Account of the Paper given at the Meeting 11 March 1869.]

Two figures are rationally transformable each into the other (or, say, there is a rational transformation between the two figures) when to a variable point of each of them there corresponds a single variable point of the other. The figures may be either loci in a space, or *locus in quo* of any number of dimensions; or they may be such spaces themselves. Thus the figures may be each a line (or space of one dimension), each a plane (or space of two dimensions), or each a space of three dimensions; these last are the cases intended to be considered in the present Memoir, which is accordingly entitled, "On the Rational Transformation between Two Spaces." I observe in explanation (to fix the ideas, attending to the case of two planes), that any rational transformation between two planes gives rise to a rational transformation between curves in these planes respectively (one of these curves being any curve whatever): but *non constat,* and it is not in fact the case, that every rational transformation between two plane curves thus arises out of a rational transformation between two planes. The problem of the rational transformation between two planes (or generally between two spaces) is thus a distinct problem from that of the rational transformation between two plane curves (or loci in the two spaces respectively).

I consider in the Memoir, (1) the rational transformation between two lines; this is simply the homographic transformation: (2) the rational transformation between two planes; and here there is little added to what has been done by Prof. Cremona in his memoirs, "Sulle Trasformazioni Geometriche delle Figure Piane," (*Mem. di Bologna*, t. II., 1863, and t. V., 1865; see also "On the Geometrical Transformation of Plane Curves," *British Assoc. Report*, 1864): (3) the rational transformation between two spaces; in regard hereto I examine the general theory, but attend mainly to what I call the lineo-linear transformation; viz., it is assumed that the coordinates of a point in the one space, and the coordinates of the corresponding point in the other space are connected by three lineo-linear equations (that is, each equation is linear in the two sets of coordinates respectively). The lineo-linear transformation presents itself in the preceding two cases; viz., between two lines, the homographic transformation (which, as already mentioned, is the only rational transformation) is lineo-linear; and between two planes, the lineo-linear transformation is in fact the well-known inverse transformation $\left(x' : y' : z' = \frac{1}{x} : \frac{1}{y} : \frac{1}{z}\right)$. As regards two spaces, the lineo-linear transformation has not, I think, been discussed in a general manner, and it gives rise to a theory of some complexity, and of great interest.

The General Principle of the Rational Transformation between Two Spaces.

1. In all that follows, the two spaces (lines, planes, or three dimensional spaces, as the case may be), or any corresponding loci in the two spaces respectively, are referred to as the first and second figures respectively. The two figures are in general considered, not as superimposed or situate in a common space, but as existing, each independently of the other, as a separate *locus in quo* or figure in such locus. The unaccented coordinates (x, y), (x, y, z), or (x, y, z, w), as the case may be, refer throughout to a point of the first figure; the accented coordinates refer in like manner to the corresponding point of the second figure (1). Moreover $X, Y, \ldots$ are used to denote functions of the same order, say n, of the coordinates $(x, y, \ldots)$; viz., (X, Y) are each of them of the form $(*\between x, y)^n$; (X, Y, Z) each of the form $(*\between x, y, z)^n$, (X, Y, Z, W) each of the form $(*\between x, y, z, w)^n$, as the case may be; and in like manner $X', Y', \ldots$ are used to denote functions of the same order, say n', of the coordinates $(x', y', \ldots)$. This being so:

The condition of a rational transformation is that we have simultaneously

$$x' : y', \ldots = X : Y, \ldots; \qquad x : y, \ldots = X' : Y', \ldots$$

viz., these equations must be such that either set shall imply the other set.

2. If, to fix the ideas, we attend to the case of two planes, or take the sets to be

$$x' : y' : z' = X : Y : Z; \qquad x : y : z = X' : Y' : Z',$$

1 The coordinates (x, y) of a point in a line may be conceived as proportional to given multiples (α times, β times) of the distances of the point from two fixed points on the line; similarly the coordinates (x, y, z) of a point in a plane as proportional to given multiples (α times, β times, γ times) of the perpendicular distances of the point from three fixed lines in the plane; and the coordinates (x, y, z, w) of a point in a space as proportional to given multiples (α times, β times, γ times, δ times) of the perpendicular distances of the point from four fixed planes in the space. Observe that even if the coordinates (x, y) and (x', y') refer to the same line, and to the same two fixed points in this line, they are not of necessity the same coordinates; viz., the factors for x, y may be α, β, and those for x', y' may be α', β'. If these are proportional (viz., if $\alpha : \beta = \alpha' : \beta'$), then (x', y') will be the same coordinates of P' that (x, y) are of P; and in this case, but not otherwise, the equation $xy' - x'y = 0$ will imply the coincidence of the points P, P'. The like remarks apply to the coordinates (x, y, z) and (x, y, z, w).

then starting with the set $x' : y' : z' = X : Y : Z$, for any given point (x, y, z) whatever in the first figure, we have a single corresponding point (x', y', z') in the second figure; but for any given point (x', y', z') in the second figure, we have *primâ facie* a system of n^2 points in the second figure, viz., these are the common points of intersection of the curves $x' : y' : z' = X : Y : Z$ (in which equations x', y', z' are regarded as given parameters, x, y, z as current coordinates, and the equations therefore represent curves of the order n in the first figure). The curves may however have only a single variable point of intersection; viz., this will be the case if each of the curves passes through the same n^2-1 fixed points (points, that is, the positions of which are independent of x', y', z'); and in order that the curves in question may each pass through the n^2-1 points, it is necessary and sufficient that these shall be common points of intersection of the curves $X=0$, $Y=0$, $Z=0$. {Observe that the condition thus imposed upon the curves $X=0$, $Y=0$, $Z=0$ will in certain cases imply that the curves have n^2 common intersections; or, what is the same thing, that the functions X, Y, Z are connected by an identical equation, or syzygy, $\alpha X+\beta Y+\gamma Z=0$. This must not happen; for if it did, not only there will be no variable point of intersection, and the transformation will on this account fail; but there would also arise a relation $\alpha x'+\beta y'+\gamma z'=0$ between (x', y', z'), contrary to the hypothesis that (x', y', z') are the coordinates of any point whatever of the second figure. It thus becomes necessary to show that there exist curves $X=0$, $Y=0$, $Z=0$, satisfying the required condition of the n^2-1 common intersections, but without a remaining common intersection, or, what is the same thing, without any syzygy $\alpha X+\beta Y+\gamma Z=0$.}

3. The curves $x' : y' : z' = X : Y : Z$ having then a single variable point of intersection, if we take (x, y, z) to be the coordinates of this point, the ratios $x : y : z$ will be determined rationally; that is, as a consequence of the first set of equations, we obtain a second set $x : y : z = X' : Y' : Z'$, where X', Y', Z' will be rational and integral functions of the same order, say n', of the coordinates (x', y', z'); that is, we have a second set of equations, and consequently a rational transformation, as mentioned above.

4. It is easy to see that we have $n'=n$; in fact, consider in the first figure a curve $\alpha X+\beta Y+\gamma Z=0$, and an arbitrary line $ax+by+cz=0$; to these respectively correspond, in the second figure, the line $\alpha x'+\beta y'+\gamma z'=0$, and the curve $aX'+bY'+cZ'=0$; the curves are of the orders n, n' respectively, or the curve and line of the first figure intersect in n points, and the line and curve of the second figure intersect in n' points; which two systems of points must correspond point to point to each other; that is, we must have $n'=n$. It will presently appear how different the analogous relation is in the transformation between two spaces.

5. Ascending to the case of two spaces, we have here the two sets

$$x' : y' : z' : w' = X : Y : Z : W; \qquad x : y : z : w = X' : Y' : Z' : W',$$

the theory is analogous; the surfaces $x' : y' : z' : w' = X : Y : Z : W$ (surfaces of the order n in the first figure) must have a single variable point of intersection, and they must therefore have a common fixed intersection equivalent to n^3-1 points of intersection: I say *equivalent* to n^3-1 points, for this fixed intersection need not be n^3-1 points, but it may be or include a curve of intersection([1]). The surfaces $X=0$, $Y=0$, $Z=0$, $W=0$ must consequently have a common intersection equivalent to n^3-1 points; there is (as in the preceding case) a cause of failure to be guarded against, viz., the condition as to the intersection must not be such as to imply one more point of intersection, that is, to imply an identical equation or syzygy $\alpha X+\beta Y+\gamma Z+\delta W=0$ between the functions X, Y, Z, W; but it is assumed that they are not thus connected. There is, then, a single variable point of intersection of the surfaces $x' : y' : z' : w' = X : Y : Z : W$; or taking the coordinates of this point to be (x, y, z, w), we have the ratios $x : y : z : w$ rationally determined; that is, we have a second set of equations $x : y : z : w = X' : Y' : Z' : W'$, where X', Y', Z', W' are rational and integral functions of the same order, say n', in the coordinates (x', y', z', w'); viz., we have the rational transformation, as above, between the two spaces.

6. Suppose that the common intersection of the surfaces $X=0$, $Y=0$, $Z=0$, $W=0$ is or includes a curve of the order ν; and consider in the first figure the two surfaces

$$\alpha X+\beta Y+\gamma Z+\delta W=0,\quad \alpha_1 X+\beta_1 Y+\gamma_1 Z+\delta_1 W=0,$$

and the arbitrary plane $ax+by+cz+dw=0$. The two surfaces intersect in the fixed curve ν, and in a residual curve of the order $n^2-\nu$; hence the two surfaces and the plane meet in ν points on the fixed curve, and in $n^2-\nu$ other points. Corresponding to the surfaces and plane in the first figure, we have in the second figure the two planes

$$\alpha x'+\beta y'+\gamma z'+\delta w'=0,\quad \alpha_1 x'+\beta_1 y'+\gamma_1 z'+\delta_1 w'=0,$$

and the surface $aX'+bY'+cZ'+dW'=0$ of the order n': these intersect in n' points, being a system corresponding point to point with the $n^2-\nu$ points of the first figure; that is, we must have $n'=n^2-\nu$. And conversely, it follows that in the second figure the common intersection of the surfaces $X'=0$, $Y'=0$, $Z'=0$, $W'=0$ will be or include a curve of the order ν'; and that we shall have $n=n'^2-\nu$. Hence also

$$\nu-\nu'=(n-n')(n+n'+1).$$

7. The principle of the rational transformation comes out more clearly in the foregoing two cases than in the case of two lines, which from its very simplicity fails to exhibit the principle so well; and I have accordingly postponed the consideration of it: but the theory is similar to that of the foregoing cases. We must have the two sets (each a single equation) $x' : y' = X : Y$, and $x : y = X' : Y'$. The equation $x' : y' = X : Y$ must give for the ratio $x : y$ a single variable value; viz., there must be $n-1$ constant values (values, that is, independent of x', y'); this can only be the case by reason of the functions having a common factor M of the order $n-1$; but this being so, the common factor divides out, and the equation assumes the form $x' : y' = X : Y$, where X, Y are linear functions of (x, y): and we have then reciprocally

[1] The curve of intersection may consist of distinct curves, each or any of which may be a singular curve of any kind in regard to the several surfaces.

$x : y = X' : Y'$, where X', Y' are linear functions of (x', y'). Thus in the present case, instead of an infinity of transformations for different values of n, n', we have only the well-known homographic transformation wherein $n = n' = 1$.

8. In the discussion of the foregoing cases of the transformation between two planes and two spaces, it was tacitly assumed that n was greater than 1, and the transformations considered were thus different from the homographic transformation; but it is hardly necessary to remark that the homographic transformation applies to these cases also; viz., for two planes we may have $x' : y' : z' = X : Y : Z$, and $x : y : z = X' : Y' : Z'$, where (X, Y, Z), (X', Y', Z') are linear functions of the two sets of coordinates respectively; and similarly for two spaces $x' : y' : z' : w' = X : Y : Z : W$ and $x : y : z : w = X' : Y' : Z' : W'$, where (X, Y, Z, W), (X', Y', Z', W') are linear functions of the two sets of coordinates respectively. We may, if we please, separate off the homographic transformation (as between two lines, planes, and spaces respectively), and restrict the notion of the rational transformation to the higher or non-linear transformations; in this point of view, the case of two lines would not be considered at all, but the theory of the rational transformation would begin with the case of the two planes. Such severance of the theory is, however, somewhat arbitrary; and moreover the homographic transformation between two lines (being, as mentioned, the only rational transformation) is analogous not only to the homographic transformation between two planes, and to the homographic transformation between two spaces, but it is also analogous to the lineo-linear (or quadric) transformation between two planes, and to the lineo-linear (which is a cubic) transformation between two spaces.

9. For the sake of bringing out this analogy, I shall consider in some detail the homographic transformation between two lines; but as regards the homographic transformations between two planes and between two spaces respectively (although there is room for a like discussion) the theories may be considered as substantially known, and I do not propose to go into them.

The Homographic Transformation between Two Lines.

10. By what precedes, it appears that we have $x' : y' = X : Y$, where (X, Y) are linear functions of (x, y); and conversely, $x : y = X' : Y'$, where X', Y' are linear functions of (x', y'); or what is the same thing, the relation is expressed by a single equation

$$(ax + by)\, x' + (cx + dy)\, y' = 0\,;$$

or, as this may conveniently be written,

$$\begin{pmatrix} a, & b \\ c, & d \end{pmatrix}\!\!\Big(x,\ y\Big)(x',\ y') = 0,$$

or, when the expression of the actual values of the coefficients is unnecessary,

$$(*\,\big)\!\!(x,\ y)(x',\ y') = 0.$$

We thus see that the rational transformation between two lines is in fact the homographic transformation; and also that it is the lineo-linear transformation.

11. A special case is when

$$ad - bc = 0.$$

Writing here

$$\frac{c}{a} = \frac{d}{b} = m, \quad = \frac{b'}{a'},$$

the equation is

$$(ax + by)(\ \ x' + my') = 0,$$

that is

$$(ax + by)(a'x' + b'y') = 0;$$

viz., either $ax + by = 0$, without any relation between x', y'; or else $a'x' + b'y' = 0$, without any relation between x, y; that is, to the single point $ax + by = 0$ of the first figure there corresponds any point whatever of the second figure; and to the single point $a'x' + b'y' = 0$ of the second figure there corresponds any point whatever of the first figure.

12. In the general case where $ad - bc \neq 0$, we may either by a linear transformation ($ax + by$, $cx + dy$ into y, $-x$ or into x, $-y$) of the coordinates of a point of the first figure, or by a linear transformation ($ax' + cy'$, $bx' + dy'$ into y', $-x'$ or into x', $-y'$) of the coordinates of a point in the second figure (or in a variety of ways by simultaneous linear transformations of the two sets of coordinates) transform the relation indifferently into either of the forms $xy' - x'y = 0$, $xx' - yy' = 0$; the former of these, or $x' : y' = x : y$, is the most simple expression of the homographic transformation; the latter, or $x' : y' = \frac{1}{x} : \frac{1}{y}$, is its expression as an inverse transformation.

13. If, to fix the precise signification of the coordinates (x, y), we employ the distances from a fixed point O in the line; taking the distances of the two fixed points (say A, B) to be α, β, and that of the variable point P to be ρ, then we have x, y proportional to given multiples $p(\rho - \alpha)$, $q(\rho - \beta)$ of the distances from the two fixed points; or writing $\frac{p}{q} = n$, we may say that the coordinate $\frac{x}{y}$ of the point P is $= n\frac{\rho - \alpha}{\rho - \beta}$; or in particular, if $n = 1$, then the coordinate is $= \frac{\rho - \alpha}{\rho - \beta}$. If for $n\frac{\rho - \alpha}{\rho - \beta}$ we write $\frac{-\beta\rho + \lambda}{\rho - \beta}$, and then take $\beta = \infty$, we see that in a particular system of coordinates, A at O, B at ∞, the coordinate $\frac{x}{y}$ is $= \rho$. Proceeding in the same manner in regard to the coordinates (x', y'), for a particular system of coordinates, A' at O', B' at ∞, the coordinate $\frac{x'}{y'}$ of P' will be $= \rho'$. And the correspondence of the points P, P' will be given by an equation

$$a\rho\rho' + b\rho' + c\rho + d = 0.$$

14. The equation just mentioned is often convenient for obtaining a precise statement of theorems. Thus taking A, B at pleasure on the first line, A', B' the corresponding points on the second line, we have

$$\rho' = -\frac{c\rho + d}{a\rho + b},$$

and thence

$$\alpha' = -\frac{c\alpha + d}{a\alpha + b},$$

$$\beta' = -\frac{c\beta + d}{a\beta + b},$$

$$\rho' - \alpha' = \frac{(ad - bc)(\rho - \alpha)}{(a\alpha + b)(a\rho + b)},$$

$$\rho' - \beta' = \frac{(ad - bc)(\rho - \beta)}{(a\beta + b)(a\rho + b)};$$

and consequently

$$\frac{\rho' - \alpha'}{\rho' - \beta'} = \frac{a\beta + b}{a\alpha + b}\,\frac{\rho - \alpha}{\rho - \beta},$$

which is of the form

$$\frac{\rho' - \alpha'}{\rho' - \beta'} = m\,\frac{\rho - \alpha}{\rho - \beta},$$

where (the correspondence $a\rho\rho' + b\rho' + c\rho + d = 0$ being given, and also the fixed points A, B) m has a determinate value not assumable at pleasure. If, however, the fixed points A, B be not given, then we may determine a relation between them, such that m shall have any given value not being $=1$; we have in fact only to write

$$a\beta + b = m(a\alpha + b),$$

that is

$$a(\beta - m\alpha) + b(1 - m) = 0,$$

($m = 1$ would give $\alpha = \beta$ and the transformation would fail). In particular we may write $m = -1$, we have then

$$a(\alpha + \beta) + 2b = 0;$$

or the sum of the two distances OA, OB has a given value $= -\dfrac{2b}{a}$ dependent on the transformation; one of these points being assumed at pleasure, the other is known; the points A', B' are also known, and the equation of correspondence is

$$\frac{\rho' - \alpha'}{\rho' - \beta'} + \frac{\rho - \alpha}{\rho - \beta} = 0;$$

it is moreover easy to show that we have

$$a(\alpha' + \beta') + 2c = 0.$$

15. In what precedes, the two lines are considered as distinct lines, not of necessity existing in a common space. But they may be considered, not only as existing in the common space, but as superimposed the one on the other. Suppose this is so, and moreover that the fixed points A', B' coincide with A, B respectively, and that the coordinates (x, y) and (x', y') are the same coordinates; so that the equation $xy' - x'y = 0$ will imply the coincidence of the points P, P'.

16. If $ad - bc = 0$, the equation of correspondence becomes

$$(ax + by)(a'x' + b'y') = 0,$$

and as before, to a single given point $ax + by = 0$, considered as belonging to the first figure, there corresponds every point whatever of the line, or second figure: to a single given point $a'x' + b'y' = 0$ (the same as, or different from, the first point), considered as belonging to the second figure, there corresponds every point whatever of the line, or first figure.

17. Excluding the foregoing case, or assuming $ad - bc \neq 0$, there are in general on the line two points such that to each of them considered as belonging to either figure there corresponds the same point considered as belonging to the other figure, or say there are two united points: in fact, writing $x' : y' = x : y$, we find $ax^2 + (b + c)xy + dy^2 = 0$, a quadric equation for the determination of the points in question. Unless $4ad - (b + c)^2 = 0$, this equation will have two unequal roots; and taking the two points so determined for the fixed points $A = A'$, $B = B'$, the equation of correspondence will assume the form $xy' - kx'y = 0$. In this equation k cannot be $= 1$; for if it were so, the equation would be $xy' - x'y = 0$; that is, the points P, P' would be always one and the same point. The equation may, however, be $xy' + x'y = 0$; the points P, P' are then harmonics in regard to the fixed points A, B. It is to be observed, that if the equation $xy' - kx'y = 0$ be unaltered by the interchange of (x, y) and (x', y') we must have $k^2 - 1 = 0$, or since $= 1$ is excluded, we must have $k = -1$.

18. The original equation $(ax + by)x' + (cx + dy)y' = 0$ is unaltered by the interchange, only if $b - c = 0$; the equation $4ad - (b + c)^2 = 0$ becomes in this case $ad - bc = 0$, which by hypothesis is not satisfied; the two distinct points $A = A'$, $B = B'$ consequently exist. That is, if the correspondence between the two points P, P' is such that whether P be considered as belonging to the first figure or to the second figure, there corresponds to it in the other figure the same point P'—or say if the correspondence between the points P, P' is a symmetrical correspondence—then as united points in the superimposed figures we have the two distinct points A, B: and the correspondence of the points P, P' is given by the condition that these are harmonics in regard to the points A, B.

19. There is still the case to be considered where $4ad - (b + c)^2 = 0$; the equation $ax^2 + (b + c)xy + dy^2 = 0$ has here equal roots, or the two united points coincide together, or form a single point. Taking this point to be the point A, the coordinate whereof is $x : y = 0 : 1$, we must, it is clear, have $d = 0$, and therefore also $b + c = 0$: the relation between the coordinates (x, y) and (x', y') is then $axx' + b(xy' - x'y) = 0$; viz., this is the form assumed by the equation of correspondence when instead of two united points there is a double united point, and this is taken to be the fixed point A.

20. It is to be observed, that we cannot have either $b = 0$, for this would give $xx' = 0$, which belongs to the excluded case $ad - bc = 0$; nor $a = 0$, for this would give $xy' - x'y = 0$: excluding these cases, the equation is of necessity altered by the inter-

change of (x, y) and (x', y'); that is, in the case of a double united point, the transformation is essentially unsymmetrical.

By what precedes, if the other fixed point be taken to be at infinity, the coordinates $x : y$ and $x' : y'$ may be taken to be ρ, ρ' respectively; viz., ρ, ρ' will denote the distances of the points P, P' from the double united point A; and the equation of correspondence then becomes $\rho\rho' + b(\rho - \rho') = 0$; that is, $(\rho - b)(\rho' + b) + b^2 = 0$.

21. The original equation $axx' + byx' + cxy' + dyy' = 0$ can be reduced to the inverse form $xx' - yy' = 0$ only (it is clear) in the symmetrical case $b = c$; here, transforming to the united points, the equation is, by what precedes (*ante*, No. 17) $xy' + x'y = 0$. This equation can be written $(lx + my)(lx' + my') - (lx - my)(lx' - my') = 0$, where $l : m$ is arbitrary; viz., we have thus an equation of the required form.

22. In further explanation, start from the equation $a\rho\rho' + b(\rho + \rho') + d = 0$; that is, $(a\rho + b)(a\rho' + b) + ad - b^2 = 0$, or say $(\rho - \alpha)(\rho' - \alpha) - k^2 = 0$; this may be reduced to $\rho\rho' - 1 = 0$; viz., the point O from which are measured the distances ρ, ρ' is here the mid-point between the two united points A, B; and the unit of distance is $\tfrac{1}{2}AB$; the equation expresses that the points P, P', harmonics in regard to the two points A, B, are the images one of the other in regard to the circle described upon AB as diameter. Take any two corresponding points L, L'; if the distances of these be λ, λ', we have $\lambda\lambda' = 1$; and hence

$$(\rho - \lambda)(\rho' - \lambda) = 1 - \lambda(\rho + \rho') + \lambda^2 = \lambda(\lambda + \lambda' - \rho - \rho'),$$
$$(\rho - \lambda')(\rho' - \lambda') = 1 - \lambda'(\rho + \rho') + \lambda'^2 = \lambda'(\lambda + \lambda' - \rho - \rho');$$

and consequently

$$\frac{\rho - \lambda}{\rho - \lambda'} \cdot \frac{\rho' - \lambda}{\rho' - \lambda'} = \frac{\lambda}{\lambda'};$$

which, writing

$$\frac{x}{y} = \frac{k(\rho - \lambda)}{\rho - \lambda'}, \quad \frac{x'}{y'} = \frac{k(\rho' - \lambda)}{\rho' - \lambda'},$$
$$k^2 = \frac{\lambda}{\lambda'} \text{ (so that } k^2 \neq 1\text{)}; \text{ or, } k = \lambda = \frac{1}{\lambda'},$$

becomes $xx' - yy' = 0$; that is, the correspondence of the points P, P' being symmetrical, if the coordinate $\dfrac{x}{y}$ of P be taken to be a multiple of the ratio of the distances PL, PL' of P from any two corresponding points L, L' (and of course the coordinate $\dfrac{x'}{y'}$ of P' to be the same multiple of the ratio of the distances $P'L, P'L'$), the equation of correspondence is obtained in the inverse form $xx' - yy' = 0$.

The Rational Transformation between Two Planes.

23. Starting from the equations $x' : y' : z' = X : Y : Z$, where $X = 0$, $Y = 0$, $Z = 0$ are curves in the first plane, of the same order n, it has been seen that in order that we may thence have a rational transformation between the two planes, the curves

$X=0$, $Y=0$, $Z=0$ must have a common intersection of n^2-1 points, and no more; that is, they must not have a complete common intersection of n^2 points. In the case $n=2$, taking the n^2-1 points in the first plane to be any three points whatever, the condition that the curves shall be conics passing through the three points does not in anywise imply that the conics shall have a common fourth point of intersection; and we have thus a rational transformation as required; viz., the first set of equations is $x' : y' : z' = X : Y : Z$, where $X=0$, $Y=0$, $Z=0$ are conics passing through the same three points of the first plane; and as it is easy to see (but which will be subsequently shown more in detail), the second set is the similar one $x : y : z = X' : Y' : Z'$, where $X'=0$, $Y'=0$, $Z'=0$ are conics passing through the same three points in the second plane; this may be called the quadric transformation between the two planes.

24. But the like theory would not apply to the case $n=3$; if the n^2-1 points in the first plane were any eight points whatever, the cubics $X=0$, $Y=0$, $Z=0$, intersecting in these eight points, would have a common ninth point of intersection, and the transformation would fail; and so for any higher value of n, taking at pleasure any $\frac{1}{2}n(n+3)-1$ of the n^2-1 points of the first plane, the curves $X=0$, $Y=0$, $Z=0$ of the order n passing through these $\frac{1}{2}n(n+3)-1$ points, would have in common all their remaining points of intersection, and the transformation would fail. A transformation can only be obtained by taking the n^2-1 points in such wise that these can be made to be the common intersection of the curves, and at the same time that the number of conditions imposed upon each of the curves $X=0$, $Y=0$, $Z=0$ shall be at most $=\frac{1}{2}n(n+3)-1$.

25. And this requirement may be satisfied; viz., the number of conditions may be made to be $=\frac{1}{2}n(n+3)-1$, by assuming that certain of the n^2-1 points of intersections are multiple intersections of the curves. For if we have a given point which is an α-tuple point on each of the curves $X=0$, $Y=0$, $Z=0$, then this counts for α^2 points of intersection of any two of the curves, and thus for α^2 points of the n^2-1 points: but the condition that the given point shall be on any one of the curves, say the curve $X=0$, an α-tuple point, imposes on the curve, not α^2, but only $\frac{1}{2}\alpha(\alpha+1)$ conditions: and we have in this way a reduction whereby the number of conditions for passing through the n^2-1 points can be lowered from n^2-1 to the required number $\frac{1}{2}n(n+3)-1$.

26. In particular, for $n=3$, we may for the n^2-1 points of the first plane take a point as a double point on each of the cubic curves $X=0$, $Y=0$, $Z=0$ (which therefore reckons as four points), and take any other four points. Each of the curves is determined by the conditions of having a given point for double point, and of passing through the same four other given points; that is, by $3+4=7$ conditions; and the three cubic curves $X=0$, $Y=0$, $Z=0$ have for the common intersection the double point reckoning as four points, and the given other four points; that is, they have a common intersection of $4+4=8$ points; but this does not imply that they have a common ninth point of intersection; we have therefore a rational transformation as required; viz., the first set of equations is $x' : y' : z' = X : Y : Z$; where $X=0$, $Y=0$, $Z=0$ are cubics in the first plane having each of them a double point at the same given point and

also each passing through the same four given points; the second set of equations is $x : y : z = X' : Y' : Z'$, where $X'=0$, $Y'=0$, $Z'=0$ are like cubics in the second plane.

27. Generally suppose that the n^2-1 points in the first plane are made up of α_1 points, which are simple points; α_2 points, which are double points; α_3 points, which are triple points, ... α_{n-1} points, which are $(n-1)$tuple points ($\alpha_{n-1}=1$ or 0), on each of the three curves; these will represent a system of n^2-1 points if only

$$\alpha_1 + 4\alpha_2 + 9\alpha_3 \ldots \quad + \quad (n-1)^2 \alpha_{n-1} = n^2 - 1.$$

The number of conditions imposed on each of the curves $X=0$, $Y=0$, $Z=0$ will be $\alpha_1 + 3\alpha_2 + 6\alpha_3 \ldots + \tfrac{1}{2}n(n-1)\alpha_{n-1}$; for the reason presently appearing, I exclude the case of this being $< \tfrac{1}{2}n(n+3)-2$; and therefore assume it to be $= \tfrac{1}{2}n(n+3)-2$. In fact, writing

$$\alpha_1 + 3\alpha_2 + 6\alpha_3 \ldots \quad + \tfrac{1}{2}n(n-1) \; \alpha_{n-1} = \tfrac{1}{2}n(n+3) - 2,$$

this combined with the former equation gives

$$\alpha_2 + 3\alpha_3 \ldots + \tfrac{1}{2}(n-1)(n-2) \; \alpha_{n-1} = \tfrac{1}{2}(n-1)(n-2);$$

viz., the singularities are equivalent to $\tfrac{1}{2}(n-1)(n-2)$ double points, that is, to the maximum number of double points of a curve of the order n; or say each of the curves $X=0$, $Y=0$, $Z=0$ is a curve of the order n having a deficiency $=0$; that is, it is a unicursal curve of the order n. Hence also, taking (a, b, c) any constant factors whatever, the curve $aX+bY+cZ=0$ is unicursal.

28. It is important to remark that the conclusion follows directly from the general notion of the rational transformation; in fact, the equation $aX+bY+cZ=0$ is satisfied if $x : y : z = X' : Y' : Z'$; $ax'+by'+cz'=0$. The last of these equations determines the ratios $x' : y' : z'$ in terms of a single parameter (e.g. the ratio $x' : y'$), and we have then $x : y : z$ expressed as rational functions of this parameter; that is, the curve is unicursal.

29. Suppose for a moment that it was possible to have

$$\alpha_1 + 3\alpha_2 + 6\alpha_3 \ldots + \quad \tfrac{1}{2}n(n-1)\alpha_{n-1} < \tfrac{1}{2}n(n+3) - 2.$$

Combining in the same way with the first equation, it would follow that

$$\alpha_2 + 3\alpha_3 \ldots + \tfrac{1}{2}(n-1)(n-2)\alpha_{n-1} > \tfrac{1}{2}(n-1)(n-2),$$

which would imply that the curves $X=0$, $Y=0$, $Z=0$ break up each of them into inferior curves: but more than this, the coefficients a, b, c being arbitrary, it would imply that the curve $aX+bY+cZ=0$ breaks up into inferior curves; this can only be the case if X, Y, Z have a common factor, say M; that is, if $X, Y, Z = MX_1, MY_1, MZ_1$; but we could then omit the common factor, and in place of $x' : y' : z' = X : Y : Z$ write $x' : y' : z' = X_1 : Y_1 : Z_1$, where $X_1=0$, $Y_1=0$, $Z_1=0$, are proper curves, not breaking up; the above supposition may therefore be excluded from consideration.

30. We have thus a transformation in which the first set of equations is $x' : y' : z' = X : Y : Z$, where $X = 0$, $Y = 0$, $Z = 0$ are curves in the first plane, of the same order n, having in common $\alpha_1, \alpha_2 \dots \alpha_{n-1}$ points which are simple points, double points,... $(n-1)$tuple points respectively on each of the curves; these numbers satisfy the conditions

$$\alpha_1 + 4\alpha_2 + 9\alpha_3 \dots + \qquad (n-1)^2 \alpha_{n-1} = n^2 - 1,$$

$$\alpha_1 + 3\alpha_2 + 6\alpha_3 \dots + \qquad \tfrac{1}{2}n(n-1)\ \alpha_{n-1} = \tfrac{1}{2}(n^2 + 3n) - 2;$$

conditions which give, as above,

$$\alpha_2 + 3\alpha_3 \dots + \tfrac{1}{2}(n-1)(n-2)\,\alpha_{n-1} = \tfrac{1}{2}(n-1)(n-2),$$

and also

$$\alpha_1 + 2\alpha_2 + 3\alpha_3 \dots + \qquad (n-1)\,\alpha_{n-1} = 3n - 3;$$

so that the relations between $\alpha_1, \alpha_2 \dots \alpha_{n-1}$ are given by any two of these four equations.

31. The second set of equations then is $x : y : z = X' : Y' : Z'$, where $X' = 0$, $Y' = 0$, $Z' = 0$ are curves in the second plane, of the same order n; and it is clear that these must be curves such as those in the first plane; viz., they must have in common $\alpha_1', \alpha_2', \dots \alpha'_{n-1}$ points, which are simple points, double points, ... $(n-1)$tuple points respectively on each of the curves, the relations between these numbers being expressed by any two of the four equations

$$\alpha_1' + 4\alpha_2' + 9\alpha_3' \dots + \qquad (n-1)^2\, \alpha'_{n-1} = n^2 - 1,$$

$$\alpha_1' + 3\alpha_2' + 6\alpha_3' \dots + \qquad \tfrac{1}{2}n(n-1)\ \alpha'_{n-1} = \tfrac{1}{2}n(n+3) - 2,$$

$$\alpha'_2 + 3\alpha_3' \dots + \tfrac{1}{2}(n-1)(n-2)\ \alpha'_{n-1} = \tfrac{1}{2}(n-1)(n-2),$$

$$\alpha_1' + 2\alpha_2' + 3\alpha_3' \dots + \qquad (n-1)\ \alpha'_{n-1} = 3n - 3.$$

32. To any line $ax' + by' + cz' = 0$ in the second plane there corresponds in the first plane a curve $aX + bY + cZ$ of the order n; and to any line $a'x + b'y + c'z = 0$ in the first plane there corresponds in the second plane a curve $a'X' + b'Y' + c'Z' = 0$ of the same order n; the curves $aX + bY + cZ = 0$ in the first plane are, it is clear, a system, and the entire system, of curves each satisfying the conditions which have been stated in regard to the individual curves $X = 0$, $Y = 0$, $Z = 0$, and being as already mentioned unicursal; and similarly the curves $a'X' + b'Y' + c'Z' = 0$ in the second plane are a system, and the entire system, of curves each satisfying the conditions which have been stated in regard to the individual curves $X' = 0$, $Y' = 0$, $Z' = 0$; and being also unicursal. We may say that to the lines of the second plane there corresponds in the first plane the *réseau* of curves $aX + bY + cZ = 0$; and to the lines of the first plane there corresponds in the second plane the *réseau* of curves $a'X' + b'Y' + c'Z' = 0$; these *réseau* being systems satisfying respectively the conditions just referred to.

33. We have next to enquire what are the curves in the second plane which correspond to the $\alpha_1 + \alpha_2 \dots + \alpha_{n-1}$ points of the first plane. I remark that the $\alpha_1 + \alpha_2 \dots + \alpha_{n-1}$ points are termed by Cremona the *principal points* of the first plane, and the corresponding curves the *principal curves* of the second plane. But it will be

more convenient to say that the $\alpha_1 + \alpha_2 \ldots + \alpha_{n-1}$ points are the *principal system* of the first plane, and the corresponding curves the *principal counter-system* of the second plane. And of course the $\alpha_1' + \alpha_2' \ldots + \alpha'_{n-1}$ points will be the principal system of the second plane, and the corresponding curves the principal counter-system of the first plane.

34. The Jacobian (curve) of the curves $X=0$, $Y=0$, $Z=0$ is, of course, the Jacobian of any three curves $aX+bY+cZ=0$ of the first plane, or it may be called the Jacobian of the reseau of the first plane; and similarly, the Jacobian of the curves $X'=0$, $Y'=0$, $Z'=0$ is the Jacobian of the reseau of the second plane.

35. I say that to each point α_1 of the first figure there corresponds in the second figure a line; to each point α_2 a conic; to each point α_3 a nodal cubic; ... and generally, to each point α_r a unicursal r-thic curve; the entire system of the curves corresponding to the $\alpha_1 + \alpha_2 + \alpha_3 \ldots + \alpha_{n-1}$ points, that is, the principal counter-system of the second plane, is thus made up of α_1 lines, α_2 conics, α_3 nodal cubics, ... α_r unicursal r-thics, ... α_{n-1} unicursal $(n-1)$thics. It is thus a curve of the aggregate order $\alpha_1 + 2\alpha_2 + 3\alpha_3 \ldots + (n-1)\,\alpha_{n-1}$, $=3n-3$; and it is in fact the Jacobian of the reseau of the second plane; as such, it passes through each point α_1' two times, each point α_2' five times, ... each point α_r' $3r-1$ times, ... each point α'_{n-1} $3n-4$ times.

36. The reciprocal theorem is of course true. The Jacobian of the reseau of the first plane is thus made up of α_1' lines, α_2' conics, α_3' nodal cubics, ... α_r' unicursal r-thics, ... α'_{n-1} unicursal $(n-1)$thics. Calculating the Jacobian of the reseau of the first plane, we have thus the numbers $\alpha_1', \alpha_2', \ldots \alpha'_{n-1}$, which determine the nature of the principal system of the second plane.

37. I indicate as follows the analytical proof of the theorem that to a principal point α_r of the first plane there corresponds in the second plane a unicursal r-thic. Consider the simplest case, $r=1$; if in the equations $x' : y' : z' = X : Y : Z$ the coordinates (x, y, z) are considered as belonging to a point α_1, these values give identically $X=0$, $Y=0$, $Z=0$; hence for the consecutive point $x+\delta x$, $y+\delta y$, $z+\delta z$, if (A, B, C) denote the derived functions of X, (A_1, B_1, C_1) those of Y, (A_2, B_2, C_2) those of Z, we have

$$\begin{aligned} x' : y' : z' = \;& A\,\delta x + B\,\delta y + C\,\delta z \\ :\;& A_1\delta x + B_1\delta y + C_1\delta z \\ :\;& A_2\delta x + B_2\delta y + C_2\delta z. \end{aligned}$$

We have $\begin{vmatrix} A, & B, & C \\ A_1, & B_1, & C_1 \\ A_2, & B_2, & C_2 \end{vmatrix} = 0$, for the determinant is the value, at the point α_1 in question, of the Jacobian of the reseau of the first plane; and the Jacobian curve passing through α_1 (in fact, having there a double point), the value is $=0$.

38. Hence x', y', z', considered as corresponding to a point indefinitely near to α_1, are connected by a linear equation. Corresponding to α_1 we have in the second figure

a line. But it is to be observed, further, that the equation of the line is that obtained by writing in the foregoing equations, say $\delta z = 0$, and eliminating the remaining quantities δx, δy; or, what is the same thing, we may consider the equation of the line as given by the equations

$$\begin{aligned} x' : y' : z' = \ & A\,\delta x + B\,\delta y \\ & : A_1\delta x + B_1\delta y \\ & : A_2\delta x + B_2\delta y, \end{aligned}$$

where δx, δy are indeterminate parameters to be eliminated.

39. In the case of a point α_r we have in like manner

$$\begin{aligned} x' : y' : z' = \ & (a\ , \ldots \between \delta x,\ \delta y,\ \delta z)^r \\ & : (a_1, \ldots \between \delta x,\ \delta y,\ \delta z)^r \\ & : (a_2, \ldots \between \delta x,\ \delta y,\ \delta z)^r, \end{aligned}$$

where $(a, \ldots)$, $(a_1, \ldots)$, $(a_2, \ldots)$ are the r-th derived functions of X, Y, Z respectively. In virtue of the relation of the point α_r to the curves $X = 0$, $Y = 0$, $Z = 0$, the coefficients will be such as to allow of the simultaneous elimination from these equations of the three quantities δx, δy, δz. The result of the elimination will be the same as if, writing say $\delta z = 0$, we eliminate δx, δy; or, what is the same thing, the relation of x', y', z' may be regarded as given by the equations

$$\begin{aligned} x' : y' : z' = \ & (a\ , \ldots \between \delta x,\ \delta y)^r \\ & : (a_1, \ldots \between \delta x,\ \delta y)^r \\ & : (a_2, \ldots \between \delta x,\ \delta y)^r, \end{aligned}$$

where δx, δy are indeterminate parameters. These equations obviously express that the point (x', y', z') is situate on a unicursal curve of the order r.

40. It is further to be shown that the r-thic curve thus corresponding to α_r is part of the Jacobian of the reseau of the second plane. The Jacobian in question is the locus of the new double point of those curves of the reseau which have a new double point; that is, a double point not included among the $\alpha_2' + \alpha_3' \ldots + \alpha'_{n-1}$ singular points of the principal system of the second plane. But a curve of the reseau being unicursal, can only acquire a new double point by breaking up into inferior curves. Consider, in the first figure, any line through α_r, the corresponding curve in the second figure is made up of the unicursal r-thic curve, which corresponds to the point α_r, together with a residual curve variable with the line through α_r; this is a unicursal curve of the order $n - r$. The aggregate curve of the order $r + (n - r)$ has singular points equivalent to $\tfrac{1}{2}(n-1)(n-2)+1$ double points[1]; that is, the singularities are those belonging to the principal system of the second plane, together with a new double

[1] In general, if $r + r' = n$, and the curves r, r' are each unicursal, then the aggregate singularity arising from the singularities of the two curves and from their intersections, is equivalent to $\tfrac{1}{2}(r-1)(r-2) + \tfrac{1}{2}(r'-1)(r'-2) + rr'$, that is, to $\tfrac{1}{2}(r+r'-1)(r+r'-2)+1$, or $\tfrac{1}{2}(n-1)(n-2)+1$ double points.

point constituted by an intersection of the curves r, $n-r$. {Observe that the two curves have only this single intersection; viz., the remaining $r(n-r)-1$ intersections are at points $\alpha_2'+\alpha_3' \ldots +\alpha'_{n-1}$ of the principal system of the second plane.} We have thus, in the second plane, a series of curves, each of them having a new double point; viz., these are the several curves which correspond to the lines through α_r in the first figure. Each of the curves is a fixed curve r together with a variable curve $n-r$. The new double point is an intersection of the two curves; that is, it is a variable point on the curve r. The locus of the new double point is thus the curve r; therefore the curve r is part of the Jacobian of the reseau of the second plane. Since each point α_r gives a curve r, the curves in question form an aggregate curve of the order $\alpha_1+2\alpha_2 \ldots +(n-1)\alpha_{n-1}$, $=3n-3$; viz., this is the order of the Jacobian; or, as stated, the curves r (that is, the principal counter-system of the second plane) constitute the Jacobian of the reseau of this plane.

41. The numerical systems $(\alpha_1, \alpha_2 \ldots \alpha_{n-1})$ and $(\alpha_1', \alpha_2' \ldots \alpha'_{n-1})$ are each of them a solution of the same two indeterminate equations

$$\Sigma r^2\alpha_r = n^2-1, \quad \Sigma r\alpha_r = 3n-3,$$

but not every solution of these equations is admissible; for instance, if $r > \frac{1}{2}n$, then α_r is $=0$ or 1, for $\alpha_r = 2$ would imply a curve of the order n with two r-tuple points, and the line joining these would meet the curve in more than r points; similarly, $r > \frac{2}{5}n$, α_r is $=4$ at most, for $\alpha_r = 5$ would imply a curve of the order n with five r-tuple points, and the conic through these would meet the curve in more than $2n$ points; and there are of course other like restrictions. The different admissible systems up to $n=10$ are tabulated in Cremona's Memoir; and he has also given systems belonging to certain specified forms of n: these results are as follows:

n	2	3	4		5			6				7				
a_1	3	4	6	3	8	3	0	10	1	4	3	12	2	0	5	3
a_2		1	0	3	0	3	6	0	4	1	4	0	3	3	0	5
a_3			1	0	0	1	0	0	2	3	0	0	2	4	3	0
a_4					1	0	0	0	0	0	1	0	1	0	1	0
a_5								1	0	0	0	0	0	0	0	1
a_6												1	0	0	0	0
a_7																
a_8																
a_9																
	1	1	1	2	1	2	3	1	2	3	4	1	2	3	4	5

n	8									9									
a_1	14	3	1	0	3	6	0	2	3	16	4	2	0	3	7	1	3	0	1
a_2	0	2	3	0	6	0	5	0	3	0	1	3	4	7	0	4	0	3	1
a_3	0	3	2	7	0	1	2	5	0	0	4	1	0	0	0	3	4	3	3
a_4	0	0	2	0	0	3	0	1	3	0	0	2	4	0	3	0	1	1	3
a_5	0	1	0	0	0	0	1	0	0	0	0	1	0	0	1	0	1	1	0
a_6	0	0	0	0	1	0	0	0	0	0	1	0	0	0	0	1	0	0	0
a_7	1	0	0	0	0	0	0	0	0	0	0	0	0	0	0	0	0	0	0
a_8										1	0	0	0	0	0	0	0	0	0
a_9																			
	1	2	3	4	5	6	7	8	9([1])	1	2	3	4	5	6	7	8	9	10

n	10																
a_1	18	5	1	0	0	3	8	2	4	1	2	3	3	3	0	0	1
a_2	0	0	4	2	0	8	0	3	0	3	1	3	3	0	6	1	0
a_3	0	5	0	2	7	0	0	4	3	2	3	0	1	0	0	5	2
a_4	0	0	2	3	0	0	1	0	2	2	1	3	0	6	0	0	5
a_5	0	0	2	1	0	0	3	0	0	0	2	0	3	0	3	2	0
a_6	0	0	0	0	1	0	0	0	1	1	0	1	0	0	0	0	0
a_7	0	1	0	0	0	0	0	1	0	0	0	0	0	0	0	0	0
a_8	0	0	0	0	0	1	0	0	0	0	0	0	0	0	0	0	0
a_9	1	0	0	0	0	0	0	0	0	0	0	0	0	0	0	0	0
	1	2	3	4	5	6	7	8	9	10	11	12	13	14	15	16	17

[1] Omitted by Cremona.

$n = p$

a_1	$= 2p - 2$
a_{p-1}	$= 1$

$n = 2p$

a_1	$= 3$	$2p-2$
a_2	$= 2p-2$	0
a_{p-1}	$= 0$	1
a_p	$= 0$	3
a_{2p-2}	$= 1$	0

$n = 2p + 1$

a_1	$= 3$	$2p-1$
a_2	$= 2p-1$	0
a_p	$= 0$	3
a_{p+1}	$= 0$	1
a_{2p-1}	$= 1$	0

$n = 3p$

a_1	$= 1$	$2p-3$
a_2	$= 4$	0
a_3	$= 2p-3$	0
a_p	$= 0$	4
a_{p+1}	$= 0$	1
a_{2p-1}	$= 0$	1
a_{3p-3}	$= 1$	0

a_1	$= 4$	$2p-2$
a_2	$= 1$	0
a_3	$= 2p-2$	0
a_{p-1}	$= 0$	1
a_p	$= 0$	4
a_{2p}	$= 0$	1
a_{3p-3}	$= 1$	0

$n = 3p + 1$

a_1	$= 2$	$2p-2$
a_2	$= 3$	0
a_3	$= 2p-2$	0
a_p	$= 0$	3
a_{p+1}	$= 0$	2
a_{2p}	$= 0$	1
a_{3p-2}	$= 1$	0

a_1	$= 5$	$2p-1$
a_3	$= 2p-1$	0
a_p	$= 0$	5
a_{2p+1}	$= 0$	1
a_{3p-2}	$= 1$	0

$n = 3p + 2$

a_1	$= 3$	$2p-1$
a_2	$= 2$	0
a_3	$= 2p-1$	0
a_p	$= 0$	2
a_{p+1}	$= 0$	3
a_{2p+1}	$= 0$	1
a_{3p-1}	$= 1$	0

a_1	$= 0$	$2p-2$
a_2	$= 5$	0
a_3	$= 2p-2$	0
a_{p+1}	$= 0$	5
a_{2p}	$= 0$	1
a_{3p-1}	$= 1$	0

$n = 4p$

a_1	$= 1$	$2p-3$
a_2	$= 3$	0
a_3	$= 2$	0
a_4	$= 2p-3$	0
a_p	$= 0$	3
a_{p+1}	$= 0$	1
a_{2p-1}	$= 0$	1
a_{2p}	$= 0$	2
a_{4p-1}	$= 1$	0

a_1	$= 2$	$2p-4$
a_3	$= 5$	0
a_4	$= 2p-4$	0
a_p	$= 0$	5
a_{p+1}	$= 0$	2
a_{3p-1}	$= 0$	1
a_{4p-4}	$= 1$	0

a_1	$= 3$	$2p-2$
a_2	$= 3$	0
a_4	$= 2p-2$	0
a_{p-1}	$= 0$	1
a_p	$= 0$	3
a_{2p}	$= 0$	3
a_{4p-4}	$= 1$	0

a_1	$= 6$	$2p-2$
a_3	$= 1$	0
a_4	$= 2p-2$	0
a_{p-1}	$= 0$	1
a_p	$= 0$	6
a_{3p}	$= 0$	1
a_{4p-4}	$= 1$	0

$n = 4p + 1$

α_1	$= 0$	$2p-3$
α_2	$= 3$	0
α_3	$= 3$	0
α_4	$= 2p-3$	0
α_p	$= 0$	1
α_{p+1}	$= 0$	3
α_{2p}	$= 0$	3
α_{4p-3}	$= 1$	0

α_1	$= 2$	$2p-2$
α_2	$= 3$	0
α_3	$= 1$	0
α_4	$= 2p-2$	0
α_p	$= 0$	3
α_{p+1}	$= 0$	1
α_{2p}	$= 0$	2
α_{2p+1}	$= 0$	1
α_{4p-3}	$= 1$	0

α_1	$= 3$	$2p-3$
α_3	$= 4$	0
α_4	$= 2p-3$	0
α_p	$= 0$	4
α_{p+1}	$= 0$	3
α_{3p}	$= 0$	1
α_{4p-3}	$= 1$	0

α_1	$= 7$	$2p-1$
α_4	$= 2p-1$	0
α_p	$= 0$	7
α_{3p+1}	$= 0$	1
α_{4p-3}	$= 1$	0

$n = 4p + 2$

α_1	$= 0$	$2p-4$
α_3	$= 7$	0
α_4	$= 2p-4$	0
α_{p+1}	$= 0$	7
α_{3p}	$= 0$	1
α_{4p-2}	$= 1$	0

α_1	$= 1$	$2p-2$
α_2	$= 3$	0
α_3	$= 2$	0
α_4	$= 2p-2$	0
α_p	$= 0$	1
α_{p+1}	$= 0$	3
α_{2p}	$= 0$	1
α_{2p+1}	$= 0$	2
α_{4p-2}	$= 1$	0

α_1	$= 3$	$2p-1$
α_2	$= 3$	0
α_4	$= 2p-1$	0
α_p	$= 0$	3
α_{p+1}	$= 0$	1
α_{2p+1}	$= 0$	3
α_{4p-2}	$= 1$	0

α_1	$= 4$	$2p-2$
α_3	$= 3$	0
α_4	$= 2p-2$	0
α_p	$= 0$	3
α_{p+1}	$= 0$	4
α_{3p}	$= 0$	1
α_{4p-2}	$= 1$	0

$n = 4p + 3$

α_1	$= 0$	$2p-2$
α_2	$= 3$	0
α_3	$= 3$	0
α_4	$= 2p-2$	0
α_{p+1}	$= 0$	3
α_{p+2}	$= 0$	1
α_{2p+1}	$= 0$	3
α_{4p-1}	$= 1$	0

α_1	$= 9$	$2p-3$
α_3	$= 6$	0
α_4	$= 2p-3$	0
α_{p+1}	$= 0$	6
α_{p+2}	$= 0$	1
α_{3p+1}	$= 0$	1
α_{4p-1}	$= 1$	0

α_1	$= 2$	$2p-1$
α_2	$= 3$	0
α_3	$= 1$	0
α_4	$= 2p-1$	0
α_p	$= 0$	1
α_{p+1}	$= 0$	3
α_{2p+1}	$= 0$	2
α_{2p+2}	$= 0$	1
α_{4p-1}	$= 1$	0

α_1	$= 5$	$2p-1$
α_3	$= 2$	0
α_4	$= 2p-1$	0
α_p	$= 0$	2
α_{p+1}	$= 0$	5
α_{3p+2}	$= 0$	1
α_{4p-1}	$= 1$	0

42. The system $(\alpha_1, \alpha_2 \dots \alpha_{n-1})$ geometrically determines completely the system $(\alpha_1', \alpha_2' \dots \alpha'_{n-1})$; it ought therefore to determine it arithmetically; that is, given the one series of numbers, we ought to be able to determine, or at least to select from the table, the other series of numbers. Cremona has shown that the two series *consist of the same numbers in the same or a different order.* By examination of the tables, it appears that there are certain columns which are single (that is, no other column contains in a different order the same numbers), others that occur in pairs, the two columns of a pair containing the same numbers in a different order. Where the column is single, it is clear that this must give as well the values of $(\alpha_1', \alpha_2' \dots \alpha'_{n-1})$ as of $(\alpha_1, \alpha_2 \dots \alpha_{n-1})$. Where there is a pair of columns, as far as Cremona has examined, if the one column is taken to be $(\alpha_1, \alpha_2 \dots \alpha_{n-1})$ the other column is $(\alpha_1', \alpha_2' \dots \alpha'_{n-1})$; it appears, however, not to be shown that this is universally the case; viz., it is not shown but that the two columns, instead of being reckoned as a pair, might be reckoned as two separate columns, each by itself representing the values as well of $(\alpha_1, \alpha_2 \dots \alpha_{n-1})$ as of $(\alpha_1', \alpha_2' \dots \alpha'_{n-1})$; neither is it shown that there are not, in any case, more than two columns having the same numbers in different orders. It seems, however, natural to suppose that the law, as exhibited in the tables, holds good generally; viz., that the tables contain only single columns, each giving the values as well of $(\alpha_1, \alpha_2 \dots \alpha_{n-1})$ as of $(\alpha_1', \alpha_2' \dots \alpha'_{n-1})$; or else pairs of columns, one giving the values of $(\alpha_1, \alpha_2 \dots \alpha_{n-1})$, and the other those of $(\alpha_1', \alpha_2' \dots \alpha'_{n-1})$; or, say, that the partitions are either *sibi-reciprocal*, or else *conjugate*.

43. Assuming that the two systems $(\alpha_1, \alpha_2 \dots \alpha_{n-1})$ and $(\alpha_1', \alpha_2' \dots \alpha'_{n-1})$ are each known, there is still a question of grouping to be settled; viz., the Jacobian of the first plane consists of α_1' lines, α_2' conics, ... α'_{n-1} unicursal $(n-1)$-thics; each line, each conic, &c., passes a certain number of times through *certain* of the points $\alpha_1, \alpha_2 \dots \alpha_{n-1}$: but through which of them? For instance, each of the α_1' lines will pass through two of the points $\alpha_1, \alpha_2, \dots \alpha_{n-1}$: will these be points α_1, or points α_2, &c., or a point α_1 and a point α_2, &c.? The mere symmetry of the different groups of points determines certain conditions of the solution[1]; for instance, if any particular one of the α_1' lines passes through two points α_r, then each of the α_1' lines must pass through two points α_r; and since the points α_r are symmetrical, we must in this way use all the pairs of points α_r; that is, if $\alpha_1' = \frac{1}{2}\alpha_r(\alpha_r+1)$, but not otherwise, *it may be* that each of the α_1' lines passes through two of the points α_r. In the case of an equality $\alpha_r = \alpha_s$ we could not hereby decide whether the line passed through two points α_r or through two points α_s. So, again, if any one of the α_1' lines pass through a point α_r and a point α_s, then each of the α_1' lines must do so likewise, and we must hereby exhaust the combinations of a point α_r with a point α_s; viz., the assumed relation can only hold good if $\alpha_1' = \alpha_r\alpha_s$. Similarly, each of the α_2' conics will pass through five of the points $\alpha_1, \alpha_2 \dots \alpha_{n-1}$; each of the α_3' nodal cubics will pass twice through one (have a double point there) and through six others of the points $\alpha_1, \alpha_2 \dots \alpha_{n-1}$: which are the points so passed through? I do not know how a general solution is to be obtained, but most of the cases within the limits of the foregoing table have

[1] It is by such considerations of symmetry that Cremona has demonstrated the before mentioned theorem of the identity of the numbers $(\alpha_1, \alpha_2 \dots \alpha_{n-1})$ and $(\alpha_1', \alpha_2' \dots \alpha'_{n-1})$.

been investigated by Cremona. The results may conveniently be stated in a tabular form; the tables exhibit in the outside upper line the values of $\alpha_1, \alpha_2 \ldots \alpha_{n-1}$, and in the outside left-hand line the values of $\alpha_1', \alpha_2' \ldots \alpha'_{n-1}$, and they are to be read as follows: Each of the $\left\{\begin{matrix} \alpha_1' \text{ lines} \\ \alpha_2' \text{ conics} \\ \text{\&c.} \end{matrix}\right\}$ passes ( ) times through ( ) of the points $\alpha_1, \alpha_2 \ldots \alpha_{n-1}$ respectively; the numbers in the table being those of the points passed through, and the indices in the table (index $=1$ when no index is expressed) showing the number of times of passage, that is, showing whether the point is a simple, double, triple, &c., point on the curve referred to.

44. Thus (in the tables which follow) the last of the tables $n=6$ gives the constitution of the Jacobian of the first plane, where the principal system is (3, 4, 0, 1, 0); and it is to be read:

Each of the 4 lines passes through 1 of the points α_1 and through the point α_4;
The 1 conic " " 4 of the points α_2 and through the point α_4;
Each of the 3 cubics " " 2 of the points α_1, 4 of the points α_2, and twice through the point α_4 (that is, α_4 is a double point on each cubic).

It is hardly necessary to remark that the tables are sibi-reciprocal, or else conjugate, as appears by the outer lines of each table.

TABLE $n=2$.

	$\alpha_1 = 3$
$\alpha_1' = 3$	2

[was originally printed, 3.]

TABLE $n=3$.

	$\alpha_1 = 4$	$\alpha_2 = 1$
$\alpha_1' = 4$	1	1
$\alpha_2' = 1$	4	1

TABLES $n=4$.

	$\alpha_1 = 6$	$\alpha_2 = 0$	$\alpha_3 = 1$
$\alpha_1' = 6$	1		1
$\alpha_2' = 0$			
$\alpha_3' = 1$	6		1^2

	$\alpha_1 = 3$	$\alpha_2 = 3$	$\alpha_3 = 0$
3		2	
3	2	3	
0			

TABLES $n=5$.

	$a_1 = 8$	$a_2 = 0$	$a_3 = 0$	$a_4 = 1$
$a_1' = 8$	1			1
$a_2' = 0$				
$a_3' = 0$				
$a_4' = 1$	8			1^3

	$a_1 = 3$	$a_2 = 3$	$a_3 = 1$	$a_4 = 0$
3		1	1	
3	1	3	1	
1	3	3	1^2	
0				

	$a_1 = 0$	$a_2 = 6$	$a_3 = 0$	$a_4 = 0$
0				
6		5		
0				
0				

TABLES $n=6$.

	$a_1 = 10$	$a_2 = 0$	$a_3 = 0$	$a_4 = 0$	$a_5 = 1$
$a_1' = 10$	1				1
$a_2' = 0$					
$a_3' = 0$					
$a_4' = 0$					
$a_5' = 1$	10				1^4

	$a_1 = 1$	$a_2 = 4$	$a_3 = 2$	$a_4 = 0$	$a_5 = 0$
1			2		
4		3	2		
2	1	4	1^2,1*		
0					
0					

	$a_1 = 4$	$a_2 = 1$	$a_3 = 3$	$a_4 = 0$	$a_5 = 0$
$a_1' = 3$			2		
$a_2' = 4$	1	1	3		
$a_3' = 0$					
$a_4' = 1$	4	1	3^2		
$a_5' = 0$					

	$a_1 = 3$	$a_2 = 4$	$a_3 = 0$	$a_4 = 1$	$a_5 = 0$
4	1			1	
1		4		1	
3	2	4		1^2	
0					
0					

* Read, "Each of the two cubics passes through the point a_1, the four points a_2, and, (1^2, 1), twice through one and once through the other of the points a_2."

45. It is to be remarked upon the tables—first, as regards the lines: if we add the numbers in each line, reckoning m^p as mp, (that is, each multiple point, according to the number of branches through it,) the sums for the successive lines are 2, 5, 8, 11, 14, &c.; that is, each line passes through 2 points, each conic through 5 points, each cubic through 8 points, each quartic through 11 points, &c. But if we add the numbers reckoning m^p as $m\,.\,\frac{1}{2}p\,(p+1)$, (that is, each multiple point according to its effect in the determination of the curve,) then the sums are 2, 5, 9, 14, 20, &c., that is, all the curves are completely determined, viz., the line by 2 conditions, the conic by 5 conditions, the cubic by 9 conditions, &c. Secondly, as regards the columns, if for any column, reckoning m^p as mp, we multiply each number by the corresponding outside left-hand number, add, and divide the sum by the outside number at the head of the column, the successive results are 2, 5, 8, 11, 14, &c.; this merely expresses the known circumstance that the Jacobian passes $3r-1$ times through each point α_r.

46. The analogous tables showing the passage of the Jacobian through the principal system, in the solutions belonging to certain special forms of n, are

TABLE $n=p$.

	$\alpha_1 = 2p-2$	$\alpha_{p-1} = 1$
$\alpha_1' = 2p-2$	1	1
$\alpha'_{p-1} = 1$	$2p-2$	1^{p-2}

TABLES $n=2p$.

	$\alpha_1 = 3$	$\alpha_2 = 2p-2$	$\alpha_{p-1} = 0$	$\alpha_p = 0$	$\alpha_{2p-2} = 1$
$\alpha_1' = 2p-2$	1				1
$\alpha_2' = 0$					
$\alpha'_{p-1} = 1$		$2p-2$			1^{p-2}
$\alpha_p' = 3$	2	$2p-2$			1^{p-1}
$\alpha'_{2p-2} = 0$					

	$\alpha_1 = 2p-2$	$\alpha_2 = 0$	$\alpha_{p-1} = 1$	$\alpha_p = 3$	$\alpha_{2p-2} = 0$
$\alpha_1' = 3$				2	
$\alpha_2' = 2p-2$	1		1	3	
$\alpha'_{p-1} = 0$					
$\alpha_p' = 0$					
$\alpha'_{2p-2} = 1$	$2p-2$		1^{p-2}	3^{p-1}	

TABLES $n = 2p+1$.

	$a_1 = 3$	$a_2 = 2p-1$	$a_p = 0$	$a_{p+1} = 0$	$a_{2p-1} = 1$
$a_1' = 2p-1$		1			1
$a_2' = 0$					
$a_p' = 3$	1	$2p-1$			1^{p-1}
$a'_{p+1} = 1$	3	$2p-1$			1^p
$a'_{2p-1} = 0$					

	$a_1 = 2p-1$	$a_2 = 0$	$a_p = 3$	$a_{p+1} = 1$	$a_{2p-1} = 0$
$a_1' = 3$	1			1	
$a_2' = 2p-1$	1		3	1	
$a_p' = 0$					
$a'_{p+1} = 0$					
$a'_{2p-1} = 1$	$2p-1$		3^{p-1}	1^p	

TABLES $n = 3p$.

	$a_1 = 1$	$a_2 = 4$	$a_3 = 2p-3$	$a_p = 0$	$a_{p+1} = 0$	$a_{2p-1} = 0$	$a_{3p-3} = 1$
$a_1' = 2p-3$			1				1
$a_2' = 0$							
$a_3' = 0$							
$a_p' = 4$		3	$2p-3$				1^{p-1}
$a'_{p+1} = 1$	1	4	$2p-3$				1^p
$a'_{2p-1} = 1$	1	4	$(2p-3)^2$				1^{2p-3}
$a'_{3p-3} = 0$							

	$a_1 = 2p-3$	$a_2 = 0$	$a_3 = 0$	$a_p = 4$	$a_{p+1} = 1$	$a_{2p-1} = 1$	$a_{3p-3} = 0$
$a_1' = 1$					1	1	
$a_2' = 4$				3	1	1	
$a_3' = 2p-3$	1			4	1	1^2	
$a_p' = 0$							
$a'_{p+1} = 0$							
$a'_{2p-1} = 0$							
$a'_{3p-3} = 1$	$2p-3$			4^{p-1}	1^p	1^{2p-3}	

	$a_1 = 4$	$a_2 = 1$	$a_3 = 2p-2$	$a_{p-1} = 0$	$a_p = 0$	$a_{2p} = 0$	$a_{3p-3} = 1$
$a_1' = 2p-2$			1				1
$a_2' = 0$							
$a_3' = 0$							
$a'_{p-1} = 1$			$2p-2$				1^{p-2}
$a_p' = 4$	1	1	$2p-2$				1^{p-1}
$a'_{2p} = 1$	4	1	$(2p-2)^2$				1^{2p-2}
$a'_{3p-3} = 0$							

	$\alpha_1 = 2p-2$	$\alpha_2 = 0$	$\alpha_3 = 0$	$\alpha_{p-1} = 1$	$\alpha_p = 4$	$\alpha_{2p} = 1$	$\alpha_{3p-3} = 0$
$\alpha_1' = 4$					1	1	
$\alpha_2' = 1$					4	1	
$\alpha_3' = 2p-2$	1			1	4	1^2	
$\alpha'_{p-1} = 0$							
$\alpha_p' = 0$							
$\alpha'_{2p} = 0$							
$\alpha'_{3p-3} = 1$	$2p-2$			1^{p-2}	4^{p-1}	1^{2p-2}	

47. The before mentioned theorem, that $(\alpha_1, \alpha_2 \ldots \alpha_{n-1})$ and $(\alpha_1', \alpha_2' \ldots \alpha'_{n-1})$ are the same series of numbers, of course implies $\Sigma\alpha_r = \Sigma\alpha_r'$; this relation Cremona demonstrates independently, by consideration of the pencil of curves $(aX + bY + cZ) + \theta(a_1X + b_1Y + c_1Z) = 0$, ($\theta$ a variable parameter,) which corresponds in the first plane to the pencil of lines $(ax' + by' + cz') + \theta\,(a_1x' + b_1y' + c_1z') = 0$, which pass through a fixed point $(ax' + by' + cz' = 0,$ $a_1x' + b_1y' + c_1z' = 0)$ in the second figure. In general, in the pencil $U + \theta V = 0$ (U, V given functions of the order n) there are $3\,(n-1)^2$ values of θ, each giving a nodal curve. But in the present case each of the curves $U = 0$, $V = 0$ has multiple points at the principal points α_r of the first plane: the question is to obtain the number of values which give a curve having one new double point; and this is found to be $= 3\,(n-1)^2 - \Sigma\,(r-1)\,(3r+1)\,\alpha_r$. We have $\Sigma r^2\alpha_r = n^2 - 1$, $\Sigma r\alpha_r = 3n - 3$; or, substituting, the value of θ is $= \Sigma\alpha_r$. But the curves which have an additional double point are those which correspond to the lines which in the second figure pass through one of the principal points α_r'; viz., these are the lines drawn from the point $(ax' + by' + cz' = 0,$ $a_1x' + b_1y' + c_1z' = 0)$ to the several principal points α_r'; and the number of them is $= \Sigma\alpha_r'$. We have thus the required relation $\Sigma\alpha_r = \Sigma\alpha_r'$.

The Quadric Transformation between Two Planes.

48. This is of course given by what precedes. The principal system in each plane is a set of three points; and the Jacobian of the same plane is the set of three lines joining each pair of points; that is, the three lines of either plane are the principal counter-system of the other plane. But to give the analytical investigation

directly: taking the coordinates (x, y, z) to refer to the principal system of the first plane (viz., taking the three points to be the vertices of the triangle formed by the lines $x=0$, $y=0$, $z=0$), then $X=0$, $Y=0$, $Z=0$ being conics through the three points, the functions X, Y, Z will be each of them of the form $fyz+gzx+hxy$; x', y', z' being proportional to three such functions, there will be linear functions of x', y', z' proportional to yz, zx, xy; or taking these linear functions of the original (x', y', z') for the coordinates (x', y', z') of a point in the second plane, the formulæ of transformation will be $x' : y' : z' = yz : zx : xy$, and we have then conversely $x : y : z = y'z' : z'x' : x'y'$; that is, the formulæ for the transformation in question are

$$x' : y' : z' = yz : zx : xy, \text{ and } x : y : z = y'z' : z'x' : x'y'.$$

We at once verify *à posteriori* that the Jacobian in the first plane is $xyz=0$, and that in the second plane is $x'y'z'=0$.

The equations may be written

$$x' : y' : z' = \frac{1}{x} : \frac{1}{y} : \frac{1}{z}, \text{ and } x : y : z = \frac{1}{x'} : \frac{1}{y'} : \frac{1}{z'},$$

or, if we please, $xx'=yy'=zz'$; the transformation is thus given as an inverse transformation.

{49. With respect to the metrical interpretation and actual construction of the transformation, it is to be observed that if x, y, z be taken to be proportional (not to given multiples of the perpendicular distances, but) to the perpendicular distances of P from the sides of the triangle in the first plane, and similarly x', y', z' to be proportional to the perpendicular distances of P' from the sides of the triangle in the second plane, then in general the equations of transformation must be written, not as above, but in the form $\frac{xx'}{f}=\frac{yy'}{g}=\frac{zz'}{h}$, involving arbitrary multipliers $f : g : h$. We may imagine in the second plane a point P'' determined by coordinates (x'', y'', z''),—the same coordinates as (x', y', z'), that is, proportional to the perpendicular distances of P'' from the sides of the triangle in the second plane,—which point P'' corresponds homographically to P in such wise that $\frac{x}{f} : \frac{y}{g} : \frac{z}{h} = x'' : y'' : z''$. We have then, in the second plane, the two points P', P'' corresponding to each other in such wise that $x'x''=y'y''=z'z''$; and either of these points being given, the other can at once be constructed; viz., it is obvious that, joining P', P'' with any vertex, say A', of the triangle $A'B'C'$, the lines $A'P'$, $A'P''$ are equally inclined to the bisectors of the angle A'; and consequently, P' being given, we have the three lines $A'P''$, $B'P''$, $C'P''$ intersecting in a common point P'', which is therefore determined by means of any two of these lines. We have thus a geometrical construction of the transformation between P and P'.}

50. The analysis assumes that the principal points A, B, C of the first figure are three distinct points; but they may two of them, or all three, coincide. In the first case, say if B, C coincide, the line BC is still to be regarded as having a definite direction; and taking $x=0$ for this line, $y=0$ for the line joining A with

(BC), and $z=0$ an arbitrary line through A, the functions X, Y, Z will be each of them of the form $by^2+2gzx+2hxy$; and replacing, as before, the original x', y', z' by linear functions of these quantities, these linear functions being taken for the coordinates (x', y', z'), we may write $x' : y' : z' = y^2 : xy : xz$. Forming the converse system, the equations for the transformation are

$$x' : y' : z' = y^2 : xy : xz, \text{ and } x : y : z = y'^2 : x'y' : x'z',$$

so that the points A', B', C' in the second plane are related as the points in the first plane; viz., B', C' coincide, the line $B'C'$ being definite.

It is easy to verify that the Jacobian in the first plane is $xy^2=0$, and the Jacobian in the second plane is $x'y'^2=0$.

51. Secondly, if A, B, C all coincide, these being however consecutive points on a curve of finite curvature, or say on a conic; then, taking $x=0$ for the tangent at (ABC), $z=0$ for any other tangent, and $y=0$ for the chord of contact, the functions X, Y, Z will be of the form $ax^2+b(y^2-zx)+2hxy$; whence we may write $x' : y' : z' = x^2 : xy : y^2-xz$. Forming the converse equations, the equations of transformation are

$$x' : y' : z' = x^2 : xy : y^2-xz, \text{ and } x : y : z = x'^2 : x'y' : y'^2-x'z';$$

so that the points A', B', C' in the second plane are related as those of the first plane; viz., they are the consecutive points of a curve of continuous curvature.

We may verify that the Jacobian of the first plane is $x^3=0$, and the Jacobian of the second plane $x'^3=0$.

The Lineo-linear Transformation between Two Planes.

52. We have two equations of the form

$$(a, \ldots \between x, y, z \between x', y', z') = 0,$$
$$(a_1, \ldots \between x, y, z \between x', y', z') = 0;$$

writing these in the form

$$P\,x' + Q\,y' + R\,z' = 0,$$
$$P_1x' + Q_1y' + R_1z' = 0,$$

where (P, Q, R, P_1, Q_1, R_1) are linear functions of (x, y, z), we have

$$x', y', z' \text{ proportional to } \left\| \begin{matrix} P, & Q, & R \\ P_1, & Q_1, & R_1 \end{matrix} \right\|,$$

that is to $X : Y : Z$, where $X=0$, $Y=0$, $Z=0$ are conics each passing through the same three points in the first plane.

And conversely, writing the equations in the form

$$P'\,x + Q'\,y + R'\,z = 0,$$
$$P_1'x + Q_1'y + R_1'z = 0,$$

where $(P', Q', R', P_1', Q_1', R_1')$ are linear functions of (x', y', z'), we have

$$x,\ y,\ z \text{ proportional to } \left\| \begin{matrix} P', & Q', & R' \\ P_1', & Q_1', & R_1' \end{matrix} \right\|,$$

that is to X', Y', Z', where $X'=0$, $Y'=0$, $Z'=0$ are conics each passing through the same three points in the second plane.

53. The lineo-linear transformation is thus the same thing as the quadric transformation. It is, moreover, clear that the equations must, by linear transformations on the two sets of variables respectively, and by linear combination of the two equations, be reducible into forms giving the before-mentioned values of $x : y : z$ and $x' : y' : z'$ respectively. Thus, in the general case, where in each plane the three points are distinct points, the lineo-linear equations will be reducible to

$$xx' - yy' = 0, \quad xx' - zz' = 0\,;$$

in the case where B, C in the first plane, and B', C' in the second plane respectively coincide, the forms will be

$$xx' - yy' = 0, \quad yz' - y'z = 0\,;$$

and in the case where A, B, C in the first plane, and A', B', C' in the second plane respectively coincide, the forms will be

$$xy' - yx' = 0, \quad xz' - yy' + zx' = 0.$$

The determination of the actual formulæ for these reductions would, it is probable, give rise to investigations of considerable interest.

The General Rational Transformation between Two Planes (resumed).

54. Consider, as above, the first plane or figure with a principal system $(\alpha_1, \alpha_2 \ldots \alpha_{n-1})$, and the second plane or figure with a principal system $(\alpha_1', \alpha_2' \ldots \alpha'_{n-1})$. To a line in the second plane there corresponds in the first plane a curve of the order n passing 1 time through each of the points α_1, 2 times through each of the points α_2, 3 times through each of the points α_3, &c.; or, as we may write this:

First figure.

Points	α_1	α_2	$\alpha_3 \ldots$	α_{n-1}	
	1	2	3	$n-1$	curve order n
	1	2	3	$n-1$	
	1	⋮	3	$n-1$	
	⋮		⋮	⋮	

Second figure.

Points	α_1'	α_2'	$\alpha_3' \ldots$	α'_{n-1}	
	0	0	0	$n-1$	curve order 1
	0	0	0	$n-1$	
	⋮	⋮	0	⋮	
			⋮		

viz., the 1's denote the number of times which the curve of the order n passes through the several points α_1 respectively; the 2's the number of times which the curve passes through the several points α_2 respectively; and so on.

55. We may, in the second figure, in the place of a line consider a curve of the order k'. If the equation hereof is $(*\between x', y', z')^{k'}=0$, then the corresponding curve in the first figure is $(*\between X, Y, Z)^{k'}=0$; viz., this is a curve of the order $k=nk'$. If, however, the curve in the second figure passes once or more times through all or any of the points $\alpha_1', \alpha_2', \dots \alpha'_{n-1}$, then there will be a depression in the order of the corresponding curve in the first figure; and, moreover, this curve will pass a certain number of times through all or some of the points $\alpha_1, \alpha_2, \alpha_3, \dots \alpha_{n-1}$. The diagram of the correspondence will be:

First figure.					Second figure.				
$\alpha_1,$	$\alpha_2,$	$\alpha_3, \dots$	α_{n-1}		$\alpha_1',$	$\alpha_2',$	$\alpha_3', \dots$	α'_{n-1}	
a_1	a_2	a_3	a_{n-1}		a_1'	a_2'	a_3'	a'_{n-1}	
b_1	b_2	b_3	b_{n-1}	curve order k	b_1'	b_2'	b_3'	b'_{n-1}	curve order k'
c_1	c_2	$\vdots$	$\vdots$		c_1'	$\vdots$	$\vdots$	c'_{n-1}	
$\vdots$	$\vdots$				$\vdots$			$\vdots$	

where $a_1, b_1, c_1 \dots$ denote the number of times that the curve of the order k passes through the several points α_1 respectively, (viz., the number of the letters $a_1, b_1, c_1 \dots$ is $=\alpha_1$, any or all of them being zeros,) $a_2, b_2, c_2 \dots$ the number of times that the curve passes through the several points α_2 respectively, (viz., the number of the letters $a_2, b_2, c_2 \dots$ is $=\alpha_2$, any or all of them being zeros,) and so on; and the like for the curve in the second figure.

56. By what precedes, it is easy to see that, if the curve k' passes through a point α_1', then the curve k throws off a line, and the depression of order is $=1$; so, if the curve passes 2 times, 3 times, ... or a_1' times through the point in question, then the curve throws off the line repeated 2 times, 3 times, ... a_1' times, or the depression of order is $=2, 3, \dots$ or a_1'; and the like for each of the points α_1'; so that, writing for shortness $a_1'+b_1'+c_1'+\dots=\Sigma a_1'$, the depression of order on account of the passages through the several points α_1 is $=\Sigma a_1'$. Similarly, for each time of passage through a point α_2', there is thrown off a conic; or if $a_2'+b_2'+\dots=\Sigma a_2'$, then the depression of order is $=2\Sigma a_2'$, and so on; and the like for the figure in the other plane; and we thus arrive at the equations

$$k = k'n - \Sigma(a_1' + 2a_2' + 3a_3' \dots + \overline{n-1}\, a'_{n-1})$$

$$k' = kn - \Sigma(a_1 + 2a_2 + 3a_3 \dots + \overline{n-1}\, a_{n-1}).$$

57. The simplest case is when the curve k' does not pass through any of the points $\alpha_1', \alpha_2', \dots \alpha'_{n-1}$. We have then

$$a_1'=b_1'=c_1' \dots =0,\ a_2'=b_2' \dots =0,\ \dots\dots\ a'_{n-1}=b'_{n-1} \dots =0;$$

consequently $k=k'n$. And, moreover, it is easy to see that

$$a_1=b_1 \dots =k',\quad a_2=b_2 \dots =2k',\ \dots\dots\ a_{n-1}=b_{n-1} \dots =(n-1)\,k';$$

so that the correspondence is:

First Figure.					Second Figure.				
$\alpha_1,$	$\alpha_2,$	$\alpha_3, \ldots$	α_{n-1}		$\alpha_1',$	$\alpha_2',$	$\alpha_3', \ldots$	α'_{n-1}	
k'	$2k'$	$3k'$	$(n-1)k'$	curve order $k=nk'$	0	0	0	0	curve order k'.
k'	$2k'$	$3k'$	$(n-1)k'$		0	0	0	0	
⋮	⋮	⋮	⋮		⋮	⋮	⋮	⋮	

We have

$$\Sigma a_1 = k'\alpha_1, \quad \Sigma a_2 = 2k'\alpha_2, \;\ldots\ldots\; \Sigma a_1(a_1-1) = k'(k'-1)\alpha_1, \;\&c.,$$

and the formulæ for k, k' become

$$k = k'n, \quad k' = kn - k'\{\alpha_1 + 4\alpha_2 \ldots + (n-1)^2 \alpha_{n-1}\};$$

viz., the second equation is here $k' = kn - k'(n^2-1)$; that is, $k'n^2 = kn$, agreeing, as it should do, with the first equation.

58. Moreover, the deficiency-relation is

$$\tfrac{1}{2}(k-1)(k-2) - \Sigma\tfrac{1}{2}\{k'(k'-1) + 2k'(2k'-1) \ldots + \overline{n-1}\,k'(\overline{n-1}\,k'-1)\} = \tfrac{1}{2}(k'-1)(k'-2);$$

or, what is the same thing, this is

$$\begin{aligned}(nk'-1)(nk'-2) - (k'-1)(k'-2) = \;& k'^2\{\alpha_1 + 4\alpha_2 \ldots + (n-1)^2\alpha_{n-1}\}. \\ & - k'\{\alpha_1 + 2\alpha_2 \ldots + (n-1)\,\alpha_{n-1}\}.\end{aligned}$$

The right-hand side is

$$k'^2(n^2-1) - k'(3n-3) = (n-1)\{(\overline{n+1}\,k'^2 - 3k')\},$$

and we have thus the identical equation

$$(nk'-1)(nk'-2) - (k'-1)(k'-2) = (n-1)k'\{(n+1)k' - 3\}.$$

59. It should be possible, when the nature of the correspondence between the two planes is completely given, to express each of the numbers $a_1, b_1, c_1, \ldots a_{n-1}, b_{n-1}, \ldots$ in terms of $k', a_1', b_1', c_1', \ldots a'_{n-1}, b'_{n-1}, \ldots$; and reciprocally each of the numbers $a_1', b_1', c_1', \ldots a'_{n-1}, b'_{n-1}, \ldots$ in terms of $k, a_1, b_1, c_1, \ldots a_{n-1}, b_{n-1}, \ldots$; thus completing a system of relations between the two sets

$$(k, a_1, b_1, \ldots a_{n-1}, b_{n-1}, \ldots), \qquad (k', a_1', b_1', \ldots a'_{n-1}, b'_{n-1}, \ldots);$$

but even if the theory was known, there would be considerable difficulty in forming a proper algorithm for the expression of these relations.

60. The two curves must have each of them the same deficiency. It is to be noticed, that if the curve in the first plane passes any number of times through a point P, which is not one of the points $\alpha_1, \alpha_2, \alpha_3, \ldots$ or α_{n-1}, then the corresponding curve in the second plane will pass the same number of times through the corresponding point P', which point will not be one of the points $\alpha_1', \alpha_2', \ldots \alpha'_{n-1}$. The

points P, P' will therefore contribute equal values to the deficiencies of the two curves respectively; so that, in equating the two deficiencies, we may disregard P, P', and attend only to the points $\alpha_1, \alpha_2, \ldots \alpha_{n-1}$ of the first plane, and $\alpha_1', \alpha_2', \ldots \alpha'_{n-1}$ of the second plane. The required relation thus is

$$\tfrac{1}{2}(k-1)(k-2) - \Sigma\, \tfrac{1}{2}\{a_1(a_1-1) + a_2(a_2-1) \ldots + a_{n-1}(a_{n-1}-1)\}$$
$$= \tfrac{1}{2}(k'-1)(k'-2) - \Sigma\, \tfrac{1}{2}\{a_1'(a_1'-1) + a_2'(a_2'-1) \ldots + a'_{n-1}(a'_{n-1}-1)\}.$$

61. In the case of the quadric transformation $n=2$, we have in the first plane the three points α_1, say these are A, B, C; and in the second plane the three points α_1', say these are A', B', C'. And if in the first plane the curve of the order k passes a, b, c times through the three points respectively, and in the second plane the corresponding curve of the order k' passes a', b', c' times through the three points respectively, then it is easy to obtain

$$\begin{aligned} k' &= 2k - a - b - c, \\ a' &= k - b - c, \\ b' &= k - c - a, \\ c' &= k - a - b. \end{aligned} \qquad \Bigg\| \qquad \begin{aligned} k &= 2k' - a' - b' - c', \\ a &= k' - b' - c', \\ b &= k' - c' - a', \\ c &= k' - a - b'. \end{aligned}$$

The Quadric Transformation any number of times repeated.

62. We may successively repeat the quadric transformation according to the type:

First Fig.	Second Fig.	Third Fig.	Fourth Fig.
A, B, C	A', B', C'		
	D', E', F'	D'', E'', F''	
		G'', H'', I''	G''', H''', I'''

viz., in the transformation between the first and second figures, the principal systems are ABC and $A'B'C'$ respectively; in that between the second and third figures, they are $D'E'F'$ and $D''E''F''$ respectively; in that between the third and fourth figures, they are $G''H''I''$ and $G'''H'''I'''$; and so on. And it is then easy to see that between the first and any subsequent figure we have a rational transformation of the order 2 for the second figure, 4 for the third figure, 8 for the fourth figure, and so on.

63. But to further explain the relation, we may complete the diagram, by taking, in the transformation between the second and third figures, A'', B'', C'' to correspond to A', B', C'; similarly, in that between the third and fourth, A''', B''', C''' to correspond to A'', B'', C''; and D''', E''', F''' to correspond to D'', E'', F''. And so in the transformation between the second and third figure, we may make G', H', I'

correspond to G'', H'', I'', and between the first and second figures make D, E, F correspond to D', E', F', and G, H, I to G', H', I', the diagram being thus:

First Fig.	Second Fig.	Third Fig.	Fourth Fig.
A, B, C	A', B', C'	A'', B'', C''	A''', B''', C'''
D, E, F	D', E', F'	D'', E'', F''	D''', E''', F'''
G, H, I	G', H', I'	G'', H'', I''	G''', H''', I'''

Observe that in the principal systems (for instance, A, B, C and A', B', C') the points A, B, C correspond, not to the points A', B', C', but to the lines $B'C'$, $C'A'$, $A'B'$ respectively; and so in the other case.

64. Consider now a line in the first figure: there corresponds hereto in the second figure a conic through the points A', B', C'; and to this conic there corresponds in the third figure a quartic curve passing through each of the points A'', B'', C'' once, and through each of the points D'', E'', F'' twice. And conversely, to a line in the third figure corresponds in the second figure a conic through the points D', E', F'; and hereto in the first figure a quartic through the points D, E, F, once and through the points A, B, C twice; that is, we have between the first and third figures a quartic transformation wherein $\alpha_1 = \alpha_2 = 3$ and $\alpha_1' = \alpha_2' = 3$, or say a quartic transformation $3_1 3_2$ and $3_1 3_2$. In like manner, passing to the fourth figure, to a line in the first figure corresponds in the fourth figure an octic curve passing through A''', B''', C''' once, through D''', E''', F''' twice, and through G''', H''', I''' four times; and conversely, to a line in the fourth figure there corresponds in the first figure an octic curve passing through the points G, H, I once, the points D, E, F twice, and the points A, B, C four times; that is, between the first and fourth figures we have an octic transformation, wherein $\alpha_1 = \alpha_2 = \alpha_4 = 3$, $\alpha_1' = \alpha_2' = \alpha_4' = 3$, or say a transformation, order 8, of the form $3_1 3_2 3_4$ and $3_1 3_2 3_4$. And so between the first and fifth figures there is a transformation, order 16, of the form $3_1 3_2 3_4 3_8$ and $3_1 3_2 3_4 3_8$.

65. It is, moreover, easy to find the Jacobians or counter-systems in the several transformations respectively. Thus, in the transformation between the first and second figures, in the second figure the Jacobian consists of 3 lines such as $B'C'$ (viz., these are, of course, the lines $B'C'$, $C'A'$, $A'B'$). Hence, in the transformation between the first and third figures, the Jacobian in the third figure consists of

$$\begin{array}{ll} 3 \text{ conics} & B''C''\,(D''E''F''), \\ 3 \text{ lines} & D''E''; \end{array}$$

viz., one of the conics is that through the five points B'', C'', D'', E'', F'', one of the lines that through the two points D'', E''. And so in the fourth figure, the Jacobian consists of

$$\begin{array}{llll} 3 \text{ quartics} & B'''C''' & (D'''E'''F''')_1 & (G'''H'''I''')_2, \\ 3 \text{ conics} & & D'''E''' & (G'''H'''I''')_1, \\ 3 \text{ lines} & & & G'''H'''; \end{array}$$

viz., one of the quartics passes through B''', C'''; through D''', E''', F''' each once; and through G''', H''', I''' each twice. And so in the fifth figure the Jacobian consists of

$$\begin{array}{llll}
3 \text{ octics} & B''''C''''\,(D''''E''''F'''')_1 & (G''''H''''I'''')_2 & (J''''K''''L'''')_4, \\
3 \text{ quartics} & D''''E'''' & (G''''H''''I'''')_1 & (J''''K''''L'''')_2, \\
3 \text{ conics} & & G''''H'''' & (J''''K''''L'''')_1, \\
3 \text{ lines} & & & J''''K'''',
\end{array}$$

and so on.

66. The conditions are in each case sufficient for the determination of the curve. This depends on the numerical relation

$$4+3\,\{1\,.\,2+2\,.\,3+4\,.\,5+8\,.\,9\,\ldots+2^\theta\,(2^\theta+1)\}=2^{\theta+1}\,(2^{\theta+1}+3).$$

The term in { } is

$$\begin{array}{r}1+4+16\,\ldots+2^{2\theta}\\ +\,1+2+\;\;4\,\ldots+2^\theta,\end{array}$$

that is

$$\frac{2^{2\theta+2}-1}{2^2-1}+\frac{2^{\theta+1}-1}{2-1},$$

which is

$$\begin{aligned}&=\tfrac{1}{3}\,[2^{2\theta+2}-1+3\,(2^{\theta+1}-1)],\\ &=\tfrac{1}{3}\,[2^{2\theta+2}+3\,.\,2^{\theta+1}-4];\end{aligned}$$

and the relation is thus identically true.

67. Conversely, in the transformation between the first figure and the several other figures respectively, the Jacobian of the first figure is

3 lines	AB; and so	for order 2, between first and second figures;
3 conics	$DE\,(ABC)_1$	for order 4, between first and third figures;
3 lines	AB	
3 quartics	$GH\,(DEF)_1\,(ABC)_2$	for order 8, between first and fourth figures;
3 conics	$DE\quad (ABC)_1$	
3 lines	AB	
3 octics	$JK\,(GHI)_1\,(DEF)_2\,(ABC)_4$	for order 16, between first and fifth figures;
3 quartics	$GH\quad (DEF)_1\,(ABC)_2$	
3 conics	$DE\quad (ABC)_1$	
3 lines	AB	

and so on.

Special Cases—Reduction of the General Rational Transformation to a Series of Quadric Transformations.

68. It was remarked by Mr Clifford that any Cremona-transformation whatever may be obtained by this method of repeated quadric transformations, if only the principal systems, instead of being completely arbitrary, are properly related to each other. To take the simplest instance; suppose that we have

First figure.	Second figure.	Third figure.
$A,\ B,\ C$	$A',\ B',\ C'$	$B'',\ C''$
	$\Vert$	
$E,\ F$	$D',\ E',\ F'$	$D'',\ E'',\ F''$

viz., in the transformation between the first and second figures, we have the principal systems ABC and $A'B'C'$ (arbitrary as before); but in the transformation between the second and third figures, the principal systems are $D'E'F'$ and $D''E''F''$, where D', instead of being arbitrary, coincides with A'. And we then have B'', C'' in the third figure corresponding to B', C' in the second figure, and E, F in the first figure corresponding to E', F' in the second figure. This being so, to a line in the first figure corresponds in the second figure a conic through A', B', C'. But $A'=D'$; viz., this conic passes through a point D' of the principal system of the second figure, in regard to the transformation between the second and third figures. That is, (k, a, b, c referring to the second figure, and k', a', b', c' to the third figure, $k=2$, $a=1$, $b=0$, $c=0$, and therefore $k'=3$, $a'=2$, $b'=1$, $c'=1$,) corresponding to the conic we have in the third figure a curve, order 3 (cubic curve), passing twice through D'', but once through E'' and F'' respectively; this cubic curve passes also through the points B'', C'' which correspond to B', C' respectively; that is,

cubic passes through E'', F'', B'', C'' each 1 time
" " D'' 2 times;

or, corresponding to a line in the first figure, we have in the third figure a curve, order 3, passing through four fixed points each 1 time, and through one fixed point 2 times. That is, we have $n=3$, $\alpha_1'=4$, $\alpha_2'=1$. And in the same manner, to a line in the third figure there corresponds in the first figure a cubic through four fixed points (viz., B, C, E, F) each 1 time, and through one fixed point, A, 2 times; so that also $\alpha_1=4$, $\alpha_2=1$. The transformation is thus of the order 3, and the form $4_1\,1_2$ and $4_1\,1_2$ (this is in fact the only cubic transformation; see the Tables, *ante*, No. 41).

69. Mr Clifford has also devised a very convenient algorithm for this decomposition of a transformation of any order into quadric transformations. The quadric transformation is denoted by [3], the cubic transformation by [41], the quartic transformations by [601], [330], the quintic ones by [8001], [3310], [0600], and so on; see the Tables just referred to. (This is substantially the same as a notation employed above, the zeros enabling the omission of the suffixes; viz., $[8001]=8_1\,1_4$; and so in other cases.)

70. The foregoing result is represented thus [4, 1] = [3⦅0, 0, 1), which I proceed to explain. Consider in the first figure a line; the symbol [3] denotes that in the second figure we have a conic with three points (α_1'). We are about to apply to this a quadric transformation; (0, 0, 0) would denote that the three points of the principal system in the second figure were all of them arbitrary; (0, 0, 1) that one of these points was a point α_1'; (0, 1, 1) that two of them were points α_1'; (1, 1, 1) that all three of them were points α_1; (0, 0, 2) would denote that one of the points was a point α_2'; only in the present case we can have no such symbol, by reason that there are no points α_2'. Hence [3⦅001) denotes that the conic has applied to it a quadric transformation such that, in the transformation thereof, one point of the principal system coincides with one of the points (α_1') on the conic. To [3], *quà* quadric transformation, belongs the number 2: and from 2, (001) we derive 3, (112), {in general k, (a, b, c) gives k', (a', b', c'), where $k' = 2k - a - b - c$, $a' = k - b - c$, $b' = k - c - a$, $c' = k - a - b$}. $k = 2$ corresponds to a symbol [3] of one number, $k' = 3$ to a symbol of two numbers; viz., we change [3] into [30]; we then, in the symbols (112) and (001), consider the frequencies of the several numbers 1, 2, ... taking those in the first symbol as positive, and those in the second symbol as negative; or, what is the same thing, representing the frequency as an index, we have $1^2\,2^1$, 1^{-1}; or, combining, $1^{2-1}\,2^1$; these indices are then added on to the numbers of [30]; viz., the index of 1 to the first number, the index of 2 to the second number (and, in the case of more numbers, so on): [30] is thus converted into [41], and we have the required equation

[41] = [3⦅001),

where the *rationale* of this algorithmic process appears by the explanation, *ante*, No. 68.

71. As another example take

[8001] = [601⦅003).

To [601], *quà* quartic transformation, belongs the number 4; and from 4, (003) we form 5, (114); where the 5 indicates that [601] is to be changed into [6010]; then (114), (003), writing them in the form $1^2 2^0 3^{-1} 4^1$, show that to the numbers of [6010] we are to add 2, 0, − 1, 1; thus changing the symbol into [8001], so that we have the required relation.

72. Mr Clifford calculated in this way the following table, showing how any transformation of an order not exceeding 8 can be expressed by means of a series of quadric transformations; the symbols Cr. 3, Cr. 4 . 1; 4 . 2, &c., refer to the order and number of Cremona's tables, *ante*, No. 41.

Cr. 3 . = [41] = [3⦅001),

Cr. 4 . 1 = [601] = [41⦅002) = [3⦅001⦆002),

4 . 2 = [330] = [3⦅000) = [41⦅011) = [3⦅001⦆011),

Cr. 5 . 1 = [8001] = [601)(003) = [3)(001)(002)(003),
5 . 2 = [3310] = [41)(001) = [3)(001)(001),
5 . 3 = [0600] = [330)(111) = [3)(000)(111) = [3)(001)(011)(111),

Cr. 6 . 1 = [10,0001] = [8001)(004) = [3)(001)(002)(003)(004),
6 . 2 = [14200] = [330)(011) = [3)(000)(011) = [3)(001)(011)(011),
6 . 3 = [41300] = [601)(011) = [3)(001)(002)(011)} = [3)(001)(000),
= [3310)(022) = [3)(001)(001)(022)}
6 . 4 = [34010] = [330)(002) = [3)(000)(002),
= [3310)(013) = [3)(001)(001)(013),

Cr. 7 . 1 = [12,00001] = [10,0001)(005) = [3)(001)(002)(003)(004)(005),
7 . 2 = [330)(001) = [3)(000)(001) = [232100],
7 . 3 = [034000] = [3310)(111) = [3)(001)(001)(111),
7 . 4 = [503100] = [601)(001) = [3)(001)(002)(001),
7 . 5 = [350010] = [3310)(003) = [3)(001)(001)(003),

Cr. 8 . 1 = [14,000001] = [3)(001)(002)(003)(004)(005)(006),
8 . 2 = [3230100] = [3310)(002) = [3)(001)(001)(002),
8 . 3 = [1322000] = [3310)(011) = [3)(001)(001)(011),
8 . 4 = [0070000] = [034000)(222) = [3)(001)(001)(111)(222),
8 . 5 = [3600010] = [34010)(004) = [330)(002)(004) = [3)(000)(002)(004),
8 . 6 = [6013000] = [601)(000) = [3)(001)(002)(000),
8 . 7 = [0520100] = [0600)(002) = [3)(000)(111)(222),
8 . 8 = [2051000] = [41300)(112) = [3)(001)(000)(112),
8 . 9 = [3303000] = [3)(000)(000)(000).

73. The reduction as above of a transformation to a series of quadric transformations, enables the determination of the reciprocal transformation; or, what is the same thing, the determination of the Jacobian of the first figure; see the example, *ante*, No. 67, where it appears that the reciprocal transformation of [41] is [41]. But I do not see any easy algorithmic process for the determination of the reciprocal transformation, or still less any general form in which the result can be expressed; and I do not at present pursue the inquiry.

The Rational Transformation between Two Spaces.

74. The general principles have been already explained: the two systems $x' : y' : z' : w' = X : Y : Z : W$ and $x : y : z : w = X' : Y' : Z' : W'$ must be derivable the one from the other; and starting with the first system, this will be the case if

only the surfaces $X=0$, $Y=0$, $Z=0$, $W=0$ have a common intersection equivalent to n^3-1 points of intersection, but not equivalent to a complete common intersection of n^3 points. The last-mentioned circumstance would arise, if the condition of the common intersection should impose upon the surface more than $\frac{1}{6}(n+1)(n+2)(n+3)-4$ conditions; viz., the surfaces would then be connected by an identical equation or syzygy $\alpha X+\beta Y+\gamma Z+\delta W=0$. The common intersection is a figure composed of points and curves: say it is the principal system in the first space; the problem is, to determine a principal system equivalent to n^3-1 points of intersection but such that the number of conditions to be satisfied by a surface passing through it is not more than

$$\tfrac{1}{6}(n+1)(n+2)(n+3)-4.$$

75. The following locutions are convenient. We may say that the number of conditions imposed upon a surface of the order n which passes through the common intersection is the *Postulation* of this intersection; and that the number of points represented by the common intersection (in regard to the points of intersection of any three surfaces each of the order n which pass through it) is the *Equivalence* of this intersection. The conditions above referred to are thus

$$\begin{aligned}&\text{Equivalence} = n^3-1,\\ &\text{Postulation} \ngtr \tfrac{1}{6}(n+1)(n+2)(n+3)-4.\end{aligned}$$

76. It would appear by the analogy of the rational transformation between two planes, that the only cases to be considered are those for which

$$\text{Postulation} = \tfrac{1}{6}(n+1)(n+2)(n+3)-4;$$

but I cannot say whether this is so.

77. In the transformation between two planes, the two conditions lead, as was seen, to the result that the curve $aX+bY+cZ=0$ is unicursal. I do not see that in the present case of two spaces, the two conditions lead to the corresponding result that the surface $aX+bY+cZ+dW=0$ is unicursal; that this is so, appears, however, at once from the general notion of the rational transformation. In fact, the equation in question $aX+bY+cZ+dW=0$ is satisfied by $x:y:z:w=X':Y':Z':W'$ and $ax'+by'+cz'+dw'=0$; the last equation determines the ratios $x':y':z':w'$ in terms of two arbitrary parameters (say these are $x':y'$ and $x':z'$), and we have then $x:y:z:w$ proportional to rational functions of these two parameters; that is, the surface $aX+bY+cZ+dW=0$ is unicursal. And similarly the surface $aX'+bY'+cZ'+dW'=0$ is unicursal.

78. In the most general point of view, the principal system will contain a given number of points which are simple points, a given number which are quadriconical points, a given number which are cubiconical points, &c. &c., on the surfaces; and similarly a given number of curves which are simple curves, a given number which are double curves, &c. &c., on the surfaces. But, to simplify, I will consider that it includes only points which are simple points, and a curve which is a simple curve

on the surfaces: this curve may, however, break up into separate curves, and we thus, in fact, admit the case where there are any number of separate curves each of them a simple curve on the surfaces. It is right to remark that we cannot assert *à priori*—and it is not in fact the case—that the principal system in the second space will be subject to the like restrictions: starting with such a principal system in the first space, we may be led in the second space to a principal system including a curve which is a double curve on the surfaces; an instance of this will in fact occur.

79. It is shown (Salmon's *Solid Geometry*, 2nd ed., p. 283, [Ed. 4, p. 321]), that in the intersection of three surfaces of the orders μ, ν, ρ respectively, a curve of intersection of the order m and class r counts as $m(\mu+\nu+\rho-2)-r$ points of intersection. For a curve without actual double points or stationary points, we have $r=m(m-1)-2h$, where h is the number of apparent double points; or, substituting, we have the curve counting for $m(\mu+\nu+\rho-2)-m(m-1)+2h$ points of intersection; this is in fact a more general form of the formula, inasmuch as it extends to the case of a curve with actual double points and stationary points. Or, what is the same thing, the three surfaces intersecting in the curve of the order m with h apparent double points, will besides intersect in $\mu\nu\rho-m(\mu+\nu+\rho-2)+m(m-1)-2h$ points; viz., the curve may, besides the apparent double points, have actual double points and stationary points; but these do not affect the formula.

80. Some caution is necessary in the application of the theorem. For instance, to consider cases that will present themselves in the sequel: let the surfaces be cubics ($\mu=\nu=\rho=3$); the number of remaining intersections is given as $=27-7m+m(m-1)-2h$. Suppose that the curve consists of four non-intersecting lines, $m=4$, $h=6$, the number is given as $=-1$. But observe in this case there are two lines each meeting the four given lines; that is, any cubic surface passing through the four given lines meets these two lines each of them in four points, that is, the cubic passes also through each of the two lines; the complete *curve*-intersection of the surfaces is made up of the six lines $m=6$, $h=7$ (since each of the two lines, as intersecting the four lines, gives actual double points, but the two lines together give one apparent double point), and the expression for the number of the remaining points of intersection becomes $=27-42+30-14=1$, which is correct.

81. Similarly, if the given curve of intersection be a conic and two non-intersecting lines, there is here in the plane of the conic a line meeting each of the two given lines, and therefore meeting the cubic surface, in four points, that is, lying wholly in the cubic surface: the complete *curve*-intersection consists of the conic, the two given lines, and the last-mentioned line, $m=5$, $h=5$, and the number of points of intersection is $=27-35+20-10$, $=2$, which is correct. Again, if the given curve of intersection be two conics, here the line of intersection of the planes of the conics lies in the cubic surface; or, for the complete *curve*-intersection we have $m=5$, $h=4$; and the number of points is $27-35+20-8$, $=4$. If in this last case each or either of the conics become a pair of intersecting lines, or if in the preceding case the conic becomes a pair of intersecting lines, the results remain unaltered.

82. If a surface of the order μ pass through a curve of the order m and class r without stationary points or actual double points, this imposes on the surface a number of conditions $=(\mu+1)m-\frac{1}{2}r$. In the case in question, the value of r is $=m(m-1)-2h$; or, substituting, the number of conditions is $=(\mu+1)m-\frac{1}{2}m(m-1)+h$; and the formula in this form holds good even in the case where the curve has stationary points and actual double points. Thus $\mu=3$, the number of conditions is $=4m-\frac{1}{2}m(m-1)+h$. If the curve be a line, $m=1$, $h=0$, number of conditions is $=4$; if the curve be a pair of non-intersecting lines, $m=2$, $h=1$, number of conditions is $=8$. And so in general, if the curve consist of k non-intersecting lines ($k=4$ at most), then $m=k$, $h=\frac{1}{2}k(k-1)$, and the number of conditions is $=4k$. If the curve be a conic, or a pair of intersecting lines, $m=2$, $h=1$, and the number of conditions is $=7$. If the curve consist of k lines, such that there are θ pairs of intersecting lines, then $m=k$, $h=\frac{1}{2}k(k-1)-\theta$, and the number of conditions is $=4k-\theta$. It is obvious that, the number of conditions for a line being $=4$, that for the k lines with θ intersecting pairs must have the foregoing value $4k-\theta$. In fact, when the lines do not intersect, we take on each line 4 points, and the cubic surface passing through any such 4 points will contain the line; but for two lines which intersect, taking this point, and on each of the intersecting lines 3 other points, the cubic surface through the 7 points will pass through the two lines; and so in other cases.

83. The formula must, in some instances, be applied with caution. Thus, given five non-intersecting lines $k=5$, $\theta=0$, and the number of conditions is $=20$; and a cubic surface cannot be, in general, made to pass through the lines. But if the five lines are met by any other line, then a cubic surface, if it pass through the five lines, will pass through this sixth line; for the six lines $k=6$, $\theta=5$, and the number of conditions is $24-5=19$; so that there is a determinate cubic surface through the six lines, and therefore through the five lines related in the manner just referred to.

84. Recurring to the problem of transformation, it appears by what precedes, that if the principal system in the first plane consists of α_1 points, and of a curve of the order m_1 with h_1 apparent double points (the α_1 points being simple points, and the curve a simple curve on the surfaces), then the conditions for a transformation are

$$(3n-2)\,m_1-\ \ m_1(m_1-1)+2h_1+\alpha_1=n^3-1,$$
$$(\ n+1)\,m_1-\tfrac{1}{2}m_1(m_1-1)+\ h_1+\alpha_1=\tfrac{1}{6}(n+1)(n+2)(n+3)-4,$$

where, in the second line, instead of $\ngtr$ I have written $=$. I remark, in passing, that I have ascertained that an actual triple point counts as an apparent double point; or, what is the same thing, that if the curve has t_1 actual triple points, then we may, instead of h_1, write h_1+t_1. The equations give

$$m_1(4n-5-m_1)=\tfrac{1}{3}(n-1)(5n^2-\ n-12)-2h_1,$$
$$(n-4)\,m_1-\alpha=\tfrac{1}{3}(n-1)(2n^2-4n-15),$$

to which may be joined

$$(3n+8)\,m_1-2m_1(m_1-1)+4h_1+5\alpha_1=(n-1)(6n+17).$$

The first two equations for the successive values of n give

$$\begin{array}{lll} n=2, & m_1(\;3-m_1)=\;\;2-2h_1, & 2m_1+\alpha_1=\;\;5;\\ n=3, & m_1(\;7-m_1)=\;20-2h_1, & m_1+\alpha_1=\;\;6;\\ n=4, & m_1(11-m_1)=\;64-2h_1, & \alpha_1=-\;1;\\ n=5, & m_1(15-m_1)=144-2h_1, & -\;m_1+\alpha_1=-20;\\ n=6, & m_1(19-m_1)=270-2h_1, & -2m_1+\alpha_1=-55;\\ \&c. & \&c. & \&c. \end{array}$$

85. It is remarkable that for $n=4$ there is no solution of the geometrical problem; in fact, $\alpha_1=-1$, a negative value of α_1, shows that this is so. For the higher values of n, there seem to be solutions with large values of m_1, h_1, α_1; for example, $n=5$, we have $m_1=20+\alpha_1$, is $=20$ at least. Writing $m_1=20$, we have $-100=144-2h_1$, or $2h_1=244$. The highest value of $2h_1$ is $=(m_1-1)(m_1-2)$, which for $m_1=20$ is $=342$; or the foregoing value $2h_1=244$ is admissible. Thus $m_1=20$, $h_1=122$, $\alpha_1=0$ gives a solution; and, moreover, any larger value of m_1, say $m_1=20+\alpha$, gives an admissible solution, $m_1=20+\alpha$, $h_1=122+\frac{1}{2}\alpha(\alpha+25)$, $\alpha_1=\alpha$. And so for $n=6$, &c.; but I have not further examined any of these cases, and do not understand them.

There remain the cases $n=2$, $n=3$. For $n=2$, since $2m_1+\alpha_1=5$, we have $m_1=0$, 1, or 2; $m_1=0$ gives $h_1=0$, which is not admissible. The remaining solutions are $m_1=1$, $h_1=0$, $\alpha_1=3$; and $m_1=2$, $h_1=0$, $\alpha_1=1$.

For $n=3$, since $m_1+\alpha_1=6$, we have $m_1=0$, 1, 2, 3, 4, 5, or 6. $m_1=0$ gives $h_1=10$; $m_1=1$ gives $h_1=7$; $m_1=2$ gives $h_1=5$; $m_1=3$ gives $h_1=4$: these values are not geometrically admissible. The remaining cases are $m_1=4$, $h_1=4$, $\alpha_1=2$; $m_1=5$, $h_1=5$, $\alpha_1=1$; $m_1=6$, $h_1=7$, $\alpha_1=0$.

86. The reciprocal transformation is in every case of the order $n'=n^2-m_1$. Hence considering the quadric transformations:

First, the case $n=2$, $m_1=1$, $h_1=0$, $\alpha_1=3$; the reciprocal transformation is of the order $n'=3$. Suppose for a moment that the principal system in the second space is of the same nature as that above considered in the first space, consisting of α_1' points, and a curve of the order m_1' with h_1' apparent double points (the α_1' points each a simple point, and the curve a simple curve on the surfaces $X'=0$, &c.). Passing back to the original transformation, we should have $2=9-m_1'$, that is, $m_1'=7$. But it has just been seen that, for $n=3$, the only values of m_1 are 4, 5, 6; hence for $n'=3$ we cannot have $m_1'=7$. The explanation is, that the principal system in the second space *is not of the form in question;* it, in fact, consists (as will appear) of three lines each a simple line, and of another line which is a double line on the surfaces $X'=0$, &c. In the intersection of any two of these surfaces, the three lines count each once, the double line four times, and the order of the curve of intersection is thus $3+4=7$, as it should be. The principal system may be characterized $\alpha_1'=0$, $m_1'=3$, $h_1'=3$, $m_2'=1$, $h_2'=0$.

Next, the case $n=2$, $m_1=2$, $h_1=0$, $\alpha_1=1$: the reciprocal transformation is of the order $n'=2$; it is evidently not of the form above considered (for this would make the original transformation to be of the order 3). Hence, assuming (as it seems allowable to do) that the principal system does not contain any multiple point or curve, the reciprocal transformation will be of the same form as the original one; viz., we shall have $n'=2$, $m_1'=2$, $h_1'=0$, $\alpha_1'=1$.

87. Considering next the cubic transformations, or those belonging to $n=3$; in the case $m_1=4$, $h_1=4$, $\alpha_1=2$, the reciprocal transformation is of the order $9-4$, $=5$; and in the case $m_1=5$, $h_1=5$, $\alpha_1=1$, the reciprocal transformation is of the order $9-5$, $=4$: I do not consider these cases. But $m_1=6$, $h_1=7$, $\alpha_1=0$, the reciprocal transformation is of the order $9-6$, $=3$; and assuming (as seems allowable) that the principal system does not contain any multiple point or curve, it must be of the same form as the original transformation, that is, we must have $n'=3$, $m_1'=6$, $h_1'=7$, $\alpha_1'=0$.

88. The transformations to be studied are thus,—1° The quadri-quadric transformation $n=2$, $m_1=2$, $h_1=0$, $\alpha_1=1$, and $n'=2$, $m_1'=2$, $h_1'=0$, $\alpha_1'=1$; the principal system in each space consists of a point and of a conic (which may be a pair of intersecting lines); and the surfaces are quadrics. 2° The quadri-cubic transformation $n=2$, $m_1=1$, $h_1=0$, $\alpha_1=3$, and $n'=3$, $\alpha_1'=0$, $m_1'=3$, $h_1'=3$, $m_2'=1$, $h_2'=0$: in the first space the principal system consists of three points and a line, and the surfaces are quadrics: in the second figure, the principal system consists of three simple lines and a double line; and the surfaces are cubic surfaces passing through this principal system, that is, they are cubic scrolls. 3° The cubo-cubic transformation $n=3$, $\alpha_1=0$, $m_1=6$, $h_1=7$, and $n'=3$, $\alpha_1'=0$, $m_1'=6$, $h_1'=7$; in each space the principal system is a sextic curve with seven apparent double points (but there are different cases to be considered according as the sextic curve does or does not break up into inferior curves), and the surfaces are cubic surfaces through the sextic curve.

The Quadri-quadric Transformation between Two Spaces.

89. Starting from the equations $x' : y' : z' : w' = X : Y : Z : W$, we have here $X=0$, &c., quadric surfaces passing through a given point and a given conic (which may be a pair of intersecting lines). Take $x=0$, $y=0$, $z=0$ for the coordinates of the given point; $w=0$ for the equation of the plane of the conic; the conic is then given as the intersection of this plane by a cone having the given point for its vertex; or say the equations of the conic are $w=0$, $(a, \ldots \between x, y, z)^2=0$; the general equation of a quadric through the point and conic is $w(\alpha x+\beta y+\gamma z)+\delta(a, \ldots \between x, y, z)^2=0$; and it hence appears that the equations of the transformation may be taken to be

$$x' : y' : z' : w' = xw : yw : zw : (a, \ldots \between x, y, z)^2;$$

these give at once a reciprocal system of the same form; viz., the two sets are

$$x' : y' : z' : w' = xw : yw : zw : (a, \ldots \between x, y, z)^2,$$

and

$$x : y : z : w = x'w' : y'w' : z'w' : (a, \ldots \between x', y', z')^2.$$

90. The Jacobian of the first space is at once found to be

$$w^2 (a, \ldots \between x, y, z)^2 = 0\,;$$

that of the second space is of course

$$w'^2 (a, \ldots \between x', y', z')^2 = 0.$$

The two spaces are similar to each other; we may say that there is in each of them a principal point and a principal conic; that the plane of the conic is the principal plane, and the cone having its vertex at the point and passing through the conic is the principal cone. To the principal point of either space corresponds any point whatever in the principal plane of the other space; and conversely. More definitely, the points of the one principal plane and the infinitesimal elements of direction through the principal point of the other space correspond according to the equations $x : y : z = x' : y' : z'$. To any point on the principal conic of either space corresponds in the other space, not a mere element of direction through the principal point of the other space, but a line of the principal cone; that is, to the points of the principal conic of the one space correspond the lines of the principal cone of the other space. The Jacobian of either space, consisting of the principal plane twice, and of the principal cone, is thus the principal counter-system of the other space.

91. {Writing $(a, \ldots \between x, y, z)^2 = x^2 + y^2 + z^2$, $w = w' = 1$, the equations of transformation become

$$x' : y' : z' : 1 = x : y : z : x^2 + y^2 + z^2,$$

and

$$x : y : z : 1 = x' : y' : z' : x'^2 + y'^2 + z'^2,$$

or, what is the same thing, if for shortness

$$x^2 + y^2 + z^2 = r^2, \quad x'^2 + y'^2 + z'^2 = r'^2,$$

the equations are

$$x' = \frac{x}{r^2}, \quad y' = \frac{y}{r^2}, \quad z' = \frac{z}{r^2}; \text{ and } x = \frac{x'}{r'^2}, \quad y = \frac{y'}{r'^2}, \quad z = \frac{z'}{r'^2},$$

whence also $rr' = 1$; this is the well known transformation by reciprocal radius vectors.}

92. The principal conic may be a pair of intersecting lines; taking its equations to be $w = 0$, $xy = 0$, the equations of transformation here become

$$x' : y' : z' : w' = xw : yw : zw : xy,$$

and

$$x : y : z : w = x'w' : y'w' : z'w' : x'y'.$$

There is no difficulty in the further development of the theory.

The Quadri-cubic Transformation between Two Spaces.

93. It will be convenient to have the unaccented letters (x, y, z, w) referring to the cubic surfaces. I will therefore take the quadric surfaces in the second figure; viz., I will start from the equations $x : y : z : w = X' : Y' : Z' : W'$, where $X' = 0$,

$Y'=0$, $Z'=0$, $W'=0$ are quadric surfaces passing through three fixed points (say the principal points) and through a fixed line (say the principal line) in the second figure. Taking $x'=0$, $y'=0$ for the planes passing through the principal line and through two of the principal points respectively; $z'=0$ for the plane passing through the three principal points, $w'=0$ for an arbitrary plane passing through the first mentioned two principal points, the implicit factors of x', y', w' may be so determined that for the third principal point $x'=y'=-w'$. That is, we shall have

$$\begin{array}{ll} \text{for principal line} & x'=0,\ y'=0, \\ \text{for principal points} & (x'=0,\ z'=0,\ w'=0), \\ \quad\text{,,} & (y'=0,\ z'=0,\ w'=0), \\ \quad\text{,,} & (x'=y'=-w',\ z'=0), \end{array}$$

and this being so, the equation of a quadric surface through the principal points and line will be

$$(\alpha x'+\beta y')\,z'+\gamma x'\,(y'+w')+\delta y'\,(x'+w'),$$

and the equations of transformation may be taken to be

$$x : y : z : w \qquad = x'z' : y'z' : x'\,(y'+w') : y'\,(x'+w').$$

94. Writing these in the extended form

$$x : y : z : w : x-y : z-w = x'z' : y'z' : x'\,(y'+w') : y'\,(x'+w') : z'\,(x'-y') : w'\,(x'-y')$$

and forming also the equation

$$xy : (xw-yz) = z' : x'-y',$$

we at once derive the reciprocal system of equations

$$x' : y' : z' : w' \qquad = x\,(xw-yz) : y\,(xw-yz) : (x-y)\,xy : (z-w)\,xy,$$

so that this is a cubic transformation. And the cubic surface in the first space (corresponding to an arbitrary plane $ax'+by'+cz'+dw'=0$ of the second space) is $(ax+by)(xw-yz)+c\,(x-y)\,xy+d\,(z-w)\,xy=0$; viz., this is a cubic surface having the fixed double line $(x=0,\ y=0)$, the fixed simple lines $(x=0,\ z=0)$, $(y=0,\ w=0)$, and $(x-y=0,\ z-w=0)$; it has also the variable simple line $(dz+cx=0,\ dw+cy=0)$. The principal figure of the first space thus consists of the three simple lines $(x=0,$ $z=0)$, $(y=0,\ w=0)$, $(x-y=0,\ z-w=0)$, and of the line $(x=0,\ y=0)$, a double line counting four times in the intersection of two of the cubic surfaces.

95. The cubic surface as having the double line $(x=0,\ y=0)$ is a cubic scroll, and this line is the nodal directrix thereof; the line $(dz+cx=0,\ dw+cy=0)$ is the simple directrix; the lines $(x=0,\ z=0)$, $(y=0,\ w=0)$, $(x-y=0,\ z-w=0)$ are at once seen to be lines meeting each of these directrix lines; and they are generating lines of the scroll. To explain the generation of the scroll, observe that the section by any plane is a cubic curve having a given double point (viz., the intersection of the plane with the nodal directrix); and three other given points (viz., the intersections of

the plane with the three generating lines respectively); this cubic also passes through the intersection of the plane with the simple directrix. Conversely, if the plane be assumed at pleasure, and if, taking for the simple directrix any line which meets the given generating lines, we draw a cubic as above, then the scroll is the scroll generated by a line which meets each of the directrix lines, and also the cubic.

If the plane be taken to pass through any generating line, then the cubic section breaks up into this line, and a conic; the conic does not meet the simple directrix, but it meets the nodal directrix; and any such conic will serve as a directrix; viz., the scroll is generated by the lines which meet the two directrix lines and the conic.

96. Any two scrolls as above meet in the three fixed generating lines, and in the nodal directrix counting four times; they consequently meet besides in a curve of the second order, which is a conic (one of the conics just referred to). In order to further explain the theory, suppose for a moment that the two scrolls had only a common nodal directrix; they would besides meet in a quintic curve; this curve would meet the nodal directrix in four points, viz., the points at which the two scrolls have a common tangent plane. Now if at any point of the nodal directrix the two scrolls have a common generating line, then the plane through this line and the nodal line is one of the two tangent planes of each scroll; that is, the scrolls have this plane for a common tangent plane. Hence, in the case of the common three generating lines, the points where these meet the nodal line are three of the four points just referred to; there remains therefore one point, which is the point where the conic meets the nodal line; through this point there are for each of the scrolls two generating lines; one of these for the first scroll, and one for the second scroll, lie in a plane with the nodal line; the other two determine the plane of the conic; and the tangent to the conic at its intersection with the nodal line is the intersection of the plane of the conic with the plane of the first-mentioned two generating lines.

97. Analytically we have the two equations

$$c\ (x-y)\,xy+(ax\ +by\ )\,(xw-yz)+d\ (z-w)\,xy=0,$$
$$c'\,(x-y)\,xy+(a'x+b'y)\,(xw-yz)+d'\,(z-w)\,xy=0\,;$$

or, combining these equations so as to eliminate successively the terms in $x\,(xw-yz)$ and $y\,(xw-yz)$, and for this purpose writing

$$(bc'-b'c,\ ca'-c'a,\ ab'-a'b,\ ad'-a'd,\ bd'-b'd,\ cd'-c'd)=(\mathrm{a},\ \mathrm{b},\ \mathrm{c},\ \mathrm{f},\ \mathrm{g},\ \mathrm{h}),$$

and therefore

$$\mathrm{af}+\mathrm{bg}+\mathrm{ch}=0,$$

we have

$$\mathrm{b}\,(x-y)\,x-\mathrm{c}\,(xw-yz)-\mathrm{f}\ (z-w)\,x=0,$$
$$-\,\mathrm{a}\ (x-y)\,y+\mathrm{c}\,(xw-yz)-\mathrm{g}\,(z-w)\,y=0,$$

and multiplying the first of these by $\mathrm{c}+\mathrm{g}$ and the second by $\mathrm{c}-\mathrm{f}$, and adding, the whole divides by $x-y$, and the final result is

$$(\mathrm{c}+\mathrm{g})\,(\mathrm{b}x-\mathrm{f}z)-(\mathrm{c}-\mathrm{f})\,(\mathrm{a}y+gw)=0\,;$$

viz., this is the equation of the plane of the conic.

98. Any two scrolls as above meeting in a conic, a third scroll will meet the conic in six points; but these include the point on the nodal directrix twice, and the points on the three fixed generating lines each once; there is left a single point of intersection, viz., this is the *one variable* point of intersection of the three scrolls; which is in accordance with the theory.

99. For the Jacobian of the second space, we have

$$\begin{vmatrix} z', & 0, & x', & 0 \\ 0, & z', & y', & 0 \\ y'+w', & x', & 0, & 0 \\ y', & x'+w', & 0, & y' \end{vmatrix} = 0;$$

that is, $2x'y'z'(x'-y')=0$; viz., $z'=0$ is the plane containing the three principal points; and $x'=0$, $y'=0$, $x'-y'=0$ are the planes which pass through the principal line and the three principal points respectively.

100. For the Jacobian of the first space, we have

$$\begin{vmatrix} 2xw-yz, & -xz, & -xy, & x^2 \\ yw, & xw-2yz, & -y^2, & xy \\ 2xy-y^2, & x^2-2xy, & 0, & 0 \\ (z-w)\,y, & (z-w)\,x, & xy, & -xy \end{vmatrix} = 0;$$

that is, $3x^2y^2(x-y)^2(xw-yz)=0$; viz., $x=0$, $y=0$, $x-y=0$ are the planes through the nodal directrix and the three fixed generators respectively (each plane therefore occurring twice); and $xw-yz=0$ is the quadric scroll generated by the lines which meet each of the three generators ($x=0$, $z=0$), ($y=0$, $w=0$), ($x-y=0$, $z-w=0$); this scroll passing also through the nodal directrix $x=0$, $y=0$.

The Cubo-cubic Transformation between Two Spaces.

101. The principal system in the first space is a sextic curve with 7 apparent double points; but this curve may be either a single curve, or it may break up into inferior curves. I have not examined all the cases which may arise; but the two extreme cases are—(A) The sextic curve breaks up into six lines, viz., two non-intersecting lines, and four other lines each meeting each of the two lines (this implies that no two of the four lines meet each other): here the two lines give 1 apparent double point, and the four lines give 6 apparent double points; total number is $=7$, as it should be. (B) The curve is a proper sextic curve, with 7 apparent double points: this gives, as will be shown, the general lineo-linear transformation. The two cases are each of them symmetrical.

(A) *The Principal System consists of Six Lines.*

102. Taking in the first space, for the equations of the two lines, $(x=0,\ y=0)$ and $(z=0,\ w=0)$, and for the equations of the four lines, $(x=0,\ z=0)$, $(y=0,\ w=0)$, $(x-y=0,\ z-w=0)$, $(x-py=0,\ z-qw=0)$, then, if the equations of transformation are taken to be

$$\begin{aligned} x'-py' : x'-y' : z'-qw' : z'-w' = \ & (x-py)(xw-yz) \\ : & (x-y)(qxw-pyz) \\ : & (z-qw)(xw-yz) \\ : & (z-w)(qxw-pyz); \end{aligned}$$

these lead conversely (see *post*, No. 104) to a like system,

$$\begin{aligned} x-py : x-y : z-qw : z-w = \ & (x'-py')M' \\ : & (x'-y')N' \\ : & (z'-qw')M' \\ : & (z'-w')N', \end{aligned}$$

where for shortness

$$\begin{aligned} M' &= p(q-1)^2 x'w' - q(p-1)^2 y'z' + (pq-1)(p-q)y'w', \\ N' &= (q-1)^2 x'w' - (p-1)^2 y'z' + (pq-1)(p-q)y'w'; \end{aligned}$$

or, as these are better written,

$$\begin{aligned} M' &= -q(p-1)y'\{(p-1)z' - (pq-1)w'\} + p(q-1)w'\{(q-1)x' - (pq-1)y'\}, \\ N' &= -(p-1)y'\{(p-1)z' - (pq-1)w'\} + (q-1)w'\{(q-1)x' - (pq-1)y'\}. \end{aligned}$$

Hence the principal system in the second plane is composed of the two non-intersecting lines $(x'=0,\ y'=0)$, $(z'=0,\ w'=0)$ and the four lines $\{(p-1)z'-(pq-1)w'=0,\ (q-1)x'-(pq-1)y'=0\}$, $(y'=0,\ w'=0)$, $(x'-y'=0,\ z'-w'=0)$, $(x'-py'=0,\ z'-qw'=0)$, each meeting each of the two lines.

103. The Jacobian of the first space is

$$\begin{vmatrix} 2xw-yz-pyw, & -xz-pxw+2pyz, & -y(x-py), & x(x-py) \\ 2qxw-pyz-qyw, & -pxz-qxw+2pyz, & -py(x-y), & qx(x-y) \\ w(z-qw), & -z(z-qw), & xw-2yz+qyw, & xz-2qxw+qyz \\ qw(z-w), & -pz(z-w), & qxw-2pyz+pyw, & qxz-2qxw+pyz \end{vmatrix} = 0,$$

viz., this is $xyzw(x-y)(x-py)(z-w)(z-qw)=0$, the equation of the planes each passing through one of the four lines and one of the two lines.

Similarly, the Jacobian of the second space is

$$\{(q-1)x'-(pq-1)y'\}\{(p-1)z'-(pq-1)w'\}(x'-y')(z'-w')(x'-py')(z'-qw')y'w'=0;$$

viz., this is the equation of the eight planes each passing through one of the four lines and one of the two lines.

104. To effect the foregoing transformation, writing

$$\begin{aligned} x' : y' : z' : w' = &\ (x-py)(\ xw - \ yz) \\ &: (x-\ y)(qxw-pyz) \\ &: (z-qw)(\ xw-\ yz) \\ &: (z-\ w)(qxw-pyz); \end{aligned}$$

or what will ultimately be the same thing, but it is more convenient for working with,

$$\begin{aligned} x' &= (x-py)(\ xw-\ yz), \\ y' &= (x-\ y)(qxw-pyz), \\ z' &= (z-qw)(\ xw-\ yz), \\ w' &= (z-\ w)(qxw-pyz); \end{aligned}$$

these give

$$\begin{aligned} x-py &= M'x', \\ x-\ y &= N'y', \\ z-qw &= M'z', \\ z-\ w &= N'w', \end{aligned}$$

where M', N' are quantities which have to be determined; and thence

$$\begin{aligned} (1-p)\,x &= M'x' - pN'y', \\ (1-p)\,y &= M'x' - \ N'y', \\ (1-q)\,z &= M'z' - qN'w', \\ (1-q)\,w &= M'z' - \ N'w'; \end{aligned}$$

whence also

$$\begin{aligned} (1-p)(1-q)(\ xw-\ yz) &= N'[\ \{(q-1)\,x'w' - (p-1)\,y'z'\}\,M' + (p-q)\,y'w'N'], \\ (1-p)(1-q)(qxw-pyz) &= M'[-(p-q)\,x'z'M' + \{(pq-q)\,x'w' - (pq-p)\,y'z'\}\,N']; \end{aligned}$$

but we have

$$\frac{xw-yz}{qxw-pyz} = \frac{x'}{y'} \div \frac{x-py}{x-y} = \frac{x'}{y'} \div \frac{M'x'}{N'y'} = \frac{N'}{M'};$$

or, substituting,

$$\begin{aligned} &M'\{\ \ (q-1)\,x'w' - (p-1)\,y'z'\} + N'(p-q)\,y'w' \\ &= M'\{-(p-q)\,x'z'\} + N'\{(pq-q)\,x'w' - (pq-p)\,y'z'\}; \end{aligned}$$

that is

$$M'\{(q-1)\,x'w' - (p-1)\,y'z' + (p-q)\,x'z'\} = N'\{(pq-q)\,x'w' - (pq-p)\,y'z' - (p-q)\,y'w'\};$$

or, what is the same thing,

$$\begin{aligned} M' &= (pq-q)\,x'w' - (pq-p)\,y'z' - (p-q)\,y'w', \\ N' &= (\ q-1)\,x'w' - (\ p-1)\,y'z' + (p-q)\,x'z'; \end{aligned}$$

viz., M', N' having these values, the original equations

$$\begin{aligned} x' : y' : z' : w' = &\ (x - py)(\ xw - \ yz) \\ &: (x - \ y)(pxw - qyz) \\ &: (z - qw)(\ xw - \ yz) \\ &: (z - \ w)(pxw - qyz), \end{aligned}$$

give

$$x - py : x - y : z - qw : z - w = M'x' : N'y' : M'z' : N'w'.$$

If, in these equations, in place of (x', y', z', w') we write $(x' - py', x' - y', z' - qw', z' - w')$, the new values of M', N' are found to be

$$\begin{aligned} M' &= p(q-1)^2 x'w' - q(p-1)^2 y'z' + (pq-1)(p-q) y'w', \\ N' &= \ \ (q-1)^2 x'w' - \ \ (p-1)^2 y'z' + (pq-1)(p-q) y'w', \end{aligned}$$

and we have the formulæ of No. 102.

(B) *The Principal System of a Proper Sextic Curve; the Lineo-linear Transformation between Two Spaces.*

105. I start with the lineo-linear transformation, and show that this is in fact a transformation such that the principal system in either space is a sextic curve with seven apparent double points. I do not attempt any formal proof, but assume that the lineo-linear transformation is the most general one which gives rise to such a principal system.

We have between (x, y, z, w), (x', y', z', w') three lineo-linear equations; writing these first under the form

$$\begin{aligned} (P_1, Q_1, R_1, S_1 \between x', y', z', w') &= 0, \\ (P_2, Q_2, R_2, S_2 \between x', y', z', w') &= 0, \\ (P_3, Q_3, R_3, S_3 \between x', y', z', w') &= 0, \end{aligned}$$

we have $x' : y' : z' : w' = X : Y : Z : W$, where X, Y, Z, W are the determinants (each with its proper sign) formed out of the matrix

$$\left| \begin{matrix} P_1, & Q_1, & R_1, & S_1 \\ P_2, & Q_2, & R_2, & S_2 \\ P_3, & Q_3, & R_3, & S_3 \end{matrix} \right| .$$

106. Each of the surfaces $X = 0$, $Y = 0$, $Z = 0$, $W = 0$, or generally any surface $aX + bY + cZ + dW = 0$, is thus a cubic surface passing through the curve

$$\left\| \begin{matrix} P_1, & Q_1, & R_1, & S_1 \\ P_2, & Q_2, & R_2, & S_2 \\ P_3, & Q_3, & R_3, & S_3 \end{matrix} \right\| = 0,$$

which is at once seen to be of the order 6. In fact any two of these surfaces, for instance

$$\begin{vmatrix} P_1, & Q_1, & R_1 \\ P_2, & Q_2, & R_2 \\ P_3, & Q_3, & R_3 \end{vmatrix} = 0 \text{ and } \begin{vmatrix} P_1, & Q_1, & S_1 \\ P_2, & Q_2, & S_2 \\ P_3, & Q_3, & S_3 \end{vmatrix} = 0,$$

have in common a curve

$$\left\| \begin{matrix} P_1, & P_2, & P_3 \\ Q_1, & Q_2, & Q_3 \end{matrix} \right\| = 0,$$

which is of the order 3; they consequently besides intersect in a curve of the order 6, which is the before mentioned curve of intersection of all the surfaces. And it further appears that the number of the apparent double points is $=7$; in fact the formula in the case of two surfaces of the orders μ, ν, the complete intersection of which consists of a curve of the order m with h apparent double points, and of a curve of the order m' with h' apparent double points, the numbers m, m', h, h' are connected by the equation $2(h-h')=(m-m')(\mu-1)(\nu-1)$. (Salmon's *Solid Geometry*, 2nd Ed., p. 273 [Ed. 4, p. 311]). Hence, in the case of the two cubic surfaces intersecting as above (since for the cubic curve we have $m'=3$, $h'=1$, and for the sextic $m=6$), the formula becomes $2(h-1)=12$, that is $h=1+6=7$; or the number of apparent double points is $=7$.

107. It thus appears that the principal system in the first plane is a curve of the order 6, with seven apparent double points: it is to be added that there are not in general any actual double points or stationary points, so that the class of the curve is $6.5-2.7$, $=16$, and its deficiency is $\frac{1}{2}5.4-7$, $=3$. For convenience I will refer to this as the curve Σ.

The transformation is obviously a symmetrical one; hence the principal system in the second space is in like manner a curve of the order 6, with seven apparent double points; say it is the curve Σ'.

108. Consider in the first space any point P on the curve Σ; for this point the three equations

$$(P_1, Q_1, R_1, S_1 \between x', y', z', w') = 0,$$
$$(P_2, Q_2, R_2, S_2 \between \quad " \quad) = 0,$$
$$(P_3, Q_3, R_3, S_3 \between \quad " \quad) = 0,$$

are not independent, but are equivalent to two linear equations in (x', y', z', w'); that is, to the point P on the curve Σ there corresponds in the second space, not a determinate point P', but any point whatever on a certain line L'; or say to the point P on Σ there corresponds a line L'; and as P describes the curve Σ, L' describes a scroll Π'; that is, to the curve Σ there corresponds a scroll Π', the principal counter-system in the second space. Similarly to the curve Σ' there corresponds a scroll Π, the principal counter-system of the first space.

109. The scroll Π is the Jacobian of the first space; and as such it is of the order 8, having the curve Σ for a triple line—and it thus appears that the Jacobian

of the first space is a scroll (a theorem the analytical verification of which seems by no means easy). But without assuming the identity of the scroll Π with this Jacobian, or taking the order of the scroll to be known, I proceed to show that the scroll Π is the scroll generated by the lines each of which meets the curve Σ three times; it will thereby appear that the order is $=8$, and that the curve is a triple line on the scroll.

Consider a point P' on Σ', and the corresponding line L of the first space: take Θ' a plane in the second space; corresponding to it the cubic surface Θ in the first space. By imposing a single relation on the coefficients (a, b, c, d) in the equation $ax'+by'+cz'+dw'=0$ of the plane Θ', we make it pass through the point P'; therefore by imposing this same *single* relation on the coefficients (a, b, c, d) of the cubic surface Θ, we make it pass through the line L; Θ is a cubic surface through Σ; and it is easy to see that the effect will be as above only if the line L cuts the curve Σ three times; this being so, the general cubic surface Θ meets L in three points (viz., the three intersections of L with Σ), and if Θ be made to pass through a fourth point on the line L, it will pass through the line L; it thus appears that the line L meets Σ three times, and consequently that the scroll Π is generated by the lines which meet Σ three times.

110. The theory of a scroll so generated is considered in my "Memoir on Skew Surfaces, otherwise Scrolls"(1). Writing $m=6$, $h=7$ and therefore $M\,[=-\tfrac{1}{2}m(m-1)+h], =-8$, the order of the scroll is $\left(\tfrac{1}{3}[m]^3+(m-2)M=40-32\right)=8$; but calculating the values of

$NG(m^3)=\tfrac{1}{2}[m]^4+6m+M(3[m]^2-12m+33)+M^2.3,$

$NR(m^3)=\tfrac{1}{18}[m]^6+\tfrac{3}{8}[m]^5-\tfrac{1}{2}[m]^3-3m+M(\tfrac{1}{3}[m]^4-\tfrac{1}{6}[m]^3-\tfrac{5}{2}m^2+8m-20)+M^2(\tfrac{1}{2}[m]^2-2m);$

these are found to be respectively $=0$; viz., there are no nodal generators, and no nodal residue; the sextic curve Σ is a triple curve on the surface, and there is not any other multiple line.

111. It may be remarked that any plane Θ' meets the sextic curve Σ' in six points; hence the corresponding cubic surface Θ contains six lines, generatrices of Π, and, therefore, each meeting the curve Σ three times; say six lines L. Through one of these lines L, draw to the cubic surface Θ a triple tangent plane meeting it in the line L and in two other lines, say M, N; this plane must meet Σ in three new points which must lie on the lines M, N; viz., one of these lines must pass through two of the points, and the other line through the third point.

Addition—September, 1870.

[Some corrections have been made in accordance with the concluding paragraph of a paper "Note on the Rational Transformation and on Special Systems of Points," 450.]

The formulæ of No. 84 are included in the following more general formulæ; viz., if the principal system consist of α_1 points, each a simple point, α_2 points each a

(1) *Phil. Trans.* vol. CLIII. 1863, pp. 453—483, [339]. See the Table $S(m^3)$ &c., p. 457; in the value of $NR(m^3)$ instead of term $+3m$ read $-3m$. [This correction should have been made in the present Reprint.]

quadri-conical point, α_3 points each a cubi-conical point, &c., and of a simple curve order m_1 with h_1 apparent double points, a double curve order m_2 with h_2 apparent double points, and so on; and if moreover, the curves m_1, m_2 intersect in $k_{1,2}$ points, the curves m_1, m_3 in $k_{1,3}$ points, &c.; then writing in general $\rho = \frac{1}{2}m(m-1) - h$; that is, $\rho_1 = \frac{1}{2}m_1(m_1 - 1) - h_1$, $\rho_2 = \frac{1}{2}m_2(m_2 - 1) - h_2$, &c., I find that the general condition of equivalence is

$$\left.\begin{array}{l} \alpha_1 + (\ 3n - \ 2)\, m_1 - \ 2\rho_1 \\ + 8\alpha_2 + (12n - 16)\, m_2 - 16\rho_2 \\ \quad\vdots \\ + r^3\alpha_r + (3r^2 n - 2r^3)\, m_r - 2r^3\rho_r \\ - 5k_{1,2} - 8k_{1,3} \ldots - (3r-1)\, k_{1,r} \\ \qquad - 28k_{2,3} \\ \qquad\qquad \ldots - s^2(3r - s)\, k_{s,r}\ (s < r) \end{array}\right\} = n^3 - 1;$$

and that the general condition of postulation is

$$\left.\begin{array}{l} \alpha_1 + (\ n + 1)\, m_1 - \ \rho_1 \\ + 4\alpha_2 + (3n + 1)\, m_2 - 5\rho_2 \\ \quad\vdots \\ + \frac{1}{6}r(r+1)(r+2)\,\alpha_r \\ + [\frac{1}{2}r(r+1)\, n - \frac{1}{6}r(r+1)(2r-5)]\, m_r \\ \qquad - \frac{1}{24}[(r-1)(r-2)(r-3)(r-4) \\ \qquad\qquad + 4r(r+1)(2r+1)]\,\rho_r \\ - 2k_{1,2} - 3k_{1,3} \ldots - rk_{1,r} \\ \quad - 8k_{2,3} \\ \qquad\vdots \\ \quad - \frac{1}{2}s(s+1)\{r + 1 - \frac{1}{3}(s+2)\}\, k_{s,r}\ (s < r) \end{array}\right\} = \frac{1}{6}(n+1)(n+2)(n+3) - 4:$$

in which formulæ it is however assumed that the curves have not any actual multiple points. This implies that if any one of the curves, say m_r, break up into two or more curves, the component curves do not intersect each other; for, of course, any such point of intersection would be an actual double point on the curve m_r. I believe, however, that the formulæ will extend to this case by admitting for s the value $s = r$; viz., if we suppose the curve m_r to be the aggregate of the two curves m_r', m_r'' intersecting in K_r points, then that the corresponding terms in the equivalence-equation are

$$(3r^2 n - 2r^3)(m_r' + m_r'') - 2r^3(\rho_r' + \rho_r'') - 2r^3 K_r,$$

and that those in the postulation-equation are

$$\begin{array}{l} [\frac{1}{2}r(r+1)\, n - \frac{1}{6}r(r+1)(2r-5)]\,(m_r' + m_r'') \\ \qquad - \frac{1}{24}[(r-1)(r-2)(r-3)(r-4) + 4r(r+1)(2r+1)]\,(\rho_r' + \rho_r'') \\ \qquad - \frac{1}{6}r(r+1)(2r+1)\, K_r. \end{array}$$

Let the r-tuple curve consist of three right lines meeting in a point: this is an actual triple point, and the formulæ do not apply. But calculating the postulation-terms by the formula, we have $m_r = 3$, $\rho_r = \frac{1}{2}3.2 - 0, = 3$; and the terms are

$$[\tfrac{1}{2}r(r+1)n - \tfrac{1}{6}r(r+1)(2r-5)]\,3 - \tfrac{1}{8}[(r-1)(r-2)(r-3)(r-4) + 4r(r+1)(2r+1)],$$

which are

$$= \tfrac{1}{2}r(r+1)(3n-4r+4) - \tfrac{1}{8}(r-1)(r-2)(r-3)(r-4),$$

or say

$$= \tfrac{1}{2}r(r+1)(3n-4r+4) + \tfrac{1}{8}(-r^4+10r^3-35r^2+50r-24).$$

I have found by an independent investigation that this value requires the correction

$$+\tfrac{1}{8}[r^4-8r^3+30r^2-56r+24+\tfrac{1}{2}\{1-(-)^r 1\}],$$

and that the true value of the postulation is

$$= \tfrac{1}{2}r(r+1)(3n-4r+6) + \tfrac{1}{8}[\quad 2r^3-5r^2\ -6r \qquad +\tfrac{1}{2}\{1-(-)^r 1\}],$$

viz., that this is the number of the conditions to be satisfied that a surface of the order n may have for an r-tuple curve three given right lines meeting in a point.

448.

NOTE ON THE CARTESIAN WITH TWO IMAGINARY AXIAL FOCI.

[From the *Proceedings of the London Mathematical Society*, vol. III. (1869—1871), pp. 181, 182. Read June 9, 1870.]

LET A, A', B, B' be a pair of points and antipoints; viz.,

(A, A') the two imaginary points, coordinates $(\pm\beta i,\ 0)$,

(B, B') the two real points, coordinates $(0,\ \pm\beta)$;

and write ρ, ρ', σ, σ' for the distances of a point (x, y) from the four points respectively; say

$$\rho\ = \sqrt{(x+\beta i)^2+y^2},\quad \sigma\ = \sqrt{x^2+(y+\beta)^2},$$

$$\rho' = \sqrt{(x-\beta i)^2+y^2},\quad \sigma' = \sqrt{x^2+(y-\beta)^2}.$$

We have

$$\rho^2+\rho'^2 = 2x^2+2y^2-2\beta^2 = \sigma^2+\sigma'^2-4\beta^2,$$

$$\rho\rho' = \sqrt{(x+\beta i+yi)(x+\beta i-yi)(x-\beta i+yi)(x-\beta i-yi)} = \sigma\sigma';$$

and thence

$$(\rho+\rho')^2 = (\sigma+\sigma')^2-4\beta^2,$$

$$(\rho-\rho')^2 = (\sigma-\sigma')^2-4\beta^2;$$

or say

$$\rho+\rho' \ = \sqrt{(\sigma+\sigma')^2-4\beta^2},$$

$$i(\rho-\rho') = \sqrt{4\beta^2-(\sigma-\sigma')^2}.$$

The equation of a Cartesian having the two imaginary axial foci A, A' is

$$(p+qi)\,\rho+(p-qi)\,\rho'+2k^2=0;$$

viz., this is

$$p(\rho+\rho')+qi(\rho-\rho')+2k^2=0;$$

or, what is the same thing, it is

$$p\sqrt{(\sigma+\sigma')^2+4\beta^2}+q\sqrt{4\beta^2-(\sigma-\sigma')^2}+2k^2=0,$$

which is the equation expressed in terms of the distances σ, σ' from the non-axial real foci B, B'. Of course, the radicals are to be taken with the signs $\pm$. This equation gives, however, the Cartesian in combination with an equal curve situate symmetrically therewith in regard to the axis of y.

The distances σ, σ' may conveniently be expressed in terms of a single variable parameter θ; in fact, we may write

$$\pm p\sqrt{(\sigma+\sigma')^2-4\beta^2}=-k^2-k\theta,$$

$$\pm q\sqrt{4\beta^2-(\sigma-\sigma')^2}=-k^2+k\theta;$$

that is

$$(\sigma+\sigma')^2-4\beta^2=\frac{k^2}{p^2}(k+\theta)^2,$$

$$4\beta^2-(\sigma-\sigma')^2=\frac{k^2}{q^2}(k-\theta)^2;$$

and therefore

$$\sigma+\sigma'=\sqrt{4\beta^2+\frac{k^2}{p^2}(k+\theta)^2},$$

$$\sigma-\sigma'=\pm\sqrt{4\beta^2-\frac{k^2}{q^2}(k-\theta)^2}:$$

so that, assigning to θ any given value, we have σ, σ', and thence the position of the point on the curve. We may draw the hyperbola $y^2=4\beta^2+\frac{k^2}{p^2}x^2$, and the ellipse $y^2=4\beta^2-\frac{k^2}{q^2}x^2$; and then measuring off in these two curves respectively the ordinates which belong to the abscissæ $k+\theta$ for the hyperbola, $k-\theta$ for the ellipse, we have

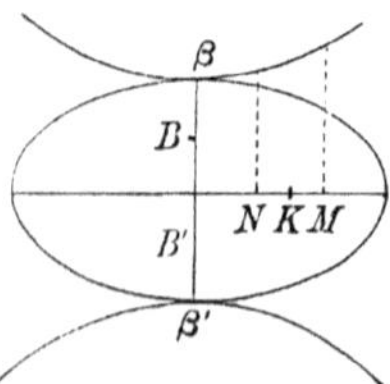

the values $\sigma+\sigma'$ and $\sigma-\sigma'$, which determine the point on the curve. Considering k, p, q, β as disposable quantities, the conics may be any ellipse and hyperbola whatever, having a pair of vertices in common; and the complete construction is,—From the fixed point K in the axis of x, measure off in opposite directions the equal distances KM, KN, and take

$\sigma+\sigma'$ the ordinate at M in the hyperbola,

$\pm(\sigma-\sigma')$ „ „ N „ ellipse;

where σ, σ' denote the distances of the required point from the fixed points B and B' respectively, the distance of each of these from the origin being $=\frac{1}{2}$ the common semi-axis. We may imagine N travelling from one extremity of the x-axis of the ellipse to the other, the value of $\sigma+\sigma'$ will be real and greater than BB', that of $\sigma-\sigma'$ real and less than BB', and the point (σ, σ') will be real. The construction gives, it will be observed, the two symmetrically situated curves.

The x-semi-axis of the ellipse is $\frac{q}{k}2\beta$, and the form of the curve depends chiefly on the value of the ratio $k : \frac{q}{k}2\beta$; or, what is the same thing, $k^2 : 2\beta q$. We see, for instance, that, in order that the curve may meet the axis of y in two real points between the foci, the value $\theta=-k$ must give a real value of $\sigma-\sigma'$; viz., that we must have $4\beta^2>\frac{4k^4}{q^2}$; that is, $\beta^2q^2>k^4$, or $k^2<\beta q$. If k has this value, viz., $k=\frac{1}{2}\frac{q}{k}2\beta=$ $\frac{1}{2}$ semi-axis, the curve *touches* the axis of y at the origin; if $k<\frac{1}{2}$ semi-axis, the curve cuts the axis of y in two real points between the foci; if $k>\frac{1}{2}$ semi-axis, the curve does not cut the axis of y between the foci.

449.

SKETCH OF RECENT RESEARCHES UPON QUARTIC AND QUINTIC SURFACES.

[From the *Proceedings of the London Mathematical Society*, vol. III. (1869—1871), pp. 186—195. Read Nov. 10, 1870.]

THE classification of quartic surfaces is even as to its highest divisions incomplete; and it is by no means easy to make it at once exhaustive and precise; an enumeration of all the *primâ facie* possible cases would include forms which do not really exist. Thus the singular curve (if any) is of the order 1, 2, or 3—but in the case where the order is $=3$, the curve, as is at once evident, cannot be a plane cubic, nor (among other excluded forms) a system of three non-intersecting lines. And certain forms of the singular curve, e.g. all but one of the admissible forms of a curve of the order 3, make the surface to be a scroll, so that, if (as is convenient) we wish to separate the scrolls, certain forms otherwise admissible must be excluded. The expression "singular" means double or cuspidal, or refers to a higher singularity, but the cases of higher singularity are very special. I will, at the cost of some inaccuracy, use the expression "nodal" as meaning, in general, double, but as including the signification "cuspidal"; and, if there are any cases of higher singularity, as extending to cases of higher singularity: and I provisionally arrange the non-scrolar quartic surfaces as follows:

1. Without a nodal curve.

2. With a nodal line.

3. With a nodal conic, or line-pair (pair of intersecting lines).

4. With three nodal lines (not in the same plane) meeting in a point.

(Observe that the omitted cases are cases which, as I believe, ought to be omitted; thus the case of a nodal skew cubic is omitted, because the surface is then of necessity a scroll.) And to these I join:

5. The quartic scrolls;

omitting altogether the torse and cones.

The references, by the name of the author, and number (if any) of his paper, are to the subjoined list of Memoirs.

As to the scrolls, we have Cayley (3) and (4), and Cremona; the division into 12 species is, I believe, complete: see *post*, the remarks upon Schwarz's paper on quintic scrolls.

As regards the non-scrolar surfaces:

1. Without a singular curve. The surface may be without a cnicnode (conical point), or it may have any number of cnicnodes up to 16, Cayley (7): the cases of singularity higher than a cnicnode are probably very numerous, but they have been scarcely at all examined. The memoir just referred to relates chiefly to the several cases of not more than 10 nodes; the cases of 11, 12, 13, 14, 15, 16 nodes are considered incidentally, Kummer (2), but it was not the object of his paper to make an enumeration, and there *may be* cases which are not considered; the discussion of the cases considered is very full and interesting. The case of 16 nodes is also considered, Kummer (1). As to the surface with 16 nodes, it is to be remarked that the wave-surface, or generally the surface obtained by the homographic deformation of the wave-surface—called, Cayley (1), the "tetrahedroid"—is a special form of surface with 16 nodes: its relation to the general surface is explained, Cayley (2).

2. Quartic surface with nodal line: considered incidentally, Clebsch (2) and (3). There are through the nodal line 8 planes, each meeting the surface in a line-pair: considering any 7 of these, and taking out of each of them a line, the 7 lines are met by a conic which also meets a determinate line out of the remaining line-pair; there are thus on the surface 2^7, $=128$, conics; viz., these form 64 pairs, each pair lying in a plane, and being the complete intersection of the surface by such plane; the number of these planes is of course $=64$.

Although not properly included in the present case, I mention the quartic surface which is the reciprocal of the cubic surface $\text{XIX} = 12 - B_6 - C_2$, Cayley (5): the nodal curve is here an oscnodal line counting as three nodal lines.

3. Quartic surfaces with nodal conic. Such a surface may be without cnicnodes, or it may have 1, 2, 3, or 4 cnicnodes; the cases, other than that of 3 cnicnodes, are mentioned, Kummer (3); but the question is examined, and the remaining case of 3 cnicnodes established, Cayley (6).

The general case of the nodal conic without cnicnodes is elaborately considered, Clebsch (1): it is shown that there are on the surface 16 lines, each meeting the conic, and which in their arrangement are strikingly analogous to the 27 lines on a cubic surface; viz., if on a cubic surface we select at pleasure any one of the 27 lines, and through this line draw a plane which besides meets the cubic surface in a conic; then, disregarding the line in question and the 10 lines which meet it, the remaining 16 lines each meet the conic, and are related to it and to each other in the same manner that the 16 lines of the quartic surface are related to the nodal conic and to each other. And the ground hereof appears, Geiser; viz., it is shown that the quartic surface with the nodal conic, is rationally transformable into a cubic surface, the 16 lines and the nodal conic corresponding respectively to the 16 lines and the conic of the cubic surface.

The several cases of 1, 2, 3, and 4 cnicnodes are considered, Korndörfer.

In the case where the nodal conic is the circle at infinity, the surfaces have been termed "anallagmatic" (perhaps "bicircular" would be a more convenient name), and a great deal has been written upon these surfaces by Moutard, Clifford, and others. Such a surface may of course have 1, 2, 3, or 4 cnicnodes; these surfaces, viz. the cnicnodal anallagmatics, in fact arise from the inversion of a quadric surface by the method of reciprocal radius vectors $\left(\text{that is, by the change of } x, y, z \text{ into } \frac{x}{r^2}, \frac{y}{r^2}, \frac{z}{r^2}\right)$: the centre of inversion is a node on the quartic surface. If the quadric surface is a cone, there is another node, the inverse point of the vertex; if the quadric surface is one of revolution, there are two other nodes; and if it is a cone of revolution, there are three other nodes—viz., in all, four nodes. The last-mentioned surface is, or includes, the Cyclide; viz., this is a quartic surface having the circle at infinity for a nodal curve, and having besides four nodes, which are a system of skew antipoints. The surface was first considered by Dupin (*Applications de Géométrie &c.*, 1822) as the envelope of a sphere touching three given spheres—its lines of curvature are thus circles; and the surface has been very frequently considered in reference to this property and otherwise: see Maxwell, where a classification (not quite complete) is made of the different forms of the surface, and also stereographic drawings given. It is to be observed, that one interesting form, the parabolic cyclide, is not a quartic but a cubic surface.

In the class of surfaces which have been under consideration, the cnicnodes have been points not on the nodal conic—in fact, a point on a nodal curve cannot be, properly speaking, a cnicnode, though it may be a point of higher singularity in the nature of a cnicnode; viz., there may be on the nodal curve points which, in the classification of the surfaces, must be counted as cnicnodes. Such a case presents itself in the "Conic Torus," or surface generated by the rotation of a conic about a line whether not in or in the plane of the conic. The surface has been considered, De la Gournerie (1), although more in reference to the constructions of descriptive geometry than as a theory of pure geometry, and Cayley (6). The surface has a nodal circle, and upon it two singular points, the circular points at infinity; so that it belongs to the case of a nodal conic with two cnicnodes. In the particular case where the axis of rotation is in the plane of the conic, then there are on the axis two cnicnodes; so that the case is that of a nodal conic and four cnicnodes; and when the generating conic is a circle, viz., when the surface is the ordinary torus, or anchor ring, generated by the rotation of a circle about a line in its own plane, then the nodal conic is the circle at infinity having upon it two cnicnodal points (its intersections by the planes at right angles to the axis) and the surface has also two cnicnodes on the axis: the surface, although presenting considerable peculiarity, may be regarded as a particular case of the cyclide. In reference to the plane sections of the conic torus and its various particular cases, see De la Gournerie (2); the ordinary torus has been the subject of numerous papers by Darboux and others, and possesses very interesting properties.

In connexion with the foregoing, I speak of the surfaces having a cuspidal conic: the general case is briefly referred to, Cayley (6); viz., this is the surface (AA) the equation of which is $V^2 - x^3y = 0$, and which it is shown has a reciprocal of the order 6. A special case is the surface (AB) having a reciprocal of the order 3; viz., the quartic surface is here the reciprocal of the cubic surface $\text{XX} = 12 - U_8$, Cayley (5). And it appears from the memoir last referred to, that there is another cubic surface, $\text{XVII} = 12 - 2B_3 - C_2$, the reciprocal of which is a quartic surface having a cuspidal conic. But the theory of the quartic surfaces with a cuspidal conic has been hardly at all considered.

I do not know that anything has been done in regard to the quartic surfaces where the nodal conic becomes a line-pair; that is, where we have two intersecting nodal lines. Although not properly belonging to the case in question, I mention here the quartic surface which is the reciprocal of the cubic surface $\text{XVIII} = 12 - B_4 - 2C_2$, Cayley (5); the nodal curve consists of two intersecting lines, but one of them is tacnodal, counting as two nodal lines.

4. Quartic surface with three nodal lines (not in the same plane) meeting in a point. This is, in fact, Steiner's quartic surface; and it has been the subject of numerous investigations.

The equation of the surface may be taken to be $\sqrt{x} + \sqrt{y} + \sqrt{z} + \sqrt{w} = 0$; and the surface thus presents itself as the reciprocal of the cubic surface $\frac{1}{x} + \frac{1}{y} + \frac{1}{z} + \frac{1}{w} = 0$, $(\text{XVI} = 12 - 4C_2,)$ with four cnicnodes [1].

It was convenient to make the foregoing enumeration before speaking of Kummer's paper (3), and of the several memoirs which relate to the Abbildung of certain quartic and quintic surfaces.

As regards Kummer's paper, the object appears by the title, viz., he considers in what cases a quartic surface has upon it a system of conics; or, what is the same thing, in what cases there is a system of planes each intersecting the surface in two conics. It is, in the first place, remarked that there is no proper quartic surface cut by every plane in a pair of conics, or even a proper quartic surface cut in a pair of conics by every plane through a fixed point. The cases considered are—I., where the planes are non-tangent planes; II., where they are single tangent planes; and III., where they are double tangent planes. The case I. is—(1) when there is a nodal conic and two cnicnodes; viz., any plane through the 2 cnicnodes gives a section with 4 nodes, therefore a pair of conics, (and the special case of 4 cnicnodes is noticed

[1] The Author exhibited, and pointed out some of the properties of, a model of Steiner's surface.

incidentally);—(2) when there is a nodal line; any plane through the nodal line besides meets the surface in a conic;—(3) when the surface has two "Selbstberührungspuncte"; viz., either of these is a point where the tangent plane is replaced by two coincident planes, and which, when a plane passes through it, gives in the section a tacnode, $=2$ nodes; the section by a plane through two such points, consists of two conics touching each other at the point in question: the equation of such a surface is $\phi^2=(*\between p, q)^4$, where ϕ is a quadric function, and p, q linear functions of the coordinates. In all the cases the planes pass through a fixed line, and the surface may be considered as the locus of a variable conic, the plane of which always passes through such line. II. is—(1) Steiner's surface, where every tangent plane meets the surface in a pair of conics; and (2) surface with a nodal conic and one cnicnode, where every tangent plane through the cnicnode meets the surface in a pair of conics. And III. is (1) the surface with a nodal conic, where every double tangent plane meets the surface in a pair of conics: it is shown that there are 5 quadric cones, such that a tangent plane of any one of these cones is always a double tangent plane of the surface. Or the surface is (2) a quartic scroll; any plane through two intersecting lines of the surface besides meets the surface in a conic.

It is in the paper, Cayley (6), remarked, that the quartic surface $(*\between U, V, W)^2=0$ can also be expressed in the form $UW-V^2=0$; under which form the surface is seen to be the envelope of the series of quadric surfaces $(U, V, W\between\theta, 1)^2=0$; and by reason of this property it is very easy to find the equations of the reciprocal surfaces, or plane-equations of the quartic surfaces in question. And, in the same paper, it is noticed that the surfaces of the form in question include the reciprocals of several interesting surfaces of the orders 6, 8, 9, 10, and 12; viz., order 6, parabolic ring: order 8, elliptic ring: order 9, centro-surface of paraboloid: order 10, parallel surface of paraboloid; envelope of planes through the points of an ellipsoid at right angles to the radius vectors from the centre: order 12, centro-surface of ellipsoid; parallel surface of ellipsoid.

It will be noticed that several of the papers by Clebsch and others refer in their titles to the "Abbildung" of a surface; viz., they show that a (1, 1) correspondence exists between the points of the surface and the points of a plane. The most simple instance is the quadric surface; here, taking any fixed point O on the surface, the line OP drawn to any point P on the surface meets a plane in a point P', and the points P, P' have, it is clear, a (1, 1) correspondence. And, of course, to any curve on the quadric surface there corresponds a curve on the plane, and the discussion of the nature of the plane curves which correspond to the different curves on the quadric surface would constitute a theory of the Abbildung of the quadric surface.

Similarly, as remarked, Clebsch (2), for a cubic surface, taking upon it any two lines which do not meet, if from a point P on the surface we draw, meeting each of the two lines, a line to meet the plane in P', then the point P on the cubic surface and the point P' on the plane will have a (1, 1) correspondence; and we have thus a like theory for the cubic surface. The Abbildung of a cubic surface had been, however, previously effected by Clebsch {in the paper "Die Geometrie auf den Flächen dritter Ordnung," *Crelle*, t. LXV. (1866), pp. 359—380}, and by Cremona, in a different and really the most simple manner [1], but having a less obvious geometrical signification.

For surfaces of the higher orders, it is only certain surfaces which admit of an Abbildung, or (1, 1) correspondence of the points thereof with the points of a plane; viz. (in the same way as a plane curve, in order to its being unicursal, must have a sufficient number of nodes or cusps) a surface, in order that it may thus correspond with the plane (or say, in order that it may be unicursal), must have a sufficient singularity in the way of a nodal or cuspidal curve. The quartic and quintic surfaces considered in the enumerated memoirs are there considered for the sake of the Abbildung theory which they give rise to; whereas, in the present sketch, the Abbildung theory is considered only for the sake of the quartic and quintic surfaces to which the theory has been applied. But the methods of the theory furnish results in relation to these surfaces; and it is proper to give some account of them.

Clebsch's memoirs (2) and (3) relate to the same subject, which is elaborately treated in the latter of them: the former of them contains, however, some valuable remarks which are not reproduced in the other. In these memoirs (2) and (3), after explaining the above method of the transformation of a cubic surface by means of two of the lines thereof, the author goes on to notice that the like method is applicable to certain quartic and quintic surfaces; viz., (1) quartic surface with a nodal conic: there are here, as already mentioned, 16 lines, each meeting the conic; if, selecting any one of these, from a point P on the surface we draw, meeting the line and the conic, a line to cut the plane in P', then the points P on the surface and P' on the plane have a (1, 1) correspondence. (2) Quartic surface with a nodal line: as already mentioned, there are on the surface 128 conics, each meeting the nodal line; selecting any one of these, if from a point P of the surface we draw, meeting the nodal line and the conic, a line cutting the plane in P', then the point P on the surface, and the point P' on the plane, have a (1, 1) correspondence.

Similarly, (3), for a quintic surface having a nodal skew cubic; then if from a point P on the surface we draw, meeting the skew cubic twice, a line to cut the plane in P', the point P on the surface and the point P' on the plane have a (1, 1) correspondence. The nodal skew cubic may break up into a conic and line which meets it, or into three lines, two of them not meeting each other, but each met by the third line; and the like theory applies to these quintic surfaces.

It is to be noticed that (as for the cubic surface) the above methods of Abbildung, although they have the most obvious geometrical significations, are (as explained in the foregoing foot-note) not the most simple ones; but for each of the foregoing cases (1), (2), (3), the most simple transformation is established in the memoirs now under consideration. The memoir [Memoirs (1) and (2)] of Korndörfer, as indicated by its title, relates to the Abbildung of a quartic surface having a nodal conic and 1, 2, 3, or 4 cnicnodes.

[1] Any method of transformation leads to an expression of the coordinates of a point on the surface as proportional to rational and integral functions of a given degree ν of the coordinates (x, y, z) of a point on the plane, and that transformation is the more simple for which ν has the smaller value: for the method of the text, the value is $\nu=3$, but for the methods previously given by Clebsch and Cremona, it is $\nu=2$.

Clebsch's paper (4) relates to the Abbildung of a quartic scroll.

As regards quintic surfaces (not being scrolls), we have, so far as I am aware, only the before-mentioned paper, Clebsch (3), relating to quintic surfaces with a nodal skew cubic; and the paper, Clebsch (5), which relates to the Abbildung of a quintic surface having a nodal quadriquadric. The method employed is that of a preliminary Abbildung upon a twofold plane (2-blättrige Ebene); that is, it consists in establishing, in the first instance, a (1, 2) correspondence between the surface and the plane; and by means hereof it is shown that there exist on the surface the conics K and C presently referred to, and which give, ultimately, an ordinary Abbildung or (1, 1) correspondence of the points of the surface with those of the plane; viz., this final result is as follows:

There is on the surface a system of conics K, such that their planes pass through a point and envelope a quadricone; and also 64 conics C each meeting each of the conics K in a single point: we select one of these and call it the conic C.

Take now the plane B_1 of a conic K_1 of the series of conics K, which plane B_1 passes, of course, through the vertex V of the cone enveloped by the planes of the conics K; viz., these planes intersect the plane B_1 in a series of lines passing through the point V.

Take a point P on the surface; this lies on a conic K meeting the conic C in a point ξ; and if we draw the line ξP to meet the plane B_1 in P', then P on the quintic surface, and P' on the plane B_1, will have a (1, 1) correspondence; in fact, it appears that, given P, there exists a single position of P'; and conversely, given P', this lies on a line VP' through which there passes the plane B_1 and one other tangent plane of the cone: this tangent plane contains a conic K meeting the conic C in a point ξ; and joining $\xi P'$, this meets the conic K in one other point P; viz., given P', there is a single position of P; and there is thus a (1, 1) correspondence.

There are, as originally shown by Schläfli, and as further appears by my memoir on cubic surfaces, Cayley (5), 3 kinds of cubic surfaces of the class 5; viz., these are the surfaces $\text{XIII} = 12 - B_3 - 2C_2$, $\text{XIV} = 12 - B_5 - C_2$, and $\text{XV} = 12 - U_7$; for each of these the reciprocal surface is a quintic surface of the class 3, having a nodal line and a cuspidal quartic curve. For the reciprocal of XIII, the cuspidal curve is a quadriquadric; for that of XIV, the cuspidal curve breaks up into the nodal line (viz., this is a cuspnodal line) and into a skew cubic; for that of XV, the cuspidal curve is a cuspidal quadriquadric, or curve of intersection of two quadric surfaces with singular contact.

It only remains to speak of Schwarz's memoir on quintic scrolls: it is to be remarked that the theory of scrolls is allied more closely with that of plane curves than with that of surfaces; viz., considering any plane section of the scroll, the lines of the scroll have, in general, a (1, 1) correspondence with the points of the plane section, and the scrolls of any given order are properly arranged according to the deficiency of the plane section. This is what is done by Cremona in the memoir on quartic scrolls above referred to; viz., for a quartic scroll the deficiency is either 0 or 1; and of the 12 species, there are 10 for which the deficiency is $=0$ (or which are unicursal), and 2 for which the deficiency is $=1$. And this is the principle of classification in Schwarz's memoir; viz., for a quintic scroll the deficiency is $=0$, 1, or 2; the number of species established being 10, 4, and 1 for these deficiencies respectively.

List of Memoirs.

Cayley. 1. Sur la surface des ondes. *Liouv.* t. XI. (1846), pp. 291—296, [47].

——— 2. Sur un cas particulier de la surface du quatrième ordre avec seize points singuliers. *Crelle,* t. LXV. (1866), pp. 284—290, [356].

——— 3. Second Memoir on Skew Surfaces, otherwise Scrolls. *Phil. Trans.,* vol. CLIV. (1864), pp. 559—577, [340].

——— 4. Third Memoir on Skew Surfaces, otherwise Scrolls. *Phil. Trans.,* vol. CLIX. (1869), pp. 111—126, [410].

——— 5. Memoir on Cubic Surfaces. *Phil. Trans.,* vol. CLIX. (1869), pp. 231—326, [412].

——— 6. On the Quartic Surfaces $(*\between U, V, W)^2=0$. *Quart. Math. Journ.* vol. X. (1868), pp. 24—34; and vol. XI. (1870), pp. 15—25, and pp. 111—113.

——— 7. A Memoir on Quartic Surfaces. *Proc. Lond. Math. Soc.,* vol. III. (1870), pp. 19—69, [445].

Clebsch. 1. Ueber die Flächen vierter Ordnung welche eine Doppelcurve zweiten Grades besitzen. *Crelle,* t. LXIX. (1868), pp. 355—358.

——— 2. Intorno alla rappresentazione di superficie algebriche sopra un piano. *Atti del R. Ist. Lomb.* (12 Nov. 1868), 13 p.

——— 3. Ueber die Abbildung algebraischer Flächen, insbesondere der vierten und fünften Ordnung. *Math. Ann.,* t. I. (1869), pp. 253—316.

——— 4. Ueber die ebene Abbildung der geradlinigen Flächen vierter Ordnung, welche eine Doppelcurve dritten Grades besitzen. *Math. Ann.,* t. II. (1870), pp. 445—466.

——— 5. Ueber die Abbildung einer Classe von Flächen fünfter Ordnung. *Gött. Abh.,* t. XV. 64 p.

Cremona. Sulle superficie gobbe di quarto grado. *Mem. di Bologna,* t. VIII. (30 April, 1868), 15 p.

De la Gournerie. 1. Mémoire sur la surface engendrée par la révolution d'une conique autour d'une droite située d'une manière quelconque dans l'espace. *Jour. de l'Éc. Polyt.,* t. XXIII. (1863), pp. 1—74.

——— 2. Mémoire sur les lignes spiriques. *Liouv.,* t. XIV. (1869), 92 p.

Geiser. Ueber die Flächen vierten Grades welche eine Doppelcurve zweiten Grades haben. *Crelle,* t. LXX. (1869), pp. 249—257.

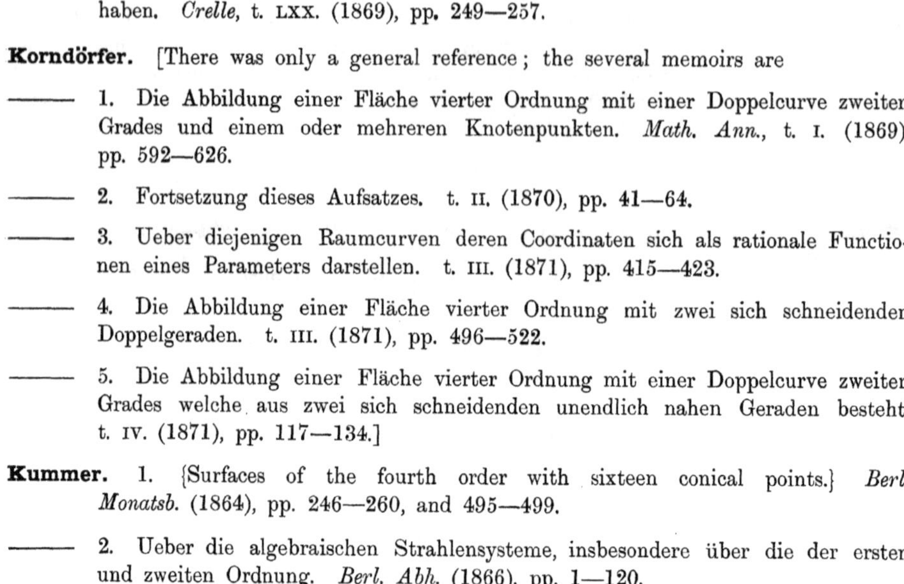

Korndörfer. [There was only a general reference; the several memoirs are

—— 1. Die Abbildung einer Fläche vierter Ordnung mit einer Doppelcurve zweiten Grades und einem oder mehreren Knotenpunkten. *Math. Ann.,* t. I. (1869), pp. 592—626.

—— 2. Fortsetzung dieses Aufsatzes. t. II. (1870), pp. 41—64.

—— 3. Ueber diejenigen Raumcurven deren Coordinaten sich als rationale Functionen eines Parameters darstellen. t. III. (1871), pp. 415—423.

—— 4. Die Abbildung einer Fläche vierter Ordnung mit zwei sich schneidenden Doppelgeraden. t. III. (1871), pp. 496—522.

—— 5. Die Abbildung einer Fläche vierter Ordnung mit einer Doppelcurve zweiten Grades welche aus zwei sich schneidenden unendlich nahen Geraden besteht. t. IV. (1871), pp. 117—134.]

Kummer. 1. {Surfaces of the fourth order with sixteen conical points.} *Berl. Monatsb.* (1864), pp. 246—260, and 495—499.

—— 2. Ueber die algebraischen Strahlensysteme, insbesondere über die der ersten und zweiten Ordnung. *Berl. Abh.* (1866), pp. 1—120.

—— 3. Ueber die Flächen vierten Grades auf welchen Schaaren von Kegelschnitten liegen. *Berl. Monatsb.* (July, 1863). *Crelle,* t. LXIV. (1864), pp. 66—76.

Maxwell. On the Cyclide. *Quart. Math. Journ.,* t. IX. (1867), pp. 111—126.

Schwarz. Ueber die geradlinigen Flächen fünften Grades. *Crelle,* t. LXVII. (1867), pp. 23—57.

450.

NOTE ON THE THEORY OF THE RATIONAL TRANSFORMATION BETWEEN TWO PLANES, AND ON SPECIAL SYSTEMS OF POINTS.

[From the *Proceedings of the London Mathematical Society*, vol. III. (1869—1871), pp. 196—198. Read December 8, 1870.]

In Prof. Cremona's theory of the transformation of plane curves, the fundamental equations are taken to be

$$\alpha_1 + 4\alpha_2 + 9\alpha_3 + \ldots = n^2 - 1 \qquad (1),$$

$$\alpha_1 + 3\alpha_2 + 6\alpha_3 + \ldots = \tfrac{1}{2}(n^2 + 3n) - 2 \qquad (2);$$

and from these we have as a consequence

$$\alpha_2 + 3\alpha_3 + \ldots = \tfrac{1}{2}(n-1)(n-1) \qquad (3);$$

viz., the first equation expresses that any two curves of the system intersect in a single variable point; the second, that the curves form a *réseau*, or system containing two arbitrary parameters; and the third, that the curves are unicursal.

In the equivalent theory of the rational transformation between two planes, as given in my "Memoir on the Rational Transformation between Two Spaces," [447], we have the equation (1); but instead of the equation (2), it would *primâ facie* appear to be sufficient if we had the inequation

$$\alpha_1 + 3\alpha_2 + 6\alpha_3 + \ldots \stackrel{=}{<} \tfrac{1}{2}(n^2 + 3n) - 2\,;$$

but on the ground there explained, the case

$$\alpha_1 + 3\alpha_2 + 6\alpha_3 + \ldots < \tfrac{1}{2}(n^2 + 3n) - 2$$

is excluded, and we thus have the equation (2), giving with (1) the equation (3).

I believe the better course is to assume (1) and (3) as the fundamental equations, from them deducing (2); and we thus also get over a difficulty presently referred to, but which did not occur to me when the memoir was written.

In fact, starting with the equations $x' : y' : z' = X : Y : Z$ (which are to give $x : y : z = X' : Y' : Z'$), we have in the first instance the equation (1). Moreover, establishing for x', y', z' a linear equation $ax' + by' + cz' = 0$, we have corresponding hereto a curve $aX + bY + cZ = 0$, and the coordinates x, y, z of a point on this curve are proportional to $X' : Y' : Z'$; that is, substituting for z' the value $-\frac{1}{c}(ax' + by')$, they are proportional to rational and integral (homogeneous) functions of (x', y'), that is, to rational and integral functions of the single parameter $x' : y'$; wherefore the curve $aX + bY + cZ = 0$ is unicursal; whence the equation (3). The like change may be made in the theory of the rational transformation between two spaces; and it is in this case a more important one.

The difficulty is as follows: It is not self-evident that we are at liberty to assume

$$\alpha_1 + 3\alpha_2 + 6\alpha_3 \ldots \mathrel{\substack{=\\<}} \tfrac{1}{2}(n^2 + 3n) - 2;$$

for imagine that we had a system of $(\alpha_1, \alpha_2, \alpha_3, \ldots)$ points, such that $\alpha_1 + 4\alpha_2 + \ldots$ being $= n^2 - 1$, and $\alpha_1 + 3\alpha_2 + \ldots$ being $> \frac{1}{2}(n^2 + 3n) - 2$, the points were such that the conditions in question (viz., the condition that the curve passes once through each of the points α_1, twice through each of the points $\alpha_2 \ldots$) should be *less* than $\alpha_1 + 3\alpha_2 + \ldots$, and in fact $=$ or $< \frac{1}{2}(n^2 + 3n) - 2$; then the functions X, Y, Z would not of necessity be connected by a linear relation $\lambda X + \mu Y + \nu Z = 0$, and the ground for the assumption in question, $\alpha_1 + 3\alpha_2 + \ldots \mathrel{\substack{=\\<}} \frac{1}{2}(n^2 + 3n) - 2$, would no longer exist. And except by the process now adopted of deriving the equation (2) from the equations (1) and (3), I do not know how the impossibility of such a system is to be established; viz., I do not know how we are to prove the following theorem:—There is not any system of $(\alpha_1, \alpha_2, \alpha_3 \ldots)$ points, where

$$\alpha_1 + 4\alpha_2 + 9\alpha_3 \ldots = n^2 - 1,$$

$$\alpha_1 + 3\alpha_2 + 6\alpha_3 \ldots > \tfrac{1}{2}(n^2 + 3n) - 2,$$

such that (for a curve of the order n passing once through each point α_1, twice through each point $\alpha_2, \ldots$) the number of conditions actually imposed on the curve is $=$ or $< \frac{1}{2}(n^2 + 3n) - 2$.

A system of $(\alpha_1, \alpha_2 \ldots)$ points such that the number of actually imposed conditions is less than $\alpha_1 + 3\alpha_2 + \ldots$, may be termed a special system; we have, of course, the well-known case $(\alpha_1 = n^2)$ of a system of n^2 points, such that any curve of the order n passing through $\frac{1}{2}(n^2 + 3n) - 1$ of these passes through all the remaining points {or what is the same thing, where the number of conditions actually imposed is $= \frac{1}{2}(n^2 + 3n) - 1$}; and we have the following special system, which presented itself to Dr Clebsch, in his researches on the Abbildung of a quintic surface with two non-intersecting nodal lines; viz., "$\alpha_1 = 12$, $\alpha_2 = 2$. We may have 12 points and 2 points such that, for a quintic curve passing once through each of the 12 points and twice through each of the 2 points, the number of conditions actually imposed (instead of being $12 + 3.2$, $= 18$) is $= 17$." The construction is as follows: viz., starting with the 2 points and any 10 points, we may draw a quartic passing twice through the first of the 2 points, once through the second of them, and through the 10 points; and another quartic passing twice through the second of the 2 points, once through the first of them, and through the 10 points: the two quartics intersect in the 2 points each twice, in the 10 points, and in 2 new points, forming, with the 10 points, a system of 12 points; and the first-mentioned 2 points and the 12 points form the system in question.

A more complicated case, $\alpha_1 = 10$, $\alpha_2 = 6$, $\alpha_3 = 1$, occurs in Dr Nöther's paper, "Ueber Flächen, welche Schaaren rationaler Curven besitzen," [*Math. Ann.*, t. III. (1871), pp. 161—227]. Except these two, I do not know any other case of a special system for which α_2, $\alpha_3 \ldots$ are not all $= 0$; the investigation of such systems would, I think, be very interesting.

[A concluding paragraph of seven lines gave some corrections to the "Memoir on the Rational Transformation between Two Spaces," 447, which corrections are made in the present reprint of that paper.]

451.

A SECOND MEMOIR ON QUARTIC SURFACES.

[From the *Proceedings of the London Mathematical Society*, vol. III. (1869—1871), pp. 198—202. Read December 8, 1870.]

IN my Memoir on Quartic Surfaces, *ante* pp. 19—69, [445], although remarking (see No. 79) that the identification was not completely made out, I tacitly assumed that the symmetroid and the decadianome (each of them a quartic surface with 10 nodes) were in fact identical. There is yet a good deal which I cannot completely explain; but the truth appears to be, that the decadianome includes two cases of coordinate generality, say the sextic decadianome, and the bicubic decadianome = symmetroid: viz., in the first of these the circumscribed cone, having for vertex any one of the 10 nodes, is a proper sextic cone with 9 double lines; in the second it is a system of two cubic cones, intersecting, of course, in 9 lines, which are double lines of the aggregate sextic cone: or, in the notation of the Table No. 11, in the case of the sextic decadianome, the circumscribed cones are each of them 6_9; in that of the bicubic decadianome = symmetroid, they are each of them (3, 3). We thus arrive at a very remarkable system of 10 points in space, viz., giving the name "ennead" to any 9 points *in plano*, which are the intersections of two cubic curves, or to any 9 lines through a point which are the intersections of two cubic cones; the 10 points in space are such that, taking any one whatever of them as vertex, and joining it with the remaining points, the 9 lines form an ennead. I purpose in the present short Memoir to consider the theories in question: the paragraphs are numbered consecutively with those of the Memoir on Quartic Surfaces.

Plane Sextic Curve with 9 Nodes.

110. A sextic curve contains 27 constants; and the number of conditions to be satisfied in order that a given point may be a node is = 3. Hence it would at first sight appear that the curve could be found so as to have 9 given nodes; this would

be 9×3, $=27$ conditions, or the curve would be completely determinate. But observe that through the 9 given points we have a determinate cubic curve $U=0$; we have therefore $U^2=0$ a sextic curve, and the only sextic curve with the 9 given nodes; that is, there is not in a proper sense any sextic curve with the 9 given nodes. The number of given nodes is thus $=8$ at most.

111. The sextic curve with 8 given nodes should contain $27-3.8=3$ constants. We may through the 8 given points draw the two cubics $P=0$, $Q=0$; and we have then $(a, b, c \between P, Q)^2=0$, a bicubic, or improper sextic curve having the 8 nodes, and also a ninth node, viz., the remaining point of intersection of the two cubic curves, or say the remaining point of the ennead. Hence if $\nabla=0$ be any particular sextic curve having the 8 given nodes, we have

$$(a, b, c \between P, Q)^2 + \theta\nabla = 0$$

a proper sextic curve having the 8 given nodes; and this, as containing the right number $(=3)$ of constants, will be the general sextic curve having the 8 given nodes.

112. There will be a ninth node if $\theta=0$; viz., the curve is then $(a, b, c \between P, Q)^2=0$, a bicubic, or improper sextic curve, having for nodes the 9 points of the ennead. Observe that the ninth node is here a point completely and uniquely determined by means of the given 8 nodes. Moreover the number of constants is $=2$, so that we have here a general (improper) solution of the question of finding a sextic curve with 9 nodes, 8 of them given.

113. But if θ is not $=0$, then the ninth node must be a point on the curve $J(P, Q, \nabla)=0$; viz., this is a curve of the order 9, determined by means of the given 8 points; say it is the "dianodal curve" of these 8 points, and, as is easy to see, it has each of these 8 points for a node. The ninth node of the sextic may be any point whatever on the dianodal curve; and regarding it as a given point, the sextic will still contain 1 constant; that is, we have the general solution of the problem of finding a sextic curve with 9 nodes, 8 of them given, and the 9th a given point on the dianodal curve.

114. So long as the 8 points are arbitrary, the dianodal curve does not pass through the 9th point of the ennead, and the two cases above considered are mutually exclusive. It will be noticed how closely analogous this theory of the plane sextic with 9 nodes, is to that of the quartic surface with 8 nodes.

115. Of course, instead of the plane sextic curve, we may have a sextic cone; such a cone has at most 8 given double lines; and if there be a 9th double line, then there are the two cases of coordinate generality; viz., (1), the new double line is the ninth line of the ennead, the cone being in this case not a proper sextic cone, but a bicubic cone; (2), the new double line may be any line whatever on the dianodal cone, (cone of the order 9 determined by the 8 given lines, and having each of these for a double line,) and regarding it as a given line on the dianodal cone, the sextic cone contains 1 constant.

Each circumscribed cone of the Symmetroid is (3, 3).

116. Using (x, y, z, w) as current coordinates of a point of the symmetroid, I take S, T, U, V to be quadric functions of the coordinates $(\alpha, \beta, \gamma, \delta)$; the equation of the symmetroid is therefore given by

$$xS + yT + zU + wV = \text{cone},$$

and the nodes thereof are determined by

$$xS + yT + zU + wV = \text{plane-pair}.$$

Suppose that a node is $(x=0,\ y=0,\ w=0)$; the condition for this is $V=$ plane-pair; and we may without loss of generality write $V=\gamma^2+\delta^2$. Hence, putting for shortness $\Theta = xS + yT + zU$, that is Θ a quadric function

$$(a, b, c, d, f, g, h, l, m, n \between \alpha, \beta, \gamma, \delta)^2,$$

wherein the coefficients $a, b, \ldots$ are arbitrary linear functions of (x, y, z), but not containing w, the equation of the symmetroid is given by

$$\Theta + w(\gamma^2+\delta^2) = \text{cone}.$$

117. It follows that the equation is

$$\begin{vmatrix} a, & h, & g, & l \\ h, & b, & f, & m \\ g, & f, & c+w, & n \\ l, & m, & n, & d+w \end{vmatrix} = 0;$$

viz., this is

$$\nabla + w(\delta_c + \delta_d)\nabla + \tfrac{1}{2}w^2(\delta_c+\delta_d)^2\nabla = 0,$$

where ∇ denotes the foregoing determinant, writing therein $w=0$. Or, observing that ∇ contains c and d each only linearly, the equation may be written

$$\nabla + w(\delta_c+\delta_d)\nabla + w^2\delta_c\delta_d\nabla = 0,$$

which is a quartic surface having, as it should have, the point $(0, 0, 0)$ for one of its ten nodes.

118. The equation of the circumscribed cone is

$$(\delta_c\nabla + \delta_d\nabla)^2 - 4\quad \nabla\,.\,\delta_c\delta_d\nabla = 0;$$

or, what is the same thing, it is

$$(\delta_c\nabla - \delta_d\nabla)^2 + 4(\delta_c\nabla\,.\,\delta_d\nabla - \nabla\,.\,\delta_c\delta_d\nabla) = 0.$$

But we have identically

$$\delta_c\nabla\,.\,\delta_d\nabla - (\tfrac{1}{2}\delta_n\nabla)^2 = \nabla\,.\,\delta_c\delta_d\nabla;$$

so that the equation is

$$(\delta_c \nabla - \delta_d \nabla)^2 + (\delta_n \nabla)^2 = 0 ;$$

a sextic cone breaking up into the two cubic cones

$$\delta_c \nabla - \delta_d \nabla \pm i\delta_n \nabla = 0,$$

so that the cone is (3, 3). And since clearly the point (0, 0, 0) may be regarded as representing any one whatever of the 10 nodes, it follows that for any node whatever of the symmetroid, the circumscribed cone is (3, 3), so that, as stated above, bicubic decadianome = symmetroid.

Deductions from the foregoing theory.

119. Referring to No. 85 of the original memoir, it appears that, with 6 given points as nodes, we can actually find for the symmetroid an equation containing 6 constants. I cannot discover any ground for doubting that 3 of these may be determined so as to give to the symmetroid a seventh given node; and I therefore assume that with 7 given points as nodes, an equation can be found with 3 constants. The symmetroid is certainly not octadic, hence the eighth node must lie on the dianodal surface of the 7 given points. I can discover no ground for doubting but that two of the constants may be determined so that the eighth node shall be any given point whatever on the dianodal surface of the 7 points; and (this being so) that further the remaining constant may be determined so that the ninth node shall be any given point whatever on the dianodal curve of the 8 points. But if all this be so, the consequence is very remarkable; the tenth node is not any one whatever of the 22 dianodal centres of the 9 points, but it is a uniquely determinate "enneadic centre," viz., we must have the following theorem:

120. "Take any 7 points; an eighth point at pleasure on the dianodal surface of the 7 points; a ninth point at pleasure on the dianodal curve of the 8 points. In the system of 9 points so determined, take any one as vertex, and joining it with the remaining 8, construct the ninth line of the ennead. Performing this construction with each of the 9 points successively as vertex, we obtain 9 lines passing through the 9 points respectively. These 9 lines meet in a point which is the 'enneadic centre' of the 9 points: and further, the 10 points form a completely symmetrical system, so that each one of them is the enneadic centre of the remaining 9."

121. Assuming that the 9 lines do intersect so as to give rise to an enneadic centre, there is no difficulty in conceiving that the loci, which by their intersection determine the dianodal centres, do each of them pass through the enneadic centre; so that this enneadic centre counts once or more among the dianodal centres, and the number of proper dianodal centres, instead of being $=22$, will be suppose $=22-\omega$, and if, further, the 9 points, together with the enneadic centre, are the nodes of a symmetroid, but the 9 points together with any one of the $22-\omega$ dianodal centres are the nodes of a sextic decadianome, then we must also have as follows:

122. "Considering any 9 points as above; taking any one as vertex, and joining it with the remaining 8, these 8 lines determine a dianodal ninthic cone. We have thus 9 dianodal cones, which cones pass all of them through the same $22 - \omega$ points."

123. I am not able to verify these theorems *à posteriori*. It appears to me that the theorem in regard to the enneadic centre subsists for a system of 9 points such as referred to in the statement; but that if by possibility the statement be too general, the theorem must, at all events, subsist for a more special system of 9 points; and that there certainly exist systems of 10 points, such that each 9 of the points have as an enneadic centre the tenth point. {I have since ascertained that if a quartic surface with 10 nodes has a single node (3, 3), the surface is a symmetroid; whence, by what precedes, the remaining nine nodes are each of them (3, 3). Added 25 March, 1871.}

124. I notice, as a subject of investigation, the following system of correspondence viz., given any 8 points in space: then to every point in space corresponds a line through this point, viz., the ninth line of the ennead obtained by joining the point with the 8 given points respectively; and to each line in space a point or points on the line, viz., the point or points for each of which the line is the ninth line of the ennead obtained by joining the point with the eight given points respectively.

452.

ON AN ANALYTICAL THEOREM FROM A NEW POINT OF VIEW.

[From the *Proceedings of the London Mathematical Society,* vol. III. (1869—1871), pp. 220, 221. Read February 9, 1871.]

THE theorem is a well-known one, derived from the equation

$$(az^2 + 2bz + c)\, w^2 + 2\,(a'z^2 + 2b'z + c')\, w + a''z^2 + 2b''z + c'' = 0\,;$$

viz., considering this equation as establishing a relation between the variables z and w, and writing it in the forms

$$2u = Aw^2 + 2Bw + C = A'z^2 + 2B'z + C' = 0,$$

(where, of course, A, B, C are quadric functions of z, and A', B', C' quadric functions of w,) we have

$$0 = \frac{du}{dw}\, dw + \frac{du}{dz}\, dz, \; = (Aw + B)\, dw + (A'z + B')\, dz;$$

but in virtue of the equation $u = 0$, we have $Aw + B = \sqrt{B^2 - AC}$, and $A'z + B' = \sqrt{B'^2 - A'C'}$, and the differential equation thus becomes

$$\frac{dw}{\sqrt{B'^2 - A'C'}} + \frac{dz}{\sqrt{B^2 - AC}} = 0,$$

where $B'^2 - A'C'$ and $B^2 - AC$ are quartic functions of w and z respectively. This is, of course, integrable (viz., the integral is the original equation $u = 0$); and it follows, from the theory of elliptic functions, that the two quartic functions must be linearly transformable into each other; viz., they must have the same absolute invariant $I^3 \div J^2$. It is, in fact, easy to verify, not only that this is so, but that the two functions have the same quadrinvariant I, and the same cubinvariant J.

The new point of view is, that we take the coefficients a, b, &c., to be homogeneous functions of (x, y), their degrees being such that the equation $u=0$ is a quartic equation $(* \between x, y, z, w)^4=0$; viz., this equation now represents a quartic surface having a node (conical point) at the point $(x=0, y=0, z=0)$, and also a node at the point $(x=0, y=0, w=0)$, say, these points are O, O' respectively. The equation $B'^2-A'C'=0$ gives the circumscribed sextic cone having O for its vertex, and the equation $B^2-AC=0$ the circumscribed sextic cone having O' for its vertex; each of these cones has the line OO' $(x=0, y=0)$ for a nodal line, as appears geometrically, and also by the equations containing z, w respectively in the degree 4. Considering $B'^2-A'C'$ as a quartic function of z, its quadrinvariant is a function $(x, y)^8$, and its cubinvariant a function $(x, y)^{12}$; and similarly, considering B^2-AC as a quartic function of w, its invariants are functions $(x, y)^8$ and $(x, y)^{12}$. We have thus, between the two cones, a geometrical relation answering to the analytical one of the identity of the invariants; but the nature of this geometrical relation is not obvious; and it presents itself as an interesting subject of investigation.

453.

ON A PROBLEM IN THE CALCULUS OF VARIATIONS.

[From the *Proceedings of the London Mathematical Society*, vol. III. (1869—1871), pp. 221, 222. Read February 9, 1871.]

THE problem is, $z = \frac{1}{3}(3x - y^2)y$, to find y a function of x such that $\int z dx = \text{max.}$ or min., subject to a given condition $\int y dx = c$ (the limits of each integral being x_1, x_0, where these quantities are each positive, and $x_1 > x_0$). The ordinary method of solution gives $y^2 = x + \lambda$, where $(x_1 + \lambda)^{\frac{3}{2}} - (x_0 + \lambda)^{\frac{3}{2}} = \frac{3}{2}c$; so long as c is not less than $(x_1 - x_0)^{\frac{3}{2}}$, there is a real value of λ, but for a smaller value of c there is no real value. The difficulty arising in this last case is somewhat illustrated by replacing the original problem by a like problem of ordinary maxima and minima; viz., $x_1, x_2 \ldots x_n$ being given positive values of x, in the order of increasing magnitude; and if, in general, $z_i = (3x_i - y_i^2) y_i$, then the problem is to find y_i a function of x_i, such that $\Sigma z_i = \text{max.}$ or min., subject to the condition $\Sigma y_i = c$. We have here $y_i^2 = x_i + \lambda$, where λ is to be determined by the condition $\Sigma y_i = c$; the remainder of the investigation turns on the question of the sign $y_i = +\sqrt{x_i + \lambda}$ or $y_i = -\sqrt{x_i + \lambda}$, to be taken for the several values of i respectively.

454.

A THIRD MEMOIR ON QUARTIC SURFACES.

[From the *Proceedings of the London Mathematical Society*, vol. III. (1869—1871), pp. 234—266. Read April 13, 1871.]

THE present Memoir is a continuation of my former researches on Nodal Quartic Surfaces, [445, 451]. The leading idea is, that for a quartic surface with k-nodes, given the nature of the circumscribed, $(k-1)$ nodal, sextic cone belonging to any one node of the surface {for instance, $k=10$, that it is a cone (3, 3) composed of two cubic cones}, we thereby determine the equation of the quartic surface, and consequently the nature of the remaining $(k-1)$ nodes thereof. By means of this general theory I complete, in an essential point, the theory of the *Symmetroid;* viz., I show that a 10-nodal quartic surface having a single node (3, 3) is a Symmetroid; whence, as appears by my second Memoir, [451], each of the remaining nine nodes is also a node (3, 3); and we have the theory of the remarkable system of ten points in space such that, joining any one of them with the remaining nine, the nine lines thus obtained are the intersections of two cubic cones. A large part of the Memoir is devoted to the consideration of the surfaces with 16, 15, 14, and 13 nodes: this is substantially a reproduction of the results obtained by Kummer in the Memoir "Ueber die algebraischen Strahlensysteme, &c.," already referred to; but the results in question are brought into connexion with the theory of the present Memoir, and they are, by a change of the constants, exhibited in a form of much greater symmetry and elegance. I attach importance also to the square diagrams by means of which I have exhibited, in a compendious form, the relation between the several nodes and circumscribed sextic cones.

The paragraphs are numbered consecutively with those of the first and second Memoirs.

Preliminary Considerations and Classification.

125. I call to mind that if a quartic surface has a node (conical point), then there is for this node a tangent quadricone and a circumscribed sextic cone; viz., if the surface has $(k-1)$ other nodes, or in all k nodes, then the sextic cone has $(k-1)$

nodal lines (passing through the other nodes respectively), and we have thus for the different forms of the sextic the table No. 11; viz., this is

CIRCUMSCRIBED SEXTIC CONE.

Nodes of Surface.	Nodal Lines of Cones.								
1	0	6							
2	1	6_1							
3	2	6_2							
4	3	6_3							
5	4	6_4							
6	5	6_5	5 , 1						
7	6	6_6	5_1, 1						
8	7	6_7	5_2, 1						
9	8	6_8	5_3, 1	4 , 2					
10	9	6_9	5_4, 1	4_1, 2	4 , 1, 1	3 , 3			
11	10	6_{10}	5_5, 1	4_2, 2	4_1, 1, 1	3_1, 3			
12	11	...	5_6, 1	4_3, 2	4_2, 1, 1	3_1, 3_1	3 , 2, 1		
13	12	...	...	...	4_3, 1, 1	...	3_1, 2, 1	3 , 1, 1, 1	2 , 2, 2
14	13	...	...	...	...	...	...	3_1, 1, 1, 1	2 , 2, 1, 1
15	14	...	...	...	...	...	...	...	2_1, 1, 1, 1, 1
16	15	...	...	...	...	...	...	...	1 , 1, 1, 1, 1, 1

viz., 6 denotes a proper sextic cone without nodal lines; 6_1 a proper sextic cone with one nodal line; 5, 1 a proper quintic cone and a plane, &c.

We may distinguish the nodes according to the sextic cones; thus, a node 6 means a node for which the circumscribed cone is a proper sextic cone, (1, 1, 1, 1, 1, 1) a node where the circumscribed cone breaks up into six planes, &c.

126. A 16-nodal surface has 16 nodes (1, 1, 1, 1, 1, 1), and a 15-nodal surface has 15 nodes (2, 1, 1, 1, 1); but, for a 14-nodal surface, the question arises how many nodes are (3_1, 1, 1, 1), and how many (2, 2, 1, 1). It was remarked, No. 13, that the only possible cases were 14, 0; 8, 6; or 2, 12; and that we might, in like manner, limit the number of possible cases for other values of k; but that the inquiry was not then further pursued. I resume this inquiry, but without obtaining as yet a complete answer.

127. It is to be observed that a line joining any two nodes is not, in general, a line on the surface, but that it may be so; the surfaces for which this is so (viz., any surface which contains upon it a line through two nodes) form, however,

a class by themselves, which at present I *altogether exclude from consideration.* This being so, it will appear in the sequel that there is but one kind of surface having a node (2, 2, 1, 1), and but one kind of surface having a node (3_1, 1, 1, 1). Now there is a surface, Kummer's 14-nodal, the nodes of which are 8 (3_1, 1, 1, 1) + 6 (2, 2, 1, 1); wherefore the two kinds are identical, and are each of them Kummer's 14-nodal surface. Similarly, for the 13-nodal surfaces, there is but one kind having a node (4_3, 1, 1), but one kind having a node (3, 1, 1, 1), and moreover but one kind having a node (3_1, 2, 1); and we have Kummer's 13-nodal surface with the nodes 3 (4_3, 1, 1) + 1 (3, 1, 1, 1) + 9 (3_1, 2, 1); hence the three kinds are identical with each other and with Kummer's. Moreover, there is but one kind having a node (2, 2, 2); hence all the other nodes must be (2, 2, 2), and we have a surface 13 (2, 2, 2) not given by Kummer. And in like manner for the 12-nodal surfaces, we have the two kinds given by Kummer, and a third kind 12 (4_2, 1, 1) not given by him; the arrangement thus far being

No. of Nodes.	Character of Surface.
16	16 (1, 1, 1, 1, 1, 1),
15	15 (2, 1, 1, 1, 1),
14	8 (3_1, 1, 1, 1) + 6 (2, 2, 1, 1),
13 (α)	3 (4_3, 1, 1) + 1 (3, 1, 1, 1) + 9 (3_1, 2, 1),
„ (β)	13 (2, 2, 2),
12 (α)	12 (4_3, 2),
„ (β)	2 (5_6, 1) + 6 (3_1, 3_1) + 4 (3, 2, 1),
„ (γ)	12 (4_2, 1, 1).

128. But in the next following case we have Kummer's surface, viz.

11 (α) 1 (6_{10}) + 10 (3_1, 3),

and I do not know whether one, two, or three kinds of surface having nodes (4_1, 1, 1), (4_2, 2), and (5_5, 1). And in the next case we have (as will appear) the Symmetroid, viz.,

10 (α) 10 (3, 3),

and I do not know how many kinds of surfaces having a node or nodes 6_9, (5_4, 1), (4_1, 2), (4, 1, 1).

It will be observed that the present division has nothing to do with the octadic and dianodal division in the former Memoir.

129. I consider a conic $A=0$, and any six tangents thereof, $t_1=0$, $t_2=0$, $t_3=0$, $t_4=0$, $t_5=0$, $t_6=0$; we have an identical equation which might be written $AC-B^2=t_1t_2t_3t_4t_5t_6$, but it will be convenient, introducing a constant factor K, to write it

$$AC-B^2=Kt_1t_2t_3t_4t_5t_6,$$

B being a cubic function and C a quartic function of the coordinates.

Consider now the series of factors, such as

$$\begin{aligned} &t_1, \\ &lA \quad + m\,t_1 t_2, \\ &sA \quad + m\,t_1 t_2 t_3, \\ &UA \quad + m\,t_1 t_2 t_3 t_4, \\ &VA \quad + m\,t_1 t_2 t_3 t_4 t_5, \\ &WA \quad + m\,t_1 t_2 t_3 t_4 t_5 t_6, \end{aligned}$$

where m is a constant, l a constant, s a linear function, U, V, W functions of the degrees 2, 3, 4 respectively; and compose with one or more such factors an expression involving the term $Kt_1t_2t_3t_4t_5t_6$; for instance, such an expression is

$$\frac{K}{mm'}(sA + mt_1t_2t_3)(l'A + m't_4t_5)\,t_6;$$

this is, of the form $A\Omega + Kt_1t_2t_3t_4t_5t_6$, viz., $A\Omega + (AC - B^2)$, or $A(\Omega + C) - B^2$, say $A\Gamma - B^2$; or what is the same thing, introducing a new coordinate w, we have a quadric function

$$Aw^2 + 2Bw + \Gamma,$$

the discriminant of which, $A\Gamma - B^2$, is equal to the expression in question.

130. In the sequel (x, y, z, w) are considered as the coordinates of a point in space; $A = 0$ is thus a quadric cone, $t_1 = 0$, $t_2 = 0 \ldots t_6 = 0$, any six tangent planes thereof; and hence $Aw^2 + 2Bw + \Gamma = 0$ a quartic surface, having the point $(x = 0, y = 0, z = 0)$ for a node, whereof the circumscribed cone $A\Gamma - B^2 = 0$ breaks up in the assumed manner.

Thus, in order that the circumscribed cone may be as above

$$(sA + mt_1t_2t_3)(l'A + m't_4t_5)\,t_6,$$

we have only to assume

$$\Gamma = C + \frac{K}{mm'}(sl'A + sm't_4t_5 + l'mt_1t_2t_3)\,t_6,$$

and so in other cases. Observe that $sA + mt_1t_2t_3 = 0$ is a cubic cone, which, so long as s, m are arbitrary, has no nodal line; but establishing a single relation (say s remains arbitrary, but a proper value is assigned to m) it will be a cubic cone having a nodal line. And so $UA + mt_1t_2t_3t_4 = 0$ is a quartic cone without any nodal line, but by particularising the constants it may be made to have one, two, or three nodal lines. Such nodal determinations are obviously required in order that the formula may extend to all the before-mentioned forms of the circumscribed cone. The foregoing analysis is the foundation of the whole theory: I have given it, as above, apart from the theory, in order that the nature of it may be the better perceived; but I have now to bring it into connexion with the theory.

On the Sextic Curves, $A_2B_4 - C_3^2 = 0$.

131. I revert to the consideration of plane curves. The equation of a sextic curve $(*\between x, y, z)^6 = 0$ cannot be in general expressed in the form $AC - B^2 = 0$, where

the degrees of A, B, C are 2, 3, 4 respectively; in fact, the existence of such a form implies that there is a conic $A=0$ touching the sextic 6 times; and since a conic can only be made to satisfy 5 conditions, there is not in general any such conic.

132. Such conic, when it exists, is said to be inscribed in the sextic, and the sextic to be circumscribed about the conic, or to be an "amphigram;" and then, $A=0$ being the equation of the conic, that of the sextic is expressible in the form in question $AC-B^2=0$. It is clear that $B=0$ is a cubic curve passing through the 6 points of contact of the conic with the sextic, and that any such curve may be taken for the curve B; in fact if a particular cubic through the 6 points is $B'=0$, and the equation of the sextic is $AC'-B'^2=0$, then taking p an arbitrary linear function of the coordinates, the equation of the general cubic is $B=B'+pA=0$; and then writing

$$A=A,$$
$$B=B'+pA,$$
$$C=C'+2B'p+Ap^2,$$

we have $AC-B^2=AC'-B'^2$; so that the original form $AC'-B'^2=0$ becomes $AC-B^2=0$. But the cubic $B=0$ being assumed at pleasure, the quartic $C=0$ is a determinate curve.

133. It is to be observed that a sextic curve may be an amphigram in more than one way: certainly in two, three, or four, and possibly in a greater number of ways. For the equation of the curve contains 27 constants, and hence determining the sextic so as to touch 4 given conics each of them 6 times, there are still 3 constants; and the curve will be an amphigram in regard to each of the 4 conics; say it is a quadruple amphigram. But in the sequel we are only concerned with a sextic curve considered as an amphigram in regard to a given conic $A=0$ (no attention being paid to the other inscribed conics, if any); and then, by what precedes, taking $B=0$ any cubic whatever through the 6 points of contact, we have a determinate quartic curve $C=0$, and the equation of the sextic curve assumes the form $AC-B^2=0$.

134. The curves $A=0$, $B=0$, $C=0$ contain respectively 5, 9, 14 constants; whence considering the function B as containing an arbitrary constant factor, for the curve $AC-B^2=0$, the number of constants is *primâ facie* $5+9+14+1=29$; but on account of the arbitrary linear function p, the real number is $29-3=26$: this is right, for a sextic curve contains 27 constants; and the curve being an amphigram, there is one relation between the constants, $27-1=26$.

135. Suppose now that the sextic curve $AC-B^2=0$ breaks up into two or more separate curves, say into the two curves $P=0$, $Q=0$ of the orders f, g respectively; $f+g=6$. We have

$$AC-B^2=PQ=0;$$

and the conic $A=0$ touching the sextic six times, must, it is clear, touch the curves $P=0$, $Q=0$, f and g times respectively. And so when the sextic breaks up into any number of curves, each component curve $P=0$ of the order f must touch the sextic g times.

136. It follows that if the sextic break up into six lines, say $AC-B^2=t_1t_2t_3t_4t_5t_6=0$, then that each of the lines $t_1=0$, $t_2=0, \ldots t_6=0$ is a tangent to the conic. And conversely, starting with the conic $A=0$ and any six tangents thereof $t_1=0$, $t_2=0, \ldots t_6=0$, we have an identity of the form in question. In fact, taking any two of the tangents, say $t_1=0$ and $t_2=0$, then, if $p=0$ be the equation of the line joining their points of intersection, the equation of the conic will be of the form $t_1t_2+p^2=0$, that is, we may write $A=t_1t_2+p^2$, or what is the same thing, $t_1t_2=A-p^2$. (Considering A as a given quadric function of the coordinates, this of course implies that the implicit constant factors of t_1, t_2, p are properly determined.) Similarly, $q=0$ being the line through the points of contact of t_3, t_4, and $r=0$ that through the points of contact of t_5, t_6, we have $t_3t_4=A-q^2$ and $t_5t_6=A-r^2$; whence, to satisfy the equation

$$AC-B^2=t_1t_2t_3t_4t_5t_6,$$

we have only to assume $B=lA+pqr$, l an arbitrary linear function of the coordinates, and the equation then gives

$$C=A^2-A(p^2+q^2+r^2)+(q^2r^2+r^2p^2+p^2q^2)+l^2A+2lpqr.$$

137. It will be observed that the grouping of the six tangents into pairs is arbitrary. By altering this grouping, we merely alter the linear function l, but do not obtain any new solution. Thus, say that the new form is $B=l'A+p'q'r'$, then, by properly determining the linear function l, we can reduce this to the original form $B=lA+pqr$; viz., we can satisfy identically the equation $(l-l')A+pqr-p'q'r'=0$; or what is the same thing, $\lambda A+pqr-p'q'r'=0$, where λ is a linear function of the coordinates. We have, in fact, the conic $A=0$ and the cubic $pqr=0$ intersecting in the six points of contact any other cubic through these six points; and consequently the cubic $p'q'r'=0$ must be expressible in the form $\lambda A+pqr=0$, and we have thus the identity in question.

138. We have just seen that the value of B is necessarily of the form $B=lA+pqr$, but we are not concerned with its expression in this particular form. What we require in the sequel is a value of B, and thence one of C, satisfying the identity in question, $AC-B^2=t_1t_2t_3t_4t_5t_6$; or what is the same thing, introducing for convenience a constant factor K, the identity

$$AC-B^2=Kt_1t_2t_3t_4t_5t_6.$$

139. Instead of C, I write Γ, and consider the sextic amphigram $A\Gamma-B^2=0$ touched by the conic $A=0$ in the points of contact of the conic with the six tangents $t_1=0$, $t_2=0$, $\ldots t_6=0$. Suppose the sextic curve breaks up into factors; if one of these factors is a line, it is one of the six tangents, say the tangent $t_1=0$. If there is a conic factor, this is a conic touching the conic $A=0$ at its points of contact with two of the tangents, say the equation is $lA+mt_1t_2=0$. Similarly, if there is a cubic, quartic, or quintic factor, then the equation hereof is $sA+mt_1t_2t_3=0$, $UA+mt_1t_2t_3t_4=0$, or $VA+m_1t_2t_3t_4t_5=0$. Or going on to the next case of a sextic factor (being of course the whole curve), we may say that this is $WA+mt_1t_2t_3t_4t_5t_6=0$. (Observe that since $AC-B^2=Kt_1t_2t_3t_4t_5t_6$, this means only that the equation of the sextic amphigram is of the assumed form $A\Gamma-B^2=0$.)

140. By what precedes we can, for a sextic amphigram which breaks up in any assigned manner, determine the value of Γ. For instance, let the amphigram break up into two cubic curves; say these are $sA + mt_1t_2t_3 = 0$, $s'A + m't_4t_5t_6 = 0$. Assume

$$A\Gamma - B^2 = \frac{K}{mm'}(sA + mt_1t_2t_3)(s'A + m't_4t_5t_6),$$

then this equation is

$$A\Gamma - B^2 = \frac{K}{mm'}\{ss'A^2 + (sm't_4t_5t_6 + s'mt_1t_2t_3)A\} + AC - B^2;$$

that is, we have

$$\Gamma = C + \frac{K}{mm'}(ss'A + sm't_4t_5t_6 + s'mt_1t_2t_3),$$

and so in any other case.

I have already adverted to the question of the "nodal determination" of the formulæ, and it might be properly here considered; viz., the question is as to the determination of the constants in such manner that, for instance, $sA + mt_1t_2t_3 = 0$ may be a nodal cubic, $UA + mt_1t_2t_3t_4 = 0$ a nodal, binodal, or trinodal quartic, &c.; but I defer it for the moment in order first to apply the theory to the quartic surfaces.

Application to Quartic Surfaces.

141. If a quartic surface has a node or nodes, we may take for a node the point $x = 0$, $y = 0$, $z = 0$; the equation of the quartic surface is then of the form

$$Aw^2 + 2Bw + \Gamma = 0,$$

where A, B, Γ are functions of x, y, z of the degrees 2, 3, 4 respectively. $A = 0$ is the tangent quadricone at the node in question; and the circumscribed cone is $A\Gamma - B^2 = 0$. By what precedes, this is an amphigram touching the quadricone along six generating lines thereof; say the tangent planes of the cone $A = 0$ along these six lines respectively are $t_1 = 0$, $t_2 = 0, \ldots t_6 = 0$. We have then an identical equation

$$AC - B^2 = Kt_1t_2t_3t_4t_5t_6,$$

viz., regarding for a moment this equation as an equation for the determination of B, and B' as any particular solution thereof, then its general solution is $B' + tA$, where t is an arbitrary linear function of (x, y, z), and the B in the equation of the surface is properly $= B' + tA$. But by the substituting $w - t$ in place of w, the B of the equation of the surface would then be made $= B'$; and it thus appears that we may, without loss of generality, take the B of this equation to be *any particular value* satisfying the identity in question; and then, B having such particular value, C is a quartic function of (x, y, z) completely determined by the same identity. And we then, by what precedes, at once determine Γ so that the circumscribed cone $A\Gamma - B^2 = 0$ may be a cone breaking up in any assigned manner; for instance, if it be a cone

(3, 3), then, as just mentioned, the two cubic cones are $sA + mt_1t_2t_3 = 0$, $s'A + m't_4t_5t_6 = 0$; and Γ has the value

$$\Gamma = C + \frac{K}{mm'}(ss'A + sm't_4t_5t_6 + s'mt_1t_2t_3),$$

above obtained.

On the Nodal Determination.

142. I am not able to discuss with much completeness the question of nodal determination. We have to consider a cubic curve $sA + mt_1t_2t_3 = 0$, a quartic curve $UA + mt_1t_2t_3t_4 = 0$, &c., as the case may be, and to determine the constants so that this shall have a node or nodes. Consider for a moment the form $PA + t_1Q = 0$, where Q denotes the product $mt_2t_3\ldots$ of all or any of the tangents $t_2, \ldots t_5$; the orders of PA, t_1Q are of course equal, that is, the order of P is less by unity than that of Q. I say that, by establishing a single relation between the constants, this may be made to have a node *at the point of contact* $A = 0$, $t_1 = 0$. In fact, writing $\Delta = \lambda\delta_x + \mu\delta_y + \nu\delta_z$, where λ, μ, ν are arbitrary, there will be a node at any point if for that point $\Delta(PA + t_1Q) = 0$. But for the point $A = 0$, $t_1 = 0$ this becomes $P\Delta A + Q\Delta t_1 = 0$; moreover, if $t = 0$ be any other tangent of the conic $A = 0$, and if $p = 0$ be the line joining the points of contact of the tangents t, t_1, then we may write $A = tt_1 - p^2$, and thence (since at the point in question, $A = 0$, $t_1 = 0$, we have also $p = 0$) we find $\Delta A = t\Delta t_1$, and the foregoing equation thus becomes $(tP + Q)\Delta t_1 = 0$; viz., this equation is satisfied irrespectively of the values of λ, μ, ν, if only at the point in question (that is, for the values of the coordinates which belong to the point $t_1 = 0$, $A = 0$) we have $tP + Q = 0$, which is a single relation between the constants.

143. In particular the cubic curve $sA + mt_1t_2t_3 = 0$ may be made to have a node at the point of contact of any one of the three tangents; the quartic curve $UA + mt_1t_2t_3t_4 = 0$, a node, or two or three nodes, at the point or points of contact of any one, two, or three of the four tangents; and so in other cases. These are not the only solutions, and they are in fact solutions which (as afterwards explained) I propose to reject, attending in each case only to the remaining or proper solutions of the problem.

144. To obtain in a different manner the foregoing result, consider again the cubic curve $sA + mt_1t_2t_3 = 0$; regarding this as a given curve, the conic $A = 0$ is a conic determined (not of course completely) as a conic having therewith 3 points of 2-pointic intersection; viz., if the cubic has a node, then the cone $A = 0$ is either a conic passing through the node and besides touching the curve twice, or else it is a conic touching the curve 3 times; the former is of course the above mentioned case where there is a node at one of the points of contact on the conic $A = 0$; the latter is regarded as the proper solution. So in the case of a quartic curve $UA + mt_1t_2t_3t_4 = 0$, regarding this as a given curve, the conic $A = 0$ is a conic having therewith four points of 2-pointic intersection; viz., if the quartic curve has one, two, or three nodes, then the conic is either a conic passing through one, two, or three nodes, and besides touching the quartic thrice, twice, or once; or else it is a conic touching the quartic

four times. The former is the above mentioned case where there is a node or nodes at a point or points of contact with the conic $A=0$; the latter is regarded as the proper solution.

145. To fix the ideas, and at the same time obtain a result which will be afterwards useful, I work out the formulæ for the cubic curve $sA+mt_1t_2t_3=0$, taking this equation under the form

$$\left(\frac{x}{\lambda}+\frac{y}{\mu}+\frac{z}{\nu}\right)(x^2+y^2+z^2-2yz-2zx-2xy)+m\,xyz=0.$$

This may have a node in two different ways; viz.,

1°. At the point of contact of one of the tangents $x=0$, $y=0$, $z=0$ with the conic $A=0$; say at the point of contact of $x=0$, that is, the point $x=0$, $y-z=0$. The value of m is $=\frac{4}{\mu}+\frac{4}{\nu}$; hence $\frac{1}{\nu}=-\frac{1}{\mu}+\tfrac{1}{4}m$; and, substituting, the equation of the curve becomes

$$\left(\frac{x}{\lambda}+\frac{y-z}{\lambda}\right)\{x^2-2x(y+z)+(y-z)^2\}+\tfrac{1}{4}mz(x+y-z)^2=0,$$

which has obviously a node at the point in question.

2°. The node may be at a point not on the conic $A=0$, viz. the value of m is $=\frac{(\lambda+\mu+\nu)^2}{\lambda\mu\nu}$, the equation is

$$\left(\frac{x}{\lambda}+\frac{y}{\mu}+\frac{z}{\nu}\right)(x^2+y^2+z^2-2yz-2zx-2xy)+\frac{(\lambda+\mu+\nu)^2}{\lambda\mu\nu}xyz=0.$$

In fact, writing for shortness

$$\begin{aligned}-\lambda+\mu+\nu&=L,\\ \lambda-\mu+\nu&=M,\\ \lambda+\mu-\nu&=N\\ \lambda+\mu+\nu&=P,\end{aligned}$$

the node is at the point $x:y:z=L\lambda:M\mu:N\nu$; which is at once verified, if we remark that, writing for convenience $x, y, z=L\lambda, M\mu, N\nu$, then we have

$$\begin{aligned}-x+y+z&=MN,\\ x-y+z&=NL,\\ x+y-z&=LM,\end{aligned}$$

$$\frac{x}{\lambda}+\frac{y}{\mu}+\frac{z}{\nu}=P,\quad x^2+y^2+z^2-2yz-2zx-2xy=-LMNP\,(=A).$$

For, of the three equations for the coordinates of a node, the first is

$$\frac{1}{\lambda}A+\left(\frac{x}{\lambda}+\frac{y}{\mu}+\frac{z}{\nu}\right)2(x-y-z)+\frac{(\lambda+\mu+\nu)^2}{\lambda\mu\nu}yz=0,$$

that is, for the values in question,

$$-\frac{1}{\lambda}LMNP + P(-2MN) + \frac{P^2}{\lambda\mu\nu}MN\mu\nu = 0,$$

that is

$$-LMNP - 2\lambda MNP + P^2MN = 0,$$

or, finally, $-L - 2\lambda + P = 0$, which is satisfied; and similarly the other two equations are satisfied.

Quartic Surfaces resumed.

146. Passing now from curves to cones, and to the theory of the quartic surface, suppose that there is a component cone having a nodal line, say the cubic cone $sA + mt_1t_2t_3 = 0$: if the remaining factor is $t_4t_5t_6$, then we have

$$A\Gamma - B^2 = \frac{K}{m}(sA + mt_1t_2t_3)\,t_4t_5t_6.$$

Suppose the nodal line is a line of contact with the cone $A = 0$, say its equations are $t_1 = 0$, $p = 0$ (p a linear function), then $sA + mt_1t_2t_3$ is a quadric function $(*\between t_1, p^2)$, (of course with variable coefficients); hence $A\Gamma - B^2$ is a quadric function; and A being a linear function $(*\between t_1, p)$, it follows that B is a linear function, and thence that Γ is also a linear function; that is, A, B, Γ are each of them a linear function $(*\between t_1, p)$, or the line in question (viz. the line of contact $A = 0$, $t_1 = 0$) is a line on the quartic surface $Aw^2 + 2Bw + \Gamma = 0$. As already mentioned, *I exclude from consideration* the surfaces which have upon them a line through two nodes; that is, I exclude from consideration the case in question where any component cone, or say where the sextic cone, has a nodal line which is a line on the tangent cone $A = 0$.

147. Now, excluding the case just referred to, *I assume as a postulate* that there is but one way in which the cubic cone $sA + mt_1t_2t_3 = 0$ can be made to have a nodal line, or the quartic cone $UA + mt_1t_2t_3t_4 = 0$ one, two, or three nodal lines &c., as the case may be. It is to be understood that this does not mean that the constants are in any of these cases completely determined, but that there is between them a relation or relations constituting a general solution which includes in itself every particular solution whatever. I have no doubt that as regards the cubic cone at least the assumption is correct. This being so, the character of a single node determines the nature of the surface; for instance, if there is a node $(3_1, 3)$, then taking this as the point ($x = 0$, $y = 0$, $z = 0$) the equation of the surface is $Az^2 + 2Bw + \Gamma = 0$, where

$$\Gamma = C + \frac{K}{mm'}(ss'A + l's t_4t_5t_6 + ls't_1t_2t_3),$$

a surface of a determinate nature; so that the character of all the remaining nodes is completely determined.

148. The point to be attended to is, that if for instance there were two essentially distinct ways of giving the cubic cone $sA + mt_1t_2t_3 = 0$ a nodal line (such as there would be if the excluded case were considered admissible), then the foregoing equation

of the surface would or might include two distinct forms of equation applying to different kinds of surface. The conclusion is that there is but one kind of quartic surface having a node $(3_1, 3)$. Admitting this, and similarly that there is but one kind of quartic surface having a node 6_{10}, it follows that if (as the fact is) there is a surface having the nodes $1(6_{10}) + 10(3_1, 3)$ (Kummer's 11-nodal surface), then that the two first-mentioned kinds are in fact each of them this last-mentioned kind of surface; and it was in this manner that I arrived at the enumeration given near the beginning of the present Memoir.

149. The reasoning is, of course, in place of a direct demonstration which would consist in showing that a surface having a node $(3_1, 3)$ has 9 other like nodes, and also a node 6_{10}; and that a surface having a node 6_{10} has 10 other nodes $(3_1, 3)$; and that, starting from either form of equation, we could, by passing to a node of the other kind, obtain the other form of equation.

Enumeration of the Cases.

150. I collect the results as follows: I call to mind that we have always the identical equation $AC - B^2 = Kt_1t_2t_3t_4t_5t_6$, that the equation of the surface is $Aw^2 + 2Bw + \Gamma = 0$, and that the circumscribed cone is $A\Gamma - B^2 = 0$. The equation of a surface having different kinds of nodes will assume different forms according as the origin (or point $x = 0$, $y = 0$, $z = 0$) is taken to be at a node of one or other of these kinds; these forms of the equations are distinguished as "node-forms,"—viz., we speak of the node-form $(3_1, 3)$ when the origin is a node $(3_1, 3)$, and so in other cases.

The 16-nodal surface

$$16\,(1,\ 1,\ 1,\ 1,\ 1,\ 1),$$

node-form

$$(1,\ 1,\ 1,\ 1,\ 1,\ 1),$$

cone is

$$t_1t_2t_3t_4t_5t_6 = 0,$$

and

$$\Gamma = C;$$

viz., equation is

$$Aw^2 + 2Bw + C = 0.$$

The 15-nodal surface

$$15\,(2,\ 1,\ 1,\ 1,\ 1),$$

node-form

$$(2,\ 1,\ 1,\ 1,\ 1)$$

cone is

$$(lA + mt_1t_2)\,t_3t_4t_5t_6 = 0,$$

and

$$\Gamma = C + \frac{Kl}{m}\,t_3t_4t_5t_6 = 0.$$

The 14-nodal surface

$$8(3_1,\ 1,\ 1,\ 1)+6(2,\ 2,\ 1,\ 1),$$

node-form

$$(3_1,\ 1,\ 1,\ 1),$$

cone is

$$(sA+mt_1t_2t_3)\,t_4t_5t_6=0,$$

where

$$sA+mt_1t_2t_3=0 \text{ is a nodal cubic } 3_1,$$

and

$$\Gamma=C+\frac{Ks}{m}t_4t_5t_6\,;$$

node-form

$$(2,\ 2,\ 1,\ 1),$$

cone is

$$(lA+mt_1t_2)(l'A+m't_3t_4)\,t_5t_6=0,$$

and

$$\Gamma=C+\frac{K}{mm'}(ll'A+lm't_3t_4+l'mt_1t_2)\,t_5t_6.$$

The 13 (α)-nodal surface

$$3(4_3,\ 1,\ 1)+1(3,\ 1,\ 1,\ 1)+9(3_1,\ 2,\ 1),$$

node-form

$$(4_3,\ 1,\ 1),$$

cone is

$$(UA+mt_1t_2t_3t_4)\,t_5t_6=0,$$

where

$$UA+mt_1t_2t_3t_4=0 \text{ is a trinodal quartic } 4_3,$$

and

$$\Gamma=C+\frac{K}{m}Ut_5t_6\,;$$

node-form

$$(3,\ 1,\ 1,\ 1),$$

cone is

$$(sA+mt_1t_2t_3)\,t_4t_5t_6=0,$$

and

$$\Gamma=C+\frac{K}{m}st_4t_5t_6\,;$$

node-form

$$(3_1,\ 2,\ 1),$$

cone is

$$(sA+mt_1t_2t_3)(l'A+m't_4t_5)\,t_6=0,$$

where

$$sA+mt_1t_2t_3=0 \text{ is a nodal cubic } 3_1,$$

and

$$\Gamma=C+\frac{K}{mm'}(sl'A+m'st_4t_5+ml't_1t_2t_3)\,t_6.$$

The 13 (β)-nodal surface

$$13(2,\ 2,\ 2),$$

node-form

$$(2,\ 2,\ 2),$$

cone is

$$(lA + mt_1t_2)(l'A + m't_3t_4)(l''A + m''t_5t_6) = 0,$$

and

$$\Gamma = C + \frac{K}{mm'm''}\{ll'l''A^2 + (l'l''mt_1t_2 + l''lm't_3t_4 + ll'm''t_5t_6)A$$
$$+ lm'm''t_3t_4t_5t_6 + l'm''mt_5t_6t_1t_2 + l''mm't_1t_2t_3t_4\}.$$

The 12 (α)-nodal surface

$$12(4_3,\ 2),$$

node-form

$$(4_3,\ 2),$$

cone is

$$(UA + mt_1t_2t_3t_4)(l'A + m't_5t_6) = 0,$$

where

$$UA + mt_1t_2t_3t_4 = 0 \text{ is a trinodal quartic } 4_3,$$

and

$$\Gamma = C + \frac{K}{mm'}(l'UA + m'Ut_5t_6 + l'mt_1t_2t_3t_4).$$

The 12 (β)-nodal surface

$$2(5_6,\ 1) + 6(3_1,\ 3_1) + 4(3,\ 2,\ 1),$$

node-form

$$(5_6,\ 1),$$

cone is

$$(VA + mt_1t_2t_3t_4t_5)\,t_6 = 0,$$

where

$$VA + mt_1t_2t_3t_4t_5 = 0 \text{ is a 6-nodal quintic } 5_6,$$

and

$$\Gamma = C + \frac{K}{m}Vt_6;$$

node-form

$$(3_1,\ 3_1),$$

cone is

$$(sA + mt_1t_2t_3)(s'A + m't_4t_5t_6) = 0,$$

where

$sA + mt_1t_2t_3 = 0$ and $s'A + m't_4t_5t_6 = 0$ are each of them a nodal cubic 3_1,

and

$$\Gamma = C + \frac{K}{mm'}(ss'A + m'st_4t_5t_6 + ms't_1t_2t_3);$$

node-form

$$(3,\ 2,\ 1),$$

cone is

$$(sA + mt_1t_2t_3)(l'A + m't_4t_5)\,t_6 = 0,$$

and

$$\Gamma = C + \frac{K}{mm'}(l'sA + m'st_4t_5 + l'mt_1t_2t_3)\,t_6.$$

The 12 (γ)-nodal surface,

$$12\,(4_2,\ 1,\ 1),$$

node-form

$$(4_2,\ 1,\ 1),$$

cone is

$$(UA + mt_1t_2t_3t_4)\,t_5t_6 = 0,$$

where

$$UA + mt_1t_2t_3t_4 = 0 \text{ is a binodal quartic } 4_2,$$

and

$$\Gamma = C + \frac{K}{m}\,Ut_5t_6.$$

The 11 (α)-nodal surface,

$$1\,(6_{10}) + 10\,(3_1,\ 3)$$

node-form

$$(6_{10}),$$

cone is

$$WA + mt_1t_2t_3t_4t_5t_6 = 0,$$

where this is a 10-nodal sextic 6_{10},

and

$$\Gamma = C + \frac{K}{m}\,W;$$

node-form

$$(3_1,\ 3),$$

cone is

$$(sA + mt_1t_2t_3)\,(s'A + m't_4t_5t_6) = 0,$$

where

$$sA + mt_1t_2t_3 = 0 \text{ is a nodal cubic } 3_1,$$

and

$$\Gamma = C + \frac{K}{mm'}\,(ss'A + m'st_4t_5t_6 + ms't_1t_2t_3).$$

Other 11-nodal surfaces,

node-form

$$(5_5,\ 1),$$

cone is

$$(VA + mt_1t_2t_3t_4t_5)\,t_6 = 0,$$

where

$$VA + mt_1t_2t_3t_4t_5 = 0 \text{ is a 5-nodal quintic } 5_5,$$

and

$$C = \Gamma + \frac{K}{m}\,Vt_6 = 0;$$

node-form

$$(4_2,\ 2),$$

cone is

$$(UA + mt_1t_2t_3t_4)\,(l'A + m't_5t_6) = 0,$$

where

$$UA + mt_1t_2t_3t_4 = 0 \text{ is a binodal quartic } 4_2,$$

and

$$\Gamma = C + \frac{K}{mm'}\,(l'UA + m'Ut_5t_6 + l'mt_1t_2t_3t_4);$$

node-form

$$(4,\ 1,\ 1),$$

cone is

$$(UA + mt_1t_2t_3t_4)\,t_5t_6 = 0,$$

where

$$UA + mt_1t_2t_3t_4 \text{ is a nodal quartic } 4_1,$$

and

$$\Gamma = C + \frac{K}{m}\,Ut_5t_6\,;$$

but whether these node-forms belong to the same or to different surfaces is not ascertained.

The enumeration is not extended to the 10-nodal surfaces, but I consider one case of these surfaces.

The 10 (α)-*nodal surface* 10 (3, 3).

151. I assume only that there is a single node (3, 3): taking the cone to be

$$(sA + mt_1t_2t_3)(s'A + m't_4t_5t_6) = 0\,;$$

then for the equation of the surface, in the node-form (3, 3) in question, we have

$$\Gamma = C + \frac{K}{mm'}(ss'A + sm't_4t_5t_6 + s'mt_1t_2t_3).$$

But I present this result under a different form, as follows: I write

$$A = p^2 + ft_1t_2 = q^2 + gt_3t_4 = r^2 + ht_5t_6,$$

where f, g, h are constants, and, as before, $p=0$, $q=0$, $r=0$ are the lines joining the points of contact of t_1, t_2; t_3, t_4; and t_5, t_6 respectively: we have

$$sA + mt_1t_2t_3 = sA + mt_3\left(\frac{A-p^2}{f}\right), \text{ and } s'A + m't_4t_5t_6 = s'A + m't_4\left(\frac{A-r^2}{h}\right);$$

or in place of s, s' introducing new linear functions σ, σ', the cubic curves may be taken to be $\sigma A - \frac{m}{f}p^2t_3$, $\sigma'A - \frac{m'}{h}r^2t_4$, so that we have

$$\begin{aligned} A\Gamma - B^2 &= \frac{K}{mm'}\left(\sigma A - \frac{m}{f}p^2t_3\right)\left(\sigma'A - \frac{m'}{h}r^2t_4\right), \\ &= \frac{K}{mm'}\left(\sigma\sigma'A^2 - \sigma A\frac{m'}{h}r^2t_4 - \sigma'A\frac{m}{f}p^2t_3 + \frac{mm'}{fh}p^2r^2\frac{A-q^2}{g}\right), \\ &= \frac{K}{mm'}\left(\sigma\sigma'A^2 - \sigma A\frac{m'}{h}r^2t_4 - \sigma'A\frac{m}{f}p^2t_3 + \frac{mm'}{fgh}p^2r^2A\right) - \frac{K}{fgh}p^2q^2r^2\,; \end{aligned}$$

whence $B = \left(\frac{K}{fgh}\right)^{\frac{1}{2}}(pqr + tA)$, where t is a linear function of the coordinates; and we then have

$$\Gamma = \frac{K}{mm'}\left(\sigma\sigma'A - \sigma\frac{m'}{h}r^2t_4 - \sigma'\frac{m}{f}p^2t_3 + \frac{mm'}{fgh}p^2r^2\right) + \frac{K}{fgh}(t^2A + 2tpqr),$$

where A may be considered as standing for $q^2+gt_3t_4$. The equation $Aw^2+2Bw+\Gamma=0$ of the surface, substituting throughout for A its value, is therefore

$$(q^2+gt_3t_4)\,w^2+2\left(\frac{K}{fgh}\right)^{\frac{1}{2}}\{pqr+t\,(q^2+gt_3t_4)\}\,w$$

$$+\frac{K}{mm'}\left[\sigma\sigma'\,(q^2+gt_3t_4)-\sigma\frac{m'}{h}\,r^2t_4-\sigma'\frac{m}{f}\,p^2t_3+\frac{mm'}{fgh}\,p^2r^2\right]+\frac{K}{fgh}\,[t^2\,(q^2+gt_3t_4)+2tpqr]=0,$$

where the cone is

$$\left\{\sigma\,(q^2+gt_3t_4)-\frac{m}{f}\,p^2t_3\right\}\left\{\sigma'\,(q^2+gt_3t_4)-\frac{m'}{h}\,r^2t_4\right\}=0.$$

152. Writing in the equation of the surface $w\left(\frac{K}{fgh}\right)^{\frac{1}{2}}$ instead of w, it becomes

$$(q^2+gt_3t_4)\,w^2+2\,[pqr+t\,(q^2+gt_3t_4)]\,w$$

$$+\frac{fgh}{mm'}\left[\sigma\sigma'\,(q^2+gt_3t_4)-\sigma\frac{m'}{h}\,r^2t_4-\sigma'\frac{m}{f}\,p^2t_3+\frac{mm'}{fgh}\,p^2r^2\right]+t^2\,(q^2+gt_3t_4)+2tpqr=0\,;$$

and then writing $\frac{m}{f}\,\sigma$ and $\frac{m'}{h}\,\sigma'$ for σ and σ' respectively, this is

$$(q^2+gt_3t_4)\,w^2+2pqrw+2tw\,(q^2+gt_3t_4)$$

$$+g\,[\sigma\sigma'\,(q^2+gt_3t_4)-\sigma r^2t_4-\sigma' p^2t_3+\frac{1}{g}\,p^2r^2]+t^2\,(q^2+gt_3t_4)+2tpqr=0.$$

We may consider t_3, t_4 as denoting not the functions originally so represented, but these functions each multiplied by a suitable constant, and thereupon write $g=-1$; viz., $t_3=0$, $t_4=0$, will now denote any two tangents to the conic $A=0$, the implicit factors being so determined that $A=q^2-t_3t_4$. The equation of the surface is

$$(q^2-t_3t_4)\,w^2+2pqrw+2tw\,(q^2-t_3t_4)$$

$$-\sigma\sigma'\,(q^2-t_3t_4)+\sigma r^2t_4+\sigma' p^2t_3+p^2r^2+t^2\,(q^2-t_3t_4)+2pqrt=0\,;$$

viz., this is

$$(q^2-t_3t_4)\,[(w+t)^2-\sigma\sigma']+2pqr\,(w+t)+\sigma r^2t_4+\sigma' p^2t_3+p^2r^2=0,$$

the sextic cone being

$$\{\sigma\,(q^2-t_3t_4)-p^2t_3\}\,\{\sigma'\,(q^2-t_3t_4)-r^2t_4\}=0.$$

153. But the foregoing equation of the surface is

$$\begin{vmatrix} -\sigma', & w+t, & ., & r \\ w+t, & -\sigma, & p, & . \\ ., & p, & t_4, & -q \\ r, & ., & -q, & t_3 \end{vmatrix}=0,$$

as is at once seen by developing the determinant; the functions $w+t$, σ, σ', p, q, r, t_3, t_4 are all of them linear; and the determinant is thus a symmetrical quartic determinant the terms whereof are linear functions of the coordinates; viz. the surface is a symmetroid. That is, a surface having a single node (3, 3) is a symmetroid; but I have shown (*Second Memoir*, No. 116) that a symmetroid has each of its ten nodes (3, 3); wherefore the surface having a single node (3, 3) is the 10 (α)-nodal surface, nodes 10 (3, 3).

154. Start from two cubic cones $U=0$, $V=0$, having each the same vertex ($x=0$, $y=0$, $z=0$); we may in a variety of ways determine the two cones $\alpha U+\beta V=0$, $\gamma U+\delta V=0$, having a common inscribed quadric cone $A=0$ (viz., $\alpha:\beta$ being assumed at pleasure, then $\gamma:\delta$ will be determined; not, I believe, uniquely, but I do not know what the multiplicity is). This being so, the quadric cone $A=0$ is uniquely determined; and then, assuming at pleasure the plane $w=0$, the 10 (α)-nodal surface $Aw^2+2Bw+\Gamma=0$ is uniquely determined: consequently the remaining nine nodes are determinate points on the nine lines $U=0$, $V=0$ respectively. And we have thus a system of ten points in space such that, joining any one of them with the remaining nine, the nine lines so obtained are the intersections of two cubic cones, or say that they are an ennead of lines.

Notation for the Cases afterwards considered.

155. I proceed to further develope the theory of some of the different surfaces. The same node-form of equation will, of course, assume different shapes according to the actual expressions in terms of the coordinates (x, y, z) of the several functions A, &c., which enter into it. I have found it convenient to attribute to A and B certain specific values which are not in every case those of the coefficients of w^2, w in the equation of the surface: this means that we must, in the equation of the surface, substitute new symbols for these coefficients, and write the equation say in the form $A'w^2+2B'w+\Gamma=0$; the change of notation, when it occurs, will be duly explained.

156. It is in general (but not always) convenient to take the equation of the tangent cone to be $x^2+y^2+z^2-2yz-2zx-2xy=0$; for then any plane $\frac{x}{\alpha}+\frac{y}{\beta}+\frac{z}{\gamma}=0$, where $\alpha+\beta+\gamma=0$, will be a tangent plane; so that six tangent planes may be represented by $x=0$, $y=0$, $z=0$, and by three equations of the form just referred to. And in reference to this assumed form of the equation of the tangent cone, and to what follows, I write

$$\alpha+\beta+\gamma=0,$$

$$\alpha'+\beta'+\gamma'=0,$$

$$\alpha''+\beta''+\gamma''=0,$$

$$P=\frac{x}{\alpha}+\frac{y}{\beta}+\frac{z}{\gamma},$$

$$P'=\frac{x}{\alpha'}+\frac{y}{\beta'}+\frac{z}{\gamma'},$$

$$P''=\frac{x}{\alpha''}+\frac{y}{\beta''}+\frac{z}{\gamma''},$$

$$\begin{aligned}
X &= \alpha\ (\gamma'\gamma''\, y - \beta'\beta''z),\\
Y &= \beta\ (\alpha'\alpha''\, z - \gamma'\gamma''x),\\
Z &= \gamma\ (\beta'\beta''x - \alpha'\alpha''y),
\end{aligned}$$

$$\begin{aligned}
X' &= \alpha'\ (\gamma''\gamma\ y - \beta''\beta\ z),\\
Y' &= \beta'\ (\alpha''\alpha\ z - \gamma''\gamma\ x),\\
Z' &= \gamma'\ (\beta''\beta\ x - \alpha''\alpha\ y),
\end{aligned}$$

$$\begin{aligned}
X'' &= \alpha''\ (\gamma\ \gamma'\ y - \beta\ \beta'\ z),\\
Y'' &= \beta''\ (\alpha\ \alpha'\ z - \gamma\ \gamma'\ x),\\
Z'' &= \gamma''\ (\beta\ \beta'\ x - \alpha\ \alpha'\ y),
\end{aligned}$$

$$\begin{aligned}
A &= x^2 + y^2 + z^2 - 2yz - 2zx - 2xy,\\
B &= \alpha\alpha'\alpha''\,(y^2z - yz^2) + \beta\beta'\beta''\,(z^2x - zx^2) + \gamma\gamma'\gamma''\,(x^2y - xy^2) + Mxyz,\\
C &= (\alpha\alpha'\alpha''yz + \beta\beta'\beta''zx + \gamma\gamma'\gamma''xy)^2;
\end{aligned}$$

where

$$\begin{aligned}
M &= (\beta\ -\gamma\ )\,\alpha'\alpha'' + (\gamma\ -\alpha\ )\,\beta'\beta'' + (\alpha\ -\beta\ )\,\gamma'\gamma'',\\
&= (\beta' - \gamma'\ )\,\alpha''\alpha + (\gamma' - \alpha'\ )\,\beta''\beta + (\alpha' - \beta'\ )\,\gamma''\gamma,\\
&= (\beta'' - \gamma'')\,\alpha\alpha' + (\gamma'' - \alpha'')\,\beta\beta' + (\alpha'' - \beta'')\,\gamma\ \gamma',\\
&= -\tfrac{1}{8}\,\{(\beta-\gamma)(\beta'-\gamma')(\beta''-\gamma'') + (\gamma-\alpha)(\gamma'-\alpha')(\gamma''-\alpha'') + (\alpha-\beta)(\alpha'-\beta')(\alpha''-\beta'')\};
\end{aligned}$$

also

$$K = 4\alpha\alpha'\alpha''\beta\beta'\beta''\gamma\gamma'\gamma'':$$

and we have identically

$$AC - B^2 = Kxyz\left(\frac{x}{\alpha} + \frac{y}{\beta} + \frac{z}{\gamma}\right)\left(\frac{x}{\alpha'} + \frac{y}{\beta'} + \frac{z}{\gamma'}\right)\left(\frac{x}{\alpha''} + \frac{y}{\beta''} + \frac{z}{\gamma''}\right).$$

The 16-*nodal Surface* 16 (1, 1, 1, 1, 1, 1).

157. Kummer starts from an irrational equation, which is readily converted into the following

$$\sqrt{x(X-w)} + \sqrt{y(Y-w)} + \sqrt{z(Z-w)} = 0,$$

and then, rationalizing, we have

$$Aw^2 + 2Bw + C = 0,$$

where as above

$$AC - B^2 = Kxyz\, PP'P''.$$

This agrees with the foregoing theory; viz., the point ($x=0$, $y=0$, $z=0$) being a node, the rationalized equation must, of course, be in the node-form (1, 1, 1, 1, 1, 1), (being the only node-form); and the symmetry of the formulæ enables us at once to write down the equations of the 16 singular planes, and thence to deduce the coordinates of the 16 nodes; viz.,

the singular planes are		and the nodes are	
(1)	$x=0$,	(1)	$(0, -\beta, \gamma, \alpha'\alpha''\beta\gamma)$,
(2)	$y=0$,	(2)	$(\alpha, 0, -\gamma, \beta'\beta''\gamma\alpha)$,
(3)	$z=0$,	(3)	$(-\alpha, \beta, 0, \gamma'\gamma''\alpha\beta)$,
(4)	$w=0$,	(4)	$(\alpha'\alpha'', \beta'\beta'', \gamma'\gamma'', 0)$,
(5)	$X-w=0$,	(5)	$(1, 0, 0, 0)$,
(6)	$Y-w=0$,	(6)	$(0, 1, 0, 0)$,
(7)	$Z-w=0$,	(7)	$(0, 0, 1, 0)$,
(8)	$P=0$,	(8)	$(0, 0, 0, 1)$,
(9)	$X'-w=0$,	(9)	$(0, -\beta', \gamma', \alpha''\alpha\beta'\gamma')$,
(10)	$Y'-w=0$,	(10)	$(\alpha', 0, -\gamma', \beta''\beta\gamma'\alpha')$,
(11)	$Z'-w=0$,	(11)	$(-\alpha', \beta', 0, \gamma''\gamma\alpha'\beta')$,
(12)	$P'=0$,	(12)	$(\alpha''\alpha, \beta''\beta, \gamma''\gamma, 0)$,
(13)	$X''-w=0$,	(13)	$(0, -\beta'', \gamma'', \alpha\alpha'\beta''\gamma'')$,
(14)	$Y''-w=0$,	(14)	$(\alpha'', 0, -\gamma'', \beta\beta'\gamma''\alpha'')$,
(15)	$Z''-w=0$,	(15)	$(-\alpha'', \beta'', 0, \gamma\gamma'\alpha''\beta'')$,
(16)	$P''=0$,	(16)	$(\alpha\alpha', \beta\beta', \gamma\gamma', 0)$,

where the nodes and planes are numbered as by Kummer; and by means of his (differently arranged) diagram of the relation between the several nodes and planes, I was enabled to form the following square diagram, which exhibits this relation in, I think, the most convenient form. To explain this, observe that in the upper and left-hand margins, the numbers refer to the nodes; in the body of the table, and in the right-hand margin to the planes, the table shows that for the node 1, the circumscribed cone is made up of the planes 1, 6, 7, 8, 9, 13; and that the remaining 15 nodes are situate on the nodal lines of this cone, the node 2 on the intersection of the planes 7, 8; the node 3 on the intersection of the planes 6, 8, and so on; and the like as regards the other lines of the table.

NODES	1	2	3	4	5	6	7	8	9	10	11	12	13	14	15	16	CONES
1	*	7, 8	6, 8	6, 7	9, 13	1, 6	1, 7	1, 8	1, 9	6, 13	7, 13	8, 13	1, 13	6, 9	7, 9	8, 9	1, 9, 13, 8, 7, 6
2	7, 8	*	5, 8	5, 7	2, 5	10, 14	2, 7	2, 8	5, 14	2, 10	7, 14	8, 14	5, 10	2, 14	7, 10	8, 10	2, 10, 14, 7, 8, 5
3	6, 8	5, 8	*	5, 6	3, 5	3, 6	11, 15	3, 8	5, 15	6, 15	3, 11	8, 15	5, 11	6, 11	3, 15	8, 11	3, 11, 15, 6, 5, 8
4	6, 7	5, 7	5, 6	*	4, 5	4, 6	4, 7	12, 16	5, 16	6, 16	7, 16	4, 12	5, 12	6, 12	7, 12	4, 16	4, 12, 16, 5, 6, 7
5	9, 13	2, 5	3, 5	4, 5	*	3, 4	2, 4	2, 3	5, 9	2, 13	3, 13	4, 13	5, 13	2, 9	3, 9	4, 9	5, 13, 9, 4, 3, 2
6	1, 6	10, 14	3, 6	4, 6	3, 4	*	1, 4	1, 3	1, 14	6, 10	3, 14	4, 14	1, 10	6, 14	3, 10	4, 10	6, 14, 10, 3, 4, 1
7	1, 7	2, 7	11, 15	4, 7	2, 4	1, 4	*	1, 2	1, 15	2, 15	7, 11	4, 15	1, 11	2, 11	7, 15	4, 11	7, 15, 11, 2, 1, 4
8	1, 8	2, 8	3, 8	12, 16	2, 3	1, 3	1, 2	*	1, 16	2, 16	3, 16	8, 12	1, 12	2, 12	3, 12	8, 16	8, 16, 12, 1, 2, 3
9	1, 9	5, 14	5, 15	5, 16	5, 9	1, 14	1, 15	1, 16	*	15, 16	14, 16	14, 15	1, 5	9, 14	9, 15	9, 16	9, 1, 5, 16, 15, 4
10	6, 13	2, 10	6, 15	6, 16	2, 13	6, 10	2, 15	2, 16	15, 16	*	13, 16	15, 13	10, 13	2, 6	10, 15	10, 16	10, 2, 6, 15, 16, 13
11	7, 13	7, 14	3, 11	7, 16	3, 13	3, 14	7, 11	3, 16	14, 16	13, 16	*	13, 14	11, 13	11, 14	3, 7	11, 16	11, 3, 7, 14, 13, 16
12	8, 13	8, 14	8, 15	4, 12	4, 13	4, 14	4, 15	8, 12	14, 15	13, 15	13, 14	*	12, 13	12, 14	12, 15	4, 8	12, 4, 8, 13, 14, 15
13	1, 13	5, 10	5, 11	5, 12	5, 13	1, 10	1, 11	1, 12	1, 5	10, 13	11, 13	12, 13	*	11, 12	10, 12	10, 11	13, 5, 1, 12, 11, 10
14	6, 9	2, 14	6, 11	6, 12	2, 9	6, 14	2, 11	2, 12	9, 14	2, 6	11, 14	12, 14	11, 12	*	9, 12	9, 11	14, 6, 2, 11, 12, 9
15	7, 9	7, 10	3, 15	7, 12	3, 9	3, 10	7, 15	3, 12	9, 15	10, 15	3, 7	12, 15	10, 12	9, 12	*	9, 10	15, 7, 3, 10, 9, 12
16	8, 9	8, 10	8, 11	4, 16	4, 9	4, 10	4, 11	8, 16	9, 16	10, 16	11, 16	4, 8	10, 11	9, 11	9, 10	*	16, 8, 4, 9, 10, 11

158. The before mentioned irrational equation may be written

$$\sqrt{1\,.\,5}+\sqrt{2\,.\,6}+\sqrt{3\,.\,7}=0,$$

and by symmetry we see that also

$$\sqrt{1\,.\,9}+\sqrt{2\,.\,10}+\sqrt{3\,.\,11}=0,$$

$$\sqrt{1\,.\,13}+\sqrt{2\,.\,14}+\sqrt{3\,.\,15}=0;$$

viz., these are three equations each containing the planes 1, 2, 3, which are three of the planes belonging to the node 1; the other three planes in any such equation (for instance, the planes 5, 6, 7, in the first equation) being three planes belonging to another node. Instead of the planes 1, 2, 3, we may have any other three planes belonging to the node 1; and instead of the node 1, any other node; but each equation belongs to two nodes: the number of equations is thus

$$\frac{6\,.\,5\,.\,4}{1\,.\,2\,.\,3}\times 16\times 3\div 2,\ =480.$$

159. To obtain the planes belonging to any such equation, combine any two of the outside right-hand lines of the diagram, these contain in common two numbers the places of which are interchanged; striking these out, we have four columns, and taking out of these any three columns, we have the corresponding sets of planes. For instance, lines 1 and 2 contain 78 and 87 respectively; striking these out, the lines are

1, 9, 13, 6;

2, 10, 14, 5;

whence we have the sets (1, 9, 13) and (2, 10, 14); viz., there is an irrational equation of the form

$$\sqrt{1\,.\,2}+\sqrt{9\,.\,10}+\sqrt{13\,.\,14}=0,$$

but it is probably necessary to introduce constant factors along with the products 1.2, 9.10, and 13.14 respectively. There are $\frac{1}{2}$ 16.15, =120 pairs of lines, and each line gives 4 equations; in all 120 × 4, = 480 equations, as above.

160. I stop to remark that Kummer gives for his 13-nodal surface an equation containing three arbitrary constants, say λ, μ, ν, such that, putting one of these = 0, we have the 14-nodal surface; putting two of them each = 0, the 15-nodal surface; putting all three of them = 0, the 16-nodal surface. The equations for the 16-nodal surface and that for the 14-nodal surface, made use of by Kummer, are, in fact, those deduced as above from the equation of the 13 (α)-nodal surface; and the like form might have been used for the 15-nodal surface. But the form actually used by Kummer, as presently appearing, is an equivalent form not thus deducible from the equation of the 13 (α)-nodal surface.

The 15-*nodal Surface* 15 (2, 1, 1, 1, 1).

161. Kummer's equation is readily converted into the following:

$$Aw^2 + 2Bw + C + \frac{Kl}{m} xyz\, P = 0,$$

the circumscribed cone being thus

$$(lA + mP'P'')\, xyz\, P = 0,$$

and the equation being in the node-form (2, 1, 1, 1, 1).

The formulæ for the 15 nodes and the 10 singular planes depend upon a quadric equation, for the symmetrical expression of which I write

$$\alpha'\alpha'' - \beta''\gamma' = \beta'\beta'' - \gamma''\alpha' = \gamma'\gamma'' - \alpha''\beta' = \omega,$$
$$\alpha'\alpha'' - \beta'\gamma'' = \beta'\beta'' - \gamma'\alpha'' = \gamma'\gamma'' - \alpha'\beta'' = \varpi;$$

so that

$$\omega - \varpi = \beta'\gamma'' - \beta''\gamma' = \gamma'\alpha'' - \gamma''\alpha' = \alpha'\beta'' - \alpha''\beta',$$
$$\omega + \varpi = \alpha'\alpha'' + \beta'\beta'' + \gamma'\gamma'':$$

the equation in question then is

$$(\rho - \omega)(\rho - \varpi) + \frac{Kl}{4\alpha\beta\gamma m} = 0;$$

so that, calling the roots of it ρ_1, ρ_2, we have

$$\rho_1 + \rho_2 = \omega + \varpi, \quad \rho_1\rho_2 = \omega\varpi + \frac{Kl}{4\alpha\beta\gamma m};$$

or we may write

$$\rho_1 = \tfrac{1}{2}(\omega + \varpi + \sqrt{\Omega}), \quad = \tfrac{1}{2}(\alpha'\alpha'' + \beta'\beta'' + \gamma'\gamma'' + \sqrt{\Omega}),$$
$$\rho_2 = \tfrac{1}{2}(\omega + \varpi - \sqrt{\Omega}), \quad = \tfrac{1}{2}(\alpha'\alpha'' + \beta'\beta'' + \gamma'\gamma'' - \sqrt{\Omega}),$$

if for shortness

$$\Omega = (\omega - \varpi)^2 - \frac{Kl}{\alpha\beta\gamma m}.$$

162. I write also for shortness

$$\mathrm{a} = (\beta - \gamma)\,\alpha'\alpha'' + \alpha\,(\beta'\beta'' - \gamma'\gamma''),$$
$$\mathrm{b} = (\gamma - \alpha)\,\beta'\beta'' + \beta\,(\gamma'\gamma'' - \alpha'\alpha''),$$
$$\mathrm{c} = (\alpha - \beta)\,\gamma'\gamma'' + \gamma\,(\alpha'\alpha'' - \beta'\beta'');$$

and I say that the singular planes are

(1) (1) $x = 0$,

(2) (2) $y = 0$,

(3) (3) $z = 0$,

(4) (9) $-\dfrac{1}{\beta\gamma}(X'-w)+(\rho_1-\omega)\left(\dfrac{y}{\beta}+\dfrac{z}{\gamma}\right)=0,$

(5) (10) $-\dfrac{1}{\gamma\alpha}(Y'-w)+(\rho_1-\omega)\left(\dfrac{z}{\gamma}+\dfrac{x}{\alpha}\right)=0,$

(6) (11) $-\dfrac{1}{\alpha\beta}(Z'-w)+(\rho_1-\omega)\left(\dfrac{x}{\alpha}+\dfrac{y}{\beta}\right)=0,$

(7) (13) $-\dfrac{1}{\beta\gamma}(X''-w)+(\rho_2-\varpi)\left(\dfrac{y}{\beta}+\dfrac{z}{\gamma}\right)=0,$

(8) (14) $-\dfrac{1}{\gamma\alpha}(Y''-w)+(\rho_2-\varpi)\left(\dfrac{z}{\gamma}+\dfrac{x}{\alpha}\right)=0,$

(9) (15) $-\dfrac{1}{\alpha\beta}(Z''-w)+(\rho_2-\varpi)\left(\dfrac{x}{\alpha}+\dfrac{y}{\beta}\right)=0,$

(10) (8) $P=0$;

and that the nodes are

(1) (1) $(0,\ -\beta,\ \gamma,\ \alpha'\alpha''\beta\gamma),$

(2) (2) $(\alpha,\ 0,\ -\gamma,\ \beta'\beta''\gamma\alpha),$

(3) (3) $(-\alpha,\ \beta,\ 0,\ \gamma'\gamma''\alpha\beta),$

(4) (9) $\{0,\ \rho_1-\gamma'\gamma'',\ -(\rho_2-\beta'\beta''),\ \alpha(\rho_1-\gamma'\gamma'')(\rho_2-\beta'\beta'')\},$

(5) (10) $\{-(\rho_2-\gamma'\gamma''),\ 0,\ \rho_1-\alpha'\alpha'',\ \beta(\rho_1-\alpha'\alpha'')(\rho_2-\gamma'\gamma'')\},$

(6) (11) $\{\rho_1-\beta'\beta'',\ -(\rho_2-\alpha'\alpha''),\ 0,\ \gamma(\rho_1-\beta'\beta'')(\rho_2-\alpha'\alpha'')\},$

(7) (13) $\{0,\ \rho_2-\gamma'\gamma'',\ -(\rho_1-\beta'\beta''),\ \alpha(\rho_2-\gamma'\gamma'')(\rho_1-\beta'\beta'')\},$

(8) (14) $\{-(\rho_1-\gamma'\gamma''),\ 0,\ \rho_2-\alpha'\alpha'',\ \beta(\rho_2-\alpha'\alpha'')(\rho_1-\gamma'\gamma'')\},$

(9) (15) $\{\rho_2-\beta'\beta'',\ -(\rho_1-\alpha'\alpha''),\ 0,\ \gamma(\rho_2-\beta'\beta'')(\rho_1-\alpha'\alpha'')\},$

(10) (8) $(0,\ 0,\ 0,\ 1),$

(11) (5) $(1,\ 0,\ 0,\ 0),$

(12) (6) $(0,\ 1,\ 0,\ 0),$

(13) (7) $(0,\ 0,\ 1,\ 0),$

(14) (16) $\left(\tfrac{1}{2}\alpha(\mathrm{a}-\alpha\sqrt{\Omega}),\ \tfrac{1}{2}\beta(\mathrm{b}-\beta\sqrt{\Omega}),\ \tfrac{1}{2}\gamma(\mathrm{c}-\gamma\sqrt{\Omega}),\ \dfrac{kl}{4m}\right),$

(15) (12) $\left(\tfrac{1}{2}\alpha(\mathrm{a}+\alpha\sqrt{\Omega}),\ \tfrac{1}{2}\beta(\mathrm{b}+\beta\sqrt{\Omega}),\ \tfrac{1}{2}\gamma(\mathrm{c}+\gamma\sqrt{\Omega}),\ \dfrac{kl}{4m}\right).$

163. The small reference numbers are those used by Kummer. It is, I think, better to retain the reference numbers belonging to the case of the 16-nodal surface; viz., there are here given, large, 1, 2, 3, 8, 9, 10, 11, 13, 14, 15 for the planes, and 1, 2, 3, 5, 6, ... to 16 for the nodes. Belonging to each node (that is, with the node as vertex) there is a quadric cone passing through 8 other nodes; and each node lies (exclusively of the cone whose vertex it is) in 8 such cones. We have thus the following square diagram:

NODES	1	2	3	5	6	7	8	9	10	11	12	13	14	15	16	CONES
1	*	*C* 8	*C* 8	9, 13	*C* 1	*C* 1	1, 8	1, 9	*C* 13	*C* 13	8, 13	1, 13	*C* 9	*C* 9	8, 9	1, 8, 9, 13 (6, 7)
2	*C* 8	*	*C* 8	*C* 2	10, 14	*C* 2	2, 8	*C* 14	2, 10	*C* 14	8, 14	*C* 10	2, 14	*C* 10	8, 10	2, 8, 13, 14 (5, 7)
3	*C* 8	*C* 8	*	*C* 3	*C* 3	11, 15	3, 8	*C* 15	*C* 15	3, 11	8, 15	*C* 11	*C* 11	3, 15	8, 11	3, 8, 11, 15 (5, 6)
5	9, 13	*C* 2	*C* 3	*	*C* 3	*C* 2	2, 3	*C* 9	2, 13	3, 13	*C* 13	*C* 13	2, 9	3, 9	*C* 9	2, 3, 9, 13 (4, 5)
6	*C* 1	10, 14	*C* 3	*C* 3	*	*C* 1	1, 3	1, 14	*C* 10	3, 14	*C* 14	1, 10	*C* 14	3, 10	*C* 10	1, 3, 10, 14 (4, 6)
7	*C* 1	*C* 2	11, 15	*C* 2	*C* 1	*	1, 2	1, 15	2, 15	*C* 11	*C* 15	1, 11	2, 11	*C* 15	*C* 11	1, 2, 11, 15 (4, 7)
8	1, 8	2, 8	3, 8	2, 3	1, 3	1, 2	*	*C* 1	*C* 2	*C* 3	*C* 8	*C* 1	*C* 2	*C* 3	*C* 8	1, 2, 3, 8 (12, 16)
9	1, 9	*C* 14	*C* 15	*C* 9	1, 14	1, 15	*C* 1	*	*C* 15	*C* 14	14, 15	*C* 1	9, 14	9, 15	*C* 9	1, 9, 14, 15 (5, 16)
10	*C* 13	2, 10	*C* 15	2, 13	*C* 10	2, 15	*C* 2	*C* 15	*	*C* 13	13, 15	10, 13	*C* 2	10, 15	*C* 10	2, 10, 13, 15 (6, 16)
11	*C* 13	*C* 14	3, 11	3, 13	3, 14	*C* 11	*C* 3	*C* 14	*C* 13	*	13, 14	11, 13	11, 14	*C* 3	*C* 11	3, 11, 13, 14 (7, 16)
12	8, 13	8, 14	8, 15	*C* 13	*C* 14	*C* 15	*C* 8	14, 15	13, 15	13, 14	*	*C* 13	*C* 14	*C* 15	*C* 8	8, 13, 14, 15 (4, 12)
13	1, 13	*C* 10	*C* 11	*C* 13	1, 10	1, 11	*C* 1	*C* 1	10, 13	11, 13	*C* 13	*	*C* 11	*C* 10	10, 11	1, 10, 11, 13 (5, 12)
14	*C* 9	2, 14	*C* 11	2, 9	*C* 14	2, 11	*C* 2	9, 14	*C* 2	11, 14	*C* 14	*C* 11	*	*C* 9	9, 11	2, 9, 11, 14 (6, 12)
15	*C* 9	*C* 10	3, 15	3, 9	3, 10	*C* 15	*C* 3	9, 15	10, 15	*C* 3	*C* 15	*C* 10	*C* 9	*	9, 10	3, 9, 10, 15 (7, 12)
16	8, 9	8, 10	8, 11	*C* 9	*C* 10	*C* 11	*C* 8	*C* 9	*C* 10	*C* 11	*C* 8	10, 11	9, 11	9, 10	*	8, 9, 10, 11 (4, 16)

164. The arrangement is the same as in the 16-nodal square diagram; only, in the right-hand margin, a bracket (6, 7) denotes that instead of the planes (6, 7) we have a quadric cone; which cones are, in the body of the table, denoted by C. Thus for the node 1 the sextic cone is made up of the planes 1, 8, 9, 13 and of a quadric cone (6, 7), $=C$: the remaining 14 nodes lie on the nodal lines of the sextic cone, viz., the node 2 on an intersection of the cone C with the plane 8, the node 3 on an intersection of the same cone and plane, the node 5 on the intersection of the planes 9, 13, and so on.

The Equation of the 15-*nodal Surface, as deduced from that of the* 13 (α)-*nodal.*

165. If, in the equation hereafter given for the 13-nodal surface, we write $\nu=0$, $\mu=0$, or (what is the same thing) in that of the 14-nodal surface we write $\mu=0$, the form is

$$\begin{aligned} &w^2(A+4\lambda yz)\\ +\,&2w\ (B-2\lambda yz\,X)\\ +\,&\qquad C=0. \end{aligned}$$

The circumscribed sextic cone is thus (2, 1, 1, 1, 1),

$$(\lambda\alpha^2 yz+\beta\gamma x^2+\gamma\alpha xy+\alpha\beta xz)\,yz\,(P'-\rho_1 P'')\,(P'-\rho_2 P'')=0,$$

where ρ_1, ρ_2 now denote the roots of the equation

$$\lambda(\rho\alpha-\alpha'')^2-\frac{1}{\beta'\beta''\gamma'\gamma''}(\omega-\varpi)^2\,\rho\alpha'\alpha''=0.$$

The singular planes are

(4)	$w=0,$
(12)	$P'-\rho_1 P''=0,$
(16)	$P'-\rho_2 P''=0,$
(3)	$z=0,$
(2)	$y=0,$
(15)	$\dfrac{\gamma''\alpha'\beta'}{\gamma'\alpha''\beta''}\rho_1(Z'-w)-(Z''-w)=0,$
(11)	$\dfrac{\gamma''\alpha'\beta'}{\gamma'\alpha''\beta''}\rho_2(Z'-w)-(Z''-w)=0,$
(14)	$\dfrac{\beta''\gamma'\alpha'}{\beta'\gamma''\alpha''}\rho_1(Y'-w)-(Y''-w)=0,$
(10)	$\dfrac{\beta''\gamma'\alpha'}{\beta'\gamma''\alpha''}\rho_2(Y'-w)-(Y''-w)=0,$
(5)	$X-w=0,$

and we have then the same square table as before: the coordinates of the 15 nodes may be obtained without difficulty.

166. The form is really equivalent to that first considered in regard to the 15-nodal surface. To show that this is so, we have only to arrange according to powers of x; viz., the equation thus becomes

$$\begin{aligned} & x^2\{(\gamma\gamma'\gamma'')^2 y^2 + (\beta\beta'\beta'')^2 z^2 + w^2 - 2\beta\beta'\beta''zw + 2\gamma\gamma'\gamma''wy + 2\beta\beta'\beta''\gamma\gamma'\gamma''yz\} \\ & + 2x\{\gamma\gamma'\gamma''\alpha\alpha'\alpha''y^2z + \alpha\alpha'\alpha''\beta\beta'\beta''yz^2 + \beta\beta'\beta''z^2w - zw^2 - w^2y - \gamma\gamma'\gamma''wy^2 + Myzw\} \\ & + \quad (\alpha\alpha'\alpha''yz + wy - wz)^2 - 4\lambda yzw(X - w) = 0, \end{aligned}$$

where, if for a moment A denotes the coefficient of x^2, we have $y = 0$, $z = 0$, $w = 0$, $X - w = 0$, four tangent planes of the quadric cone $A = 0$.

14-*nodal Surface* $8(3_1, 1, 1, 1) + 6(2, 2, 1, 1)$, *Node-form* $(3_1, 1, 1, 1)$.

167. In the equation hereafter given for the 13-nodal surface, writing $\nu = 0$, the sextic cone becomes

$$4z(\lambda\alpha^2y^2z + \mu\beta^2zx^2 + \beta\gamma x^2y + \gamma\alpha xy^2 + \alpha\beta xyz)(P' - \rho_1P'')(P' - \rho_2P'') = 0;$$

viz., this is of the form in question $(3_1, 1, 1, 1)$; and the equation of the surface is

$$\begin{aligned} & w^2A + 4(\lambda y + \mu x)z - 4\lambda\mu z^2 \\ & + 2w\{B - 2(\lambda yX + \mu xY)z\} \\ & + \quad C = 0. \end{aligned}$$

The singular planes are

$$\begin{aligned} & w = 0, && (4) \quad (4) \\ & P' - \rho_1 P'' = 0, && (12) \quad (2) \\ & P' - \rho_2 P'' = 0, && (16) \quad (3) \\ & z = 0, && (3) \quad (1) \\ & \frac{\gamma''\alpha'\beta'}{\gamma'\alpha''\beta''}\rho_1(Z' - w) - (Z'' - w) = 0, && (15) \quad (6) \\ & \frac{\gamma''\alpha'\beta'}{\gamma'\alpha''\beta''}\rho_2(Z' - w) - (Z'' - w) = 0, && (11) \quad (5) \end{aligned}$$

and the nodes are

$$\begin{aligned} & (0,\ 0,\ 0,\ 1), && (8) \quad (1) \\ & (1,\ 0,\ 0,\ 0), && (5) \quad (13) \\ & (0,\ 1,\ 0,\ 0), && (6) \quad (14) \\ & (0,\ 0,\ 1,\ 0), && (7) \quad (5) \\ & \left(-\frac{\beta'' - \beta'\rho_1}{\beta'\beta''},\ \frac{\alpha'' - \alpha'\rho_1}{\alpha'\alpha''},\ 0,\ \frac{\gamma\gamma'\gamma''(\beta'' - \beta'\rho_1)(\alpha'' - \alpha'\rho_1)}{\alpha''\beta''\gamma' - \alpha'\beta'\gamma''\rho_1}\right), && (11) \quad (4) \\ & \left(-\frac{\beta'' - \beta'\rho_2}{\beta'\beta''},\ \frac{\alpha'' - \alpha'\rho_2}{\alpha'\alpha''},\ 0,\ \frac{\gamma\gamma'\gamma''(\beta'' - \beta'\rho_2)(\alpha'' - \alpha'\rho_2)}{\alpha''\beta''\gamma' - \alpha'\beta'\gamma''\rho_2}\right), && (15) \quad (3) \end{aligned}$$

$$\left(\frac{\alpha\alpha'\alpha''}{\alpha''-\alpha'\rho_1},\ \frac{\beta\beta'\beta''}{\beta''-\beta'\rho_1},\ \frac{\gamma\gamma'\gamma''}{\gamma''-\gamma'\rho_1},\ 0\right), \qquad (16)\ \ (7)$$

$$\left(\frac{\alpha\alpha'\alpha''}{\alpha''-\alpha'\rho_2},\ \frac{\beta\beta'\beta''}{\beta''-\beta'\rho_2},\ \frac{\gamma\gamma'\gamma''}{\gamma''-\gamma'\rho_2},\ 0\right), \qquad (12)\ \ (8)$$

$$(\ \alpha'\alpha'',\ \beta'\beta'',\ \gamma'\gamma'',\ 0\ ), \qquad (4)\ \ (6)$$

$$(\ -\alpha,\ \beta,\ 0,\ \gamma'\gamma''\alpha\beta), \qquad (3)\ \ (2)$$

two nodes on line

$$\frac{\gamma''\alpha'\beta'}{\gamma'\alpha''\beta''}\rho_2(Z'-w)-(Z''-w)=0,\quad P'-\rho_1P''=0, \qquad (13,\ 14)\ \ (9,\ 10)$$

and two nodes on line

$$\frac{\gamma''\alpha'\beta'}{\gamma'\alpha''\beta''}\rho_1(Z'-w)-(Z''-w)=0,\quad P'-\rho_2P''=0, \qquad (9,\ 10)\ \ (11,\ 12)$$

where the large numbers are those for the 16-nodal surface, the small numbers are Kummer's.

168. In these formulæ, ρ_1, ρ_2 denote the roots of the equation

$$\begin{aligned}&(\lambda\beta''^2+\mu\alpha''^2)(\rho\alpha'\beta')^2\\ &+(\lambda\beta'^2+\mu\alpha'^2)(\alpha''\beta'')^2\\ &-2\left[\lambda\beta'\beta''+\mu\alpha'\alpha''+\frac{1}{2\gamma'\gamma''}(\omega-\varpi)^2\right]\rho\alpha'\beta'\alpha''\beta''=0.\end{aligned}$$

169. The relation of the nodes and sextic cones is given by means of the square diagram on the opposite page:

NODES	3	4	5	6	7	8	9	10	11	12	13	14	15	16	CONES
3	*	×	C 3	C 3	11, 15	C 3	C 15	C 15	3, 11	C 15	C 11	C 11	3, 15	C 11	3, 11, 15, (5, 6, 8)
4	×	*	C 4	C 4	C 4	12, 16	C 16	C 16	C 16	4, 12	C 12	C 12	C 12	4, 16	4, 12, 16, (5, 6, 7)
5	D 3	D 4	*	3, 4	D 4	D 3			D' 3	D' 4			D' 3	D' 4	3, 4, (2, 5), (9, 13)
6	D 3	D 4	3, 4	*	D 4	D 3			D' 3	D' 4			D' 3	D' 4	3, 4, (1, 6), (10, 14)
7	11, 15	C 4	C 4	C 4	*	×	C 15	C 15	C 11	4, 15	C 11	C 11	C 15	4, 11	4, 11, 15, (1, 2, 7)
8	C 3	12, 16	C 3	C 3	×	*	C 16	C 16	3, 16	C 12	C 12	C 12	3, 12	C 16	3, 12, 16, (1, 2, 8)
9	D 15	D 16			D' 15	D' 16	*	15, 16	D 16	D 15			D' 15	D' 16	15, 16, (5, 14), (1, 9)
10	D 15	D 16			D' 15	D' 16	15, 16	*	D 16	D 15			D' 15	D' 16	15, 16, (6, 13), (2, 10)
11	3, 11	C 16	C 3	C 3	C 11	3, 16	C 16	C 16	*	×	C 11	C 11	C 3	11, 16	3, 11, 16, (7, 13, 14)
12	C 15	4, 12	C 4	C 4	4, 15	C 12	C 15	C 15	×	*	C 12	C 12	12, 15	C 4	4, 12, 15, (8, 13, 14)
13	D 11	D 12			D' 11	D' 12			D' 11	D' 12	*	11, 12	D 12	D 11	11, 12, (5, 10), (1, 13)
14	D 11	D 12			D' 11	D' 12			D' 11	D' 12	11, 12	*	D 12	D 11	11, 12, (6, 9), (2, 14)
15	3, 15	C 12	C 3	C 3	C 15	3, 12	C 15	C 15	C 3	12, 15	C 12	C 12	*	×	3, 12, 15, (7, 9, 10)
16	C 11	4, 16	C 4	C 4	4, 11	C 16	C 16	C 16	11, 16	C 4	C 11	C 11	×	*	4, 11, 16, (8, 9, 10)

where the arrangement is the same as before; only in the right-hand margin, a bracket (5, 6, 8) denotes that, instead of the planes 5, 6, 8, we have a nodal cubic cone (5, 6, 8), which is in the body of the table referred to as C, and the nodal line thereof by $\times$; and the brackets (2, 5), (9, 13) denote that, instead of the planes 2, 5, we have a quadric cone (2, 5), and instead of the planes 9, 13, a quadric cone (9, 13); which cones are in the body of the table referred to as D, D' respectively. And it is to be understood that each vacant square of the table should contain the symbol (D, D'), this being omitted only for the purpose of better exhibiting the form of the table.

170. The equation of the surface may be presented in the irrational form

$$\begin{aligned}&\sqrt{(P'-\rho_1 P'')\,[\gamma''\alpha'\beta'\rho_2\,(Z'-w)-\gamma'\alpha''\beta''\,(Z''-w)]}\\&+\sqrt{(P'-\rho_2 P'')\,[\gamma''\alpha'\beta'\rho_1\,(Z'-w)-\gamma'\alpha''\beta''\,(Z''-w)]}\\&+\quad(\rho_1-\rho_2)\sqrt{\alpha'\beta'\alpha''\beta''\left(\frac{\lambda}{\alpha''^2}+\frac{\mu}{\beta''^2}\right)zw}=0.\end{aligned}$$

In fact the norm of the left-hand side is

$$\begin{aligned}=(\rho_1-\rho_2)^2\,(\omega-\varpi)^2\,\{&w^2\,[A+4z\,(\lambda y+\mu x)-4\lambda\mu z^2]\\&+2w\,[B-2z\,(\lambda yX+\mu xY)]\\&+C\qquad\qquad\qquad\qquad\}.\end{aligned}$$

To partially verify this, observe that, writing the equation under the form $\sqrt{R}+\sqrt{S}+\sqrt{T}=0$, on writing therein $w=0$, we ought to have

$$(R-S)^2=(\rho_1-\rho_2)^2\,(\omega-\varpi)^2\,(\alpha\alpha'\alpha''yz+\beta\beta'\beta''zx+\gamma\gamma'\gamma''xy)^2.$$

But writing $w=0$, we have

$$\begin{aligned}R-S=&\quad(P'-\rho_1P'')\,(\gamma''\alpha'\beta'\rho_2Z'-\gamma'\alpha''\beta''Z'')\\&-(P'-\rho_2P'')\,(\gamma''\alpha'\beta'\rho_1Z'-\gamma'\alpha''\beta''Z''),\\=&\quad(\rho_2-\rho_1)\,(\gamma''\alpha'\beta'P'Z'-\gamma'\alpha''\beta''P''Z''),\\=&\quad(\rho_2-\rho_1)\,\{\gamma''\,(\beta'\gamma'x+\gamma'\alpha'y+\alpha'\beta'z)\,(\beta\beta''x-\alpha\alpha''y)\\&\qquad\qquad-\gamma'\,(\beta''\gamma''x+\gamma''\alpha''y+\alpha''\beta''z)\,(\beta\beta'x-\alpha\alpha'y)\},\end{aligned}$$

which is easily found to be

$$=-(\rho_2-\rho_1)\,(\omega-\varpi)\,(\alpha\alpha'\alpha''yz+\beta\beta'\beta''zx+\gamma\gamma'\gamma''xy);$$

and thus $(R-S)^2$ has the value in question.

171. In further verification, observe that, writing $x, y, z=\alpha'\alpha'', \beta'\beta'', \gamma'\gamma''$, and therefore $P'=0$, $P''=0$, we ought to have

$$(\rho_1-\rho_2)^2\,(\alpha'\beta'\alpha''\beta'')^2\left(\frac{\lambda}{\alpha''^2}+\frac{\mu}{\beta''^2}\right)z^2w^2=(\rho_1-\rho_2)^2\,(\omega-\varpi)^2\,w^2\,[A+4z\,(\lambda y+\mu x)-4\lambda\mu z^2],$$

observing that, for the values in question, B, X, Y, C all vanish.

This is

$$(\rho_1-\rho_2)^2(\alpha'\beta'\alpha''\beta'')^2\left(\frac{\lambda}{\alpha''^2}+\frac{\mu}{\beta''^2}\right)^2(\gamma'\gamma'')^2=(\omega-\varpi)^2\{(\omega-\varpi)^2+4\gamma'\gamma''(\lambda\beta'\beta''+\mu\alpha'\alpha''-\lambda\mu\gamma'\gamma'')\},$$

which is in fact the value of $(\rho_1-\rho_2)^2$ obtained from the equation in ρ.

The 13 (α)-*nodal Surface* $3(4_3, 1, 1)+1(3, 1, 1, 1)+9(3_1, 2, 1)$.

172. The equation, node-form $(4_3, 1, 1)$, is

$$\begin{aligned}&w^2\{A+4(\lambda yz+\mu zx+\nu xy)-4(\mu\nu x^2+\nu\lambda y^2+\lambda\mu z^2)\}\\&+2w\{B-2(\lambda yzX+\mu zxY+\nu xyZ)\}\\&+\quad C=0;\end{aligned}$$

viz., for the circumscribed cone we have

$$\begin{aligned}&\{A+4(\lambda yz+\mu zx+\nu xy)-4(\mu\nu x^2+\nu\lambda y^2+\lambda\mu z^2)\}\,C\\&-\{B-2(\lambda yzX+\mu zxY+\nu xyZ)\}^2\\&\quad=4\{\lambda\alpha^2y^2z^2+\mu\beta^2z^2x^2+\nu\gamma^2x^2y^2+\beta\gamma x^2yz+\gamma\alpha y^2zx+\alpha\beta z^2xy\}\\&\qquad\times\left\{-\lambda\frac{X^2}{\alpha^2}-\mu\frac{Y^2}{\beta^2}-\nu\frac{Z^2}{\gamma^2}+\alpha'\alpha''\beta'\beta''\gamma'\gamma''P'P''\right\},\end{aligned}$$

where, on the right-hand side, the first factor, equated to zero, represents a trinodal quartic cone, the nodal lines whereof are $(y=0,\ z=0)$, $(z=0,\ x=0)$, $(x=0,\ y=0)$.

173. As regards the second factor, it is to be observed that, writing as above

$$\omega-\varpi=\beta'\gamma''-\beta''\gamma',\quad=\gamma'\alpha''-\gamma''\alpha',\quad=\alpha'\beta''-\alpha''\beta',$$

we have identically

$$\begin{aligned}\alpha'P'-\alpha''P''&=\frac{\omega-\varpi}{\beta'\beta''\gamma'\gamma''}\,\frac{X}{\alpha},\\\beta'P'-\beta''P''&=\frac{\omega-\varpi}{\gamma'\gamma''\alpha'\alpha''}\,\frac{Y}{\beta},\\\gamma'P'-\gamma''P''&=\frac{\omega-\varpi}{\alpha'\alpha''\beta'\beta''}\,\frac{Z}{\gamma};\end{aligned}$$

so that the second factor is

$$\begin{aligned}=\frac{1}{(\omega-\varpi)^2}\{&-\lambda(\beta'\beta''\gamma'\gamma'')^2(\alpha'P'-\alpha''P'')^2-\mu(\gamma'\gamma''\alpha'\alpha'')^2(\beta'P'-\beta''P'')^2\\&-\nu(\alpha'\alpha''\beta'\beta'')^2(\gamma'P'-\gamma''P'')^2+(\omega-\varpi)^2\alpha'\alpha''\beta'\beta''\gamma'\gamma''P'P''\};\end{aligned}$$

or, what is the same thing,

$$\frac{1}{(\omega-\varpi)^2}\{-\ [\lambda(\beta''\gamma'')^2+\mu(\gamma''\alpha'')^2+\nu(\alpha''\beta'')^2](\alpha'\beta'\gamma'P')^2$$
$$-\ [\lambda(\beta'\gamma')^2+\mu(\gamma'\alpha')^2+\nu(\alpha'\beta')^2](\alpha''\beta''\gamma''P'')^2$$
$$+2[\lambda(\beta'\gamma'\beta''\gamma'')+\mu(\gamma'\alpha'\gamma''\alpha'')+\nu(\alpha'\beta'\alpha''\beta'')+\tfrac{1}{2}(\omega-\varpi)^2]\times\alpha'\beta'\gamma'P'\alpha''\beta''\gamma''P''\},$$

so that, equating to zero, we have a pair of planes, each passing through the line $P'=0$, $P''=0$, and which it is clear must be the tangent planes from this line to the quadric cone $A'=0$. I will presently return to this, but I consider first the foregoing identical equation in regard to the circumscribed cone.

174. In verification hereof, observe first that, if $x=0$, the equation becomes

$$\{(y-z)^2+4\lambda yz-4\lambda(\nu y^2+\mu z^2)\}(\alpha\alpha'\alpha'')^2y^2z^2-\{\alpha\alpha'\alpha''(y^2z-yz^2)-2\lambda yz\alpha(\gamma'\gamma''y-\beta'\beta''z)\}^2$$
$$=4\lambda y^2z^2\{-\lambda\alpha^2(\gamma'\gamma''y-\beta'\beta''z)^2-(\alpha\alpha'\alpha'')^2(\nu y^2+\mu z^2)-\alpha^2\alpha'\alpha''(\gamma'y+\beta'z)(\gamma''y+\beta''z)\},$$

viz., omitting terms which destroy each other, this is

$$[(y-z)^2+4\lambda yz](\alpha\alpha'\alpha'')^2y^2z^2-[\alpha\alpha'\alpha''(y^2z-yz^2)-2\lambda yz\alpha(\gamma'\gamma''y-\beta'\beta''z)]^2$$
$$=4\lambda y^2z^2[-\lambda\alpha^2(\gamma'\gamma''y-\beta'\beta''z)^2-\alpha^2\alpha'\alpha''(\gamma'y+\beta'z)(\gamma''y+\beta''z)].$$

Or again, this is

$$4\lambda yz(\alpha\alpha'\alpha'')^2y^2z^2+4\lambda yz\,\alpha^2\alpha'\alpha''(y^2z-yz^2)(\gamma'\gamma''y-\beta'\beta''z)$$
$$=-4\lambda y^2z^2\alpha^2\alpha'\alpha''(\gamma'y+\beta'z)(\gamma''y+\beta''z);$$

viz., this is

$$\alpha'\alpha''yz+(y-z)(\gamma'\gamma''y-\beta'\beta''z)=(\gamma'y+\beta'z)(\gamma''y+\beta''z),$$

which is at once seen to be true.

175. Again, compare the terms which contain x^4yz. On the right-hand side, we have

$$4\beta\gamma\,x^2yz\times\text{term in }x^2\text{ of }\left(-\lambda\frac{X^2}{\alpha^2}-\mu\frac{Y^2}{\beta^2}-\nu\frac{Z^2}{\gamma^2}+\alpha'\alpha''\beta'\beta''\gamma'\gamma''P'P''\right);$$

viz., the coefficient is

$$=4\beta\gamma\{-\mu(\gamma'\gamma'')^2-\nu(\beta'\beta'')^2+\beta'\beta''\gamma'\gamma''\}.$$

On the left-hand side, that is in $A'C-B'^2$, the only terms which give rise to the terms in question are

$$\text{in }A',\ (1-4\mu\nu)x^2;\ \text{ in }C,\ 2\beta\beta'\beta''\gamma\gamma'\gamma''x^2yz;$$

and in B'

$$(\gamma\gamma'\gamma''-2\nu\gamma\beta'\beta'')x^2y,\quad-(\beta\beta'\beta''-2\mu\beta\gamma'\gamma'')x^2z;$$

whence the coefficient is

$$2\beta\beta'\beta''\gamma\gamma'\gamma''(1-4\mu\nu)+2(\gamma\gamma'\gamma''-2\nu\gamma\beta'\beta'')(\beta\beta'\beta''-2\mu\beta\gamma'\gamma''),$$

which is in fact

$$= 4\beta\gamma\,\{-\mu\,(\gamma'\gamma'')^2 - \nu\,(\beta'\beta'')^2 + \beta'\beta''\gamma'\gamma''\},$$

which is right; and the verification may be completed without difficulty.

176. The singular planes are

$$\begin{array}{lr} w = 0, & (4) \\ P' - \rho_1 P'' = 0, & (12) \\ P' - \rho_2 P'' = 0, & (16); \end{array}$$

and the nodes are

$$\begin{array}{lll}
(0,\ 0,\ 0,\ 1), & (8) & (1) \\
(1,\ 0,\ 0,\ 0), & (5) & (5) \\
(0,\ 1,\ 0,\ 0), & (6) & (6) \\
(0,\ 0,\ 1,\ 0), & (7) & (7) \\
\left(\dfrac{\alpha\alpha'\alpha''}{\alpha''-\alpha'\rho_1},\ \dfrac{\beta\beta'\beta''}{\beta''-\beta'\rho_1},\ \dfrac{\gamma\gamma'\gamma''}{\gamma''-\gamma'\rho_1},\ 0\right), & (16) & (2) \\
\left(\dfrac{\alpha\alpha'\alpha''}{\alpha''-\alpha'\rho_2},\ \dfrac{\beta\beta'\beta''}{\beta''-\beta'\rho_2},\ \dfrac{\gamma\gamma'\gamma''}{\gamma''-\gamma'\rho_2},\ 0\right), & (12) & (3) \\
(\ \alpha'\alpha'',\ \beta'\beta'',\ \gamma'\gamma'',\ 0), & (4) & (4) \\
\text{three nodes in } P' - \rho_1 P'' = 0 & (13,\ 14,\ 15) & (8,\ 9,\ 10) \\
\qquad ,, \quad ,, \quad P' - \rho_2 P'' = 0 & (9,\ 10,\ 11) & (11,\ 12,\ 13);
\end{array}$$

where the small numbers are those used by Kummer, the large ones are those referring to the 16-nodal surface, and are here adopted. In the foregoing formulæ ρ_1, ρ_2 are the roots of

$$\begin{aligned}
&[\lambda\,(\beta''\gamma'')^2 + \mu\,(\gamma''\alpha'')^2 + \nu\,(\alpha''\beta'')^2]\,(\rho\alpha'\beta'\gamma')^2 \\
&+ [\lambda\,(\beta'\gamma')^2 + \mu\,(\gamma'\alpha')^2 + \nu\,(\alpha'\beta')^2]\,(\alpha''\beta''\gamma'')^2 \\
&- 2\,[\lambda\beta'\gamma'\beta''\gamma'' + \mu\gamma'\alpha'\gamma''\alpha'' + \nu\alpha'\beta'\alpha''\beta'' + \tfrac{1}{2}(\omega - \varpi)^2]\,\rho\alpha'\beta'\gamma'\alpha''\beta''\gamma'' = 0.
\end{aligned}$$

177. We have the square diagram in the following page:

NODES	4	5	6	7	8	9	10	11	12	13	14	15	16	CONES
4	*	*C* 4	*C* 4	*C* 4	12, 16	*C* 16	*C* 16	*C* 16	4, 12	*C* 12	*C* 12	*C* 12	4, 16	4, 12, 16, (5, 6, 7)
5	*C* 4	*	*C* 4	*C* 4	×				*C′* 4				*C′* 4	4, (2, 3; 5) nodal, (9, 13)
6	*C* 4	*C* 4	*	*C* 4	×				*C′* 4				*C′* 4	4, (1, 3; 6) nodal, (10, 14)
7	*C* 4	*C* 4	*C* 4	*	×				*C′* 4				*C′* 4	4, (1, 2; 7) nodal, (11, 15)
8	12, 16	×	×	×	*	*C* 16	*C* 16	*C* 16	*C* 16	*C* 12	*C* 12	*C* 12	*C* 12	12, 16 (8; 1, 2, 3) trinodal
9	*C* 16				*C″* 16	*	*C* 16	*C* 16	×				*C′* 16	16, (5; 14, 15) nodal, (1, 9)
10	*C* 16				*C″* 16	*C* 16	*	*C* 16	×				*C′* 16	16, (6; 13, 15) nodal, (2, 10)
11	*C* 16				*C″* 16	*C* 16	*C* 16	*	×				*C′* 16	16, (7; 13, 14) nodal, (3, 11)
12	4, 12	*C* 4	*C* 4	*C* 4	*C* 12	×	×	×	*	*C* 12	*C* 12	*C* 12	*C* 4	4, 12 (8; 13, 14, 15) trinodal
13	*C* 12				*C″* 12				*C′* 12	*	*C* 12	*C* 12	×	12, (5; 10, 11) nodal, (1, 13)
14	*C* 12				*C″* 12				*C′* 12	*C* 12	*	*C* 12	×	12, (6; 9, 11) nodal, (2, 14)
15	*C* 12				*C″* 12				*C′* 12	*C* 12	*C* 12	*	×	12, (7; 9, 10) nodal, (3, 15)
16	4, 16	*C* 4	*C* 4	*C* 4	*C* 16	*C* 16	*C* 16	*C* 16	*C* 4	×	×	×	*	4, 16, (8; 9, 10, 11) trinodal

where for greater clearness I have omitted the symbol CC', which is to be understood as occupying each of the vacant squares.

The arrangement is the same as before; the right-hand margin shows the sextic cone; viz., for the node 4 this is made up of the singular planes 4, 12, 16 and of a cubic cone represented by (5, 6, 7) (as replacing the planes 5, 6, 7 in the case of the 16-nodal surface). Similarly for the node 5, the sextic cone is made up of the singular plane 4, the nodal cubic cone (2, 3; 5), and the quadric cone (9, 13) (the numbers in these last symbols indicating the planes in the case of the 16-nodal surface, which are here replaced by cones). So for the node 8, the sextic cone is made up of the singular planes 12, 16 and of the trinodal quartic cone (8; 1, 2, 3). As regards a nodal cubic cone, for example (2, 3; 5), the semicolon is used to indicate that the nodal line replaces the intersection of the planes 2, 3; the other intersections 2, 5 and 3, 5 having disappeared. And so for a trinodal quartic cone (8; 1, 2, 3), the semicolon is used to indicate that the nodal lines replace the intersection (1, 2), the intersection (1, 3), and the intersection (2, 3) respectively; the other intersections 1, 2; 2, 8; and 3, 8 having disappeared. Finally, in the body of the table, C is used to denote the cubic or the quartic cone (as the case may be); $\times$ to denote a nodal line of either of these cones; and C' the quadric cone; as already mentioned, the vacant places are considered to be CC'.

The reading of the table is then as follows; viz., for the node 4, the remaining twelve nodes lie on the nodal lines of the sextic cone 4, 12, 16, (5, 6, 7), as shown; viz., 5, 6, 7 are each of them on the intersection of the cubic cone with the plane 4; 8 is on the intersection of the planes 12 and 16; and so on.

I reserve for another Memoir the discussion of the 13 (β)-nodal surface, and the surfaces with less than 13 nodes.

455.

ON PLÜCKER'S MODELS OF CERTAIN QUARTIC SURFACES.

[From the *Proceedings of the London Mathematical Society*, vol. III. (1869—1871), pp. 281—285. Read June 8, 1871.]

THE Society possesses a series of 14 wooden models of surfaces, constructed under the direction of the late Prof. Plücker, in illustration of the theory developed in his posthumous work, "Neue Geometrie des Raumes gegründet auf die Betrachtung der geraden Linie als Raumelemente," Leipzig, 1869. These all of them represent, I believe, Equatorial Surfaces; viz., models 1 to 8 represent cases of the 78 forms of equatorial surfaces "deren Breiten-Curven eine feste Axenrichtung besitzen," vol. II. pp. 352—363, the remaining models, Nos. 9—14, I have not completely identified. I propose to go into the theory only so far as is required for the explanation of the models.

In a "Complex," or triply infinite system of lines, there is in any plane whatever a singly infinite system of lines enveloping a curve; and if we attend only to the curves the planes of which pass through a given fixed line, the locus of these curves is a "complex surface." Similarly, there is through any point whatever a single infinite series of lines generating a cone; and if we attend only to the cones which have their vertices in the given fixed line, then the envelope of these cones is the same complex surface. In the case considered of a complex of the second degree, the curves and cones are each of them of the second order; the fixed line is a double line on the surface, so that (attending to the first mode of generation) the complete section by any plane through the fixed line is made up of this line twice, and of a conic; the surface is thus of the order 4: it is also of the class 4; the surface has, in fact, the nodal line, and also 8 nodes (conical points), and we have thus a reduction $=32$ in the class of the surface.

In the particular case where the nodal line is at infinity, the complex surface becomes an equatorial surface; viz., (attending to the first mode of generation) we have here a series of parallel planes each containing a conic, and the locus of these conics is the equatorial surface.

It is convenient to remark that, taking a, b, h to be homogeneous functions of (x, w) of the order 2; f, g of the order 1; and c of the order 0 (a constant); then the equation of a complex surface (see vol. I. p. 162) is

$$\begin{vmatrix} & y, & z, & 1 \\ y, & a, & h, & g \\ z, & h, & b, & f \\ 1, & g, & f, & c \end{vmatrix} = 0;$$

and that, writing $w = 1$, or considering a, h, b; f, g; c as functions of x of the orders 2, 1, 0 respectively, we have an equatorial surface.

A particular form of equatorial surface is thus $bcy^2 + caz^2 + ab = 0$, or taking $c = 1$, this is $by^2 + az^2 + ab = 0$, where a, b are quadric functions of x.

The surface is still, in general, of the fourth order: it may however degenerate into a cubic surface, or even into a quadric surface; the last case is however excluded from the enumeration. The section by any plane parallel to that of yz is a conic; the section by the plane $y = 0$ is made up of the pair of lines $a = 0$, and of the conic $z^2 + b = 0$; that by the plane $z = 0$ is made up of the pair of lines $b = 0$, and of the conic $y^2 + a = 0$; the last-mentioned planes may be called the principal planes, and the conics contained in them principal conics. The surface is thus the locus of a variable conic, the plane of which is parallel to that of yz, and which has for its vertices the intersections of its plane with the two principal conics respectively. And we have thus the particular equatorial surfaces considered by Plücker, vol. II. pp. 346—363 (as already mentioned), under the form

$$\frac{y^2}{Ex^2 + 2Ux + C} + \frac{z^2}{Fx^2 - 2Rx + B} + 1 = 0,$$

and of which he enumerates 78 kinds; viz., these are

1 to 17	Principal conics each proper.
18 „ 29	One of them a line-pair.
30 „ 32	Each a line-pair.
33 „ 39	Principal conics, each proper, but having a common point.
40 „ 43	One of them a line-pair, its centre on the other principal conic.
44 „ 61	One principal conic a parabola.
62 „ 73	One principal conic a pair of parallel lines.
74 „ 76	Principal conics each a parabola.
77 and 78	Principal conics, one of them a parabola, the other a pair of parallel lines.

The models 1 to 8 correspond hereto, as follows:

		is		i.e.		is		
Mod.	1	is	13,	i.e.	2	is	Mod.	5
„	2		9		3		„	6
„	3		40		4		„	7
„	4		34		9		„	2
„	5		2		13		„	1
„	6		3		32		„	8
„	7		4		34		„	4
„	8		32		40		„	3.

Mod. 5 is 2: the form of the equation is here

$$\frac{y^2}{l^2[(x-\alpha)^2+\beta^2]}-\frac{z^2}{l'^2[(x-\alpha')^2+\beta'^2]}=1;$$

viz., the principal conics are one of them a hyperbola, the other imaginary; hence the generating conic has always two, and only two, real vertices, viz., it is always a hyperbola: there are no real lines.

Mod. 6 is 3: the form of the equation is

$$\frac{y^2}{l^2[(x-\alpha)^2+\beta^2]}+\frac{z^2}{l'^2[(x-\alpha')^2+\beta'^2]}=1;$$

viz., the principal conics are each of them a hyperbola; the generating conic has four real vertices, viz., it is always an ellipse: there are no real lines.

Mod. 7 is 4: the form of the equation is

$$\frac{y^2}{l^2(x-\gamma)(x-\delta)}+\frac{z^2}{l'^2[(x-\alpha')^2+\beta'^2]}+1=0.$$

The principal conics are one of them an ellipse, the other imaginary; for values of x between γ and δ, the variable conic has two real vertices or it is a hyperbola; for any other values it is imaginary, so that the surface lies wholly between the planes $x=\gamma$, $x=\delta$: the surface contains the real lines $y=0$, $x=\gamma$ and $y=0$, $x=\delta$.

Mod. 2 is 9: the form of the equation is

$$\frac{y^2}{l^2(x-\gamma)(x-\delta)}+\frac{z^2}{l'^2(x-\gamma')(x-\delta')}+1=0,$$

where, say the values γ, δ lie between the values γ', δ': the principal conics are each of them an ellipse, the vertices (on the axis or line $y=0$, $z=0$) of the one ellipse lying between those of the other ellipse. The variable conic for values of x between γ, δ has four real vertices, or it is an ellipse; for values beyond these limits, but within the limits γ', δ'—say from γ to γ', and from δ to δ'—there are two real vertices, or the conic is a hyperbola; and for values beyond the limits γ', δ', the variable conic is imaginary.

There are four real lines ($y=0$, $x=\gamma$), ($y=0$, $x=\delta$), ($z=0$, $x=\gamma'$), ($z=0$, $x=\delta'$). The surface consists of a central pillow-like portion, joined on by two conical points to an upper portion, and by two conical points to an under portion, the whole being included between the planes $x=\gamma'$, $x=\delta'$.

Mod. 1 is 13: the form of the equation is

$$\frac{y^2}{l^2(x-\gamma)(x-\delta)}-\frac{z^2}{l'^2(x-\gamma')(x-\delta')}+1=0\,;$$

the values γ', δ' lying between γ, δ; the principal conics are one of them a hyperbola, the other an ellipse, the vertices (on the axis or line $y=0$, $z=0$) of the hyperbola lying between those of the ellipse.

The variable conic, for values of x between γ', δ', has two real vertices, or it is a hyperbola; for the values, say, from γ' to γ, and δ' to δ, there are four real vertices, or the conic is an ellipse; for values beyond the limits γ, δ, there are two real vertices, and the conic is a hyperbola. There are the four real lines ($y=0$, $x=\gamma$), ($y=0$, $x=\delta$), and ($z=0$, $x=\gamma'$), ($z=0$, $x=\delta'$). The surface consists of 8 portions joined to each other by 8 conical points, but the form can scarcely be explained by a description.

Mod. 8 is 32: the form of the equation is

$$\frac{y^2}{l^2(x-\gamma)^2}+\frac{z^2}{l'^2(x-\gamma')^2}=1\,;$$

viz., the principal conics are each of them a line-pair, the variable conic is always an ellipse.

There are the two real nodal lines ($y=0$, $x=\gamma$) and ($z=0$, $x=\gamma'$), each of these being in the neighbourhood of the axis crunodal, and beyond certain limits acnodal; the surface is a scroll, being, in fact, the well-known surface which is the boundary of a small circular pencil of rays obliquely reflected, and consequently passing through two focal lines.

Mod. 4 is 34: the equation is

$$\frac{y^2}{l^2(x-\gamma)(x-\delta)}+\frac{z^2}{l'^2(x-\gamma')(x-\delta)}+1=0,$$

where $x=\delta$ is not intermediate between the values $x=\gamma$ and $x=\gamma'$; say the order is δ, γ, γ'. The surface is thus a *cubic* surface; the principal conics are ellipses having on the axis a common vertex at the point $x=\delta$, and the remaining two vertices on the same side of the last-mentioned one. The variable conic for values between δ and γ has four real vertices, or it is an ellipse; for values between γ and γ' two real vertices, or it is a hyperbola; and for values beyond the limits δ, γ' it is imaginary. There are on the surface the two real lines ($y=0$, $x=\gamma$) and ($z=0$, $x=\gamma'$). The surface consists of a finite portion joined on by two conical points to the remaining portion.

Mod. 3 is 40: the form of equation is

$$\frac{y^2}{l^2(x-\gamma)(x-\delta)}+\frac{z^2}{l'^2(x-\delta)^2}+1=0.$$

The surface is thus a *cubic* surface; the principal conics are, one of them an ellipse, the other a pair of imaginary lines intersecting on the ellipse; for values of x between γ and δ, the variable conic has thus two real vertices, and it is a hyperbola; for values beyond these limits it is imaginary, and the whole surface is thus included between the planes $x=\gamma$ and $x=\delta$. There are the two real lines $(y=0, x=\gamma)$ and $(z=0, x=\delta)$.

Taking $l^2=l'^2=1$, the surface is

$$\frac{y^2}{(x-\gamma)(x-\delta)}+\frac{z^2}{(x-\delta)^2}+1=0,$$

which is a *particular* case of the parabolic cyclide.

As already mentioned, I have not completely identified the remaining models 9 to 14, but I will say a few words about them.

The equatorial surfaces, not included in the preceding 78 cases, Plücker distinguishes (vol. II. p. 363) as "gedrehte" or "tordirte," say as twisted equatorial surfaces; the equation of such a surface is

$$by^2+2hyz+az^2+ab-h^2=0,$$

where

$$\begin{aligned} b&=Fx^2-2Rx+B,\\ a&=Ex^2+2Ux+C,\\ h&=Kx^2-\ Ox-G \text{ (or in particular } =-Ox-G). \end{aligned}$$

Mod. 13 is such a surface, being a twisted form of model 2.

Mod. 9 and Mod. 14 belong, I think, to the case $a=0$; viz., the form of the equation is $by^2+2hyz-h^2=0$. The variable conic is a hyperbola, the direction of one of the asymptotes being constant (vol. II. p. 368).

There are moreover (p. 372) equatorial surfaces in which the variable conic is always a parabola, and where there are on the surface four real or imaginary lines.

Mods. 10, 11, and 12 seem to represent such surfaces.

456.

NOTE ON THE DISCRIMINANT OF A BINARY QUANTIC.

[From the *Quarterly Journal of Pure and Applied Mathematics*, vol. X. (1870), p. 23.]

IT is well known that the discriminant of a binary quantic $(a, b, c, d, \ldots \between t, 1)^n$ is of the form

$$Ma + Nb^2,$$

but it is further to be remarked that if $b = 0$, then the form is

$$a\ (Ma + Nc^3),$$

if $b = 0$, $c = 0$, the form is

$$a^2\,(Ma + Nd^4),$$

and so on, until only the lowest two coefficients are not put $= 0$. Or, what is the same thing, if in the discriminant of the original function we put $a = 0$, then the discriminant divides by b^2; if $b = 0$, the discriminant divides by a, and, omitting this factor, if we then write $a = 0$, it divides by c^3; if $b = 0$, $c = 0$, the discriminant divides by a^2, and omitting this factor, if we then write $a = 0$, it divides by d^4; and so on, until as before.

Thus if $b = 0$, the discriminant of $(a, 0, c, d, e \between t, 1)^4$, divides by a, and omitting this factor it is

$$\begin{array}{l} a^2e^3 \\ - 18\,ac^2e^2 \\ + 54\,acd^2e \\ - 27\,ad^4 \\ + 81\,c^4e \\ - 54\,c^3d^2 \end{array}$$

which for $a = 0$ has the factor c^3; if $b = 0$, $c = 0$, the discriminant of $(a, 0, 0, d, e \between t, 1)^4$ has the factor a^2, and omitting this factor it is

$$\begin{array}{l} ae^2 \\ - 27\,d^4, \end{array}$$

which for $a = 0$ has the factor d^4; the series of theorems here terminates, since the lowest two coefficients d, e are not to be put $= 0$.

457.

ON THE QUARTIC SURFACES $(*\between U, V, W)^2=0$.

[From the *Quarterly Journal of Pure and Applied Mathematics*, vol. x. (1870), pp. 24—34.]

I PROPOSE to myself for investigation the quartic surfaces represented by an equation

$$(*\between U, V, W)^2=0,$$

where U, V, W are quadric functions of the coordinates.

Such a surface has 8 nodes (conical points), viz., these are the points of intersection of the quadric surfaces $U=0$, $V=0$, $W=0$. It is to be observed, that not every quartic surface with 8 nodes is included under the above form; in fact the equation of a quartic surface contains (homogeneously) 35 coefficients, or say 34 arbitrary parameters; in order that a given point may be a node, 4 conditions must be satisfied, and it is consequently possible to find a quartic surface having 8 given points as nodes (and having in its equation $(34-8\,.\,4=)$ 2 arbitrary parameters): but 8 given points are not in general the intersections of three quadric surfaces, and such a quartic surface is therefore not in general included under the above form. I think, however, that it may be assumed that the above form includes all the quartic surfaces having 8 nodes, points of intersection of three quadric surfaces. It will presently appear that, included in the form, we have surfaces where (instead of the 8 nodes) there is a nodal or cuspidal conic; and that these are the most general forms of such quartic surfaces. A quartic surface has at most 16 nodes, and the general form with 8 nodes must admit of being particularised so that the surface shall acquire any number not exceeding 8 of additional nodes. This does not show, but it is probable, that the above special form with 8 nodes can be particularised so that the surface shall in like manner acquire any number not exceeding 8 of additional nodes. Similarly, a quartic surface with a nodal conic may have besides 1, 2, 3, or 4 nodes; and it will be shown in the sequel how the form, particularised so as to give a nodal conic, may be further particularised so as to give the 1, 2, 3, or 4 nodes. So a quartic surface with a cuspidal conic may besides have 1 node, and it will be shown how the form, particularised so as to give a cuspidal conic, may be further particularised so as to give 1 node.

Starting from the equation $(*\between U, V, W)^2=0$, we may, by substituting for U, V, W linear functions of these expressions, transform the equation precisely in the manner of a conic, and therefore into any of the forms under which the equation of a conic can be exhibited; for instance, in the forms $aU^2+bV^2+cW^2=0$, $fVW+gWU+hUV=0$, $UW-V^2=0$, &c. I attend at present only to the last-mentioned form $UW-V^2=0$, which, it thus appears, is equally general with the original form $(*\between U, V, W)^2=0$.

The quartic surface

$$UW-V^2=0$$

where U, V, W are any quadric functions of the coordinates, may be considered as the envelope of the quadric surface

$$(U, V, W\between\theta, 1)^2=0,$$

where θ is an arbitrary parameter. And it thus appears that it is very easy to reciprocate (in regard to any given quadric surface) the quartic surface. For the reciprocal of the quartic surface is clearly the envelope of the reciprocal of the variable quadric surface; this reciprocal is itself a quadric surface, and the reciprocal of the quartic surface is thus given in the same form as the original surface, viz., as the envelope of a quadric surface the equation whereof contains rationally the variable parameter θ; the equation of the reciprocal surface is consequently obtained by equating to zero the discriminant in regard to θ, of the equation of the reciprocal quadric surface.

It is to be observed that, inasmuch as the equation of the reciprocal quadric surface is of the third degree in the coefficients of the original quadric, it is in general of the degree 6 in the parameter θ; we have thus a sextic function of θ, the coefficients whereof are quadric functions of the coordinates; and the discriminant is a function of the order 10 in these coefficients, that is, of the order 20 in the coordinates. The reciprocal of the quartic surface is thus a surface of the order 20; this is right, for in a general quartic surface the order of the reciprocal surface is $=36$, and the 8 nodes reduce the order by 16; $36-16=20$.

In the equation $UW-V^2=0$, or say $V^2-UW=0$; if U reduce itself to the square of a linear function, $U=P^2$, the equation becomes $V^2-P^2W=0$, which is the general form of the quartic surface having the nodal conic $V=0$, $P=0$. And if, moreover, W be the product of this same linear function P by another linear function Q, $W=PQ$, then the form is $V^2-P^3Q=0$, which is the general form of the quartic surface having the cuspidal conic $V=0$, $P=0$.

Writing for greater convenience x, y in the place of P, Q respectively, we have the quartic

$$(AA) \qquad V^2-x^3y=0,$$

having the cuspidal conic $V=0$, $x=0$; and which has besides the conic of plane contact $V=0$, $y=0$. In virtue of the cuspidal conic the reciprocal surface should be of the order 6; and by the foregoing method of obtaining the equation of the reciprocal surface, I will verify that this is so. To effect this as simply as possible, I fix the remaining coordinates z, w as follows. The line $x=0$, $y=0$ is not in general a tangent to the surface $V=0$; it therefore meets this surface in two points, and we may take $z=0$, $w=0$ to be the equations of the tangent planes at these two points respectively; we have thus $V = ax^2 + 2hxy + by^2 + 2nzw$. Introducing for convenience the numerical factor 2, and taking the equation of the surface to be

$$(ax^2 + 2hxy + by^2 + 2nzw)^2 - 2x^3y = 0,$$

this is the envelope of the quadric surface

$$\theta^2x^2 + 2\theta\,(ax^2 + 2hxy + by^2 + 2nzw) + 2xy = 0,$$

which is a surface $(a, b, c, d, f, g, h, l, m, n \between x, y, z, w)^2 = 0$, where $a = \theta^2 + 2a\theta$, $b = 2\theta b$, $h = 2\theta h + 1$, $n = 2\theta n$, and where all the other coefficients vanish. Assuming, as usual, that the reciprocation is effected in regard to the surface $x^2 + y^2 + z^2 + w^2 = 0$, the general equation is

$$\begin{aligned}
&x^2\,.\,d\,(bc - f^2) - cm^2 - bn^2 + 2fmn\\
+\;&y^2\,.\,d\,(ca - g^2) - an^2 - cl^2 + 2gnl\\
+\;&z^2\,.\,d\,(ab - h^2) - bl^2 - am^2 + 2hlm\\
+\;&w^2\,.\,abc - af^2 - bg^2 - ch^2 + 2fgh\\
+\;&2yz\,.\,d\,(gh - af) + l^2f + amn - hnl - glm\\
+\;&2zx\,.\,d\,(hf - bg) + m^2g + bnl - flm - hmn\\
+\;&2xy\,.\,d(fg - ch) + n^2h + clm - gmn - fnl\\
+\;&2xw\,.\, - l\;(bc - f^2) - n\,(hf - bg) - m\,(fg - ch)\\
+\;&2yw\,.\, - m\,(ca - g^2) - l\;(fg - ch) - n\;(gh - af)\\
+\;&2zw\,.\, - n\;(ab - h^2) - m\,(gh - af) - l\;(hf - bg),
\end{aligned}$$

(I write down this general result as it will be useful for reference in other cases); in the present case this becomes simply

$$x^2.\, - bn^2 + y^2.\, - an^2 + 2xy\,.\,nh^2 + 2zw\,(-nab + nh^2) = 0,$$

where a, b, n, h have the foregoing values; the equation is thus only of the order 4 in regard to θ; but it in fact divides by $n\,(=2\theta n)$ and thus reduces itself to the third order, viz. it becomes

$$n\,(bx^2 + ay^2) - 2h^2xy + 2\,(ab - h^2)\,zw = 0,$$

or, substituting for a, b, n, h their values, this is

$$x^2.\,4bn\theta^2 + y^2\,(2n\theta^3 + 4an\theta^2) + 2zw\,(2b\theta^3 + 4ab\theta^2) - 2\,(xy + zw)\,(2\theta h + 1)^2 = 0,$$

say this is

$$(A, B, C, D\between\theta, 1)^3=0,$$

where A, B, C, D are each of them a quadric function of the coordinates; it being observed that C and D are respectively numerical multiples of the same function $xy+zw$. Hence equating the discriminant to zero, we have

$$A^2D^2+4AC^3+4B^3D-3B^2C^2-6ABCD=0,$$

which equation, inasmuch as every term contains either C or D as a factor, divides by $xy+zw$, and thus becomes an equation of the order 6 in the coordinates: that is the order of the reciprocal surface is $=6$. Multiplying by $\frac{3}{2}$ to avoid fractions, the actual values of A, B, C, D are

$$\left\{\begin{aligned} A &= \;\;\,3\;(ny^2\;+2bzw),\\ B &= \;\;\,2\;\{bnx^2+any^2+2abzw-2h^2(xy+zw)\},\\ C &= -4h\,(xy\;+zw),\\ D &= -3\;(xy\;+zw),\end{aligned}\right.$$

or say $A=3\alpha$, $B=2\beta$, $C=-4h\gamma$, $D=-3\gamma$; where $\gamma=xy+zw$; substituting these values and omitting the factor 3γ, the equation is

$$27\alpha^2\gamma-256h^3\alpha\gamma^2-32\beta^3-64h^2\beta^2\gamma-144h\alpha\beta\gamma=0,$$

which is an equation of the form $(*\between\alpha, \beta, \gamma)^3=0$. The sextic surface has thus singular points $\alpha=0$, $\beta=0$, $\gamma=0$, viz. these are the two points $(x=0, y=0, z=0)$, $(x=0, y=0, w=0)$ each four times. The further discussion of the sextic surface is reserved for another occasion.

I do not at present attempt to enumerate the particular cases of the surface $V^2-x^3y=0$, but content myself with the discussion of a particular case in which the order of the reciprocal surface is $=3$. Suppose that $V=0$ is a cone, $y=0$ a tangent plane to the cone (so that the conic $y=0$, $V=0$ breaks up into a line twice repeated), $x=0$ an arbitrary plane (so that we have still the proper cuspidal conic $x=0$, $V=0$). Any other tangent plane of the cone may be taken for the plane $z=0$; the plane containing the lines of contact of the two tangent planes for the plane $w=0$; the equation of the conic then is $V=dw^2+2fyz=0$; and the equation of the surface is

$$(AB)\qquad (dw^2+2fyz)^2-x^3y=0.$$

For convenience of comparison, I change x, y, w, z into y, w, z, x, and assign numerical values to the coefficients, writing the equation under the form

$$(AB)\qquad 27\,(4xw+z^2)^2-64y^3w=0.$$

The quartic is here the envelope of the quadric surface

$$\theta^2\,.\,4yw+\theta\,.\,9\,(4xw+z^2)+12y^2=0,$$

viz., comparing with the general form

$$(a, b, c, d, f, g, h, l, m, n\between x, y, z, w)^2=0,$$

we have $b=12$, $c=9\theta$, $l=18\theta$, $m=2\theta^2$, all the other coefficients vanishing. The reciprocal equation is

$$x^2(-cm^2)+y^2(-cl^2)+z^2(-bl^2)+2xy\,.\,clm+2xw(-lbc)=0,$$

or substituting for b, c, l, m their values, this is found to be

$$\theta(\theta x-9y)^2+108(z^2+xw)=0.$$

Representing this by $(A, B, C, D\between\theta, 1)^3=0$, the discriminant in regard to θ would, in virtue of the values of A, B, C, contain D as a factor; the reason of this appears from the original form; in fact, forming the derived equation in regard to θ, this is found to be $(\theta x-9y)(\theta x-3y)=0$; the value $\theta x-9y=0$ gives as a factor of the discriminant z^2+xw; the value $\theta x-3y=0$ gives $3y(-6y)^2+108x(z^2+xw)$, that is the factor $y^3+x(z^2+xw)$; the complete value of the discriminant as obtained by substitution of the values of A, B, C, D being $x^3(z^2+xy)\{y^3+x(z^2+xw)\}$; the equation of the reciprocal surface is

$$y^3+x(z^2+xw)=0,$$

viz. this is a cubic surface, Prof. Schläfli's Case XX., having a uniplanar point $x=0$, $y=0$, $z=0$ reducing the class by 8, and so giving a reciprocal surface of the order $(12-8=)\,4$, viz. the surface $27(4xw+z^2)^2-64y^3w=0$. See the Memoir, Schläfli, "On the distribution of surfaces of the third order into species in reference to the absence or presence of singular points and the reality of their lines," *Phil. Trans.*, vol. CLIII. (1853), pp. 193—241.

I pass to the case of a surface

$$V^2-P^2U=0$$

having a nodal conic $V=0$, $P=0$, but not having in general any nodes. And I propose to show how the constants may be determined so that the surface shall have 1, 2, 3, or 4 nodes. It is to be remarked that in the above equation the plane $P=0$ is a determinate plane, but the quadric surface $V=0$ is not a determinate quadric, we may in fact substitute for it the quadric $V+\lambda P^2=0$, writing the equation under the form

$$(V+\lambda P^2)^2-P^2(U+2\lambda V+\lambda^2P^2),$$

so that we may without loss of generality, by means of the disposable constant λ, subject the surface $V=0$ to any single condition; for instance, take it to be a cone, or to pass through a given point, &c.

Taking the planes $x=0$, $y=0$, $z=0$, $w=0$ to be arbitrary planes, the implicit constant factors in these equations may be determined in such wise that the equation of the given plane $P=0$ shall be $x+y+z+w=0$. The equation of the surface will then be

$$\{(a, b, c, d, f, g, h, l, m, n\between x, y, z, w)^2\}^2$$
$$=(x+y+z+w)^2.(a', b', c', d', f', g', h', l', m', n'\between x, y, z, w)^2,$$

and we may assume that the node or nodes (if any) lie at a vertex or vertices of the tetrahedron $x=0$, $y=0$, $z=0$, $w=0$, say at the points A, B, C, D. The conditions for a node at each of these points are at once found to be

Node at A,	Node at B,	Node at C,	Node at D,
$2a^2 = a' + a'$	$2bh = h' + b'$	$2cg = g' + c'$	$2dl = l' + d'$
$2ah = h' + a'$	$2b^2 = b' + b'$	$2cf = f' + c'$	$2dm = m' + d'$
$2ag = g' + a'$	$2bf = f' + b'$	$2c^2 = c' + c'$	$2dn = n' + d'$
$2al = l' + a'$	$2bm = m' + b'$	$2cn = n' + c'$	$2d^2 = d' + d'$

The first set of equations gives

$$a' = a^2, \quad h' = 2ah - a^2, \quad g' = 2ag - g^2, \quad l' = 2al - a^2.$$

If the first and second sets are satisfied simultaneously, we have $2(a-b)h = a^2 - b^2$, that is $a=b$, or else $h = \frac{1}{2}(a+b)$; that is, the two sets may be satisfied in two different ways according as a and b are equal or unequal. Similarly the first, second, and third sets may be satisfied in three different ways and the four sets in five different ways according as there are or are not any equalities between a, b, c, and between a, b, c and d respectively. The several solutions are shown in the annexed table, viz., in the line I no set is satisfied; in the line II only the first set; in the lines III and IV the first and second sets; in the lines V to VII the first, second, and third sets; and in the lines VIII to XII the four sets.

[See next page for this Table, which should come in here.]

I is the general case, $V^2 = P^2 U$, of a quartic and a nodal conic but without nodes. (AC).

II is the case of a single node; writing, as without loss of generality we may do, $a=0$, the equation is

$$[(0, b, c, d, f, g, h, l, m, n \between x, y, z, w)^2]^2 = (x+y+z+w)^2 . (b, c, d, n, m, f \between y, z, w)^2,$$

viz. the quadric $U=0$ is here a cone having its vertex on the quadric $V=0$. (AD).

III and IV are two cases each of them with two nodes, viz. III, the equation is

$$\begin{aligned}\{(ax+by)(x+y) + cz^2 + 2nzw + dw^2 + 2x(gz+lw) + 2y(fz+mw)\}^2 \\ = (x+y+z+w)^2 [(ax+by)^2 + c'z^2 + 2n'zw + d'w^2 \qquad (AE) \\ + 2x\{(2ag-a^2)z + (2al-a^2)w\} \\ + 2y\{(2bf-b^2)z + (2bm-b^2)w\}],\end{aligned}$$

where it is to be observed that the line $z=0$, $w=0$ joining the two nodes $(y=0, z=0, w=0)$ and $(w=0, z=0, x=0)$ is a line on the surface. Writing, as we may do, $a=0$, the equation assumes the more simple form

$$\begin{aligned}\{by(x+y) + cz^2 + 2nzw + dw^2 + 2x(gz+lw) + 2y(fz+mw)\}^2 \qquad (AE) \\ = (x+y+z+w)^2 [b^2y^2 + c'z^2 + 2n'zw + d'w^2 + 2y\{(2bf-b^2)z + (2bm-b^2)w\}].\end{aligned}$$

I	a	b	c	d	f	g	h	l	m	n	a'	b'	c'	d'	f'	g'	h'	l'	m'	n'
II	a	b	c	d	f	g	h	l	m	n	a^2	b'	c'	d'	f'	$2ag-a^2$	$2ah-a^2$	$2al-a^2$	m'	n'
III	a	b	c	d	f	g	$\frac{1}{2}(a+b)$	l	m	n	a^2	b^2	c'	d'	$2bf-b^2$	$2ag-a^2$	ab	$2al-a^2$	$2bm-b^2$	n'
IV	a	a	c	d	f	g	h	l	m	n	a^2	a^2	c'	d'	$2af-a^2$	$2ag-a^2$	$2ah-a^2$	$2al-a^2$	$2am-a^2$	n'
V	a	b	c	d	$\frac{1}{2}(b+c)$	$\frac{1}{2}(a+c)$	$\frac{1}{2}(a+b)$	l	m	n	a^2	b^2	c^2	d'	bc	ac	ab	$2al-a^2$	$2bm-b^2$	$2cn-c^2$
VI	a	a	c	d	$\frac{1}{2}(a+c)$	$\frac{1}{2}(a+c)$	h	l	m	n	a^2	a^2	c^2	d'	ac	ac	$2ah-a^2$	$2al-a^2$	$2am-a^2$	$2cn-c^2$
VII	a	a	a	d	f	g	h	l	m	n	a^2	a^2	a^2	d'	$2af-a^2$	$2ag-a^2$	$2ah-a^2$	$2al-a^2$	$2am-a^2$	$2an-a^2$
VIII	a	b	c	d	$\frac{1}{2}(b+c)$	$\frac{1}{2}(a+c)$	$\frac{1}{2}(a+b)$	$\frac{1}{2}(a+d)$	$\frac{1}{2}(b+d)$	$\frac{1}{2}(c+d)$	a^2	b^2	c^2	d^2	bc	ac	ab	ad	bd	cd
IX	a	a	c	d	$\frac{1}{2}(a+c)$	$\frac{1}{2}(a+c)$	h	$\frac{1}{2}(a+d)$	$\frac{1}{2}(a+d)$	$\frac{1}{2}(c+d)$	a^2	a^2	c^2	d^2	ac	ac	$2ah-a^2$	ad	ad	cd
X	a	a	a	d	f	g	h	$\frac{1}{2}(a+d)$	$\frac{1}{2}(a+d)$	$\frac{1}{2}(a+d)$	a^2	a^2	a^2	d^2	$2af-a^2$	$2ag-a^2$	$2ah-a^2$	ad	ad	ad
XI	a	a	c	c	$\frac{1}{2}(a+c)$	$\frac{1}{2}(a+c)$	h	$\frac{1}{2}(a+c)$	$\frac{1}{2}(a+c)$	n	a^2	a^2	c^2	c^2	ac	ac	$2ah-a^2$	ac	ac	$2cn-c^2$
XII	a	a	a	a	f	g	h	l	m	n	a^2	a^2	a^2	a^2	$2af-a^2$	$2ag-a^2$	$2ah-a^2$	$2al-a^2$	$2am-a^2$	$2an-a^2$

In IV the equation is

$$\begin{aligned}\{a(x^2+y^2)&+2hxy+cz^2+dw^2+2nzw+2x(gz+lw)+2y(fz+mw)\}^2\\ &=(x+y+z+w)^2[a^2(x^2+y^2)+2(2af-a^2)xy+c'z^2+2n'zw+d'w^2 \qquad (AF)\\ &\qquad+2x\{(2ag-a^2)z+(2al-a^2)w\}\\ &\qquad+2y\{(2af-a^2)z+(2am-a^2)w\}],\end{aligned}$$

where it will be observed that the line $z=0$, $w=0$ joining the two nodes is *not* a line on the surface.

Writing, as we may do, $a=0$, the equation becomes

$$\begin{aligned}\{2hxy&+cz^2+2nzw+dw^2+2x(gz+lw)+2y(fz+mw)\}^2\\ &=(x+y+z+w)^2(c'z^2+2n'zw+d'w^2), \qquad (AF)\end{aligned}$$

viz. the form is $V^2=P^2QR$, the quadric surface $U=0$ breaking up into the two planes $Q=0$, $R=0$; and the nodes being situate at the intersections of the line $Q=0$, $R=0$ with the surface $V=0$.

V, VI, VII are apparently cases with three nodes, but in fact VI is the only case of a proper quartic surface with three nodes. For in V the equation is

$$\begin{aligned}\{(ax&+by+cz)(x+y+z)+dw^2+2w(lx+my+nz)\}^2\\ &=(x+y+z+w)^2[(ax+by+cz)^2+d'w^2+2w\{(2al-a^2)x+(2bm-b^2)y+(2cn-c^2)z\}],\end{aligned}$$

which is satisfied by $w=0$, and the surface thus breaks up into the plane $w=0$ and a cubic surface.

And in VII the equation is

$$\begin{aligned}\{a(x^2&+y^2+z^2)+dw^2+2fyz+2gzx+2hxy+2lxw+2myw+2nzw\}^2\\ &=(x+y+z+w)^2[a^2(x^2+y^2+z^2)+d'w^2\\ &\qquad+2\{(2af-a^2)yz+(2ag-a^2)zx+(2ah-a^2)xy\\ &\qquad+(2al-a^2)xw+(2am-a^2)yw+(2an-a^2)zw\}],\end{aligned}$$

which putting $a=0$ is

$$(dw^2+2fyz+2gzx+2hxy+2lxw+2myw+2nzw)^2=(x+y+z+w)^2d'w^2,$$

viz. this is a pair of quadric surfaces.

In the remaining case VI the equation is

$$\begin{aligned}\{a(x^2&+y^2)+2hxy+cz^2+dw^2+(a+c)(yz+zx)+2w(lx+my+nz)\}^2 \qquad (AG)\\ &=(x+y+z+w)^2[a^2(x^2+y^2)+c^2z^2+d'w^2+2acz(x+y)+2(2ah-a^2)xy\\ &\qquad+2w\{(2al-a^2)x+(2am-a^2)y+(2an-a^2)z\}],\end{aligned}$$

which putting therein $a=0$ is

$$\begin{aligned}\{2hxy&+dw^2+cz(x+y+z)+2w(lx+my+nz)\}^2\\ &=(x+y+z+w)^2\{c^2z^2+d'w^2+2(2cn-c^2)zw\}, \qquad (AG)\end{aligned}$$

which is a surface having the nodes A, B, C; and it is to be observed that the lines CA, CB, but *not* the line AB, are lines on the surface.

IX, X, XI, XII are apparently cases with four nodes, but it is only XI which is a proper quartic with four nodes. In fact IX is

$$\{(ax+ay+cz+dw)(x+y+z+w)+2(h-a)xy\}^2$$
$$=(x+y+z+w)^2\{(ax+ay+cz+dw)^2+4a(h-a)xy\},$$

which is satisfied if $x=0$ or if $y=0$; that is, the surface breaks up into the two planes $x=0$, $y=0$, and a quadric surface.

X is

$$\{a(x^2+y^2+z^2)+dw^2+2fyz+2gzx+2hxy+(a+d)(xw+yw+zw)\}^2$$
$$=(x+y+z+w)^2\{a^2(x^2+y^2+z^2)+d^2w^2$$
$$+2(2af-a^2)yz+2(2ag-a^2)zx+2(2af-a^2)xy+2ad(xw+yw+zw)\},$$

which putting therein $a=0$ is

$$\{dw(x+y+z+w)+2fyz+2gzx+2hxy\}^2=(x+y+z+w)^2d^2w^2,$$

and thus breaks up into two quadrics.

And XII is

$$\{a^2(x^2+y^2+z^2+w^2)+2fyz+2gzx+2hxy+2lxw+2myw+2nw\}^2$$
$$=(x+y+z+w)^2\{a^2(x^2+y^2+z^2+w^2)$$
$$+2(2af-a^2)yz+2(2ag-a^2)zx+2(2ah-a^2)xy$$
$$+2(2al-a^2)xw+2(2am-a^2)yw+2(2ah-a^2)zw\},$$

which putting $a=0$ is

$$(2fyz+2gzx+2hxy+2lxw+2myw+2nzw)^2=0,$$

and is thus a quadric surface twice repeated.

There remains XI, and here the equation is

$$\{(ax+ay+cz+cw)(x+y+z+w)+2(h-a)xy+2(n-c)zw\}^2$$
$$=(x+y+z+w)^2\{(ax+ay+cz+cw)^2+4a(h-a)xy+4c(n-c)zw\},$$

or writing $h+a$, $n+c$ in place of a, c respectively, this is

$$\{(ax+ay+cz+cw)(x+y+z+w)+2hxy+2nw\}^2$$
$$=(x+y+z+w)^2\}(ax+ay+cz+cw)^2+4ahxy+4cnzw\}^2, \qquad (AH)$$

or putting herein $a=0$ it is

$$\{c(z+w)(x+y+z+w)+2hxy+2nzw\}^2=c(x+y+z+w)^2\{c(z+w)^2+4nzw\}, \quad (AH)$$

which may also be written

$$c(x+y+z+w)\{hxy(z+w) - nzw(x+y)\} + (hxy+nzw)^2 = 0, \qquad (AH)$$

the equation of a quartic surface with the four nodes A, B, C, D; it is to be observed that the lines AC, AD, BC, BD are, the lines AB and CD are *not*, lines on the surface.

A more simple form may be given to the equation as follows; using the second of the above forms, multiplying the equation by 4, and writing therein

$$p = \sqrt{(c)}\,(x+y+z+w),$$
$$qr = c\,(z+w)^2 + 4nzw,$$
$$st = c\,(x+y)^2 - 4hxw,$$

q, r, and s, t being the linear factors of the two quadric functions respectively, we have

$$qr - st = c\,(x+y+z+w)\,(-x-y+z+w) + 4hxy + 4nzw,$$

and thence

$$p^2 + qr - st = 2c\,(z+w)\,(x+y+z+w) + 4hxy + 4nzw,$$

wherefore the equation is

$$(p^2 + qr - st)^2 = 4p^2qr, \qquad (AH)$$

or, what is the same thing,

$$p + \sqrt{(qr)} + \sqrt{(st)} = 0, \qquad (AH)$$

where p, q, r, s, t are any linear functions of the coordinates; this is the equation of a quartic surface having the nodal conic $p=0$, $qr-st=0$; and the four nodes $(q=0, r=0, p^2-st=0)$ and $(s=0, t=0, p^2-qr=0)$. It includes the Cyclide, the equation of which may be written

$$b^2 = \sqrt{\{(ax-ek)^2 + b^2y^2\}} + \sqrt{\{(ex-ak)^2 - b^2z^2\}}.$$

I remark that Prof. Kummer in his most valuable Memoir, "Ueber die Flächen vierten Grades auf welchen Schaaren von Kegelschnitten liegen," *Crelle*, t. LXVI. (1864), pp. 66—76, has considered several of the cases of a quartic surface with a nodal conic, viz. no node, (AC); a single node, (AD); two nodes (the case AF); and four nodes, (AH); but he has not considered two nodes, the case (AE); nor three nodes, (AG).

In reference to the general case of a quartic surface with a nodal conic, some most interesting properties have recently been obtained by Prof. Clebsch, see *Berl. Monatsb.*, April 30, 1868, where it is shown that there are on the surface 16 right lines forming 20 systems of double-fours, analogous in some respect to the 27 lines and 36 systems of double-sixes of a cubic surface.

458.

ON THE ANHARMONIC-RATIO SEXTIC.

[From the *Quarterly Journal of Pure and Applied Mathematics*, vol. x. (1870), pp. 56, 57.]

Mr Walker's equation is $\Delta(\lambda^2-\lambda+1)^3+I^3(\lambda^2-\lambda)^2=0$; changing the sign of λ, and also the numerical multipliers of I, Δ (so as to convert the discriminant equation into its standard form $\Delta=I^3-27J^2$), the equation is

$$4\Delta(\lambda^2+\lambda+1)^3-27I^3(\lambda^2+\lambda)^2=0.$$

I remark that this is most readily obtained as follows; writing

$$\begin{aligned} A &= (a-d)(b-c),\\ B &= (b-d)(c-a),\\ C &= (c-d)(a-b), \end{aligned}$$

then we have $A+B+C=0$,

$$\begin{aligned} I &= \tfrac{1}{24}(A^2+B^2+C^2)=-\tfrac{1}{12}(BC+CA+AB),\\ J &= \tfrac{1}{432}(B-C)(C-A)(A-B),\\ \sqrt{(\Delta)} &= \tfrac{1}{16}ABC, \end{aligned}$$

see my Fifth Memoir on Quantics, *Phil. Trans.*, vol. CXLVIII. (1858), pp. 429—460, [156]. And observe also, that in virtue of the relation $A+B+C=0$, we have

$$12I=A^2+AB+B^2=A^2+AC+C^2=B^2+BC+C^2.$$

Hence writing

$$u=\frac{4\sqrt{(\Delta)}}{3I}\left(\lambda+\frac{1}{\lambda}+1\right),$$

when λ has any one of the values $\frac{A}{B}, \frac{B}{A}, \frac{A}{C}, \frac{C}{A}, \frac{B}{C}, \frac{C}{B}$, we see that u assumes only the values A, B, C, and u is thus determined by the equation

$$u^3 - 12Iu - 16\sqrt{(\Delta)} = 0.$$

Eliminating u, we obtain

$$16\sqrt{(\Delta)}\left\{\frac{4\Delta}{27I^3}\left(\lambda+\frac{1}{\lambda}+1\right)^3 - \left(\lambda+\frac{1}{\lambda}+1\right) - 1\right\} = 0,$$

or, what is the same thing,

$$4\Delta\left(\lambda+\frac{1}{\lambda}+1\right)^3 - 27I^3\left(\lambda+\frac{1}{\lambda}+2\right) = 0,$$

that is

$$4\Delta(\lambda^2+\lambda+1)^3 - 27I^3\lambda^2(\lambda+1)^2 = 0,$$

the required equation.

459.

ON THE DOUBLE-SIXERS OF A CUBIC SURFACE.

[From the *Quarterly Journal of Pure and Applied Mathematics*, vol. X. (1870), pp. 58—71.]

THE 27 lines on a cubic surface include, and that in 36 different ways, *a double-sixer;* viz. a system of two sets of six lines 1, 2, 3, 4, 5, 6; 1′, 2′, 3′, 4′, 5′, 6′, such that every line of the one set intersects all the non-corresponding lines of the other set, thus

	1	2	3	4	5	6
1′		.	.	.	.	.
2′	.		.	.	.	.
3′	.	.		.	.	.
4′	.	.	.		.	.
5′	.	.	.	.		.
6′	.	.	.	.	.	

there being in all 30 intersections.

Any line say 4, of the one set, intersects five lines 1′, 2′, 3′, 5′, 6′ of the other set; and these six lines being given the double-sixer may be constructed; viz. (besides the line 4) we have a line 1 meeting the lines 2′, 3′, 5′, 6′; a line 2 meeting the lines 3′, 5′, 6′, 1′; a line 3 meeting the lines 5′, 6′, 1′, 2′; a line 5 meeting the lines 6′, 1′, 2′, 3′; and a line 6′ meeting the lines 1′, 2′, 3′, 5′; and then the lines 1, 2, 3, 5, 6 are all of them met by a single line 4′, which completes the system.

We may, if we please, consider the lines 4, 2 as given, and then 1′, 3′, 5′, 6′ will be any four lines each of them meeting the two given lines 4, 2; 2′ will be any line meeting 4; and we have to determine a line 4′ meeting 2, such that there may exist the lines 1, 3, 5, 6, completing the system as above. Or what is the same thing, we have a skew quadrilateral 1′, 2, 3′, 4; 5′ and 6′ meet 2 and 4; 2′ meets

4, and 4′ meets 2: 5 and 6 meet 1′ and 3′; 1 meets 3′ and 3 meets 1′; and the two sets 2′, 4′, 5′, 6′ and 1, 3, 5, 6 meet thus

	1	3	5	6
2′	.	.	.	.
4′	.	.	.	.
5′	.	.		.
6′	.	.	.	

Hence, starting with the skew quadrilateral 1′23′4, and taking $x=0$, $y=0$, $z=0$, $w=0$ for the equations of the four planes 41′, 1′2, 23′, 3′4 respectively; or what is the same thing $x=0$, $y=0$ for the equations of the line 1′; $y=0$, $z=0$ for those of the line 2; $z=0$, $w=0$ for those of the line 3′; and $w=0$, $x=0$ for those of the line 4; the several lines may be determined, each of them by means of its six coordinates, as follows:

	a	b	c	f	g	h
1′	0	0	0	0	0	1
2	0	0	0	1	0	0
3′	0	0	1	0	0	0
4	1	0	0	0	0	0
2′	A_2	B_2	C_2	0	G_2	H_2
4′	0	B_4	C_4	F_4	G_4	H_4
5′	0	B_5	C_5	0	G_5	H_5
6′	0	B_6	C_6	0	G_6	H_6
1	a_1	b_1	c_1	f_1	g_1	0
3	a_3	b_3	0	f_3	g_3	h_3
5	a_5	b_5	0	f_5	g_5	0
6	a_6	b_6	0	f_6	g_6	0

where

$$B_2G_2 + C_2H_2 = 0,$$
$$B_4G_4 + C_4H_4 = 0,$$
$$B_5G_5 + C_5H_5 = 0,$$
$$B_6G_6 + C_6H_6 = 0,$$
$$a_1f_1 + b_1g_1 = 0,$$
$$a_3f_3 + b_3g_3 = 0,$$
$$a_5f_5 + b_5g_5 = 0,$$
$$a_6f_6 + b_6g_6 = 0.$$

The conditions in regard to the intersections of the lines 2′, 4′, 5′, 6′ and 1, 3, 5, 6, are formed by means of the diagram

$$\begin{array}{c|cccccc||cccccc|c}
1 & f_1 & g_1 & 0 & a_1 & b_1 & c_1 & A_2 & B_2 & C_2 & 0 & G_2 & H_2 & 2', \\
3 & f_3 & g_3 & h_3 & a_3 & b_3 & 0 & 0 & B_4 & C_4 & F_1 & G_4 & H_4 & 4', \\
5 & f_5 & g_5 & 0 & a_5 & b_5 & 0 & 0 & B_5 & C_5 & 0 & G_5 & H_5 & 5', \\
6 & f_6 & g_6 & 0 & a_6 & b_6 & 0 & 0 & B_6 & C_6 & 0 & G_6 & H_6 & 6',
\end{array}$$

viz. we have the equations

first set,

$$\begin{aligned}
f_1A_2+g_1B_2 \qquad\quad &+b_1G_2+c_1H_2=0,\\
g_1B_4+a_1F_4&+b_1G_4+c_1H_4=0,\\
g_1B_5 \qquad\quad &+b_1G_5+c_1H_5=0,\\
g_1B_6 \qquad\quad &+b_1G_6+c_1H_6=0;
\end{aligned}$$

second set,

$$\begin{aligned}
f_3A_2+g_3B_2+h_3C_2 \qquad\quad &+b_3G_2=0,\\
g_3B_4+h_3C_4+a_3F_3&+b_3G_4=0,\\
g_3B_5+h_3C_5 \qquad\quad &+b_3G_5=0,\\
g_3B_6+h_3C_6 \qquad\quad &+b_3G_6=0;
\end{aligned}$$

third set,

$$\begin{aligned}
f_5A_2+g_5B_2 \qquad\quad &+b_5G_2=0,\\
g_5B_4 \qquad +a_5F_4&+b_5G_4=0,\\
g_5B_6 \qquad\quad &+b_5G_6=0;
\end{aligned}$$

fourth set,

$$\begin{aligned}
f_6A_2+g_6B_2 \qquad\quad &+b_6G_2=0,\\
g_6B_4 \qquad +a_6F_4&+b_6G_4=0,\\
g_6B_5 \qquad\quad &+b_6G_5=0;
\end{aligned}$$

and it is to be shown, that taking as given the coordinates of 2′, 5′, 6′, that is $(A_2, B_2, C_2, G_2, H_2)$, (B_5, C_5, G_5, H_5) and (B_6, C_6, G_6, H_6), we can find the coordinates of the remaining lines 4′, 1, 3, 5, 6.

The first set of equations gives

$$g_1,\ b_1,\ c_1 = \left\|\begin{matrix} B_5, & G_5, & H_5 \\ B_6, & G_6, & H_6 \end{matrix}\right\|,$$

viz. g_1, b_1, c_1 are proportional, but as only the ratios are material, they may be taken equal, to the determinants $G_5H_6-G_6H_5$, $H_5B_6-H_6B_5$, $B_5G_6-B_6G_5$. And then retaining g_1, b_1, c_1 to signify these values respectively, the first equation gives f_1A_2, and the second equation gives a_1F_4; multiplying together these values, and writing $a_1f_1=-b_1g_1$, we find

$$(B_2B_4,\ G_2G_4,\ H_2H_4,\ G_2H_4+G_4H_2,\ H_2B_4+H_4B_2,\ B_2G_4+B_4G_2+A_2F_4 \between g_1,\ b_1,\ c_1)^2=0.$$

Proceeding in a similar manner with the second set of equations, we have first

$$g_3,\ h_3,\ b_3 = \left\| \begin{matrix} B_5, & C_5, & G_5 \\ B_6, & C_6, & G_6 \end{matrix} \right\|,$$

$$\text{(observe } h_3 = G_5B_6 - G_6B_5,\ = -c_1),$$

and then

$$(B_2B_4,\ C_2C_4,\ G_2G_4,\ C_2G_4 + C_4G_2,\ G_2B_4 + G_4B_2 + A_2F_4,\ B_2C_4 + B_4C_2 \between g_3,\ h_3,\ b_3)^2 = 0.$$

The third set gives more simply

$$g_5^2B_2B_4 + g_5b_5(B_2G_4 + B_4G_2 + A_2F_4) + b_5^2G_2G_4 = 0,$$

or since

$$g_5 : b_5 = G_6 : -B_6,$$

this is

$$G_6^2B_2B_4 - G_6B_6(B_2G_4 + B_4G_2 + A_2F_4) + B_6^2G_2G_4 = 0,$$

and similarly, the fourth set gives

$$g_6^2B_2B_4 + g_6b_6(B_2G_4 + B_4G_2 + A_2F_4) + b_6^2G_2G_4 = 0,$$

or since $g_6 : b_6 = G_5 : -B_5$, this is

$$G_5^2B_2B_4 - G_5B_5(B_2G_4 + B_4G_2 + A_2F_4) + B_5^2G_2G_4 = 0:$$

and these last two results lead to the values of the ratios of B_2B_4, $B_2G_4 + B_4G_2 + A_2F_4$, G_2G_4; viz. these are proportional to expressions containing the common factor $B_5G_6 - B_6G_5$, and omitting this common factor, and taking them equal instead of merely proportional to the resulting expressions (which is allowable, since the absolute values are not material), we have

$$B_2B_4,\ B_2G_4 + B_4G_2 + A_2F_4,\ G_2G_4 = B_5B_6,\ B_5G_6 + B_6G_5,\ G_5G_6.$$

Returning to the result obtained from the first set of equations, this now becomes

$$(B_5B_6,\ G_5G_6,\ H_2H_4,\ G_2H_4 + G_4H_2,\ H_2B_4 + H_4B_2,\ B_5G_6 + B_6G_5 \between g_1,\ b_1,\ c_1)^2 = 0:$$

but the terms containing g_1, b_1 are $(B_5g_1 + G_5b_1)(B_6g_1 + G_6b_1)$, viz. this is $= -H_5c_1 . -H_6c_1$, that is $H_5H_6c_1^2$; the whole equation is thus divisible by c_1, and omitting this factor, it becomes

$$g_1(H_2B_4 + H_4B_2) + b_1(G_2H_4 + G_4H_2) + c_1(H_2H_4 + H_5H_6) = 0.$$

Proceeding in like manner with the result obtained from the second set of equations, this becomes

$$(B_5B_6,\ C_2C_4,\ G_5G_6,\ C_2G_4 + C_4G_2,\ B_5G_6 + B_6G_5,\ B_2G_4 + B_4C_2 \between g_3,\ h_3,\ b_3)^2 = 0,$$

where the terms containing g_3, b_3 are $(Bg_3 + G_5b_3)(Bg_3 + G_3b)$, viz. this is $-h_3C_5 . -h_3C_6 = h_3^2C_5C_6$; the whole equation divides by h_3, and it then becomes

$$g_3(B_2C_4 + B_4C_2) + h_3(C_2C_4 + C_5C_6) + b_3(C_2G_4 + C_4G_2) = 0.$$

Considering B_4, G_4, F_4 as given by the equations

$$B_2B_4 = B_5B_6,\ G_2G_4 = G_5G_6,\ B_2G_4 + G_2B_4 + A_2F_4 = B_5G_6 + B_6G_5,$$

the equations last obtained determine the values of H_4 and C_4, viz. these equations may be written

$$(g_1B_2+b_1G_2+c_1H_2)\,H_4+H_2\,(g_1B_4+b_1G_4)+c_1H_5H_6=0,$$
$$(g_3B_2+h_3C_2+b_3G_2)\,C_4+C_2\,(g_3B_4+b_3G_4)+h_3C_5\,C_6=0\,;$$

but in order that the values $(B_4,\ C_4,\ F_4,\ G_4,\ H_4)$ given by these five equations may belong to a line $(0,\ B_4,\ C_4,\ F_4,\ G_4,\ H_4)$, they must satisfy the equation

$$B_4G_4+C_4H_4=0,$$

viz. in order to the existence of the line 4, this equation must be satisfied identically by the foregoing values; and I proceed to show that it is in fact thus satisfied. Multiplying the values of C_4, H_4, and writing $C_4H_4=-B_4G_4$, the identity to be verified is

$$\begin{aligned}&(g_1B_2+b_1G_2+c_1H_2)(g_3B_2+h_3C_2+b_3G_2)\,B_4G_4\\&\qquad+[H_2\,(g_1B_4+b_1G_4)+c_1H_5H_6]\,[C_2\,(g_3B_4+b_3G_4)+h_3C_5C_6]=0.\end{aligned}$$

The first line includes the terms

$$\{g_1g_3B_2^2+b_1b_3G_2^2+(b_1g_3+b_3g_1)\,B_2G_4+c_1h_3C_2H_2\}\,B_4G_4,$$

which, writing $C_2H_2=-B_2G_2$ and $B_2B_4=B_5B_6$, $G_2G_4=G_5G_6$, are

$$=g_1g_3B_2G_4B_5B_6+b_1b_3G_2B_4G_5G_6+(b_1g_3+g_1b_3-c_1h_3)\,B_5B_6G_5G_6.$$

The second line includes the terms

$$C_2H_2\,(g_1B_4+b_1G_4)\,(g_3B_4+b_3G_4)+c_1h_3C_5H_5C_6H_6,$$

which, reducing in like manner, are

$$=-g_1g_3G_2B_4B_5B_6-b_1b_3B_2G_4G_5G_6-(b_1g_3+g_1b_3-c_1h_3)\,B_5B_6G_5G_6,$$

and these are together

$$=\ (g_1g_3B_5B_6-b_1b_3G_5G_6)\,(B_2G_4-B_4G_2).$$

The remaining terms from the first line are at once reduced to

$$(g_1h_3C_2G_4+c_1g_3H_2G_4)\,B_5B_6+(b_1h_3C_2B_4+c_1b_3H_2B_4)\,G_5G_6,$$

and those from the second line are

$$C_5C_6\,H_2\,(g_1h_3B_4+b_1h_3G_4)+H_5H_6C_2\,(c_1g_3B_4+c_1b_3G_4).$$

Hence, attending to the relation $c_1=-h_3$, and collecting and arranging, the equation to be verified is

$$\begin{aligned}&(g_1g_3B_5B_6-b_1b_3G_5G_6)\,(B_2G_4-B_4G_2)\\&\quad+h_3C_2\,G_4\,(g_1B_5B_6-b_3H_5H_6)\\&\quad+h_3H_2B_4\,(g_1C_5C_6-b_3G_5\,G_6)\\&\quad+h_3C_2\,B_4\,(b_1\,G_5G_6-g_3H_5H_6)\\&\quad+h_3H_2G_4\,(b_1\,C_5C_6-g_3B_5\,B_6)=0.\end{aligned}$$

But we have

$$g_1B_5B_6 - b_3H_5H_6 = B_5B_6(G_5H_6 - G_6H_5) - H_5H_6(B_5C_6 - B_6C_5)$$
$$= B_6H_6(B_5G_5 + C_5H_5) - B_5H_5(B_6G_6 + C_6H_6) = 0,$$

and similarly

$$g_1C_5C_6 - b_3G_5G_6 = 0,$$
$$b_1G_5G_6 - g_3H_5H_6 = 0,$$
$$b_1C_5C_6 - g_3B_5B_6 = 0.$$

Moreover, writing $g_1B_5B_6 = b_3H_5H_6$, we have

$$g_1g_3B_5B_6 - b_1b_3G_5G_6$$
$$= b_3(g_3H_5H_6 - b_1G_5G_4) = 0,$$

and the five terms of the equation in question thus separately vanish; and the equation is consequently verified.

We may collect the results as follows:

Data are lines 1′, 2, 3′, 4, 2′, 5′, 6′;

and then, for the remaining lines, 1, 3, 5, 6, 4′ the coordinates are as follows:

For 4′,

$$B_2B_4 = B_5B_6,\quad G_2G_4 = G_5G_6,\quad A_2F_4 = B_5G_6 + B_6G_5 - B_2G_4 - B_4G_2,\quad A_4 = 0,$$

$$\begin{vmatrix} B_2, & G_2, & H_2 \\ B_5, & G_5, & H_5 \\ B_6, & G_6, & H_6 \end{vmatrix} H_4 + \begin{vmatrix} H_2B_4, & H_2G_4, & H_5H_6 \\ B_5, & G_5, & H_5 \\ B_6, & G_6, & H_6 \end{vmatrix} = 0,$$

$$\begin{vmatrix} B_2, & C_2, & G_2 \\ B_5, & C_5, & G_5 \\ B_6, & C_6, & G_6 \end{vmatrix} C_4 + \begin{vmatrix} C_2B_4, & C_5C_6, & C_2G_4 \\ B_5, & C_5, & G_5 \\ B_6, & C_6, & G_6 \end{vmatrix} = 0,$$

$(B_4G_4 + C_4H_4 = 0$, identity).

For 1,

$$(g_1, b_1, c_1) = \begin{Vmatrix} B_5, & G_5, & H_5 \\ B_6, & G_6, & H_6 \end{Vmatrix},\quad A_2f_2 = -\begin{vmatrix} B_2, & G_2, & H_2 \\ B_5, & G_5, & H_5 \\ B_6, & G_6, & H_6 \end{vmatrix},\quad F_4a_1 = -\begin{vmatrix} B_4, & G_4, & H_4 \\ B_5, & G_5, & H_5 \\ B_6, & G_6, & H_6 \end{vmatrix},\quad h_1 = 0,$$

$(a_1f_1 + b_1g_1 = 0$, identity).

For 3,

$$(g_3, h_3, b_3) = \begin{Vmatrix} B_5, & C_5, & G_5 \\ B_6, & C_6, & G_6 \end{Vmatrix},\quad A_2f_3 = -\begin{vmatrix} B_2, & C_2, & G_2 \\ B_5, & C_5, & G_5 \\ B_6, & C_6, & G_6 \end{vmatrix},\quad F_4a_3 = -\begin{vmatrix} B_4, & C_4, & G_4 \\ B_5, & C_5, & G_5 \\ B_6, & C_6, & G_6 \end{vmatrix},\quad c_3 = 0,$$

$(a_3f_3 + b_3g_3 = 0$, identity).

For 5,

$$g_5 = G_6,\quad b_5 = -B_6,\quad A_2 f_5 = B_6 G_2 - B_2 G_6,\quad F_4 a_5 = B_6 G_4 - B_4 G_6,\quad c_5 = 0,\quad h_5 = 0,$$

$$(a_5 f_5 + b_5 g_5 = 0,\ \text{identity}).$$

For 6,

$$g_6 = G_5,\quad b_6 = -B_5,\quad A_2 f_6 = B_5 G_2 - B_2 G_5,\quad F_4 a_6 = B_5 G_4 - B_4 G_5,\quad c_6 = 0,\quad h_6 = 0;$$

and, for actual calculation, it is convenient to remark that as only the ratios are material, a set of six coordinates may be multiplied or divided by any common number at pleasure.

But these results may be further reduced. Writing

$$(g_3,\ h_3,\ b_3) = \left\| \begin{matrix} B_5, & C_5, & G_5 \\ B_6, & C_6, & G_6 \end{matrix} \right\|,$$

we have

$$-(g_3 B_2 + h_3 C_2 + b_3 G_2)\, C_4 = C_2 \left(g_3 \frac{B_5 B_6}{B_2} + b_3 \frac{G_5 G_6}{C_2} \right) + h_3 C_5 C_6 = \frac{C_2}{B_2 G_2} (g_3 B_5 B_6 G_2 + b_3 G_5 G_6 B_2) + h_3 C_5 C_6.$$

But

$$g_3 B_5 B_6 G_2 + b_3 G_5 G_6 B_2 = \begin{Bmatrix} B_5 B_6 G_2 (C_5 G_6 - C_6 G_5) \\ + G_5 G_6 B_2 (B_5 C_6 - C_5 B_6) \end{Bmatrix} = \begin{Bmatrix} B_6 G_6 C_5 (B_5 G_2 - B_2 G_5) \\ + B_5 G_5 C_6 (G_6 B_2 - G_2 B_6) \end{Bmatrix}$$

$$= C_5 C_6 \begin{Bmatrix} - H_6 (B_5 G_2 - B_2 G_5) \\ - H_5 (G_6 B_2 - G_2 B_6) \end{Bmatrix}$$

$$= C_5 C_6 \{ B_2 (G_5 H_6 - G_6 H_5) + G_2 (B_6 H_5 - B_5 H_6) \}$$

$$= C_5 C_6 (B_2 g_1 + G_2 b_1),$$

since

$$(g_1,\ b_1,\ c_1) = \left\| \begin{matrix} B_5, & G_5, & H_5 \\ B_6, & G_6, & H_6 \end{matrix} \right\|;$$

the equation obtained is thus

$$g_3 B_4 + b_3 G_4 = \frac{C_5 C_6}{B_2 G_2} (g_1 B_2 + b_1 G_2).$$

We then have

$$-(g_3 B_2 + h_3 C_2 + b_3 G_2)\, C_4 = \frac{C_5 C_6}{B_2 G_2} [C_2 (g_1 B_2 + b_1 G_2) + h_3 B_2 G_2]$$

$$= \frac{-C_5 C_6}{C_2 H_2} [C_2 (g_1 B_2 + b_1 G_2) + c_1 C_2 H_2]$$

$$= -\frac{C_5 C_6}{H_2} (g_1 B_2 + b_1 G_2 + c_1 H_2),$$

that is

$$C_4 = \frac{C_5 C_6}{H_2} \frac{g_1 B_2 + b_1 G_2 + c_1 H_2}{g_3 B_2 + h_3 C_2 + b_3 G_2};$$

and in like manner

$$-(g_1B_2+b_1G_2+c_1H_2)H_4=\frac{H_2}{B_2G_2}(g_1G_2B_5B_6+b_1B_2G_5G_6)+c_1H_5H_6,$$

$$g_1G_2B_5B_6+b_1B_2G_5G_6=\left\{\begin{matrix}G_2B_5B_6(G_5H_6-G_5H_5)\\+B_2G_5G_6(H_5B_6-H_6G_5)\end{matrix}\right\}=\left\{\begin{matrix}B_5G_5H_6(G_2B_6-G_6B_2)\\+B_6G_6H_5(B_2G_5-B_5G_2)\end{matrix}\right\}$$

$$=H_5H_6\left\{\begin{matrix}-C_5(G_2B_6-G_6B_2)\\-C_6(B_2G_5-B_5G_2)\end{matrix}\right\}$$

$$=H_5H_6\{B_2(C_5C_6=C_6G_5)+G_2(B_5C_6-B_6C_5)\}$$

$$=H_5H_3(B_2g_3+G_2b_3);$$

the equation obtained is thus

$$g_1B_4+b_1G_4=\frac{H_5H_6}{B_2G_2}(B_2g_3+G_2b_3);$$

and then

$$-(g_1B_2+b_1G_2+c_1H_2)H_4=\frac{H_5H_6}{B_2G_2}\{H_2(B_2g_3+G_2b_3)+c_1B_2G_2\}$$

$$=-\frac{H_5H_6}{C_2H_2}\{H_2(B_2g_3+G_2b_3)+h_3C_2H_2\}$$

$$=-\frac{H_5H_6}{C_2}(B_2g_3+C_2h_3+G_2b_3),$$

that is

$$H_4=\frac{H_5H_6}{C_2}\frac{g_3B_2+h_3C_2+b_3G_2}{g_1B_2+b_1G_2+c_1H_2},$$

which values of C_4, H_4 satisfy, as they should do, the relation

$$C_4H_4=-B_4G_4.$$

We have also

$$A_2F_4=B_5G_6+B_6G_5-\frac{B_2G_5G_6}{G_2}-\frac{G_2B_5C_6}{B_2}$$

$$=-\frac{1}{B_2G_2}(G_2B_5-B_2G_5)(G_2B_6-B_2G_6)$$

$$=\frac{1}{C_2H_2}(G_2B_5-B_2G_5)(G_2B_6-B_2G_6),$$

which gives F_4.

Moreover

$$-F_4a_1=g_1B_4+b_1G_4+c_1H_4=-\frac{H_5H_6}{C_2H_2}(g_3B_2+b_2G_3)-\frac{h_3H_5H_6}{C_2}\frac{g_3B_2+b_3G_2+h_3C_2}{g_1B_2+b_1G_2+c_1H_2}$$

$$=\frac{-H_5H_6}{C_2H_2(g_1B_2+b_1G_2+c_1H_2)}\left\{\begin{matrix}(g_3B_2+b_3G_2)(g_1B_2+b_1G_2+c_1H_2)\\+(g_3B_2+b_3G_2+h_3C_2)h_3H_2\end{matrix}\right\}$$

$$=\frac{-H_5H_6}{C_2H_2(g_1B_2+b_1G_2+c_1H_2)}\{(g_3B_2+b_3G_2)(g_1B_2+b_1G_2)+c_1h_3B_2G_2\},$$

which combined with the foregoing value of F_4, gives a_1.

Again,

$$-F_4a_3 = g_3B_4 + h_3C_4 + b_3G_4 = -\frac{C_5C_6}{C_2H_2}(g_1B_2 + b_1G_2) = \frac{c_1C_5C_6}{H_2}\,\frac{g_1B_2 + b_1G_2 + c_1H_2}{g_3B_2 + h_3C_2 + b_3G_2}$$

$$= -\frac{C_5C_6}{C_2H_2(g_3B_2 + h_3C_2 + b_3G_2)}\left\{\begin{array}{l}(g_1B_2 + b_1G_2)(g_3B_2 + h_3C_2 + b_3G_2)\\ + (g_1B_2 + b_1G_2 + c_1H_2)\,c_1C_2\end{array}\right\}$$

$$= \frac{-C_5C_6}{C_2H_2(g_3B_2 + h_3C_2 + b_3G_2)}\{(g_1B_2 + b_1G_2)(g_1B_2 + b_3G_2) + c_1h_3B_2C_2\},$$

which combined with the foregoing value of F_4, gives a_3.

Write

$$\omega = B_5G_6 + B_6G_5 + C_5H_6 + C_3H_5,$$

we have

$$\begin{aligned} b_1g_1 &= (H_5B_6 - H_6B_5)(G_5H_6 - G_6H_5) \\ &= -H_5^2B_6G_6 - H_6^2B_5G_5 + H_5H_6(B_5G_6 + B_6G_5) \\ &= H_5H_6(C_6H_5 + C_5H_6 + B_5G_6 + B_6G_5), \end{aligned}$$

that is

$$b_1g_1 = H_5H_3\omega,$$

and similarly

$$b_3g_3 = C_5C_6\,\omega,$$

$$b_1b_3 = B_5B_6\,\omega,$$

$$g_1g_3 = G_5G_6\,\omega,$$

$$b_1g_3 + b_3g_1 + c_1h_3 = -(B_5G_6 + B_3G_5)\,\omega,$$

$$\begin{aligned}(b_1G_2 + g_1B_2)(b_3G_2 + g_3B_2) + c_1h_2B_2G_2 &= \{G_2^2B_5B_6 + B_2^2G_5G_6 - B_2G_2(B_5G_6 + B_6G_5)\}\,\omega \\ &= (G_2B_5 - B_2G_5)(G_2B_6 - B_2G_6)\,\omega,\end{aligned}$$

which last value is to be substituted for the left-hand function in the formulæ for a_1 and a_3 respectively.

Whence, finally recollecting that

$$B_2G_2 + C_2H_2 = 0, \quad B_5G_5 + C_5H_5 = 0, \quad B_6G_6 + C_6H_6 = 0,$$

and

$$\omega = C_6H_5 + C_5H_6 + B_6G_5 + B_5G_6,$$

we have

$$\text{For 1}\left\{\begin{array}{ll} (g_1,\ b_1,\ c_1) = \left\|\begin{array}{ccc} B_5, & G_5, & H_5 \\ B_6, & G_6, & H_6 \end{array}\right\|, & f_1 = -\dfrac{1}{A_2}(g_1B_2 + b_1G_2 + c_1H_2), \\ b_1g_1 = H_5H_6\omega. & a_1 = \dfrac{A_2H_5H_6\omega}{g_1B_2 + h_1G_2 + c_1H_2}, \quad h_1 = 0. \end{array}\right.$$

For 3
$$\left\{\begin{array}{ll} (g_3,\ h_3,\ b_3) = \left\Vert \begin{array}{ccc} B_5, & C_5, & G_6 \\ B_6, & C_6, & G_6 \end{array} \right\Vert, & f_3 = -\dfrac{1}{A_2}(g_3B_2 + h_3C_2 + b_3G_2), \\ b_3g_3 = C_5C_6\omega. & a_3 = \dfrac{A_2C_5C_6\omega}{g_3B_2 + h_3C_2 + b_3G_2}, \quad h_3 = 0, \end{array}\right.$$

where observe that $c_1 + h_3 = 0$.

For 5
$$\left\{\begin{array}{ll} g_5 = G_6, \quad b_5 = -B_6, & f_5 = \dfrac{1}{A_2}(G_2B_6 - B_2G_6), \\ c_5 = 0\ , \quad h_5 = \quad 0, & a_5 = \dfrac{A_2B_6G_6}{G_2B_6 - B_2G_6}. \end{array}\right.$$

For 6
$$\left\{\begin{array}{ll} g_6 = G_5, \quad b_6 = -B_5, & f_6 = \dfrac{1}{A_2}(G_2B_5 - B_2G_5), \\ c_6 = 0\ , \quad h_6 = \quad 0, & a_6 = \dfrac{A_2B_5G_5}{G_2B_5 - B_2G_5}. \end{array}\right.$$

For 4′
$$\left\{\begin{array}{ll} A_4 = 0, & B_4 = \dfrac{B_5B_6}{B_2}, \\ F_4 = \dfrac{(G_2B_5 - G_5B_2)(G_2B_6 - G_6B_2)}{A_2C_2H_2}, & G_4 = \dfrac{G_5G_6}{G_2}, \\ C_4 = \dfrac{C_5C_6}{H_2}\ \dfrac{g_1B_2 + b_1G_2 + c_1H_2}{g_3B_2 + h_3C_2 + b_3G_2}, & \\ H_4 = \dfrac{H_5H_6}{C_2}\ \dfrac{g_3B_2 + h_3C_2 + b_3G_2}{g_1B_2 + b_1G_2 + c_1H_2}. & \end{array}\right.$$

I have thought it worth while to effect the numerical calculations for enabling the construction of a drawing or model. For this purpose taking X, Y, Z as ordinary rectangular coordinates, I write

$$\begin{aligned} x &= \quad X + Y + Z - 10, \\ y &= \qquad\qquad\ \ Z, \\ z &= -X + Y + Z - 10, \\ w &= \qquad\quad Y, \end{aligned}$$

that is, I take 1′ and 2 to be lines in the plane of XY, defined by the equations $X + Y = 10$ and $X - Y = -10$ respectively, and 3′ and 4 to be lines in the plane of XZ defined by the equations $X - Z = -10$, $X + Z = 10$ respectively. And I take 5′ to be the line joining the points (2, 0, 8) and (−9, 1, 0); 6′ the line joining the points (3, 0, 7) and (−8, 2, 0); 2′ the line joining the points (9, 0, 1) and (−3, 7, 4). We calculate then for each of the lines the $xyzw$ coordinates of two points thereof; and thence the six coordinates of the line, viz.:

	x y z w	x y z w	A B C F G H	reduced A B C F G H
5′	0, 8, − 4, 0	− 18, 0, 0, 1	0, 72, 144, 0, 8, − 4	0, 18, 36, 0, 2, − 1
6′	0, 7, − 6, 0	− 16, 0, 0, 2	0, 96, 112, 0, 14, − 12	0, 48, 56, 0, 7, − 6
2′	0, 1, − 18, 0	− 2, 4, 4, 7	76, 36, 2, 0, 7, −126	76, 36, 2, 0, 7, −126

and effecting the calculations for the remaining lines, we have

	A	B	C	F	G	H
4′	0	24	− 944	$-\frac{9}{38}$	2	$\frac{3}{59}$
1	$\frac{380}{59}$	60	30	$\frac{385}{19}$	− 5	0
3	127680	− 720	0	$\frac{15}{19}$	140	− 30
5	304	− 48	0	$\frac{21}{19}$	7	0
6	$\frac{152}{3}$	− 18	0	$\frac{27}{38}$	0	0
	a	b	c	f	g	h

or reducing to integers, the values are

	A	B	C	F	G	H
5′	0	18	36	0	2	− 1
6′	0	48	56	0	7	− 6
2′	76	36	2	0	7	− 126
4′	0	53808	− 2116448	− 531	4484	114
1	1444	13452	6726	10443	− 1121	0
3	485184	− 2736	0	3	532	− 114
5	5776	− 912	0	21	133	0
6	5776	− 2052	0	81	228	0
	a	b	c	f	g	h

The line (a, b, c, f, g, h) is given as the intersection of any two of the four planes

$$\left(\begin{array}{cccc} & h, & -g, & a \\ -h, & & f, & b \\ g, & -f, & & c \\ -a, & -b, & -c, & \end{array}\right\between x, y, z, w) = 0,$$

or substituting for x, y, z, w the values $X+Y+Z-10$, Z, $-X+Y+Z-10$, Y, these become

$$\left(\begin{array}{cccc} g\;, & a-g\;, & h-g\;, & g \\ -f-h, & b+f-h, & f-h\;, & h-g \\ g\;, & c+g\;, & g-f\;, & -g \\ c-a, & -c-a\;, & -c-a-b, & c+a \end{array}\right\between X, Y, Z, 10) = 0,$$

or, what is the same thing,

$$\left(\begin{array}{cccc} \cdot\;, & 2g-a+c\;, & 2g-f-h\;, & -2g \\ -2g+a-c, & \cdot\;, & a+b+c+f-h, & -a-c \\ -2g+f+h, & -a-b-c-f+h, & \cdot\;, & -f+h \\ 2g\;, & a+c\;, & -f+h\;, & \cdot \end{array}\right\between X, Y, Z, 10) = 0.$$

And substituting, we have the equations of the several lines, viz.:

(1′) $X+Y=10,\quad Z=0,$

(2) $-X+Y=10,\quad Z=0,$

(3′) $-X+Z=10,\quad Y=0,$

(4) $X+Z=10,\quad Y=0,$

(5′) $$\left(\begin{array}{cccc} \cdot\;, & 40, & 5, & -4 \\ -40, & \cdot\;, & 55, & -36 \\ -5, & -55, & \cdot\;, & 1 \\ 4, & 36, & -1, & \cdot \end{array}\right\between X, Y, Z, 10) = 0,$$

(6′) $$\left(\begin{array}{cccc} \cdot\;, & -35, & 10, & -7 \\ -35, & \cdot\;, & 55, & -28 \\ -10, & -55, & \cdot\;, & 3 \\ 7, & 28, & -3, & \cdot \end{array}\right\between X, Y, Z, 10) = 0,$$

$$(2')\qquad \left(\begin{array}{cccc} \cdot & -30 & 70 & -7 \\ 30 & \cdot & 120 & -39 \\ -70 & -120 & \cdot & 63 \\ 7 & 39 & -63 & \cdot \end{array}\right)\!\!\left(X,\ Y,\ Z,\ 10\right) = 0,$$

$$(4')\qquad \left(\begin{array}{cccc} \cdot & -2107480 & 9385 & -8968 \\ 2107480 & \cdot & -2063285 & 2116448 \\ -9385 & 2063285 & \cdot & -645 \\ 8968 & -2116448 & 645 & \cdot \end{array}\right)\!\!\left(X,\ Y,\ Z,\ 10\right) = 0,$$

$$(1)\qquad \left(\begin{array}{cccc} \cdot & 3040 & -12685 & 2242 \\ -3040 & \cdot & 32065 & -8170 \\ 12685 & -32065 & \cdot & 10443 \\ -2242 & 8170 & -10443 & \cdot \end{array}\right)\!\!\left(X,\ Y,\ Z,\ 10\right) = 0,$$

$$(3)\qquad \left(\begin{array}{cccc} \cdot & -484120 & 1175 & -1064 \\ 484120 & \cdot & 482565 & -485184 \\ -1175 & -482565 & \cdot & 117 \\ 1064 & 485184 & -117 & \cdot \end{array}\right)\!\!\left(X,\ Y,\ Z,\ 10\right) = 0,$$

$$(5)\qquad \left(\begin{array}{cccc} \cdot & 5510 & 245 & -266 \\ -5510 & \cdot & 4885 & -5776 \\ -245 & -4885 & \cdot & 21 \\ 266 & 5776 & -21 & \cdot \end{array}\right)\!\!\left(X,\ Y,\ Z,\ 10\right) = 0,$$

$$(6)\qquad \left(\begin{array}{cccc} \cdot & -5320 & 375 & -456 \\ 5320 & \cdot & 3805 & -5776 \\ -3751 & -3805 & \cdot & 81 \\ 456 & 5776 & -81 & \cdot \end{array}\right)\!\!\left(X,\ Y,\ Z,\ 10\right) = 0.$$

The several lines intersect as they should do, the coordinates of the points of intersection being as follows:

	1′	2′	3′	4′	5′	6′
1	*	621 1239 } ÷ 305 836	−354 0 } ÷ 43 76	42417 −177 } ÷ 5098 8968	−493 59 } ÷ 78 152	−1188 354 } ÷ 253 532
2	0 10 0	*	−10 0 0	−472 −2 } ÷ 47 0	−9 1 0	−8 2 0
3	912 −2 } ÷ 91 0	9693 −21 } ÷ 1073 1064	*	−2484 66 } ÷ 24727 251464	807 −1 } ÷ 398 3192	3672 −6 } ÷ 1213 8512
4	10 0 0	9 0 1	0 0 10	*	2 0 8	3 0 7
5	304 −14 } ÷ 29 0	459 −21 } ÷ 47 38	6 0 } ÷ 7 76	4752 42 } ÷ 6521 71744	*	1080 −42 } ÷ 283 2128
6	76 −6 } ÷ 7 0	2421 −189 } ÷ 233 152	54 0 } ÷ 25 304	1479 9 } ÷ 727 8968	65 −3 } ÷ 16 154	*

viz. the coordinates of 12′ (intersection of lines 1 and 2′) are $\left(\frac{621}{305}, \frac{1239}{305}, \frac{836}{305}\right)$, and so in other cases; where there is no divisor the coordinates are integers. I find however, on laying down the figure, that the lines 3 and 4, 3′ and 4′ come so close together, that the figure cannot be obtained with any accuracy.

460.

NOTE ON MR FROST'S PAPER ON THE DIRECTION OF LINES OF CURVATURE IN THE NEIGHBOURHOOD OF AN UMBILICUS.

[From the *Quarterly Journal of Pure and Applied Mathematics*, vol. x. (1870), pp. 111—113.]

I REMARK as follows:

1. In regard to a quadric surface, it is not, I think, correct to say that the generatrices through an umbilic are curves of curvature; notwithstanding that, as shown p. 80, the normals at every point of such generatrix lie in one plane and consequently intersect. The way in which these generatrices as *quasi*-curves-of-curvature present themselves is as follows:

The curves of curvature satisfy a certain differential equation, the complete integral of which gives these curves as the intersections of the given quadric surface by the series of confocal surfaces $\frac{x^2}{a^2+h}+\frac{y^2}{b^2+h}+\frac{z^2}{c^2+h}=1$, h being the constant of integration of the differential equation. The singular solution of the differential equation, or envelope of the curves of curvature determined as above, gives the umbilicar generatrices.

2. In regard to a surface in general, I think it must be considered, not that there pass through the umbilic three distinct curves, but that the umbilicar curve of curvature is a curve having at the umbilic a triple point, or rather a point at which there are in general three distinct directions of the curve. The umbilicar curve of curvature in fact presents itself as the curve belonging to a certain value of the constant of integration h; in order that the curve of curvature may pass through a given point on the surface, h must satisfy a certain quadratic equation, that is for a given point of the surface there are two values of h, and therefore two curves of

curvature; but an umbilic is a point for which (as in effect shown, p. 81, for the particular case of a quadric surface) the two values of h become equal; that is, there is through the umbilic only a singular curve of curvature; but $\frac{dy}{dx}$ is determined by a cubic equation, and the umbilic is thus (as just mentioned) a point at which there are in general three distinct directions of the curve.

3. Some researches on the subject are contained in my paper "On Differential Equations and Umbilici," *Phil. Mag.*, vol. XXVI. (1863), pp. 373—379 and 441—452, [330]. It is noticeable that in the integral equations which I have there obtained for the differential equations $cy(p^2-1)+(a-c)xp=0$, and the more general form $(bx+cy)(p^2-1)+2(fx+gy)=0$, which belong to the neighbourhood of an umbilic, the curve through the umbilic *does* break up into three distinct curves; and the same is the case with the umbilic on the surface $xyz=1$ presently referred to.

4. In the paper "Mémoire sur les surfaces orthogonales," *Liouv.*, t. XII. (1847), pp. 241—254, M. Serret has given two very remarkable cases of three systems of surfaces intersecting each other at right angles, and consequently in the curves of curvature of the surfaces of each system. It was only on referring to this paper, in connexion with that of Mr Frost, that I perceived an obvious enough simplification of M. Serret's formulæ, whereby it appears that the curves of curvature on the surface $xyz=1$ are given as the intersection of this surface with the series of surfaces

$$h=(x^2+\omega y^2+\omega^2 z^2)^{\frac{3}{2}}+(x^2+\omega^2 y^2+\omega z^2)^{\frac{3}{2}},$$

where ω is an imaginary cube root of unity; the rationalised equation is of the twelfth order in (x, y, z), and for the particular value $h=0$, reduces itself as is easily seen to $0=(y^2-z^2)^2(z^2-x^2)^2(x^2-y^2)^2$. The point $x=y=z=1$ is obviously an umbilic on the surface $xyz=1$, and the corresponding value of h being $h=0$, the equation just obtained determines the umbilicar curves of curvature, viz. combining therewith the equation $xyz=1$ of the surface, we have the three hyperbolic curves

$$(y=z,\ xy^2=1),\quad (z=x,\ yz^2=1),\quad (x=y,\ zx^2=1).$$

461.

ON THE GEOMETRICAL INTERPRETATION OF THE COVARIANTS OF A BINARY CUBIC.

[From the *Quarterly Journal of Pure and Applied Mathematics*, vol. x. (1870), pp. 148—149.]

CONSIDER the binary cubic $U = (a, b, c, d \between x, y)^3$, and its covariants, viz. the discriminant (invariant)

$$\nabla = a^2d^2 - 6abcd + 4ac^3 + 4b^3d - 3b^2c^2,$$

the Hessian

$$H = (ac - b^2)\, x^2 + (ad - bc)\, xy + (bd - c^2)\, y^2,$$

and the cubicovariant

$$\begin{aligned}\Phi = \;&(\quad a^2d - 3abc + 2b^3\;)\, x^3\\ &-(-3abd + 6ac^2 - 3b^2c)\, x^2y\\ &+(-3acd + 6bd^2 - 3bc^2)\, xy^2\\ &-(\quad ad^2 - 3bcd + 2c^3\;)\, y^3,\end{aligned}$$

connected by the identical equation

$$\Phi^2 - \nabla\, U^2 = -4H^3.$$

Then if we regard (a, b, c, d) as the coordinates of a point in space, but (x, y) as variable parameters, the equation

$$\nabla = 0$$

represents a quartic torse, having for its cuspidal curve the skew cubic $ac - b^2 = 0$, $ad - bc = 0$, $bd - c^2 = 0$; the equation

$$U = 0$$

is that of the tangent plane to the torse along the line $ax^2+2bxy+cy^2=0$, $bx^2+2cxy+dy^2=0$: this line meets the cuspidal curve in the point whose coordinates are $a:b:c:d=y^3:-xy^2:x^2y:-y^3$. The equation

$$H=0$$

is that of a quadric cone having the last mentioned point for its vertex, and passing through the cuspidal curve: and the equation

$$\Phi=0$$

is that of the cubic surface which is the first polar of the same point in regard to the torse.

The equation $\Phi^2-\nabla U^2=-4H^3$, writing therein $U=0$, gives $\Phi^2=-4H^3$, a result which implies that $U=0$, $H=0$ is a certain curve repeated twice, and that $U=0$, $\Phi=0$ is the same curve repeated three times. The curve in question is at once seen to be the line of contact $\delta_x U=0$, $\delta_y U=0$; it thus appears that the tangent plane $U=0$ meets the cubic surface $\Phi=0$ in this line taken three times. This can only be the case if the equation $\Phi=0$ be expressible in the form $MU+(\delta_x U)^3=0$, or, what is the same thing,

$$MU+(\alpha\delta_x U+\beta\delta_y U)^3=0,$$

α and β constants, M a quadric function of (a, b, c, d); that is, Φ must be equal to a function of the form

$$MU+(\alpha\delta_x U+\beta\delta_y U)^2.$$

Seeking for this expression of Φ, and writing the symbols out at length, I find that the required identical equation is

$$-(\beta x-\alpha y)^3\left\{\begin{array}{l}(\quad a^2d-3abc+2b^3\ )\,x^3\\ -(-3abd+6ac^2-3b^2c)\,x^2y\\ +(-3acd+6bd^2-3bc^2)\,xy^2\\ -(-\quad ad^2-3bcd+2c^3\ )\,y^3\end{array}\right\}+2\,\{\alpha(ax^2+2bxy+cy^2)+\beta(bx^2+2cxy+dy^2)\}^3=$$

$$(a, b, c, d\between x, y)^3.\left(\begin{array}{llll}2a^2, & 6ab, & 6b^2, & -ad+3bc\\ 6ab, & 12ac+6b^2, & 3ad+15bc, & 6c^2\\ 6b^2, & 3ad+15bc, & 12bd+6c^2, & 6cd\\ -ad+3bc, & 6c^2, & 6cd, & 2d^2\end{array}\right)\dagger(x, y)^3(\alpha, \beta)^3,$$

(where the $\dagger$ indicates that the binomial coefficients are *not* to be inserted, viz. the function on the right hand is $\{2a^2x^3+6abx^2y+6b^2xy^3+(-ad+3bc)\,y^3\}\,\alpha^3+$&c.). As a verification remark that for $x=\alpha$, $y=\beta$, the equation becomes simply $2U^3=U.2U^2$.

462.

A NINTH MEMOIR ON QUANTICS.

[From the *Philosophical Transactions of the Royal Society of London*, vol. CLXI. (for the year 1871), pp. 17—50. Received April 7,—Read May 19, 1870.]

It was shown not long ago by Professor Gordan that the number of the irreducible covariants of a binary quantic of any order is finite (see his memoir "Beweis dass jede Covariante und Invariante einer binären Form eine ganze Function mit numerischen Coefficienten einer endlichen Anzahl solcher Formen ist," *Crelle*, t. LXIX. (1869), Memoir dated 8 June 1868), and in particular that for a binary quintic the number of irreducible covariants (including the quintic and the invariants) is $=23$, and that for a binary sextic the number is $=26$. From the theory given in my "Second Memoir on Quantics," *Phil. Trans.*, 1856, [141], I derived the conclusion, which, as it now appears, was erroneous, that for a binary quintic the number of irreducible covariants was infinite. The theory requires, in fact, a modification, by reason that certain linear relations, which I had assumed to be independent, are really not independent, but, on the contrary, linearly connected together: the interconnexion in question does not occur in regard to the quadric, cubic, or quartic; and for these cases respectively the theory is true as it stands; for the quintic the interconnexion first presents itself in regard to the degree 8 in the coefficients and order 14 in the variables, viz. the theory gives correctly the number of covariants of any degree not exceeding 7, and also those of the degree 8 and order less than 14; but for the order 14 the theory as it stands gives a non-existent irreducible covariant $(a,..)^8(x, y)^{14}$, viz. we have, according to the theory, $5=(10-6)+1$, that is, of the form in question there are 10 composite covariants connected by 6 syzygies, and therefore equivalent to $10-6$, $=4$ asyzygetic covariants; but the number of asyzygetic covariants being $=5$, there is left, according to the theory, 1 irreducible covariant of the form in question. The fact is that the 6 syzygies being interconnected and equivalent to 5 independent syzygies only, the composite covariants are equivalent to $10-5$, $=5$, the full number of the asyzygetic covariants. And similarly the theory as it stands gives a non-existent irreducible covariant $(a, ..)^8 (x, y)^{20}$. The theory being thus in error, by reason that it omits to take account of the interconnexion of the syzygies, there is no difficulty in conceiving that the effect is the introduction of an infinite series of non-existent irreducible covariants, which, when the error is corrected, will disappear, and there will be left only a finite series of irreducible covariants.

Although I am not able to make this correction in a general manner so as to show from the theory that the number of the irreducible covariants is finite, and so to present the theory in a complete form, it nevertheless appears that the theory can be made to accord with the facts; and I reproduce the theory, as well to show that this is so as to exhibit certain new formulæ which appear to me to place the theory in its true light. I remark that although I have in my Second Memoir considered the question of finding the number of irreducible covariants of a given degree θ in the coefficients but of any order whatever in the variables, the better course is to separate these according to their order in the variables, and so consider the question of finding the number of the irreducible covariants of a given degree θ in the coefficients, and of a given order μ in the variables. (This is, of course, what has to be done for the enumeration of the irreducible covariants of a given quantic; and what is done completely for the quadric, the cubic, and the quartic, and for the quintic up to the degree 6 in my Eighth Memoir, *Phil. Trans.* 1867, [405].) The new formulæ exhibit this separation; thus (Second Memoir, No. 49), writing a instead of x, we have for the quadric the expression $\frac{1}{(1-a)(1-a^2)}$, showing that we have irreducible covariants of the degrees 1 and 2 respectively, viz. the quadric itself and the discriminant: the new expression is $\frac{1}{(1-ax^2)(1-a^2)}$, showing that the covariants in question are of the actual forms $(a, ..\between x, y)^2$ and $(a, ..)^2$ respectively. Similarly for the cubic, instead of the expression No. 55, $\frac{1-a^6}{(1-a)(1-a^2)(1-a^3)(1-a^4)}$, we have $\frac{1-a^6x^6}{(1-ax^3)(1-a^2x^2)(1-a^3x^3)(1-a^4)}$, exhibiting the irreducible covariants of the forms $(a, ..\between x, y)^3$, $(a, ..)^2 (x, y)^2$, $(a ..)^3 (x, y)^3$, and $(a, ..)^4$, connected by a syzygy of the form $(a, ..)^6 (x, y)^6$; and the like for quantics of a higher order.

In the present Ninth Memoir I give the last-mentioned formulæ; I carry on the theory of the quintic, extending the Table No. 82 of the Eighth Memoir up to the degree 8, calculating all the syzygies, and thus establishing the interconnexions in virtue of which it appears that there are really no irreducible covariants of the forms $(a, ..)^8 (x, y)^{14}$, and $(a, ..^8\between x, y)^{20}$. I reproduce in part Gordan's theory so far as it applies to the quintic, and I give the expressions of such of the 23 covariants as are not given in my former memoirs; these last were calculated for me by Mr W. Barrett Davis, by the aid of a grant from the Donation Fund at the disposal of the Royal Society. [The expressions referred to are in fact printed, 143.] The paragraphs of the present memoir are numbered consecutively with those of the former memoirs on Quantics.

Article Nos. 328 to 332. *Reproduction of my original Theory as to the Number of the Irreducible Covariants.*

328. I reproduce to some extent the considerations by which, in my Second Memoir on Quantics, I endeavoured to obtain the number of the irreducible covariants of a given binary quantic $(a, b, \ldots \between x, y)^n$.

Considering in the first instance the covariants as functions of the coefficients $(a, b, c, \ldots)$, without regarding the variables (x, y), and attending only to the following properties—1°, a covariant is a rational and integral homogeneous function of the coefficients; 2°, if P, Q, $R, \ldots$ are covariants, any rational and integral function $F(P, Q, R, \ldots)$, homogeneous in regard to the coefficients, is also a covariant,—we say that the covariants X, $Y, \ldots$ of the same degree in regard to the coefficients, and not connected by any identical equation $\alpha X + \beta Y \ldots = 0$ (where α, $\beta, \ldots$ are quantities independent of the coefficients $(a, b, c, \ldots)$), are *asyzygetic* covariants, and that a covariant not expressible as a rational and integral function of covariants of lower degrees is an *irreducible* covariant; and it is assumed that we know the number of the asyzygetic covariants of the degrees 1, 2, 3,....; say, these are A_1, A_2, $A_3, \ldots$, or, what is the same thing, that the number of the asyzygetic covariants of the degree θ, or form $(a, b, \ldots)^\theta$, is equal to the coefficient of a^θ in a given function

$$\phi(a) = 1 + A_1 a + A_2 a^2 \ldots + A_\theta a^\theta + \ldots,$$

where I have purposely written a, as a representative of the coefficients $(a, b, c, \ldots)$, in place of the x of my Second Memoir.

329. The theory was, that determining α_1, $\alpha_2, \ldots$ by the conditions

$$A_1 = \alpha_1,$$
$$A_2 = \tfrac{1}{2}\alpha_1(\alpha_1 + 1) + \alpha_2,$$
$$A_3 = \tfrac{1}{6}\alpha_1(\alpha_1 + 1)(\alpha_1 + 2) + \alpha_1\alpha_2 + \alpha_3,$$
$$\vdots$$

that is, throwing

$$1 + A_1 a + A_2 a^2 + A_3 a^3 + \ldots$$

into the form

$$(1-a)^{-\alpha_1}(1-a^2)^{-\alpha_2}(1-a^3)^{-\alpha_3}\ldots,$$

the index α_r would express the number of irreducible covariants of the degree r *less* the number of the (irreducible) linear relations, or syzygies, between the composite or non-irreducible covariants of the same degree. Thus $A_1 = \alpha_1$, there would be α_1 covariants of the degree 1[1]; these give rise to $\tfrac{1}{2}\alpha_1(\alpha_1 + 1)$ composite covariants of the degree 2; or, assuming that these are connected by k_2 syzygies, the number of asyzygetic composite covariants of the degree 2 would be $\tfrac{1}{2}\alpha_1(\alpha_1+1) - k_2$; and thence there would be $A_2 - \tfrac{1}{2}\alpha_1(\alpha_1+1) + k_2$, that is, $\alpha_2 + k_2$ irreducible covariants of the same degree; so that (irreducible invariants *less* syzygies) $(\alpha_2 + k_2) - k_2$ is $= \alpha_2$.

[1] For the case of covariants, α_1 is of course $= 1$; but in the investigation the term covariant properly stands for any function satisfying the conditions 1° and 2°.

330. The k_2 syzygies are here irreducible syzygies; for, calling P, Q, R,... the covariants of the degree 1, there is no identical relations between the terms P^2, Q^2, PQ,...: imagine for a moment that we could have l_2 such identical relations (viz. this might very well be the case if instead of the $\frac{1}{2}\alpha_1(\alpha_1+1)$ functions P^2, Q^2, PQ,..., we were dealing with the same number of other quadric functions of these quantities), that is, relations not establishing any relation between P^2, Q^2, PQ,..., and besides these k_2 non-identical relations as above; then the number of irreducible invariants would be $\alpha_2+k_2+l_2$, and the number of irreducible syzygies being as before k_2, the difference would be not α_2 but α_2+l_2. The l_2 identical relations are here relations between composite covariants, and the effect (if any such relation could subsist) would, it appears, be to increase α_2; between syzygies such identical relations do actually exist, and the effect is contrariwise to diminish the α; we may, for instance, for the degree s have irreducible covariants *less* irreducible syzygies $=\alpha_s-l_s$.

331. Assume for a moment that, for a given value of s, α_s is positive; but for the term l_s it would of course follow that there was for the degree in question a certain number of irreducible covariants; and it was in this manner that I was led to infer that the number of the covariants of a quintic was infinite—viz. the transformed expression for the number of asyzygetic covariants is

$$=\text{coeff. } a^\theta \text{ in } (1-a^4)^{-1}(1-a^8)^{-3}(1-a^{12})^{-6}(1-a^{14})^{-4}\ldots,$$

a product which does not terminate, and as to which it is also assumed that the series of negative indices does not terminate.

332. The principle is the same, but the discussion as to the number of the irreducible covariants becomes more precise, if we attend to the covariants as involving not only the coefficients $(a, b, \ldots)$ but also the variables (x, y); we have then to consider the covariants of the form $(a, b, \ldots)^\theta (x, y)^\mu$, or, say, of the form $a^\theta x^\mu$ (degree θ and order μ), and the number of the asyzygetic covariants of this form is given as the coefficient of $a^\theta x^\mu$ in a given function of (a, x), (I write a instead of the z of my Second Memoir in the formulæ which contain x and z): by taking account of the composite covariants and syzygies, we successively determine, from the given number of asyzygetic covariants for each value of θ and μ, the number of the irreducible covariants for the same values of θ and μ. This is, in fact, done for the quintic in my Eighth Memoir up to the covariants and syzygies of the degree 6. But before resuming the discussion for the quintic, I will consider the preceding cases of the quadric, the cubic, and the quartic.

Article Nos. 333 to 336. *New formulæ for the number of Asyzygetic Covariants.*

333. For the quadric $(a, b, c \between x, y)^2$, the number of asyzygetic covariants $a^\theta x^\mu$

$$=\text{coeff. } a^\theta x^{\theta-\frac{1}{2}\mu} \text{ in } \frac{1-x}{(1-a)(1-ax)(1-ax^2)},$$

(see Second Memoir, No. 35, observing that q is there $=\theta-\frac{1}{2}\mu$, and that the subtraction of successive coefficients is effected by means of the factor $1-x$ in the numerator. See also Eighth Memoir, No. 251, where a like form is used for the quintic). Writing ax^2 for a, and $\frac{1}{x^2}$ for x, this is

$$= \text{coeff. } a^\theta x^\mu \text{ in } \frac{1-\frac{1}{x^2}}{(1-ax^2)(1-a)\left(1-\frac{a}{x^2}\right)}.$$

The development is

$$\begin{array}{ll|l} 1 & -\frac{1}{x^2} & 1 \\ +ax^2 & & +a\left(\frac{1}{x^2}\right) \\ +a^2(x^4+1) & & +a^2\left(\frac{1}{x^4}+1\right) \\ +a^3(x^6+x^2) & & +a^3\left(\frac{1}{x^6}+\frac{1}{x^2}\right) \\ +a^4(x^8+x^4+1) & & +a^4\left(\frac{1}{x^8}+\frac{1}{x^4}+1\right) \\ \vdots & & \vdots \end{array}$$

which is

$$= A(x) - \frac{1}{x^2} A\left(\frac{1}{x}\right),$$

where

$$A(x) = \frac{1}{(1-ax^2)(1-a^2)};$$

and, since $\frac{1}{x^2} A\left(\frac{1}{x}\right)$ contains only negative powers, the required number is

$$= \text{coeff. } a^\theta x^\mu \text{ in } \frac{1}{(1-ax^2)(1-a^2)},$$

indicating that the covariants are powers and products of (ax^2 and a^2), the quadric itself, and the discriminant. Compare Second Memoir, No. 49, according to which, writing therein a for x, the number of asyzygetic covariants is

$$= \text{coeff. } a^\theta \text{ in } \frac{1}{(1-a)(1-a^2)}.$$

334. For the cubic $(a, b, c, d \between x, y)^3$ the number of asyzygetic covariants $a^\theta x^\mu$ is

$$= \text{coeff. } a^\theta x^{\theta - \frac{1}{2}\mu} \text{ in } \frac{1-x}{(1-a)(1-ax)(1-ax^2)(1-ax^3)};$$

or transforming as before, this is

$$= \text{coeff. } a^\theta x^\mu \text{ in } \frac{1-\frac{1}{x^2}}{(1-ax^3)(1-ax)(1-ax^{-1})(1-ax^{-3})}:$$

the function is here

$$= A(x) - \frac{1}{x^2} A\left(\frac{1}{x}\right),$$

where

$$A(x) = \frac{1 - a^6x^6}{(1 - ax^3)(1 - a^2x^2)(1 - a^3x^3)(1 - a^4)}$$

(that this is so may be easily verified); and since the second term contains only negative powers, the required number is = coeff. $a^\theta x^\mu$ in $A(x)$. The formula, in fact, indicates that the covariants are made up of $(ax^3, a^2x^2, a^3x^3, a^4)$, the cubic itself, the Hessian, the cubicovariant, and the discriminant, these being connected by a syzygy (a^6x^6) of the degree 6 and order 6. Compare Second Memoir, No. 50, according to which the number of covariants of degree θ is

$$= \text{coeff. } a^\theta \text{ in } \frac{1 - a^6}{(1 - a)(1 - a^2)(1 - a^3)(1 - a^4)}.$$

335. For the quartic $(a, b, c, d, e \between x, y)^4$ the number of asyzygetic covariants $a^\theta x^\mu$ is

$$= \text{coeff. } a^\theta x^{\theta - \frac{1}{2}\mu} \text{ in } \frac{1 - x}{(1 - a)(1 - ax)(1 - ax^2)(1 - ax^3)(1 - ax^4)};$$

or transforming as before, this is

$$= \text{coeff. } a^\theta x^\mu \text{ in } \frac{1 - x^{-2}}{(1 - ax^4)(1 - ax^2)(1 - a)(1 - ax^{-2})(1 - ax^{-4})}:$$

the function is here

$$= A(x) - \frac{1}{x^2} A\left(\frac{1}{x}\right),$$

where

$$A(x) = \frac{1 - a^6x^{12}}{(1 - ax^4)(1 - a^2x^4)(1 - a^2)(1 - a^3)(1 - a^3x^6)};$$

and the second term containing only negative powers, the required number is = coeff. $a^\theta x^\mu$ in $A(x)$. The formula indicates that the covariants are made up of $(ax^4, a^2x^4, a^2, a^3, a^3x^6)$, the quartic itself, the Hessian, the quadrinvariant, the cubinvariant, and the cubicovariant, these being connected by a syzygy (a^6x^{12}) of the degree 6 and order 12. Compare Second Memoir, No. 51, according to which the number of covariants of degree θ is

$$= \text{coeff. } a^\theta \text{ in } \frac{1 - a^6}{(1 - a)(1 - a^2)^2(1 - a^3)^2}.$$

336. For the quintic $(a, b, c, d, e, f \between x, y)^5$ the number of asyzygetic covariants $a^\theta x^\mu$ is

$$= \text{coeff. } a^\theta x^{\theta - \frac{1}{2}\mu} \text{ in } \frac{1 - x}{(1 - a)(1 - ax)(1 - ax^2)(1 - ax^3)(1 - ax^4)(1 - ax^5)};$$

or transforming as before, this is

$$= \text{coeff. } a^\theta x^\mu \text{ in } \frac{1-x^{-2}}{(1-ax^5)(1-ax^3)(1-ax)(1-ax^{-1})(1-ax^{-3})(1-ax^{-5})}.$$

The developed expression is

$$\begin{array}{ll|l} 1 & -\dfrac{1}{x^2} & 1 \\ +ax^5 & & +ax^{-5} \\ +a^2(x^{10}+x^6+x^2) & & +a^2(x^{-10}+x^{-6}+x^{-2}) \\ \vdots & & \vdots \end{array}$$

but here there is not any *finite* function $A(x)$ such that this development is

$$= A(x) \qquad -\frac{1}{x^2}A\left(\frac{1}{x}\right).$$

The numerical coefficients are of course the same as those in the development of the untransformed function; viz. they are the numbers given in the third column of Table No. 82 (Eighth Memoir), and also (carried further) in the third column of the following Table, No. 87. And we can, from the discussion of these coefficients, deduce the form of $A(x)$, viz. this is

$1-a^5x^{11}$	$1-a^6x^{18}$	$1-a^7x^{15}$	$(1-a^8x^{12})^3$	...
	14	13	$(10)^3$	
	12	11	$(8)^2$	
	10	$(9)^3$	$(6)^2$	
	8	7		
	6			

=

$1-ax^5$	$1-a^2x^6$	$1-a^3x^9$	$1-a^4x^6$	$1-a^5x^7$	$1-a^6x^4$	$1-a^7x^3$	$1-a^8x^3$	$1-a^{18}$	...
	2	5	4	3	2	1	0		
		3	0	1			20		
							14		

where, for shortness, I have written $\underset{2}{1-a^2x^6}$ to stand for $(1-a^2x^6)(1-a^2x^2)$, and so in other cases: moreover in the third column of the numerator the $(9)^3$ shows that the factor is $(1-a^7x^9)^3$, and so in other cases: this will be further explained presently. Compare herewith the form, Second Memoir, No. 52, viz. the number of asyzygetic covariants of the degree θ is

$$= \text{coeff. } a^\theta \text{ in } (1-a)^{-1}(1-a^2)^{-2}(1-a^3)^{-3}(1-a^4)^{-3}(1-a^5)^{-2}(1-a^6)^4(1-a^7)^5(1-a^8)^6\ldots$$

each index being, it will be observed, equal to the number of factors in the numerator, less the number of factors in the denominator, in the corresponding column of the new formula.

Article Nos. 337 to 346. *The 23 Fundamental Covariants.*

337. Gordan's result is that the entire number of the irreducible covariants of the binary quintic is $=23$. I represent these by the letters $A, B, C, \ldots, W$, identifying such of them as were given in my former Memoirs on Quantics with the Tables of these Memoirs, and the new ones, O, P, R, S, T, V, with the Tables Nos. 90, 91, 92, 93, 94, 95 of the present Memoir.

Table No. 87. Identification of the 23 irreducible covariants of the binary quintic.

			Table No.
A	$(a, b, c, d, e, f \between x, y)^5$	f	13
$B = \frac{1}{28000}(A, A)^4$	$(\quad)^2 (\quad)^2$	$\iota = (ff)^4$	14
$C = \frac{1}{800}(A, A)^2$	$(\quad)^2 (\quad)^6$	$\phi = (ff)^2$	15
$D = -\frac{1}{3}(A, B)^2$	$(\quad)^3 (\quad)^3$	$j = (f\iota)^2$	16
$E = \frac{1}{5}(A, B)$	$(\quad)^3 (\quad)^5$	$(f\iota)$	17
$F = \frac{1}{15}(A, C)$	$(\quad)^3 (\quad)^9$	$(f\phi)$	18
$G = -\frac{1}{2}(B, B)^2$	$(\quad)^4 (\quad)^0$	$(\iota\,\iota)^2$	19
$H = -\frac{1}{5}(B, C)^2 + \frac{2}{5}B^2$	$(\quad)^4 (\quad)^4$	$p = (\phi\,\iota)^2$	20
$I = -\frac{1}{6}(B, C)$	$(\quad)^4 (\quad)^6$	$(\phi\,\iota)$	21
$J = -\frac{1}{4}(B, D)^2$	$(\quad)^5 (\quad)^1$	$\alpha = (j\,\iota)^2$	22
$K = -(B, D)$	$(\quad)^5 (\quad)^3$	$(j\,\iota)$	23
$L = -\frac{1}{20}(A, H) + \frac{1}{2}BE$	$(\quad)^5 (\quad)^7$	(fp)	24
$M = -\frac{1}{48}(B, H)^2 - \frac{1}{6}BG$	$(\quad)^6 (\quad)^2$	$\tau = (p\iota)^2$	83
$N = \frac{1}{4}(B, H)$	$(\quad)^6 (\quad)^4$	$(p\iota)$	84
$O = -(B, J)$	$(\quad)^7 (\quad)^1$	$(\iota\alpha)$	*90
$P = -\frac{1}{5}(A, M) - BK$	$(\quad)^7 (\quad)^5$	$(f\tau)$	*91
$Q = \frac{1}{2}(B, M)^2$	$(\quad)^8 (\quad)^0$	$(\iota\tau)^2$	25
$R = -\frac{1}{2}(B, M)$	$(\quad)^8 (\quad)^2$	$(\tau\iota)$	*92
$S = -96(D, M) + 16BO - 7GK$	$(\quad)^9 (\quad)^3$	$(j\tau)$	*93
$T = -(J, M)$	$(\quad)^{11} (\quad)^1$	$\gamma = (\tau\alpha)$	*94
$U = \frac{1}{18}(J, O) + \frac{1}{9}GQ$	$(\quad)^{12} (\quad)^0$	$((\iota\alpha), \alpha)$	29
$V = -(B, T)$	$(\quad)^{13} (\quad)^1$	$(\iota\gamma)$	*95
$W = -\frac{1}{6}(O, T)$	$(\quad)^{18} (\quad)^0$	$((\iota\alpha), \gamma)$	29A

338. The Table exhibits the generation of the several covariants; viz. (A, B) denotes $\partial_x A \,.\, \partial_y B - \partial_y A \,.\, \partial_x B$, $(A, B)^2$ denotes $\partial_x{}^2 A \,.\, \partial_y{}^2 B - 2\partial_x\partial_y A \,.\, \partial_x\partial_y B + \partial_y{}^2 A \,.\, \partial_x{}^2 B$, &c. (see *post*, No. 348). The column f, $\iota = (ff)^4$, &c. shows Gordan's notation, and the generation of his 23 forms $\big((ff)^4$ written as with him for $(f, f)^4$, &c.$\big)$: it will be observed that the forms are not identical; if the calculations had been made *de novo*, I should have adopted his values, simply omitting numerical factors of the several forms (thus every term of ι, $=(ff)^4$ contains the factor $2.(120)^2$, $=28800$): of course the presence of these numerical factors renders the f, ι, ϕ, &c. as they stand inconvenient for the expression of results; and the numerical fixation of the values was no part of Gordan's object. But by reason of the existing Tables the change of notation is in fact more than this; thus H instead of being a submultiple of $(B, C)^2$, that is, of p, is in fact $= -\frac{1}{5}(B, C)^2 + \frac{2}{5}B^2$; and so in other cases. If the occasion for it arises, there is no difficulty in expressing any one of the forms f, ι, ϕ, &c. in terms of the $(A, B, C \,..\, V, W)$; thus in the instance just referred to, $p = (\phi\iota)^2$, we have

$$\phi = (ff)^2 = (A,\ A)^2 = 800\, C,$$

and

$$\iota\ = (ff)^4 = (A,\ A)^4 = 28800\, B,$$

whence $p = 2304000\,(B, C)^2$; also $(B, C)^2 = -5H + 2B^2$; and therefore, finally,

$$p = -11520000\, H + 4608000\, B^2.$$

339. I remark upon the value $S = -96\,(D, M) + 16BO - 7GK$, that S is the complete value of a covariant $(\quad)^9 (\quad)^3$, the leading coefficient of which is given in Table No. 86 of my Eighth Memoir; the form (D, M), omitting a numerical factor (if any), would have had smaller numerical coefficients, but there is in the form actually adopted the advantage that it vanishes for $a = 0$, $b = 0$, that is, when the quintic has two equal roots, [see *post*, No. 346].

340. I now form the following Table No. 88, viz. this is the Table No. 82 of my Eighth Memoir, carried as far as a^8, but with the composite covariants expressed by means of the foregoing letters $A, B, C, \ldots, W$; instead of giving the syzygies as in Table No. 82, I transfer them to a separate Table, No. 89. In all other respects the arrangement is as explained, Eighth Memoir, No. 253; but in place of N, S, S' I have written $*$, Σ, Σ' to denote new covariant, new syzygy, derived syzygy, respectively; and I have, as to the terms a^8x^{14}, a^8x^{20} respectively, introduced the new symbol σ to denote an interconnexion of syzygies, as appearing by the Table No. 89, and as will be further explained.

Table No. 88.

[In this Table and the subsequent Table 89, I have for convenience used, instead of capitals, the small italic letters a, b, c, ... w to denote the 23 irreducible covariants of the quintic.]

Ind. a.	Ind. x.	Coeff.		
1	5	1	a	*
	3	0		
	1	0		
2	10	1	a^2	
	8	0		
	6	1	c	*
	4	0		
	2	1	b	*
	0	0		
3	15	1	a^3	
	13	0		
	11	1	ac	
	9	1	f	*
	7	1	ab	
	5	1	e	*
	3	1	d	*
	1	0		
4	20	1	a^5	
	18	0		
	16	1	a^2c	
	14	1	af	
	12	2	a^2b, c^2	
	10	1	ae	
	8	2	ad, bc	
	6	1	i	*
	4	2	b^2, h	*
	2	0		
	0	1	g	*

Table No. 88 (continued).

Ind. a.	Ind. x.	Coeff.		
5	25	1	a^5	
	23	0		
	21	1	a^3c	
	19	1	a^2f	
	17	2	$a^3b,\ ac^2$	
	15	2	$a^2e,\ cf$	
	13	2	$a^2d,\ abc$	
	11	2	$ai,\ bf,\ ce$	Σ
	9	3	$ab^2,\ ah,\ cd$	
	7	2	$be,\ l$	*
	5	2	$ag,\ bd$	
	3	1	k	*
	1	1	j	*
6	30	1	a^6	
	28	0		
	26	1	a^4c	
	24	1	a^3f	
	22	2	$a^4b,\ a^2c^2$	
	20	2	$a^3e,\ acf$	
	18	3	$a^3d,\ a^2bc,\ c^3,\ f^2$	Σ
	16	2	$a^2i,\ abf,\ ace$	Σ′
	14	4	$a^2b^2,\ a^2h,\ acd,\ bc^2,\ ef$	Σ
	12	3	$abe,\ al,\ ci,\ df$	Σ
	10	4	$a^2g,\ abd,\ b^2c,\ ch,\ e^2$	Σ
	8	2	$ak,\ bi,\ de$	Σ
	6	4	$aj,\ b^3,\ bh,\ cg,\ d^2$	Σ
	4	1	n	*
	2	2	$bg,\ m$	*
	0	0		
7	35	1	a^7	
	33	0		
	31	1	a^5c	
	29	1	a^4f	
	27	2	$a^5b,\ a^3c^2$	
	25	2	$a^4e,\ a^2cf$	
	23	3	$a^4d,\ a^3bc,\ ac^3,\ af^2$	Σ′
	21	3	$a^3i,\ a^2bf,\ a^2ce,\ c^2f$	Σ′
	19	4	$a^3b^2,\ a^3h,\ a^2cd,\ abc^2,\ aef$	Σ′
	17	4	$a^2be,\ a^2l,\ aci,\ adf,\ bcf,\ c^2e$	2Σ′
	15	5	$a^2g,\ a^2bd,\ ab^2c,\ ach,\ ae^2,\ c^2d,\ fi$	Σ′, Σ
	13	4	$a^2k,\ abi,\ ade,\ b^2f,\ bce,\ cl,\ fh$	Σ
	11	5	$a^2j,\ ab^3,\ abh,\ acg,\ ad^2,\ bcd,\ ei$	Σ
	9	4	$an,\ b^2e,\ bl,\ ck,\ di,\ eh,\ fg$	3Σ
	7	4	$abg,\ am,\ b^2d,\ cj,\ dh$	Σ
	5	3	$bk,\ p,\ eg$	*
	3	2	$bj,\ dg$	
	1	1	o	*

Table No. 88 (concluded).

Ind. a.	Ind. x.	Coeff.		
8	40	1	a^8 .	
	38	0	.	
	36	1	a^6c .	
	34	1	a^5f .	
	32	2	$a^6b,\ a^4c^2$.	
	30	2	$a^5e,\ a^3cf$.	
	28	3	$a^5d,\ a^4bc,\ a^2c^3,\ a^2f^2$.	Σ'
	26	3	$a^4i,\ a^3bf,\ a^3ce,\ ac^2f$.	Σ'
	24	5	$a^4b^3,\ a^4h,\ a^3cd,\ a^2bc^2,\ a^2ef,\ cf^2,\ c^4$.	Σ'
	22	4	$a^3be,\ a^3l,\ a^2ci,\ a^2df,\ abcf,\ ac^2e$.	$2\Sigma'$
	20	6	$a^4g,\ a^3bd,\ a^2b^2c,\ a^2ch,\ a^2e^2,\ ac^2d,\ afi,\ bc^3,\ bf^2,\ cef$.	σ
	18	5	$a^3k,\ a^2bi,\ a^2de,\ ab^2f,\ abce,\ acl,\ afh,\ c^2i,\ cdf$.	$4\Sigma'$
	16	7	$a^3j,\ a^2b^3,\ a^2bh,\ a^2cg,\ a^2d^2,\ abcd,\ aei,\ b^2c^2,\ bef,\ c^2h,\ ce^2,\ fl$.	$5\Sigma'$
	14	5	$a^2n,\ ab^2e,\ abl,\ ack,\ adi,\ aeh,\ afg,\ bci,\ bdf,\ cde$.	σ
	12	7	$a^2bg,\ a^2m,\ ab^2d,\ acj,\ adh,\ b^3c,\ bch,\ be^2,\ c^2g,\ cd^2,\ el,\ fk,\ i^2$.	3Σ
	10	5	$abk,\ aeg,\ ap,\ b^2i,\ bde,\ cn,\ dl,\ fj,\ hi$.	3Σ
	8	6	$abj,\ adg,\ b^4,\ b^2h,\ bcg,\ bd^2,\ cm,\ ek,\ h^2$.	2Σ
	6	3	$ao,\ bn,\ dk,\ ej,\ gi$.	2Σ
	4	4	$b^2g,\ bm,\ dj,\ gh$.	
	2	1	r .	*
	0	2	$g^2,\ q$.	*

341. The syzygies and interconnexions of syzygies are given in

Table No. 89.

[See *ante* Table No. 88.]

(5, 11)	$ai + bf - ce = 0$
(6, 18)	$a^3d - a^2bc + 4c^3 + f^2 = 0$
(6, 14)	$a^2h - 6acd - 4bc^2 - ef = 0$
(6, 12)	$al - 2ci + 3df = 0$
(6, 10)	$a^2g - 12abd - 4b^2c - e^2 = 0$
(6, 8)	$ak + 2bi - 3de = 0$
(6, 6)	$aj - b^3 + 2bh - cg - 9d^2 = 0$
(7, 15)	$a^2bd - ab^2c + ach - 6c^2d - fi = 0$
(7, 13)	$a^2k - abi - 3b^2f + 6cl + 3fh = 0$
(7, 11)	$a^2j - ab^3 + abh - 9ad^2 - 6bcd - ei = 0$
(7, 9)	$an - b^2e - 6di + 2eh - fg = 0$
	$2bl + 6di - eh + fg = 0$
	$2ck - 12di + eh - fg = 0$
(7, 7)	$am + 2b^2d + cj - 3dh = 0$

Table No. 89 (continued).

σ, (8, 20)	$0 . a^2 (a^2g - 12abd - 4b^2c - e^2)$	suprà (6, 10)
	$- a (a^2bd - abc^2 + ach - 6c^2d - fi)$	„ (7, 15)
	$+ b (a^3d - a^2bc + 4c^3 + f^2)$	„ (6, 18)
	$+ c (a^2h - 6acd - 4bc^2 - ef)$	„ (6, 14)
	$- f (ai + bf - ce) = 0$	„ (5, 11)
σ, (8, 14)	$0 . a (an - b^2e - 6di + 2eh - fg)$	suprà (7, 9)
	$+ a (2bl + 6di - eh + fg)$	„ („)
	$+ a (2ck - 12di + eh - fg)$	„ („)
	$- 2b (al - 2ci + 3df)$	„ (6, 12)
	$- 2c (ak + 2bi - 3de)$	„ (6, 8)
	$+ 6d (ai + bf - ce) = 0$	„ (5, 11)
(8, 12)	$ab^2d - b^3c + 2bch - c^2g + i^2 = 0$	
	$- 3adh - 2bch + 2c^2g + 18cd^2 + fk - 2i^2 = 0$	
	$el + fk - 2i^2 = 0$	
(8, 10)	$abk - cn - 6dl - 2fj + hi = 0$	
	$ap + 2cn + fj = 0$	
	$b^2i + cn + 3dl + fj - 2hi = 0$	
(8, 8)	$abj - b^4 + 4b^2h - 9bd^2 + 12cm - ek - 3h^2 = 0$	
	$adg + 2b^2h - 12bd^2 + 8cm - ek - 2h^2 = 0$	
(8, 6)	$ao + 6dk - 3ej + 2gi = 0$	
	$bn + 3dk - ej + gi = 0$	

342. In illustration take any one of the lines of Table No. 88, for instance [resuming the notation by capital letters] the line

$$(7,\ 17)\ |\ 4\ |\quad A^2BE,\ A^2L,\ ACI,\ ADF,\ BCF,\ C^2E\quad |\ 2\Sigma'\ |$$

there are here 6 composite covariants, but the number of asyzygetic covariants is $=4$; there must therefore be $6-4$, $=2$ syzygies; we have however (see Table No. 89) two derived syzygies of the right form, viz. these are

$$A (AL - 2CI + 3DF) = 0,$$
$$C (AI + BF - CE) = 0,$$

which are designated as $2\Sigma'$, and there is consequently no new syzygy Σ.

But in the line

$$(7,\ 15)\ |\ 5\ |\quad A^3G,\ A^2BD,\ AB^2C,\ ACH,\ AE^2,\ C^2D,\ FI\quad |\ \Sigma',\ \Sigma\ |$$

there are 7 composite covariants, but the number of asyzygetic covariants is $=5$; there must therefore be $7-5$, $=2$ syzygies. One of these is the derived syzygy

$$A (A^2G - E^2 - 12ABD - 4B^2C) = 0,$$

which is designated by Σ'; the other is a new syzygy (see Table No. 89),

$$A^2BD - ABC^2 + ACH - 6C^2D - FI = 0,$$

designated by Σ.

343. Take now the line

(8, 20) | 6 | A^4G, A^3BD, A^2B^2C, A^2CH, A^2E^2, AC^2D, AFI, BC^3, BF^2, CEF | $5\Sigma'$, σ | ;

there are here 10 composite covariants, but the number of irreducible covariants is $= 6$; there should therefore be $10 - 6$, $= 4$ syzygies. There are, however, the 5 derived syzygies

$$A^2(A^2G - 12ABD - 4B^2C - E^2) = 0, \text{ \&c. (see Table No. 89)}$$

designated by $5\Sigma'$; since these are equivalent to 4 syzygies only there must be 1 identical relation between them (designated by σ), viz. this is the equation $0 = 0$ obtained by adding the several syzygies, multiplied each by the proper numerical factor as shown Table No. 89.

344. Again, for the line

(8, 14) | 5 | A^2N, AB^2E, ABL, ACK, ADI, AEH, AFG, BCI, BDF, CDE | $6\Sigma'$, σ |

there are here 10 composite covariants, but only 5 irreducible covariants; there should therefore be $10 - 5$, $= 5$ syzygies; we have in fact the 6 derived syzygies

$$A(AN - B^2E - 6DI + 2EH - FG) = 0 \text{ \&c. (see Table No. 89)}$$

designated by $6\Sigma'$; these must therefore be connected by 1 identical relation (designated by σ), viz. this is the equation $0 = 0$ obtained by adding the several syzygies, each multiplied by the proper numerical factor as shown Table No. 89.

345. These two cases (σ) are in fact the instances which present themselves where a correction is required to my original theory. The two identical relations in question were disregarded in my original theory, and this accordingly gave the two non-existent irreducible covariants $(a,..)^8(x, y)^{14}$ and $(a,..)^8(x, y)^{20}$. And reverting to No. 336, these give in the denominator of $A(x)$ the factors $(1 - a^8x^{20})(1 - a^8x^{14})$. In virtue hereof, writing $x = 1$, we have in $A(x)$ the factor $\dfrac{(1 - a^8)^{10}}{(1 - a^8)^4}$, $= (1 - a^8)^6$, agreeing with the function $(1 -)^{-1}(1 - a)^{-2} \ldots (1 - a^8)^6 \ldots$. And we thus see that the denominator factors of $A(x)$ do not all of them refer to irreducible covariants; viz. we have

$$ax^5,\ a^2x^6,\ a^2x^2,\ a^3x^9,\ a^3x^5,\ a^3x^3,\ a^4x^6,\ a^4x^4,\ a^4,\ a^5x^7,\ a^5x^3,\ a^5x,\ a^6x^4,\ a^6x^2,\ a^7x^3,\ a^7x,\ a^8x^2,\ a^8,$$

each referring to an irreducible covariant, but a^8x^{20} and a^8x^{14} each referring to an identical relation (σ) or interconnexion of syzygies. And we thus understand how, consistently with the number of the irreducible covariants being finite, the expression for $A(x)$ may be as above the quotient of two infinite products; viz. there will be in the denominator a finite number of factors each referring to an irreducible covariant, but the remaining infinite series of denominator factors will refer each factor to an identical relation or interconnexion of syzygies. But I do not see how we can by the theory distinguish between the two classes of factors, so as to determine the number of the irreducible covariants, or even to make out affirmatively that the number of them is finite.

346. The new covariants O, P, R, S, T, V are as follows:

[Table No. 90 (Covariant O),
Table No. 91 (Covariant P),
Table No. 92 (Covariant R),
Table No. 93 (Covariant S),
Table No. 94 (Covariant V),

printed in the paper 143, "Tables of the Covariants M to W of the Binary Quintic; from the second, third, fifth, eighth, ninth and tenth Memoirs on Quantics" with the insertion as therein mentioned of the terms with zero coefficients. The covariant $S, = -96(D, M) + 16BO - 7GK$, of the present Memoir is there called S', and there is given the more simple form $S = (D, M)$, of this covariant.]

Article Nos. 347 to 365. *Sketch of Professor* GORDAN'S *proof for the finite Number,* $= 23$, *of the Covariants of a Binary Quintic.*

347. I propose to reproduce the leading points of Professor Gordan's proof that the binary quintic $(a, b, c, d, e, f \between x, y)^5$ has a finite system of 23 covariants, viz. a system such that every other covariant whatever is a rational and integral function of these 23 covariants.

348. DERIVATION. Consider for a moment any two binary quantics ϕ, ψ of the same or different orders, and which may be either independent quantics, or they may be both or one of them covariants, or a covariant, of a binary quantic f. We may form the series of *derivatives*

$$(\phi, \psi)^0 = \phi\psi,$$
$$(\phi, \psi)^1 = \overline{12}\ \phi_1\psi_2 = \partial_x\phi \,.\, \partial_y\psi - \partial_y\phi \,.\, \partial_x\psi,$$
$$(\phi, \psi)^2 = \overline{12}^2\ \phi_1\psi_2 = \partial_x^2\phi \,.\, \partial_y^2\psi - 2\partial_x\partial_y\phi \,.\, \partial_x\partial_y\psi + \partial_y^2\phi \,.\, \partial_x^2\psi,$$
$$\vdots$$

where, however, there is no occasion to use the notation $(\phi, \psi)^0$ (as this is simply the product $\phi\psi$), and the succeeding derivatives may (when there is no risk of ambiguity) be written more shortly $(\phi\psi)$, $(\phi\psi)^2$, $(\phi\psi)^3$, &c.; in all that follows the word "derivative" (Gordan's *Uebereinanderschiebung*) is to be understood in this special sense.

349. The degree of the derivative $(\phi\psi)^k$ is the sum of the degrees of the constituents ϕ, ψ; the order of the derivative is the sum of the orders *less* $2k$; it being understood throughout that the word degree refers to the coefficients, and the

word order to the variables. In speaking generally of the covariants or of all the covariants of a quantic f, or of the covariants or all the covariants of a given degree or order, we of course exclude from consideration covariants linearly connected with other covariants (for otherwise the number of terms would be infinite); but unless it is expressly so stated, we do not carry this out rigorously so as to make the system to consist of asyzygetic covariants; viz. it is assumed that the system is complete, but not that it is divested of superfluous terms.

350. THEOREM A. The covariants of a quantic f of a given degree m can be all of them obtained by derivation from f and the covariants of the next inferior degree $(m-1)$.

In particular for the degree 1 the only covariant is the quantic f itself; for the degree 2 the covariants are $(ff)^0$, $(ff)^2$, $(ff)^4, \ldots$: using for a moment β to denote each of these in succession, the covariants of the third degree are $(\beta f)^0$, $(\beta f)^1$, $(\beta f)^2, \ldots$; and so on.

351. Suppose that the covariants of the second degree $(ff)^0$, $(ff)^2$, $(ff)^4 \ldots$ are in this order represented by β_1, β_2, $\beta_3 \ldots$, then the covariants of the third degree written in the order

$$(\beta_1 f)^0,\ (\beta_1 f),\ (\beta_1 f)^2,\ \ldots (\beta_2 f)^0,\ (\beta_2 f),\ (\beta_2 f)^2,\ \ldots (\beta_3 f)^0,\ (\beta_3 f),\ (\beta_3 f)^2 \ldots$$

may be represented by γ_1, γ_2, $\gamma_3, \ldots$, the covariants of the fourth degree written in the order

$$(\gamma_1 f)^0,\ (\gamma_1 f),\ (\gamma_1 f)^2,\ \ldots (\gamma_2 f)^0,\ (\gamma_2 f),\ (\gamma_2 f)^2,\ \ldots (\gamma_3 f)^0,\ (\gamma_3 f),\ (\gamma_3 f)^2 \ldots$$

may be represented by δ_1, δ_2, $\delta_3 \ldots$, and so on: we thus obtain in a definite order the covariants of a given degree m; say, these are μ_1, μ_2, μ_3, $\mu_4, \ldots$: any term μ_3 is said to be a *later* term than the preceding terms μ_1, μ_2, and an *earlier* term than the following ones, μ_5, μ_6, &c.

Observe that each term μ_r is a derivative $(\lambda_q f)^k$, the derivatives of an earlier λ are earlier than those of a later λ; and as regards the derivatives of the same λ, the derivative with a less index of derivation is earlier than that with a greater index of derivation, or, what is the same thing, those are earlier which are of the higher order.

352. The series μ_1, μ_2, μ_3, $\mu_4 \ldots$ is not asyzygetic; we make it so, by considering in succession whether the several terms μ_2, $\mu_3, \ldots$ respectively are expressible as linear functions of the earlier terms, and by omitting every term which is so expressible. The reduced series thus obtained is called T_1, T_2, $T_3, \ldots$. Observe that not every μ is a T, but that every T is a μ; every T therefore arises from a derivation upon f and a certain term λ; which term λ (supposing the λ series reduced in like manner to S_1, S_2, $S_3, \ldots$) is a linear function of certain of the S's. Each later T is derived from later S's, or it may be from the same S's as an earlier T; viz. if the later T is derived from $(S_1, S_2, \ldots S_\theta)$, then the earlier T is derived, it may be, from $(S_1, S_2, \ldots S_\theta)$, or from $(S_1, S_2, \ldots S_{\theta-k})$, but so that there is not in the series any term later than S_θ.

And if, considering any T as thus derived from certain of the S's, and in like manner each of these S's as derived from certain of the R's, and so on, we descend to any preceding series,

$$M_1,\ M_2,\ M_3 \dots$$

it will appear that the T is derived from a certain number $(M_1,\ M_2, \dots M_\phi)$ of the terms of this series.

353. The quadricovariants $(ff)^0$, $(ff)^2$, $(ff)^4, \dots$ are of different orders, and consequently asyzygetic. They form therefore a series such as the T-series, and they may be represented by

$$B_1,\ B_2,\ B_3, \dots .$$

Supposing f to be of the order n, B_1 is of the order $2n$, B_2 of the order $2n-4$, B_3 of the order $2n-8$, and so on. Those terms which are of an order greater than n, are said to be of the form W (agreeing with a subsequent more general definition of W); those which are of an order equal to or less than n, are said to be of the form χ; so that the earlier terms of the B series are W, and the later terms are χ; viz. the χ terms taken in order, beginning with the earliest, are $\chi_1, \chi_2, \chi_3, \dots$.

354. By what precedes any particular T is derived from certain terms $B_1, B_2, \dots B_\theta$, of the B series. This series, $B_1, B_2, \dots B_\theta$, may stop short of the terms χ, or it may include a certain number of them, say $\chi_1, \chi_2, \dots \chi_r$. The terms derived from the χ's are in the sequel denoted by P_χ.

355. Every covariant whatever is a form or sum of forms such as

$$\overline{12}^\alpha\, \overline{13}^\beta\, \overline{23}^\gamma \dots f_1 f_2, \dots f_m;$$

writing in regard to any such expression

$$\Sigma \text{ ind. } 1 = i,\ \Sigma \text{ ind. } 2 = j, \dots$$

(viz. i is the sum of all those indices α, β, &c. which belong to a term containing the symbolic number 1, j the sum of all the indices α, γ, &c. which belong to a term containing the symbolic number 2, and so on) then each of the numbers $i, j, \dots$ is at most $=n$, that is $n-i$, $n-j, \dots$ may be any of them $=0$, but they cannot be any of them negative; the degree of the function is $=m$, and its order is $=mn-i-j\dots$ It is to be further observed that the form is a function of the differential coefficients of f of the orders $n-i$, $n-j$, &c. respectively. It follows that if $n-i$, $n-j, \dots$ are none of them $=0$, the form in question may be obtained from a like form belonging to a quantic f' of the next inferior order $n-1$ by replacing therein the coefficients $a', b', \dots$ by $ax+by$, $bx+cy$, &c. respectively: for example, if f denote the cubic function $(a, b, c, d \between x, y)^3$, then the Hessian hereof is $\overline{12}^2 f_1 f_2$; the like form in regard to the quadric $f' = (a', b', c' \between x, y)^2$ is $\overline{12}^2 f_1' f_2'$, which is $= a'c' - b'^2$; and substituting herein $ax+by$, $bx+cy$, $cx+dy$ for a', b', c' respectively, we have the Hessian $\overline{12}^2 f_1 f_2$ of the cubic. A covariant of f derivable in this manner from a covariant of the next inferior quantic f' is said to be a special covariant.

356. Reverting to the form

$$\overline{12}^{\alpha}\,\overline{13}^{\beta}\,\overline{23}^{\gamma}\ldots f_1 f_2 \ldots f_m;$$

if, as before, $n-1$, $n-j$, &c. are each of them >0; if there is at least one index i which is $=$ or $<\frac{1}{2}n$ (that is, for which $n-i>\frac{1}{2}n$), and if the order $mn-i-j\ldots$ be $>n$, then the form, or any sum of such forms, is said to be a form or covariant W. Every covariant W is thus a special covariant, but not conversely. In the particular case $m=2$, the form is

$$\overline{12}^{\alpha} f_1 f_2,$$

which will be a form W if $n-\alpha>\frac{1}{2}n$, or, what is the same thing, $2n-2\alpha>n$, that is if the order be $>n$. Hence, as already mentioned, the covariants T of the degree 2 are W, or else χ, according as the order is greater than n, or as it is equal to or less than n.

357. *Theorem B.* If any covariant T be expressible as the sum of a form W and of earlier T's than itself, then forming the derivative $(Tf)^k$, either this is not a form T, or being a form T, it is expressible as the sum of a form W and of earlier T's than itself; or, what is the same thing, $(Tf)^k$, if it be a form T, is (like the original T) the sum of a form W and of earlier T's than itself.

Hence also every form T is the sum of a form W, and of forms derived from the functions $\chi_1, \chi_2, \ldots$, say

$$T=W+P_\chi,$$

or, what is the same thing, every covariant whatever is of the form $W+P_\chi$.

358. The proof that for a form f of the order n the number of covariants is finite, depends on the assumption that the number is finite for a form f' of the next inferior order $n-1$: this being so, the number of the special covariants of f will be finite; say these are $A_1, A_2, A_3, \ldots$ (f is itself one of the series, but we may separate it, and speak of the form f and its special covariants): the forms W are functions of the special covariants, and hence every covariant whatever of f is of the form $F(A)+P_\chi$; but it requires still a long investigation to pass from this to the theorem of the existence of a finite number of forms V such that every covariant whatever is $F(V)$. I pass this over, and reproduce only the investigation for the case of the quintic.

359. Starting from the assumed system of forms,

$$\begin{aligned}&f,\ \phi=(ff)^2,\ i=(ff)^4,\ j=(fi)^2,\ \alpha=(ji)^2,\ p=(\phi i)^2,\ \tau=(pi)^2,\ \gamma=(\tau\alpha),\\ &(f\phi),\ (fp),\ (f\tau),\ (j\tau),\\ &(fi),\ (\phi i),\ (ji),\ (pi),\ (\tau i),\\ &(i\alpha),\ (i\gamma),\ (ii)^2,\ \big((i\alpha),\ \alpha\big),\ (i\tau)^2,\ \big((i\alpha),\ \gamma\big),\end{aligned}$$

say, the 23 forms U, it is to be shown that every other covariant whatever of the quintic is of the form $F(U)$.

The special covariants are f, ϕ, $(f\phi)$, i, j, which are forms U; the only form χ is i, so that instead of P_χ writing P_i, every covariant whatever of f is

$$=F(U)+P_i;$$

and it remains to show that every form P_i is $F(U)$; or, what is the same thing, that if H be any form $F(U)$ whatever, then that (Hi) and $(Hi)^2$ are each of them $F(U)$.

360. In order to show that every covariant of a degree not exceeding m is $F(U)$, it will be sufficient to show that the several forms (Hi) and $(Hi)^2$ of a degree not exceeding m are each of them $F(U)$: and if for this purpose we assume that it is shown that every covariant of a degree not exceeding $m-1$ is $F(U)$, then in regard to the forms (Hi) and $(Hi)^2$ of the degree m, it will be sufficient to show that any such form is a function of covariants of a degree inferior to m.

361. First for the form (Hi): we have $(PQ,\ i)=P(Qi)+Q(Pi)$; and hence we see that (Hi) will be $F(U)$ if only (Ui) is always $F(U)$.

In forming the derivative of i with the several covariants U, we may omit i itself, and also the four invariants $(ii)^2$, $(i\tau)^2$, $\big((i\alpha),\ \alpha\big)$, $\big((i\alpha),\ \gamma\big)$, since in each of these cases the derivative is $=0$. We have therefore to consider the derivative of i with

$$f,\ \phi,\ j,\ \alpha,\ p,\ \tau,\ \gamma,\ (f\phi),\ (fp),\ (f\tau),\ (j\tau),\ (fi),\ (\phi i),\ (ji),\ (pi),\ (\tau i),\ (i\alpha),\ (i\gamma),$$

respectively: the first seven of these are each of them U; the remaining eleven are each of them of the form $\big((PQ),\ i\big)$. Now $\big((PQ),\ i\big)$ is a linear function of $P(Qi)^2$, $Q(Pi)^2$, and $i(PQ)^2$, that is $\big((PQ),\ i\big)$ is a function of covariants of a lower degree than itself.

362. Next for the form $(Hi)^2$, we have $(PQ,\ i)^2$, a linear function of $P(Qi)^2$, $Q(Pi)^2$, $i(PQ)^2$; and we hence see that $(Hi)^2$ will be $F(U)$ if only $(Ui)^2$ is always $F(U)$.

In forming the second derivative of i with the several covariants U, we may omit as before the four invariants, and also omit the four linear covariants α, $i\alpha$, γ, $i\gamma$; we have therefore to consider the second derivatives of i with

$$f,\ \phi,\ i,\ j,\ p,\ \tau,\ (f\phi),\ (fp),\ (f\tau),\ (j\tau),\ (fi),\ (\phi i),\ (ji),\ (pi),\ (\tau i),$$

respectively: the first six of these are each of them U; the remaining nine are each of the form $\big((PQ),\ i\big)^2$. Now $\big((PQ),\ i\big)^2$ is a linear function of $\big((Pi)^2,\ Q\big)$, $\big((Qi)^2,\ P\big)$, $P(Qi)^3$, and $Q(Pi)^3$. The first two of these are terms of the same form; $(Pi)^2$, as a covariant of a lower degree than $\big((PQ),\ i\big)^2$, is $F(U)$, and hence $\big((Pi)^2,\ Q\big)$ will be $F(U)$ if only $(U,\ Q)$ is $F(U)$; Q being here any one of the functions f, ϕ, i, j, p, τ, and U being any one of the functions

$$f,\ \phi,\ i,\ j,\ p,\ \tau,\ \alpha,\ \gamma,\ (f\phi),\ (fp),\ (f\tau),\ (j\tau),\ (fi),\ (\phi i)\,(ji)\,(pi)\,(\tau i)\,(i\alpha)\,(i\gamma).$$

363. For U equal to any one of the last eleven values, the form is $(Q, S) R$ which is $= R(QS) + S(QR)$, and is thus a function of covariants of a lower degree; there remains only the derivatives formed with two of the functions f, ϕ, i, j, p, τ, or of one of these with α or γ. But these are all U other than the derivatives

$$(fj),\ (\phi j),\ (\phi p),\ (\phi\tau),\ (p\tau);\ (f\alpha),\ (\phi\alpha),\ (j\alpha),\ (p\alpha);\ (f\gamma),\ (\phi\gamma),\ (j\gamma),\ (p\gamma),\ (\tau\gamma),$$

and since $\gamma = (\tau\alpha)$, the derivatives containing γ will depend upon covariants of a lower degree; there remain therefore only (fj), (ϕj), (ϕp), $(\phi\tau)$, $(p\tau)$; $(f\alpha)$, $(\phi\alpha)$, $(j\alpha)$, $(p\alpha)$: each of these can be actually calculated in the form $F(U)$.

Hence finally, assuming that every covariant of a degree inferior to m is $F(U)$, it follows that every covariant of the degree m is $F(U)$; whence every covariant whatever is $F(U)$, viz. it is a rational and integral function of the 23 covariants U.

364. It will be observed that, writing A, B, C for P, Q, i, the proof depends on the theorems

$((AB), C)$,	a linear function of	$A(BC)^2$, $B(CA)^2$, $C(AB)^2$,
$(AB, C)^2$	" "	do. do. do.
$((AB), C)^2$	" "	$((AC)^2, B)$, $((BC)^2, A)$, $B(AC)^3$, $C(AB)^3$,

which are theorems relating to any three functions A, B, C whatever.

365. I remark upon the proof that the really fundamental theorem seems to be that which I have called theorem A. As to the forms W it is difficult to see *à priori* why such forms are to be considered, or what the essential property involved in their definition is; and in fact in a more recent paper, "Die simultanen Systeme binären Formen" (*Math. Annalen*, t. II. (1869), see p. 256), Professor Gordan has modified the definition of the forms W by omitting the condition that the order of the function shall exceed n; if it were possible further to omit the condition of at least one index being $=$ or $< \frac{1}{2}n$, and so only retain the conditions $n - i$, $n - j$, &c., each of them > 0, then the essential property of the forms W would be that any such form was a rational and integral function of the special covariants formed, as above, by means of the quantic of the next inferior order. And moreover, as regards the theorem B, there seems something indirect and artificial in the employment of such a property; one sees no reason why, when a system of irreducible covariants is once written down, it should not be possible to show that the derivatives of $F(U)$ with the original quantic f are each of them $F(U)$, instead of having to show this in regard to the derivatives of $F(U)$ with the several covariants χ: as regards the quintic, where there is a single covariant χ, the *quadric* function i, there is obviously a great abbreviation in this employment of i in place of f; but for the higher orders, assuming that the proof could be conducted by means of the quantic f itself, it does not appear that there would be even an abbreviation in the employment in its stead of the several covariants χ. The like remarks apply to the proof in the last-mentioned paper. I cannot but hope that a more simple proof of Professor Gordan's theorem will be obtained—a theorem the importance of which, in reference to the whole theory of forms, it is impossible to estimate too highly.

463.

NOTE ON A DIFFERENTIAL EQUATION.

[From the *Memoirs of the Literary and Philosophical Society of Manchester*, vol. II. (1865), pp. 111—114. Read February 18, 1862.]

THE following investigation was suggested to me by Mr Harley's "Remarks on the Theory of the Transcendental Solution of Algebraic Equations," communicated to the Society at the Meeting of the 4th of February.

Mr Harley's equation

$$y^n - ny + (n-1)x = 0$$

may be written

$$y = \frac{n-1}{n}x + \frac{1}{n}y^n;$$

or putting

$$\frac{n-1}{n}x = u,\ \frac{1}{n} = a,$$

it becomes

$$y = u + ay^n,$$

which equation may be considered instead of the original equation; and it is to be shown that y, regarded as a function of u, satisfies a certain linear differential equation of the order $n-1$. In fact, expanding y by Lagrange's theorem, we have

$$y = u + au^n + \frac{a^2}{1.2}(u^{2n})' + \frac{a^3}{1.2.3}(u^{3n})'' + \&\text{c.},$$

$$= u + au^n + \frac{a^2}{1.2}2n.u^{2n-1} + \frac{a^3}{1.2.3}3n(3n-1)u^{3n-2} + \&\text{c.},$$

the law whereof is obvious, and using the ordinary notation of factorials, viz. $[n]^r = n(n-1)\dots(n-r+1)$, we may write

$$y = S_\theta \cdot \frac{[n\theta]^{\theta-1}}{[\theta]^\theta} a^\theta u^{(n-1)\theta+1},$$

where θ extends from 0 to ∞.

It is now very easy to show that y satisfies the differential equation

$$\left[u\frac{d}{du}\right]^{n-1} y = na\left[\frac{n}{n-1}u\frac{d}{du} - \frac{2n-1}{n-1}\right]^{n-1} u^{n-1}y.$$

In fact, using on the left-hand side the foregoing value of y, and on the right-hand side the following value of $u^{n-1}y$, obtained from that of y by writing $\theta-1$ in the place of θ, viz.

$$u^{n-1}y = S_\theta \frac{[n\theta-n]^{\theta-2}}{[\theta-1]^{\theta-1}} a^{\theta-1} u^{(n-1)\theta+1},$$

and observing that in general the symbol $u\dfrac{d}{du}$, as regards u^m, is $=m$, the equation in question will be satisfied, if only

$$\frac{[n\theta]^{\theta-1}}{[\theta]^\theta}[(n-1)\theta+1]^{n-1} = \frac{n[n\theta-n]^{\theta-2}}{[\theta-1]^{\theta-1}}\left[\frac{n}{n-1}\left((n-1)\theta+1\right) - \frac{2n-1}{n-1}\right]^{n-1},$$

where the right-hand side is

$$= \frac{n[n\theta-n]^{\theta-2}}{[\theta-1]^{\theta-1}}[n\theta-1]^{n-1};$$

and the equation may be written

$$\frac{n\theta[n\theta-1]^{\theta-2}}{\theta[\theta-1]^{\theta-1}}[(n-1)\theta+1]^{n-1} = \frac{n[n\theta-n]^{\theta-2}}{[\theta-1]^{\theta-1}}[n\theta-1]^{n-1},$$

that is,

$$[n\theta-1]^{\theta-2}[(n-1)\theta+1]^{n-1} = [n\theta-1]^{n-1}[n\theta-n]^{\theta-2},$$

which, since each side of the equation is $=[n\theta-1]^{\theta+n-3}$, is obviously true.

The foregoing differential equation is developable in the form

$$\left\{\alpha_0 + \alpha_1 u\frac{d}{du} + \alpha_2 u^2\left(\frac{d}{du}\right)^2 \dots + \alpha_{n-1}u^{n-1}\left(\frac{d}{du}\right)^{n-1}\right\} y = \frac{1}{na}\left(\frac{d}{du}\right)^{n-1} y;$$

but to find the coefficients $\alpha_0, \alpha_1, \dots \alpha_{n-1}$, I start from this form, and proceed to substitute in the equation the value of y, which on the left-hand side I use in the original form, and on the right-hand side in the form obtained by writing $\theta+1$ in the place of θ, viz.

$$y = S_\theta \frac{[n(\theta+1)]^\theta}{[\theta+1]^{\theta+1}} a^{\theta+1} u^{(n-1)\theta+n}.$$

The equation to be satisfied is

$$\frac{[n\theta]^{\theta-1}}{[\theta]^{\theta}}\left(\alpha_0+\alpha_1[(n-1)\theta+1]^1+\alpha_2[(n-1)\theta+1]^2\ldots+\alpha_{n-1}[(n-1)\theta+1]^{n-1}\right)$$
$$=\frac{1}{n}\frac{[n(\theta+1)]^{\theta}}{[\theta+1]^{\theta+1}}[(n-1)\theta+n]^{n-1},$$

or, what is the same thing,

$$\frac{1}{[\theta]^{\theta}}\left(\alpha_0[n\theta]^{\theta-1}+\alpha_1[n\theta]^{\theta}+\alpha_2[n\theta]^{\theta+1}\ldots+\alpha_{n-1}[n\theta]^{\theta+n-2}\right)=\frac{1}{n}\frac{[n(\theta+1)]^{\theta+n-1}}{[\theta+1]^{\theta+1}}.$$

Observing that the right-hand side may be written

$$\frac{1}{n}\cdot\frac{n(\theta+1)[n\theta+n-1]^{\theta+n-2}}{(\theta+1)[\theta]^{\theta}},$$

the equation becomes

$$\alpha_0[n\theta]^{\theta-1}+\alpha_1[n\theta]^{\theta}+\alpha_2[n\theta]^{\theta+1}\ldots+\alpha_{n-1}[n\theta]^{\theta+n-2}=[n\theta+n-1]^{\theta+n-2},$$

or, what is the same thing,

$$\alpha_0+\alpha_1[(n-1)\theta+1]^1+\alpha_2[(n-1)\theta+1]^2\ldots+\alpha_{n-1}[(n-1)\theta+1]^{n-1}=[n\theta+n-1]^{n-1};$$

so that $\alpha_0, \alpha_1, \ldots \alpha_{n-1}$ are the coefficients of the expansion of $[n\theta+n-1]^{n-1}$ (which is a rational and integral function of θ, of the degree $n-1$) in a factorial series, as shown by the left-hand side of the equation.

To determine the actual values, write

$$(n-1)\theta+1=\phi,$$

this gives

$$n\theta+n-1=\frac{n\phi+n^2-3n+1}{n-1};$$

and we have therefore

$$\left[\frac{n\phi+n^2-3n+1}{n-1}\right]^{n-1}=\alpha_0+\alpha_1[\phi]^1+\alpha_2[\phi]^2\ldots+\alpha_{n-1}[\phi]^{n-1};$$

and thus the general expression is

$$\alpha_s=\frac{1}{[s]^s}\Delta^s\left(\frac{n\phi+n^2-3n+1}{n-1}\right),$$

where Δ denotes the difference in regard to ϕ ($\Delta U_\phi=U_{\phi+1}-U_\phi$), and, after the operation Δ^s is performed, ϕ is to be put equal to zero.

464.

NOTE ON PLANA'S LUNAR THEORY.

[From the *Monthly Notices of the Royal Astronomical Society*, vol. XXIII. (1862—1863), pp. 211—215.]

I HAVE been much surprised to find that there is an error of the order $m^2\gamma^4$, arising from the omission of a factor $(1+\gamma^2)^{-1}$, in the expression for $\dfrac{d^2\delta u}{dv^2}+\delta u$, as given by the equation (II.)' (*Théorie de la Lune*, t. I., p. 267), being the equation made use of in the theory for the determination of δu, the perturbation of the reciprocal of the radius vector. This error may probably be the cause of some of the discrepancies in the terms of the fourth and higher orders, between Plana's results and those of Pontécoulant and Delaunay.

Plana's equation (6), t. I., p. 260, is

$$\begin{aligned}\frac{d^2\delta u}{dv^2}+\delta u = \ & aR''\\ & + f(e,\ \gamma)\,Q'e\cos\left(cv-\int\varpi dv\right)\\ & - \left\{f(e,\ \gamma)(1+\gamma^2)(1+s_,^2)^{-\frac{3}{2}}-\frac{a(1+s^2)^{-\frac{3}{2}}}{a_,\psi(e,\ \gamma)}\right\}\\ & + f(e,\ \gamma)\,P\gamma^2(1+s_,^2)^{-\frac{3}{2}}\,\Theta,\end{aligned}$$

if for shortness

$$\Theta=\tfrac{3}{8}\gamma^2-(1+\tfrac{1}{2}\gamma^2)\cos\left(2gv-2\int\theta dv\right)+\tfrac{1}{8}\gamma^2\cos\left(4gv-4\int\theta dv\right).$$

R'' (p. 256) should be

$$R''=\frac{1}{\sigma a_,\psi(e,\ \gamma)}\,\frac{\Omega_2-\dfrac{du}{dv}\dfrac{u^2dv}{d\Omega}-\sigma(1+s^2)^{-\frac{3}{2}}\,2\displaystyle\int Udv}{1+2\displaystyle\int Udv},$$

but, by an error which is implicitly corrected, the σ which multiplies $(1+s^2)^{-\frac{3}{2}}\, 2\int U dv$ is omitted. Hence the equation (6) becomes

$$\begin{aligned}(1+2\int U dv)\left(\frac{d^2\delta u}{dv^2}+\delta u\right) = {} & \frac{\mathrm{a}}{\sigma \mathrm{a}_{,}\, \psi(e,\gamma)}\left\{\Omega_2 - \frac{du}{dv}\,\frac{d\Omega}{u^2 dv} - \sigma(1+s^2)^{-\frac{3}{2}}\, 2\int U dv\right\} \\ & + (1+2\int U dv) f(e,\gamma)\, Q'e \cos\left(cv - \int \varpi dv\right) \\ & - (1+2\int U dv)\left\{f(e,\gamma)(1+\gamma^2)(1+s_{,}^2)^{-\frac{3}{2}} - \frac{\mathrm{a}(1+s^2)^{-\frac{3}{2}}}{\mathrm{a}_{,}\psi(e,\gamma)}\right\} \\ & + (1+2\int U dv) f(e,\gamma)\, P\gamma^2 (1+s_{,}^2)^{-\frac{3}{2}}\, \Theta,\end{aligned}$$

in which equation

$$\Omega_2 = \frac{d\Omega}{du} + \frac{s}{u}\,\frac{d\Omega}{ds}, \text{ pp. 26, 245,} \quad \frac{d\Omega}{u^2 dv} = \frac{\sigma \mathrm{a}_{,}}{\lambda^{\frac{2}{3}}(1+\gamma^2)^{\frac{4}{3}}}\, U, \text{ p. 265,}$$

$$\psi(e,\gamma) = \lambda^{-\frac{2}{3}}(1+\gamma^2)^{-\frac{4}{3}}, \quad f(e,\gamma) = \lambda^{\frac{2}{3}}(1+\gamma^2)^{\frac{4}{3}}, \text{ p. 261.}$$

But retaining for greater convenience the function $f(e, \gamma)$ in two of the terms, we have

$$\begin{aligned}(1+2\int U dv)\left(\frac{d^2\delta u}{dv^2}+\delta u\right) = {} & \\ & \frac{\mathrm{a}}{\sigma \mathrm{a}_{,}}\lambda^{\frac{2}{3}}(1+\gamma^2)^{\frac{4}{3}}\left\{\frac{d\Omega}{du} + \frac{s}{u}\,\frac{d\Omega}{ds} - \frac{\sigma \mathrm{a}_{,}}{\lambda^{\frac{2}{3}}(1+\gamma^2)^{\frac{4}{3}}}\, U\frac{du}{dv} - \sigma(1+s^2)^{-\frac{3}{2}}\, 2\int U dv\right\} \\ & + (1+2\int U dv) f(e,\gamma)\, Q'e\cos\left(cv - \int\varpi dv\right) \\ & - (1+2\int U dv)\,\lambda^{\frac{2}{3}}(1+\gamma^2)^{\frac{4}{3}}\left\{(1+s_{,}^2)^{-\frac{3}{2}} - \frac{\mathrm{a}}{\mathrm{a}_{,}}(1+s^2)^{-\frac{3}{2}}\right\} \\ & + (1+2\int U dv) f(e,\gamma)\, P\gamma^2(1+s_{,}^2)^{-\frac{3}{2}}\,\Theta \\ = {} & \frac{\mathrm{a}}{\sigma \mathrm{a}_{,}}\lambda^{\frac{2}{3}}(1+\gamma^2)^{\frac{4}{3}}\left(\frac{d\Omega}{du} + \frac{s}{u}\,\frac{d\Omega}{ds}\right) \\ & - \mathrm{a} U\frac{du}{dv} \\ & - \frac{\mathrm{a}}{\mathrm{a}_{,}}\lambda^{\frac{2}{3}}(1+\gamma^2)^{\frac{4}{3}}(1+s^2)^{-\frac{3}{2}}\, 2\int U dv \\ & + (1+2\int U dv) f(e,\gamma)\, Q'e\cos\left(cv - \int\varpi dv\right) \\ & - (1+2\int U dv)\,\lambda^{\frac{2}{3}}(1+\gamma^2)^{\frac{4}{3}}\left\{(1+s_{,}^2)^{-\frac{3}{2}} - \frac{\mathrm{a}}{\mathrm{a}_{,}}(1+s^2)^{-\frac{3}{2}}\right\} \\ & + (1+2\int U dv) f(e,\gamma)\, P\gamma^2(1+s_{,}^2)^{-\frac{3}{2}}\,\Theta,\end{aligned}$$

which is

$$
\begin{aligned}
= \; & \frac{\mathrm{a}}{\sigma \mathrm{a}_{,}} \lambda^{\frac{2}{3}} (1+\gamma^2)^{\frac{4}{3}} \left(\frac{d\Omega}{du} + \frac{s}{u} \frac{d\Omega}{dv} \right) \\
& - \mathrm{a} U \frac{du}{dv} \\
& - \frac{\mathrm{a}}{\mathrm{a}_{,}} \lambda^{\frac{2}{3}} (1+\gamma^2)^{\frac{4}{3}} (1+s^2)^{-\frac{3}{2}} \, 2 \int U dv \text{ (destroyed by term } \textit{infrà}) \\
& + \left(1 + 2 \int U dv\right) f(e, \gamma) \, Q'e \cos \left(cv - \int \varpi dv\right) \\
& - \lambda^{\frac{2}{3}} (1+\gamma^2)^{\frac{4}{3}} \left\{ (1+s_{,}^2)^{-\frac{3}{2}} - \frac{\mathrm{a}}{\mathrm{a}_{,}} (1+s^2)^{-\frac{3}{2}} \right\} \\
& - \lambda^{\frac{2}{3}} (1+\gamma^2)^{\frac{4}{3}} (1+s_{,}^2)^{-\frac{3}{2}} \, 2 \int U dv \\
& + \frac{\mathrm{a}}{\mathrm{a}_{,}} \lambda^{\frac{2}{3}} (1+\gamma^2)^{\frac{4}{3}} (1+s^2)^{-\frac{3}{2}} \, 2 \int U dv \text{ (destroyed by term } \textit{suprà}) \\
& + \left(1 + 2 \int U dv\right) f(e, \gamma) \, P \gamma^2 (1+s_{,}^2)^{-\frac{3}{2}} \Theta ;
\end{aligned}
$$

or, putting $u = \frac{1}{\mathrm{a}} (u_{,} + \delta u)$, this becomes

$$
\begin{aligned}
\left(1 + 2 \int U dv\right) \left(\frac{d^2 \delta u}{dv^2} + \delta u \right) = \; & \frac{\mathrm{a}}{\sigma \mathrm{a}_{,}} \lambda^{\frac{2}{3}} (1+\gamma^2)^{\frac{4}{3}} \left(\frac{d\Omega}{du} + \frac{s}{u} \frac{d\Omega}{dv} \right) \\
& - U \left(\frac{du_{,}}{dv} + \frac{d\delta u}{dv} \right) \\
& - \lambda^{\frac{2}{3}} (1+\gamma)^{\frac{4}{3}} (1+s_{,}^2)^{-\frac{3}{2}} \, 2 \int U dv \\
& + \left(1 + 2 \int U dv\right) f(e, \gamma) \, Q'e \cos \left(cv - \int \varpi dv\right) \\
& - \lambda^{\frac{2}{3}} (1+\gamma^2)^{\frac{4}{3}} \left\{ (1+s_{,}^2)^{-\frac{3}{2}} - \frac{\mathrm{a}}{\mathrm{a}_{,}} (1+s^2)^{-\frac{3}{2}} \right\} \\
& + \left(1 + 2 \int U dv\right) f(e, \gamma) \, P \gamma^2 (1+s_{,}^2)^{-\frac{3}{2}} \Theta ,
\end{aligned}
$$

agreeing with the Formula II. p. 265, except that in Plana's last term, instead of the factor $f(e, \gamma) \left(= \lambda^{\frac{2}{3}} (1+\gamma^2)^{\frac{1}{3}}\right)$, we have the factor $\lambda^{\frac{2}{3}} (1+\gamma^2)^{\frac{4}{3}}$. That is, the last term, as given by Plana, should be divided by $1+\gamma^2$. And this error is introduced from the formula (II.) into the formula (II.)′, p. 267, viz. the incorrect factor $\lambda^{\frac{2}{3}}(1+\gamma^2)^{\frac{4}{3}}$ is there replaced by its value q; whereas, the true value being $\lambda^{\frac{2}{3}}(1+\gamma^2)^{\frac{1}{3}}$, the factor in (II.)′ should be $=\dfrac{q}{1+\gamma^2}$.

The corrected formula (II.)′ is

$$
\begin{aligned}
-\frac{d^2\delta u}{dv^2} - \delta u = &- Q'\frac{qe}{1+\gamma^2}\cos\left(cv - \int \varpi dv\right) \\
&+ q\left\{(1+s_{,}^{2})^{-\frac{3}{2}} - \frac{\mathrm{a}}{\mathrm{a}_{,}}(1+s^2)^{-\frac{3}{2}}\right\} \\
&+ \mu^2(R_4+R_5) - \mu^2\left(\frac{du_{,}}{dv} + \frac{d\delta u}{dv}\right)R_{,} \\
&- 2\mu^2\left\{\frac{d^2\delta u}{dv^2} + \delta u + q(1+s_{,}^{2})^{-\frac{3}{2}} - \frac{Q'qe}{1+\gamma^2}\cos\left(cv - \int \varpi dv\right)\right\}\int R_{,} dv \\
&- Pq\gamma^3(1+\gamma^2)^{-1}(1+s_{,}^{2})^{-\frac{3}{2}}\left(1 - 2\mu^2\int R_{,} dv\right)\times \\
&\quad \left\{\tfrac{3}{8}\gamma^2 - (1+\tfrac{1}{2}\gamma^2)\cos\left(2gv - 2\int\theta dv\right) + \tfrac{1}{8}\gamma^2\cos\left(4gv - 4\int\theta dv\right)\right\}.
\end{aligned}
$$

Observing that P is of the order m^2, and that q is approximately equal to unity, the error in $\dfrac{d^2\delta u}{dv^2} + \delta u$ is of the order $m^2\gamma^4$, as noticed above. It may be right to mention that I obtained the correction in the first instance by starting from the fundamental equations, and not as here from the intermediate equation (6), so that there is not in that equation any error afterwards implicitly corrected in the transformation to (II.)′.

465.

NOTE ON THE LUNAR THEORY.

[From the *Monthly Notices of the Royal Astronomical Society*, vol. xxv. (1864—1865), pp. 182—189.]

I ATTEND, in the expressions for the lunar coordinates, only to the coefficients independent of m. Plana's values, taken to the fourth order only, are as follows; for greater simplicity I write $a=1$; and, instead of $nt+$ constant, $cnt+$ constant, $gnt+$ constant, I write l, c, g respectively; viz., l is the mean longitude, c the mean anomaly, g the mean distance from node: this being so, then r, v, y, denoting the radius vector longitude and latitude respectively, we have

$$\frac{1}{r}\,(\text{Plana}) =$$

$$\begin{array}{llll}
 & e - \tfrac{1}{8}e^3 - \tfrac{1}{4}\gamma^2 e & \cos & c \\
+ & e^2 - \tfrac{1}{3}e^4 - \tfrac{1}{2}\gamma^2 e^2 & ,, & 2c \\
+ & \tfrac{9}{8}e^3 & ,, & 3c \\
+ & \tfrac{4}{3}e^4 & ,, & 4c \\
- & \tfrac{5}{4}\gamma^2 e^2 & ,, & 2g \\
- & \tfrac{5}{8}\gamma^2 e & ,, & c-2g
\end{array}$$

(but I omit Plana's term $+\tfrac{1}{8}\gamma^2 e^2 \quad \cos \quad 2c+2g$ which should be $=0$).

$$v\,(\text{Plana}) = l +$$

$$\begin{array}{llll}
+ & 2e - \tfrac{1}{4}e^3 - \tfrac{1}{2}\gamma^2 e & \sin & c \\
+ & \tfrac{5}{4}e^2 - \tfrac{11}{24}e^4 - \tfrac{15}{16}\gamma^2 e^2 & ,, & 2c \\
+ & \tfrac{13}{12}e^3 & ,, & 3c \\
+ & \tfrac{103}{96}e^4 & ,, & 4c \\
- & \tfrac{1}{4}\gamma^2 - \tfrac{9}{16}\gamma^2 e^2 + \tfrac{1}{8}\gamma^4 & ,, & 2g \\
+ & \tfrac{3}{4}\gamma^2 e & ,, & c-2g
\end{array}$$

$-\frac{1}{2}\gamma^2 e$	sin	$c+2g$
$-\frac{1}{8}\gamma^2 e^2$	„	$2c-2g$
$-\frac{13}{16}\gamma^2 e^2$	„	$2c+2g$
$+\frac{1}{32}\gamma^4$	„	$4g.$

y (Plana) =

$\gamma - \gamma e^2 - \frac{3}{8}\gamma^3$	sin	g
$+ \gamma e - \frac{5}{8}\gamma e^3$	„	$c-g$
$+ \gamma e - \frac{5}{4}\gamma e^3 - \frac{5}{8}\gamma^3 e$	„	$c+g$
$+\frac{3}{4}\gamma e^2$	„	$2c-g$
$+\frac{9}{8}\gamma e^2$	„	$2c+g$
$+\frac{17}{24}\gamma e^3$	„	$3c-g$
$+\frac{4}{3}\gamma e^3$	„	$3c+g$
$-\frac{1}{24}\gamma^3$	„	$3g$
$+\frac{1}{2}\gamma^3 e$	„	$c-3g$
$-\frac{1}{8}\gamma^3 e$	„	$c+3g.$

To compare these with the elliptic values, it is necessary to write $e(1+\frac{1}{4}\gamma^2)$ in place of e. Making this change, or say reducing Plana's (e, γ) to the elliptic (e, γ), I write down in a first column the transformed coefficients, and in a second column the elliptic coefficients, as follows:

Plana, with Elliptic e, γ	Elliptic		
$\frac{1}{r}=$	$\frac{1}{r}=$		
1	1		
$+ e - \frac{1}{8}e^3$	$+ e - \frac{1}{8}e^3$	cos	c
$+ e^2 - \frac{1}{3}e^4$	$+ e^2 - \frac{1}{3}e^4$	„	$2c$
$+\frac{9}{8}e^3$	$+\frac{9}{8}e^3$	„	$3c$
$+\frac{4}{3}e^4$	$+\frac{4}{3}e^4$	„	$4c$
$-\frac{5}{4}\gamma^2 e^2$	0	„	$2g$
$-\frac{5}{8}\gamma^2 e$	0	„	$c-2g.$

Plana, with Elliptic e, γ	Elliptic		
$v=$	$v=$		
l	l		
$+ 2e - \frac{1}{4}e^3$	$+ 2e - \frac{1}{4}e^3$	sin	c
$+\frac{5}{4}e^2 - \frac{11}{24}e^4 - \frac{5}{16}\gamma^2 e^2$	$+\frac{5}{4}e^2 - \frac{11}{24}e^4$	„	$2c$
$+\frac{13}{12}e^3$	$+\frac{13}{12}e^3$	„	$3c$
$+\frac{103}{96}e^4$	$+\frac{103}{96}e^4$	„	$4c$
$-\frac{1}{4}\gamma^2 - \frac{9}{16}\gamma^2 e^2 + \frac{1}{8}\gamma^4$	$-\frac{1}{4}\gamma^2 + \gamma^2 e^2 + \frac{1}{8}\gamma^4$	„	$2g$
$+\frac{3}{4}\gamma^2 e$	$-\frac{1}{2}\gamma^2 e$	„	$c-2g$
$-\frac{1}{2}\gamma^2 e$	$-\frac{1}{2}\gamma^2 e$	„	$c+2g$
$-\frac{1}{8}\gamma^2 e^2$	$+\frac{3}{16}\gamma^2 e^2$	„	$2c-2g$
$-\frac{13}{16}\gamma^2 e^2$	$-\frac{13}{16}\gamma^2 e^2$	„	$2c+2g$
$+\frac{1}{32}\gamma^4$	$+\frac{1}{32}\gamma^4$	„	$4g.$

Plana, with Elliptic e, γ	Elliptic		
$y=$	$y=$		
$\gamma - \gamma e^2 - \frac{3}{8}\gamma^3$	$+ \gamma e - \gamma e^2 - \frac{3}{8}\gamma^3$	sin	g
$+ \gamma e - \frac{5}{4}\gamma e^3 - \frac{3}{8}\gamma^3 e$	$+ \gamma e - \frac{5}{4}\gamma e^3 - \frac{3}{8}\gamma^3 e$	,,	$c+g$
$+ \gamma e - \frac{5}{8}\gamma e^3 + \frac{1}{4}\gamma^3 e$	$+ \gamma e \quad - \frac{3}{8}\gamma^3 e$	,,	$c-g$
$+ \frac{3}{4}\gamma e^2$	$+ \frac{1}{8}\gamma e^2$	,,	$2c-g$
$+ \frac{9}{8}\gamma e^2$	$+ \frac{9}{8}\gamma e^2$	,,	$2c+g$
$+ \frac{17}{24}\gamma e^3$	$+ \frac{1}{12}\gamma e^3$	,,	$3c-g$
$+ \frac{4}{3}\gamma e^3$	$+ \frac{4}{3}\gamma e^3$	,,	$3c+g$
$- \frac{1}{24}\gamma^3$	$- \frac{1}{24}\gamma^3$	,,	$3g$
$+ \frac{1}{2}\gamma^3 e$	$- \frac{1}{8}\gamma^3 e$	,,	$c-3g$
$- \frac{1}{8}\gamma^3 e$	$- \frac{1}{8}\gamma^3 e$	,,	$c+3g$,

where, for greater clearness, I remark that the values called "elliptic" of e, γ, c, g, refer to an ellipse, such that the longitude of the node, and the longitude (in orbit) of the pericentre, vary uniformly with the time,—viz., we have mean distance $=1$, excentricity $=e$, tangent of inclination $=\gamma$, mean longitude $=l$, mean anomaly $=c$, distance from node $=g$.

We have therefore

$$
\begin{array}{rlll}
\delta\dfrac{1}{r} = & -\frac{5}{4}\gamma^2 e^2 & \cos & 2g \\
& -\frac{5}{8}\gamma^2 e & \text{,,} & c-2g \\
\delta v = & -\frac{5}{16}\gamma^2 e^2 & \sin & 2c \\
& -\frac{25}{16}\gamma^2 e^2 & \text{,,} & 2g \\
& +\frac{5}{4}\gamma^2 e & \text{,,} & c-2g \\
& -\frac{5}{16}\gamma^2 e^2 & \text{,,} & 2c-2g \\
\delta y = & -\frac{5}{8}\gamma e^3 + \frac{5}{8}\gamma^3 e & \text{,,} & c-g \\
& +\frac{5}{8}\gamma e^2 & \text{,,} & 2c-g \\
& +\frac{5}{8}\gamma e^3 & \text{,,} & 3c-g \\
& +\frac{5}{8}\gamma^3 e & \text{,,} & c-3g,
\end{array}
$$

viz., these are the increments to be added to the elliptic values of $\frac{1}{r}$, v, y, respectively, in order to obtain the disturbed values of $\frac{1}{r}$, v, y, attending only to the coefficients independent of m; *they represent, in fact, the lunar inequalities which rise two orders by integration.*

The elliptic values of $\frac{1}{r}$ and y are functions, and that of v, is equal $l+$, a function, of e, γ, c, g, and the foregoing disturbed values may be obtained by affecting each of the quantities e, γ, c, g, and l, with an inequality depending on the argument $2c-2g$, viz., these inequalities are

$$\begin{aligned}\delta e &= -\tfrac{5}{8}\gamma^2 e \cos 2c-2g\\ \delta c &= \tfrac{5}{8}\gamma^2 \sin 2c-2g\\ \delta\gamma &= \tfrac{5}{8}\gamma e^2 \cos 2c-2g\\ \delta g &= \tfrac{5}{8} e^2 \sin 2c-2g\\ \delta l &= -\tfrac{5}{16}\gamma^2 e^2 \sin 2c-2g.\end{aligned}$$

The verification may be effected without difficulty; thus, for instance, starting from the elliptic value of $\frac{1}{r}$, we have to the fourth order

$$\begin{aligned}\delta\frac{1}{r} = \delta\begin{pmatrix} e^2\cos c\\ +e\cos 2c\end{pmatrix} &= \begin{pmatrix}-e\sin c\\ -2e^2\sin 2c\end{pmatrix}\delta c + \begin{pmatrix}\cos c\\ +2e\cos 2c\end{pmatrix}\delta e\\ &= \tfrac{5}{8}\gamma^2 e\,(-\sin c\sin 2c-2g-\cos c\cos 2c-2g)\\ &\quad+\tfrac{5}{4}\gamma^2 e^2(-\sin 2c\sin 2c-2g-\cos 2c\cos 2c-2g)\\ &= -\tfrac{5}{8}\gamma^2 e\cos c-g\\ &\quad-\tfrac{5}{4}\gamma^2 e^2\cos 2g,\end{aligned}$$

which is right; and the verification of the values of δv, δy, may be effected in a similar manner.

I have, in order to fix the ideas, preferred to give in the first instance the foregoing *à posteriori* proof; but I now inquire generally as to the form of the values of $\frac{1}{r}$, v, y, or say of r, v, y, taking account only of coefficients independent of m; and I proceed to show that these may be obtained from the elliptic values expressed as above in terms of l, e, γ, c, g, by affecting l, e, γ, c, g, each with an inequality depending on the multiple sines or cosines of $c-g$.

Writing for greater simplicity $n=1$, we have $l=t+L$, $c=\mathrm{c}t+C$, $g=\mathrm{g}t+G$, where $\mathrm{c}=1-\frac{3}{4}m^2+\&c.$, $\mathrm{g}=1+\frac{3}{4}m^2+\&c.$; viz., c, g, are constants which differ from unity by terms involving m^2.

The required values of r, v, y, satisfy the *undisturbed* equations of motion, if after the differentiations we write in the coefficients (which coefficients are functions of m through c, g) $m=0$; that is, if we write in the coefficients $\mathrm{c}=1$, $\mathrm{g}=1$. In fact, the required values of r, v, y, are what the complete values become, upon writing in the coefficients of the complete values $m=0$; that is, the required values of r, v, y, differ from the complete values by terms the coefficients whereof contain m as a factor; and the disturbed equations differ from the undisturbed equations in that they contain the differential coefficients of the disturbing function; that is, terms the coefficients whereof have the factor m^2. Imagine the complete values of r, v, y, substituted in the disturbed equations of motion; the resulting equations are satisfied identically; and, therefore, whatever be the value of m; that is, they are satisfied if in these equations respectively we write $m=0$: it requires a little consideration to see that this is so, if *in the coefficients only* we write $m=0$; but recollecting that c, g, stand for functions $\mathrm{c}t+C$, $\mathrm{g}t+G$, so that, for example, $c-g$, $=(\mathrm{c}-\mathrm{g})t+C-G$, upon writing therein $m=0$, becomes equal, not to zero, but to the constant value $C-G$, the identity subsists in regard to the coefficient of the sine or cosine of each separate argument $\alpha c+\beta g$, and, consequently, it subsists notwithstanding that in the arguments c and g, instead of being each put $=1$, are left indeterminate. And granting this (viz. that the equations are satisfied if *in the coefficients only* we write $m=0$), then it is clear that, as above stated, the required values of r, v, y, satisfy the undisturbed equations of motion, if after the differentiations we write in the coefficients $\mathrm{c}=1$, $\mathrm{g}=1$.

The required values of r, v, y, are of the form $r=\phi(c, g)$, $y=\psi(c, g)$, $v=l+\chi(c, g)$, but writing $w=v+c-l$, $=c+\chi(c, g)$, the last mentioned property will equally subsist in regard to the functions r, w, y: in fact, v enters into the differential equations only through its differential coefficient $\frac{dv}{dt}$, and the differential coefficients of v and w, that is, of $l+\chi(c, g)$ and $c+\chi(c, g)$, differ only by the quantity $\mathrm{c}-1$, which becomes $=0$, in virtue of the assumed relations $\mathrm{c}=1$, $\mathrm{g}=1$.

Hence the undisturbed equations are satisfied by the values $r=\phi(c, g)$, $y=\psi(c, g)$, $w=c+\chi(c, g)$, when after the differentiations we write in the coefficients $\mathrm{c}=1$, $\mathrm{g}=1$; the foregoing values contain t through the quantities c, g, only; and we have, therefore, $\frac{d}{dt}=\mathrm{c}\frac{d}{dc}+\mathrm{g}\frac{d}{dg}$.

Hence, writing in the coefficients $\mathrm{c}=1$, $\mathrm{g}=1$, we have $\frac{d}{dt}=\frac{d}{dc}+\frac{d}{dg}$; that is, the values $r=\phi(c, g)$, $y=\psi(c, g)$, $w=\chi(c, g)$, regarding r, v, y, as functions of c, g, satisfy the partial differential equations obtained from the undisturbed equations of motion by writing therein $\frac{d}{dc}+\frac{d}{dg}$ in place of $\frac{d}{dt}$. Hence also, considering r, w, y, as functions of c and $c-g$, then observing that $\left(\frac{d}{dc}+\frac{d}{dg}\right)(c-g)$ is $=0$, the values of r, v, y, satisfy the partial differential equations obtained by writing $\frac{d}{dc}$ in place of $\frac{d}{dt}$; and inasmuch as these partial differential equations do not contain $\frac{d}{dg}$, they are to be integrated as ordinary differential equations in regard to c as the independent variable, the constants of integration being replaced by arbitrary functions of $c-g$.

Consider the pure elliptic values of r, v, y, in an elliptic orbit with the following elements: A, the mean distance; N, the mean motion ($N^2A^3=1$ and therefore $A=N^{-\frac{2}{3}}$); E, the excentricity; $Nt+D$, the mean anomaly; $Nt+H$, the mean distance from node; $Nt+K$, the mean longitude; then writing c in place of t, we have

$$\begin{aligned} r &= N^{-\frac{2}{3}}\,\mathrm{elqr}\,(E,\ Nc+D), \\ v(=l-c+w) &= l-c+Nc+K+P\,(E,\ \Gamma,\ Nc+D,\ Nc+H), \\ y &= Q\,(E,\ \Gamma,\ Nc+D,\ Nc+H), \end{aligned}$$

where N, E, Γ, D, H, K, are arbitrary functions of $c-g$: P and Q denote given functional expressions. But, in order that r, v, y, considered as functions of c and g may be of the proper form, it is necessary as regards N to write simply $N=1$; we have then

$$\begin{aligned} r &= \text{elqr}\,(E,\ c+D), \\ v &= l+K+P\,(E,\ \Gamma,\ c+D,\ c+H), \\ y &= \qquad\quad Q\,(E,\ \Gamma,\ c+D,\ c+H), \end{aligned}$$

where E, Γ, D, H, K, are arbitrary functions of $c-g$; or, what is the same thing, writing for these quantities respectively $e+\delta e$, $\gamma+\delta\gamma$, δc, $g-c+\delta g$, δl, where δe, $\delta\gamma$, δc, δg, δl are arbitrary functions of $c-g$, we have

$$\begin{aligned} r &= \text{elqr}\,(e+\delta e,\ c+\delta c), \\ v &= l+\delta l+P\,(e+\delta e,\ \gamma+\delta\gamma,\ c+\delta c,\ g+\delta g), \\ y &= \qquad\quad Q\,(e+\delta e,\ \gamma+\delta\gamma,\ c+\delta c,\ g+\delta g), \end{aligned}$$

that is, the values of r, v, y, are obtained from the elliptic values

$$\begin{aligned} r &= \text{elqr}\,(e,\ c), \\ v &= l+P\,(e,\ \gamma,\ c,\ g), \\ y &= \quad\ Q\,(e,\ \gamma,\ c,\ g), \end{aligned}$$

by affecting each of the quantities e, γ, c, g, l, with an inequality which is a function of $c-g$.

466.

SECOND NOTE ON THE LUNAR THEORY.

[From the *Monthly Notices of the Royal Astronomical Society*, vol. XXV. (1864—1865), pp. 203—207.]

THE elliptic values of

r, the radius vector,
v, the longitude,
y, the latitude,

are functions of

a, the mean distance,
e, the excentricity,
γ, the tangent of the inclination,
l, the mean longitude,
c, the mean anomaly,
g, the mean distance from node;

see my Note in the last *Monthly Notice*, p. 182, [465], where, for the present purpose, $\frac{a}{r}$ should be written instead of $\frac{1}{r}$; and it is there shown that the disturbed values, attending only to the coefficients independent of m, are obtained by affecting a, e, γ, c, g, l, with the inequalities

$$\begin{array}{llll}
\delta a = & 0 & & \\
\delta e = & -\tfrac{5}{8}\gamma^2 e & \cos & 2c-2g \\
\delta\gamma = & +\tfrac{5}{8}\gamma e^2 & \text{,,} & 2c-2g \\
\delta c = & +\tfrac{5}{8}\gamma^2 & \sin & 2c-2g \\
\delta g = & +\tfrac{5}{8}e^2 & \text{,,} & 2c-2g \\
\delta l = & -\tfrac{5}{16}\gamma^2 e^2 & \text{,,} & 2c-2g,
\end{array}$$

or, what is the same thing, adding to the elliptic values the inequalities

$$\begin{array}{lll}
\delta \frac{a}{r} = -\tfrac{5}{4}\gamma^2 e^2 & \cos & 2g \\
\quad -\tfrac{5}{8}\gamma^2 e & ,, & c-2g, \\
\delta v = -\tfrac{5}{16}\gamma^2 e^2 & \sin & 2c \\
\quad -\tfrac{25}{16}\gamma^2 e^2 & ,, & 2g \\
\quad +\tfrac{5}{4}\gamma^2 e & ,, & c-2g \\
\quad -\tfrac{5}{16}\gamma^2 e^2 & ,, & 2c-2g, \\
\delta y = -\tfrac{5}{8}\gamma e^3 + \tfrac{5}{8}\gamma^3 e & \sin & c-g \\
\quad +\tfrac{5}{8}\gamma e^2 & ,, & 2c-g \\
\quad +\tfrac{5}{8}\gamma e^3 & ,, & 3c-g \\
\quad +\tfrac{5}{8}\gamma^3 e & ,, & c-3g.
\end{array}$$

I propose to show how these results may be obtained by the method of the variation of the elements. For this purpose, treating a, e, γ, c, g, l, as elements, the proper formulæ are obtained very readily from those given in my "Memoir on the Problem of Disturbed Elliptic Motion," *Mem. R. Ast. Soc.*, vol. XXVII. (1859), pp. 1—29, [212]; viz., writing c in place of g, the formulæ, p. 25, give the variations of a, e, c, ϖ, θ, ϕ; we have then

$$g = c + \varpi$$
$$l = c + \varpi + \theta$$
$$\gamma = \tan\phi,$$

and therefore

$$dg = dc + d\varpi$$
$$dl = dc + d\varpi + d\theta$$
$$d\gamma = (1+\gamma^2)\, d\phi,$$

which give for the transformation of the differential coefficients of Ω,

$$\frac{d\Omega}{dc} = \frac{d\Omega}{dc} + \frac{d\Omega}{dg} + \frac{d\Omega}{dl}$$

$$\frac{d\Omega}{d\varpi} = \frac{d\Omega}{dg} + \frac{d\Omega}{dl}$$

$$\frac{d\Omega}{d\theta} = \frac{d\Omega}{dl}$$

$$\frac{d\Omega}{d\phi} = (1+\gamma^2)\frac{d\Omega}{d\gamma}$$

and the formulæ finally become

$$\frac{da}{dt} = \frac{2}{na}\frac{d\Omega}{dc} + \frac{2}{na}\frac{d\Omega}{dg} + \frac{2}{na}\frac{d\Omega}{dl},$$

$$\frac{de}{dt} = \frac{1-e^2}{na^2e}\frac{d\Omega}{dc} + \frac{1-e^2-\sqrt{1-e^2}}{na^2e}\frac{d\Omega}{dg} + \frac{1-e^2-\sqrt{1-e^2}}{na^2e}\frac{d\Omega}{dl},$$

$$\frac{d\gamma}{dt} = \frac{1+\gamma^2}{na^2\sqrt{1-e^2}\,\gamma}\,\frac{d\Omega}{dg} + \frac{(1+\gamma^2)(1-\sqrt{1+\gamma^2})}{na^2\sqrt{1-e^2}\,\gamma}\,\frac{d\Omega}{dl},$$

$$\frac{dc}{dt} = -\frac{2}{na}\,\frac{d\Omega}{da} - \frac{1-e^2}{na^2e}\,\frac{d\Omega}{de},$$

$$\frac{dg}{dt} = -\frac{2}{na}\,\frac{d\Omega}{da} - \frac{1-e^2-\sqrt{1-e^2}}{na^2e}\,\frac{d\Omega}{de} - \frac{1+\gamma^2}{na^2\sqrt{1-e^2}\,\gamma}\,\frac{d\Omega}{d\gamma},$$

$$\frac{dl}{dt} = -\frac{2}{na}\,\frac{d\Omega}{da} - \frac{1-e^2-\sqrt{1+e^2}}{na^2e}\,\frac{d\Omega}{de} - \frac{(1+\gamma^2)(1-\sqrt{1+\gamma^2})}{na^2\sqrt{1-e^2}\,\gamma}\,\frac{d\Omega}{d\gamma}.$$

The disturbing function contains the term

$$m^2n^2a^2\left(+\tfrac{15}{16}e^2\gamma^2\right)\quad \cos\quad 2c-2g.$$

If after the differentiations we write for greater simplicity $a=1$, $n=1$, we have

$$\begin{aligned}
\frac{d\Omega}{da} &= +\tfrac{15}{8}m^2e^2\gamma^2 \quad \cos \quad 2c-2g,\\
\frac{d\Omega}{de} &= +\tfrac{15}{8}m^2e\,\gamma^2 \quad ,, \quad 2c-2g,\\
\frac{d\Omega}{d\gamma} &= +\tfrac{15}{8}m^2e^2\gamma \quad ,, \quad 2c-2g,\\
\frac{d\Omega}{dc} &= -\tfrac{15}{8}m^2e^2\gamma^2 \quad \sin \quad 2c-2g,\\
\frac{d\Omega}{dg} &= -\tfrac{15}{8}m^2e^2\gamma^2 \quad ,, \quad 2c-2g,\\
\frac{d\Omega}{dl} &= 0,
\end{aligned}$$

and the formulæ for the variations give

$$\begin{aligned}
\frac{da}{dt} &= 2\left(\frac{d\Omega}{dc}+\frac{d\Omega}{dg}\right) &&= 0\\
\frac{de}{dt} &= \frac{1}{e}\,\frac{d\Omega}{dc} &&= -\tfrac{15}{8}m^2e\gamma^2 \quad \sin \quad 2c-2g,\\
\frac{d\gamma}{dt} &= \frac{1}{\gamma}\,\frac{d\Omega}{dg} &&= -\tfrac{15}{8}m^2e^2\gamma \quad ,, \quad 2c-2g,\\
\frac{dc}{dt} &= -\frac{1}{e}\,\frac{d\Omega}{de} &&= -\tfrac{15}{8}m^2\gamma^2 \quad \cos \quad 2c-2g,\\
\frac{dg}{dt} &= -\frac{1}{\gamma}\,\frac{d\Omega}{d\gamma} &&= -\tfrac{15}{8}m^2e^2 \quad ,, \quad 2c-2g,\\
\frac{dl}{dt} &= -2\,\frac{d\Omega}{da}+\tfrac{1}{2}e\,\frac{d\Omega}{de}+\tfrac{1}{2}\gamma\,\frac{d\Omega}{d\gamma} &&= \left(-\tfrac{15}{4}+\tfrac{15}{16}+\tfrac{15}{16}=\right)-\tfrac{15}{8}m^2e^2\gamma^2 \quad ,, \quad 2c-2g,
\end{aligned}$$

but this value of $\frac{dl}{dt}$ is, as will presently be seen, incomplete.

Writing $a+\delta a$, $e+\delta e$, &c., in place of a, e, &c., and observing that the divisor for the integration of the term in $2c-2g$ is $2(c-g)$, $=-3m^2$, the first five equations give respectively

$$\begin{aligned}
\delta a &= 0, \\
\delta e &= -\tfrac{5}{8}\gamma^2 e \quad \cos \quad 2c-2g, \\
\delta\gamma &= +\tfrac{5}{8}\gamma e^2 \quad ,, \quad 2c-2g, \\
\delta c &= +\tfrac{5}{8}\gamma^2 \quad \sin \quad 2c-2g, \\
\delta g &= +\tfrac{5}{8}c^2 \quad ,, \quad 2c-2g.
\end{aligned}$$

The constant term in Ω is

$$= m^2n^2a^2\left(\tfrac{1}{4}+\tfrac{3}{8}e^2-\tfrac{3}{8}\gamma^2\right),$$

and this gives in

$$\frac{dl}{dt}, \ = -2\frac{d\Omega}{da}+\tfrac{1}{2}e\frac{d\Omega}{de}+\tfrac{1}{2}\gamma\frac{d\Omega}{d\gamma},$$

a term

$$\begin{aligned}
m^2(-1-\tfrac{3}{2}e^2+\tfrac{3}{2}\gamma^2 \\
+\tfrac{3}{8}e^2-\tfrac{3}{8}\gamma^2),
\end{aligned}$$

which is

$$= m^2(-1-\tfrac{9}{8}e^2+\tfrac{9}{8}\gamma^2).$$

Substituting for e, γ, their correct values $e+\delta e$, $\gamma+\delta\gamma$, it appears that $\frac{dl}{dt}$ contains the term

$$m^2(-\tfrac{9}{4}e\delta e+\tfrac{9}{4}\gamma\delta\gamma),$$

which is

$$\begin{aligned}
&= m^2\left(\tfrac{45}{32}+\tfrac{45}{32}=\right)\tfrac{45}{16} \quad e^2\gamma^2 \quad \cos \quad 2c-2g, \\
&= \qquad \tfrac{45}{16}m^2e^2\gamma^2 \quad ,, \quad 2c-2g,
\end{aligned}$$

and joining to this the before-mentioned term

$$= \qquad -\tfrac{15}{8}m^2e^2\gamma^2 \quad ,, \quad 2c-2g,$$

we find

$$\frac{dl}{dt} = \left(\tfrac{45}{16}-\tfrac{15}{8}=\right)\tfrac{15}{16}m^2e^2\gamma^2 \quad ,, \quad 2c-2g,$$

whence, writing as above $l+\delta l$ for l, and integrating, we have

$$\delta l = \qquad -\tfrac{5}{16} \quad e^2\gamma^2 \quad \sin \quad 2c-2g,$$

and it thus appears that the values of δa, δe, $\delta\gamma$, δc, δg, δl, agree with those obtained in my former Note.

467.

EXPRESSIONS FOR PLANA'S e, γ IN TERMS OF THE ELLIPTIC e, γ.

[From the *Monthly Notices of the Royal Astronomical Society,* vol. XXV. (1864—1865), pp. 265—271.]

THE coefficient of $\sin cnt$ in Plana's expression for the true longitude v (see Plana, t. I. p. 574), putting therein $E' = \epsilon' = e'$, that is, neglecting the terms which depend on the variation of the solar excentricity, is

$$
\begin{aligned}
= \; & e \; \left(2 + \tfrac{3}{2} m^2 - \tfrac{75}{64} m^3 - \tfrac{6659}{256} m^4 - \tfrac{4884375}{36864} m^5 - \tfrac{65756819}{147456} m^6 \right) \\
& + e^3 \; \left(- \tfrac{1}{4} - 17 m^2 - \tfrac{3195}{32} m^3 - \tfrac{4635997}{7680} m^4 \right) \\
& + e^5 \; \left(\tfrac{5}{96} + \tfrac{66863}{6144} m^2 \right) \\
& + e^7 \; \left(\tfrac{5921}{161280} \right) \\
& + e\gamma^2 \; \left(- \tfrac{1}{2} - \tfrac{63}{32} m^2 + \tfrac{1467}{256} m^3 + \tfrac{22857}{512} m^4 \right) \\
& + e^3\gamma^2 \; \left(\tfrac{23}{16} - \tfrac{405}{128} m + \tfrac{16029}{512} m^2 \right) \\
& + e^5\gamma^2 \; \left(- \tfrac{195}{128} \right) \\
& + e\gamma^4 \; \left(- \tfrac{3}{8} + \tfrac{135}{256} m + \tfrac{3749}{2048} m^2 \right) \\
& + e^3\gamma^4 \; \left(\tfrac{1761}{1280} \right) \\
& + e\gamma^6 \; \left(- \tfrac{5}{16} \right) \\
& + ee'^2 \; \Big(\left(- \tfrac{45}{4} + \tfrac{9}{4} = \right) - 9m^2 + \left(- \tfrac{6455}{64} + \tfrac{165}{42} = \right) - \tfrac{6125}{64} m^3 \\
& \qquad\qquad + \left(- \tfrac{281095}{512} - \tfrac{147}{256} = \right) - \tfrac{281389}{512} m^4 \Big) \\
& + e^3e'^2 \; \Big(\left(- \tfrac{12831}{480} - \tfrac{33}{2} = \right) - \tfrac{20751}{480} m^2 \Big) \\
& + ee^2\gamma^2 \; \Big(\left(\tfrac{5171}{128} - \tfrac{453}{128} = \right) + \tfrac{2359}{64} m^2 \Big) \\
& + ee'^4 \; \Big(\left(\tfrac{45}{16} - \tfrac{2025}{64} = \right) - \tfrac{1845}{64} m^2 \Big) \\
& + eb^4 \; \left(- \tfrac{135}{32} m^2 \right) \\
& + ee'^2b^4 \; \left(- \tfrac{75}{16} \right).
\end{aligned}
$$

Taking this to the fifth order only, and comparing it with the coefficient in the elliptic theory, we have

$$\begin{array}{ll} \textbf{Plana.} & \textbf{Elliptic.} \\ = e\left(2 - \tfrac{3}{2}m^2 - \tfrac{75}{64}m^3 - \tfrac{6659}{256}m^4\right) & = e\left(2\right) \\ + e^3\left(-\tfrac{1}{4} - 17m^2\right) & + e^3\left(-\tfrac{1}{4}\right) \\ + e^5\left(\tfrac{5}{96}\right) & + e^5\left(\tfrac{5}{96}\right) \\ + e\gamma^2\left(-\tfrac{1}{2} - \tfrac{63}{32}m^2\right) & \\ + e^3\gamma^2\left(\tfrac{23}{16}\right) & \\ + e\gamma^4\left(-\tfrac{3}{8}\right) & \\ + ee'^2\left(-9m^2\right). & \end{array}$$

The coefficient of $\sin gnt$ in Plana's expression for the latitude (see t. I. p. 704) is

$$\begin{aligned} = \gamma &\left(1 + \tfrac{33}{128}m^3 + \tfrac{241}{512}m^4 - \tfrac{82495}{24576}m^5\right) \\ + \gamma e^2 &\left(-1 - \tfrac{31}{512}m^2 - \tfrac{7977}{256}m^3\right) \\ + \gamma e^4 &\left(\tfrac{5}{64} + \tfrac{945}{512}m\right) \\ + \gamma^3 &\left(-\tfrac{3}{8} + \tfrac{5}{128}m^2 + \tfrac{69}{256}m^3\right) \\ + \gamma^3 e^2 &\left(\tfrac{7}{32} - \tfrac{405}{256}m\right) \\ + \gamma^5 &\left(\tfrac{1}{4}\right) \\ + \gamma e'^2 &\left(\tfrac{27}{8}m^2 - \tfrac{113}{128}m^3\right). \end{aligned}$$

But according to the calculation of Prof. Adams (quoted by M. Delaunay, *Comptes Rendus*, t. LIV. (1862), this should be

$$\begin{aligned} = \gamma &\left(1 + \tfrac{33}{128}m^3 - \tfrac{1}{512}m^4 - \tfrac{82497}{24576}m^5 - \tfrac{4801697}{294012}m^6\right) \\ + \gamma e^2 &\left(-1 - \tfrac{1111}{256}m^2 - \tfrac{7977}{256}m^3\right) \\ + \gamma e^4 &\left(\tfrac{3}{16} - \tfrac{135}{512}m\right) \\ + \gamma^3 &\left(-\tfrac{3}{8} + \tfrac{5}{128}m^2 - \tfrac{15}{128}m^3\right) \\ + \gamma^3 e^2 &\left(\tfrac{23}{32} + \tfrac{135}{256}m\right) \\ + \gamma^5 &\left(\tfrac{15}{64}\right) \\ + \gamma e'^2 &\left(\tfrac{9}{8}m^2 - \tfrac{113}{128}m^3 + \tfrac{3521}{1024}m^4\right). \end{aligned}$$

Adopting this as the true expression according to Plana's theory, taking it to the fifth order only, and comparing with the elliptic value of the same coefficient, we have

$$\begin{array}{ll} \textbf{Plana.} & \textbf{Elliptic.} \\ \gamma\left(1 + \tfrac{33}{128}m^3 - \tfrac{1}{512}m^4\right) & = \gamma\left(1\right) \\ + \gamma e^2\left(-1 - \tfrac{1111}{256}m^2\right) & + \gamma e^2\left(-1\right) \\ + \gamma e^4\left(\tfrac{3}{16}\right) & + \gamma e^4\left(\tfrac{7}{64}\right) \\ + \gamma^3\left(-\tfrac{3}{8} + \tfrac{5}{128}m^2\right) & + \gamma^3\left(-\tfrac{3}{8}\right) \\ + \gamma^3 e^2\left(\tfrac{23}{32}\right) & + \gamma^3 e^2\left(\tfrac{3}{8}\right) \\ + \gamma^5\left(\tfrac{15}{64}\right) & + \gamma^5\left(\tfrac{55}{64}\right). \\ + \gamma e'^2\left(\tfrac{9}{8}m^2\right) & \end{array}$$

We have thus two equations for the determination of Plana's e, γ in terms of the elliptic e, γ. And the solution of these equations give

Elliptic.

$$
\begin{aligned}
e\text{ (Plana)} = \;& e\,\left(1 - \tfrac{3}{4} m^2 + \tfrac{75}{128} m^3 + \tfrac{6947}{512} m^4\right) \\
&+ e^5 \left(\tfrac{263}{32} m^2\right) \\
&+ \gamma^2 e \left(\tfrac{1}{4} + \tfrac{39}{64} m^2\right) \\
&+ \gamma^2 e^3 \left(- \tfrac{5}{8}\right) \\
&+ \gamma^4 e \left(\tfrac{1}{4}\right) \\
&+ ee'^2 \left(\tfrac{9}{2} m^2\right),
\end{aligned}
$$

$$
\begin{aligned}
\gamma\text{ (Plana)} = \;& \gamma\,\left(1 - \tfrac{33}{128} m^2 + \tfrac{1}{512} m^4\right) \\
&+ \gamma e^2 \left(\tfrac{727}{256} m^2\right) \\
&+ \gamma e^4 \left(- \tfrac{5}{64}\right) \\
&+ \gamma^3 \left(- \tfrac{5}{128} m^2\right) \\
&+ \gamma^3 e^2 \left(\tfrac{5}{32}\right) \\
&+ \gamma^5 \left(\tfrac{5}{8}\right) \\
&+ \gamma e'^2 \left(- \tfrac{9}{8} m^2\right).
\end{aligned}
$$

I annex the verification of these expressions; we have

Plana. Elliptic.

$$
\begin{aligned}
e\,\left(2 + \tfrac{3}{2} m^2 - \tfrac{75}{64} m^3 - \tfrac{6659}{256} m^4\right) = \;& e\,\Big(2 - \tfrac{3}{2} m^2 + \tfrac{75}{64} m^3 + \tfrac{6947}{256} m^4 \\
& \quad + \tfrac{3}{2} m^2 \qquad\qquad - \tfrac{9}{8} m^4 \\
& \quad\qquad - \tfrac{75}{64} m^3 - \tfrac{6659}{256} m^4\Big) \\
&+ e^3 \left(\tfrac{263}{32} m^2\right) \\
&+ e\gamma^2 \Big(\tfrac{1}{2} + \tfrac{39}{32} m^2 \\
& \qquad + \tfrac{3}{8} m^2\Big) \\
&+ e^3\gamma^2 \left(- \tfrac{5}{4}\right) \\
&+ e\gamma^4 \left(\tfrac{1}{2}\right) \\
&+ ee'^2 \left(9m^2\right), \\
e^3 \left(- \tfrac{1}{4} - 17m^2\right) = \;& e^3 \Big(- \tfrac{1}{4} + \tfrac{9}{16} m^2 \\
& \qquad - 17\, m^2\Big) \\
&+ e^3\gamma^2 \left(- \tfrac{3}{16}\right), \\
e^5 \left(\tfrac{5}{96}\right) = \;& e^5 \left(\tfrac{5}{96}\right), \\
e\gamma^2 \left(- \tfrac{1}{2} - \tfrac{63}{32} m^2\right) = \;& e\gamma^2 \Big(- \tfrac{1}{2} - \tfrac{63}{32} m^2 \\
& \qquad + \tfrac{3}{8} m^2\Big) \\
&+ e\gamma^4 \left(- \tfrac{1}{8}\right) \\
e^3\gamma^2 \left(\tfrac{23}{16}\right) = \;& e^3\gamma^2 \left(\tfrac{23}{16}\right) \\
e\gamma^4 \left(- \tfrac{3}{8}\right) = \;& e\gamma^4 \left(- \tfrac{3}{8}\right) \\
ee'^2 \left(- 9m^2\right) = \;& ee'^2 \left(- 9m^3\right),
\end{aligned}
$$

whence, adding, we have the first equation.

And, moreover,

$$
\begin{aligned}
\gamma\ (1+\tfrac{33}{128}m^3-\tfrac{1}{512}m^4) &= \gamma\ (\ 1-\tfrac{33}{128}m^3+\tfrac{1}{512}m^4\\
&\qquad\qquad +\tfrac{33}{128}m^3-\tfrac{1}{512}m^4)\\
&\quad +\gamma e^2\ (\tfrac{727}{256}m^2)\\
&\quad +\gamma^4 e\ (-\tfrac{5}{64})\\
&\quad +\gamma^3\ (-\tfrac{5}{128}m^2)\\
&\quad +\gamma^3 e^2\ (\tfrac{5}{32})\\
&\quad +\gamma^5\ (\tfrac{5}{8})\\
&\quad +\gamma e'^2\ (-\tfrac{9}{8}m^2),\\
\gamma e^2\ (-1-\tfrac{1111}{256}m^2) &= \gamma e^2\ (-1+\tfrac{3}{2}m^2\\
&\qquad\qquad -\tfrac{1111}{256}m^2)\\
&\quad +\gamma^3 e^2\ (-\tfrac{1}{2}),\\
\gamma e^4\ (\tfrac{3}{16}) &= \gamma e^4\ (\tfrac{3}{16}),\\
\gamma^3\ (-\tfrac{3}{8}+\tfrac{5}{128}m^2) &= \gamma^3\ (-\tfrac{3}{8}+\tfrac{5}{128}m^2)\\
\gamma^3 e^2\ (\tfrac{23}{32}) &= \gamma^3 e^2\ (\tfrac{23}{32})\\
\gamma^5\ (\tfrac{15}{64}) &= \gamma^5\ (\tfrac{15}{64})\\
\gamma e'^2\ (\tfrac{9}{8}m^2) &= \gamma e'^2\ (\tfrac{9}{8}m^2),
\end{aligned}
$$

whence, adding, we have the second equation.

It may be noticed that, taking the foregoing expressions only as far as the third order, we have

Plana.		Elliptic.
e	=	$e\ (1+\frac{1}{4}\gamma^2-\frac{3}{4}m^2)$,
γ	=	γ.

And moreover that, attending only to the terms which are independent of m, we have

$$
\begin{aligned}
e &= e\ (1+\tfrac{1}{4}\gamma^2-\tfrac{5}{8}\gamma^2+\tfrac{1}{4}\gamma^4),\\
\gamma &= \gamma\,(1-\tfrac{5}{64}e^4+\tfrac{5}{32}e^2\gamma^2-\tfrac{5}{8}\gamma^4).
\end{aligned}
$$

which are formulæ that may be found useful.

468.

ADDITION TO SECOND NOTE ON THE LUNAR THEORY.

[From the *Monthly Notices of the Royal Astronomical Society,* vol. XXVII. (1866—1867), pp. 267—269.]

WRITING as in my Second Note, *Monthly Notices,* Vol. XXV., pp. 203—207 (May 1865), [466], for the Moon,

a, the mean distance,

e, the excentricity,

γ, the tangent of the inclination,

l, the mean longitude,

c, the mean anomaly,

g, the mean distance from node,

I obtained by the ordinary method of the variation of the elements, from the constant term of R and the term involving $\cos(2c-2g)$, the following expressions of the variations,

$$\begin{array}{lll} \delta a = \ \ 0, & & \\ \delta e = - \tfrac{5}{8}\gamma^2 e & \cos & 2c-2g, \\ \delta\gamma = + \tfrac{5}{8}\gamma e^2 & ,, & 2c-2g, \\ \delta c = + \tfrac{5}{8}\gamma^2 & \sin & 2c-2g, \\ \delta g = + \tfrac{5}{8}e^2 & ,, & 2c-2g, \\ \delta l = + \tfrac{5}{16}\gamma^2 e^2 & ,, & 2c-2g, \end{array}$$

viz. if in the elliptic expressions of the radius vector, longitude, and latitude, we apply to a, e, γ, c, g, l, the foregoing increments, we obtain to the fourth order in (e, γ) the portions independent of m in the expressions of the radius vector, latitude, and longitude. I wish to notice that the results, to the very limited extent to which they go, agree with those obtained by M. Delaunay in his "Théorie du Mouvement de la Lune," from his 49th operation, the object of which is to take away the term (63) of R, that is the term involving $\cos(2c-2g)$. The formulæ (see vol. I. p. 788), taken only to the necessary degree of approximation are

$$\begin{array}{rcllll} a & \text{replaced by} & a, & & & \\ e^2 & ,, & e^2 - 5\gamma^2e^2 & \cos & 2g, \\ \gamma^2 & ,, & \gamma^2 + \tfrac{5}{4}\gamma^2e^2 & ,, & 2g, \\ l & ,, & l - \tfrac{5}{2}\gamma^2 & \sin & 2g, \\ h+g+l & ,, & h+g+l + \tfrac{5}{4}\gamma^2e^2 & ,, & 2g, \\ h & ,, & h + \tfrac{5}{8}e^2 & ,, & 2g, \end{array}$$

which, observing that

$$\begin{aligned} \gamma\ (\text{Del.}) &= \tfrac{1}{2}\gamma\ (\text{for present purpose}), \\ l &= c, \\ g+l &= g, \\ h+g+l &= l, \end{aligned}$$

and therefore $\qquad g = -(c-g),$

become

$$\begin{array}{rcllll} a & \text{replaced by} & a, & & \\ e^2 & ,, & e^2 - \tfrac{5}{4}\gamma^2e^2 & \cos & 2c-2g, \\ \gamma^2 & ,, & \gamma^2 + \tfrac{5}{4}\gamma^2e^2 & ,, & 2c-2g, \\ c & ,, & c + \tfrac{5}{8}\gamma^2 & \sin & 2c-2g, \\ l & ,, & l - \tfrac{5}{16}\gamma^2e^2 & ,, & 2c-2g, \\ l-g & ,, & l-g - \tfrac{5}{8}e^2 & ,, & 2c-2g, \end{array}$$

the last of which may be changed into

$$g \quad ,, \quad g + \tfrac{5}{8}e^2 \quad ,, \quad 2c-2g,$$

or if the new values of a, e, γ, c, g, l, are called $a+\delta a$, $e+\delta e$, $\gamma+\delta\gamma$, $c+\delta c$, $g+\delta g$, $l+\delta l$, then the increments δa, δe, $\delta\gamma$, δc, δg, δl, have the values given above. The process of my Second Note, taken as a first transformation, has in fact the object of removing the term $\cos(2c-2g)$, and to the degree of approximation regarded, the result is not affected by the previous transformations, or by the substitution, t. II. p. 800, introducing for a, e, γ, their standard elliptic values.

469.

ON AN EXPRESSION FOR THE ANGULAR DISTANCE OF TWO PLANETS.

[From the *Monthly Notices of the Royal Astronomical Society*, vol. XXVII. (1866—1867), pp. 312—315.]

If for the planet m, referred to any fixed plane and origin of longitudes, we have

v, the longitude in orbit,

θ, the longitude of node,

ϕ, the inclination,

and similarly for the planet m' referred to the same fixed plane and origin of longitudes, if the corresponding quantities are v', θ', ϕ'; then the angular distance of the two planets will of course be expressible in terms of v, θ, ϕ, v', θ', ϕ', but I am not aware that the actual expression has been given. To obtain it in the most simple manner, I write further for the planet m:

$\theta+x$, the reduced longitude,

y, the latitude,

z, the distance from node,

so that z $(=v-\theta)$, x, y, are the hypothenuse, base, and perpendicular of a right-angled spherical triangle, the base angle of which is $=\phi$. And similarly $\theta'+x'$, y', z', have the like significations for the planet m'. I write also r, r', for the distances of the two planets respectively.

This being so, the rectangular coordinates of the planet m are

$$r \cos y \cos(\theta + x),$$
$$r \cos y \sin(\theta + x),$$
$$r \sin y.$$

But observing that from the right-angled triangle we have

$$\begin{aligned}\cos z &= \cos x \cos y,\\ \cos\phi &= \tan x \cot z,\\ \sin x &= \cot\phi \tan y,\\ \sin y &= \sin\phi \sin z,\end{aligned}$$

and therefore also

$$\sin x \cos y = \cot\phi \sin y, = \cos\phi \sin z,$$

the expressions for the coordinates become

$$\begin{aligned}&r(\cos z\cos\theta - \sin z\sin\theta\cos\phi),\\ &r(\cos z\sin\theta + \sin z\cos\theta\cos\phi),\\ &r(\qquad\qquad \sin z\sin\phi).\end{aligned}$$

Forming the analogous expressions for the coordinates of m', then if H be the angular distance of the two planets, we deduce at once the expression for $\cos H$, viz. this is

$$\begin{aligned}\text{Cos } H = &\ (\cos z\cos\theta - \sin z\sin\theta\cos\phi)(\cos z'\cos\theta' - \sin z'\sin\theta'\cos\phi')\\ &+(\cos z\sin\theta + \sin z\cos\theta\cos\phi)(\cos z'\sin\theta' + \sin z'\cos\theta'\cos\phi')\\ &+(\qquad \sin z\sin\phi \qquad)(\qquad \sin z'\sin\phi' \qquad),\end{aligned}$$

or, multiplying out, this is

$$\begin{aligned}\text{Cos } H = &\ \cos z\cos z' \ \cos(\theta-\theta')\\ &+\cos z\sin z' \ \sin(\theta-\theta')\cos\phi'\\ &-\sin z\cos z' \ \sin(\theta-\theta')\cos\phi\\ &+\sin z\sin z' \big(\cos(\theta-\theta')\cos\phi\cos\phi' + \sin\phi\sin\phi'\big),\end{aligned}$$

say this is

$$\begin{aligned}= &\ A\cos z\cos z'\\ &+B\cos z\sin z'\\ &+C\sin z\cos z'\\ &+D\sin z\sin z',\end{aligned}$$

viz. it is

$$\begin{aligned}= &\ \cos(z-z').\ \ \tfrac{1}{2}A + \tfrac{1}{2}D\\ &+\sin(z-z'). -\tfrac{1}{2}B + \tfrac{1}{2}C\\ &+\cos(z+z').\ \ \tfrac{1}{2}A - \tfrac{1}{2}D\\ &+\sin(z+z').\ \ \tfrac{1}{2}B + \tfrac{1}{2}C.\end{aligned}$$

But we have

$$z - z' = v - v' - \theta + \theta', \quad z + z' = v + v' - \theta - \theta',$$

whence the expression becomes

$$\begin{aligned}\operatorname{Cos} H = \;& \cos(v-v') \,.\, (\tfrac{1}{2}A+\tfrac{1}{2}D)\cos(\theta-\theta') - (-\tfrac{1}{2}B+\tfrac{1}{2}C)\sin(\theta-\theta')\\ &+\sin(v-v') \,.\, (\tfrac{1}{2}A+\tfrac{1}{2}D)\sin(\theta-\theta') + (-\tfrac{1}{2}B+\tfrac{1}{2}C)\cos(\theta-\theta')\\ &+\cos(v+v') \,.\, (\tfrac{1}{2}A-\tfrac{1}{2}D)\cos(\theta+\theta') - (\;\tfrac{1}{2}B+\tfrac{1}{2}C)\sin(\theta+\theta')\\ &+\sin(v+v') \,.\, (\tfrac{1}{2}A-\tfrac{1}{2}D)\sin(\theta+\theta') + (\;\tfrac{1}{2}B+\tfrac{1}{2}C)\cos(\theta+\theta'),\end{aligned}$$

or substituting for A, B, C, D, their values, and after a few easy reductions, we find

$$\begin{aligned}\operatorname{Cos} H = \;& \cos(v-v')\left\{\begin{array}{l}\tfrac{1}{2}+\tfrac{1}{2}\cos\phi\cos\phi' - \tfrac{1}{2}(1-\cos\phi)(1-\cos\phi')\sin^2(\theta-\theta')\\ \qquad + \tfrac{1}{2}\sin\phi\sin\phi' \qquad \cos(\theta-\theta')\end{array}\right\}\\[1ex] &+\sin(v-v')\left\{\begin{array}{l}\tfrac{1}{2}(1-\cos\phi)(1-\cos\phi')\sin(\theta-\theta')\cos(\theta-\theta')\\ +\tfrac{1}{2}\sin\phi\sin\phi' \qquad \sin(\theta-\theta')\end{array}\right\}\\[1ex] &+\cos(v+v')\left\{\begin{array}{l}\tfrac{1}{2}(1-\cos\phi\cos\phi')\cos(\theta-\theta')\cos(\theta+\theta')\\ +\tfrac{1}{2}(\cos\phi-\cos\phi') \quad \sin(\theta-\theta')\sin(\theta+\theta')\\ -\tfrac{1}{2}\sin\phi\sin\phi' \qquad \cos(\theta+\theta')\end{array}\right\}\\[1ex] &+\sin(v+v')\left\{\begin{array}{l}\tfrac{1}{2}(1-\cos\phi\cos\phi')\cos(\theta-\theta')\sin(\theta+\theta')\\ -\tfrac{1}{2}(\cos\phi-\cos\phi') \quad \sin(\theta-\theta')\cos(\theta+\theta')\\ -\tfrac{1}{2}\sin\phi\sin\phi' \qquad \sin(\theta+\theta')\end{array}\right\}\end{aligned}$$

For $\phi=\phi'=0$, the formula becomes, as of course it should do,

$$\operatorname{Cos} H = \cos(v-v').$$

It may be added, that if f, f' are the true anomalies, ω, ω' the longitudes of pericentre in orbit, then $v=\omega+f$, $v'=\omega'+f'$; and we thence have for $\cos H$, formulæ of the like form, containing $\cos f\cos f'$, $\cos f\sin f'$, $\sin f\cos f'$, $\sin f\sin f'$, or containing $\cos(f-f')$, $\sin(f-f')$, $\cos(f+f')$, $\sin(f+f')$, respectively, in place of the like functions of z, z', but with of course altered values of the coefficients.

470.

NOTE ON THE ATTRACTION OF ELLIPSOIDS.

[From the *Monthly Notices of the Royal Astronomical Society*, vol. XXIX. (1868—1869), pp. 254—257.]

IF an indefinitely thin shell of uniform density, bounded by two similar and similarly-situated ellipsoids, attracts a point P on its outer surface, it has been shown geometrically by M. Chasles that the attraction is in the direction of the normal at P, and is equal to twice the attraction of an infinite plate, the thickness of which is equal to the normal thickness at P. Assuming that the attraction is in the direction of the normal, the proof is in fact as follows:—with P as vertex, circumscribe to the interior surface a cone; this divides the shell into three parts; the one, $D+E+F$, exterior to the cone, the other two, $A+B$ and C, interior to the cone. It is shown that in the direction of the normal the attraction of C is equal to that of $A+B$;

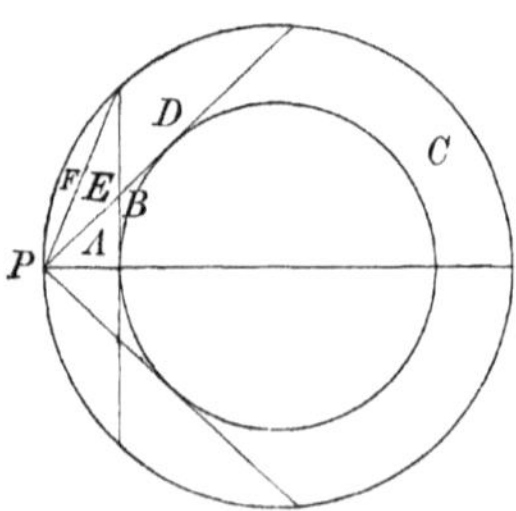

and it is assumed that in comparison with these the attraction of $D+E+F$ may be neglected; the whole attraction is thus equal to twice that of the portion $A+B$. At the point where the normal at P meets the internal surface draw the tangent plane to the internal surface, thus dividing the portion $A+B$ into the solid cone A and a remaining portion B; it is assumed that in comparison with that of A the attraction of B may be neglected; the whole attraction is thus equal to twice the attraction of the solid cone A; and the attraction of this solid cone is in the limit (the aperture or solid angle then becoming $=2\pi$) equal to the attraction of an infinite plate whose thickness is equal to the altitude of the solid cone, that is, to the normal thickness at P. And the attraction of the whole ellipsoidal shell is thus ultimately (that is, when the shell is indefinitely thin) equal to twice the attraction of the infinite plate.

It is interesting to ascertain the orders of magnitude of the attractions of the several portions of the shell, which attractions are compared in the foregoing investigation; and this can be done very easily, when, instead of the ellipsoidal shell, we have a spherical shell (bounded by two concentric spherical surfaces). The tangent plane to the inner surface divides the portion $D+E+F$ into two portions D and $E+F$; and if with P as vertex we describe a cone standing on the circle in which the tangent plane meets the outer surface, the last-mentioned portion is hereby divided into the portions E and F; the whole shell is thus divided into the portions A, B, C, D, E, F, each of them symmetrical in regard to the normal or radius at P, and consequently attracting in the direction of this radius. I proceed to find the attractions of each of these portions; it will appear, in accordance with the assumptions of the foregoing investigation, that, taking the radii to be 1 and $1+\alpha$, that is, α the thickness of the shell, and supposing ultimately α to become indefinitely small, the attractions of A and C are each ultimately $=2\pi\alpha$, that is $=$ to the attraction of the infinite plate, while the attractions of the other portions are of the order $\alpha^{\frac{3}{2}}$, and thus vanish in comparison with that of A or C.

The attraction of an indefinitely thin cone or frustum of a cone, length r and solid angle $d\omega$ is $=rd\omega$; considering any such cone having P for its vertex, if the inclination of r to the radius through P is $=\theta$, and if the azimuth of the plane through r and the radius is $=\phi$, then we have $d\omega = \sin\theta\, d\theta\, d\phi$, the attraction $rd\omega$ is $=r\sin\theta d\theta d\phi$, and this attraction resolved in the direction of the radius is $=r\sin\theta\cos\theta d\theta d\phi$. For the several cases which have to be considered, the value of r is independent of ϕ, and the integration in regard to ϕ is always from $\phi=0$ to $\phi=2\pi$;—the attraction is thus in each case $=2\pi\int r\sin\theta\cos\theta\, d\theta$, the expression of r in the terms of θ, and the limits of θ being known for each of the several portions of the shell. Taking θ_1 for the semi-angle of the tangent cone, we have it is clear

$$\sin\theta_1 = \frac{1}{1+\alpha}, \quad \cos\theta_1 = \frac{\sqrt{2\alpha+\alpha^2}}{1+\alpha};$$

and taking θ_2 for the semi-angle of the cone which divides the portions E, F,

$$\tan\theta_2 = \sqrt{\frac{2+\alpha}{\alpha}}, \quad \sin\theta_2 = \frac{\sqrt{2+\alpha}}{\sqrt{2(1+\alpha)}}, \quad \cos\theta_2 = \frac{\sqrt{\alpha}}{\sqrt{2(1+\alpha)}}.$$

For F we have

$$r = 2(1+\alpha)\cos\theta,\quad \theta = \theta_2 \text{ to } \theta = \tfrac{1}{2}\pi,$$

Integral is

$$= 2(1+\alpha)\int \sin\theta\cos^2\theta\, d\theta, = \tfrac{2}{3}(1+\alpha)\cos^3\theta_2.$$

For $D+E$ we have

$$r = 2(1+\alpha)\cos\theta,\quad \theta = \theta_1 \text{ to } \theta = \theta_2,$$

Integral is

$$= 2(1+\alpha)\int \sin\theta\cos^2\theta\, d\theta, = \tfrac{2}{3}(1+\alpha)(\cos^3\theta_1 - \cos^3\theta_2).$$

For E we have

$$r = \frac{\alpha}{\cos\theta},\quad \theta = \theta_1 \text{ to } \theta = \theta_2,$$

Integral is

$$= \alpha\int \sin\theta\, d\theta, = \alpha(\cos\theta_1 - \cos\theta_2).$$

For A we have

$$r = \frac{\alpha}{\cos\theta},\quad \theta = 0 \text{ to } \theta = \theta_1,$$

Integral is

$$= \alpha\int \sin\theta\, d\theta, = \alpha(1 - \cos\theta_1).$$

For $A+B$ we have

$$r = (1+\alpha)\cos\theta - \sqrt{1-(1+\alpha)^2\sin^2\theta},\quad \theta = 0 \text{ to } \theta = \theta_1,$$

Integral is

$$= \int \{(1+\alpha)\cos\theta - \sqrt{1-(1+\alpha)^2\sin^2\theta}\}\sin\theta\cos\theta\, d\theta,$$

$$= (1+\alpha)(-\tfrac{1}{3}\cos^3\theta) + \frac{1}{3(1+\alpha)^2}\{1-(1+\alpha)^2\sin^2\theta\}^{\frac{3}{2}}, \text{ between the limits,}$$

$$= \tfrac{1}{3}\left\{(1+\alpha)(1-\cos^3\theta_1) - \frac{1}{(1+\alpha)^2}\right\},$$

and subtracting the above value of the integral for A, it at once appears that, for B, the integral is

$$= 2\pi\left\{\alpha(-1+\cos\theta_1) + \tfrac{1}{3}(1+\alpha)(1-\cos^2\theta_1) - \tfrac{1}{3}\frac{1}{(1+\alpha)^2}\right\}.$$

Hence, calculating the approximate values, and restoring in each case the omitted factor, 2π, we have

$$\begin{aligned} \text{Attraction}\quad A &= 2\pi\alpha - 2\sqrt{2}\,\pi\alpha^{\frac{3}{2}}, \\ \text{,,}\quad B &= \tfrac{2}{3}\sqrt{2}\,\pi\alpha^{\frac{3}{2}}, \\ \text{,,}\quad C &= 2\pi\alpha - \tfrac{4}{3}\sqrt{2}\,\pi\alpha^{\frac{3}{2}}, \\ \text{,,}\quad D &= \tfrac{4}{3}\sqrt{2}\,\pi\alpha^{\frac{3}{2}}, \\ \text{,,}\quad E &= 1\sqrt{2}\,\pi\alpha^{\frac{3}{2}}, \\ \text{,,}\quad F &= \tfrac{1}{3}\sqrt{2}\,\pi\alpha^{\frac{3}{2}}; \end{aligned}$$

or, if we please,

$$\begin{aligned} \text{Attraction}\quad A+B &= 2\pi\alpha - \tfrac{4}{3}\sqrt{2}\,\pi\alpha^{\frac{3}{2}}, \\ \text{,,}\quad C &= 2\pi\alpha - \tfrac{4}{3}\sqrt{2}\,\pi\alpha^{\frac{3}{2}}, \\ \text{,,}\quad D+E+F &= \tfrac{8}{3}\sqrt{2}\,\pi\alpha^{\frac{3}{2}}; \end{aligned}$$

so that ultimately the attraction of the portion $D+E+F$ vanishes in comparison with those of the portions $A+B$ and C; and the attraction of these last, that is, of the whole shell, is $=4\pi\alpha$, twice the attraction of an infinite plate of the thickness α.

471.

NOTE ON THE PROBLEM OF THE DETERMINATION OF A PLANET'S ORBIT FROM THREE OBSERVATIONS.

[From the *Monthly Notices of the Royal Astronomical Society*, vol. XXIX. (1868—1869), pp. 257—259.]

THE principle of the solution given in the *Theoria Motus* may be explained very simply as follows:

Consider three successive positions of C, C', C'', of a planet revolving about the focus S; let n, n', n'', denote the doubles of the triangular areas $C'SC''$, CSC', and CSC'' respectively (viz. the triangular area means the area of the triangle included between the two radius vectors and the chord joining their extremities), r' the radius

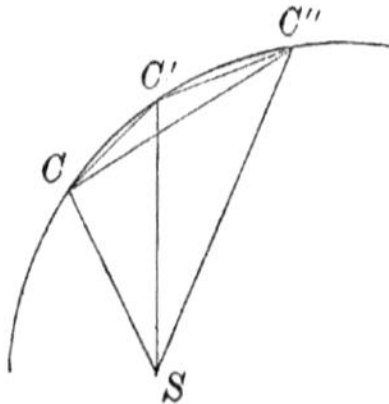

vector SC'; θ'', θ, the times of describing the arcs CC' and $C'C''$ respectively, the units of time and distance being such that the time is equal to the double area divided by the square root of the half latus rectum ($t = 2\pi a^{\frac{3}{2}}$ for the Period in a circular or elliptic orbit).

Then writing

$$P = \frac{n''}{n}, \quad Q = 2\left(\frac{n + n''}{n'} - 1\right) r'^3,$$

(observe that $n+n''-n'$ is $=$ twice the triangle $CC'C''$), for neighbouring positions of the planet, the values of P and Q are *approximately* $=\frac{\theta''}{\theta}$ and $\theta\,\theta''$ respectively: the solution consists in the determination of an orbit for which P and Q have these approximate values; then, by means of such approximate orbit, the values of P and Q are more accurately determined, and by means of these new values of P and Q, a new determination is effected of the orbit: and so on, to the requisite accuracy of approximation.

The foregoing approximate values of P and Q respectively are deduced from the accurate values

$$P=\frac{\theta''\eta}{\theta\eta''},\quad Q=\frac{\theta\theta''}{\eta\eta''}\,\frac{r'^2}{r'r''}\,\frac{1}{\cos f\cos f'\cos f''};$$

where r, r', r'' are the radius vectors SC, SC', SC''; $2f$, $2f'$, $2f''$ are the angular distances $C'SC''$, CSC'', CSC' $(f'=f+f'')$ and η, η', η'' are the ratios of the sectorial areas $C'SC''$, CSC'', $C'SC''$, to the triangular areas represented by the same letters respectively: the doubles of the sectorial areas are thus $n\eta$, $n'\eta'$, and $n''\eta''$, and if the half latus rectum be denoted by p, then we have

$$\sqrt{p}=\frac{n\eta}{\theta}=\frac{n'\eta'}{\theta'}=\frac{n''\eta''}{\theta''};$$

and it thus at once appears that the accurate value of P is $=\frac{\theta''\eta}{\theta\eta''}$, as above. To obtain the expression for Q, taking ϕ, ϕ', ϕ'' for the true anomalies (and, for greater symmetry, writing for the moment ν, $-\nu'$, ν'', g, $-g'$, g'' in place of n, n', n'', f, f', f'' respectively), we have

$$r=\frac{p}{1+e\cos\phi},\quad 2g=\phi''-\phi',$$

$$r'=\frac{p}{1+e\cos\phi'},\quad 2g'=\phi-\phi'',$$

$$r''=\frac{p}{1+e\cos\phi''},\quad 2g''=\phi'-\phi,$$

$$(g+g'+g''=0);$$

whence identically

$$\frac{\sin 2g}{r}+\frac{\sin 2g'}{r'}+\frac{\sin 2g''}{r''}=-\frac{4\sin g\sin g'\sin g''}{p};$$

or writing

$$\nu=r'r''\sin 2g,\quad \nu'=r''r\sin 2g',\quad \nu''=rr'\sin 2g'',$$

this is

$$\nu+\nu'+\nu''=-\frac{4rr'r''\sin g\sin g'\sin g''}{p},$$

$$=-\frac{(rr'r'')^2\sin 2g\sin 2g'\sin 2g''}{2prr'r''\cos g\cos g'\cos g''}$$

$$=-\frac{\nu\nu'\nu''}{2prr'r''\cos g\cos g'\cos g''}.$$

This is, in fact,

$$n - n' + n'' = \frac{nn'n''}{2prr'r'' \cos f \cos f' \cos f''},$$

or since

$$\frac{nn''}{p} = \frac{\theta\theta''}{\eta\eta''},$$

it is

$$2\left(\frac{n+n''}{n'} - 1\right) = \frac{\theta\theta''}{\eta\eta'' rr'r'' \cos f \cos f' \cos f''},$$

viz. multiplying by r'^3, it is

$$Q \quad = \frac{\theta\theta''}{\eta\eta''} \frac{r'^2}{rr''} \frac{1}{\cos f \cos f' \cos f''},$$

the above-mentioned value of Q.

472.

NOTE ON LAMBERT'S THEOREM FOR ELLIPTIC MOTION.

[From the *Monthly Notices of the Royal Astronomical Society*, vol. XXIX. (1868—1869), pp. 318—320.]

CONSIDER any two positions, A, B, in an elliptic orbit, focus S, and semi-axis major $=a$; then if ρ, ρ', c denote the radius vectors SA, SB, and the chord AB respectively, and if P, $=\dfrac{2\pi a^{\frac{3}{2}}}{\sqrt{\mu}}$, be the periodic time, the time of passage from A to B is given by the formula

$$\text{Time } AB = \frac{P}{2\pi}(\chi - \chi' - \sin\chi + \sin\chi')$$

where

$$2a\cos\chi = 2a - \rho - \rho' - c, \quad 2a\cos\chi' = 2a - \rho - \rho' + c.$$

To fix the ideas we may consider the time of passage as being in every case positive; and, for Time AB, the motion from A as being towards the apocentre; Time BA will, of course, in like manner denote that the motion from B is towards the apocentre; and we thus have according to the positions of A, B, either Time $AB=$ Time BA; or else Time $AB+$ Time $BA=P$.

This being so (see the *Theoria Motus*, p. 120), χ will be always a positive arc between 0 and 360°; χ' a positive or negative arc between 0 and $\pm 180°$; and moreover χ' will be positive or negative according as the described focal angle is $<180°$ or $>180°$; whence, $\cos\chi'$ being known, the arc χ' is determined without ambiguity.

But as noticed in the place referred to, there is when only ρ, ρ', c, a, are known, a real ambiguity as regards the arc χ; viz. χ may be either the arc $>180°$ or the arc $<180°$, having for its cosine the given value of $\cos\chi$. For, given the points S, A, B, and the semi-axis major a, there exist two elliptic orbits determined by these data; and the two values of χ correspond to the times of passage between A and B, in these two orbits respectively. If, however, the actual orbit be given, there is no longer any real ambiguity; and it must be possible to decide between the two values of χ: the criterion is, in fact, a very simple one, viz. drawing a chord from A through the other focus H of the ellipse, this either separates, or it does not separate, B from the force-focus S; and I say that in the expression of Time AB, in the former case (viz. when chord A is a separator) we have $\chi < 180°$; in the latter case (viz. when chord A is not a separator) we have $\chi > 180°$.

It of course follows that, in the case of transition, when the line AB passes through H, we must have $\chi = 180°$: this is at once seen to be so; for $\chi = 180°$ gives the condition $4a = \rho + \rho' + c$; but if σ, σ', are the distances of AB from the focus H, then $2a = \rho + \sigma$, $2a = \rho' + \sigma'$, and the condition becomes $\sigma + \sigma' = c$; that is AB must pass through H.

As a verification of the new criterion, I consider the point A as having a fixed position on the orbit, but the point B as having successively different positions; and writing down the two formulæ

$$\text{Time } AB = \chi - \chi' - \sin\chi + \sin\chi',$$
$$\text{Time } BA = \omega - \omega' - \sin\omega + \sin\omega',$$

(where for simplicity the constant factor $P \div 2\pi$ is omitted) I proceed to compare these for different positions of the point B. We have, in every case, $\cos\omega = \cos\chi$, and $\cos\omega' = \cos\chi'$; whence (χ, ω being each positive and less than 360°) $\omega = \chi$ or else $\omega + \chi = 360°$, viz. the former equation subsists if ω, χ, are each less or each greater than 180°, the latter if the one is greater, the other less than 180°. And again (χ', ω' being each less than $\pm 180°$) we have $\omega' = \chi'$, or else $\omega' = -\chi'$, according as ω', χ' have the same or opposite signs.

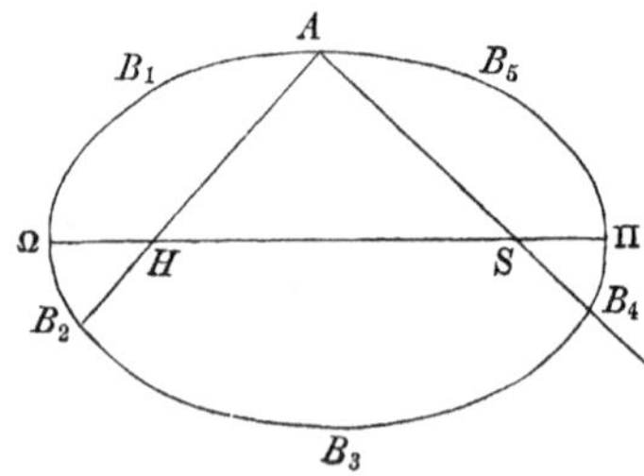

Now in the figure, suppose that B occupies successively the different positions $B_1, B_2, \ldots B_5$, the criteria for χ, χ' (or ω, ω') give as follows,

	Ch. A.	Ch. B.	therefore			$\angle AS$	$\angle BSA$			
1	sep.	not	$\chi < 180°$	$\omega > 180°$	or $\omega + \chi = 2\pi$	$< 180°$	$> 180°$	$\chi' = +$	$\omega' = -$	or $\omega' = -\chi'$,
2	sep.	sep.	$<$	$<$	„ $\omega = \chi$	$<$	$<$	$+$	$+$	„ $\omega' = \chi'$,
3	not	not	$>$	$>$	„ $\omega = \chi$	$>$	$>$	$+$	$+$	„ $\omega' = \chi'$,
4	not	not	$>$	$>$	„ $\omega = \chi$	$>$	$>$	$-$	$-$	„ $\omega' = \chi'$,
5	not	sep.	$>$	$<$	„ $\omega + \chi = 2\pi$	$>$	$<$	$-$	$-$	„ $\omega' = -\chi'$.

Hence substituting for ω, ω' their values in terms of χ, χ', we have

$$\begin{aligned}
\text{Time } AB_{1 \ldots 5} &= \chi - \chi' - \sin\chi + \sin\chi', \\
\text{,, } BA_1 &= 2\pi - \chi + \chi' + \sin\chi - \sin\chi', \\
\text{,, } BA_2 &= \chi - \chi' - \sin\chi + \sin\chi', \\
\text{,, } BA_3 &= \chi - \chi' - \sin\chi + \sin\chi', \\
\text{,, } BA_4 &= \chi - \chi' - \sin\chi + \sin\chi', \\
\text{,, } BA_5 &= 2\pi - \chi + \chi' + \sin\chi - \sin\chi';
\end{aligned}$$

and thence (restoring the omitted factor $P \div 2\pi$)

$$\begin{aligned}
\text{Time } AB_1 + \text{Time } BA_1 &= P, \\
\text{,, } AB_2 - \text{ ,, } BA_2 &= 0, \\
\text{,, } AB_3 - \text{ ,, } BA_3 &= 0, \\
\text{,, } AB_4 - \text{ ,, } BA_4 &= 0, \\
\text{,, } AB_5 + \text{ ,, } BA_5 &= P,
\end{aligned}$$

which are the relations which in fact subsist between the times AB_1 and BA_1 &c.

473.

ON THE GRAPHICAL CONSTRUCTION OF THE UMBRAL OR PENUMBRAL CURVE AT ANY INSTANT DURING A SOLAR ECLIPSE.

[From the *Monthly Notices of the Royal Astronomical Society*, vol. XXX. (1869—1870), pp. 162—164.]

THE curve in question, say the penumbral curve, is the intersection of a sphere by a right cone,—I wish to show that the stereographic projection of this curve may be constructed as the envelope of a variable circle, having its centre on a given conic, and cutting at right angles a fixed circle; this fixed circle being in fact the projection of the circle which is the section of the sphere by the plane through the centre and the axis of the cone, or say by the axial plane. The construction thus arrived at is Mr Casey's construction for a bicircular quartic; and it would not be difficult to show that the stereographic projection of the penumbral curve is in fact a bicircular quartic.

The construction depends on the remark that a right cone is the envelope of a variable sphere, having its centre on a given line and its radius proportional to the distance of the centre from a given point on this line; and on the following theorem of plane geometry:

Imagine a fixed circle, and a variable circle having its centre on a given line and its radius proportional to the distance of the centre from a given point on the line (or, what is the same thing, the variable circle always touches a given line); then the locus of the pole in regard to the fixed circle, of the common chord of the two circles (or, what is the same thing, the locus of the centre of a new variable circle which cuts the fixed circle at right angles in the points where it is met by the first-mentioned variable circle) is a conic.

To fix the ideas, say that P is the centre of the first variable circle; AB its common chord with the fixed circle; Q the centre of the circle which cuts the fixed circle at right angles in the points A and B; then the locus of Q is a conic.

To prove this, take $x^2+y^2=1$ for the equation of the fixed circle, $(x-\alpha)^2+(y-\beta)^2=\gamma^2$ for that of the variable circle; the foregoing law of variation being in fact such that α, β, γ, are linear functions of a variable parameter θ; the equation of the common chord AB is $-2\alpha x-2\beta y+1+\alpha^2+\beta^2-\gamma^2=0$; viz., this equation contains θ quadratically; hence the envelope of the common chord is a conic; and thence (reciprocating in regard to the fixed circle) the locus of the pole of AB, that is, of the point Q, is also a conic.

Consider now a solid figure in which the circles are replaced by spheres; viz. we have a fixed sphere, and a variable sphere having its centre on a given line and its radius proportional to the distance of the centre from a given point on the line. The envelope of the variable sphere is a right cone; the intersection of the cone with the fixed sphere is the envelope of the small circle of the sphere, say the circle AB, which is the intersection of the fixed sphere by the variable sphere. This circle AB is also the intersection of the fixed sphere by a sphere, centre Q, which cuts the fixed sphere at right angles; and by what precedes the locus of Q is a conic. Hence the penumbral curve is given as the envelope of the circle AB which is the intersection of the fixed sphere by a sphere which has its centre Q on a conic, and which cuts the fixed sphere at right angles. It is obvious that the circle AB always cuts at right angles the great circle which is the section of the fixed sphere by the axial plane, or say the axial circle. Project the whole figure stereographically; the projection of the circle AB is a variable circle which cuts at right angles the circle which is the projection of the axial circle, and which has for its centre the point Q' which is the projection of Q. But the locus of Q being a conic, the locus of its projection Q' is also a conic; and we have thus the projection of the penumbral curve as the envelope of a variable circle which has its centre on a conic, and which cuts at right angles a fixed circle.

We may in the axial plane construct five points of the conic which is the locus of Q, by means of any five assumed positions of the variable circle, and somewhat simplify the construction by a proper choice of the five positions of the variable circle. This is not a convenient construction, and even if it were accomplished we should still have to construct the projection of the conic so obtained, in order to find, in the figure of the stereographic projection, the conic which is the locus of Q'. I do not at present perceive any direct construction for the last-mentioned conic; but assuming that a tolerably simple construction can be obtained, the construction of the projection of the penumbral curve as the envelope of the variable circle is as easy and rapid as possible. Probably the easiest course would be (without using the conic at all) to calculate numerically, for a given position of the variable sphere, the terrestrial latitude and longitude of the two points of intersection of the variable sphere by the axial circle; laying these down on the projection, we have then a position of the variable circle; and a small number of properly selected positions would give the penumbral curve with tolerable accuracy.

I have throughout spoken of the penumbral curve, as it is in regard hereto that a graphical construction is most needed; but the theory is applicable, without any alteration, to the umbral curve.

474.

ON THE GEOMETRICAL THEORY OF SOLAR ECLIPSES.

[From the *Monthly Notices of the Royal Astronomical Society*, vol. XXX. (1869—1870), pp. 164—168.]

THE fundamental equation in a solar eclipse is, I think, most readily established as follows:

Take the centre of the Earth for origin, and consider a set of axes fixed in the Earth and moveable with it; viz., the axis of z directed towards the North Pole; those of x, y, in the plane of the Equator; the axis of x directed towards the point longitude 0°; that of y towards the point longitude 90° W. of Greenwich. Take a, b, c, for the coordinates of the Moon; k for its radius (assuming it to be spherical); a', b', c', for the coordinates of the Sun; k' for its radius (assuming it to be spherical); then, writing $\theta+\phi=1$, the equation

$$\{\theta(x-a)+\phi(x-a')\}^2+\{\theta(y-b)+\phi(y-b')\}^2+\{\theta(z-c)+\phi(z-c')\}^2=(\theta k\pm\phi k')^2$$

is the equation of the surface of the Sun or Moon, according as $\theta,\ \phi=1,\ 0$ or $=0,\ 1$: and for any values whatever of θ, ϕ, it is that of a variable sphere, such that the whole series of spheres have a common tangent cone. Writing the equation in the form

$$\begin{aligned}&\theta^2\{(x-a)^2+(y-b)^2+(z-c)^2-k^2\}\\+&2\theta\phi\{(x-a)(x-a')+(y-b)(y-b')+(z-c)(z-c')\mp kk'\}\\+&\phi^2\{(x-a')^2+(y-b')^2+(z-c')^2-k'^2\}=0,\end{aligned}$$

or, putting for shortness,

$$\begin{aligned}\rho&=a^2+b^2+c^2-k^2\\ \rho'&=a'^2+b'^2+c'^2-k'^2\\ \sigma&=aa'+bb'+cc'\mp kk'\\ P&=ax+by+cz\\ P'&=a'x+b'y+c'z,\end{aligned}$$

the equation is

$$\begin{aligned}&\theta^2\ (x^2+y^2+z^2-2P+\rho)\\ +\,&2\theta\phi\ (x^2+y^2+z^2-P-P'+\sigma)\\ +\,&\phi^2\,(x^2+y^2+z^2-2P'+\rho')=0,\end{aligned}$$

and the equation of the envelope consequently is

$$(x^2+y^2+z^2-2P+\rho)(x^2+y^2+z^2-2P'+\rho')-(x^2+y^2+z^2-P-P'+\sigma)^2=0,$$

that is

$$(x^2+y^2+z^2)(\rho+\rho'-2\sigma)-(P-P')^2-2(\rho'-\sigma)P-2(\rho-\sigma)P'+\rho\rho'-\sigma^2=0,$$

which is the equation of the cone in question.

Observe that one sphere of the series is a *point*, viz., taking first the upper signs if we have $\theta k+\phi k'=0$, that is

$$\theta=\frac{k'}{k'-k},\quad \phi=\frac{-k}{k'-k},$$

then the sphere in question is the point the coordinates whereof are

$$x=\frac{k'a-ka'}{k'-k},\quad y=\frac{k'b-kb'}{k'-k},\quad z=\frac{k'c-kc'}{k'-k},$$

which point is the vertex of the cone: it hence appears that, taking the upper signs, the cone is the *umbral* cone, having its vertex on this side of the Moon; and similarly taking the lower signs, then if we have $\theta k-\phi k'=0$, that is

$$\theta=\frac{k'}{k'+k},\quad \phi=\frac{k}{k'+k},$$

then the variable sphere will be the point the coordinates of which are

$$\frac{k'a+ka'}{k'+k},\quad \frac{k'b+kb'}{k'+k},\quad \frac{k'c+kc'}{k'+k},$$

which point is the vertex of the cone; viz. the cone is here the penumbral cone having its vertex between the Sun and Moon.

Taking as unity the Earth's equatorial radius, if p, p' are the parallaxes, κ, κ' the angular semi-diameters of the Moon and Sun respectively, then the distances are $\dfrac{1}{\sin p}$, $\dfrac{1}{\sin p'}$ and the radii are $\dfrac{\sin\kappa}{\sin p}$, $\dfrac{\sin\kappa'}{\sin p'}$ respectively; hence, if h, h' are the hour-angles west from Greenwich, Δ, Δ' the N.P.D.'s of the Moon and Sun respectively, we have

$$\begin{aligned} a&=\frac{1}{\sin p}\sin\Delta\cos h, & a'&=\frac{1}{\sin p'}\sin\Delta'\cos h',\\ b&=\frac{1}{\sin p}\sin\Delta\sin h, & b'&=\frac{1}{\sin p'}\sin\Delta'\sin h',\\ c&=\frac{1}{\sin p}\cos\Delta\quad, & c'&=\frac{1}{\sin p'}\cos\Delta',\\ k&=\frac{\sin\kappa}{\sin p}\quad, & k'&=\frac{\sin\kappa'}{\sin p'};\end{aligned}$$

and thence

$$\rho = \frac{1}{\sin^2 p}(1 - \sin^2 \kappa'),$$

$$\rho' = \frac{1}{\sin^2 p'}(1 - \sin^2 \kappa'),$$

$$\sigma = \frac{1}{\sin p \sin p'}[\cos \Delta \cos \Delta' + \sin \Delta \sin \Delta' \cos (h' - h) \mp \sin \kappa \sin \kappa'],$$

$$P = \frac{1}{\sin p} \quad \{\sin \Delta \ (x \cos h + y \sin h) + z \cos \Delta\},$$

$$P' = \frac{1}{\sin p'} \quad \{\sin \Delta' (x \cos h' + y \sin h') + z \cos \Delta'\}.$$

Moreover, if the right ascensions of the Moon and Sun are α, α' respectively, and if the R.A. of the meridian of Greenwich (or sidereal time in angular measure) be $= \Sigma$, then we have

$$h = \Sigma - \alpha, \quad h' = \Sigma - \alpha'.$$

It is to be observed that $h - h'$, Δ, Δ' are slowly varying quantities, viz., their variation depends upon the variation of the celestial positions of the Sun and Moon; but h and h' depend on the diurnal motion, thus varying about 15° per hour; to put in evidence the rate of variation of the several angles h, h', Δ, Δ' during the continuance of the eclipse, instead of the foregoing values of h, h', I write

$$h' = \left\{E + \left(1 + \frac{E_1 - E}{24}\right) t\right\} 15^\circ,$$

where t is the Greenwich mean time, E, E_1 are the values (reckoned in parts of an hour) of the Equation of Time at the preceding and following mean noons respectively, taken positively or negatively, so that E, E_1 are the mean times of the two successive apparent noons respectively; whence also

$$h = \left\{E + \left(1 + \frac{E_1 - E}{24}\right) t\right\} 15^\circ - \alpha + \alpha';$$

and moreover

$$\begin{aligned}
\alpha &= A + m \ (t - T), \\
\alpha' &= A + m' (t - T), \\
\Delta &= D + n \ (t - T), \\
\Delta' &= D' + n' \ (t - T),
\end{aligned}$$

if T be the time of conjunction, A, A, D, D' the values at that instant of the R.A.'s and N.P.D.'s; m, m' and n, n' the horary motions in R.A. and N.P.D. respectively.

It appears to me not impossible but that the foregoing form of equation,

$$(x^2+y^2+z^2)(\rho+\rho'-2\sigma)-(P-P')^2-2(\rho'-\sigma)P-2(\rho-\sigma)P'+\rho\rho'-\sigma^2=0,$$

for the umbral or penumbral cone might present some advantage in reference to the calculation of the phenomena of an eclipse over the Earth generally: but in order to obtain in the most simple manner the equation of the same cone referred to a set of principal axes, I proceed as follows:

Writing

$$\begin{aligned} \mathrm{a} &= bc'-b'c, & \mathrm{f} &= a-a', \\ \mathrm{b} &= ca'-c'a, & \mathrm{g} &= b-b', \\ \mathrm{c} &= ab'-a'b, & \mathrm{h} &= c-c', \end{aligned}$$

$$(\text{and therefore}\quad \mathrm{af}+\mathrm{bg}+\mathrm{ch}=0).$$

Then, if

$$X=\frac{(\mathrm{bh}-\mathrm{cg})\,x+(\mathrm{cf}-\mathrm{ah})\,y+(\mathrm{ag}-\mathrm{bf})\,z}{\sqrt{\mathrm{a}^2+\mathrm{b}^2+\mathrm{c}^2}\,\sqrt{\mathrm{f}^2+\mathrm{g}^2+\mathrm{h}^2}},$$

$$Y=\frac{\mathrm{a}x+\mathrm{b}y+\mathrm{c}z}{\sqrt{\mathrm{a}^2+\mathrm{b}^2+\mathrm{c}^2}},$$

$$Z=\frac{\mathrm{f}x+\mathrm{g}y+\mathrm{h}z}{\sqrt{\mathrm{f}^2+\mathrm{g}^2+\mathrm{h}^2}},$$

X, Y, Z, will be coordinates referring to a new set of rectangular axes; viz., the origin is, as before, at the centre of the Earth, the axis of Z is parallel to the line joining the centres of the Sun and Moon; the axis of X cuts at right angles the last-mentioned line; and the axis of Y is perpendicular to the plane of the other two axes; or, what is the same thing, to the plane through the centres of the Earth, Sun, and Moon.

The coordinates of the vertex of the cone are therefore X_0, Y_0, Z_0, where these denote what the foregoing values of X, Y, Z, become on substituting therein for x, y, z, the values

$$\frac{k'a\mp ka'}{k'\mp k},\quad \frac{k'b\mp kb'}{k'\mp k},\quad \frac{k'c\mp kc'}{k'\mp k},$$

and the equation of the cone therefore is

$$(X-X_0)^2+(Y-Y_0)^2=\tan^2\lambda\,(Z-Z_0)^2,$$

where

$$\sin\lambda=\frac{k'\mp k}{G},$$

if for a moment G denotes the distance between the centres of the Sun and Moon. We have therefore

$$\tan\lambda = \frac{k' \mp k}{\sqrt{G^2 - (k' \mp k)^2}},$$

or since

$$G^2 = (a' - a)^2 + (b' - b)^2 + (c' - c)^2,$$

this is in fact

$$\tan\lambda = \frac{k' \mp k}{\sqrt{\rho + \rho' - 2\sigma}},$$

where ρ, ρ', σ signify as before; and thus X_0, Y_0, Z_0, $\tan\lambda$ are all of them given functions of a, b, c, k, a', b', c', k', and consequently of the before-mentioned astronomical data of the problem. The form is substantially the same as Bessel's equation (3), *Ast. Nach.* No. 321 (1837), (but the direction of the axes of X, Y is not identical with those of his x, y); and it is therefore unnecessary to consider here the application of it to the calculation of the eclipse for a given point on the Earth.

475.

ON A PROPERTY OF THE STEREOGRAPHIC PROJECTION.

[From the *Monthly Notices of the Royal Astronomical Society*, vol. XXX. (1869—1870), pp. 205—207.]

I AM not aware whether it has been noticed that the very same circles which in the direct stereographic projection of a hemisphere (viz., that wherein the projection is on the plane of a meridian) represent the meridians and parallels respectively,—represent also in the oblique projection of the hemisphere meridians and parallels respectively. In fact, in the direct projection where the poles N, S, are in the horizon-meridian, or bounding circle of the projection, if we take a chord AB at right angles to NS, and on AB as diameter describe a circle, the original (meridian and parallel) circles will, as the appearance of the figure at once suggests, represent meridians and parallels in the oblique projection in which the horizon or bounding circle of the projection is the circle diameter AB, and where consequently the North Pole N is brought into view, the South Pole S being beyond the limits of the projection. That this really is so, is clear from the consideration that in any stereographic projection whatever, the meridians will be circles passing through two fixed points N, S, and the parallels be circles cutting the meridians at right angles. (Or, what is the same thing, the parallels also pass each of them through two fixed imaginary points, the antipoints of N, S, but this in passing.) And moreover since in the oblique, as well as in the direct, projection, the longitude of any meridian, as reckoned from the central meridian NS, is the angle at N between the two meridians, the longitude for a given meridian is the same in the two projections respectively. But the co-latitudes are not the same in the two projections respectively; viz., a circle which in the direct projection represents the parallel co-latitude c, will in the oblique projection represent the parallel of a different co-latitude c'. The relation between the values of c, c', will of course depend upon the position of the bounding circle AB of the oblique direction: to define this position, we may use either the arc NM which in the direct projection determines the co-latitude of the centre M of the oblique projection (say $NM=\Delta$, that is, $NV=\Delta$), or by the arc NM which in the oblique projection determines the distance of N from the centre, or co-latitude of the

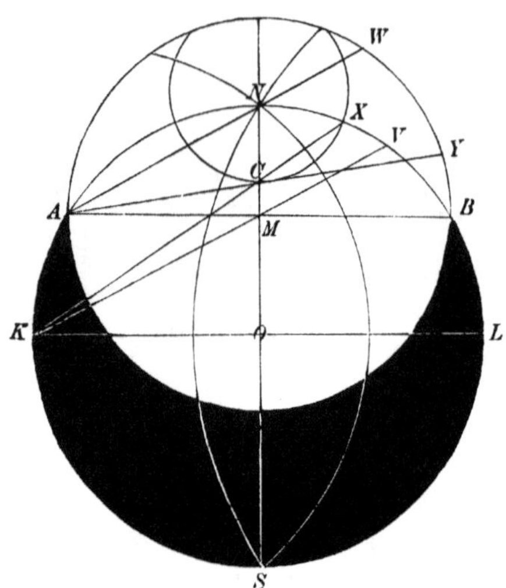

centre (say $NM=\Delta'$, that is, $BW=\Delta'$). The obliquity in the oblique projection is thus $90°-\Delta'$, viz., this is the inclination of the plane of projection to that of the horizon-meridian in the direct projection. We have also $c=NX$, $c'=WY$. The relation between the angles Δ, Δ', is easily found to be

$$\tan\tfrac{1}{2}\Delta=\tan^2\tfrac{1}{2}\Delta',$$

viz., taking the radius in the direct projection to be $=1$, we have

$$\begin{aligned} OM &= \tan\tfrac{1}{2}(90°-\Delta),\\ MA &= \sqrt{1-\tan^2\tfrac{1}{2}(90°-\Delta)},\\ MN &= 1-\tan\tfrac{1}{2}(90°-\Delta); \end{aligned}$$

wherefore

$$\sqrt{1-\tan^2\tfrac{1}{2}(90°-\Delta)}\,.\tan\tfrac{1}{2}\Delta'=1-\tan\tfrac{1}{2}(90°-\Delta),$$

and thence

$$\tan^2\tfrac{1}{2}\Delta'=\frac{1-\tan\tfrac{1}{2}(90°-\Delta)}{1+\tan\tfrac{1}{2}(90°-\Delta)}=\tan\tfrac{1}{2}\Delta,$$

the required relation.

We have moreover

$$\begin{aligned} NC = 1 - \tan\tfrac{1}{2}(90^\circ - c) = AM & \;\{\tan\tfrac{1}{2}\Delta' - \tan\tfrac{1}{2}(\Delta' - c')\}, \\ & = \sin\Delta' \{\tan\tfrac{1}{2}\Delta' - \tan\tfrac{1}{2}(\Delta' - c')\}, \\ & = 2\sin^2\tfrac{1}{2}\Delta' - \sin\Delta' \tan\tfrac{1}{2}(\Delta' - c'), \end{aligned}$$

that is

$$\begin{aligned} \tan\tfrac{1}{2}(90^\circ - c) & = \cos\Delta' + \sin\Delta' \tan\tfrac{1}{2}(\Delta' - c'), \\ & = \frac{\cos\tfrac{1}{2}(\Delta' + c')}{\cos\tfrac{1}{2}(\Delta' - c')}, \end{aligned}$$

or, what is the same thing,

$$\frac{1 - \tan\tfrac{1}{2}c}{1 + \tan\tfrac{1}{2}c} = \frac{1 + \delta\tan\tfrac{1}{2}c' \tan\tfrac{1}{2}\Delta'}{1 + \tan\tfrac{1}{2}c' \tan\tfrac{1}{2}\Delta'},$$

that is

$$\tan\tfrac{1}{2}c = \tan\tfrac{1}{2}\Delta' \tan\tfrac{1}{2}c',$$

which is the required relation between c and c'. In the particular case $\Delta = \Delta' = 90^\circ$, the two projections coincide, and we have, as we should do, $c' = c$.

476.

ON THE DETERMINATION OF THE ORBIT OF A PLANET FROM THREE OBSERVATIONS.

[From the *Memoirs of the Royal Astronomical Society*, vol. XXXVIII. (1870), pp. 17—111. Read December 10, 1869.]

I PROPOSE to consider from a geometrical point of view the problem of the determination of the orbit of a planet from three observations. The orbit is a conic, having the Sun for a focus; and each observation shows that the planet is at the date thereof in a given line. We have thus a given point or focus S, and three given lines, say the "rays." The orbit-plane, if known, would, by its intersections with the three rays, determine the three positions of the planet; that is, we should have the focus and three points on the orbit; or (what is the same thing) three radius vectors from the focus, say a "trivector." Geometrically, through three given points, and with a given focus, there may be described four conics; but (as will be explained) there is only one of these which can be the orbit; we may therefore say that the orbit will be determined, and that uniquely, by means of a given trivector. The problem is therefore to find the orbit-plane, such that in the orbit determined by means of the trivector the times of passage between the three positions on the orbit may have the observed values; or (what is the same thing) that the orbital areas, each divided by the square root of the latus rectum, may have given values. If, instead of the orbit-plane, we consider the orbit-axis (that is, the line normal to the orbit-plane at the point S), or, what is more convenient, the orbit-pole, or intersection of the axis with a sphere about the centre S; then to a given position of the orbit-pole, there corresponds, as above, a determinate orbit; and the problem is to find the position of the orbit-pole, so that in the orbit belonging thereto the times of passage may have given values as already mentioned; and it is clear that the required position of the orbit-pole may be obtained as the intersection of two spherical curves; the one of them, the locus of those positions of the orbit-pole for which the time of passage between the first and second points on the orbit has its proper given value; the other of them, the locus of those positions for which the time of passage between the second and third points on the orbit has its proper given value: and in connexion therewith we may consider other isoparametric loci of the orbit-pole; for instance, the iseccentric lines, or loci of the orbit-pole such that along each of them the eccentricity of the orbit has a given value. It is in this point of view that the problem is considered in the present memoir, viz., the object proposed is the discussion of the configuration, &c. of these loci. I consider, in the first instance, any three given rays whatever; but in the ulterior discussion of the spherical curves, which it is difficult to carry out otherwise than numerically, I have confined myself to the case of a particular symmetrical position of the three rays; viz., these are taken to be lines each of them at an inclination of 60° to a fixed plane through S, and such that their projections on this plane form an equilateral triangle having S for its centre, and that each ray cuts the plane in the mid-point of the corresponding side of the triangle.

The general theory as above explained is further developed in the memoir; and I consider the formulæ for the determination of the orbit, &c. by means of a given trivector; those relating to the determination of the trivector obtained as above by means of a variable plane passing through a given point and intersecting three given rays; and lastly, the application to the particular system of three rays already referred to. The Plates refer to this particular system; they are as follow:

Plate 1.	General Planogram for a single ray,	See Nos. 8—10 for explanation of the terms Planogram and Spherogram.
„ 2.	Planogram for Meridian 90°—270°,	
„ 3.	Planogram for Meridian 0°—180°,	
„ 4.	Spherogram for the Eccentricity,	
„ 5.	Spherogram for the Time.	

Article Nos. 1 to 14. *Considerations on the General Theory.*

1. As explained in the introduction, we have a point or focus S, and three lines called the "rays." The orbit-plane is any plane through S; it meets the rays in three points, which are points on the orbit; and joining these with S, we have a "trivector." The orbit is for the present considered as in general uniquely determined by means of the trivector.

2. There are certain critical positions of the orbit-plane.

First, the orbit-plane may be parallel to one of the rays; or (what is the same thing) it may pass through the line through S parallel to the ray: the point on the ray is at infinity; or say that it is at an indefinitely great distance in one direction or in the other direction along the ray; and (from the particular way in which the orbit is selected as one of four conics) there is, as will appear (see *post,* No. 20), a discontinuity of orbit as the point passes from the one to the other of these positions.

3. *Secondly*, the orbit-plane may be parallel to two of the rays; or (what is the same thing) it may pass through the lines through S parallel to these two rays; the points on the two rays are each at infinity; viz. each of them is at an indefinitely great distance in one or the other direction along the ray; and there is a discontinuity of the orbit as each point passes from the one to the other of its two positions.

4. *Thirdly*, the orbit-plane may be such that the orbit is a right line. To see how this arises, observe that we may consider a system of lines meeting each of the three rays, and of course generating a hyperboloid; say these are the generating lines: there is on the hyperboloid another system of lines, say the directrix lines, in which are included the three rays; the point S is not on the hyperboloid. Then, if the orbit-plane pass through a generating line, it will meet the three rays in the points in which these are met by the generating line: and the orbit is, consequently, the generating line (described, as being a right line not passing through S, with a velocity $=\infty$). Any plane through S and a generating line also meets the hyperboloid in a directrix line; and consequently touches it at the intersection of the two lines, viz. it is a tangent plane of the hyperboloid. The planes in question thus envelope the circumscribed cone whose vertex is S; or (what is the same thing) when the orbit-plane is any tangent-plane of this cone, the orbit is a right line.

5. The only exception is, *fourthly*, when the orbit-plane passes through one of the rays. Observe that the plane then meets the hyperboloid in another line, that is, a generating line, or the case under consideration is included in the third case; it is also included in the first case. The point on the ray in question is here not a determinate point, but any point whatever of the ray; the points on the other two rays being (as in general) determinate: the orbit is consequently indeterminate; viz. to any point selected at pleasure as the intersection of the orbit-plane with the ray contained therein, there corresponds a determinate orbit (in particular, the selected point may be such that the orbit is, as in the third case, a right line); and, corresponding to the position in question of the orbit-plane, we have the entire system of such orbits.

6. Consider now the corresponding positions of the orbit-pole on a sphere described about the centre S. It will be convenient for the moment to attend to the two opposite positions of the orbit-pole belonging to any position of the orbit-plane, and thus to regard the orbit-pole as moving over the entire spherical surface. The parallel through S to a ray meets the sphere in two points, poles of a great circle which I call a "separator;" we have thus three separators, each two meeting in a pair of opposite points which I call the points B; viz., these are the intersections with the sphere of a line through S perpendicular to the plane containing the parallels of the two rays. A line through S perpendicular to the plane through a ray meets the sphere in a pair of opposite points which I call the points A; these lying on the corresponding separator; there are thus three pairs of points A. The cone reciprocal to the circumscribed cone (that is, generated by a line through S at right angles to any tangent plane of the circumscribed cone) meets the sphere in a spherical conic which I call the "regulator;" this touches each of the separators at the pair of points A on such separator.

7. I say that in the *first* of the cases above considered the locus of the orbit-pole is a separator; in the *second* case the orbit-pole is a point B; in the *third* case the locus is the regulator; and in the *fourth* case the orbit-pole is a point A.

8. In the absence of models, the spherical figure must be represented by a projection; the stereographic projection is convenient for facility of description; and it has the very great advantage that we can by means of it exhibit, no matter how large a portion of the spherical surface. In the figures called "spherograms," afterwards referred to, the representation of a hemisphere is all that is required; but, to give a more distinct general idea, I annex a figure representing a larger portion of the surface; the data are those belonging to the particular symmetrical case referred to as intended to be specially considered: and the regulator conic is accordingly a pair of opposite small circles, the points A and B being related to it symmetrically; but, disregarding these specialities, the figure is adapted to the illustration of the general

Fig. 1.

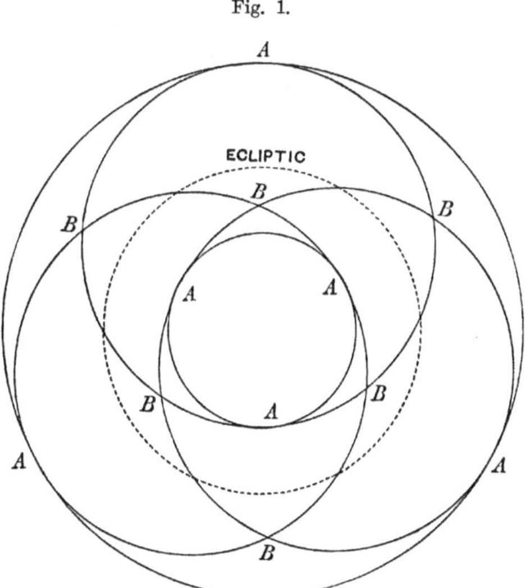

case (at least if the point S be situate *within* the hyperboloid), and it is here given for that purpose. The circle marked "Ecliptic" does not properly belong to the figure: it is added as showing the boundary of a hemisphere, so that, by omitting all that lies outside this circle, the figure would be limited to the representation of a hemisphere; and the orbit-pole be in every case represented, no longer as a pair of opposite points, but as a single point; we should have the separators each as a half circle, and the regulator as a single small circle; the separators would intersect in pairs, in the *three* points B, and would touch the regulator in the *three* points A, &c.

9. The figure constructed as above, but omitting so much of it as lies outside the ecliptic circle, is the representation of a hemisphere—say of the northern hemi-

sphere. It is readily seen that the central triangle BBB and the three circumjacent triangles BBB, represent also the half-surface of the sphere, viz., instead of the omitted portions of the northern hemisphere we have the equal opposite portions of the southern hemisphere. The adoption of this figure as the representation of the half-surface of the sphere has the great advantage that the spherical curves can be delineated without the apparent breaks which would otherwise occur at their intersections with the ecliptic circle: I accordingly adopt it, and call the figure in question (viz., that composed of the four triangles) a blank "spherogram." We wish for any given position thereon of the orbit-pole to determine the values of certain parameters (eccentricity, latus rectum, time of passage between two rays, &c., as the case may be) belonging to the orbit, with a view to the subsequent delineation of the corresponding isoparametric (iseccentric, isochronic, &c.) lines, so constructing a "spherogram" for any such parameter, or system of lines.

10. It is for this purpose convenient to consider the values of the parameter corresponding to a single series of positions of the orbit-pole, viz., we consider the orbit-pole as describing on the sphere a curve selected at pleasure. Consider for a moment the orbit-plane as a material plane rigidly connected with the orbit-axis; the motion of the orbit-pole does not absolutely determine the motion of the orbit-plane, inasmuch as the orbit-plane, occupying the same position in space, might rotate about the orbit-axis; but if we exclude any such motion by the assumption that the motion of the orbit-plane is always about an axis in the orbit-plane, then the motion of the orbit-pole determines that of the orbit-plane, viz., the orbit-plane envelopes a cone, the reciprocal to that described by the orbit-axis. If then on the orbit-plane in each position thereof we mark, as well its line of contact with the enveloped cone, as also its intersections with the three rays, we obtain a figure (which may, if we please, be regarded as drawn on the orbit-plane in some particular position thereof), such figure consisting of a series of trivectors, and (belonging to each of them) a line through S serving to fix the position of the trivector in space. The locus of each extremity of the trivector is a certain curve, and the construction establishes a point-to-point correspondence between these three curves; viz., to any point on one of them there corresponds on each of the other two a single point, the three points being the extremities of a trivector. The figure would be rendered more complete by drawing the orbit belonging to each trivector thereof. Such a figure (with or without the orbits) is termed a "planogram."

11. The most simple case is when the orbit-pole describes a great circle; the orbit-plane here rotates about a fixed line, the axis of the circle, or (what is the same thing) the enveloped cone reduces itself to this axis of rotation; and the line of contact is thus a fixed line in the orbit-plane; or (what is the same thing) the lines through S in the planogram are here a single fixed line, the axis of rotation. I say that, for each extremity of the trivector, the locus is a hyperbola, having the axis of rotation for its conjugate axis. In fact, attending to any one ray, it is the same thing whether the orbit-plane be made to revolve round the axis of rotation, so as continually to intersect the ray, or whether, considering the orbit-plane as fixed,

and the ray as rigidly connected with the axis, we make the ray to rotate about this axis, so as continually to intersect the orbit-plane. But in this last case the ray describes about the axis a hyperboloid of revolution, and the orbit-plane, as an axial plane, meets this surface in a hyperbola having the axis for its conjugate axis; which hyperbola is the required locus of the trivector-extremity. It is moreover easy to see that if the angle of position of the variable orbit-plane, or (what is the same thing) the angle of position of the orbit-pole in the great circle which it describes be $=q$ (where q is measured from any fixed plane or point), and if the coordinates x' and y' be measured from S in the direction of and perpendicular to the axis of rotation, then the coordinates of the point on the hyperbola are expressed in the form $x' = \alpha + a \tan(q+\beta)$, $y' = b \sec(q+\beta)$, where α, a, b, β, are constants depending on the position of the ray in regard to the axis of rotation: see as to this *post*, No. 49.

12. Considering the orbit-pole as describing a given curve, the value for the several positions thereof of any parameter of the orbit may be exhibited by means of a "diagram," viz., we may take for abscissa any quantity serving to fix the position of the orbit-pole on the described curve, and for ordinate the value of the parameter in question. In the particular case where the orbit-pole describes a great circle passing through the axis of the stereographic projection, and which is consequently in the spherogram represented by a diameter of the ecliptic or bounding circle, it is natural to take for the abscissa the distance (from the centre) of the representation of the orbit-pole; the diagram will then fit on to the diameter, and for any position of the orbit-pole on such diameter give at once the value of the parameter to which the diagram relates.

13. It is right to remark that the construction of planograms and diagrams is merely subsidiary to that of the spherograms; the information given by any number of planograms or diagrams would be all of it embodied in a spherogram for the same parameter. And theoretically the construction of a spherogram is a mere matter of geometry; for a given position of the orbit-pole we construct the trivector, thence the orbit, and in relation thereto any parameters which it is desired to consider; and so, for a sufficient number of points on the spherogram, determine the value of the parameter, or parameters; and lay down the isoparametric lines. The construction of the orbit from a given trivector, and in particular the selection of the orbit as one of the four conics given by the trivector, has not yet been explained: in connexion herewith we have the discontinuity of orbit which arises when the orbit-pole is upon a separator, and which is a leading circumstance in the theory; until it is gone into, there is little more to be said in the way of general explanation as to the spherogram, or the isoparametric lines thereof.

14. It may however be noticed that for any parameter whatever, the points A of the spherogram are common points, through which pass in general the lines belonging to any value whatever of the parameter; the reason of course is that the orbit-plane then passing through the ray, and the orbit itself being indeterminate, the value of any parameter belonging to the orbit is also indeterminate. Moreover, for some parameters the curve belonging to any particular value of the parameter not only

passes through the points A, but passes through each point twice, or (what is the same thing) has each of the points A for a nodal point; when this is so, then it is to be further observed that, for certain values of the parameters, they will be acnodal points, properly belonging to the curve, although there is not any real branch of the curve passing through the points A; for others they will be crunodal points, with two real branches through each; and in the transition between the two cases they will be cuspidal points on the isoparametric curve; it will appear in the sequel that this is really the case in regard to the iseccentric lines.

Article Nos. 15 to 30. *Determination of the Orbit from a given Trivector.*

15. With a given point S as focus, and through three given points, that is with a given trivector, there may be described *four* conics. This appears from the general theory according to which a given focus is equivalent to two given tangents; and also

Fig. 2.

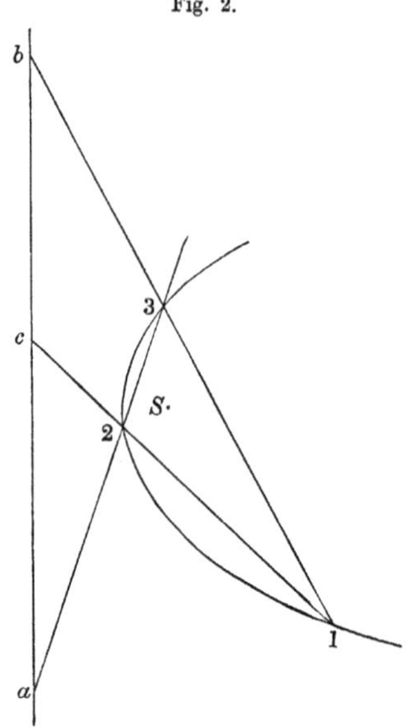

from the geometrical construction, *Principia*, book I. sect. 4; Scholium to Prop. XXI.: viz. given the focus S and the points 1, 2, 3, then if

On 23 we find a so that $a2 : a3 = S2 : S3$,

„ 31 „ b „ $b3 : b1 = S3 : S1$,

„ 12 „ c „ $c1 : c2 = S1 : S2$,

the points a, b, c, are each of them on the directrix, so that any two of them determine the directrix. In the figure (as in Newton's) the distances $S1$, $S2$, $S3$, are each regarded as positive, but the very same construction, taking two of the distances each as positive and negative successively, would lead to three other positions of the directrix; or the construction would give in all four conics.

16. In the figure the directrix lies on the same side of the three points; and the conic is thus an ellipse or parabola, or, if a hyperbola, then the three points lie in the same branch thereof; and it is consequently an orbit such that along it a body can pass through the three points successively. The construction as varied would give in each case a directrix having on one side of it one, and on the other side two, of the three points; so that the conic would be a hyperbola having the three points not on the same branch thereof; consequently it would not be an orbit such that along it a body could pass through the three points successively.

And it thus appears that though the trivector really determines four conics, yet it is only one of these in which the directrix lies on the same side of the three points; and this conic I call the "orbit:" the given trivector thus determines a single orbit.

17. It is to be noticed however that the orbit constructed as above *may* be a hyperbolic branch separated by the directrix from the focus S, and consequently convex to the focus S; viz., the three points lie here in a hyperbolic branch convex to S, and which is therefore not an orbit which can be described under the action of an *attractive* force at S: say we have a "convex orbit." I regard this as a real orbit, but the times of passage therein as imaginary, or rather as non-existent, and the case is thus excluded from consideration in the formulæ and figures which relate to the times of passage.

18. The same results are established analytically in a very similar manner, viz., taking the focus for origin and starting from the focal equation

$$r = Ax + By + C;$$

then if we take (x_1, y_1), (x_2, y_2), (x_3, y_3), as the coordinates of the three given points and write

$$r_1 = \sqrt{x_1^2 + y_1^2},\quad r_2 = \sqrt{x_2^2 + y_2^2},\quad r_3 = \sqrt{x_3^2 + y_3^2},$$

we have for the determination of the constants

$$\begin{aligned} r_1 &= Ax_1 + By_1 + C,\\ r_2 &= Ax_2 + By_2 + C,\\ r_3 &= Ax_3 + By_3 + C, \end{aligned}$$

and the equation therefore is

$$\begin{vmatrix} r, & x, & y, & 1 \\ r_1, & x_1, & y_1, & 1 \\ r_2, & x_2, & y_2, & 1 \\ r_3, & x_3, & y_3, & 1 \end{vmatrix} = 0,$$

which, attributing therein to r_1, r_2, r_3, the signs $+$, $-$ at pleasure, represents eight different equations: these however give only four conics, viz., we have the same conic whether we attribute to r_1, r_2, r_3, any particular combination of signs, or reverse all the signs simultaneously.

19. But the focal equation $r = Ax + By + C$ is precisely equivalent to the equation

$$r = \frac{p}{1 + e\cos(\theta - \varpi)},$$

and in this equation (taking as is allowable p as positive) then if $\pm e$ be $=$ or < 1, that is for an ellipse or parabola whatever be the value of $\theta - \varpi$, r is always positive; but if $\pm e$ be > 1, that is for a hyperbola, r is positive for those values of $\theta - \varpi$ which belong to one branch, negative for those which belong to the other branch, of the curve. Hence in the determinant equation, unless r_1, r_2, r_3, have the same sign, the curve will be a hyperbola with the points two of them on one branch, the third on the other branch thereof. But in the remaining case, when r_1, r_2, r_3, have all the same sign, or say when they are all positive, then the conic is an ellipse or parabola, or else it is a hyperbola with the three points on the same branch thereof; that is, the foregoing determinant equation, regarding therein r_1, r_2, r_3, as all of them positive, gives the orbit.

20. When one of the points is at infinity on a given line there is a discontinuity of orbit. To explain this, suppose that the point (x_1, y_1) is situate on the line $y = x\tan\alpha$, at an indefinitely great distance r_1 in one or the other direction along the line; viz., r_1 is an indefinitely large positive quantity, and we have in the one case $x_1, y_1 = r_1\cos\alpha, r_1\sin\alpha$; and in the other case $x_1, y_1 = -r_1\cos\alpha, -r_1\sin\alpha$: the corresponding equations of the orbit, putting therein ultimately $r_1 = +\infty$, are

$$\begin{vmatrix} r, & x, & y, & 1 \\ 1, & \cos\alpha, & \sin\alpha, & 0 \\ r_2, & x_2, & y_2, & 1 \\ r_3, & x_3, & y_3, & 1 \end{vmatrix} = 0, \qquad \begin{vmatrix} r, & x, & y, & 1 \\ 1, & -\cos\alpha, & -\sin\alpha, & 0 \\ r_2, & x_2, & y_2, & 1 \\ r_3, & x_3, & y_3, & 1 \end{vmatrix} = 0,$$

which equations belong, it is clear, to two distinct conics; or as the point (x_1, y_1) passes from a positive to a negative infinity along the given line, there is an abrupt change of orbit. It is proper to remark that the two orbits are the very same as would be obtained by writing $x_1, y_1 = r_1\cos\alpha, r_1\sin\alpha$, $r_1 = +\infty$ and $r_1 = -\infty$ in the determinant equation: that is, the orbit passes abruptly from one to another of the four conics which belong to the position (x_1, y_1), and we thus understand how the transition from $+\infty$ to $-\infty$, which is geometrically no breach of continuity, occasions in the actual problem a discontinuity.

21. The same thing appears from the geometrical construction; and we derive a further result which will be useful. Suppose first that the point 1 is at infinity in the direction shown by the arrow; then drawing $2c = 2S$ and $3b = 3S$ each in the direction opposite to $S1$, we have the points b, c on the directrix, which is thus the

line D joining these points. But if 1 is at infinity on the same line in the opposite direction, then instead of c, b we have the points c', b', and the directrix is the line D' joining these points.

Fig. 3.

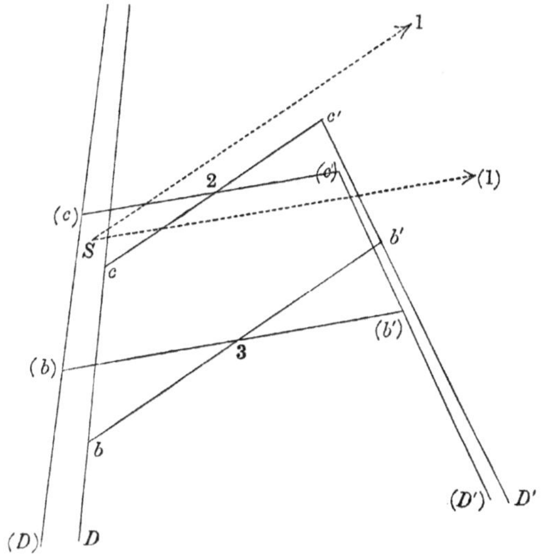

22. Observe that in the first case the focus S and the three points are on opposite sides of the directrix D, or the orbit is convex; but in the second case the focus S and the three points are on the same side of the directrix D', and the orbit is concave. That is, the line S_1 does not separate the two points 2, 3, and the orbits are the one convex, the other concave.

23. But if 1 be at infinity along the line $S(1)$ first in the direction shown by the arrow, and then in the opposite direction; in the first case the directrix is (D) not separating the focus S from the three points, and the orbit is concave; in the second case the orbit is (D'), not separating S from the three points, and the orbit is still concave; here the line $S(1)$ does separate the points 2, 3, and the orbits are both concave.

24. And we thus see in general that as the point 1 passes from a positive to a negative infinity along a line passing through S; then, according as the line through S does not or does separate the remaining two points 2, 3, the orbits corresponding to the two positions of 1 are the one convex, the other concave, or they are both concave.

25. The points 1 and 2 may be each of them at infinity along a given ray; we have here in a similar manner $x_1,\ y_1 = r_1 \cos \alpha_1,\ r_1 \sin \alpha_1$, or else $= -r_1 \cos \alpha_1,\ -r_1 \sin \alpha_1$, where r_1 is an indefinitely large positive quantity; and $x_2,\ y_2 = r_2 \cos \alpha_2,\ r_2 \sin \alpha_2$, or else

$= -r_2 \cos \alpha_2, -r_2 \sin \alpha_2$, where r_2 is an indefinitely large positive quantity. And writing ultimately $r_1 = +\infty$, $r_2 = +\infty$, the equation of the orbit is obtained in the form

$$\begin{vmatrix} r, & x, & y, & 1 \\ 1, & \pm\cos\alpha_1, & \pm\sin\alpha_1, & 0 \\ 1, & \pm\cos\alpha_2, & \pm\sin\alpha_2, & 0 \\ r_3, & x_3, & y_3, & 1 \end{vmatrix} = 0,$$

where the $\pm$ of the second line and the $\pm$ of the third line have each of them the value + or − at pleasure. There are consequently four distinct orbits, corresponding to the combinations of each of the two directions of the point 1 with each of the two directions of the point 2. And it is moreover clear that these are the very conics which are obtained from the determinant equation by writing therein $x_1, y_1 = r_1 \cos \alpha_1, r_1 \sin \alpha_1$; $x_2, y_2 = r_2 \cos \alpha_2, r_2 \sin \alpha_2$ and $r_1 = +\infty, -\infty$; $r_2 = +\infty, -\infty$ successively; viz., the orbit changes abruptly between the four conics which correspond to the given position of the points 1, 2, 3.

Fig. 4.

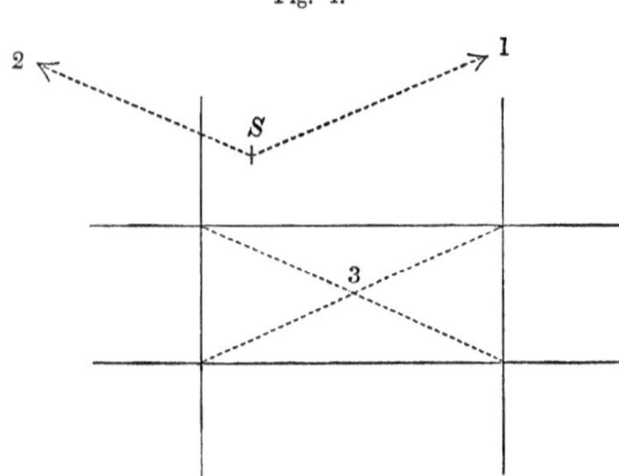

26. The geometrical construction is very simple indeed; viz., measuring off from 3 in the directions $S1$, $S2$, and in the opposite directions respectively, a distance $= S3$, we have four points, the angles of a rectangle; and joining these in pairs, we have the four positions of the directrix: the figure shows at once that the orbits are three of them concave, the remaining one convex.

27. The determinant equation obtained for the orbit is an equation of the form

$$r = Ax + By + C;$$

and it is clear that the equation of the directrix is $Ax + By + C = 0$. By what precedes, this line will lie on the same side of the three points; viz., either it does not separate them from the focus, and the orbit is then concave, or it does separate them from the focus, and the orbit is then convex. Although in general the sign of C is no criterion (for the equations $r = Ax + By + C$ and $r = -Ax - By - C$ represent the same curve) yet in the present case it is so; for, observe that, in taking r_1, r_2, r_3 each of them positive, we make r to be positive for the orbit, that is, for the entire curve if an ellipse or parabola, but for the branch containing the three points if the curve is a hyperbola. Hence, considering the radius vector through S parallel to the directrix, this is positive for a concave, negative for a convex orbit; or writing $Ax+By=0$, we have $r=C$ positive for a concave, negative for a convex orbit; wherefore the orbit is concave or convex according as C is positive or negative.

28. Comparing the equation with

$$r=e(x\cos\varpi+y\sin\varpi)\pm a(1-e^2),$$

we see that the eccentricity and semiaxis major, taken to be each of them positive, are

$$e=\sqrt{A^2+B^2},\quad a=\frac{\pm C}{1-A^2-B^2},$$

($+C$ or $-C$, according as $e<1$ or $e>1$); and inasmuch as the focus and directrix are known, there is no ambiguity as to the position of the orbit: it may be added that the coordinates of the centre are given by

$$\begin{aligned}(A^2-1)\,x+AB\quad\;\; y+AC&=0,\\ AB\quad x+(B^2-1)\,y+BC&=0,\end{aligned}$$

that is, we have for the coordinates of the centre

$$x=\frac{AC}{1-A^2-B^2},\quad y=\frac{BC}{1-A^2-B^2};$$

and thence also

$$x=\frac{2AC}{1-A^2-B^2},\quad y=\frac{2BC}{1-A^2-B^2}$$

for the coordinates of the other focus.

29. But to effect the comparison rather more precisely it is to be observed that a, e being positive, then for a concave orbit, if X be measured from the focus in the direction *away from* the directrix, we should have

$$r=eX\pm a(1-e^2)$$

($+$ for the ellipse, $-$ for the hyperbola, so that $\pm a(1-e^2)$ is positive): whence

$$e=\sqrt{A^2+B^2},\quad X=\frac{Ax+By}{\sqrt{A^2+B^2}},\quad a=\frac{\pm C}{1-A^2-B^2}$$

(by what precedes, C is $=+$, so that the formula gives as it should do $a=+$).

And similarly for a convex orbit, if X be measured in the direction *towards* the directrix, we should have

$$r=eX-a(e^2-1);$$

whence

$$e=\sqrt{A^2+B^2},\quad X=\frac{Ax+By}{\sqrt{A^2+B^2}},\quad a=\frac{-C}{A^2+B^2-1},$$

where by what precedes C is $=-$, and the formula gives as it should do $a=+$.

30. It is not necessary for the purpose of the present memoir, but I notice an elegant form of the polar equation of the orbit belonging to a given trivector; viz., taking (r, θ) as polar coordinates, and therefore (r_1, θ_1), (r_2, θ_2), (r_3, θ_3), as the coordinates of the given points, the equation of the orbit is

$$\frac{1}{r}=\Sigma\frac{1}{r_1}\cdot\frac{\sin\frac{1}{2}(\theta-\theta_2)\sin\frac{1}{2}(\theta-\theta_3)}{\sin\frac{1}{2}(\theta_1-\theta_2)\sin\frac{1}{2}(\theta_1-\theta_3)}.$$

In fact, it is clear that this is an equation of the form

$$\frac{1}{r}=(\alpha,\ \beta,\ \gamma)(\sin\tfrac{1}{2}\theta,\ \cos\tfrac{1}{2}\theta)^2;$$

that is of the form

$$\frac{1}{r}=\lambda\cos\theta+\mu\sin\theta+\nu;$$

and that it thus represents a conic with the given focus; and moreover that the equation is satisfied by writing therein (r_1, θ_1), (r_2, θ_2), or (r_3, θ_3), in place of (r, θ); that is, the conic passes through the three given points. The foregoing remarks as to the signs of r_1, r_2, r_3, apply without alteration to this polar equation.

Article Nos. 31 to 41. *Time Formulæ;* Lambert's *Equation.*

31. Suppose for a moment that the orbit is an ellipse; as the ellipse may be described in either direction, the time of passage between any two points, 1 to 2, or 2 to 1, indifferently, may be regarded as positive. With only two points 1, 2, we might pass, say from 1 to 2, in either direction along the ellipse, and the time of passage would have ambiguously either of two positive values. In the case however where we have on the ellipse three points, 1, 2, 3, this ambiguity is avoided; viz., it is assumed that the passage between any two of the points is along the elliptic arc which does not contain the third point; the three times of passage are thus all of them positive, and their sum is equal to the periodic time, or time of describing the entire ellipse.

32. But if the orbit be a parabola or concave hyperbolic branch, then, if the points taken in their order of position along the orbit be 1, 2, 3, we have in like manner a positive time of passage between 1 and 2, and also a positive time of passage between 2 and 3; but, inasmuch as there is no passage between 1 and 3 except through 2 (which mode is excluded from consideration), I say that there is no time of passage between 1 and 3; and so consider only two times of passage; viz., between 1 and 2, and between 2 and 3.

33. In the case of a convex hyperbolic branch, since this cannot be described under the action of an attractive force, there is not any time of passage to be considered.

In the transition case of a right line not passing through the focus, since, as mentioned, the velocity is infinite, if the order of the points on the line is 1, 2, 3, the times of passage from 1 to 2 and from 2 to 3 are each $=0$; and these are the only times of passage which are to be considered.

34. The preceding conventions are of course to be attended to in the application of any formula to the calculation of the times of passage between given points of the orbit; in the case of a parabolic or hyperbolic orbit we have only to ascertain which are the two times of passage to be calculated; but, in the case of an ellipse, we must take care that the time of passage between each two of the three points is calculated along the arc not containing the third point; viz., it is in some cases to be calculated through the angle $<\pi$ between the two radius vectors, and in other cases through the angle $>\pi$ between the two radius vectors; or, more simply, the time to be calculated is sometimes the longer, and at other times the shorter time of passage.

35. For the purpose of the present memoir the unit of time is so fixed that the periodic time in a circle radius 1 shall be equal 3. The period in a circle or ellipse, radius or semiaxis major $=a$, is thus $=3a^{\frac{3}{2}}$, and generally

$$\text{Time} = \frac{3}{\pi} \cdot \frac{\text{Area}}{\sqrt{\tfrac{1}{2} \text{ latus rectum}}}.$$

The time formulæ are first the ordinary ones in which the time from pericentre is expressed in terms of an angle (the eccentric anomaly for an ellipse or hyperbola, true anomaly for the parabola); secondly, Lambert's formulæ, in which the time between any two points on the orbit is expressed by means of the two radius vectors and the chord.

36. The first set of formulæ may be written:

Ellipse. u, the eccentric anomaly from pericentre, viz. $x=a(\cos u-e)$, $y=a\sqrt{1-e^2}\sin u$, if x, y, are the coordinates from the focus, x measured in the direction *towards* the directrix.

$$\text{Time from pericentre} = \frac{3}{\pi} a^{\frac{3}{2}} (u - e\sin u).$$

Parabola. θ, the true anomaly, viz., $r=p\sec^2\tfrac{1}{2}\theta$, if p be the pericentric distance or $\tfrac{1}{4}$-latus rectum.

$$\text{Time from pericentre} = \frac{3}{\pi} \frac{p^{\frac{3}{2}}}{\sqrt{2}} (\tan\tfrac{1}{2}\theta + \tfrac{1}{3}\tan^3\tfrac{1}{2}\theta).$$

Hyperbola; concave branch. u, the eccentric anomaly from pericentre, viz., $x=a(\sec u-e)$, $y=a\sqrt{e^2-1}\tan u$, if x, y are the coordinates from the focus, x measured in the direction *away* from the directrix.

$$\text{Time from pericentre} = \frac{3a^{\frac{3}{2}}}{2\pi} \{e\tan u - \text{hyp. log} \tan(\tfrac{1}{4}\pi + \tfrac{1}{2}u)\},$$

and by taking the sum or the difference of two of these expressions, we obtain the time of passage between two given points of the orbit.

37. I remark that as to the elliptic and parabolic orbits, I have preferred using Lambert's equations, and I should have done the same for the hyperbolic orbits, but for the absence of a table (see *post*, No. 39). As it is, for the few hyperbolic orbits which it was necessary to calculate, I have used the foregoing formula[1]: a table of hyp. log tan $(\frac{1}{4}\pi + \frac{1}{2}u)$, $u = 0°$ to $u = 90°$, at intervals of 30′ to 12 places of decimals, with fifth differences is given, Table IV. Legendre, *Traité des Fonctions Elliptiques*, t. II. pp. 256—259.

38. The other set of formulæ may be written:

Ellipse. r, r' the radius vectors, γ the chord.

$$2a\cos\chi = 2a - r - r' - \gamma, \qquad 2a\cos\chi' = 2a - r - r' + \gamma.$$

$$\text{Time} = \frac{3}{2\pi} a^{\frac{3}{2}} (\chi - \chi' - \sin\chi + \sin\chi').$$

Parabola. r, r', γ, *ut suprà*;

$$\text{Time} = \frac{1}{4\pi} \{(r + r' + \gamma)^{\frac{3}{2}} - (r + r' - \gamma)^{\frac{3}{2}}\}.$$

Hyperbola.

$$2a\cosh\chi = 2a + r + r' + \gamma, \qquad 2a\cosh\chi' = 2a + r + r' - \gamma.$$

$$\text{Time} = \frac{3}{2\pi} a^{\frac{3}{2}} (-\chi + \chi' + \sinh\chi - \sinh\chi'),$$

where cosh, sinh, denote the hyperbolic cosine and sine of χ, viz.:

$$\cosh\chi = \tfrac{1}{2}(e^{\chi} + e^{-\chi}), \qquad \sinh\chi = \tfrac{1}{2}(e^{\chi} - e^{-\chi}).$$

39. The logarithms (ordinary) of the functions $\cosh\chi$, $\sinh\chi$, and of $\tanh\chi$ are tabulated by Gudermann, *Crelle*, tt. VIII. and IX. from $\chi = 2{\cdot}000$ to $\chi = 8{\cdot}00$ at intervals of ·001 and subsequently of ·01 to eight places of decimals. I do not know why the tabulation was not commenced from $\chi = 0$, but the omission from them of the values 0 to 2 rendered the tables unavailing for the present purpose, and I therefore, for the hyperbolic orbits, resorted to the first set of formulæ.

40. As regards the elliptic formulæ it remains to be explained how the values of χ, χ' are to be selected from those which satisfy the required conditions

$$2a\cos\chi = 2a - r - r' - \gamma, \qquad 2a\cos\chi' = 2a - r - r' + \gamma.$$

It is remarked in Gauss' *Theoria Motûs*, p. 120, that χ is a positive angle between 0° and 360°; χ' a positive or negative angle between +180°, −180°, viz. χ' is positive

[1] I rather regret that I did not use the foregoing formulæ in all cases.

or negative according as the angle between the two radius vectors is $< 180^\circ$ or $> 180^\circ$. This determines χ', but it is said that χ is really indeterminate; viz. it is so if only the values r, r', γ, a, are given, for there are then two orbits in which these quantities have their given values, and the times in these have different values. But when, as in the case here considered the orbit is known, χ will of course have a determinate signification, and it is easy to explain how this is to be fixed. I observe, in the first place, that if $\chi=\pi$ we have $\gamma=(2a-r)+(2a-r')$, that is, the chord γ passes through the other focus of the ellipse. The criterion thus depends on the position of the two points on the ellipse in relation to the other focus, and it is easy to see that it is as follows: viz. let the time between the points 1, 2, on the ellipse be understood to mean the time of passage from 1 through apocentre to 2; then I say that, in the preceding formula

$$\text{Time} = \frac{3}{2\pi} a^{\frac{3}{2}} (\chi - \chi' - \sin\chi + \sin\chi'),$$

χ will be $< 180^\circ$ or $> 180^\circ$ according as the chord from 1 through the other focus H does not or does separate the point 2 from the focus S.

41. It is hardly necessary to remark that in the application of the formulæ, χ, χ' must be reckoned according to their lengths as circular arcs to the radius unity: a table for the conversion of degrees and minutes to such circular measure, is given in most collections of Trigonometrical Tables.

Article Nos. 42 to 45. *Formulæ for the Transformation between two sets of Rectangular Axes.*

42. Consider an arbitrary set of fixed rectangular axes, Sx, Sy, Sz, which are considered as intersecting the sphere, centre S, in the points X, Y, Z, and so the axes Sx', Sy', Sz', afterwards defined are considered as intersecting the sphere in the points X', Y', Z'. For convenience Sx is considered as an origin of longitudes, which are measured in the plane of xy in the direction towards y; and an angular distance from Sz is termed a polar distance or colatitude; so that the position of any line through S, or point on the sphere, will be determined by its longitude b and colatitude c.

43. It is wished in the sequel to made the orbit-pole revolve about an arbitrary line Sx', and for this purpose I take the new set of rectangular axes, Sx', Sy', Sz', or points on the sphere X', Y', Z', as follows,

$$X', \text{ longitude } G, \text{ colatitude } 90^\circ + N.$$

$Y'Z'$, is then a great circle, pole X', meeting ZX' in a point Π, longitude G, colatitude N, and the position of Z' in this great circle is fixed by its distance from Π, $\Pi Z' = H$, the distance of Y' being $\Pi Y' = 90^\circ + H$, and these being each of them reckoned from Π in the direction of longitude X to Y. The position of the new axes Sx', Sy', Sz', or points X', Y', Z', is thus fixed by means of the three angles G, N, H.

It is to be added that if the angle $X'ZZ'$ is called q, and if b, c, are the longitude and colatitude of Z', then we have $\sin N = \cot q \tan H$, which gives q, and then

$$b = G + q$$

$$\cos c = \cos N \cos H.$$

Fig. 5.

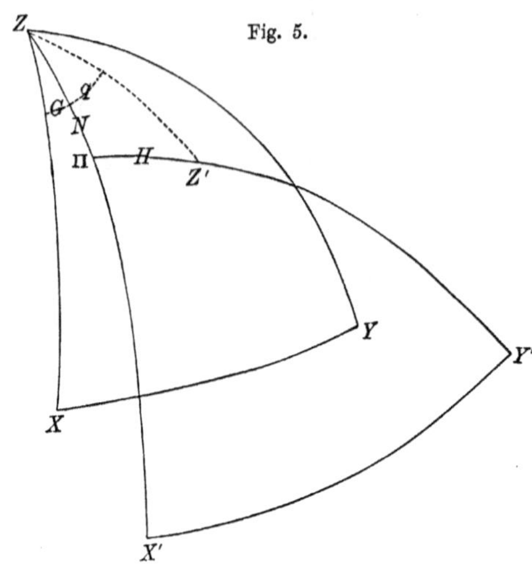

44. The transformation-formulæ between the two sets of axes are at once found to be

	X	Y	Z
X'	$\cos G \cos N$	$\sin G \cos N$	$-\sin N$
Y'	$-\sin G \cos H - \cos G \sin H \sin N$	$\cos G \cos H - \sin G \sin H \sin N$	$-\sin H \cos N$
Z'	$-\sin G \sin H + \cos G \cos H \sin N$	$\cos G \sin H + \sin G \cos H \sin N$	$\cos H \cos N$

which are for shortness represented by

	X	Y	Z
X'	α	β	γ
Y'	α'	β'	γ'
Z'	α''	β''	γ''

45 In the particular case where Sx' is in the plane of xy, $N=0$; Π coincides with Z, and the longitude and colatitude of Z' are $b=G+90°$, $c=H$. Writing accordingly in the formula $N=0$, and introducing b, c in the place of G, H, the formulæ become

	X	Y	Z
X'	$\sin b$	$-\cos b$	0
Y'	$\cos b \cos c$	$\sin b \cos c$	$-\sin c$
Z'	$\cos b \sin c$	$\sin b \sin c$	$\cos c$

and in particular if $c=0$, (Sz' here coincides with Sz, and the axes Sx', Sy', are in the plane of xy) then we have simply

	X	Y	Z
X'	$\sin b$	$-\cos b$	0
Y'	$\cos b$	$\sin b$	0
Z'	0	0	1

Article Nos. 46 to 60. *Application to finding the Intersection of the Orbit-plane by a Single Ray.*

46. The equations of the ray referred to the fixed axes are taken to be

$$\frac{x-A}{\mathrm{f}}=\frac{y-B}{\mathrm{g}}=\frac{z-C}{\mathrm{h}}, \ =R \text{ suppose},$$

or, what is the same thing,

$$x=A+R\mathrm{f},$$
$$y=B+R\mathrm{g},$$
$$z=C+R\mathrm{h},$$

and if in the foregoing formulæ the point Z' is taken to be the orbit-pole (longitude $b=G+90°$, and colatitude $c=\cos^{-1}\cos N\cos H$ as above) then the equation of the orbit-plane is $z'=0$. We have therefore merely to transform the equations of the ray to the new axes by writing for x, y, z, the values

$$\begin{aligned}&\alpha x'+\alpha' y'+\alpha'' z',\\ &\beta x'+\beta' y'+\beta'' z',\\ &\gamma x'+\gamma' y'+\gamma'' z',\end{aligned}$$

and then putting $z'=0$, we find x', y', the coordinates in the orbit-plane of its intersections with the ray.

47. The equations thus become,

$$\begin{aligned}\alpha x'+\alpha' y'-A-R\mathrm{f}&=0,\\ \beta x'+\beta' y'-B-R\mathrm{g}&=0,\\ \gamma x'+\gamma' y'-C-R\mathrm{h}&=0,\end{aligned}$$

or, what is the same thing, we have

$$x' : y' : R : 1$$

$$= \begin{vmatrix} 1 & & & \\ \alpha, & \alpha', & \mathrm{f}, & A \\ \beta, & \beta', & \mathrm{g}, & B \\ \gamma, & \gamma', & \mathrm{h}, & C \end{vmatrix} : \begin{vmatrix} & 1 & & \\ \alpha, & \alpha', & \mathrm{f}, & A \\ \beta, & \beta', & \mathrm{g}, & B \\ \gamma, & \gamma', & \mathrm{h}, & C \end{vmatrix} : \begin{vmatrix} & & -1 & \\ \alpha, & \alpha', & \mathrm{f}, & A \\ \beta, & \beta', & \mathrm{g}, & B \\ \gamma, & \gamma', & \mathrm{h}, & C \end{vmatrix} : \begin{vmatrix} & & & -1 \\ \alpha, & \alpha', & \mathrm{f}, & A \\ \beta, & \beta', & \mathrm{g}, & B \\ \gamma, & \gamma', & \mathrm{h}, & C \end{vmatrix}$$

$$= \begin{vmatrix} \alpha', & \mathrm{f}, & A \\ \beta', & \mathrm{g}, & B \\ \gamma', & \mathrm{h}, & C \end{vmatrix} : - \begin{vmatrix} \alpha, & \mathrm{f}, & A \\ \beta, & \mathrm{g}, & B \\ \gamma, & \mathrm{h}, & C \end{vmatrix} : - \begin{vmatrix} A, & \alpha, & \alpha' \\ B, & \beta, & \beta' \\ C, & \gamma, & \gamma' \end{vmatrix} : \begin{vmatrix} \mathrm{f}, & \alpha, & \alpha' \\ \mathrm{g}, & \beta, & \beta' \\ \mathrm{h}, & \gamma, & \gamma' \end{vmatrix}.$$

In these formulæ we have identically

$$\beta\gamma'-\beta'\gamma,\quad \gamma\alpha'-\gamma'\alpha,\quad \alpha\beta'-\alpha'\beta=\alpha'',\ \beta'',\ \gamma'',$$

and if we write moreover

$$\mathrm{a},\ \mathrm{b},\ \mathrm{c},\ = C\mathrm{g}-B\mathrm{h},\quad A\mathrm{h}-C\mathrm{f},\quad B\mathrm{f}-A\mathrm{g},$$

(whence identically $\mathrm{af}+\mathrm{bg}+\mathrm{ch}=0$, and where (a, b, c, f, g, h) are the "six coordinates" of the ray), then we have the very simple formulæ

$$x' : y' : R : 1$$
$$=(\mathrm{a},\ \mathrm{b},\ \mathrm{c}\between\alpha',\ \beta',\ \gamma') : -(\mathrm{a},\ \mathrm{b},\ \mathrm{c}\between\alpha,\ \beta,\ \gamma) : (A,\ B,\ C\between\alpha'',\ \beta'',\ \gamma'') : (\mathrm{f},\ \mathrm{g},\ \mathrm{h}\between\alpha'',\ \beta'',\ \gamma''),$$

or omitting (as not required for the present purpose) one of the proportional terms, we have

$$x' : y' : 1=(\mathrm{a},\ \mathrm{b},\ \mathrm{c}\between\alpha',\ \beta',\ \gamma') : -(\mathrm{a},\ \mathrm{b},\ \mathrm{c}\between\alpha,\ \beta,\ \gamma) : (\mathrm{f},\ \mathrm{g},\ \mathrm{h}\between\alpha'',\ \beta'',\ \gamma''),$$

which are the required expressions for the coordinates.

48. Consider in the equations just obtained the axis of x' as fixed but H as variable; that is, let the orbit-pole Z' describe a great circle about the fixed pole X' (longitude G, colatitude $90° + N$). We have x', y', 1, proportional to linear functions of $\sin H$, $\cos H$; viz., writing for shortness

$$\begin{aligned}
X_c &= -\mathrm{a} \sin G + \mathrm{b} \cos G,\\
X_s &= (-\mathrm{a} \cos G - \mathrm{b} \sin G) \sin N - \mathrm{c} \cos N,\\
Y_o &= (-\mathrm{a} \cos G - \mathrm{b} \sin G) \cos N + \mathrm{c} \sin N,\\
W_c &= (\ \mathrm{f} \cos G + \mathrm{g} \sin G) \sin N + \mathrm{h} \cos N,\\
W_s &= (-\mathrm{f} \sin G + \mathrm{g} \cos G),
\end{aligned}$$

we have

$$x' = \frac{X_c \cos H + X_s \sin H}{W_c \cos H + W_s \sin H},$$

$$y' = \frac{Y_o}{W_c \cos H + W_s \sin H}.$$

49. I write

$$\left\{\begin{aligned}
&\frac{W_c}{Y_o} = \frac{1}{m} \cos \Delta, \qquad \frac{W_s}{Y_o} = \frac{1}{m} \sin \Delta,\\
&\frac{X_c}{Y_o} = \frac{l}{m} \cos \Delta - \cot \delta \sin \Delta,\\
&\frac{X_s}{Y_o} = \frac{l}{m} \sin \Delta + \cot \delta \cos \Delta,
\end{aligned}\right.$$

equations which determine m, Δ, l, δ, viz., we have

$$\tan \Delta = \frac{W_s}{W_c}, \qquad m = \frac{Y_o}{\sqrt{W_c^2 + W_s^2}},$$

$$l = m \frac{X_c \cos \Delta + X_s \sin \Delta}{Y_o} = \frac{1}{W_c^2 + W_s^2} (X_c W_c + X_s W_s),$$

$$\cot \delta = \frac{-X_c \sin \Delta + X_s \cos \Delta}{Y_o} = \frac{1}{Y_o \sqrt{W_c^2 + W_s^2}} (X_s W_c - X_c W_s),$$

and we then very easily find

$$\begin{aligned}
x' &= l + m \cot \delta \tan (H - \Delta),\\
y' &= \qquad\quad m \sec (H - \Delta),
\end{aligned}$$

and thence also

$$y'^2 - (x' - l)^2 \tan^2 \delta = m^2;$$

viz. the orbit-plane revolving about the fixed axis SX', meets the ray in a series of points forming in the orbit-plane a hyperbola having the line SX' for its conjugate axis.

50. As already remarked (*ante*, No. 11), this hyperbola is nothing else than the intersection of the orbit-plane regarded as fixed, by the hyperboloid generated by the rotation of the ray about the axis SX'. And we thus see the interpretation of the constants, viz.

l is the distance from S along the axis SX' of the "arm," or shortest distance of SX' and the ray.

m is the length of this arm.

δ is the inclination of the ray to the axis SX';

and for the remaining quantity Δ, imagine parallel to the ray a line through S meeting the sphere in L (L is the pole of the separator), I say that $\Delta - H$ is the angle $LX'Z'$: or (what is the same thing) drawing $X'L$ to meet $\Pi Z'Y'$ in Λ, we have $\Pi\Lambda = \Delta = H + Z'\Lambda$, or (what is the same thing) $Z'\Lambda = \Delta - H$.

51. To verify this, observe that the cosine distances of L from X, Y, Z, are as $\mathrm{f} : \mathrm{g} : \mathrm{h}$; and thence its cosine distances from X', Y', Z', are as $(\mathrm{f}, \mathrm{g}, \mathrm{h} \between \alpha, \beta, \gamma) : (\mathrm{f}, \mathrm{g}, \mathrm{h} \between \alpha', \beta', \gamma') : (\mathrm{f}, \mathrm{g}, \mathrm{h} \between \alpha'', \beta'', \gamma'')$; say, for a moment, as $\mathrm{f}' : \mathrm{g}' : \mathrm{h}'$.

Now $L\Lambda$ is the perpendicular from L on the side $Y'Z'$ of the quadrantal spherical triangle $LY'Z'$, and we thence have

$$\frac{\mathrm{h}'}{\mathrm{g}'} = \frac{\cos \Lambda Y'}{\cos \Lambda Z'} = \tan \Lambda Z' = \tan (\Delta - H),$$

if Δ has the geometrical signification just assigned to it. But this equation is

$$\mathrm{g}' \cos (H - \Delta) + \mathrm{h}' \sin (H - \Delta) = 0,$$

that is

$$\tan \Delta = \frac{\mathrm{g}' \cos H + \mathrm{h}' \sin H}{-\mathrm{g}' \sin H + \mathrm{h}' \cos H},$$

or substituting for g', h' their values, the numerator is

$$\mathrm{f}(\alpha' \cos H + \alpha'' \sin H) + \mathrm{g}(\beta' \cos H + \beta'' \sin H) + \mathrm{h}(\gamma' \cos H + \gamma'' \sin H),$$

which is

$$= -\mathrm{f} \sin G + \mathrm{g} \cos G, \ = W_s,$$

and the denominator is

$$\mathrm{f}(-\alpha' \sin H + \alpha'' \cos H) + \mathrm{g}(-\beta' \sin H + \beta'' \cos H) + \mathrm{h}(-\gamma' \sin H + \gamma'' \cos H),$$

which is

$$= (\mathrm{f} \cos G + \mathrm{g} \sin G) \sin N + \mathrm{h} \cos N, \ = W_c,$$

so that the formula becomes

$$\tan \Delta = \frac{W_s}{W_c},$$

which is the original expression of $\tan \Delta$.

52. We might in the equations

$$x' : y' : 1 = (\mathrm{a, b, c} \between \alpha', \beta', \gamma') : -(\mathrm{a, b, c} \between \alpha, \beta, \gamma) : (\mathrm{f, g, h} \between \alpha'', \beta'', \gamma'')$$

consider for instance G or N as alone variable, and then eliminate the variable parameter so as to obtain a locus; but the results would be complicated and the geometrical interpretations not very obvious.

53. I assume (as was done before) $N=0$, $G=b-90°$, $H=c$, that is, the position of the orbit-pole Z' is longitude b, colatitude c, and the axis SX' is the line of nodes or intersection of the orbit-plane with the ecliptic, viz., the longitude of this line is $=b-90°$.

The formulæ become

$$\begin{aligned} x' : y' : 1 = \quad & (\mathrm{a} \cos b + \mathrm{b} \sin b) \cos c - \mathrm{c} \sin c \\ : \; - \; & \mathrm{a} \sin b + \mathrm{b} \cos b \\ : \quad & (\mathrm{f} \cos b + \mathrm{g} \sin b) \sin c + \mathrm{h} \cos c. \end{aligned}$$

or if these are

$$x' = \frac{X_c \cos c + X_s \sin c}{W_c \cos c + W_s \sin c},$$

$$y' = \frac{Y_0}{W_c \cos c + W_s \sin c},$$

the values now are

$$\begin{aligned} X_c &= \mathrm{a} \cos b + \mathrm{b} \sin b, \\ X_s &= -\mathrm{c}, \\ Y_0 &= -\mathrm{a} \sin b + \mathrm{b} \cos b, \\ W_c &= \mathrm{h}, \\ W_s &= \mathrm{f} \cos b + \mathrm{g} \sin b, \end{aligned}$$

and thence forming as before the values of $\tan \Delta$, l, m, $\cot \delta$, and putting for shortness

$$\sqrt{W_c^2 + W_s^2}, = \sqrt{\mathrm{h}^2 + (\mathrm{f} \cos b + \mathrm{g} \sin b)^2}, = \Omega$$

we find after some easy reductions

$$\tan \Delta = \frac{\mathrm{f}}{\mathrm{h}} \cos b + \frac{\mathrm{g}}{\mathrm{h}} \sin b,$$

$$m = \frac{1}{\Omega} (-\mathrm{a} \sin b + \mathrm{b} \cos b),$$

$$l = \frac{1}{\Omega^2} [(\mathrm{ah} - \mathrm{cf}) \cos b + (\mathrm{bh} - \mathrm{cg}) \sin b],$$

$$\begin{aligned} \cot \delta &= \frac{1}{\Omega Y_0} (-\mathrm{a} \sin b + \mathrm{b} \cos b)(-\mathrm{f} \sin b + \mathrm{g} \cos b), \\ &= \frac{1}{\Omega} (-\mathrm{f} \sin b + \mathrm{g} \cos b), \end{aligned}$$

and with these values

$$x' = l + m\cot\delta\ \tan(c-\Delta),$$

$$y' = \qquad m\sec(c-\Delta),$$

and thence

$$y'^2 - (x'-l)^2\tan^2\delta = m^2,$$

viz., this is the hyperbola obtained by rotating the orbit-plane about the line of nodes, longitude $b - 90°$.

54. Imagine the orbit-plane (having upon it the hyperbola) brought by such rotation into the plane $z=0$, or plane of the ecliptic, so that the hyperbola will be a curve in this plane, the inclination to Sx, or longitude of the axis Sx', being of course $= b - 90°$. Transforming the equation to axes Sx, Sy, we must write in the equation

$$x' = x\sin b - y\cos b,$$

$$y' = x\cos b + y\sin b,$$

and the equation thus becomes

$$(x\cos b + y\sin b)^2 - (x\sin b - y\cos b - l)^2\tan^2\delta = m^2.$$

55. It will be recollected that the equations of the ray were

$$\frac{x-A}{\mathrm{f}} = \frac{y-B}{\mathrm{g}} = \frac{z-C}{\mathrm{h}};$$

writing herein $z=0$ we find

$$x = A - \frac{\mathrm{f}}{\mathrm{h}}C, = \frac{\mathrm{b}}{\mathrm{h}},$$

$$y = B - \frac{\mathrm{f}}{\mathrm{g}}C, = -\frac{\mathrm{a}}{\mathrm{h}},$$

and it is clear that this point $\left(\frac{\mathrm{b}}{\mathrm{h}}, -\frac{\mathrm{a}}{\mathrm{h}}\right)$ should lie on the hyperbola.

Substituting for (x, y) the values in question, we have first

$$\mathrm{b}\sin b + \mathrm{a}\cos b - \mathrm{h}l$$

$$= \frac{1}{\Omega^2}\left\{\left(\mathrm{h}^2 + (\mathrm{f}\cos b + \mathrm{g}\sin b)^2\right)(\mathrm{b}\sin b + \mathrm{a}\cos b) - \mathrm{h}(\mathrm{ah}-\mathrm{cf})\left(\cos b + (\mathrm{bh}-\mathrm{cg})\sin b\right)\right\}$$

$$= \frac{1}{\Omega^2}\{(\mathrm{f}\cos b + \mathrm{g}\sin b)^2(\mathrm{b}\sin b + \mathrm{a}\cos b) + (\mathrm{f}\cos b + \mathrm{g}\sin b)\,\mathrm{ch}\}$$

$$= \frac{1}{\Omega^2}(\mathrm{f}\cos b + \mathrm{g}\sin b)\{(\mathrm{f}\cos b + \mathrm{g}\sin b)(\mathrm{b}\sin b + \mathrm{a}\cos b) + \mathrm{ch}(\cos^2 b + \sin^2 b)\}$$

$$= \frac{1}{\Omega^2}(\mathrm{f}\cos b + \mathrm{g}\sin b)(-\mathrm{a}\sin b + \mathrm{b}\cos b)(\mathrm{f}\sin b - \mathrm{g}\cos b);$$

or observing that

$$\tan\delta = \frac{-\Omega}{\mathrm{f}\sin b - \mathrm{g}\cos b},$$

we have

$$(\mathrm{b}\sin b + \mathrm{a}\cos b - \mathrm{h}l)\tan\delta = -\frac{1}{\Omega}(\mathrm{f}\cos b + \mathrm{g}\sin b)(-\mathrm{a}\sin b + \mathrm{b}\cos b);$$

and hence the result of the substitution is at once found to be

$$(-\mathrm{a}\sin b + \mathrm{b}\cos b)^2 - \frac{1}{\Omega^2}(-\mathrm{a}\sin b + \mathrm{b}\cos b)^2(\mathrm{g}\sin b + \mathrm{f}\cos b)^2$$

$$= m^2\mathrm{h}^2, \ = \frac{\mathrm{h}^2(-\mathrm{a}\sin b + \mathrm{b}\cos b)^2}{\Omega^2};$$

viz., the factor $(-\mathrm{a}\sin b + \mathrm{b}\cos b)^2$ divides out, and the equation then becomes

$$1 - \frac{1}{\Omega^2}(\mathrm{g}\sin b + \mathrm{f}\cos b)^2 = \frac{\mathrm{h}^2}{\Omega^2},$$

that is

$$\Omega^2 = \mathrm{h}^2 + (\mathrm{g}\sin b + \mathrm{f}\cos b)^2,$$

which is in fact the value of Ω^2.

56. I seek for the direction of the hyperbola at the point $\left(\frac{\mathrm{b}}{\mathrm{h}}, -\frac{\mathrm{a}}{\mathrm{h}}\right)$ in question. We have

$$\begin{aligned} dx : dy = &\ (\mathrm{b}\cos b - \mathrm{a}\sin b)\sin b + \cos b\tan^2\delta(\mathrm{b}\sin b + \mathrm{a}\cos b - \mathrm{h}l) \\ : &-(\mathrm{b}\cos b - \mathrm{a}\sin b)\cos b + \sin b\tan^2\delta(\mathrm{b}\sin b + \mathrm{a}\cos b - \mathrm{h}l), \end{aligned}$$

and from the above values of $(\mathrm{b}\sin b + \mathrm{a}\cos b - \mathrm{h}l)$ and $\tan\delta$, we have

$$\tan^2\delta(\mathrm{b}\sin b + \mathrm{a}\cos b - \mathrm{h}l) = \frac{\mathrm{g}\sin b + \mathrm{f}\cos b}{\mathrm{f}\sin b - \mathrm{g}\cos b}(-\mathrm{a}\sin b + \mathrm{b}\cos b);$$

whence

$$\begin{aligned} dx : dy = &\ (\mathrm{b}\cos b - \mathrm{a}\sin b)\sin b(\mathrm{f}\sin b - \mathrm{g}\cos b) + (\mathrm{g}\sin b + \mathrm{f}\cos b)\cos b(-\mathrm{a}\sin b + \mathrm{b}\cos b) \\ : &-(\mathrm{b}\cos b - \mathrm{a}\sin b)\cos b(\mathrm{f}\sin b - \mathrm{g}\cos b) + (\mathrm{g}\sin b + \mathrm{f}\cos b)\sin b(-\mathrm{a}\sin b + \mathrm{b}\cos b), \end{aligned}$$

which, multiplying out and reducing by means of the relation $\mathrm{af} + \mathrm{bg} + \mathrm{ch} = 0$, becomes

$$dx : dy = (-\mathrm{a}\sin b + \mathrm{b}\cos b)(\sin^2 b + \cos^2 b)\mathrm{f} : (-\mathrm{a}\sin b + \mathrm{b}\cos b)(\sin^2 b + \cos^2 b)\mathrm{g};$$

that is

$$dx : dy = \mathrm{f} : \mathrm{g}, \quad \text{or} \quad \frac{dy}{dx} = \frac{\mathrm{g}}{\mathrm{f}},$$

which shows that the hyperbola, at the point $\left(\frac{\mathrm{b}}{\mathrm{h}}, -\frac{\mathrm{a}}{\mathrm{h}}\right)$ where it meets the ray, touches the projection

$$\frac{x - A}{\mathrm{f}} = \frac{y - B}{\mathrm{g}}$$

of the ray on the plane of xy, which contains the hyperbola.

57. We may consider various particular forms of the hyperbola $y'^2 - (x' - l)^2 \tan^2 \delta = m^2$.

1°. If $\tan \delta = 0$, the hyperbola is the pair of parallel lines $y'^2 = m^2$.

This can only happen if $\mathrm{h} = 0$, $\mathrm{f} \cos b + \mathrm{g} \sin b = 0$. The first equation gives $\mathrm{af} + \mathrm{bg} = 0$, whence $\tan b = -\frac{\mathrm{f}}{\mathrm{g}} = \frac{\mathrm{b}}{\mathrm{a}}$; we have thus $m = \frac{-\mathrm{a} \sin b + \mathrm{b} \cos b}{\Omega} = \frac{0}{0}$, which is consistent with m finite. The equations show that the ray is parallel to the line of nodes.

2°. If $\tan \delta = \infty$, the hyperbola is $(x' - l)^2 = 0$, viz., the line $x' = l$ twice: the condition is $-\mathrm{f} \sin b + \mathrm{g} \cos b = 0$; viz., the ray (not in general cutting the line of nodes) is at right angles to the line of nodes.

3°. If $m = 0$, the hyperbola is the pair of intersecting lines $y'^2 = (x' - l)^2 \tan^2 \delta$. The condition is $-\mathrm{a} \sin b + \mathrm{b} \cos b = 0$, signifying that the ray cuts the line of nodes.

4°. We may have simultaneously $\tan \delta = \infty$, $m = 0$. The hyperbola (as in 2°) is $(x' - l)^2 = 0$. The conditions are $-\mathrm{f} \sin b + \mathrm{g} \cos b = 0$, $-\mathrm{a} \sin b + \mathrm{b} \cos b = 0$, whence $\tan b = \frac{\mathrm{g}}{\mathrm{f}} = \frac{\mathrm{b}}{\mathrm{a}}$, and therefore also $\mathrm{ag} - \mathrm{bf} = 0$; these signify that the ray cuts at right angles the line of nodes.

The line $x' = l$ passes through the point $\left(\frac{\mathrm{b}}{\mathrm{h}}, -\frac{\mathrm{a}}{\mathrm{h}}\right)$, that is, we ought to have $\mathrm{h}^2 l^2 = \mathrm{a}^2 + \mathrm{b}^2$. The value of l is in the first instance given in the form

$$l = \frac{1}{\Omega^2} \{(\mathrm{ah} - \mathrm{cf}) \cos b + (\mathrm{bh} - \mathrm{cg}) \sin b\},$$

where

$$\Omega^2 = \mathrm{h}^2 + (\mathrm{f} \cos b + \mathrm{g} \sin b)^2 = \mathrm{h}^2 + \mathrm{f}^2 + \mathrm{g}^2 - (-\mathrm{f} \sin b + \mathrm{g} \cos b)^2 = \mathrm{f}^2 + \mathrm{g}^2 + \mathrm{h}^2.$$

But observe that the equations

$$\mathrm{ag} - \mathrm{bf} = 0,$$

$$\mathrm{bg} + \mathrm{af} = -\mathrm{ch},$$

give

$$\mathrm{f} = \frac{-\mathrm{ch}}{\mathrm{a}^2 + \mathrm{b}^2} \mathrm{a}, \quad \mathrm{g} = \frac{-\mathrm{ch}}{\mathrm{a}^2 + \mathrm{b}^2} \mathrm{b},$$

and thence

$$\Omega^2 = \mathrm{f}^2 + \mathrm{g}^2 + \mathrm{h}^2 \quad = \mathrm{h}^2 \left(1 + \frac{\mathrm{c}^2}{\mathrm{a}^2 + \mathrm{b}^2}\right) \quad = \frac{\mathrm{h}^2 (\mathrm{a}^2 + \mathrm{b}^2 + \mathrm{c}^2)}{\mathrm{a}^2 + \mathrm{b}^2},$$

$$\mathrm{ah} - \mathrm{cf} = \mathrm{ah} \frac{\mathrm{a}^2 + \mathrm{b}^2 + \mathrm{c}^2}{\mathrm{a}^2 + \mathrm{b}^2} \quad = \frac{\mathrm{a}}{\mathrm{h}} \Omega,$$

$$\mathrm{bh} - \mathrm{cg} = \mathrm{bh} \frac{\mathrm{a}^2 + \mathrm{b}^2 + \mathrm{c}^2}{\mathrm{a}^2 + \mathrm{b}^2} \quad = \frac{\mathrm{b}}{\mathrm{h}} \Omega;$$

consequently

$$l = \frac{1}{\Omega^2 \mathrm{h}} (\mathrm{a} \cos b + \mathrm{b} \sin b) \, \Omega^2 = \frac{1}{\mathrm{h}} (\mathrm{a} \cos b + \mathrm{b} \sin b) = \frac{1}{\mathrm{h}} \sqrt{\mathrm{a}^2 + \mathrm{b}^2},$$

which is right.

58. I return to the equation of the hyperbola written in the form

$$(x\cos b + y\sin b)^2 - (x\sin b - y\cos b - l)^2\tan^2\delta = m^2;$$

being (as was shown) a hyperbola passing through the point $\left(\frac{\mathrm{b}}{\mathrm{h}}, -\frac{\mathrm{a}}{\mathrm{h}}\right)$ where its plane is met by the ray, and touching at this point the projection $\frac{x-A}{\mathrm{f}} = \frac{y-B}{\mathrm{g}}$.

If in the equation we consider b as variable, we have a series of hyperbolas, viz., these are the intersections of the plane of xy with the hyperboloids of revolution obtained by making the ray rotate successively round the several lines $x\cos b + y\sin b = 0$ through the focus S. And, as just seen, these hyperbolas all of them touch at $\left(\frac{\mathrm{b}}{\mathrm{h}}, -\frac{\mathrm{a}}{\mathrm{h}}\right)$ the projection of the ray.

59. The hyperbola to any particular angle b is the hyperbola belonging to the ray, in the planogram for an orbit-plane rotating about the axis $x\cos b + y\sin b = 0$; so that the system of hyperbolas would be useful for the construction of any such planogram. And there is another series of curves which, if they could be constructed with moderate facility, would be very useful for the same purpose; viz., reverting to the equations

$$\begin{aligned} x' : y' : 1 = &\quad (\mathrm{a}\cos b + \mathrm{b}\sin b)\cos c - \mathrm{c}\sin c \\ &: -\ \mathrm{a}\sin b + \mathrm{b}\cos b \\ &: \quad (\mathrm{f}\cos b + \mathrm{g}\sin b)\sin c + \mathrm{h}\cos c, \end{aligned}$$

which determine in the orbit-plane the coordinates x', y' of the intersection thereof with the ray: imagine as before that the point is marked on the orbit-plane, and let it by a rotation of the orbit-plane be brought into the plane of xy; so that x', y', will be the coordinates in the direction of and perpendicular to the line of nodes of a point on the hyperbola $y'^2 - (x'-l)^2\tan^2\delta = m^2$, or $(x\cos b + y\sin b)^2 - (x\sin b - y\cos b - l)^2\tan^2\delta = m^2$, viz., of the point corresponding to an orbit-pole, colatitude c. Suppose that x, y, are the coordinates of this same point referred to the fixed axes, we have

$$\begin{aligned} x &= \quad x'\sin b + y'\cos b, \\ x &= -\,x'\cos b + y'\sin b, \end{aligned}$$

and thence

$$\begin{aligned} x : y : 1 = &\quad (\mathrm{a}\cos b + \mathrm{b}\sin b)\sin b\cos c - \mathrm{c}\sin b\sin c + (-\,\mathrm{a}\sin b + \mathrm{b}\cos b)\cos b \\ &: -(\mathrm{a}\cos b + \mathrm{b}\sin b)\cos b\cos c + \mathrm{c}\cos b\sin c + (-\,\mathrm{a}\sin b + \mathrm{b}\cos b)\sin b \\ &: \quad (\mathrm{f}\cos b + \mathrm{g}\sin b)\sin c + \mathrm{h}\cos c, \end{aligned}$$

the coordinates of the point just referred to. Now, if from these equations we could eliminate b, we should have a series of curves containing the variable parameter c, intersecting the series of hyperbolas; and thus marking out on each of these hyperbolas the points which belong to the successive values of the parameter c; we should thus have in the plane of xy the point corresponding to an orbit-pole longitude b and colatitude c. The series of curves in question may be called "graduation curves," viz., they would serve for the graduation of the hyperbola in the planogram for an orbit-plane rotating round any line $x\cos b + y\sin b = 0$ in the plane of xy. But the elimination cannot be easily effected, and I am not in possession of any method of tracing the series of curves.

60. I remark that from the equations

$$\begin{aligned} x' : y' : 1 = &\quad (\mathrm{a}\cos b + \mathrm{b}\sin b)\cos c - \mathrm{c}\sin c \\ &: -\ \mathrm{a}\sin b + \mathrm{b}\cos b \\ &: \quad (\mathrm{f}\cos b + \mathrm{g}\sin b)\sin c + \mathrm{h}\cos c, \end{aligned}$$

we may without difficulty eliminate b; the result is, in fact,

$$\begin{aligned} &[x'(-\mathrm{ah}\cos c) + y'(-\mathrm{bh}\cos^2 c - \mathrm{cg}\sin^2 c) - \mathrm{ac}\sin c]^2 \\ + &[x'(\ \ \mathrm{bh}\cos c) + y'(-\mathrm{ah}\cos^2 c + \mathrm{cf}\sin^2 c) + \mathrm{bc}\sin c]^2 \\ = &[x'(\ \ \mathrm{ch}\sin c) + y'(\ \ \mathrm{ag} - \mathrm{bf})\sin c\cos c + (\mathrm{a}^2 - \mathrm{b}^2)\cos c]^2, \end{aligned}$$

a conic; but the geometrical signification of this result is not obvious, and I do not make any use of it.

Article Nos. 61 to 63. *The Trivector and the Orbit.*

61. Considering now the three rays, these are determined by their six coordinates,

$$\begin{aligned} &(\mathrm{a}_1,\ \mathrm{b}_1,\ \mathrm{c}_1,\ \mathrm{f}_1,\ \mathrm{g}_1,\ \mathrm{h}_1), \\ &(\mathrm{a}_2,\ \mathrm{b}_2,\ \mathrm{c}_2,\ \mathrm{f}_2,\ \mathrm{g}_2,\ \mathrm{h}_2), \\ &(\mathrm{a}_3,\ \mathrm{b}_3,\ \mathrm{c}_3,\ \mathrm{f}_3,\ \mathrm{g}_3,\ \mathrm{h}_3), \end{aligned}$$

respectively; and the intersections with the orbit-plane are given by

$$\begin{aligned} x_1' : y_1' : 1 &= (\mathrm{a}_1,\ \mathrm{b}_1,\ \mathrm{c}_1 \between \alpha,\ \beta,\ \gamma) : -(\mathrm{a}_1,\ \mathrm{b}_1,\ \mathrm{c}_1 \between \alpha',\ \beta',\ \gamma') : (\mathrm{f}_1,\ \mathrm{g}_1,\ \mathrm{h}_1 \between \alpha'',\ \beta'',\ \gamma''), \\ x_2' : y_2' : 1 &= (\mathrm{a}_2,\ \mathrm{b}_2,\ \mathrm{c}_2 \between\ \ \text{,,}\ \ ) : -(\mathrm{a}_2,\ \mathrm{b}_2,\ \mathrm{c}_2 \between\ \ \text{,,}\ \ ) : (\mathrm{f}_2,\ \mathrm{g}_2,\ \mathrm{h}_2 \between\ \ \text{,,}\ \ ), \\ x_3' : y_3' : 1 &= (\mathrm{a}_3,\ \mathrm{b}_3,\ \mathrm{c}_3 \between\ \ \text{,,}\ \ ) : -(\mathrm{a}_3,\ \mathrm{b}_3,\ \mathrm{c}_3 \between\ \ \text{,,}\ \ ) : (\mathrm{f}_3,\ \mathrm{g}_3,\ \mathrm{h}_3 \between\ \ \text{,,}\ \ ), \end{aligned}$$

where the axes Sx', Sy', are an arbitrary set of rectangular axes in the orbit-plane; or where, as before, the axis Sx' may be taken to be the line of nodes.

There is no difficulty in finding the equation of the orbit. Writing $r_1 = \sqrt{x_1^2 + y_1^2}$, we have

$$r_1 = \frac{u_1}{(\mathrm{f}_1,\ \mathrm{g}_1,\ \mathrm{h}_1 \between \alpha'',\ \beta'',\ \gamma'')},$$

if

$$u_1 = \pm\sqrt{[(\mathrm{a}_1,\ \mathrm{b}_1,\ \mathrm{c}_1 \between \alpha',\ \beta',\ \gamma')]^2 + [(\mathrm{a}_1,\ \mathrm{b}_1,\ \mathrm{c}_1 \between \alpha,\ \beta,\ \gamma)]^2},$$

the sign being taken in such manner that r_1 shall be positive; viz., the sign must be the same as that of $(\mathrm{f}_1, \mathrm{g}_1, \mathrm{h}_1 \between \alpha'', \beta'', \gamma'')$. And we have the like formulæ for r_2 and r_3. Substituting these values, the equation of the orbit becomes

$$\left|\begin{array}{llll} r, & x', & y', & 1 \\ u_1, & (\mathrm{a}_1, \mathrm{b}_1, \mathrm{c}_1 \between \alpha', \beta', \gamma'), & -(\mathrm{a}_1, \mathrm{b}_1, \mathrm{c}_1 \between \alpha, \beta, \gamma), & (\mathrm{f}_1, \mathrm{g}_1, \mathrm{h}_1 \between \alpha'', \beta'', \gamma'') \\ u_2, & (\mathrm{a}_2, \mathrm{b}_2, \mathrm{c}_2 \between \quad ,, \quad), & -(\mathrm{a}_2, \mathrm{b}_2, \mathrm{c}_2 \between \quad ,, \quad), & (\mathrm{f}_2, \mathrm{g}_2, \mathrm{h}_2 \between \quad ,, \quad) \\ u_3, & (\mathrm{a}_3, \mathrm{b}_3, \mathrm{c}_3 \between \quad ,, \quad), & -(\mathrm{a}_3, \mathrm{b}_3, \mathrm{c}_3 \between \quad ,, \quad), & (\mathrm{f}_3, \mathrm{g}_3, \mathrm{h}_3 \between \quad ,, \quad) \end{array}\right| = 0.$$

62. Considering the minor determinants formed with the terms under the x' and y', for instance

$$(\mathrm{a}_1, \mathrm{b}_1, \mathrm{c}_1 \between \alpha', \beta', \gamma') \,.\, -(\mathrm{a}_2, \mathrm{b}_2, \mathrm{c}_2 \between \alpha, \beta, \gamma)$$
$$+ (\mathrm{a}_1, \mathrm{b}_1, \mathrm{c}_1 \between \alpha, \beta, \gamma) \,.\, \quad (\mathrm{a}_2, \mathrm{b}_2, \mathrm{c}_2 \between \alpha', \beta', \gamma')$$

this is

$$= (\mathrm{b}_1\mathrm{c}_2 - \mathrm{b}_2\mathrm{c}_1)(\beta\gamma' - \beta'\gamma) + (\mathrm{c}_1\mathrm{a}_2 - \mathrm{c}_2\mathrm{a}_1)(\gamma\alpha' - \gamma'\alpha) + (\mathrm{a}_1\mathrm{b}_2 - \mathrm{a}_2\mathrm{b}_1)(\alpha\beta' - \alpha'\beta)$$
$$= \alpha''(\mathrm{b}_1\mathrm{c}_2 - \mathrm{b}_2\mathrm{c}_1) + \beta''(\mathrm{c}_1\mathrm{a}_2 - \mathrm{c}_2\mathrm{a}_1) + \gamma''(\mathrm{a}_1\mathrm{b}_2 - \mathrm{a}_2\mathrm{b}_1),$$

or, what is the same thing,

$$= (\mathrm{b}_1\mathrm{c}_2 - \mathrm{b}_2\mathrm{c}_1,\ \mathrm{c}_1\mathrm{a}_2 - \mathrm{c}_2\mathrm{a}_1,\ \mathrm{a}_1\mathrm{b}_2 - \mathrm{a}_2\mathrm{b}_1 \between \alpha'', \beta'', \gamma'');$$

with the like expressions for the other two minors. And we thus obtain the following developed form of the equation, viz.

$$\begin{aligned}
&\{x'(\mathrm{a}_1, \mathrm{b}_1, \mathrm{c}_1 \between \alpha, \beta, \gamma) + y'(\mathrm{a}_1, \mathrm{b}_1, \mathrm{c}_1 \between \alpha', \beta', \gamma')\}\,[-u_2(\mathrm{f}_3, \mathrm{g}_3, \mathrm{h}_3 \between \alpha'', \beta'', \gamma'') \\
&\qquad\qquad + u_3(\mathrm{f}_2, \mathrm{g}_2, \mathrm{h}_2 \between \alpha'', \beta'', \gamma'')] \\
&+ \{x'(\mathrm{a}_2, \mathrm{b}_2, \mathrm{c}_2 \between \quad ,, \quad) + y'(\mathrm{a}_2, \mathrm{b}_2, \mathrm{c}_2 \between \quad ,, \quad)\}\,[-u_3(\mathrm{f}_1, \mathrm{g}_1, \mathrm{h}_1 \between \quad ,, \quad) \\
&\qquad\qquad + u_1(\mathrm{f}_3, \mathrm{g}_3, \mathrm{h}_3 \between \quad ,, \quad)] \\
&+ \{x'(\mathrm{a}_3, \mathrm{b}_3, \mathrm{c}_3 \between \quad ,, \quad) + y'(\mathrm{a}_3, \mathrm{b}_3, \mathrm{c}_3 \between \quad ,, \quad)\}\,[-u_1(\mathrm{f}_2, \mathrm{g}_2, \mathrm{h}_2 \between \quad ,, \quad) \\
&\qquad\qquad + u_2(\mathrm{f}_1, \mathrm{g}_1, \mathrm{h}_1 \between \quad ,, \quad)] \\
&+ (\mathrm{b}_2\mathrm{c}_3 - \mathrm{b}_3\mathrm{c}_2,\ \mathrm{c}_2\mathrm{a}_3 - \mathrm{c}_3\mathrm{a}_2,\ \mathrm{a}_2\mathrm{b}_3 - \mathrm{a}_3\mathrm{b}_2 \between \alpha'', \beta'', \gamma'')\,[r(\mathrm{f}_1, \mathrm{g}_1, \mathrm{h}_1 \between \alpha'', \beta'', \gamma'') - u_1] \\
&+ (\mathrm{b}_3\mathrm{c}_1 - \mathrm{b}_1\mathrm{c}_3,\ \mathrm{c}_3\mathrm{a}_1 - \mathrm{c}_1\mathrm{a}_3,\ \mathrm{a}_3\mathrm{b}_1 - \mathrm{a}_1\mathrm{b}_3 \between \quad ,, \quad)\,[r(\mathrm{f}_2, \mathrm{g}_2, \mathrm{h}_2 \between \quad ,, \quad) - u_2] \\
&+ (\mathrm{b}_1\mathrm{c}_2 - \mathrm{b}_2\mathrm{c}_1,\ \mathrm{c}_1\mathrm{a}_2 - \mathrm{c}_2\mathrm{a}_1,\ \mathrm{a}_1\mathrm{b}_2 - \mathrm{a}_2\mathrm{b}_1 \between \quad ,, \quad)\,[r(\mathrm{f}_3, \mathrm{g}_3, \mathrm{h}_3 \between \quad ,, \quad) - u_3] = 0,
\end{aligned}$$

being an equation of the form $\Omega r = Ax' + By' + C$.

63. The coefficient of r is a quadric function of $(\alpha'', \beta'', \gamma'')$, and if this vanish the orbit is a right line. It thus appears that the orbit will be a right line provided only the orbit-axis be situate in a certain quadric cone, or (what is the same thing) the orbit-pole be situate in a certain spherical conic: agreeing with a preceding result, viz. the cone is that reciprocal to the cone, vertex S, circumscribed about the hyper-

boloid which contains the three rays. And we see that the equation of this reciprocal cone is

$$\left|\begin{array}{ll} & \alpha'',\ \beta'',\ \gamma'' \\ (f_1,\ g_1,\ h_1\between\alpha'',\ \beta'',\ \gamma''), & a_1,\ b_1,\ c_1 \\ (f_2,\ g_2,\ h_2\between\quad ,, \quad), & a_2,\ b_2,\ c_2 \\ (f_3,\ g_3,\ h_3\between\quad ,, \quad), & a_3,\ b_3,\ c_3 \end{array}\right| = 0.$$

Article Nos. 64 and 65. *The Special Symmetrical System of three Rays.*

64. In what follows I consider the three rays forming a symmetrical system as already referred to: viz. the three rays intersect the plane of the ecliptic at points equidistant from S at longitudes 0°, 120°, 240°; each of them is at right angles to

Fig. 6.

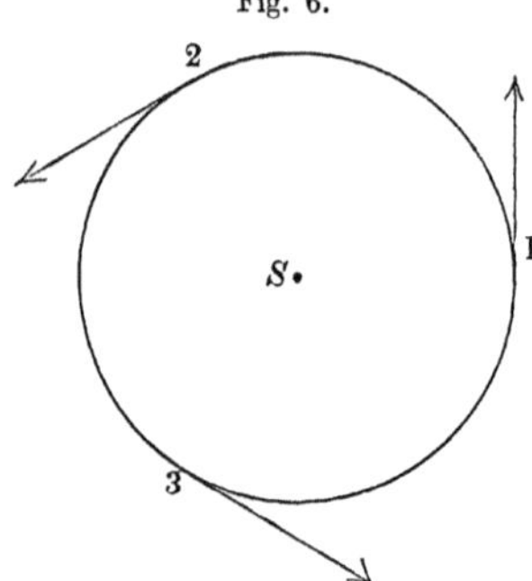

the line joining S with the intersection with the plane of the ecliptic, and at an inclination $=60°$ to this plane: the figure shows the projection on the plane of the ecliptic of the portions which lie above this plane of the three rays respectively.

The three rays lie on a hyperboloid of revolution having the line Sz for its axis; the circumscribed or asymptotic cone vertex S, is a right cone of the semi-aperture $=30°$; the reciprocal cone is therefore a right cone semi-aperture 60°, or (what is the same thing) the regulator is a small circle, angular radius 60°, and the regulator and separators have the positions shown in fig. 1, see No. 8.

Taking $S1 = S2 = S3 = 1$, and writing down the equations of the three rays in the forms

$$\frac{x-1}{0} = \frac{y}{1} = \frac{z}{\tan 60°},$$

$$\frac{x+\cos 60°}{-\sin 60°} = \frac{y-\sin 60°}{-\cos 60°} = \frac{z}{\tan 60°},$$

$$\frac{x+\cos 60°}{\sin 60°} = \frac{y+\sin 60°}{-\cos 60°} = \frac{z}{\tan 60°},$$

we obtain the six coordinates of the three rays respectively

$$(a_1, b_1, c_1, f_1, g_1, h_1) = (\ 0,\ \sqrt{3},\ -1,\ 0,\ 1,\ \sqrt{3}),$$
$$(a_2, b_2, c_2, f_2, g_2, h_2) = (\ 3,\ \sqrt{3},\ 2,\ \sqrt{3},\ 1,\ -2\sqrt{3}),$$
$$(a_3, b_3, c_3, f_3, g_3, h_3) = (-3,\ \sqrt{3},\ 2,\ -\sqrt{3},\ 1,\ -2\sqrt{3}),$$

whence the intersections with the orbit-plane are given by

$$x_1' : y_1' : 1 = \beta'\sqrt{3} - \gamma' : -\beta\sqrt{3} + \gamma : \beta'' + \gamma''\sqrt{3},$$
$$x_2' : y_2' : 1 = 3\alpha' + \beta'\sqrt{3} + 2\gamma' : -3\alpha - \beta\sqrt{3} - 2\gamma : \alpha''\sqrt{3} + \beta'' - 2\sqrt{3}\gamma'',$$
$$x_3' : y_3' : 1 = -3\alpha' + \beta'\sqrt{3} + 2\gamma' : 3\alpha - \beta\sqrt{3} - 2\gamma : -\alpha''\sqrt{3} + \beta'' - 2\sqrt{3}\gamma'',$$

where if (as before) the position of the orbit-plane be determined by means of the longitude b and colatitude c of the orbit-pole, we have

$$\alpha, \beta, \gamma = \sin b, \quad -\cos b, \quad 0,$$
$$\alpha', \beta', \gamma' = \cos b \cos c, \quad \sin b \cos c, \quad -\sin c,$$
$$\alpha'', \beta'', \gamma'' = \cos b \sin c, \quad \sin b \sin c, \quad \cos c,$$

and the passage from the coordinates x', y', to x, y, is given by

$$x' = x \sin b - y \cos b,$$
$$y' = x \cos b + y \sin b,$$

or conversely

$$x = x' \sin b + y' \cos b,$$
$$y = -x' \cos b + y' \sin b.$$

65. To develope the results, I consider the orbit-pole as passing through certain series of positions. The locus may be a meridian circle: by reason of the symmetry of the system, the results are not altered by a change of 120° in the longitude of the meridian; so that, by considering the two meridians 0°—180° and 90°—270°, we, in fact, consider twelve half meridians at the intervals of 30°. An illustration is afforded by Plate I.; the orbit-pole describes successively the meridians 0°, 30°, 60°, 90°, and the line 1, by its intersection with the orbit-plane, traces out on this plane a series of hyperbolas shown in the figure; the hyperbola for the meridian 90° is a right line, but (except for the position where the orbit-plane passes through the line 1) the locus is a determinate point on this line. Planogram No. 1 (Plate II.) refers to the meridian 90°—270°, and Planogram No. 2 (Plate III.) to the meridian 0°—180°. Next, if the orbit-pole be at one of the points A, that is, if the orbit-plane pass through a ray—though the position of the orbit-pole be here determinate, yet as there is a series of orbits, this also will give rise to a planogram: I call it Planogram No. 3. The orbit-pole may pass along a separator circle (viz. the orbit-plane be parallel to a ray), this is Planogram No. 4. And, lastly, the orbit-pole may pass along the ecliptic (or the orbit-plane may pass through the axis SZ), I call this Planogram No. 5. But the last three planograms are not considered in the like detail as the first two, and I have not, in regard to them, tabulated the results, nor given any Plates.

Article Nos. 66 to 82. *Planogram No. 1, the Meridian* 90°—270° (see Plate II.).

66. Supposing that the orbit-plane rotates about the axis $S1$ (fig. 6, see No. 64) in the plane of the ecliptic, the orbit-pole will describe the meridian 90°—270°, the position of the orbit-pole being $b=90°$, $c=0°$ to $90°$, or else $b=270°$, $c=0°$ to $90°$. But the same analytical formula extends to the two half meridians, viz., we may take $b=90°$, and extend c over $180°$, in the final results making c an arc between $0°$ and $90°$, and $b=90°$, or $=270°$, as the case requires.

67. Assuming then $b=90°$, we have

$$\begin{aligned} \alpha\,,\ \beta\,,\ \gamma\ &= 1, \quad 0\,, \quad 0\,, \\ \alpha',\ \beta',\ \gamma' &= 0, \quad \cos c, \ -\sin c, \\ \alpha'',\ \beta'',\ \gamma'' &= 0, \quad \sin c, \quad \cos c, \end{aligned}$$

and, moreover, $x', y'=x, y$: so that instead of (x_1', y_1'), &c., we may write at once (x_1, y_1), &c. The formulæ become

$$\begin{aligned} x_1 : y_1 : 1 &= \sqrt{3}\cos c + \sin c : \ \ 0 : \sin c + \sqrt{3}\cos c, \\ x_2 : y_2 : 1 &= \sqrt{3}\cos c - 2\sin c : -3 : \sin c - 2\sqrt{3}\cos c, \\ x_3 : y_3 : 1 &= \sqrt{3}\cos c - 2\sin c : \ \ 3 : \sin c - 2\sqrt{3}\cos c, \end{aligned}$$

that is

$$x_1 = 1, \quad y_1 = 0,$$

(viz. the orbit-plane, as is evident, meets the ray 1 in a fixed point, its intersection with the plane of xy);

$$x_2 = \frac{\sqrt{3}\cos c - 2\sin c}{\sin c - 2\sqrt{3}\cos c}, \quad x_3 = x_2,$$

$$y_2 = \frac{-3}{\sin c - 2\sqrt{3}\cos c}, \quad y_3 = -y_2,$$

and writing

$$\frac{2\sqrt{3}}{\sqrt{13}} = \cos\omega, \qquad \frac{1}{\sqrt{13}} = \sin\omega, \qquad \frac{1}{2\sqrt{3}} = \tan\omega,$$

(whence $\omega = 16°\,6'$) we find

$$x_2 = -\tfrac{8}{13} + \frac{3\sqrt{3}}{13}\tan(c+\omega),$$

$$y_2 = \frac{3}{\sqrt{13}}\sec(c+\omega),$$

and we thence have for the hyperbola, the locus of (x_2, y_2) and (x_3, y_3)

$$(x+\tfrac{8}{13})^2 = \tfrac{3}{13}(y^2 - \tfrac{9}{13}),$$

viz. the points (x_2, y_2) and (x_3, y_3) are situate on the hyperbola, symmetrically on opposite sides of the axis of x. For $c=0$, we have $x_2=-\frac{1}{2}$, $y_2=\frac{1}{2}\sqrt{3}$, $(x_2^2+y_2^2=1)$, and the hyperbola at this point touches the circle $x^2+y^2=1$; and similarly for x_3, y_3. The inclination of the asymptotes to the axis of y is given by $\tan\eta=\sqrt{\frac{3}{13}}$, $\eta=22^\circ\ 56'$.

68. The orbits are conics, focus S and vertex 1. It will be convenient to consider c as passing from 0° to $90^\circ-\omega$, and from 0° to $-(90^\circ+\omega)$; that is, from 0° to $73^\circ\ 54'-\epsilon$, and from 0° to $-116^\circ\ 6'+\epsilon$, if ϵ be indefinitely small: the point 2 will thus traverse the upper branch (alone shown in the Plate) of the guide-hyperbola, viz., for $c=0^\circ$ it will be at the point of contact with the circle; for $c=73^\circ\ 54'-\epsilon$ it will be at ∞, and for $c=-106^\circ\ 6'+\epsilon$ at ∞'. For $c=0^\circ$ the orbit is the circle; as c increases positively, it becomes an ellipse of increasing eccentricity and major axis, until for a certain value ($c=46^\circ\ 48'$ as will appear) it becomes a parabola; it then becomes a hyperbola (concave branch); for $c=52^\circ\ 45'$ it becomes the hyperbola Σ' subsequently referred to; and for $c=60^\circ$ (the point 2 being then on the line shown in the figure) the orbit becomes this right line. As c continues to increase, the orbit becomes a hyperbola (convex branch); and ultimately for $c=73^\circ\ 54'-\epsilon$, the point 2 goes to ∞, and the orbit becomes a hyperbola (convex) Σ, having an asymptote parallel to that of the guide-hyperbola: the inclination to the axis of x being thus $90^\circ-22^\circ\ 56'$, $=67^\circ\ 4'$.

69. Next as c increases negatively, the point 2 moves from the point of contact in the other direction to ∞': for $c=0^\circ$ the orbit is of course the circle, and as c increases negatively the orbits are at first the very same series of orbits as those belonging to the positive values [1], viz., they are first ellipses, of increasing eccentricity and major axis; then for $c=-92^\circ\ 54'$ the orbit is the parabola; the orbits are then hyperbolas (concave), and finally for $c=-106^\circ\ 6'+\epsilon$, when 2 is at ∞', the orbit is a hyperbola Σ', the asymptote of which is parallel to that of the guide-hyperbola, viz., the inclination to the axis of x is $=67^\circ\ 4'$.

70. It will be observed that the orbits from the circle to the hyperbola Σ' each intersect the guide-hyperbola (that is, the branch shown in the figure) in two points, the one corresponding to a positive, the other to a negative value of c; in the positive series, the remaining orbits from the hyperbola Σ', through the right line to the convex hyperbola Σ, each intersect the guide-hyperbola (same branch) in a single point only, for which c is positive.

71. There is, in the passage of the orbit-pole from $c=-106^\circ\ 6'+\epsilon$ to $c=73^\circ\ 54'-\epsilon$, say at $c=73^\circ\ 54'$, a discontinuity of orbit, viz., an abrupt change from the concave hyperbola Σ' to the concave hyperbola Σ; observe that the direction of the asymptotes being the same in each, the eccentricity e has the same value.

[1] Of course, as corresponding to different values of c, they are not the same orbits in space, but they are only the same curves in the planogram.

The point in question ($b=90^\circ$, $c=73^\circ\,54'$) is one of the points B of the spherogram, and the hyperbolas Σ, Σ' are two of the four orbits belonging to this point. And, by what precedes, it appears that as the orbit-pole passes through this point along a meridian downwards to the ecliptic the change is from a concave to a convex orbit.

72. On account of the symmetry in regard to the axis of x, the equation of the orbit will be of the form $r=Ax+B$; viz., the equation is at once found to be

$$r-1=\frac{r_2-1}{x_2-1}(x-1).$$

73. The eccentricity is the coefficient A taken positively ($e=\pm A$): it is in the present case proper to attend to the value of the coefficient itself,

$$A=\frac{r_2-1}{x_2-1},$$

the sign of A will then indicate the position of the centre of the orbit, viz., according as A is positive or negative the centre will be on the negative or the positive side of the focus S. To investigate the variation of A, we may express it as a function of $\tan c$, $=\lambda$ suppose. We have

$$x_2=\frac{\sqrt{3}-2\lambda}{\lambda-2\sqrt{3}},\qquad y_2=\frac{-3}{\lambda-2\sqrt{3}},$$

and thence

$$r_2=\frac{1}{\lambda-2\sqrt{3}}R_2,\qquad R_2=\pm\sqrt{12-4\sqrt{3}\lambda+13\lambda^2};$$

viz., r_2 must be positive, that is, R_2 is positive or negative according to the sign of $\lambda-2\sqrt{3}$; negative if $\lambda<2\sqrt{3}$ or $c<73^\circ\,54'$, positive if $\lambda>2\sqrt{3}$ or $c>73^\circ\,54'$. And we have then

$$A=\frac{\lambda-2\sqrt{3}-R_2}{3(\lambda-\sqrt{3})}.$$

But a more convenient formula is obtained by writing

$$\theta=-\cot c+\frac{1}{2\sqrt{3}},$$

$$a=\frac{1}{2\sqrt{3}},$$

we then have

$$\sqrt{1+\theta^2}=\theta r_2,$$

which determines the sign of the radical, viz., this must have the same sign as θ; and then for the coefficient

$$A=\frac{2}{3(\theta+a)}(-\sqrt{1+\theta^2}+\theta).$$

74. For c a small arc $=\epsilon$, θ is large and negative, and $\sqrt{1+\theta^2}$, having the same sign as θ, is $=\theta+\frac{1}{2\theta}$ nearly; we have therefore

$$A=\frac{2}{3\theta}\cdot\frac{-1}{2\theta}=-\frac{1}{3\theta^2} \text{ approximately.}$$

For c nearly $=60°$, say $c=60°\pm\epsilon$,

$$\cot c=\cot 60°\pm\epsilon\operatorname{cosec}^2 60°=\frac{1}{\sqrt{3}}\pm\frac{4\epsilon}{3},$$

$$\theta=-\frac{1}{2\sqrt{3}}\pm\frac{4\epsilon}{3}\qquad \theta+a=\pm\frac{4\epsilon}{3},\qquad \sqrt{1+\theta^2}=-\frac{\sqrt{13}}{2\sqrt{3}},$$

and thence

$$A=\frac{-1+\sqrt{13}}{2\sqrt{3}}\div\pm\frac{4\epsilon}{3}=\pm\frac{\sqrt{3}}{8}\cdot\frac{-1+\sqrt{13}}{\epsilon},$$

viz., this is $-\infty$ for $c=60°-\epsilon$, and $+\infty$ for $c=60°+\epsilon$.

For c nearly $=90°-\omega$, say first $c=73°\,54'-\epsilon$, we have

$$\cot c=\frac{1}{2\sqrt{3}}+\tfrac{13}{12}\epsilon,\qquad \theta=-\tfrac{13}{12}\epsilon,\qquad \theta+a=\frac{1}{2\sqrt{3}},\qquad \sqrt{1+\theta^2}=-1,$$

whence

$$A=\frac{4}{\sqrt{3}}=2{\cdot}30940;$$

but if $c=73°\,54'+\epsilon$, then $\theta=\frac{13}{12}\epsilon$, $\theta+a=\frac{1}{2\sqrt{3}}$, $\sqrt{1+\theta^2}=+1$, and

$$A=-\frac{4}{\sqrt{3}}=-2{\cdot}30940,$$

viz., there is an abrupt change from $A=+\frac{4}{\sqrt{3}}$ to $A=-\frac{4}{\sqrt{3}}$; corresponding to the discontinuity of orbit already referred to. We may diminish c by 180°, and consider the last-mentioned value, $A=-\frac{4}{\sqrt{3}}$, as belonging to $c=-90°-\omega+\epsilon=-(106°\,6'-\epsilon)$.

75. Consider next that c passes from 0 to $-(106°\,6'-\epsilon)$. First if c is a small negative quantity $c=-\epsilon$, θ is large and positive, and $\sqrt{1+\theta^2}$ having the same sign as θ (positive) is $=\theta+\frac{1}{2\theta}$ nearly, we have therefore $A=\frac{2}{3\theta}\cdot\frac{-1}{2\theta}=-\frac{1}{3\theta^2}$ (same as for $c=+\epsilon$). And it is easy to see that as c increases negatively, A is always increasing negatively, its value for $c=-90°$ being $A=\frac{1-\sqrt{13}}{3}=-{\cdot}8685$, and for $c=-106°\,6'+\epsilon$ being $=-2{\cdot}30940$ as above. We have a diagram of A (see next page).

76. It thus appears that from $A=0$ to $A=-\frac{4}{\sqrt{3}}$, there are always to any given value of A two values of c, or positions of the orbit-pole. In particular if A be $=-1$, the curve will be a parabola; the values of c lying between 0°, 60° and between 73° 54′, 90° respectively.

Fig. 7.

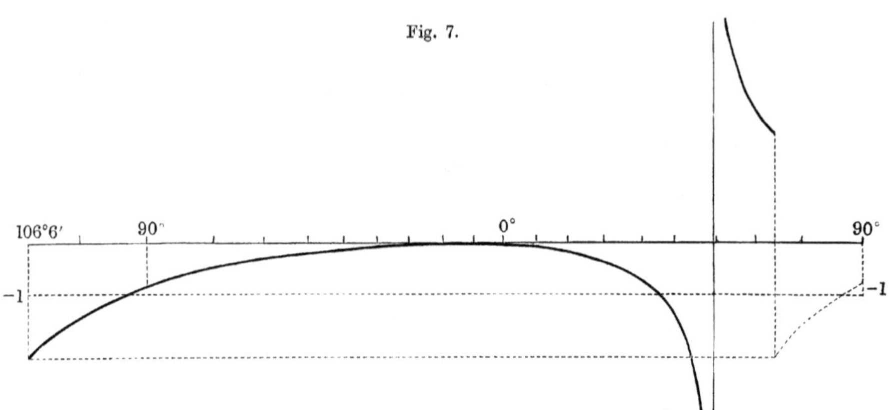

To find them, writing $A=-1$, we have

$$-3\theta-3a=2\theta-2\sqrt{1+\theta^2}, \text{ that is, } 5\theta+3a=2\sqrt{1+\theta^2},$$

or

$$21\theta^2+30a\theta+9a^2-4=0,$$

that is, substituting for a its value $=\frac{1}{2\sqrt{3}}$,

$$21\,\theta^2+5\sqrt{3}\,\theta-\tfrac{13}{4}=0, \quad (14\theta\sqrt{3}+5)^2=116,$$

or

$$\theta=\frac{-5\pm\sqrt{116}}{14\sqrt{3}},$$

that is

$$\theta=-\cdot65034, \qquad \theta=\ \cdot23797,$$

giving

$$\cot c=+\cdot93902, \qquad \cot c=+\cdot05071,$$

or

$$c=\ 46^\circ\,48', \qquad c=\ 87^\circ\,6'.$$

77. It has been seen that $c=73^\circ\,54'+\epsilon$ gives $A=-\frac{4}{\sqrt{3}}=-2\cdot30940$; there will be between 0° and 60° another value of c, giving for A this same value; to find this value write $A=-\frac{4}{\sqrt{3}}$, then we have

$$-4\sqrt{3}\left(\theta+\frac{1}{2\sqrt{3}}\right)=2\theta-2\sqrt{1+\theta^2},$$

that is

$$(1+2\sqrt{3})\,\theta+1=\sqrt{1+\theta^2},$$

or

$$(12+4\sqrt{3})\,\theta^2+(2+4\sqrt{3})\,\theta=0,$$

satisfied as it should be by $\theta=0$, and also by

$$\theta=-\frac{1+2\sqrt{3}}{2\,(3+\sqrt{3})}=-\cdot 47170,$$

giving

$$\cot c=\cdot 76038 \text{ or } c=52^\circ\,45'.$$

78. Representing the equation of the orbit by

$$r=Ax\pm a\,(1-A^2),$$

we have for the point 1,

$$1=A\ \ \pm a\,(1-A^2),$$

that is

$$a=\frac{\pm 1}{1+A},$$

where the sign is to be taken so that a shall be positive.

79. With a view to the calculation of the times of passage, I calculate a series of values of x_2, y_2, r_2, A, a, for values of c at the intervals of 5° and for a few intermediate values; we have x_3, y_3, $r_3=x_2$, y_2, r_2, so that these are known; so long as the orbit is an ellipse, the time of passage between the points 2 and 3, say T_{23}, may be calculated by Lambert's equation, the length of the chord y_2-y_3, $=2y_2$ being known without any fresh calculation. And then the times T_{12} and T_{31} being equal, and the sum $T_{12}+T_{23}+T_{31}$ being equal to the whole periodic time (reckoned as $=3a^{\frac{3}{2}}$) the times T_{12} and T_{31} are also known. But when the orbit is a concave hyperbola there is no time T_{23}, and the other two times T_{12}, $=T_{31}$ must be calculated. For the reason referred to (*ante*, No. 39) I did not use Lambert's equation,—and it was less necessary to do so, by reason that, the transverse axis coinciding with the axis of x, the other formula could be employed without difficulty.

80. The formulæ for x_2, y_2 adapted to logarithmic calculation are

$$\log(x_2+\cdot 61539)=\overline{11}\cdot 60174+\log\tan(c+16^\circ\,6'),$$

$$\log y_2\qquad\qquad=\overline{11}\cdot 92015+\log\sec(c+16^\circ\,6'),$$

where y_2 is always positive, but the sign of x_2 must be attended to. The values of r_2 and its inclination ϕ_2 to the axis of x are then to be calculated from

$$\tan\phi_2=\frac{y_2}{x_2};\quad r_2=x_2\sec\phi_2 \text{ or } =y_2\operatorname{cosec}\phi_2,$$

(viz. for r_2 it is proper to use the first or the second value, according as x_2 is greater or less than y_2).

We have then $e = (\pm A)$ and a from the foregoing formulæ

$$A = \frac{r_2 - 1}{x_2 - 1}, \quad a = \frac{\pm 1}{1 + A},$$

where a, e are each of them positive.

And then for the Times

$$T_{23} = \frac{3}{2\pi}(\chi - \chi' - \sin\chi + \sin\chi'); \quad \left(\log\frac{3}{2\pi} = \bar{1}\cdot 67894\right),$$

where

$$a\cos\chi = a - r_2 - y_2,$$
$$a\cos\chi' = a - r_2 + y_2,$$

and attention is necessary in order to the selection of the proper values of the angles χ, χ'.

And finally

$$T_{12} = T_{31} = \tfrac{1}{2}(3a^{\frac{3}{2}} - T_{23}).$$

81. I subjoin a specimen; the characteristics of the logarithms are (as in the actual calculations) omitted.

$b = 90° \quad c = 20°,$

$c + 16° 6' \quad = 36° 6'$

log sec	09259	log tan	86285
	92015		60174
	01274		46459
$y_2 = 1\cdot 0297$			61539
			29147
		$x_2 =$	$-\cdot 32392$
		log =	51044
	01274		02046
	51044		01274
	50230		03320
$\phi_2 = 72° 33'$		$r_2 =$	$1\cdot 0794$

$\cdot 0794$ log = 89982 $\qquad\qquad \cdot 94003$ log = 97314

$1\cdot 3239$ log = 12185 $\qquad\qquad$ comp = 02686

77797

$A = -\cdot 059975 \qquad a = 1\cdot 0638$

$$\begin{array}{rl}
 & 1\cdot 0638 \\
 & 1\cdot 0794 \\
\hline
A - r_2 = - & \cdot 0156 \\
y_2 = + & 1\cdot 0297 \\
\hline
 & 1\cdot 0141 = a \cos \chi' \\
- & 1\cdot 0453 = a \cos \chi
\end{array}$$

$$\begin{array}{lcl}
00608 & \qquad & 01924 \\
02686 & & 02686 \\
\hline
97922 & & 99238
\end{array}$$

$\chi' = 17^\circ\, 35' \qquad \chi\,(=\text{Supp. } 10^\circ\, 42') = 169^\circ\, 18' \qquad \chi - \chi' = 151^\circ\, 43'$

$$\begin{array}{rrrrr}
151^\circ & 2\cdot 63544 & & & \\
43' & \cdot 01250 & & & \\
-\sin\chi & & -\cdot 18566 & & \\
\sin\chi' & \cdot 30209 & & & 02686 \\
 & \overline{2\cdot 95003} & & & 01343 \\
 & \cdot 18566 & & & 47712 \\
 & \overline{} & & & \overline{51741} \\
 & 2\cdot 76437 & \log = 44160 & & \\
 & & 02686 & & 3a^{\frac{3}{2}} = 3\cdot 2916 \\
 & & 01343 & & 1\cdot 4482 \\
 & & 67894 & & \overline{} \\
 & & \overline{} & & 1\cdot 8434 \\
T_{23} = 1\cdot 4482 & & 16083 & & T_{12} = T_{31} = \cdot 9217
\end{array}$$

82. For the Time in a hyperbola, we have

$$T_{12} = T_{31} = \frac{3}{2\pi}\, a^{\frac{3}{2}} \left\{ e \tan u_2 - h\,.\,l\,.\tan\left(45^\circ + \tfrac{1}{2} u_2\right)\right\},$$

where

$$\tan u_2 = \frac{y_2}{a\sqrt{e^2 - 1}}\,.$$

Taking as a specimen the case $c=75°$, we have here

$$a = \cdot 9004 \qquad e = 2\cdot 1106 \qquad y_2 = 43\cdot 341$$
$$\log = \cdot 95444 \qquad \log = \cdot 32441 \qquad \log = 1\cdot 63690$$
$$a(e^2-1) = 3\cdot 1106$$
$$\log \quad ,, \quad = \cdot 49284$$

and then the calculation is

$$\begin{aligned}
\log a &= 95444 \\
,, \ a(e^2-1) &= 49284 \\
& 44728 \\
,, \ a\sqrt{e^2-1} &= 22364 \\
\log \ y_2 &= 63690 \\
\log \tan u &= 41326 \\
& 32441 \\
& 73767
\end{aligned}$$

$$\begin{aligned}
u &= 87°\ 47' \\
h\,.\,l \tan(45° + \tfrac{1}{2}u) &= 3\cdot 95140 \\
e \tan u &= 54\cdot 660 \\
& 3\cdot 951 \\
& 50\cdot 709 \\
\log &= 70508 \\
& 95444 \\
& 47722 \\
& 67894 \\
& 31568 \\
T_{12} = T_{31} &= 20\cdot 686
\end{aligned}$$

83. In the case of the parabola $p=1$, and the expression for the Times is

$$T_{13} = T_{21} = \frac{1}{4\pi}\{(\rho+\rho'+\gamma)^{\frac{3}{2}} - (\rho+\rho'-\gamma)^{\frac{3}{2}}\},$$

where for

$$c = 46°\ 48'$$
$$c = 87°\ \ 6'$$

we have

$$T_{12} = T_{31} = \cdot 787,$$
$$T_{12} = T_{31} = 2\cdot 588.$$

Planogram No. 1, *first part*, $b = 90°$.

	c	x_2	y_2	ϕ_2	r_2	A	a	T_{12}	T_{23}	T_{31}
Circle	0°	− ·500	+ ·866	60°	1·000	0	1·000	1·000	1·000	1·000
Ellipses	5	·461	·892	62° 39′	1·004	− ·003	1·003	. .		
	10	·420	·927	65 38	1·017	·012	1·012	·960	1·135	·960
	15	·374	·972	68 56	1·041	·030	1·031	. .		
	20	·324	1·030	72 33	1·079	·060	1·064	·922	1·448	·922
	25	·267	1·104	76 26	1·136	·107	1·120			
	30	·200	1·200	80 32	1·216	·180	1·220	·887	2·275	·887
	35	·120	1·325	84 49	1·330	·295	1·418			
	40	− ·021	1·492	90 48	1·492	·482	1·931	·838	6·371	·838
	40° 54′	·000	1·515	90	1·515	·515	2·061			
	45	+ ·109	1·722	93 37	1·725	·814	5·362			
Parab.	46° 48′	·166	1·826	95 11	1·834	1·000	∞	·787	∞	·787
Hyperbs.	50	·287	2·054	97 57	2·074	1·505	1·981	·750	~	·750
	52° 45′	·418	2·306	100 16	2·344	2·309	·764		~	
	55	·552	2·569	102 8	2·627	− 3·632	·380	·628	~	·628
Line	60	1·000	3·464	196 6	3·605	− ∞ + ∞	·000	·000	~	·000
Convex	65	1·937	5·378	109 48	5·716	+ 5·032	·166	Convex orbit.		
	70	+ 5·248	12·233	113 13	13·311	2·898	·257			
	73° 54′	+ ∞ − ∞	∞	115 39 64 21	∞	+ 2·309 − 2·309	·302 ·764	 ∞	 ~	 ∞
Hyperbs.	75	− 21·432	43·341	63 42	48·346	2·111	·900	20·68	~	20·68
	80	4·356	7·830	60 55	8·960	1·486	2·056	3·856	~	3·856
	85	2·653	4·322	58 27	5·072	1·115	8·718	2·912	~	2·912
Parab.	87° 6′	2·320	3·644	57 31	4·320	1·000	∞	2·588	∞	2·588
Ellipse	90	− 2·000	3·000	56 18	3·606	− ·869	7·622	2·255	58·62	2·255

The mark ~ in the T_{23} column shows that there is no Time T_{23}.

Planogram No. 1, *second part*, $b = 270°$.

	c	x_2	y_2	ϕ_2	r_2	A	a	T_{12}	T_{23}	T_{31}
Circ.	0°	− ·500	+ ·866	60° ′	1·000	0	1·000	1·000	1·000	1·000
All Ellipses	5	·537	·848	57 39	1·003	− ·002	1·002			
	10	·573	·837	55 37	1·014	·009	1·009	1·044	·951	1·044
	15	·608	·832	53 52	1·030	·019	1·019			
	20	·643	·834	52 23	1·053	·032	1·033	1·091	·969	1·091
	25	·678	·842	51 10	1·081	·048	1·051			
	30	·714	·857	50 11	1·116	·068	1·073	1·145	1·043	1·145
	35	·752	·879	49 27	1·157	·090	1·098			
	40	·793	·910	48 51	1·207	·115	1·130	1·207	1·192	1·207
	45	·836	·950	48 40	1·266	·145	1·169			
	50	·884	1·002	48 36	1·336	·179	1·217	1·283	1·464	1·283
	55	·938	1·069	48 45	1·442	·218	1·278			
	60	1·000	1·154	49 7	1·527	·264	1·358	1·377	1·983	1·377
	65	1·074	1·266	49 42	1·660	·318	1·466			
	70	1·164	1·412	50 31	1·830	·383	1·622	1·506	3·036	1·506
	75	1·280	1·611	51 35	2·056	·464	1·864			
	80	1·431	1·891	52 53	2·372	·564	2·295	1·771	6·888	1·771
	85	1·651	2·311	54 28	2·840	·694	3·269			
	90	− 2·000	+ 3·000	56 18	3·606	− ·869	7·622	2·255	58·62	2·255

Article Nos. 84 to 94. *Planogram No.* 2, *the Meridian* 0°—180° (see Plate III.).

84. The orbit-plane here rotates about an axis in the plane of the ecliptic at right angles to $S1$ (Fig. 6). The entire meridian is given by $b=0°$, $c=0°$ to $90°$, and $b=180°$, $c=0°$ to $90°$, but it is sufficient to consider one of these half meridians, say the latter of them, as the series of values is the same for each of them, with only an interchange of the points 2, 3. I write therefore, $b=180°$, so that we have

$$\begin{aligned}\alpha\ ,\ \beta\ ,\ \gamma\ &= \quad 0\ ,\ 1,\quad 0\ ,\\ \alpha'\ ,\ \beta'\ ,\ \gamma'\ &= -\cos c,\ 0,\ -\sin c,\\ \alpha'',\ \beta'',\ \gamma'' &= -\sin c,\ 0,\quad \cos c,\end{aligned}$$

consequently

$$\begin{aligned}x_1' : y_1' : 1 &= \sin c : -\sqrt{3} : \sqrt{3}\cos c,\\ x_2' : y_2' : 1 &= -3\cos c - 2\sin c : -\sqrt{3} : -\sqrt{3}\sin c - 2\sqrt{3}\cos c,\\ x_3' : y_3' : 1 &= 3\cos c - 2\sin c : -\sqrt{3} : \sqrt{3}\sin c - 2\sqrt{3}\cos c,\end{aligned}$$

and moreover $x'=y$, $y'=-x$; so that, introducing into the formulæ (x_1, y_1), &c., in place of the (x_1', y_1'), &c., we have

$$\begin{aligned}x_1 &= \sec c, & y_1 &= \frac{1}{\sqrt{3}}\tan c,\\ x_2 &= \frac{1}{\sin c + 2\cos c}, & y_2 &= \frac{1}{\sqrt{3}}\,\frac{2\sin c + 3\cos c}{\sin c + 2\cos c},\\ x_3 &= \frac{-1}{\sin c - 2\cos c}, & y_3 &= \frac{1}{\sqrt{3}}\,\frac{2\sin c - 3\cos c}{\sin c - 2\cos c},\end{aligned}$$

which, putting

$$\cos\vartheta = \frac{2}{\sqrt{5}},\quad \sin\vartheta = \frac{1}{\sqrt{5}},\quad \tan\vartheta = \tfrac{1}{2},\quad \vartheta = 26°\,34',$$

become

$$\begin{aligned}x_1 &= \sec c, & y_1 &= \frac{1}{\sqrt{3}}\tan c,\\ x_2 &= -\frac{1}{\sqrt{5}}\sec(c-\vartheta), & y_2 &= \frac{1}{\sqrt{3}}\left\{\ \tfrac{8}{5} + \tfrac{1}{5}\tan(c-\vartheta)\right\},\\ x_3 &= -\frac{1}{\sqrt{5}}\sec(c+\vartheta), & y_3 &= \frac{1}{\sqrt{3}}\left\{-\tfrac{8}{5} + \tfrac{1}{5}\tan(c+\vartheta)\right\};\end{aligned}$$

so that the guide-hyperbolas are

$$\begin{aligned}&x_1^2 - \ 3y_1^2 = 1, && \tfrac{1}{2}\text{ angle of asymptotes} = 30°\\ &x_2^2 = 15y_2^2 - 16\sqrt{3}\,y_2 + 13, && \text{,,}\qquad\text{,,}\qquad \tan^{-1}\frac{1}{\sqrt{15}} = 14°\,28'\\ &x_3^2 = 15y_3^2 + 16\sqrt{3}\,y_2 + 13, && \text{,,}\qquad\text{,,}\qquad\quad \text{,,}\qquad \text{,,}\end{aligned}$$

It is easy to verify that

Hyperbola 2 passes through $x_2 = -\frac{1}{2}$, $y_2 = \frac{1}{2}\sqrt{3}$, and touches there circle $x^2 + y^2 = 1$,

" 3 " $x_3 = -\frac{1}{2}$, $y_3 = -\frac{1}{2}\sqrt{3}$ " " "

and we thus have the figure in the Plate.

85. The figure shows the motion of the points 1, 2, 3, along their respective hyperbolas, viz. $c = 0°$ to $90°$, the point 1 moves from contact with the circle, along a half branch to infinity: 2 moves from contact along a small portion of the half branch; 3 moves from contact, along the half branch to infinity for $c = \tan^{-1} 2 = 63° 26'$, and then reappearing at the opposite infinity, as c increases to $90°$ describes a portion of the opposite half branch.

86. For $c = 0$, the orbit is the circle; as c increases the orbit becomes elliptic; then parabolic, $c = 51°$, and afterwards hyperbolic (concave); until for $c = 60°$, the three points are on the horizontal line of the figure, and the orbit is this right line; it is to be noticed that the arrangement of the points on these orbits is 1, 2, 3; so that for the parabola, T_{31} is $= \infty$, and for the hyperbolas and right line T_{31} does not exist.

87. For $c < 60°$ until $c = 63° 26'$ the orbit is a convex hyperbola, the arrangement of the points being still 1, 2, 3: say for $c = 63° 26' - \epsilon$, the orbit is the convex hyperbola Ω. At $c = 63° 26'$ there is an abrupt change of orbit; say for $c = 63° 26' + \epsilon$ the orbit is a concave hyperbola Ω_1; and for $c = 65° 52'$ the orbit is a parabola; the arrangement of the points on these orbits is 2, 1, 3; so that for the hyperbolas T_{23} does not exist, and T_{23} is $= \infty$ for the parabola. Observe also that for the hyperbola Ω_1, the point 3 is at infinity, or we have $T_{31} = \infty$. As c continues to increase, the orbit becomes an ellipse, the eccentricity having a minimum value = ·628 (about), for $c = 69°$ (about). For $c = 89° 20'$ the orbit is again a parabola, and then until $c = 90°$ it is a hyperbola; the order of the points on the last-mentioned parabola and hyperbolas being 1, 3, 2; so that for the parabola T_{12} is $= \infty$, and for the hyperbolas T_{12} does not exist. In the hyperbola for $c = 90°$, say the hyperbola Ω', the point 1 is at infinity, or we have $T_{12} = \infty$. The foregoing results, obtained (except as to the numerical values) by consideration of the figure, will be confirmed by means of the calculated values of e.

88. The equation of the orbit may be written

$$\begin{vmatrix} \dfrac{r}{\sqrt{3}}, & \dfrac{x}{\sqrt{3}}, & y, & \dfrac{1}{\sqrt{3}} \\ r_1 \cos c, & 1, & \sin c, & \cos c \\ r_2(\sin c + 2\cos c), & -1, & 2\sin c + 3\cos c, & \sin c + 2\cos c \\ r_3(\sin c - 2\cos c), & 1, & -2\sin c + 3\cos c, & \sin c - 2\cos c \end{vmatrix} = 0,$$

or developing, this is

$$\frac{r}{\sqrt{3}}6(\sin^2 c - 3\cos^2 c),$$

$$-\frac{x}{\sqrt{3}}\{\ 4r_1(\sin^2 c - 3\cos^2 c)\cos c$$
$$-\ r_2(\sin c + 2\cos c)(\sin^2 c - 3\cos^2 c)$$
$$+\ r_3(\sin c - 2\cos c)(\sin^2 c - 3\cos^2 c)\}$$

$$+y\ \{-\ 2r_1 \sin c\cos c$$
$$+\ r_2(-\sin^2 c + \sin c\cos c + 6\cos^2 c)$$
$$+\ r_3(\ \sin^2 c + \sin c\cos c - 6\cos^2 c)\}$$

$$-\frac{1}{\sqrt{3}}\{\ r_1 . -6\cos^2 c$$
$$+3r_2(\sin^2 c + \sin c\cos c - 2\cos^2 c)$$
$$+3r_3(\sin^2 c - \sin c\cos c - 2\cos^2 c)\} = 0;$$

(observe that the orbit will be a right line if $\sin^2 c - 3\cos^2 c = 0$, that is if $c = 60°$, which is right, since 60° is the angular radius of the regulator circle).

89. Putting in the equation $\tan c = \lambda$, and therefore $\cos c = \frac{1}{\sqrt{1+\lambda^2}}$, the equation becomes

$$r = \frac{1}{6\sqrt{1+\lambda^2}}\Big(4r_1 - (\lambda+2)r_2 + (\lambda-2)r_3\Big)x$$
$$+\frac{1}{2\sqrt{3}(\lambda^2-3)}\Big(2\lambda\, r_1 + (\lambda+2)(\lambda-3)r_2 - (\lambda+3)(\lambda-2)r_3\Big)y$$
$$+\ \frac{1}{2(\lambda^2-3)}\Big(-2r_1 + (\lambda-1)(\lambda+2)r_2 + (\lambda+1)(\lambda-2)r_3\Big).$$

We have

$$x_1 = \sqrt{1+\lambda^2},\quad x_2 = \frac{-\sqrt{1+\lambda^2}}{\lambda+2},\quad x_3 = \frac{\sqrt{1+\lambda^2}}{\lambda-2},$$

$$y_1 = \frac{1}{\sqrt{3}}\lambda\ ,\quad y_2 = \frac{1}{\sqrt{3}}\frac{2\lambda+3}{\lambda+2},\quad y_3 = -\frac{1}{\sqrt{3}}\frac{2\lambda-3}{\lambda-2},$$

and thence, writing for shortness

$$R_1 = \sqrt{1+\tfrac{4}{3}\lambda^2},$$

$$R_2 = \frac{1}{\sqrt{3}}\sqrt{7\lambda^2+12\lambda+12},$$

$$R_3 = \frac{1}{\sqrt{3}}\sqrt{7\lambda^2-12\lambda+12},$$

we have

$$\begin{aligned} r_1 &= R_1, \\ r_2(\lambda+2) &= R_2, \\ r_3(\lambda-2) &= R_3, \end{aligned}$$

where r_1, r_2, r_3 are positive, and the signs of R_1, R_2, R_3 must be determined accordingly; viz., R_1 is always positive, and ($c=0°$ to $c=90°$, as here supposed) R_2 is also positive; but R_3 has the same sign as $\lambda-2$; viz., $c=0°$ to $c=63° 26'$, R_3 is negative; and $c=63° 26'$ to $c=90°$, R_3 is positive. It is to be observed that this position, $c=\tan^{-1}2=63° 26'$, of the pole is the intersection of the meridian $b=180°$ by a separator circle, and corresponds to an intersection at infinity on the ray 3.

90. Substituting the foregoing values of r_1, r_2, r_3, the equation of the orbit becomes

$$\begin{aligned} r = &\frac{1}{6\sqrt{1+\lambda^2}}(4R_1 - R_2 + R_3)x \\ &+ \frac{1}{2\sqrt{3}(\lambda^2-3)}\{\lambda(2R_1+R_2-R_3) - 3(R_2+R_3)\}y \\ &+ \frac{1}{2(\lambda^2-3)}\{\lambda(R_2+R_3) - 2R_1 - R_2 + R_3\}, \end{aligned}$$

where $\lambda = \tan c$; and the equation of the orbit may thence be calculated for any given value of c.

91. The analytical expression for the eccentricity is

$$e = \sqrt{A^2+B^2},$$

Fig. 8.

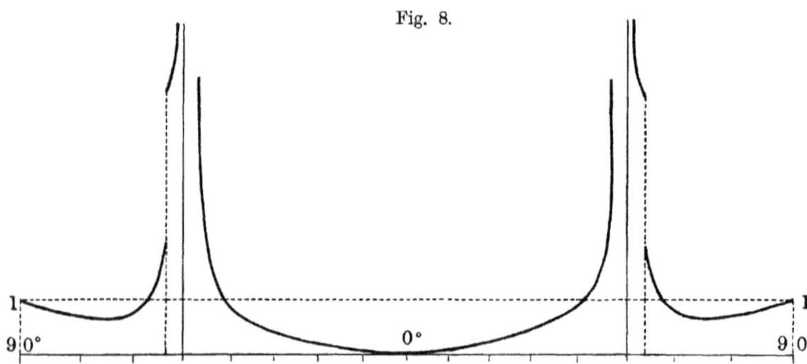

where, as above,

$$A = \frac{1}{6\sqrt{1+\lambda^2}}(4R_1 - R_2 + R_3),$$

$$B = \frac{1}{2\sqrt{3}(\lambda^2-3)}\{\lambda(2R_1+R_2-R_3) - 3(R_2+R_3)\};$$

but this expression is too complicated to allow of an analytical discussion of the series of values of e (such as was given for $A, = \pm e$, in planogram No. 1). The numerical calculation gives the results mentioned *ante* No. 87, viz., $c = 0$, $e = 0$; $c = 51°$, $e = 1$; $c = 60°$, $e = \infty$; $c = 63° \, 26' - \epsilon$, $e = 4{\cdot}912$; $c = 63° \, 26' + \epsilon$, $e = 1{\cdot}853$; $c = 69°$, $e = {\cdot}628$ (min.); $c = 89° \, 20'$, (viz. $\lambda = 86{\cdot}176$), $e = 1$; $c = 90°$, $e = 1{\cdot}018$; values which are exhibited in the diagram in the preceding page.

92. It may be further remarked, in reference to the formula

$$r = Ax + By + C,$$

that for $c = 60°$, that is $\lambda = \sqrt{3}$, we have A finite, B and C each infinite, but equal and of opposite signs; viz., the equation becomes $r = {\cdot}2242\,x \pm \infty\,(y-1)$, that is $y = 1$, orbit a right line as above.

The abrupt change at $c = 63° \, 26'$, $\lambda = 2$, arises from the change of sign of R_3; viz., $c = 63° \, 26' - \epsilon$, $R_3 = -\dfrac{4}{\sqrt{3}} = -2{\cdot}309$, but $c = 63° \, 26' + \epsilon$, $R_3 = \dfrac{4}{\sqrt{3}} = +2{\cdot}309$; the two orbits are

$$c = 63° \, 26' - \epsilon, \quad r = {\cdot}234\,x + 4{\cdot}906\,y - 3{\cdot}671, \quad e = 4{\cdot}912, \quad a = {\cdot}159,$$

$$c = 63° \, 26' + \epsilon, \quad r = {\cdot}578\,x - 1{\cdot}761\,y + 3{\cdot}257, \quad e = 1{\cdot}853, \quad a = 1{\cdot}338$$

For $c = 90°$ the equation is

$$r = \frac{4}{3\sqrt{3}}\,x + \tfrac{2}{3}\,y + \sqrt{\tfrac{7}{3}}$$

$$= {\cdot}770\,x + {\cdot}666\,y + 1{\cdot}527$$

and therefore $e = \sqrt{\tfrac{28}{27}} = 1{\cdot}018$ as above; $a = 9\sqrt{21} = 41{\cdot}243$.

It is to be added that for c nearly $= 90°$, or λ very large, we have

$$R_1 = \frac{2}{\sqrt{3}}\lambda, \quad R_2 = \sqrt{\tfrac{7}{3}}\,\lambda + 2\sqrt{\tfrac{3}{7}}, \quad R_3 = \sqrt{\tfrac{7}{3}}\,\lambda - 2\sqrt{\tfrac{3}{7}},$$

and thence

$$A = \frac{4}{3\sqrt{3}} - \frac{2}{\sqrt{21}}\,\frac{1}{\lambda} = {\cdot}770 - {\cdot}430\,\frac{1}{\lambda},$$

$$B = \tfrac{2}{3} - \frac{5}{\sqrt{7}}\,\frac{1}{\lambda} = {\cdot}666 - 1{\cdot}890\,\frac{1}{\lambda},$$

$$C = \sqrt{\tfrac{7}{3}} - \frac{1}{\sqrt{3}}\,\frac{2}{\lambda} = 1{\cdot}527 - 1{\cdot}555\,\frac{1}{\lambda}.$$

It was, in fact, by means of these expressions that the value $\lambda = 86{\cdot}176$ ($c = 89° \, 20'$) corresponding to the last-mentioned parabolic orbit was obtained.

93. For the calculation of the table we have

$$\log x_1 = \overline{10} + \log \sec c,$$

$$\log y_1 = \overline{10}\cdot 76144 + \log \tan c,$$

$$\log x_2 = \overline{10}\cdot 65052 + \log \sec (c - 26^\circ\, 34'),$$

$$\log (y_2 - \cdot 92376) = \overline{10}\cdot 06247 + \log \tan (c - 26^\circ\, 34'),$$

$$\log x_3 = \overline{10}\cdot 65052 + \log \sec (c + 26^\circ\, 34'),$$

$$\log (y_3 + \cdot 92376) = \overline{10}\cdot 06247 + \log \sec (c + 26^\circ\, 34'),$$

the values of r_1, r_2, r_3, are then calculated from

$$x_1 = r_1 \cos \phi_1, \quad y_1 = r_1 \sin \phi_1,$$

or say

$$\frac{y_1}{x_1} = \tan \phi_1, \qquad r_1 = x_1 \sec \phi_1, \text{ \&c.}$$

and those of the chords γ_{12}, γ_{23}, γ_{31}, from

$$x_1 - x_2 = \gamma_{12} \cos \theta_{12}, \quad y_1 - y_2 = \gamma_{12} \sin \theta_{12},$$

or say

$$\tan \theta_{12} = \frac{x_1 - x_2}{y_1 - y_2}, \quad \gamma_{12} = (x_1 - x_2) \sec \theta_{12}.$$

We have then to find the equation of the orbit $r = Ax + By + C$; this might be done by substituting in the determinant expression the numerical values of x_1, y_1, r_1, x_2, y_2, r_2, x_3, y_3, r_3, and so calculating the result, but I have preferred to employ the formula of No. 90, using only the calculated values of r_1, r_2, r_3; viz. we have

$$r_1 = R_1,$$

$$r_2 (\lambda + 2) = R_2,$$

$$r_3 (\lambda - 2) = R_3,$$

which gives the values of R_1, R_2, R_3. And then we have e, ϖ, a, from the equations

$$A = e \cos \varpi, \quad B = e \sin \varpi, \quad a = \frac{\pm C}{1 - e^2},$$

e and a being each regarded as positive. The times in the elliptic, and parabolic orbits are then calculated from Lambert's equation, as explained in regard to Planogram No. 1, but for the hyperbolic orbits, the other formulæ were made use of.

94. I annex a specimen; the characteristics of the logarithms are omitted.

$c = 20°$.

20°		$-6° 34'$		$+46° 34'$	
02701	56107	00286	06113	16272	02376
	76144	65052	06247	65052	06247
	32251	65338	12360	81324	08623
$x_1 = 1{\cdot}06418$	$y_1 = {\cdot}21014$	$x_2 = -{\cdot}45017$	·92367	$x_3 = -{\cdot}65049$	·92367
32251	02701		·01329		·12196
02701	00830		$y_2 = +{\cdot}91038$		$y_3 = {\cdot}80171$
29950	03531		log = ·95922		$\log y_3 =$ 90402
$\phi_1 = 11° 10'$	$r_1 = 1{\cdot}0847$	95922	95922	90402	90402
		65338	04752	81324	10980
		30584	00674	09078	01382

$\phi_2 (= 63° 41') = 116° 19'$, $r_2 = 1{\cdot}0157$ $\phi_3 (= 50° 57') = 230° 57'$, $r_3 = 1{\cdot}0323$.

The calculation of the equation of the orbit is then as follows:

$\lambda =$ ·36397	$\log R_1 = 03531$
log = 56107	$R_1 = 1{\cdot}0847$
12214	$\lambda + 2 = 2{\cdot}36397$
$\lambda^2 =$ ·13248	log = 37364
$\lambda^2 - 3 = -2{\cdot}86752$	$\log r_2 =$ 00674
log = 45750	38038
$\log \sqrt{1 + \lambda^2} = 02701$	$R_2 = +2{\cdot}4010$
77815	$\lambda - 2 = -1{\cdot}63603$
$\log 6 \sqrt{1 + \lambda^2} = 80516$ (*a*)	log = 21378
45750	$\log r_3 =$ 01382
30103	22760
$\log 2 (\lambda^2 - 3) = 75853$ (*c*)	$R_3 = -1{\cdot}6889$
23856	
$\log 2 \sqrt{3} (\lambda^2 - 3) = 99709$ (*b*)	

$4R_1 = 4{\cdot}3388$
$-R_2 \quad -2{\cdot}4010$
$+R_3 \quad -1{\cdot}6889$
$4{\cdot}0899$
$\cdot 2489$
$\log = 39602$
$(a) = 80516$
59086
$A = \cdot 038982$

$2R_1 = 2{\cdot}1694$
$+\ R_2 \quad 2{\cdot}4010$
$-\ R_3 \quad 1{\cdot}6889$
$6{\cdot}2593^*$
$\log = 79653$
$\lambda = 56107$
35760
$+\ 2{\cdot}2782$
$-\ 3R_2 \quad -7{\cdot}2032$
$-\ 3R_3 \quad 5{\cdot}0667$
$7{\cdot}3449$
$-\ 7{\cdot}2032$
$0{\cdot}1417$
$\log = 15137$
$(b) = 99709$
$\cdot 15428$
$B = -\ \cdot 014265$

$R_2 = \quad 2{\cdot}4010$
$R_3 = -\ 1{\cdot}6889$
$+\ \cdot 7121$
$\log = \quad 85254$
$\lambda = \quad 56107$
41361
$+\ \cdot 25919$
$-\ 6{\cdot}2593^*$
$-\ 6{\cdot}00011$
$\log = \quad 77818$
$(c) = \quad 75853$
$\cdot 01965$
$C = +\ 1{\cdot}0464$

$\log B = 15428 \qquad 02729$
$\log A = 59086 \qquad 59086$
$56342 \qquad 61815$
$\varpi\ (= 20^\circ\ 6') = 160^\circ\ 52' \qquad e = \cdot 04151$
23630
$e^2 = \cdot 001723 \qquad \log C = 01965$
$1 - e^2 = \cdot 998277 \qquad \log \quad = 99925$
$a = 1{\cdot}0481 \qquad 02040$

The calculation of the Times is similar to that for the first planogram, and requires no further illustration.

The Table for Planogram No. 2 is as follows:

Planogram No

	c	x_1 all +	y_1 all +	x_2 all −	y_2 all +	x_3	y_3	r_1	ϕ_1	r_2	ϕ_2	r_3	ϕ
Circle	0°	1·000	·000	·500	·866	− ·500	− ·866	1·000	0° 0′	1·000	120° 0′	1·000	240
Ellipses	5	1·004	·051	·481	·878	·525	·853	1·005	2 52	1·001	118 42	1·001	238
	10	1·015	·102	·467	·889	·557	·838	1·020	5 44	1·004	117 41	1·006	236
	15	1·035	·155	·456	·900	·598	·821	1·047	8 30	1·009	116 54	1·016	233
	20	1·064	·210	·450	·910	·650	·802	1·085	11 10	1·016	116 19	1·032	230
	25	1·103	·269	·447	·921	·719	·778	1·136	13 43	1·024	115 55	1·060	227
	30	1·155	·333	·448	·931	·812	·749	1·202	16 6	1·033	115 42	1·104	222
	35	1·221	·404	·452	·941	·939	·710	1·286	18 19	1·044	115 40	1·178	217
	40	1·305	·484	·460	·951	1·125	·657	1·392	20 21	1·056	115 48	1·302	210
	45	1·414	·577	·471	·962	1·414	·577	1·527	22 12	1·071	116 6	1·528	202
	50	1·556	·688	·487	·974	1·925	·440	1·701	23 51	1·088	116 35	1·975	192
Parab.	51° 0′	1·589	·713	·491	·976	2·077	·400	1·741	24 9	1·093	116 43	2·115	190
Hyperbs.	52	1·624	·739	·495	·978	2·256	·353	1·787	24 28	1·097	116 51	2·283	188
	54	1·701	·795	·504	·984	2·729	·229	1·878	25 2	1·105	117 7	2·738	184
	55	1·743	·824	·509	·986	3·049	− ·145	1·928	25 18	1·109	117 16	3·053	182
	56° 18′	1·802	·866	·515	·990	3·601	·000	1·999	25 39	1·116	117 30	3·601	180
	59	1·942	·961	·530	·997	5·786	+ ·566	2·166	26 20	1·129	118 59	5·813	174
Line	60	2·000	1·000	·536	1·000	7·468	1·000	2·236	26 34	1·134	118 11	7·534	172
Convex	61	2·063	1·042	·542	1·003	−10·53	+ 1·793	2·311	26 48	1·140	118 24	10·68	170
	63° 26′ − ε 63° 26′ + ε	2·236	1·155	·559	1·010	− ∞ + ∞	+ ∞ − ∞	2·517	27 19	1·155	118 57	∞	165 345
Hyperbs.	64	2·281	1·184	·563	1·012	+45·22	−12·60	2·570	27 26	1·157	119 6	46·94	344
	65	2·366	1·238	·571	1·015	16·36	5·146	2·670	27 37	1·165	119 21	17·15	342
Parab.	65° 52′	2·446	1·289	·578	1·019	10·12	3·552	2·765	27 47	1·171	119 35	10·80	340
Ellipses	66	2·459	1·297	·579	1·019	9·987	3·500	2·779	27 48	1·172	119 37	10·59	340
	68	2·669	1·429	·596	1·026	5·617	2·369	3·028	28 10	1·186	120 11	6·090	337
	70	2·924	1·586	·616	1·033	3·912	1·927	3·326	28 29	1·202	120 48	4·360	333
	72	3·237	1·777	·638	1·041	3·008	1·694	3·693	28 47	1·221	121 29	3·455	330
	75	3·864	2·155	·674	1·054	2·230	1·488	4·424	29 9	1·251	122 36	2·681	326
	80	5·759	3·274	·751	1·079	1·568	1·312	6·624	29 37	1·315	124 49	2·045	320
	85	11·47	6·599	·854	1·112	1·217	1·216	13·25	29 54	1·402	127 32	1·720	315
Parab.	89° 20′	86·41	49·79	·979	1·148	1·024	1·161	99·5	29 56	1·508	130 24	1·548	311
Hyperbs.	90 − ε	∞	∞	1·000	1·155	+ 1·000	− 1·155	∞	30 0	1·527	130 54	1·527	310

$b = 180°$, $c = 0°$ to $90°$.

Equation of Orbit. $r = Ax$	$+By$	$+C$	e	ϖ	a	γ_{23}	γ_{31}	γ_{12}	T_{23}	T_{31}	T_{12}	
·000	·000	+ 1·000	·000	ind.° ′	1·000	1·732	1·732	1·732	1·000	1·000	1·000	0°
												5
+ ·0101	− ·0015	1·0104	·010	171 19	1·010	1·729	1·832	1·678	·987	1·106	·953	10
												15
·039	·014	1·046	·041	160 52	1·048	1·724	1·991	1·668	·956	1·316	·946	20
												25
·083	·061	1·126	·103	143 48	1·138	1·718	2·244	1·710	·924	1·777	·962	30
												35
·135	·209	1·317	·248	122 54	1·404	1·740	2·684	1·826	·878	3·238	·966	40
·161	·395	1·527	·426	112 12	1·867	1·805	3·055	1·925				45
·186	·815	1·972	·836	102 50	6·554	2·016	3·659	2·063	·878	48·60	·849	50
·191	·982	2·140	1·000	101 14	∞	2·100	3·831	2·097	·879	∞	·820	51° 0′
·196	1·150	2·319	1·166	99 39	6·434							52
·203	1·719	2·898	1·720	96 44	1·481							54
·207	2·182	3·366	2·192	95 25	·885	2·781	4·890	2·258	·895	~	·665	55
·212	3·074	4·227	3·081	93 56	·498							56° 18′
·221	− 14·15	+ 15·42	14·15	90 30	·077							59
·224	± ∞ (y	− 1)	∞	90 0	·000	6·932	9·468	2·536	·000	~	·000	60
									Convex Orbits.			61
·234	+ 4·906	− 3·671	4·912	87 17	·159							63° 26′ − ε
·578	− 1·761	+ 3·257	1·853	108 11	1·338	∞	∞	2·799	~	∞	·909	63° 26′ + ε
												64
·587	·979	2·494	1·134	120 21	8·666	18·014	15·380	2·945				65
·591	·805	2·257	1·000	126 19	∞	11·633	9·072	3·036	∞	7·746	1·386	65° 52′
·593	·779	2·221	·979	127 15	53·83							66
·606	·338	1·894	·693	150 53	3·645							68
·619	− ·120	1·708	·630	169 0	2·834	5·409	3·649	3·584	5·735	6·343	2·685	70
·635	+ ·027	1·599	·636	182 25	2·674							72
·654	·185	1·497	·680	195 47	2·783	3·859	3·981	4·644	2·62	6·68	4·64	75
·692	·366	1·439	·783	207 52	3·716	3·327	6·212	6·870	1·97	10·21	9·343	80
·740	·514	1·455	·892	214 47	7·721	3·115	12·895	13·480				85
·764	·645	1·505	1·000	219 51	∞	3·055	99·43	102·7	1·200	225·4	∞	89° 20′
+ ·770	+ ·666	+ 1·527	1·018	220 6	41·000	3·055	∞	∞	1·148	∞	~	90 − ε

The mark ~ in any of the T columns shows that the Time does not exist.

Article Nos. 95 to 98. *Planogram No. 3, the Orbit-pole at one of the points A.*

95. When the orbit-pole is at one of the points A, the orbit-plane passes through one of the rays, and as there is no longer on this ray any determinate point of intersection, the orbit (as was seen) becomes indeterminate. Thus consider the point A for which $b = 270°$, $c = 60°$: we have

$$\begin{array}{llll} \alpha, \beta, \gamma = & -1, & 0, & 0, \\ \alpha', \beta', \gamma' = & 0, & -\tfrac{1}{2}, & -\tfrac{1}{2}\sqrt{3}, \\ \alpha'', \beta'', \gamma'' = & 0, & -\tfrac{1}{2}\sqrt{3}, & \tfrac{1}{2}, \end{array}$$

Fig. 9.

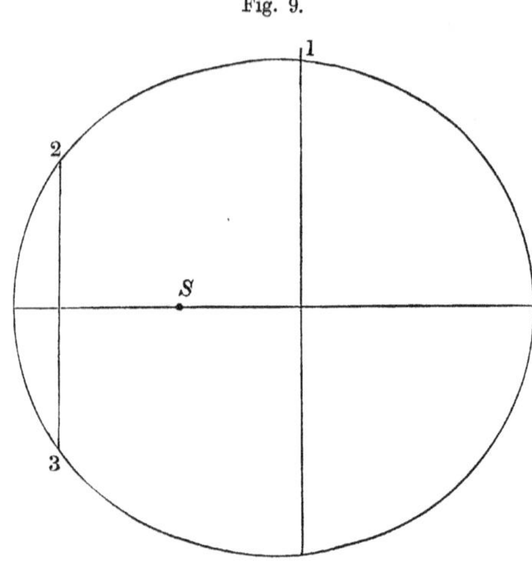

and consequently the formula gives

$$\begin{array}{l} x_1' : y_1' : 1 = 0 : 0 : 0, \\ x_2' : y_2' : 1 = -\tfrac{1}{2}\sqrt{3} - \sqrt{3} : 3 : -\tfrac{1}{2}\sqrt{3} - \sqrt{3}, \\ x_3' : y_3' : 1 = -\tfrac{1}{2}\sqrt{3} - \sqrt{3} : -3 : -\tfrac{1}{2}\sqrt{3} - \sqrt{3}, \end{array}$$

and, moreover, $x = -x'$, $y = -y'$. From the formula the value of x_1' or x_1 is given as $\tfrac{0}{0}$, but the true value is obviously $x_1 = 1$; the value of y_1 is actually indeterminate. The formulæ give the values of (x_2, y_2), (x_3, y_3), viz. the system is

$$\begin{array}{ll} x_1 = 1, & y_1 = \text{ind.} \\ x_2 = -1, & y_2 = \dfrac{2}{\sqrt{3}}, \quad \text{whence } r_2 = r_3 = \sqrt{\tfrac{7}{3}}, \\ x_3 = -1, & y_3 = -\dfrac{2}{\sqrt{3}}, \end{array}$$

so that the orbits in the planogram are the whole series of conics having a given focus, S, and passing through two fixed points, 2, 3, having the common abscissa $x=-1$, and at equal distances $\frac{2}{\sqrt{3}}$ $(=1\cdot15470)$ on opposite sides of the axis. The axis of x is obviously the common transverse axis for all the orbits; that is, the equation of the orbit will be of the form $r=Ax+B$; and writing $x=-1$, we have $\sqrt{\tfrac{7}{3}}=-A+B$, viz. the equation is $r-\sqrt{\tfrac{7}{3}}=A(x+1)$; the value of A will be determined if we assume for the point 1 a determinate position on the line $x=1$, say its ordinate is $=y_1$; for then if $r_2=\sqrt{1+y_1^2}$ we have $r_1-\sqrt{\tfrac{7}{3}}=2A$, and the equation is $r-\sqrt{\tfrac{7}{3}}=\tfrac{1}{2}(r_1-\sqrt{\tfrac{7}{3}})(x+1)$. In particular if $y_1=0$, we have $r_1=1$, and the equation of the orbit is $r-\sqrt{\tfrac{7}{3}}=\tfrac{1}{2}(1-\sqrt{\tfrac{7}{3}})(x+1)$: this is the orbit, eccentricity $\tfrac{1}{2}(\sqrt{\tfrac{7}{3}}-1)$, $=\cdot264$, belonging to the point A as a point in planogram No. 1: for the value of y, being in that planogram originally assumed $=0$, is of course $=0$ when the orbit-pole comes to be the point A.

96. We may conversely take the equation of the orbit, or say the value of $A\ (=\pm e)$ in the equation $r-\sqrt{\tfrac{7}{3}}=A(x+1)$, to be given; and then writing $x=x_1=1$, we have

$$r_1=\sqrt{\tfrac{7}{3}}+2A, \qquad \text{that is } y_1^2-(\sqrt{\tfrac{7}{3}}+2A)^2-1\,;$$

for

$$r_1=1 \text{ or } y_1=0, \qquad A=\tfrac{1}{2}(1-\sqrt{\tfrac{7}{3}})=-\cdot264,$$

and as r_1 increases to $r_1=\sqrt{\tfrac{7}{3}}$, or y_1 increases to $\pm\frac{2}{\sqrt{3}}$, A diminishes from $-\cdot264$ to 0; viz., for $r_1=\sqrt{\tfrac{7}{3}}$, or $y_1=+\frac{2}{\sqrt{3}}$, the orbit is a circle; as r_1 increases from $\sqrt{\tfrac{7}{3}}$, or y_1 from $\pm\frac{2}{\sqrt{3}}$, A increases from 0 positively; for $r_1=\sqrt{\tfrac{7}{3}}+2$, $=3\cdot527$, or $y_1=\pm\sqrt{\frac{16+2\sqrt{21}}{3}}$, $=\pm 2\cdot896$, A becomes $=1$; that is, the orbit is a parabola; and for larger positive values of r_1, or positive or negative values of y_1, the orbit is a hyperbola (concave); and ultimately for $r_1=\infty$ or $y_1=\pm\infty$, the orbit is the right line $x+1=0$. Thus A extends from $-\cdot264$ to 0, and thence from 0 positively to $+\infty$.

97. In further illustration, suppose that the orbit-pole, instead of being at A, is a point in the immediate neighbourhood of A, say that the rectangular spherical coordinates, measured from A in the direction of the meridian and perpendicular thereto, are ξ and η; the colatitude and longitude of the orbit-pole being thus $c=60^\circ+\xi$, and $b=270^\circ+\frac{2}{\sqrt{3}}\eta$; we have then, ξ, η being indefinitely small,

$$\begin{array}{llll} \alpha\,,\ \beta\,,\ \gamma & = -1\,, & -\frac{2}{\sqrt{3}}\eta, & 0, \\[2ex] \alpha',\ \beta',\ \gamma' & = \frac{1}{\sqrt{3}}\eta, & -\tfrac{1}{2}+\frac{\sqrt{3}}{2}\xi, & \frac{\sqrt{3}}{2}-\tfrac{1}{2}\xi, \\[2ex] \alpha'',\ \beta'',\ \gamma'' & = \eta\,, & -\frac{\sqrt{3}}{2}-\tfrac{1}{2}\xi, & \tfrac{1}{2}-\frac{\sqrt{3}}{2}\eta\,; \end{array}$$

and thence

$$x_1' = \left(-\tfrac{1}{2} + \frac{\sqrt{3}}{2}\,\xi\right)\sqrt{3} + \frac{\sqrt{3}}{2} + \tfrac{1}{2}\,\xi = \quad 2\xi$$

$$: y_1' \quad : -\frac{2}{\sqrt{3}}\,\eta. \quad \sqrt{3} \qquad : -2\eta$$

$$: 1 \quad : -\frac{\sqrt{3}}{2} - \tfrac{1}{2}\,\xi + \left(\tfrac{1}{2} - \frac{\sqrt{3}}{2}\right)\xi \qquad : -2\xi\,;$$

that is, $x_1' = -1$, $y_1' = \frac{\eta}{\xi}$, or what is the same thing, $x_1 = -1$, $y_1 = \frac{\eta}{\xi}$; the values of x_2, y_2, and x_3, y_3, differ from their former values only by terms in ξ, η, which may be neglected; that is, we have as before $x_2 = -1$, $y_2 = \frac{2}{\sqrt{3}}$ and $x_3 = -1$, $y_3 = -\frac{2}{\sqrt{3}}$; and we thus see that the foregoing determination of the orbit for an arbitrary value of y_1, writing therein $y_1 = -\frac{\eta}{\xi}$ $\left(\text{or what would be the same thing } y_1 = \frac{\eta}{\xi}\right)$ gives the orbit for the neighbouring position $c = 60° + \xi$, and $b = 270° + \frac{2}{\sqrt{3}}\,\eta$ of the orbit-pole. Writing for greater convenience $\xi = \rho\cos\psi$, $\eta = \rho\sin\psi$, the indefinitely small quantity ρ will denote the distance of the orbit-pole from A, and its azimuth measured from the meridian will be $=\psi$. We then have $y_1 = -\tan\psi$, and $r_1 = \sqrt{1 + y_1^2} = \pm\sec\psi$, or, if to fix the ideas, ψ be considered as $< \pm 90°$, then $r_1 = \sec\psi$: we have thus ($A = \pm e$ as before) $A = \tfrac{1}{2}(-\sqrt{\tfrac{7}{3}} + \sec\psi)$; viz., observing that $\sqrt{\tfrac{7}{3}} = 1\cdot527$, we obtain

$$\begin{array}{llr}
\psi = 0, & & A = -\tfrac{1}{2}(\sqrt{\tfrac{7}{3}} - 1) = -\cdot264 \\
\psi = \sec^{-1}\ \sqrt{\tfrac{7}{3}} & = \pm 49°\ 6', & A = \qquad 0 \\
\psi = \sec^{-1}(2\sqrt{\tfrac{7}{3}} - 1) & = \pm 60°\ 52', & A = \tfrac{1}{2}(\sqrt{\tfrac{7}{3}} - 1) = +\cdot264 \\
\psi = \sec^{-1}(\ \sqrt{\tfrac{7}{3}} + 2) & = \pm 73°\ 32', & A \qquad = \ 1 \\
\psi = & \pm(90° - \epsilon), & A \qquad = +\infty.
\end{array}$$

98. These results will have to be further considered in reference to the course of the iseccentric curves through the point A. I remark here that, although it appears that although for eccentricities less than $\cdot264$, and in particular for the eccentricity $=0$, there are real directions of passage from A to a neighbouring point, yet there are not through A any real branches of the corresponding iseccentric curves; viz., A is in regard to these curves, an isolated point with *real* tangents; that is a point in the nature of an evanescent lemniscate. As regards the eccentricity $=0$, it is obvious that this must be so; viz., there can be no real branch through A. In fact, the orbit can only be a circle when the intersection by the orbit-plane of the hyperboloid which contains the three rays is also a circle; that is, the orbit is a circle only when the orbit-plane coincides with the plane of the ecliptic.

Article Nos. 99 to 103. *Planogram No. 4, the Orbit-pole in the Ecliptic.*

99. When the orbit-pole describes the circle of the ecliptic, the orbit-plane passes through the axis of z, or polar axis. We have $c = 90°$, and consequently

$$\begin{aligned} \alpha\,,\ \beta\,,\ \gamma\ &= \sin b, \quad -\cos b, \quad 0, \\ \alpha',\ \beta',\ \gamma'\ &= 0\ , \quad 0\ , \quad -1, \\ \alpha'',\ \beta'',\ \gamma''\ &= \cos b, \quad \sin b, \quad 0. \end{aligned}$$

Reverting for a moment to the general case where the six coordinates of the ray are (a, b, c, f, g, h), the formulæ for the intersection by the orbit-plane are

$$\begin{aligned} x' : y' : 1 = &\ (\mathrm{a, b, c} \between \alpha', \beta', \gamma') = -\mathrm{c} \\ &: -(\mathrm{a, b, c} \between \alpha\,, \beta\,, \gamma\,) : -\mathrm{a}\sin b + \mathrm{b}\cos b \\ &: \ (\mathrm{f, g, h} \between \alpha'', \beta'', \gamma'') : \ \mathrm{f}\cos b + \mathrm{g}\sin b, \end{aligned}$$

that is

$$\frac{1}{x'} + \frac{\mathrm{f}}{\mathrm{c}}\cos b + \frac{\mathrm{g}}{\mathrm{c}}\sin b = 0,$$

$$\frac{y'}{x'} + \frac{\mathrm{b}}{\mathrm{c}}\cos b - \frac{\mathrm{a}}{\mathrm{c}}\sin b = 0\,;$$

and thence

$$\begin{aligned} 1 : \cos b : \sin b &= \frac{-\mathrm{af} - \mathrm{bg}}{\mathrm{c}^2} : \frac{\mathrm{g}y' + \mathrm{a}}{\mathrm{c}x'} : \frac{\mathrm{b} - \mathrm{f}y'}{\mathrm{c}x'} \\ &= \quad \mathrm{h}x' \quad : \mathrm{g}y' + \mathrm{a} : -\mathrm{f}y' + \mathrm{b}\,; \end{aligned}$$

consequently

$$\mathrm{h}x'^2 = (\mathrm{g}y' + \mathrm{a})^2 + (\mathrm{f}y' - \mathrm{b})^2,$$

or, what is the same thing,

$$\mathrm{h}x'^2 = (\mathrm{f}^2 + \mathrm{g}^2)\,y'^2 + 2\,(\mathrm{ag} - \mathrm{bf})\,y' + \mathrm{a}^2 + \mathrm{b}^2,$$

or, in particular, if (as in the special symmetrical case) $\mathrm{ag} - \mathrm{bf} = 0$, then

$$\mathrm{h}x'^2 = (\mathrm{f}^2 + \mathrm{g}^2)\,y'^2 + \mathrm{a}^2 + \mathrm{b}^2.$$

100. For the symmetrical system of rays we have as before

$$\begin{aligned} \mathrm{a}_1, \mathrm{b}_1, \mathrm{c}_1, \mathrm{f}_1, \mathrm{g}_1, \mathrm{h}_1 &= \ \ 0, \ \sqrt{3}, \ -1, \ \ 0\,, \ 1, \ \ \sqrt{3}, \\ \mathrm{a}_2, \mathrm{b}_2, \mathrm{c}_2, \mathrm{f}_2, \mathrm{g}_2, \mathrm{h}_2 &= \ \ 3, \ \sqrt{3}, \ \ 2, \ \sqrt{3}, \ 1, \ -2\sqrt{3}, \\ \mathrm{a}_3, \mathrm{b}_3, \mathrm{c}_3, \mathrm{f}_3, \mathrm{g}_3, \mathrm{h}_3 &= -3, \ \sqrt{3}, \ \ 2, \ \sqrt{3}, \ 1, \ -2\sqrt{3}, \end{aligned}$$

and thence

$$\begin{aligned} x_1' : y_1' : 1 &= \ \ 1 : \qquad \sqrt{3}\cos b : \sin b \qquad , \\ x_2' : y_2' : 1 &= -2 : -3\sin b + \sqrt{3}\cos b : \sin b + \sqrt{3}\cos b, \\ x_3' : y_3' : 1 &= -2 : \ \ 3\sin b + \sqrt{3}\cos b : \sin b - \sqrt{3}\cos b, \end{aligned}$$

or, what is the same thing,

$$x_1' = \operatorname{cosec} b \quad , \qquad y_1' = \sqrt{3}\cot b,$$

$$x_2' = \frac{-1}{\sin b + \sqrt{3}\cos b}, \qquad y_2' = \frac{\sqrt{3}(\cos b - \sqrt{3}\sin b)}{\sin b + \sqrt{3}\cos b},$$

$$x_3' = \frac{-2}{\sin b - \sqrt{3}\cos b}, \qquad y_3' = \frac{\sqrt{3}(\cos b + \sqrt{3}\sin b)}{\sin b - \sqrt{3}\cos b};$$

or as these may also be written

$$x_1' = \operatorname{cosec} b \quad , \qquad y_1' = \sqrt{3}\cot b \quad ,$$

$$x_2' = -\operatorname{cosec}(b+60^\circ), \qquad y_2' = \sqrt{3}\cot(b+60^\circ),$$

$$x_3' = -\operatorname{cosec}(b-60^\circ), \qquad y_3' = \sqrt{3}\cot(b-60^\circ),$$

so that for each of these sets we have

$$x'^2 - \tfrac{1}{3}y'^2 = 1.$$

(The curve is in fact a section of the hyperboloid of revolution, $x^2 + y^2 - \frac{1}{3}z^2 = 1$, which passes through the three rays.)

101. As regards the equation of the orbit I will first consider the particular cases $b = 90^\circ$, $b = 0^\circ$, which should agree with the orbits for $c = 90^\circ$ in the planograms 1 and 2 respectively.

For $b = 90^\circ$ we have $x' = x$, $y' = y$ and

$$x_1' = 1, \quad y_1' = 0,$$

$$x_2' = -2, \quad y_2' = -3,$$

$$x_3' = -2, \quad y_3' = 3,$$

and the orbit is at once found to be

$$r = \tfrac{1}{3}(1 - \sqrt{13})(x' - 1),$$

the eccentricity (regarded as positive) being thus $\frac{1}{3}(\sqrt{13} - 1)$, $= \cdot 7685$ as before. For $b = 0^\circ$ there is a discontinuity, and I write successively $b = +\epsilon$, and $b = -\epsilon$. For $b = +\epsilon$ we have $x' = -y$, $y' = x$, and

$$x_1' = \infty \; , \quad y_1' = \infty\sqrt{3},$$

$$x_2' = -\frac{2}{\sqrt{3}}, \quad y_2' = -1,$$

$$x_3' = \frac{2}{\sqrt{3}}, \quad y_3' = -1,$$

and the orbit is found to be

$$r = \tfrac{2}{3}x' + \frac{4}{3\sqrt{3}}y' + \frac{\sqrt{7}}{\sqrt{3}} = \quad \cdot 666\,x + \cdot 770\,y + 1\cdot 527\,;$$

and similarly for $b = -\epsilon$ the equation is

$$r = -\tfrac{2}{3}x' + \frac{4}{\sqrt{3}}\,y' + \frac{\sqrt{7}}{\sqrt{3}} = -\cdot 666\,x + \cdot 770\,y + 1\cdot 527\,;$$

hence the eccentricity is

$$e = \sqrt{\tfrac{28}{27}}, \quad = 1\cdot 018, \text{ as before.}$$

102. Considering now the general case where b has any value whatever, the equation of the orbit is

$$\begin{vmatrix} r & , & x', & y' & , & 1 \\ r_1 \sin b & , & 1, & \sqrt{3}\cos b & , & \sin b \\ r_2(\sin b + \sqrt{3}\cos b), & & -2, & -3\sin b + \sqrt{3}\cos b, & & \sin b + \sqrt{3}\cos b \\ r_3(\sin b - \sqrt{3}\cos b), & & -2, & 3\sin b + \sqrt{3}\cos b, & & \sin b - \sqrt{3}\cos b \end{vmatrix} = 0,$$

($x' = x\sin b - y\cos b$, $y' = x\cos b + y\sin b$, as before).

The coefficient of r is readily found to be $-6\sqrt{3}(\sin^2 b + \cos^2 b)$, $= -6\sqrt{3}$; hence completing the development, dividing by $6\sqrt{3}$, and transposing, the equation of the orbit is

$$\begin{aligned} r = {} & \tfrac{1}{6}\,[2r_1 \sin b - r_2(\sin b + \sqrt{3}\cos b) - r_3(\sin b - \sqrt{3}\cos b)]\,x' \\ & + \frac{1}{6\sqrt{3}}[4r_1 \sin b\cos b + r_2(-2\sin b\cos b + \sqrt{3}(\cos^2 b - \sin^2 b)) \\ & \qquad + r_3(-2\sin b\cos b - \sqrt{3}(\cos^2 b - \sin^2 b))]\,y' \\ & + \tfrac{1}{6}\,[4r_1\sin^2 b + r_2(\sin^2 b + 3\cos^2 b + 2\sqrt{3}\sin b\cos b) \\ & \qquad + r_3(\ \sin^2 b + 3\cos^2 b - 2\sqrt{3}\sin b\cos b)], \end{aligned}$$

where

$$r_1 = \frac{\sqrt{\sin^2 b + 4\cos^2 b}}{\sin b},$$

$$r_2 = \frac{\sqrt{13\sin^2 b + 7\cos^2 b - 6\sqrt{3}\sin b\cos b}}{\sin b + \sqrt{3}\cos b},$$

$$r_3 = \frac{\sqrt{13\sin^2 b + 7\cos^2 b + 6\sqrt{3}\sin b\cos b}}{\sin b - \sqrt{3}\cos b};$$

in which expressions the signs of the radicals must be such that r_1, r_2, r_3 shall be positive. Hence writing $\tan b = \eta$, ($\sec b = \sqrt{1+\eta^2}$, which determines the sign of $\sqrt{1+\eta^2}$), also

$$R_1 = \sqrt{\eta^2 + 4}, \quad R_2 = \sqrt{13\eta^2 - 6\sqrt{3}\,\eta + 7}, \quad R_3 = \sqrt{13\eta^2 + 6\sqrt{3}\,\eta + 7},$$

and therefore

$$\eta r_1 = R_1, \qquad (\eta + \sqrt{3})\, r_2 = R_2, \qquad (\eta - \sqrt{3})\, r_3 = R_3,$$

which last equations determine the signs of R_1, R_2, R_3 respectively, the equation of the orbit is

$$\begin{aligned} r = & \frac{1}{6\sqrt{1+\eta^2}} \quad (2R_1 - R_2 - R_3)\, x' \\ & + \frac{1}{6(1+\eta^2)\sqrt{3}} \left(4R_1 + R_2(1 - \eta\sqrt{3}) + R_3(1 + \eta\sqrt{3})\right) y' \\ & + \frac{1}{6(1+\eta^2)} \quad \left(4R_1\eta + R_2(\eta + \sqrt{3}) + R_3(\eta - \sqrt{3})\right). \end{aligned}$$

Thus if $b = +\epsilon$, then also $\eta = +\epsilon$,

$$\sqrt{1+\eta^2} = 1, \quad R_1 = 2, \quad R_2 = \sqrt{7}, \quad R_3 = -\sqrt{7},$$

and the equation is

$$r = \tfrac{1}{6}\, 4x' + \frac{1}{6\sqrt{3}} 8y' + \tfrac{1}{6}\, 2\sqrt{7}\sqrt{3}, \;= \tfrac{2}{3}\, x' + \frac{4}{3\sqrt{3}} y' + \frac{\sqrt{7}}{\sqrt{3}}, \;= \cdot 666\, x' + \cdot 770\, y' + 1 \cdot 527,$$

as before; and similarly if $b = 90°$.

And moreover, if $b = 30°$, then

$$\eta = \frac{1}{\sqrt{3}}, \quad R_1 = \sqrt{\tfrac{13}{3}}, \quad R_2 = \frac{4}{\sqrt{3}}, \quad R_3 = -\frac{2\sqrt{13}}{3},$$

whence the equation of the orbit is

$$\begin{aligned} r &= \tfrac{1}{3}(\sqrt{13} - 1)\, x' + 0\, y' + \tfrac{1}{3}(\sqrt{13} + 2), \\ &= \cdot 868\, x' + 0\, y' + 1 \cdot 868. \end{aligned}$$

103. The equation of the orbit should be tabulated from $b = 0$ to $b = 30°$, the equations for the remainder of the circumference will be then found by successive repetition of this interval in direct and reverse order, with however a change of sign, in the manner about to be explained,

$$\begin{aligned} & b = \epsilon, && r = + \cdot 666\, x' + \cdot 770\, y' + 1 \cdot 527, \\ & b = 30°, && r = + \cdot 868\, x' + \; 0 \;\; y' + 1 \cdot 868, \\ & b = 60° - \epsilon, && r = + \cdot 666\, x' - \cdot 770\, y' + 1 \cdot 527, \\ & b = 60° + \epsilon, && r = - \cdot 666\, x' + \cdot 770\, y' + 1 \cdot 527, \\ & b = 90°, && r = - \cdot 868\, x' + \; 0 \;\; y' + 1 \cdot 868, \\ & b = 120° - \epsilon, && r = - \cdot 666\, x' - \cdot 770\, y' + 1 \cdot 527, \end{aligned}$$

$30° + \beta$ same as $30° - \beta$, reversing sign of the y' coefficient.

$90° + \beta$ same as $90° - \beta$, reversing sign of the y' coefficient, and whole interval 60° to 120° same as interval 0° to 60°, except that the signs of the x' coefficient are reversed, and the remaining two intervals, 120° to 240° and 240° to 360°, are merely repetitions of the interval 0° to 120°.

As regards the interval 0° to 30° the only intermediate value that I have calculated is $b=15°$, viz., we then have

$$b=15°,\quad r=\cdot 811\,x'+\cdot 403\,y'+1\cdot 787.$$

Calculating for the foregoing values $b=0°$, $b=15°$, $b=30°$, the values of e, ϖ, a, these are found to be

$b=0°$,	$e=1\cdot 018$	$\varpi=220°\ 6'$	$a=41\cdot 24$
$b=15°$,	$e=\cdot 906$	$\varpi=206°\ 27'$	$a=10\cdot 008$
$b=30°$,	$e=\cdot 868$	$\varpi=180°$	$a=7\cdot 604$

Article Nos. 104 to 113. *Planogram No. 5. The Orbit-pole on a Separator.*

104. If the orbit-plane rotate round a line parallel to one of the rays, the orbit-pole will describe a separator circle, and conversely. I consider the general case of a ray the six coordinates of which are (a, b, c, f, g, h), and for which the intersections with the orbit-plane are given by

$$x' : y' : 1=(\text{a, b, c}\between\alpha', \beta', \gamma') : -(\text{a, b, c}\between\alpha, \beta, \gamma) : (\text{f, g, h}\between\alpha'', \beta'', \gamma'').$$

The axis of x' is parallel to the ray

$$\frac{x-A}{\text{f}}=\frac{y-B}{\text{g}}=\frac{z-C}{\text{h}},$$

that is, we have

$$\alpha : \beta : \gamma=\text{f} : \text{g} : \text{h},$$

whence, putting for shortness

$$\Omega=\sqrt{\text{f}^2+\text{g}^2+\text{h}^2}\ \text{ and }\ \Pi=\sqrt{\text{f}^2+\text{g}^2},$$

we have

$$\alpha=\frac{\text{f}}{\Omega}=\cos N\cos G,\quad \beta=\frac{\text{g}}{\Omega}=\cos N\sin G,\quad \gamma=\frac{\text{h}}{\Omega}=-\sin N,$$

and thence

$$\tan G=\frac{\text{g}}{\text{f}},\quad \sin G=\frac{\text{g}}{\Pi},\quad \cos G=\frac{\text{f}}{\Pi},\quad \cos N=\frac{\Pi}{\Omega},$$

and we thus obtain the values of α', β', γ'; α'', β'', γ'' in terms of f, g, h and the variable angle H, viz., these are

$$\alpha'=-\frac{\text{g}\cos H}{\Pi}+\frac{\text{hf}\sin H}{\Pi\Omega},\qquad \alpha''=\frac{-\text{g}\sin H}{\Pi}-\frac{\text{hf}\cos H}{\Pi\Omega},$$

$$\beta'=\frac{\text{f}\cos H}{\Pi}+\frac{\text{gh}\sin H}{\Pi\Omega},\qquad \beta''=\frac{\text{f}\sin H}{\Pi}-\frac{\text{hg}\cos H}{\Pi\Omega},$$

$$\gamma'=-\frac{\Pi^2\sin H}{\Pi\Omega},\qquad \gamma''=\frac{\Pi^2\cos H}{\Pi\Omega},$$

where H is the angular distance of the orbit-pole, along the separator, from the point A. The foregoing values give

$$(\mathrm{a, b, c} \between \alpha\ ,\ \beta\ ,\ \gamma\ ) = 0,$$

$$(\mathrm{a, b, c} \between \alpha',\ \beta',\ \gamma'\ ) = -\frac{1}{\Pi}\{(\mathrm{ag}-\mathrm{bf})\cos H + \mathrm{c}\Omega \sin H\},$$

$$(\mathrm{f, g, h} \between \alpha''\ \beta'',\ \gamma'') = 0,$$

so that the coordinates x', y' of the intersection with the ray are given in the form

$$x' : y' : 1 = M : 0 : 0,$$

that is

$$x' = \frac{M}{0} = \infty,\quad y' = \frac{0}{0},$$

but the value of y' is determinate, viz., this is equal to the perpendicular distance of the ray from the point S.

105. In particular when the rays are the special symmetrical system before considered, then if (a, b, c, f, g, h) refer to the ray 1, we have $\mathrm{f}=0$, $\mathrm{g}=1$, $\mathrm{h}=\sqrt{3}$, $\Pi=1$, $\Omega=2$, and thence

$$\begin{array}{llll} \alpha\ ,\ \beta\ ,\ \gamma = & 0 & ,\ \tfrac{1}{2} & ,\ \tfrac{1}{2}\sqrt{3}\ , \\ \alpha',\ \beta',\ \gamma' = & -\cos H, & \tfrac{1}{2}\sqrt{3}\sin H, & -\tfrac{1}{2}\sin H, \\ \alpha'',\ \beta'',\ \gamma'' = & -\sin H, & \tfrac{1}{2}\sqrt{3}\cos H, & \tfrac{1}{2}\cos H. \end{array}$$

For the intersection with the ray 1 we have

$$x_1' = \pm\infty,\quad y_1' = 1,$$

and for the intersections with the other two lines

$$\begin{array}{lll} x_2' : y_2' : 1 = & & \\ & (\ 3,\ \sqrt{3},\ \ 2\ \ )(-\cos H,\ \ \tfrac{1}{2}\sqrt{3}\sin H,\ -\tfrac{1}{2}\sin H) & = -3\cos H + \tfrac{1}{2}\sin H \\ : -(& 3,\ \sqrt{3},\ \ 2\ \ )(\ \ 0,\ 1,\ \ \tfrac{1}{2}\sqrt{3}) & : \tfrac{3}{2}\sqrt{3} \\ : & (\sqrt{3},\ \ 1,\ -2\sqrt{3})(-\sin H,\ -\tfrac{1}{2}\sqrt{3}\cos H,\ \ \tfrac{1}{2}\cos H) & : -\sqrt{3}\sin H - \tfrac{3}{2}\sqrt{3}\cos H, \end{array}$$

and

$$\begin{array}{lll} x_3' : y_3' : 1 = & & \\ & (-\ 3,\ \sqrt{3},\ 2\ \ )(-\cos H,\ \ \tfrac{1}{2}\sqrt{3}\sin H,\ -\tfrac{1}{2}\sin H) = & 3\cos H + \tfrac{1}{2}\sin H \\ : -(& -\ 3,\ \sqrt{3},\ 2\ \ )(\ 0,\ \tfrac{1}{2},\ \ \tfrac{1}{2}\sqrt{3}) & : \tfrac{3}{2}\sqrt{3} \\ : & (-\sqrt{3},\ \ 1,\ -2\sqrt{3})(-\sin H,\ -\tfrac{1}{2}\sqrt{3}\cos H,\ \ \tfrac{1}{2}\cos H) & : \sqrt{3}\sin H - \tfrac{3}{2}\sqrt{3}\cos H, \end{array}$$

that is, we have

$$x_2' = \frac{1}{\sqrt{3}} \frac{6\cos H - \sin H}{3\cos H + 2\sin H}, \qquad x_3' = -\frac{1}{\sqrt{3}} \frac{6\cos H + \sin H}{3\cos H - 2\sin H},$$

$$y_2' = \frac{3}{3\cos H + 2\sin H}, \qquad y_3' = \frac{3}{3\cos H - 2\sin H}.$$

106. Writing herein

$$\cos\omega = \frac{3}{\sqrt{13}}, \quad \sin\omega = \frac{2}{\sqrt{13}}, \quad \tan\omega = \tfrac{2}{3}, \quad \omega = 33^\circ\, 41'$$

the formulæ are readily converted into

$$x_2' = \frac{1}{13\sqrt{3}}\{16 - 15\tan(H-\omega)\}, \qquad x_3' = \frac{1}{13\sqrt{3}}\{-16 - 15\tan(H+\omega)\},$$

$$y_2' = \frac{3}{\sqrt{13}}\sec(H-\omega), \qquad y_3' = \frac{3}{\sqrt{13}}\sec(H+\omega),$$

where, in regard to this angle ω, it is to be observed that it represents the angular distance from the ecliptic along the separator to a point B, or what is the same thing, the complement of the angular distance on the separator, of the points A and B. We have, in fact, a right-angled spherical triangle ZAB, $\angle Z = 60^\circ$, $\angle A = 90^\circ$, $ZA = 60^\circ$ whence $\sin 60^\circ = \tan AB \cot 60^\circ$, that is, $\tan AB = \sin 60^\circ \tan 60^\circ = \frac{3}{2}$, or $AB = 90^\circ - \omega$.

Hence, $H = \pm 90^\circ$, the orbit-pole is on the ecliptic, $H = \pm(90^\circ - \omega)$, it is at a point B (the intersection of the separator by one of the other two separators), and $H = 0$, it is at the point A on the separator.

The foregoing values of (x_2', y_2') satisfy the equation

$$25y^2 = 39x^2 - 32x\sqrt{3} + 37,$$

and similarly the values of (x_3', y_3') satisfy

$$25y^2 = 39x^2 + 32x\sqrt{3} + 37,$$

results which would be useful for the delineation of the planogram.

107. As regards the equation of the orbit we have $x_1' = \pm\infty$, and consequently $x_1' = \pm r_1 = \theta r_1$ if for convenience θ be written to stand for ± 1. The equation of the orbit then is

$$0 = \begin{vmatrix} r, & x', & y', & 1 \\ 1, & \theta, & 0, & 0 \\ r_2(3\cos H + 2\sin H), & \frac{1}{\sqrt{3}}(6\cos H - \sin H), & 3, & 3\cos H + 2\sin H \\ r_3(3\cos H - 2\sin H), & \frac{1}{\sqrt{3}}(-6\cos H - \sin H), & 3, & 3\cos H - 2\sin H \end{vmatrix},$$

that is

$$-(r\theta - x')\, 12 \sin H =$$

$$y'\left\{\frac{1}{\sqrt{3}}(36\cos^2 H + 4\sin^2 H) - \theta r_2 (9\cos^2 H - 4\sin^2 H) + \theta r_3 (9\sin^2 H - 4\cos^2 H)\right\}$$

$$- 12\sqrt{3}\cos H + 3\theta(3\cos H + 2\sin H)\, r_2 - 3\theta(3\cos H - 2\sin H)\, r_3,$$

where

$$r_2 = \frac{\sqrt{21\cos^2 H - 4\cos H \sin H + \tfrac{28}{3}\sin^2 H}}{3\cos H + 2\sin H},$$

$$r_3 = \frac{\sqrt{21\cos^2 H + 4\cos H \sin H + \tfrac{28}{3}\sin^2 H}}{3\cos H - 2\sin H}.$$

Hence, writing $\tan H = \lambda$, and therefore $\sec H = \sqrt{1+\lambda^2}$, which determines the sign of $\sqrt{1+\lambda^2}$, and moreover

$$R_2 = \sqrt{21 - 4\lambda + \tfrac{28}{3}\lambda^2}, \quad R_3 = \sqrt{21 + 4\lambda + \tfrac{28}{3}\lambda^2},$$

and thence also

$$(3 + 2\lambda)\, r_2 = R_2, \quad (3 - 2\lambda)\, r_3 = R_3,$$

which last equations, since r_2, r_3 must be positive, determine the signs of the radicals R_2, R_3; the equation of the orbit is

$$r = \theta x' + \frac{y'}{12\lambda\sqrt{1+\lambda^2}}\left\{\frac{\theta}{\sqrt{3}}(36 + 4\lambda^2) - (3 - 2\lambda)\, R_2 + (3 + 2\lambda)\, R_3\right\} + \frac{-4\theta\sqrt{3} + R_2 - R_3}{4\lambda},$$

where θ it will be recollected denotes $+1$ or -1 at pleasure.

108. I remark that $\theta = +1$ and $\theta = -1$ may be considered as belonging to positions of the orbit-pole indefinitely near the separator on the opposite sides thereof

Fig. 10.

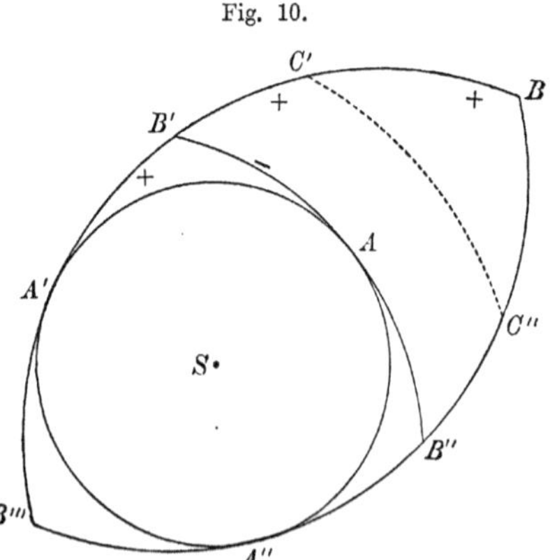

respectively; the annexed figure represents a portion of the blank spherogram, and the two sides of the half-separator $A'C'$ will be traversed by the orbit-pole, if H

extend from 0° to $90° - \omega$ ($= 56° 19'$, value at B') and thence to 90°, $\theta = +1$ belonging to the side marked + in the figure, and $\theta = -1$ to the opposite side. But the same result may be stated, more conveniently, in reference to the blank spherogram, as follows:

$H = 0°$ to $H = 56° 19'$, $\theta = +1$ belongs to the outside of AB', viz. to positions within the region of convex orbits,

$\theta = -1$, to inside of AB',

$H = 56° 19'$ to $H = 90°$, $\theta = +1$ belongs to inside of $B'C'$,

$H = 90°$ to $H = 123° 41'$, $\theta = +1$ belongs to inside of $C'B$,

the last-mentioned values being identical with those for $H = 90°$ to $H = 56° 19'$, $\theta = -1$: viz. the formula for $H = 90° + K$, $\theta = +1$ is equivalent to that for $H = 90° - K$, $\theta = -1$.

109. I consider some particular cases.

Orbit-pole at A: here $H = 0$ and therefore $\lambda = 0$, $R_2 = R_3 = \sqrt{21}$; the orbit is $r = \theta x' + \frac{\theta\sqrt{3}}{\lambda}(y' - 1)$, viz. it is the right line $y' - 1 = 0$.

Orbit-pole in the neighbourhood of B. Suppose first $H = 90° - \omega - \epsilon$, $\lambda = \cot\omega - \epsilon \operatorname{cosec}^2\omega = \frac{3}{2} - \frac{13}{4}\epsilon$, $3 - 2\lambda$, $= \frac{13}{2}\epsilon$, is positive, and therefore R_3 is positive, and we have $R_2 = 6$, $R_3 = 4\sqrt{3}$; whence the equation is

$$r = \theta x' + \frac{y'}{9\sqrt{13}}(15\theta + 24) + 1 - \frac{2}{\sqrt{3}}(\theta + 1),$$

viz. $\theta = -1$, this is

$$r = +x' + \sqrt{\tfrac{3}{13}}\,y' + 1,$$

and $\theta = +1$, it is

$$r = x' + \sqrt{\tfrac{13}{3}}\,y' + 1 - \frac{4}{\sqrt{3}},$$

and so secondly, if $H = 90° - \omega + \epsilon$, $\lambda = \frac{3}{2} + \frac{13}{4}\epsilon$, $3 - 2\lambda$, $= -\frac{13}{2}\epsilon$, is negative, or R_3 is also negative, viz. $R_2 = 6$, $R_3 = -4\sqrt{3}$, and the equation is

$$r = \theta x' - \frac{y'}{9\sqrt{13}}(15\theta - 24) + 1 - \frac{2}{\sqrt{3}}(\theta - 1),$$

viz. $\theta = +1$, this is

$$r = x' - \sqrt{\tfrac{3}{13}}\,y' + 1,$$

and $\theta = -1$, it is

$$r = -x' - \sqrt{\tfrac{13}{3}}\,y' + 1 + \frac{4}{\sqrt{3}}.$$

At the point B there are thus four orbits: viz. $H = -90° - \omega - \epsilon$, $\theta = +1$, and $H = 90° - \omega + \epsilon$, $\theta = -1$, these are orbits wherein the eccentricity is $= \sqrt{\frac{16}{3}}$, $= 2{\cdot}309$, agreeing with that found for the point B in planogram No. 1, or say for an orbit-pole near B in the direction of the meridian; whereas for $H = 90° - \omega - \epsilon$, $\theta = -1$ and $H = 90° - \omega + \epsilon$, $\theta = +1$ the eccentricity is $\sqrt{\frac{16}{13}} = 1{\cdot}101$.

Suppose again that the orbit-pole is on the ecliptic, or say $H = 90^\circ - \epsilon$, $\lambda = +\infty$, $R_2 = 2\sqrt{\tfrac{7}{3}}\lambda$, $R_3 = -2\sqrt{\tfrac{7}{3}}\lambda$, and $\sqrt{1+\lambda^2} = \lambda$, and the equation is

$$r = \theta\left(x' + \frac{y'}{3\sqrt{3}}\right) + \sqrt{\tfrac{7}{3}},$$

and similarly for $H = 90^\circ + \epsilon$, $\lambda = -\infty$, $R_2 = 2\sqrt{\tfrac{7}{3}}\lambda$, $R_3 = -2\sqrt{\tfrac{7}{3}}\lambda$, $\sqrt{1+\lambda^2} = \lambda$, and the equation still is

$$r = \theta\left(x' + \frac{y'}{3\sqrt{3}}\right) + \sqrt{\tfrac{7}{3}},$$

viz. θ retaining the same sign, there is no discontinuity in the passage through 90°.

The eccentricity, whether $\theta = +1$ or $= -1$, is $\sqrt{\tfrac{28}{27}}$, $= 1{\cdot}018$, agreeing with Planogram No. 2.

110. For the more complete discussion of the eccentricity, we have

$$e^2 = 1 + \frac{1}{144\lambda^2(1+\lambda^2)}\left(\frac{4\theta}{\sqrt{3}}(9+\lambda^2) - (3-2\lambda)R_2 + (3+2\lambda)R_3\right)^2.$$

The eccentricity cannot be less than 1, which is evidently right, for the point 3 being at infinity, the orbit cannot be an ellipse. We may have $e = 1$ (or the orbit a parabola), viz. this will be the case if

$$\frac{4\theta}{\sqrt{3}}(9+\lambda^2) - (3-2\lambda)R_2 + (3+2\lambda)R_3 = 0.$$

Proceeding to rationalize this equation, we have first

$$(3-2\lambda)^2 R_2^2 + (3+2\lambda)^2 R_3^2 - \tfrac{16}{3}(9+\lambda^2)^2 = 2(9-4\lambda^2)R_2R_3,$$

viz. substituting for R_2, R_3 their values $\sqrt{21 - 4\lambda + \tfrac{28}{3}\lambda^2}$ and $\sqrt{21 + 4\lambda + \tfrac{28}{3}\lambda^2}$, this is found to be

$$2(9-4\lambda^2)\sqrt{\left(21 + \frac{28}{3\lambda^2}\right)^2 - 16\lambda^2} = -54 + 336\lambda^2 + \tfrac{208}{3}\lambda^4;$$

or, what is the same thing,

$$(9-4\lambda^2)\sqrt{3969 + 3384\lambda^2 + 784\lambda^4} = -81 + 504\lambda^2 + 104\lambda^4$$

whence, squaring and reducing, we have

$$432(4\lambda^8 - 248\lambda^6 - 819\lambda^4 + 162\lambda^2 + 729) = 0;$$

or, what is the same thing,

$$432(\lambda^2+1)(4\lambda^6 - 252\lambda^4 - 567\lambda^2 + 729) = 0,$$

or, finally, the condition for a parabola is

$$4\lambda^6 - 252\lambda^4 - 567\lambda^2 + 729 = 0.$$

111. I stop to remark that this equation may be obtained differently, as follows. Since the point 1 is at infinity on the axis of x, this line will be the axis of the parabola; or the equation of the parabola will be

$$-y^2 + 4ax + 4a^2 = 0,$$

and we have therefore

$$-y_2^2 + 4ax_2 + 4a^2 = 0,$$

$$-y_3^2 + 4ax_3 + 4a^2 = 0,$$

that is

$$1 : 4a : 4a^2 = x_2 - x_3 : y_2^2 - y_3^2 : -y_2^2 x_3 + y_3^2 x_2,$$

and therefore

$$(y_2^2 - y_3^2)^2 = -4(x_2 - x_3)(y_2^2 x_3 - y_3^2 x_2)$$

as the condition for a parabola.

But the values of x_2, y_2; x_3, y_3, *ante* No. 104, introducing λ in the place of H, are

$$x_2 = \frac{1}{\sqrt{3}}\frac{6-\lambda}{3+2\lambda}, \quad x_3 = \frac{1}{\sqrt{3}}\frac{6+\lambda}{3-2\lambda},$$

$$y_2 = \frac{3\sqrt{1+\lambda^2}}{3+2\lambda}, \quad y_3 = \frac{3\sqrt{1+\lambda^2}}{3-2\lambda},$$

and thence

$$x_2 - x_3 = \frac{4}{\sqrt{3}}\frac{9+\lambda^2}{9-4\lambda^2},$$

$$y_2^2 - y_3^2 = -\frac{216\lambda(1+\lambda^2)}{(9-4\lambda^2)^2},$$

$$y_2^2 x_3 - y_3^2 x_2 = -\frac{36}{\sqrt{3}}\frac{(1+\lambda^2)(9-\lambda^2)}{(9-4\lambda^2)^2},$$

and substituting these values and omitting a factor $\dfrac{1+\lambda^2}{(9-4\lambda^2)^2}$, the result is

$$\frac{243\lambda^2(1+\lambda^2)}{9-4\lambda^2} = (9+\lambda^2)(9-\lambda^2),$$

viz. this is

$$(4\lambda^2 - 9)(\lambda^4 - 81) - 243\lambda^2(\lambda^2 + 1) = 0,$$

that is

$$4\lambda^6 - 252\lambda^4 - 567\lambda^2 + 729 = 0,$$

as before.

112. The equation considered as a cubic equation in λ^2 has its three roots real, but only two of them are positive; viz. there is a root not very different from 1, and which is easily approximated to by writing $\lambda^2 = 1 - x$, this gives

$$4x^3 + 240x^2 - 1068x + 86 = 0,$$

or nearly $x=\frac{86}{1088}=\cdot08$; a second approximation gives $x=\cdot0802$; or we have $\lambda^2=\cdot9198$, $\lambda=\cdot9592$, whence $H=43^\circ\,49'$. Substituting in the equation

$$\frac{4\theta}{\sqrt{3}}(9+4\lambda^2)-(3-2\lambda)R_2+(3+2\lambda)R_3=0,$$

this will be satisfied by $\theta=-1$, viz. the parabola belongs (as it obviously should do) to a point of AB' within the triangle $BB'B''$.

To obtain the other positive root we may write the equation in the form

$$\lambda^2=63+\frac{141\cdot75}{\lambda^2}-\frac{182\cdot25}{\lambda^4},$$

the approximate value $\lambda^2=63$, gives more nearly $\lambda^2=65$ and then

$$\lambda^2=63+\frac{141\cdot75}{65}-\frac{128\cdot24}{4225},\ =65\cdot177,$$

whence $\lambda^2=8\cdot073$ or $H=82^\circ\,56'$. Substituting in the equation

$$\frac{4\theta}{\sqrt{3}}(9+4\lambda^2)-(3-2\lambda)R_2+(3+2\lambda)R_3=0,$$

we have $\theta=+1$, viz. this parabola belongs to a point of $B'C'$ within the triangle $BB'B''$.

The two values of e for $\theta=+1$ and $\theta=-1$, are each infinite for $\lambda=0$, and they become equal for $\lambda=\infty$ (viz. when the orbit-pole is on the ecliptic), but not in any other case; in fact they can only do so for $9+\lambda^2=0$, or else for $(3-2\lambda)R_2=(3+2\lambda)R_3$, that is, $\lambda(288+128\lambda^2)=0$, viz., $\lambda(9+4\lambda^2)=0$.

113. In further explanation I give a diagram of the eccentricity.

Fig. 11.

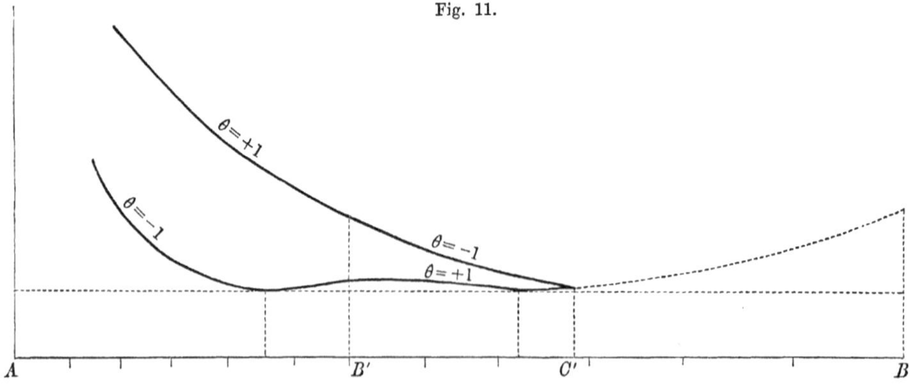

The base $AB'C'B$ is here the broken line $AB'C'B'$ of figure 10: the ordinates along the base AC' ($=90^\circ$) of the two continuous curves exhibit the values of e, as given by $\theta=+1$ and $\theta=-1$ respectively; the dotted curve on the base $C'B\,(=C'B')$ is merely the upper curve on the base $C'B'$ transferred to the base $C'B$; and the curve composed of the lower curve on the base AC' and of the dotted curve gives by its ordinates the value of the eccentricity as the orbit-pole moves along $AB'B$ within the triangle $B'BB''$: the upper curve on the base AB' gives by its ordinates the value of the eccentricity as the orbit-pole moves along AB' on the other side thereof, that is, within the convex region.

The base of the diagram is graduated not for the value of H, but for that of the angular distance (or distance in longitude) of the orbit-pole from the point A (or A'); viz. this is the angle opposite H in a right-angled spherical triangle, the sides and hypothenuse of which are 60°, H, c; writing β for the angle in question we have

$$\cos c=\tfrac{1}{2}\cos H,\quad \tan\beta=\frac{2}{\sqrt{3}}\tan H\left(=\frac{2\lambda}{\sqrt{3}}\right),$$

and any position of the orbit-pole on the separator may be conveniently laid down by means of this angle β. The values of β corresponding to the before-mentioned values $\lambda=\cdot9592$ and $\lambda=8\cdot073$ are $\beta=47°\ 54'$ and $\beta=83°\ 53'$ respectively.

Article Nos. 114 and 115. *The Spherogram and Isoparametric Lines—General Considerations.*

114. We first construct a blank spherogram, as already explained (and see also Plates IV. and V.), viz., we draw on the stereographic projection a hemisphere—say the northern hemisphere: the meridians being radii and the parallels of colatitude circles with the pole as centre; the parallel of 60° is the regulator circle, and the separators are great circles touching this at the points A, A, A, in longitudes 30°, 150°, 270° respectively; the separators intersect in the points B, B, B, in the northern hemisphere, and they are produced to meet again in the points B, B, B, of the southern hemisphere; but instead of taking the whole northern hemisphere, we omit portions thereof, and take in the opposite portions of the southern hemisphere; the spherogram being thus bounded by portions of the separator circles, and consisting of the inner spherical triangle B, B, B, and three surrounding triangles B, B, B. The inner triangle contains the regulator-circle, touching its sides at the points A, A, A respectively, and dividing it into an inner circular region and three surrounding regions A, B, A; these last are the *loci in quibus* of the orbit-poles which correspond to convex orbits; and to mark them off from the other regions, it is proper to shade them in the spherogram. Excluding them from consideration, we have the inner circular region and the outer triangular regions separated off from each other by the shaded regions, except at the points A, where these are thinned away to nothing. The points A are positions of the orbit-pole for which the orbit is indeterminate; and consequently any parameter belonging to the orbit is also indeterminate. Hence the isoparametric line for any given value of the parameter will always pass through the points A; that is, all the isoparametric lines will pass through these points, which are thus points of connexion

between the inner circular region and the three outer regions, but it must be recollected that for certain given values of the parameter, the points A may be isolated points on the isoparametric line.

115. It is sometimes necessary (more particularly as regards the Time-spherogram and isochronic lines) to distinguish from each other the several points A and B; and for this purpose I consider the several points, as situated in the spherogram, to be accented in the following manner:

$$\begin{array}{ccccc} B^{\text{IV}} & & B' & & B \\ & A' & & A & \\ B''' & & A'' & & B'' \\ & & B^{\text{V}} & & \end{array}$$

so that the inner triangle is $B'B''B'''$ and the outer triangles are $BB'B''$, $B'B^{\text{IV}}B'''$ and $B'''B^{\text{V}}B''$ respectively; this distinction has been already partially made in Fig. 10.

Article Nos. 116 to 122. *The e-spherogram and Iseccentric Lines. See Plate IV.*

116. Constructing a blank spherogram as above, we may from the tables for planograms Nos. 1 and 2 lay down numerically the values of the eccentricity at the several points of each meridian for the longitudes 0°, 30°,.. 330°, viz.

LONGITUDES 0°, 60°, 120°, 180°, 240°, 300°. Planogram No. 2 shows that e increases from 0 at the centre to ∞ at 60°, then, 60° to 63° 26′ (shaded region), it diminishes from ∞ to 4·912; on passing 63° 26′ it changes abruptly to 1·853; thence diminishes to a minimum = ·628 at 59°, and again increases to 1·018 at 90°.

LONGITUDES 10°, 210°, 330°. Planogram No. 1, part 1, shows that e increases from 0 at the centre to ∞ at 60°, then, 60° to 73° 54′ (shaded region), it diminishes from ∞ to 2·309, this last value being at a point B, the termination of the spherogram.

LONGITUDES 30°, 150°, 210°. Planogram No. 1, part 2, and for values over 90°, part 1, shows that e increases from 0 at the centre to ·264 at 60° (point A), ·869 at 90°, and 2·309 at 100° 6′, point B.

It will be recollected that, although e has the same value, 2·309 at the two opposite points B, yet there is an abrupt change of orbit, indicated by the change of sign of A $(=\pm e)$.

117. Planogram No. 3 shows the directions at the points A of the several iseccentric lines. Planogram No. 4, if the calculations were completed, would give the

value of the eccentricity at the several points of the ecliptic, but besides the already-mentioned values 1·018 at 0°, 60°, &c., and ·868 at 30°, 90°, &c., the only value calculated is ·906 at 15°, 45°, &c. It thus appears that the eccentricity $=1·018$ for longitude 0° diminishes through ·906 at 15° to ·868 at 30°, and then again increases through ·906 at 45° to 1·018 at 60°, and so on through successive intervals of 60°.

118. Planogram No. 5, if the calculations were completed, would give the value of e for the arc AB within the shaded region (but no values have been found except those given by Planograms 1 and 2, viz. $e=\infty$ at A, $=4·912$ at longitude 30° from A, and $=2·309$ at B); and it would also give the value of e for the whole bounding arc ABB within the exterior triangular region. We have $e=\infty$ at A, $=1·853$ at longitude 30° from A, $=1$ at distance $H=43°\ 49'$ from A, $=1·101$ at B, and then proceeding along the arc BB, $=1$ at distance $H=82°\ 56'$ from A, $=1·018$ on the ecliptic, and, finally, $=2·309$ at B. The two values $e=1$ are very important, as will presently appear, with regard to the parabolic curve.

119. It is now easy to trace the form of the iseccentric lines.

$e=0$, the curve is a point at the centre, and for any value less than ·264 it is a trigonoid form surrounding the centre, the maxima radii being directed towards the points A. The points A belong as isolated points to all these curves.

$e=·264$, the curve is tricuspidal, having a cusp at each of the points A. The numerical values seem to show a singularly blunt form of cusp (the points A are, in fact, not ordinary cusps, but singular points of a higher order); but the data do not enable me to draw with certainty the precise forms of the arcs between the three cusps: the wavy form was drawn purposely, but there is no sufficient evidence for its correctness.

120. It is convenient to pass at once to the case $e=1$, or say the parabolic curve, locus of the orbit-pole when the orbit is a parabola. This is a three-looped curve cutting itself (having a node) at each of the points A; and it appears from planogram No. 5 that each loop touches at four points (two points, $H=43°\,49'$, and two points, $H=82°\ 56'$), the sides of the bounding triangle BBB. The loop thus divides the triangle BBB into six regions, viz. one within the loop, two subjacent, two lateral, and one superjacent.

For any value between $e=·264$ and $e=1$, the curve is a three-looped curve intersecting itself at the points A, and such that the loops lie wholly within those of the parabolic curve, and the remaining portions between the parabolic and cuspidal curves.

121. For any value of $e>1$, we must imagine a three-looped curve intersecting itself at the points A, the loops respectively containing those of the parabolic curve, and the remaining portions within the regulator-circle lying between the regulator-circle and the parabolic curve; and we must then obliterate so much of each loop as lies in the shaded regions, or outside the spherogram; viz. instead of a continuous loop there will be thus a broken loop with detached portions thereof in the subjacent regions, the lateral regions, and the superjacent region respectively. More precisely

this is the form for any value of e from $e = 1$ to $e = 1{\cdot}101$, but for this last value the unobliterated portion for each lateral region evanesces; for any value of e between $e = 1{\cdot}101$ and $e = 2{\cdot}309$, the unobliterated portions lie wholly within the subjacent regions and the superjacent region; for $e = 2{\cdot}309$ the portion within the superjacent region evanesces; and for any greater value of e the unobliterated portion lies wholly within the subjacent regions, the loop being thus a mere fragment.

122. The iseccentric curves within the shaded regions form a distinct system: such curves belong to the values $e = 2{\cdot}309$ to $e = \infty$, and any one of them is a fragment of a three-looped curve intersecting itself at the points A, obtained by obliterating so much of the complete curve as lies outside the shaded regions. But it is perhaps better to disregard these curves altogether, thus in effect excluding the shaded regions from the spherogram.

Article Nos. 123 to 143. *The Time-spherogram and Isochronic Lines. See Plate V.*

123. We construct a blank spherogram, and lay down upon it the parabolic curve; we may then lay down (as will be explained) the numerical values, say of the times T_{13}, but in order to gain some idea of the form of the T_{13}-lines I will first consider the question in a more general manner.

124. When the orbit is a line, parabola, or hyperbola, we may distinguish it by the letters L, P, H accordingly; and by the numbers 1, 2, 3, written in the proper order, show the arrangement of the three points on the orbit; observe that if 1 be the middle point on the orbit, we may write indifferently 213 or 312, and so in other cases, the fixation of the middle number is alone material. When the orbit is a line the distances of the points are always finite; and if the orbit be, for example, $L\,123$, then T_{12} and T_{13} are each $= 0$, but T_{13} is non-existent. For the parabola and hyperbola the distances are in general finite; but it is necessary to distinguish for the parabola, e.g. the case $P\,\dot{1}23$ where an extreme point, and for the hyperbola, e.g. the cases $H\,\dot{1}23$ and $H\,\dot{1}2\dot{3}$ where one or each of the extreme points, is at infinity. We have in these cases respectively

$P\,123$,	T_{12} finite,	T_{23} finite,	$T_{31} = \infty$
$P\,\dot{1}23$,	$T_{12} = \infty$,	T_{23} finite,	$T_{31} = \infty$

and it may be added, as regards $P\,1\dot{2}3$, that, by a continuous change of the parabolic orbit the point 1 may change over to infinity on the other half-branch of the parabola, or the arrangement become $P\,23\dot{1}$. And, moreover

$H\,123$,	T_{12} finite,	T_{23} finite,	T_{31} non-existent.
$H\,\dot{1}23$,	$T_{12} = \infty$,	T_{23} finite,	T_{31} non-existent.
$H\,\dot{1}2\dot{3}$,	$T_{12} = \infty$,	$T_{23} = \infty$,	T_{31} non-existent.

Thus the proper symbol $L\,123$, $P\,\dot{1}23$, &c. as the case may be, will always at once indicate as to each of the times T_{12}, T_{23}, T_{31}, whether this is $= 0$, finite, infinite, or non-existent.

125. We may without difficulty attach to the several portions of the regulator, the separators and the parabolic curve, to each portion its proper symbol L, P, H and 123, $\dot{1}23$, &c. as the case may be.

First, as to the regulator, it is obvious that this is separated by the points A into the three portions $L\,213$, $L\,321$, $L\,132$, respectively. And inside the regulator, adjacent to these, we have portions of the parabolic curve $P\,213$, $P\,321$, $P\,132$, respectively.

Again, for one of the separators, say $B^{\text{IV}}B'AB''B^{\text{V}}$ (see here and in all that follows the notation-diagram, No. 115); since the point 2 is here at infinity this must be at every portion thereof either $H\,13\dot{2}$ or else $H\,31\dot{2}$. The point B^{IV} is $H\,\dot{1}3\dot{2}$ and the point B' is $H\,\dot{3}1\dot{2}$; consequently, as the orbit-pole passes along the separator from B^{IV} to B', the symbol is at first $H\,13\dot{2}$ and at last $H\,31\dot{2}$; the transition takes place at the point of contact of the parabolic curve which is indifferently $P\,13\dot{2}$ or $P\,\dot{2}13$. (In further explanation of the transition, consider the orbit-pole as passing from B^{IV} to B, not on the separator, but indefinitely near it; it can only do so by twice crossing the parabolic curve near the point of contact; the orbit is first $H\,13\dot{2}$, or say $H\,132$, then $P\,132$, then an ellipse, which when the orbit-pole again arrives at the parabolic curve changes into $P\,312$; and it finally becomes $H\,312$ or $H\,31\dot{2}$.)

126. Again, since, on the two separators through B^{IV}, in the portions adjacent to B^{IV}, the symbols are $H\,13\dot{2}$ and $H\,\dot{1}32$, it is clear that in the adjacent portion of the parabolic curve (terminated each way by a point of contact with these separators respectively) the symbol must be $P\,132$; at the point of contact with the first-mentioned separator $B^{\text{IV}}B'AB''B^{\text{V}}$, this becomes $P\,13\dot{2}$, $=P\,\dot{2}13$; and beyond the point of contact it becomes $P\,213$, continuing so until it arrives at the next point of contact with the separator $B'A'B''$: there is always in the symbol for the parabolic curve this change of form as we pass through a point of contact with a separator; and there is the same change, when *travelling along the loop* (that is without going inside the regulator) we pass through a point A. The foregoing considerations fully explain how the proper symbol is to be attached to each portion of the regulator, the separators, and the parabolic curve: to avoid confusion, I have abstained from attaching them in the Plate.

127. Imagining the symbols attached as above, it at once appears that, for the two portions $A'A$ and AA'' of the regulator curve, we have $T_{13}=0$; while, for the arc $A''A'$ of the parabolic curve we have $T_{13}=\infty$. Moreover, T_{13} can only be infinite on one of the separators through B''' and on the parabolic curve; and the symbols show that the curve T_{13} is made up, in a peculiar discontinuous manner, of portions of these two separators and of the parabolic curve, as shown by the strongly marked line of the figure; we have thus the boundary of certain *lightly* shaded regions within which (as well as within the shaded regions) T_{13} is non-existent; excluding these, the remaining regions (instead of a trilateral symmetry) have a symmetry about the axis BB'''; there are still four regions which may be distinguished as the inner region, the axial outer region, and the lateral outer regions; or, more shortly, as the inner, axial, and lateral regions.

128. The times T_{12}, T_{23}, T_{31} are calculated, Planogram 1, part 1, for the meridian long. 90°, and ditto part 2 for the meridian long. 270°; and in Planogram 2 for the meridian long. 180°. As regards these last values, it is easy to see that, in order to pass to the meridian long. 0°, the numbers 2, 3 must be interchanged; that is, long. 0°, the T_{12}, T_{13}, T_{23} are respectively equal to the values, long. 180°, T_{13}, T_{12}, T_{23}. Moreover, the numbers 1, 2, 3 may be changed into 2, 3, 1, or into 3, 1, 2, provided the longitude is increased by 120° and 240° in the two cases respectively; that is,

$$\begin{aligned} T_{31} \text{ long. } \alpha &= T_{31} \text{ long. } \alpha \\ &= T_{12} \text{ long. } (\alpha + 120°) \\ &= T_{23} \text{ long. } (\alpha + 240°). \end{aligned}$$

129. By means of the foregoing two relations, T_{13} for the several longitudes 0°, 30°, 60°, ... 330°, is given as equal to the T_{12}, T_{23}, or T_{31}, for long. 90°, 270°, or 180°, that is, to the T_{12}, T_{23}, or T_{31}, of Planogram No. 1, part 1 or 2, or of Planogram No. 2. For example, T_{31} long. 240° = T_{12} long. 0° = T_{13} long. 180°, that is, it is equal to the T_{31} of Planogram No. 2. We thus find

Long.	T_{13} is =		
0° . . .	T_{12} of Plan. No. 2		
30° . . .	. . . .	T_{23} of Plan. No. 1, pt. 2	
60° . . .	T_{12} „		
90° . . .	. . . .	. . . .	T_{12} of Plan. No. 1, pt. 1
120° . . .	T_{23} „		
150° . . .	. . . .	T_{31} „	
180° . . .	T_{31} „		
210° . . .	. . . .	. . . .	T_{23} „
240° . . .	T_{12} „		
270° . . .	. . . .	T_{12} „	
300° . . .	T_{23} „		
330° . . .	. . . .	. . . .	T_{31} „

and observing that for Planogram No. 1, part 1 or 2, we have $T_{12} = T_{31}$, it hence appears as above, that the meridian 30°—210° is an axis of symmetry of the spherogram. In what precedes it has been assumed that the colatitudes only extend from 0° to 90°, but in the spherogram they extend for the meridians 30°, 150°, 270°, to the colatitude 106° 6′, the values for the colatitudes above 90° are those for the omitted portions 90° to 73° 54′ of the opposite meridian.

N.B. A meridian extends from the pole *in one direction only*, unless the contrary is expressed or implied, as in speaking of a meridian 0°—180°.

130. I attend, in the first instance, to the axis of symmetry or meridian 30°—210°. Proceeding along the meridian long. 30° or towards the point A, the value of T_{13} decreases from 1 at the centre to a minimum $= \cdot 950$ at colatitude 11° (call this the point X), and it then increases to 1·983 at A, and thence to 58·62 at 90° and ∞ at the parabolic boundary of the axial region. In the opposite direction it increases from 1 at the centre to ∞ at the parabolic boundary of the inner region. The minimum value ·950 on the axis of symmetry indicates a node on the isochronic curve; that is, the point X is a node on the isochronic $T_{13} = \cdot 950$. This will consist of two branches, proceeding from A', A'', respectively, cutting the axis and each other at X, then again cutting at A, and thence passing on into the axial region, and respectively terminating on the separator boundary $B'AB''$ thereof.

131. This curve, which I call the nodal isochronic, divides the inner region into a loop, antiloop, and two side regions. On each of the meridians 0°, 60°, the value of T_{13} diminishes from 1 at the centre to a minimum which is less, and then increases to a maximum which is greater, than ·950; the value then diminishes to 0 on the regulator: on emergence of the meridian from the shaded into the axial region, the value is $= \cdot 909$, and it thence increases to ∞ at the parabolic boundary of the axial region; these data further determine the form of the nodal isochronic, viz., each of the two half meridians cuts the loop twice, and again cuts the curve in the axial region.

The nodal isochronic, at each of the points A', A'', continues its course into the lateral region, returning to the same point A or A', so as to form in each of the lateral regions a loop. Considering the loop as formed of two branches, each proceeding from A' or A'', the one which is the continuation of the course within the inner region I call the lower branch; the other, the upper branch; and I say that the upper branch *touches* the separator at A' or A''. The two branches and the entire loop lie on the left-hand side (or side away from A) of the meridians through A' or A''. As to the contact of the upper branch of this and other isochronics at A' or A'' with the separator, see *post* No. 142.

132. It is convenient at this point to consider the form of the isochronic curves within the axial region. The parabolic boundary thereof is an isochronic $T_{13} = \infty$, and it thence appears that for any large value of T_{13} the isochronic curve (portion of the curve) is a curve not meeting the parabolic boundary, and terminated each way in the separator boundary $B'AB''$. As the value of T_{13} diminishes, the curve (which is of course always symmetrical in regard to the axis) bends inwards towards the point A and for $T_{13} = 1 \cdot 983$ (value on the axis at A) the curve acquires a cusp at A. I call this the cuspidal isochronic; I remark that it intersects in the axial region each of the meridians 0° and 60°.

As T_{13} further diminishes to any value between 1·983 and ·950, the curve, commencing in the separator boundary, passes through A into the inner region, and, forming a loop within the loop of the nodal isochronic, emerges through A into the axial region, terminating again in the separator boundary.

133. On the meridians 90°, 330°, through the points B', B'', respectively, the value of T_{13} diminishes from 1 at the centre to 0 at the regulator, where these meridians are considered as terminating.

On the meridians 120°, 300° (meridian at right angles to the axis of symmetry), the value of T_{13} diminishes from 1 at the centre to a minimum less than ·878, and then increasing to a maximum of over ·895 diminishes to 0 at the regulator. On emergence of the meridian from the shaded and half-shaded region on the parabolic boundary of the lateral region the value is $=\infty$, and it thence diminishes to 1·148 on the separator boundary $B^{\text{IV}}B'$ or $B^{\text{V}}B''$.

On the meridians 150°, 270°, which pass through A', A'', respectively, the value of T_{13} increases from 1 at the centre to 1·377 at the regulator, and thence through 2·255 at 90° to ∞ at B^{IV} or B^{V}.

And finally, on the meridians 180°, 240°, the value of T_{13} increases from 1 at the centre to ∞ at the parabolic inner boundary, and then on emergence from the half-shaded and shaded region at the separator boundary $B'''A'$ or $B'''A''$, the value is $=\infty$, and it thence diminishes to a minimum under 6·343, and again increases to ∞ at the separator boundary $B'''B^{\text{IV}}$ or $B'''B^{\text{V}}$.

134. By what precedes, it appears that on the separator boundary $B^{\text{IV}}B'$ or $B^{\text{V}}B''$ of either of the lateral regions, the values of T_{13} is at each extremity $=\infty$, and at an intermediate point $=1·148$; there is consequently a minimum value less than 1·148, and therefore two points at each of which the value is $=1·983$.

Now resuming the consideration of the cuspidal isochronic ($T_{13}=1·983$) as regards the remaining portions thereof, viz., those in the lateral and inner regions; and considering first the lateral region $B'''B^{\text{IV}}B'$, there will be from each of the points just referred to on the boundary $B^{\text{IV}}B'$ a branch; one (which I call the lower branch) from the point nearer B', passes, on the right-hand side of the meridian through A', to A'; the other (which I call the upper branch) proceeding from the point nearer B^{IV}, cuts the same meridian, and then on the left-hand side thereof arrives at A', touching there the separator: at A'' in the other lateral region there are in like manner an upper and a lower branch (situate symmetrically, in regard to the axis, with the upper and lower branches at A'); and continuous with the two lower branches there is a branch from A' to A'', through the antiloop of the inner region.

135. Imagine the given value of T_{13} as continuously increasing from the value ·950, which belongs to the nodal isochronic; and attend in the first instance to the form within the lateral regions. There will be a loop of continually increasing magnitude (viz., the loop for a larger value of T_{13} will always wholly include that for a smaller value); each loop formed by an upper branch, which at A' touches the separator, and a lower branch the direction of which from A' is variable. So long as T_{13} is less than 1·377 (value at A' along the meridian) the lower branch, and consequently the whole loop, will lie on the left hand of the meridian; but when T_{13} is $=1·377$, the lower branch touches the meridian, and for any greater value of T_{13} lies on the right of the meridian; and in either of the last-mentioned cases the loop

is cut by the meridian, and thus lies partly on the left, and partly on the right thereof.

136. Now by what precedes there is on the separator boundary $B'B^{IV}$ of the lateral region a point where T_{13} has a minimum value less than 1·148, and consequently, for any given value, say for a value between this minimum and 1·377, there are on $B'B^{IV}$ two points where T_{13} has the given value. These points cannot lie on the loop of the curve belonging to the given value (for this loop is wholly on the left hand of the meridian); hence the complete curve for the given value of T_{13} will include (within the lateral region) besides the loop, a branch uniting the two points in question; say a link branch.

137. It follows that there is between $T_{13}=1·377$ and 1·983, a value (to fix the ideas, say $=1·80$?, it being understood that I do not attempt to determine this value) for which the loop and link branch will unite themselves together, the point of junction becoming as usual a node; viz., there will be a curve $T_{13}=1·80$? having in the two lateral regions respectively the nodes Y, Y'; or say the curve has in each lateral region a self-intersecting loop. For any greater value of T_{13} (as for example the value 1·983 belonging to the cuspidal curve) there are two branches inclosing the self-intersecting loop; for a less value, as has been seen, instead of the self-intersecting loop, there is a loop and link branch; at least this is the case until for the minimum value $<1·148$ of T_{13} on the separator boundary $B^{IV}B'$ the link branch disappears. For smaller values down to $T_{13}=·950$, which belongs to the nodal isochronic, there is no link branch, but only the loop; and as T_{13} diminishes below this value, there is still a continually diminishing loop, lying wholly on the left hand of the meridian, and with its upper branch always touching the separator; and ultimately for $T_{13}=0$ the loop vanishes.

138. We have attended wholly to the lateral regions; but the consideration of the axial and inner regions is very easy: for any value between the values 1·983 and ·950, there are in the axial region (between the nodal and cuspidal curves) two branches each proceeding from the separator to A, where they unite, and, crossing each other, pass into the inner region, forming a loop within the loop of the nodal isochronic; and, moreover, there is in the inner region a branch, the continuation of the lower branches of the lateral loops, uniting the points A', A'', and lying between the nodal and cuspidal isochronic. And for T_{13} less than ·950 there are in the axial region, between the nodal curve and the separator, two branches, each proceeding from the separator to A, where, crossing each other, they enter the inner region passing outside the nodal curve (or in the side regions of the inner region) to the points A', A'', where they respectively join on to the lower branches of the lateral loops. Ultimately, for $T_{13}=0$, the curve coincides with the finite portions AA', AA'' of the regulator circle.

139. We have finally to consider the case T_{13} greater than 1·983: there is in the axial region a branch lying outside the cuspidal curve, and extending from separator to separator; in each lateral region two branches (lying outside those of the cuspidal curve) each proceeding from A' (or A'') to the separator boundary $B'B^{IV}$

or $B''B^{v}$, an upper branch touching the separator at A' or A'', and a lower branch; and in the inner region, a branch (continuation of the lower branches) lying between the cuspidal curve and the parabolic boundary of the antiloop, and uniting the points A', A''. In the ultimate case $T_{13} = \infty$, the curve coincides with the before-mentioned discontinuous curve composed of portions of the parabolic curve and of two separators.

140. To obtain a comprehensive statement of the foregoing results, we may (as in the case of the iseccentric lines) imagine the curves completed and rendered continuous by the insertion of portions lying outside the spherogram, or within the half-shaded and shaded regions; which inserted portions are to be ultimately obliterated. The upper and under branches terminating in the separator boundary of a lateral region are thus completed into a loop; the link branch into a closed curve or oval; the vanishing of the link branch happens when the oval, on the point of passing outside the separator boundary of the lateral region, just touches this boundary; as T_{13} diminishes to the value for which this happens, and continues still further to diminish, I think it may be assumed that there is some value (to fix the ideas, say $T_{13} = 1{\cdot}10$?, but I do not attempt to determine it) for which the oval becomes a conjugate point, viz., for this value $T_{13} = 1{\cdot}10$? the curve will have two conjugate points (nodes) Z', Z'', outside the two lateral regions respectively.

141. We may now state the forms of the curve. The points A, A', A'', are always nodes, viz., A', A'', nodes with real branches, but A is either a conjugate point, a cusp, or a node with real branches.

$T_{13} > 1{\cdot}983$: two-looped curve, containing within it A as a conjugate point:

as T_{13} diminishes, the curve bends inwards towards A, and

$T_{13} = 1{\cdot}983$: cuspidal isochronic; A, a cusp.

$T_{13} < 1{\cdot}983$, the curve cuts itself at A, having thus acquired an internal loop: as T_{13} diminishes, changes occur first as regards the lateral loops, and afterwards as regards the internal loop; viz., each of the lateral loops is gradually pinched together until

$T_{13} = 1{\cdot}80$? there are two new nodes Y', Y'', each lateral loop being a figure of 8.

As T_{13} diminishes the figure-of-8-loop breaks up into a loop and oval, which oval continually diminishes until for

$T_{13} = 1{\cdot}10$? the ovals have each become conjugate points, or there is a curve with two conjugate points Z', Z''. As T_{13} diminishes the conjugate points have disappeared, and we have again a curve with an internal and two lateral loops; but in the meantime the internal loop and the branch $A'A''$ are continually approaching each other; and, $T_{13} = {\cdot}950$, nodal isochronic, there is a node X on the axis. The curve consists of two figures of 8, each crossing itself at one of the points A', A'', and the two crossing each other at the points A, X.

As T_{13} diminishes, the curve breaks off from X on each side of the axis so as not any longer to cross the axis (except at A), that is

$T_{13} < \cdot 950$; curve is a chain intersecting itself at A', A, A''; viz., from each loop there pass two branches, one inside, the other outside, the regulator, uniting themselves at A with the branches from the other loop outside and inside the regulator respectively; and finally

$T_{13} = 0$, the curve is the arc $A'AA''$ of the regulator circle.

142. There is not in the several curves any discontinuity of direction at the point A' or A'': the branch from A within the shaded or half-shaded region, emerges at A' or A'' into the lateral region, uniting itself with the upper branch of the loop; it can only do this in virtue of its being at A' or A'' a tangent to the separator (for otherwise it would cross the separator and regulator into the inner region); that is, the continuation thereof, or upper branch of the loop, must at the point A' or A'' *touch the separator;* it has been previously throughout assumed that this is so.

143. It is to be observed, both as regards the iseccentric and the isochronic curves, that there is a real meaning in the obliterated portions; viz., to any position of the orbit-pole on such obliterated portion of the curve there corresponds a conic determined by means of a given trivector, but which, by reason of its being a convex hyperbola, or hyperbola such that the three points do not lie on the same branch thereof, is not regarded as an orbit. The obliterated portions have been in the present Memoir considered only so far as they present themselves in continuity with the curves which are the loci of the pole of a proper orbit, and for the purpose of explaining the course of these curves; and the curves completed as above are not the complete loci which would be obtained if, instead of the selected conic called the orbit, we had considered simultaneously the four conics determined by means of any given trivector; such extension of the theory would, it is probable, be interesting geometrically; but it would be devoid of all astronomical significance.

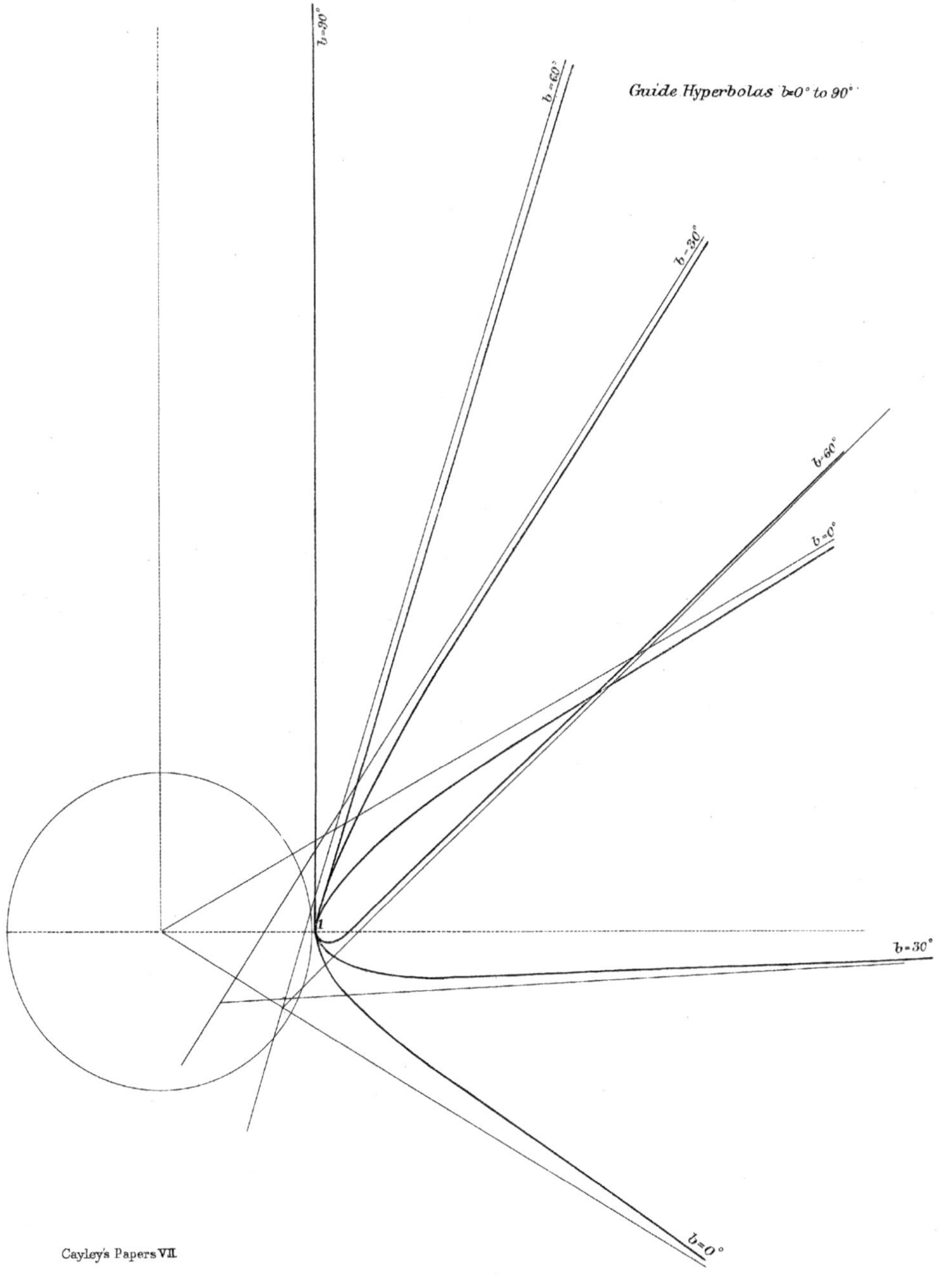
Guide Hyperbolas b=0° to 90°
b=90°
b=60°
b=30°
b=60°
b=0°
b=30°
b=0°
1

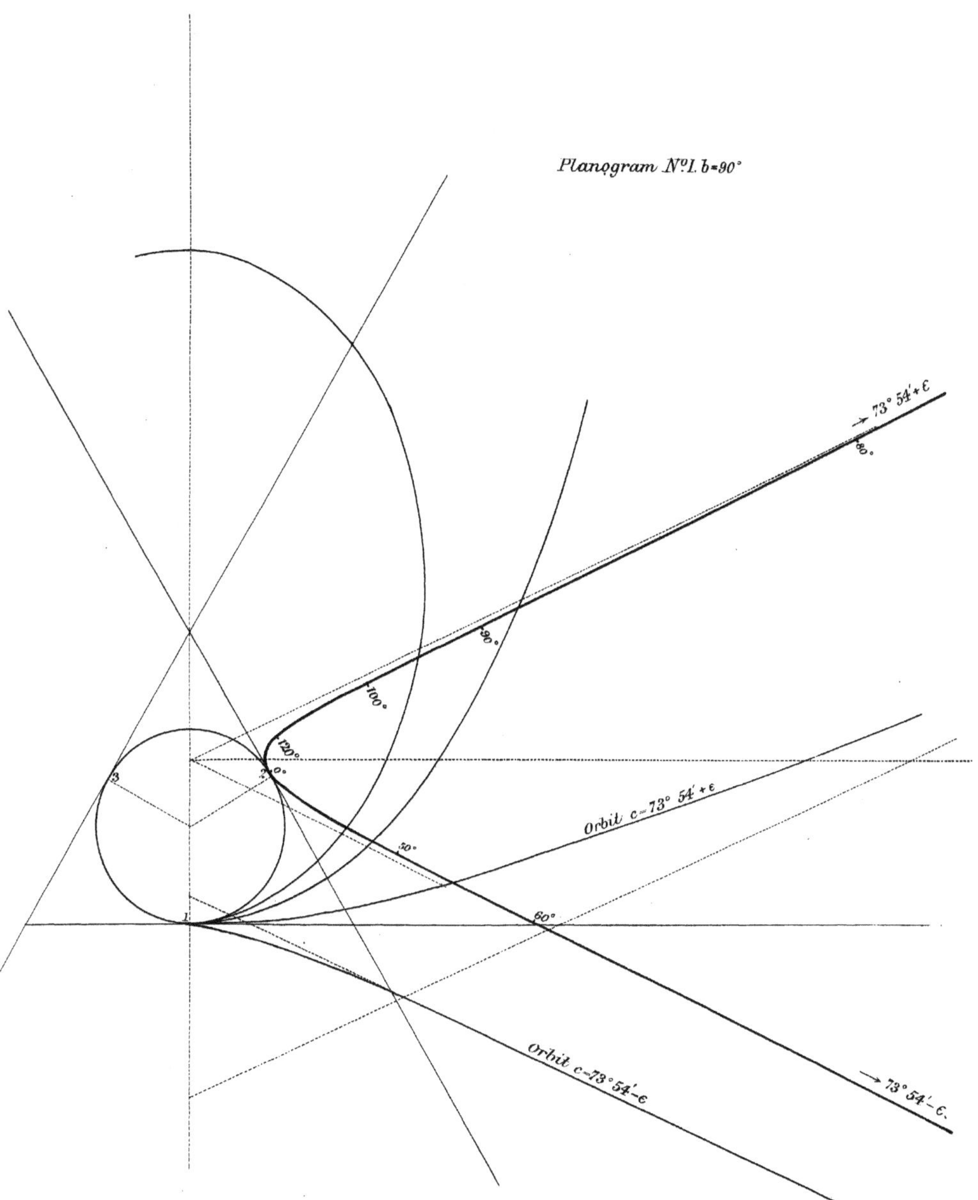
Planogram No. I. b=90°
→ 73° 54′+ε
80°
90°
100°
120°
0°
3
2
30°
Orbit c=73° 54′+ε
60°
1
Orbit c=73° 54′−ε
→ 73° 54′−ε.

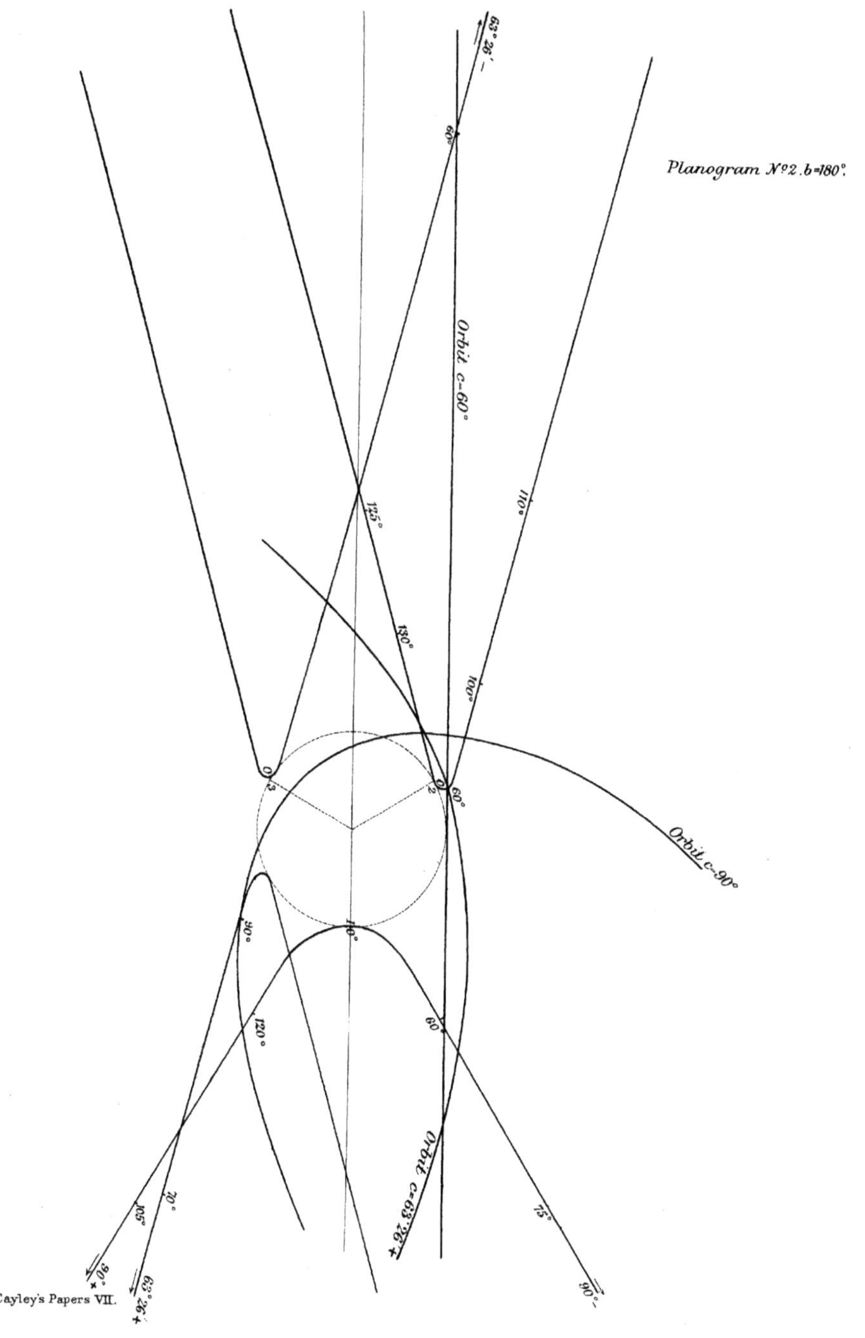
Planogram No. 2. b=180°.
Orbit c=60°
Orbit c=90°
Orbit c=63°26'+
63°26'−
60°
125°
110°
130°
100°
60°
90°
120°
60
70°
105°
90°+
63°26'+
75°
90°−

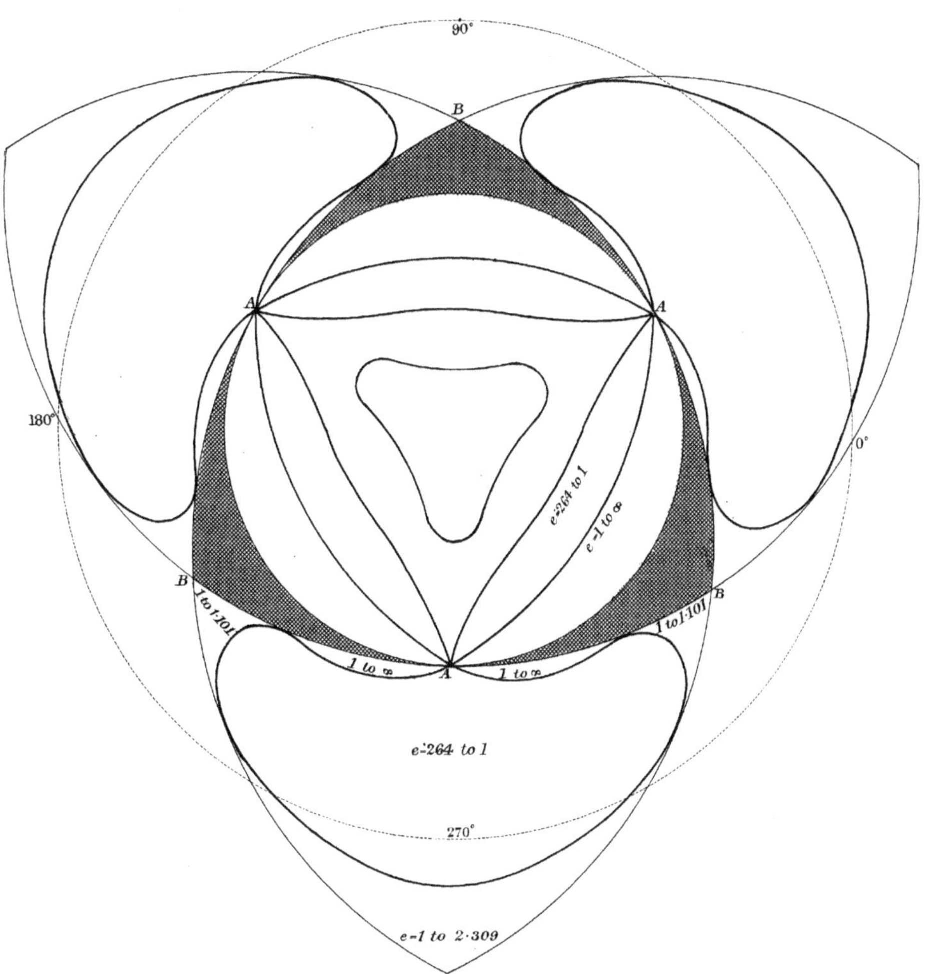

e-Spherogram.
90°
B
A
A
180°
0°
e·264 to 1
e·1 to ∞
B
B
1 to 1·101
1 to 1·101
1 to ∞
A
1 to ∞
e·264 to 1
270°
e·1 to 2·309

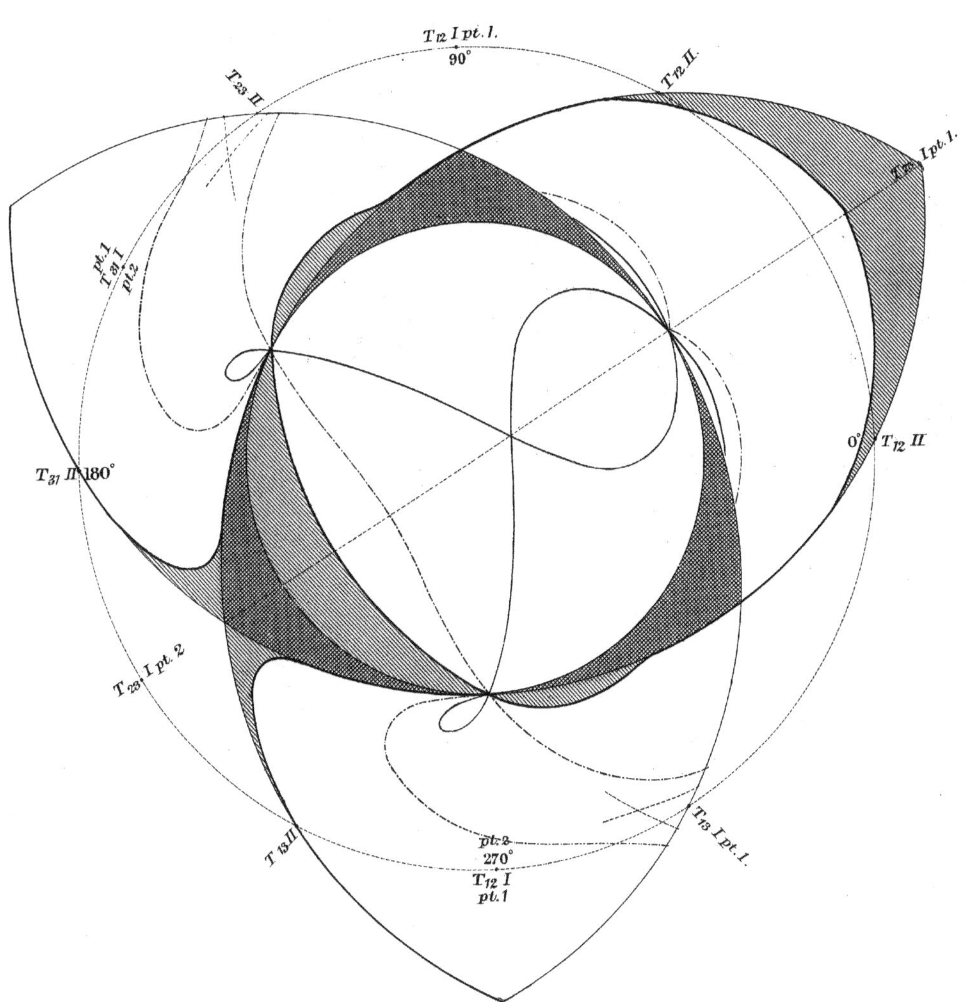
T_{13} - Spherogram.
T_{12} I pt. 1.
90°
T_{23} II
T_{12} II.
T_{12} I pt. 1.
pt. 1
T_{31} I
pt. 2
0°
T_{12} II
T_{31} II 180°
T_{23} I pt. 2
T_{13} I pt. 1.
T_{13} II
pt. 2
270°
T_{12} I
pt. 1

477.

ON THE GRAPHICAL CONSTRUCTION OF A SOLAR ECLIPSE.

[From the *Memoirs of the Royal Astronomical Society*, vol. XXXIX. (1870—1871), pp. 1—17. Read January 13, 1871.]

THE present Memoir contains the explanation of a Graphical Construction of a Solar Eclipse, which (it appears to me) is at once easy, and susceptible of considerable accuracy: I think that if made on the suggested scale (radius $=12$ inches) we might by means of it construct a diagram such as the eclipse-diagrams of the *Nautical Almanac*, with at least as much accuracy as could be exhibited in a diagram on that scale.

Article Nos. 1 to 9. *General Explanation of the Construction.*

1. We may imagine the celestial sphere as seen from the centre of the Earth stereographically projected at each instant during the eclipse—the radius of the bounding circle of the projected hemisphere being a given length, say twelve inches, which is taken as unity—in such wise that the centre of the Moon is always at the centre of the projection, say M, and the pole (suppose the north pole, say N) of the Earth on a given radius: its position on this radius will in strictness be variable, viz. distance from centre = projection of Moon's N. P. D. $=\tan\frac{1}{2}\Delta$. Suppose, for a moment, that the position at each instant of the Sun's centre were also laid down on the projection, so as to obtain the projection of the Sun's relative orbit; this will be a terminated short line $A'B'$ (fig. 1), nearly straight, and lying near the centre of the projection (this relative orbit is not to be actually laid down, but it is replaced, as will presently be explained, by a relative orbit on a very enlarged scale); if at any instant the position of the Sun on the relative orbit be denoted by S', then the straight line MS' is the projection of the arc of great circle through the centres of the Moon and Sun, so that E being the angular distance of the centres, the length of the line MS' is $=\tan\frac{1}{2}E$, or (E being small) it is $=\frac{1}{2}E$.

2. Produce $S'M$ through the centre M to a point Z, and consider Z as representing a point on the Earth's surface: to determine the geographical position of Z, we must consider the projected meridian NZ which passes through Z: the arc NZ,

Fig. 1.

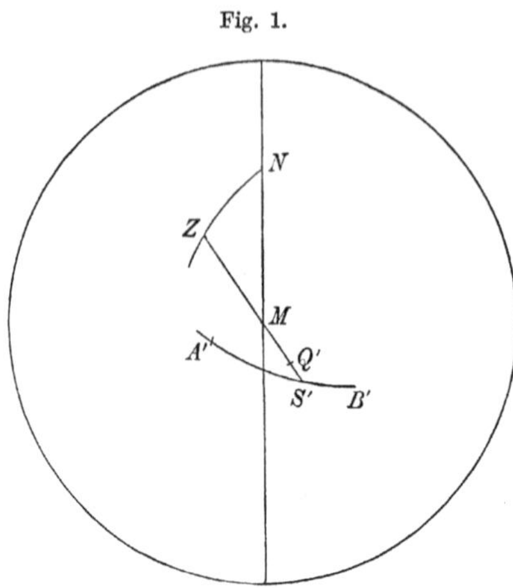

regarded as a projection, represents the N. P. D. or colatitude of Z, and the actual angle at N which the tangent of NZ makes with the line NM is equal to the celestial angle ZNM which is = Moon's hour-angle from Z, or what is the same thing = difference of Moon's hour-angle from Greenwich and of the longitude of Z (as the figure is drawn, $\angle ZNM$ = Moon's hour-angle E. of Greenwich, *less* E. longitude of Z).

3. Now, considering the Moon and Sun as seen from Z, we may disregard the parallactic depression of the Sun, and attribute to the Moon a displacement equal to the difference of the parallactic displacements of the Moon and Sun; that is, regarding the zenith distances ZM, ZS' as equal, we may consider the Moon's centre as depressed by parallax in the direction of the arc MS' through an arc MQ', $= \sin^{-1} P' \sin ZM$, where $P' = \cdot 99837 (\sigma' - \pi')$ is the quantity thus designated in the Appendix to the *Nautical Almanac* for 1836, viz. it is $= \sigma' - \pi'$, the difference of the equatoreal horizontal parallaxes at the time of the eclipse, multiplied by a factor ·99837, which answers to a distance of Z from the Earth's centre = Earth's radius for latitude 45°. And if we take Q' such that its angular distance from S' = sum of angular semidiameters of the Sun and Moon, the locus of Q' is very nearly a circle about the centre S', and the corresponding positions of Z give the positions on the Earth where the limbs are in exterior contact, or, what is the same thing, give the penumbral curve on the Earth's surface for the position S' of the Sun.

4. Instead of

$$\text{Arc } MQ' = \sin^{-1} P' \sin ZM,$$

we may write

$$\text{Arc } MQ' = P' \sin ZM,$$

or, using ρ to denote the linear distance ZM in the projection, we have $\rho = \tan \frac{1}{2} ZM$, and therefore $\sin ZM = \frac{2\rho}{\rho^2+1}$, hence

$$\text{Arc } MQ' = P' \frac{2\rho}{\rho^2+1},$$

and the linear distance MQ' in the projection is $= \tan \frac{1}{2} \text{arc } MQ'$, say this is $= \frac{1}{2} \text{arc } MQ'$, or calling this linear distance r' we have

$$r' = P' \frac{\rho}{\rho^2+1}.$$

5. Hence, if instead of the original representation of the Sun's relative orbit we consider an enlarged representation thereof and of the depressed positions Q' of the Moon, obtained by increasing the several distances from the centre of the projection in the ratio $\frac{1}{2}P'$ to 1, and if instead of A', B', S', Q', we use A, B, S, Q, as referring to this enlarged representation, then representing by r the linear distance MQ, we have $r = \frac{2}{P'} r'$, and consequently

$$r = \frac{2\rho}{\rho^2+1}.$$

We have here r representing the parallactic depression corresponding to the zenith distance ZM, where $\rho = \tan \frac{1}{2} ZM$; that is, $ZM = 90°$, $\rho = 1$, and therefore $r = 1$; but for $ZM = 90°$ the parallactic depression is $= P'$; that is, the scale of the enlarged representation of the Sun's relative orbit, or say simply the scale of the relative orbit (for on the original scale it was never actually constructed at all) is such that we have P' (= about 60′) represented by the radius of the bounding circle of the projected hemisphere, = 12 inches.

6. The process is, construct the relative orbit on the scale P' = radius of bounding circle: take S for the position at any given instant of the Sun in the relative orbit, and with centre S and radius $= s + \sigma$ (sum of the angular semidiameters, of course on the same scale) describe a circle. The positions A and B of the Sun at the beginning and end of the eclipse respectively are such that this circle just touches the bounding circle externally, viz. the distances of A and B from the centre of the projection are each = radius of bounding circle $+ s + \sigma$. At any intermediate instant the circle, radius $s + \sigma$, lies wholly or partly within the bounding circle; in the latter case we attend only to the arc thereof which lies within the bounding circle. Taking then Q any point whatever on the circle or arc in question, we join Q with the centre M of the projection, and produce this line through M to a point Z, such that the distances MQ, MZ, being r, ρ respectively, we have as above

$$r = \frac{2\rho}{\rho^2+1},$$

or, what is the same thing, writing θ in place of z, and regarding this angle θ as a variable parameter, the relation between r, ρ, may be expressed by means of the two equations, $\rho = \tan \frac{1}{2} \theta$, $r = \sin \theta$.

7. Practically the construction may be performed very easily by means of a straight edge twenty-four inches long, graduated from the centre, one half of it for the values of r, and the other half for the corresponding values of ρ (that is, the first half is graduated for $\sin\theta$, and the second half for $\tan\frac{1}{2}\theta$): we have thus, corresponding to the circle or arc of circle which is the locus of Q, a closed curve, or arc thereof terminated each way at the bounding circle, for the locus of Z: which curve or arc of a curve is the penumbral curve on the Earth's surface for the position S of the Sun in the relative orbit.

8. The north pole of the Earth occupies in the projection a given position, viz. it is situate on a given radius at a distance $=\tan\frac{1}{2}$ Moon's N.P.D.; which N.P.D. may be considered as being throughout the eclipse constant, and equal to its value at the middle of the eclipse. But in order to arrive at the geographical signification of the figure it is necessary to lay down on the projection the position of the meridian of Greenwich; which position, it will be remembered, varies according to the position of S. Supposing this done, we could of course (at least theoretically) draw the whole series of meridians and parallels, and thereby determine the latitudes and longitudes of the several points of the penumbral curve, or (if need is) transfer it to a different projection of the Earth's surface. The actual description of the meridians and parallels would, however, be very laborious, and fortunately it can be avoided by means of a single blank projection and a slight modification of the foregoing process, as will be explained.

9. But before considering how this is, it is proper to remark that constructing as above a figure of the penumbral curves corresponding to the several positions of the Sun: by what precedes these different curves may indeed be considered as belonging to the same position of the north pole in the projection, but they belong to different positions of the meridian of Greenwich; and thus they do not constitute a representation of the penumbral curves each in its proper terrestrial position, but only a representation in which the penumbral curves are affected each of them by a different displacement in longitude.

Article Nos. 10 to 13. *Modification in order to the Applicability of a Single Blank Projection.*

10. Imagine a stereographic projection of the meridians and parallels *on the plane of a meridian*, radius of this meridian, that is of the bounding circle of the projected hemisphere, being $=12$ inches as before; and the poles N, Σ being of course opposite points on the circumference of the bounding circle—the meridians and parallels are, however, to be produced outside the bounding circle; say this is the "blank projection," and let its centre be denoted by M_1. Then, if at any point M on the radius MN, we draw the chord CD at right angles to M_1N, and on CD as diameter describe a circle, this will cut out from the blank projection a new projection having the last-mentioned circle for its bounding circle, and in which N is the north pole; viz. the meridians of the blank projection will be meridians, and the parallels of the blank projection will be parallels, in this new projection. And, moreover, if the longitudes are reckoned from the meridian NMM_1, then the meridian of a given longitude in the blank projection will in the new projection be the meridian of the *same*

longitude—but the parallel of a given colatitude c in the blank projection will, in the new projection, be the parallel of a different colatitude c',—the relation of c, c' being, however, a very simple one, as presently explained.

Fig. 2.

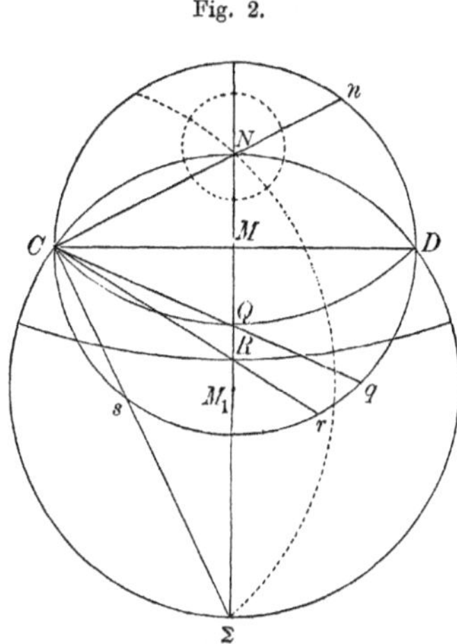

11. The blank projection thus at once gives a projection in which the north pole N has any assumed position whatever; and it is easy to see that in order that its distance MN from the centre of the projection may represent a given angle Δ, we have only to take $M_1M = \cos\Delta$ (that is $= 12$ inches $\times \cos\Delta$), the corresponding value of MC being $MC = \sin\Delta$ (that is $= 12$ inches $\times \sin\Delta$). Hence Δ denoting the Moon's N.P.D. at the middle of the eclipse, we can by means of the blank projection construct a projection such as that above referred to, only the radius of its bounding circle, instead of being unity (12 inches), is in the reduced ratio of $1 : \sin\Delta$.

12. The figure of the penumbral curves as originally constructed requires, therefore, to be reduced in the ratio $1 : \sin\Delta$, viz. each of the distances from the centre M should be reduced in this ratio; this could of course be done easily enough with a pair of proportional compasses; but by means of a different graduation of the straight edge we may, *in the first instance*, construct the penumbral curves on the proper reduced scale; viz. assuming that we have on the proper scale a proportional-scale figure such as is here shown, the line Mr ($= 12$ inches) being graduated for $\sin\theta$, and the line MA (also $= 12$ inches) for $\tan\frac{1}{2}\theta$, and a set of parallel lines being drawn through the last-mentioned graduations—then taking the distance $M\rho = \sin\Delta$, that is $= 12$ inches $\times \sin\Delta$, and drawing the line $M\rho$, this line will, it is clear, be graduated for $\sin\Delta\tan\frac{1}{2}\theta$: so that we may from the figure graduate the straight edge, the one half of it by means of the line Mr, and the other half of it by means of the line $M\rho$; and with the straight edge thus graduated, at once lay down the penumbral curve on the scale now in question. And we thus obtain a figure containing as well

the penumbral curves, as the meridians and parallels which serve to fix their terrestrial position.

Fig. 3.

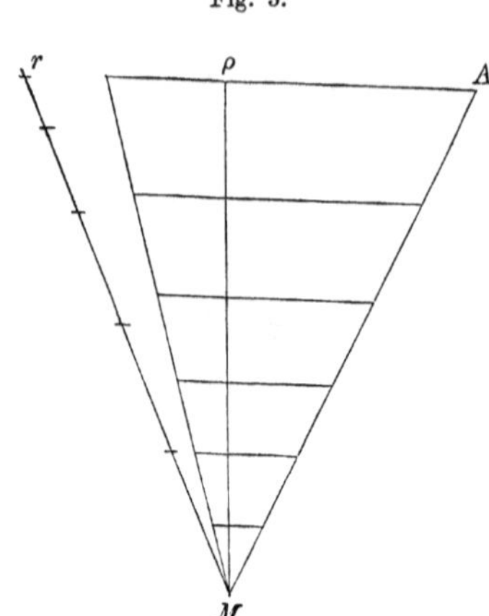

13. It remains in the new projection to find the colatitude belonging to any given parallel. Supposing that the colatitude in the blank projection is $=c'$, then it may be shown that the colatitude c of the same parallel in the reduced projection is given by means of the equation

$$\tan\tfrac{1}{2}c = \cot\tfrac{1}{2}\Delta\tan\tfrac{1}{2}c',$$

from which c might be calculated numerically: but the required value may also be obtained graphically. In fact, considering the parallel which cuts $N\Sigma$ (see fig. 2) in a point R, then, if by lines drawn from C as a centre we project N, R, Σ, on the circumference of the bounding circle of the new projection—say the projections of these points are n, r, s, respectively, the arc ns is a semicircle, and the arcs nr, sr, are respectively the N.P.D. and the S.P.D. of the parallel in question. It may be added that in the new projection the equator is represented by the parallel through the points C, D; so that if this cuts $N\Sigma$ in Q, and the point Q be in like manner projected on the bounding circle—say its projection is q, then the arcs nq, sq, will be each of them a quadrant, and the arc qr will be the latitude of the parallel in question.

Article Nos. 14 to 18. *As to the Construction of the Relative Orbits.*

14. It is convenient to notice that if e, e′, be the values of the equation of time at the preceding and following Greenwich Mean Noons (viz. e or e′ = G.M.T. of apparent Noon) then that the Sun's hour-angle E. of Greenwich at the Greenwich mean time t is

$$h' = \mathrm{e} + t\left(1 + \frac{\mathrm{e}' - \mathrm{e}}{24^{\mathrm{h}}}\right)$$

and that if a, a', are the R.A.'s of the Moon and Sun respectively, then $h' - h = a' - a$, which is also of the form $A + Bt$. In the reduced projection, the Moon is always at the centre M; by means of the values of $h' - h$ we lay down at any instant the Sun's position in R.A. and then by means of the values of h', the position of the meridian of Greenwich; and we thus at any instant read off the terrestrial longitude of any point of the reduced projection, or say, of a point on the penumbral curve.

15. With regard to the construction of the relative orbit, it is to be observed that if at any instant the hour-angle and N.P.D. of the Moon are h, Δ, and those of the Sun, h', Δ', then taking M as origin, and the axes Mx, My, in the direction

Fig. 4.

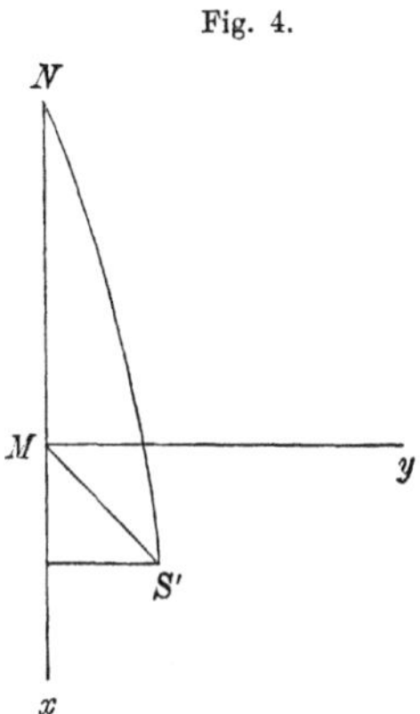

of NM produced, and perpendicular hereto to the right (or eastwards), then the rectangular coordinates of S' are approximately $x = \frac{1}{2}(\Delta' - \Delta)$, $y = \frac{1}{2}(h' - h)\sin\Delta$, where $h' - h$ is equal to the difference of R.A. of the Sun and Moon. Hence, in the adopted relative orbit, the coordinates of S would be

$$x = \frac{\Delta' - \Delta}{P'}\, 12 \text{ in.} \qquad y = \frac{h' - h}{P'} \sin\Delta \,.\, 12 \text{ in.}$$

where, P' being reckoned in minutes, $\Delta' - \Delta$ and $h' - h$ are also reckoned in minutes.

16. Moreover, Δ may be considered as constant during the eclipse: and the relative orbit, assumed to be a straight line, will be determined by means of two points thereof; viz. knowing the values of $\Delta' - \Delta$, and $h' - h$ at about the time of the beginning and at about the time of the end of the eclipse, we construct by these formulæ two points of the orbit, and joining them by a straight line, we have the orbit. Also the position at any instant of the Sun in this relative orbit will be obtained by considering its motion therein as being uniform. I think there is no advantage in the adoption of a more accurate construction: for although we may for any given instant use the accurate values of h, Δ, h', Δ', and so construct the position in the relative orbit, and the corresponding penumbral curve, yet if in the deter-

mination of the geographical significance thereof, we were to use for each curve a different value of Δ, the simplicity of the construction would disappear; and it is, moreover, doubtful whether the trifling corrections would not be within the limits of the necessary errors of the drawing.

17. But if MS' be $=E$, and $\angle xMS'=\theta$, the accurate values for the coordinates of S' are $x=\tan\frac{1}{2}E\,.\cos\theta$, $y=\tan\frac{1}{2}E\,.\sin\theta$, and the values for the coordinates of S are $x=\dfrac{2}{P'\,.\,\text{arc}\,1'}\tan\frac{1}{2}E\,.\cos\theta\,.\,12\text{ in.}$, $y=\dfrac{2}{P'\,.\,\text{arc}\,1'}\tan\frac{1}{2}E\,.\sin\theta\,.\,12\text{ in.}$, where P' is still reckoned in minutes, and of course arc $1'=\dfrac{\pi}{10800}$. As the scale is considerable, it is worth while to inquire whether the employment of the accurate formulæ would produce an appreciable difference in the position of S.

We have $\sin\theta\div\sin\Delta'=\sin(h'-h)\div\sin E$, that is, $\sin E\sin\theta=\sin(h'-h)\sin\Delta'$, and $\cos E=\cos\Delta\cos\Delta'+\sin\Delta\sin\Delta'\cos(h'-h)$; or putting for shortness $\Delta'-\Delta=\alpha$, $h'-h=\beta$, we have $\sin E\sin\theta=\sin\beta\sin\Delta'$, and $\cos E=\cos\alpha-\sin\Delta\sin\Delta'\,.\,2\sin^2\frac{1}{2}\beta$. Hence, attending to the equations $\cos^2\frac{1}{2}E=2(\cos^2\frac{1}{2}\alpha-\sin\Delta\sin\Delta'\sin^2\frac{1}{2}\beta)$ and $\sin^2\frac{1}{2}E=2(\sin^2\frac{1}{2}\alpha+\sin\Delta\sin\Delta'\sin^2\frac{1}{2}\beta)$, we find

$$\tan\tfrac{1}{2}E\sin\theta=\frac{\sin\frac{1}{2}\beta\cos\frac{1}{2}\beta\sin\Delta'}{\cos^2\frac{1}{2}\alpha-\sin\Delta\sin\Delta'\sin^2\frac{1}{2}\beta},$$

and

$$\tan\tfrac{1}{2}E\cos\theta=\sqrt{\frac{\sin^2\frac{1}{2}\alpha+\sin\Delta\sin\Delta'\sin^2\frac{1}{2}\beta}{\cos^2\frac{1}{2}\alpha-\sin\Delta\sin\Delta'\sin^2\frac{1}{2}\beta}-\frac{\sin^2\frac{1}{2}\beta\cos^2\frac{1}{2}\beta\sin^2\Delta'}{(\cos^2\frac{1}{2}\alpha-\sin\Delta\sin\Delta'\sin^2\frac{1}{2}\beta)^2}},$$

whence, considering α, β as small quantities of the same order, and neglecting terms of the third order, we have $\tan\frac{1}{2}E\sin\theta=\sin\frac{1}{2}\beta\cos\frac{1}{2}\beta\sin\Delta'$, or what is the same thing, $=\sin\frac{1}{2}\beta\sin\Delta$, or finally, $=\frac{1}{2}\beta\sin\Delta$, that is $\frac{1}{2}(h'-h)\sin\Delta$, which is the foregoing approximate value, and thus in the adopted orbit $y=\dfrac{h'-h}{P'}\,12\text{ in.}=\text{approx. value}$. As regards the expression for $\tan\frac{1}{2}E\cos\theta$, writing for a moment $\Omega=\sec^2\frac{1}{2}\alpha\sin\Delta\sin\Delta'\sin^2\frac{1}{2}\beta$, the quantity under the radical sign is

$$\frac{\tan^2\frac{1}{2}\alpha+\Omega}{1-\Omega}-\frac{\sin^2\frac{1}{2}\beta\cos^2\frac{1}{2}\beta\sin^2\Delta'}{\cos^4\frac{1}{2}\alpha\,.\,(1-\Omega)^2},$$

and, taking this to the third order, it is

$$=\tan^2\alpha+\Omega\,(1+\tan^2\tfrac{1}{2}\alpha)-\frac{\sin^2\frac{1}{2}\beta\cos^2\frac{1}{2}\beta\sin^2\Delta'}{\cos^4\frac{1}{2}\alpha},$$

which, substituting for Ω its value, is

$$=\tan^2\tfrac{1}{2}\alpha+\frac{\sin^2\frac{1}{2}\beta\sin\Delta'}{\cos^4\frac{1}{2}\alpha}(\sin\Delta-\sin\Delta'\cos^2\tfrac{1}{2}\beta),$$

where $\sin\Delta-\sin\Delta'\cos^2\frac{1}{2}\beta=\sin\Delta-\sin(\Delta+\alpha)\cos^2\frac{1}{2}\beta$, or neglecting herein terms of the second order, this is

$$= \sin \Delta - (\sin \Delta + \sin \alpha \cos \Delta) \cos^2 \tfrac{1}{2} \beta,$$

$$= - \sin \alpha \cos \Delta, \quad = - 2 \tan \tfrac{1}{2} \alpha \cos^2 \tfrac{1}{2} \alpha \cos \Delta,$$

so that to the third order the quantity under the radical sign is

$$= \tan^2 \tfrac{1}{2} \alpha - \frac{2 \tan \frac{1}{2} \alpha \sin^2 \frac{1}{2} \beta \sin \Delta \cos \Delta}{\cos^2 \frac{1}{2} \alpha};$$

and to the second order, that is finally neglecting terms of the third order,

$$\tan \tfrac{1}{2} E \cos \theta = \tan \tfrac{1}{2} \alpha - \frac{\sin^2 \frac{1}{2} \beta \sin \Delta \cos \Delta}{\cos^2 \frac{1}{2} \alpha},$$

or, what is the same thing,

$$= \tfrac{1}{2} \alpha - \sin \Delta \cos \Delta \,.\, \tfrac{1}{4} \beta^2.$$

18. Hence, writing $\alpha = (\Delta' - \Delta) \text{ arc } 1'$, $\beta = (h' - h) \text{ arc } 1'$, and passing to the adopted orbit, we have

$$x = \frac{\Delta' - \Delta}{P'} 12 \text{ in.} - \tfrac{1}{2} \sin \Delta \cos \Delta \frac{h' - h}{P'} 12 \text{ in.} \times (h' - h) \text{ arc } 1',$$

viz. putting

$$y = \frac{h' - h}{P'} \sin \Delta \,.\, 12 \text{ in.}$$

we have

$$x = \frac{\Delta' - \Delta}{P'} 12 \text{ in.} - y \,.\, \tfrac{1}{2} \cos \Delta \times (h' - h) \text{ arc } 1',$$

or say

$$= \frac{\Delta' - \Delta}{P'} 12 \text{ in.} - y \,.\, \tfrac{1}{2} \cos \Delta \, (h' - h) \frac{\pi}{10800}.$$

The value of the second term may amount to about $\frac{1}{10}$ of an inch, and thus be sensible, but there is no difficulty in taking account of it.

Article No. 19. *As to the Equation* $r = \dfrac{2\rho}{\rho^2 + 1}$.

19. It may be remarked that the equation $r = \dfrac{2\rho}{\rho^2 + 1}$, which served for the graduation of the straight edge, was in effect obtained from the equations

$$r = \frac{2}{P'} \tan \tfrac{1}{2} E, \quad P' \sin z = \sin E, \quad \rho = \tan \tfrac{1}{2} z$$

by assuming therein $\tan \frac{1}{2} E = \frac{1}{2} E$ and $\sin E = E$ respectively. But the elimination of E and z can be effected without this assumption, viz. we have $\sin E = \dfrac{2 \tan \frac{1}{2} E}{1 + \tan^2 \frac{1}{2} E} = \dfrac{P' r}{1 + \frac{1}{4} P'^2 r^2}$, and then as before, $\sin z = \frac{2\tan\frac{1}{2}z}{1+\tan^2\frac{1}{2}z} = \frac{2\rho}{1+\rho^2}$, whence the relation between r and ρ is found to be

$$\frac{r}{1+\frac{1}{4}P'^2 r^2} = \frac{2\rho}{1+\rho^2},$$

which however assumes that P' is reckoned in parts of the radius; reckoning it as before in minutes, we must, instead of P', write $P' \operatorname{arc} 1' = \frac{P'\pi}{10800}$, viz. the numerical value is about $\frac{1}{60}$, and taking it to be this number, the formula is

$$\frac{r}{1+\frac{1}{14400}r^2} = \frac{2\rho}{1+\rho^2},$$

where r, ρ are reckoned in parts of the radius (= 12 inches). Supposing that r_1 is calculated from the formula $r_1 = \frac{2\rho}{1+\rho^2}$, then we have very nearly $r = r_1\left(1+\frac{r_1^2}{14400}\right)$, and r_1 being $=1$ at most, the correction is inappreciable: if however this were not the case, the more accurate formula might have been used; the only difference being that the making of the graduation would have been more laborious.

Article No. 20. *Remark as to the Geometrical Theory of the Projection of the Penumbral Curve.*

20. The stereographic projection of the penumbral curve on the Earth's surface (assumed to be spherical) is, as I have elsewhere shown, a bicircular quartic. It may be shown that the stereographic projection, as given by the foregoing approximate method, is a bicircular quartic: we have, in fact a circle, the equation of which in the polar coordinates r, θ is

$$(r\cos\theta - \alpha)^2 + r^2\sin^2\theta = \beta^2,$$

and where (θ being unaltered) r is changed into ρ, where $r = \frac{2\rho}{\rho^2+1}$, that is $\frac{1}{r} = \frac{1}{2}\left(\rho + \frac{1}{\rho}\right)$. The equation of the circle is

$$r^2 - 2\alpha r\cos\theta + \alpha^2 - \beta^2 = 0,$$

or say

$$1 - \frac{2\alpha\cos\theta}{r} + \frac{\alpha^2-\beta^2}{r^2} = 0,$$

and the transformed equation is therefore

$$1 - \alpha\cos\theta\left(\rho+\frac{1}{\rho}\right) + \tfrac{1}{4}(\alpha^2-\beta^2)\left(\rho+\frac{1}{\rho}\right)^2 = 0,$$

that is

$$(\alpha^2-\beta^2)(\rho^2+1)^2 - 4\alpha\cos\theta\rho(\rho^2+1) + 4\rho^2 = 0,$$

or in rectangular coordinates

$$\rho^4 + 2\rho^2 + 1 - \frac{4\alpha}{\alpha^2-\beta^2}x(\rho^2+1) + \frac{4}{\alpha^2-\beta^2}\rho^2 = 0,$$

that is

$$\rho^4 - \frac{4\alpha}{\alpha^2-\beta^2} x\rho^2 + \left(2 + \frac{4}{\alpha^2-\beta^2}\right)\rho^2 - \frac{4\alpha}{\alpha^2-\beta^2} x + 1 = 0,$$

where $\rho^2 = x^2 + y^2$; the form of the equation shows that the curve is a bicircular quartic. Writing for shortness $\frac{2}{\alpha^2-\beta^2} = m$, the equation is

$$\rho^4 - 2m\alpha x\rho^2 + (2+2m)\rho^2 + 1 - 2m\alpha x = 0,$$

that is

$$\{\rho^2 - m(\alpha x - 1) + 1\}^2 - m^2(\alpha x - 1)^2 - 2m = 0,$$

or, what is the same thing,

$$\{(x - \tfrac{1}{2}m\alpha)^2 + y^2 - \tfrac{1}{4}m^2\alpha^2 + m + 1\}^2 - m^2(\alpha x - 1)^2 - 2m = 0,$$

which putting $x + \frac{1}{2}m\alpha$ for x is

$$(x^2 + y^2 - \tfrac{1}{4}m^2\alpha^2 + m + 1)^2 - m^2(\alpha x + \tfrac{1}{2}m\alpha^2 - 1)^2 - 2m = 0,$$

viz. the terms of the fourth order being $(x^2+y^2)^2$, and there being no terms of the third order, the curve represented by this equation is a bicircular quartic.

Article Nos. 21 to 30. *Practical Details and Application to Eclipse of December* 21–22, 1870.

21. There are some practical details which it is proper to explain, using to fix the ideas the eclipse of December 21–22, 1870: the constant value of Δ (see *infrà*) is taken to be $+90° + 22° 35'$[1].

I have a blank projection (radius 12 in. as above) with the meridians and parallels each at intervals of 5°. And also another blank form which has on it merely a circle, radius 12 in., graduated as to one quadrant thereof with lines about $1\frac{1}{2}$ in. long, inwards towards the centre. It contains also in a corner the foregoing proportional-scale figure.

22. On the blank projection I measure off, downwards from the centre, a distance $M_1M = 12 \sin 22° 35' \,(= 4{\cdot}61)$, distances all in inches; and then with the centre M and radius $MC = 12 \cos 22° 35' \,(= 11{\cdot}08)$, describe a circle which is the bounding circle of the reduced projection. With this same radius I describe on the second form, concentric with the 12-inch circle and above the horizontal diameter thereof, a semicircle: and, cutting out the included area, replace it with tracing cloth. The form thus prepared is placed over the blank projection, so that the semicircle shall coincide with the corresponding semicircle on the blank projection, and the two sheets are fixed together by their lower edges, and by folding down the remaining sides. We have thus the upper half of the reduced projection, represented by the semicircle, with the

[1] See Plate, which exhibits in dotted lines the blank projection under the other blank form; the part within the red semicircle, as seen through the tracing cloth, the rest really hidden.

meridians and parallels marked out thereon by lines seen through the tracing cloth. See the Plate; the dotted line shows a paper scale afterwards affixed to the second form or upper sheet. Observe that so far the only eclipse-datum made use of is the value $\Delta = 90° + 22° \, 35'$.

23. We have for the eclipse in question, taking t for the G.M.T. in hours, positive or negative according as the time is after or before G.M. Noon, Dec. 22, and h' also in hours,

$$h' = 0^{\text{h}}\cdot 02 + t \cdot 9996,$$

and then taking the values of α, α', Δ, Δ' from the N.A. we have as follows:

G.M.T. 1870, Dec.	$h'+2^{\text{h}}=$	$\alpha =$	$\alpha' =$	$\Delta = 90° +$	$\Delta' = 90° +$	$h+2^{\text{h}} = h'+2^{\text{h}}+\alpha'-\alpha$	$\Delta'-\Delta$ in Minutes of Arc.	$h'-h$ in ditto.
d h	h m s	h m s	h m s	° ′ ″	° ′ ″	h m s	′	′
21 22	0 1 13·42	17 56 1·84	18 1 48·72	22 27 8·5	23 27 17·1	0 7 0·30	60·143	− 86·620
22 3	5 1 7·37	18 9 26·74	18 2 44·27	22 43 12·5	23 27 13·9	4 54 24·90	44·023	+ 100·632

and moreover

$$\begin{array}{lrl} \text{Moon's Parallax} & \sigma' = & 60' \, 38''\cdot 6 \\ \text{Sun's ditto} & \pi' = & 9 \cdot 1 \\ \hline & \sigma' - \pi' = & 60 \;\; 29 \cdot 5 = 60'\cdot 49 \\ & P' & = 60 \cdot 39 \end{array}$$

$$\begin{array}{lrl} \text{Moon's Semidiam.} & s = & 16' \, 33''\cdot 2 \\ \text{Sun's ditto} & s' = & 16 \;\; 17 \cdot 9 \\ \hline & s + s' = & 32 \;\; 51 \cdot 2 = 32'\cdot 85 \end{array}$$

whence

$$\frac{s+s'}{P'} \, 12 \text{ in.} = \quad 6\cdot 53$$

viz. this is the radius of the circles used in the construction of the penumbral curves.

24. We have for x, y the formulæ

$$x = \frac{\Delta' - \Delta}{P'} \, 12 \text{ in.} + y \, (h' - h) \, 0\cdot 00006,$$

$$y = \frac{h' - h}{P'} \; 12 \text{ in.} \times \sin \Delta,$$

viz. I find Dec. 21, 22^{h},

$$x = 11\cdot 95 - \cdot 06 = \quad 11\cdot 89,$$

$$y \qquad\qquad\qquad -15\cdot 80,$$

and Dec. 22—3^h,

$$x = 8\cdot75 + \cdot11 = 8\cdot85,$$

$$y = +18\cdot45,$$

where I have taken account of the small corrections to the approximate values of x: it may be added that, using the conjunction-value $52' \, 9''\cdot4$ of $\Delta' - \Delta$, we have at conjunction,

$$x = 10\cdot36, \quad y = 0.$$

25. We thus lay down on the relative orbit the two points 22^h and 3^h, and the point of conjunction or intersection with the axis of x; the three points are found to be sensibly in a straight line: the distance between the extreme points is about 34 inches, representing 5 hours, so that the scale is nearly 7 inches to an hour: the line is then graduated to quarters of an hour. We then, by means of the distance $12 + 6\cdot53 = 18\cdot53$, mark off on the relative orbit, the points B, E, which correspond to the beginning and end of the eclipse respectively: the times as read off from my figure, and compared with the true times given in the N. A. are

	from figure	N. A.
Beginning	$22^h \, 12^{m}\cdot5$	22 13·6
End	2 40 ·5	2 41·1

26. With centre B describing a circle radius 6·53 this will of course just touch the 12-inch circle, and the penumbral curve will be a mere point, viz., this is the point B' on the bounding circle, opposite to the point of contact. And so with centre E describing a circle of the same radius 6·53, that will just touch the 12-inch circle, and the penumbral curve will be a mere point, viz., the point E' on the bounding circle, opposite the point of contact.

27. I draw the circles corresponding to the times $22^h \, 30^m$, 45^m, $23^h \, 0^m$, viz., so much of each as lies within the 12-inch circle. Each of these is then transformed into a penumbral curve, drawn in the upper semicircle on the tracing cloth. For this purpose we construct a straight edge of paper, the one half graduated for $12 \sin\theta$, the other half for $11\cdot08 \tan\frac{1}{2}\theta$, by means of the proportional-scale figure, as already explained: $\theta = 0°$ to $90°$ at intervals of $5°$, is quite sufficient; the points on any particular penumbral curve are laid down in pairs with the utmost facility, and the curve is traced by hand from 4 or 5 pairs of points.

28. We then graduate for latitude; viz., we see through the tracing cloth, the equator cutting the vertical radius in Q, and a parallel cutting the same radius, say in R; drawing lines from C, we refer these to the points q, r on the bounding circle, viz., on the quadrant thereof which is graduated by means of the graduation-lines of the 12-inch circle; and we thus read off the latitude of the parallel in question; this latitude is then marked for each parallel on the vertical radius from Q up to the bounding circle, viz., not on the tracing cloth, but on the paper affix; and we then on this same radius (on the paper affix) interpolate the positions where this would be intersected by the parallels for the colatitudes, $5°$, $10°$, $15°$, &c. Or

(what is perhaps better) we may without marking the latitudes of the parallels of the blank form, construct directly the last-mentioned graduations; viz., marking off on the bounding circle from the point q, equal intervals of 5°, and from any such mark drawing to C, a line to meet the vertical radius, the point of intersection is the point belonging to the parallel, latitude equal to the corresponding multiple of 5°.

29. Finally, we must (not on the tracing cloth but on the paper affix) graduate an arc of the equator for the position of the meridian of Greenwich, that is for h. We have

$$\text{At } 22^{\text{h}} \quad h = -2^{\text{h}} + 0^{\text{h}}\ 7^{\text{m}}\ 0^{\text{s}}{\cdot}30 = -1^{\text{h}}\ 52^{\text{m}}\ 52^{\text{s}}{\cdot}30 = -28^{\circ}\ 13'{\cdot}08$$

$$\text{At } 3^{\text{h}} \quad h = -2^{\text{h}} + 4\ 54\ 24{\cdot}90 = +2\ 54\ 24{\cdot}90 = +43\ 36{\cdot}20$$

The equator is already graduated in longitude by means of the meridians of the blank projection: hence we lay down the marks for 22^{h} and 3^{h} in the positions belonging to $-28^{\circ}\,13'$, and $+43^{\circ}\,36'$ respectively. And then since the interval of 5 hours answers to $71^{\circ}\,49'$, that of 1 hour will answer to $14^{\circ}\,22'$, so that, measuring off these intervals of longitude, we have the marks for the intermediate times 23^{h}, 0^{h}, 1^{h}, 2^{h}; or it might be proper to find in this way the marks corresponding to each interval of 20^{m} of time, answering to about 5° of longitude; the further subdivisions would be proportional to the intervals of time.

30. I have in this way read off the positions of the points B' and E' belonging to the beginning and end of the eclipse; the values, as compared with the true ones, are

			From Figure	N. A.
			°	° ′
B′	latitude	N.	34	35 37
	longitude	W.	47	45 44
E′	latitude	N.	26	26 5
	longitude	W.	38½	37 16

I remark that my figure, although drawn carefully, is not drawn with anything like the degree of accuracy which would be easily attainable; and I think that far better results might be obtained. I merely from a scale lay down tenths and estimate hundredths of an inch, but certainly fiftieths might be laid down from a scale.

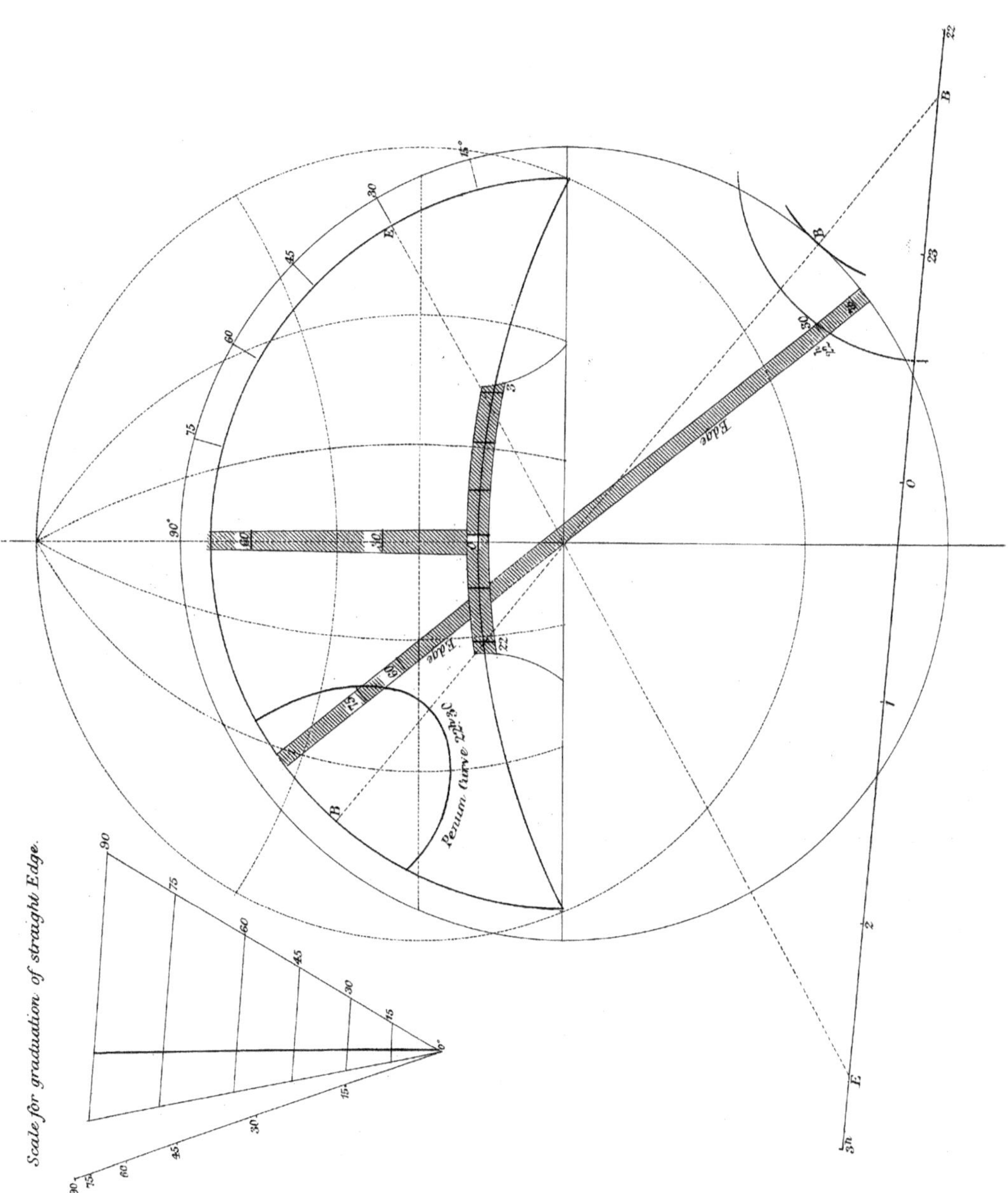
Scale for graduation of straight Edge.

478.

ON THE GEODESIC LINES ON AN ELLIPSOID.

[From the *Memoirs of the Royal Astronomical Society*, vol. XXXIX. (1872), pp. 31—53. Read January 13, 1871.]

THE fundamental equations, in regard to the geodesic lines on an ellipsoid, were established by Jacobi, viz., representing by a, b, c, the squares of the semiaxes, that is, taking the ellipsoid to be

$$\frac{x^2}{a}+\frac{y^2}{b}+\frac{z^2}{c}=1$$

(where $a>b>c$), if we introduce the elliptic coordinates h, k, and write

$$\frac{x^2}{a+h}+\frac{y^2}{b+h}+\frac{z^2}{c+h}=1,$$

$$\frac{x^2}{a+k}+\frac{y^2}{b+k}+\frac{z^2}{c+k}=1,$$

or, what is the same thing,

$$x^2=\frac{a(a+h)(a+k)}{(a-b)(a-c)},$$

$$y^2=\frac{b(b+h)(b+k)}{(b-c)(b-a)},$$

$$z^2=\frac{c(c+h)(c+k)}{(c-a)(c-b)};$$

then, if β be an arbitrary constant, the differential equation of a geodesic line is

$$(1)\quad \text{const.}=\int dh\sqrt{\frac{h}{(a+h)(b+h)(c+h)(\beta+h)}}+\int dk\sqrt{\frac{h}{(a+k)(b+k)(c+k)(\beta+k)}},$$

and the expression for the length of any arc of the curve is given by

$$(2)\quad s=\int dh\sqrt{\frac{h(\beta+h)}{(a+h)(b+h)(c+h)}}+\int dk\sqrt{\frac{k(\beta+k)}{(a+k)(b+k)(c+k)}}.$$

I propose in the present Memoir to develope the theory to the extent of showing how we can, by means of the first of these equations, explain the course of the geodesic lines; and for given numerical values of a, b, c, calculate, construct, and exhibit in a drawing the course of these lines: I attend more particularly to the series of geodesic lines through an umbilicus (which lines pass also through the opposite umbilicus), and to the case where the semiaxes are connected by the equation $ac-b^2=0$, a relation which simplifies the formulæ.

General Considerations as to the Course of the Lines.

1. It will be observed that h and k enter into the formulæ symmetrically: it will be convenient to distinguish between these coordinates by considering h as extending between the values $-a$, $-b$; and k as extending between the values $-b$, $-c$. Thus:

$h=$ const. denotes a curve of curvature of the one kind, viz.:

$h=-a$, the principal section ABA' (or major-mean section), $h=-b$, the curves UU' and $U''U'''$ (or portions of the umbilicar section $ACA'C'$); similarly,

$k=$ const. denotes a curve of curvature of the other kind, viz.:

$k=-c$, the principal section CBC' (or minor-mean section), $k=-b$, the curves UU''' and $U'U''$ (remaining portions of the umbilicar section $ACA'C'$).

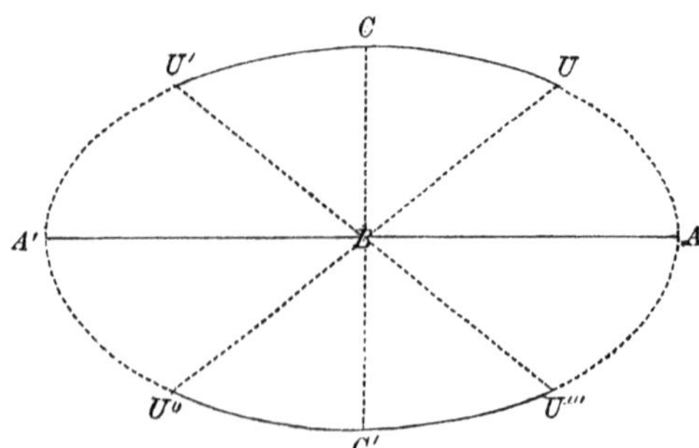

2. To any given (admissible) values of h, k there correspond eight points, situate in the eight octants of the surface respectively; but, unless the contrary is expressed, it is assumed that the coordinates x, y, z, are positive, and that the point is situate in the octant ABC.

3. The constant β may have any value from $+a$ to $+c$; viz., if it has a value between a and b, or say, if $-\beta$ has an h-value, then the geodesic lines wholly

between the two ovals of the curve of curvature $h = -\beta$ (being in general an indefinite undulating curve touching each oval an indefinite number of times). Similarly, if β has any value between b and c, or say, if $-\beta$ has a k-value, then the geodesic line lies wholly between the two ovals of the curve of curvature $k = -\beta$ (being in general an indefinite undulating curve touching each oval an indefinite number of times). The intermediate case is when $\beta = b$, or say when $-\beta$ has the umbilicar value: here the geodesic line is in general an indefinite undulating curve passing an infinite number of times through the opposite umbilici U, U'', or U', U'''; to fix the ideas, say through U, U''.

Lines through an Umbilicus.

4. I attend in particular to the last-mentioned case, and thus write $\beta = b$. We may in the formula (1) fix at pleasure a limit of each integral; and writing for convenience

$$\Pi(h) = \int_{-a}^{h} \frac{-dh}{b+h}\sqrt{\frac{h}{(a+h)(c+h)}},$$

$$\Psi(k) = \int_{k}^{-c} \frac{dk}{b+k}\sqrt{\frac{k}{(a+k)(c+k)}},$$

the equation (1) becomes

$$\text{Const.} = \Pi(h) + \Psi(k).$$

5. It is to be observed, in regard to these integrals, that writing $h = -a + u$, we have

$$\Pi(h) = \int_{0}^{u} \frac{du}{a-b-u}\sqrt{\frac{a-u}{u(a-c-u)}},$$

which, for u small, is

$$= \frac{1}{a-b}\sqrt{\frac{a}{a-c}}\int_{0}^{u}\frac{du}{\sqrt{u}}, \quad = \frac{2\sqrt{u}}{a-b}\sqrt{\frac{a}{a-c}}.$$

By the assistance of this formula the value of the integral may be calculated by quadratures; viz., the formula gives the integral for any small value of u, and we can then proceed by the method of quadratures. The integral becomes infinite for $h = -b$: suppose that we have by quadratures calculated it up to $h = -b - m$ (m small), then to calculate it up to any value $-b - m + u$ nearer to $-b$, we have

$$\Pi(h) = \Pi(-b-m) + \int_{0}^{u} \frac{du}{m-u}\sqrt{\frac{b+m-u}{(a-b-m+u)(b-c+m-u)}}$$

$$= \Pi(-b-m) + \sqrt{\frac{b}{(a-b)(b-c)}}\int_{0}^{u}\frac{du}{m-u}$$

$$= \Pi(-b-m) - \sqrt{\frac{b}{(a-b)(b-c)}}\log\left(1-\frac{u}{m}\right), \text{[1]}$$

where the second term is positive, and the value thus increases slowly with u, becoming as it should do $= \infty$ for $u = m$ or $h = -b$.

[1] Except when the contrary is stated, the symbol "log" denotes throughout the hyperbolic logarithm.

6. Similarly in the second integral writing $k=-c-v$, we have

$$\Psi(k)=\int_0^v \frac{dv}{b-c-v}\sqrt{\frac{c+v}{(a-c-v)\,v}},$$

which, for v small, is

$$=\frac{1}{b-c}\sqrt{\frac{c}{a-c}}\int\frac{dv}{\sqrt{v}},\quad =\frac{2\sqrt{v}}{b-c}\sqrt{\frac{c}{a-c}},$$

which is of the like assistance in regard to the calculation by quadratures. And if we have by quadratures calculated the integral up to $h=-b+n$ (n small), then, to calculate it up to any value $-b+n-v$ nearer to $-b$, we have

$$\Psi(k)=\Psi(-b+n)+\int_0^v \frac{dv}{n-v}\sqrt{\frac{b-n+v}{(a-b+n-v)(b-c-n+v)}}$$

$$=\Psi(-b+n)+\sqrt{\frac{b}{(a-b)(b-c)}}\int_0^v\frac{dv}{n-v}$$

$$=\Psi(-b+n)-\sqrt{\frac{b}{(a-b)(b-c)}}\log\left(1-\frac{v}{n}\right),$$

where the second term is positive, and the value thus increases slowly with v, becoming as it should do $=\infty$ for $v=n$, or $k=-b$.

7. It may be remarked that in $\Pi(h)$ and $\Psi(k)$ respectively the coefficient of the logarithmic term has in each case the same value $=\sqrt{\frac{b}{(a-b)(b-c)}}$. As regards the initial terms $\sqrt{u}$ and $\sqrt{v}$, the coefficients are $\frac{1}{a-b}\sqrt{\frac{a}{a-c}}$ and $\frac{1}{b-c}\sqrt{\frac{c}{a-c}}$ respectively, which are equal if $\frac{a}{(a-b)^2}=\frac{c}{(b-c)^2}$, or $ac-b^2=0$.

8. We may consider the two geodesic lines $\Pi(h)\pm\Psi(k)=\text{const.}$; suppose that these each of them pass through the point P, coordinates (h_0, k_0) in the ABC octant of the ellipsoid; then for one of them we have $\Pi(h)-\Psi(k)=\Pi(h_0)-\Psi(k_0)$, and for the other of them we have $\Pi(h)+\Psi(k)=\Pi(h_0)+\Psi(k_0)$: I attend first to the former of these, say $\Pi(h)-\Psi(k)=C$ (where C is $=\Pi(h_0)-\Psi(k_0)$); and I say that this denotes the curve UPU''. In fact, by reason of the equation $\Pi(h)$ and $\Psi(k)$ must both increase or both diminish; they both increase as h passes from h_0 to $-b$, and as k passes from k_0 to $-b$: we may have $h=-b+u$, $k=-b+v$ where u and v are both indefinitely small, the functions Π and Ψ being then indefinitely large, but $\Pi-\Psi=C$; and we have thus a series of points nearer and nearer to the umbilicus U; that is, we have the portion PU of the curve. Tracing the curve in the opposite direction, or considering h as passing from h_0 to $-a$, and k as passing from k_0 to $-c$, then if C be positive, k will attain the value $-c$, before h attains the value $-a$, say that we have simultaneously $h=h_1$, $k=-c$; the equation is $\Pi(h_1)-\Psi(-c)=C$, that is, $\Pi(h_1)=C$; and the geodesic line then arrives at a point P_1 on the arc CB of the minor-mean

principal section. The function Ψ then changes its sign, viz., considering it as always positive, the equation is now $\Pi(h)+\Psi(k)=C$, k passing from the value $-c$ towards $-b$, that is, $\Psi(k)$ increasing, and therefore $\Pi(h)$ diminishing, or h passing from h_1 towards the value $-a$; until at last, say for $k=k_2$, we have $h=-a$, that is, $C=\Pi(-a)+\Psi(k_2)$, or $C=\Psi(k_2)$; the geodesic line here arrives at a point P_2 on the arc BA' of the major mean principal section. The function Π then changes its sign, viz., Π denoting a positive function as before, the equation is $-\Pi(h)+\Psi(k)=C$; h passes from $-a$ towards $-b$, that is $\Pi(h)$ increases, and therefore $\Psi(k)$ must also increase, or k pass from k_2 towards $-b$: we have at length $h=-b-u$, $k=-b+v$, u and v being each indefinitely small; and therefore Π and Ψ each indefinitely large (but $-\Pi+\Psi=C$); that is, we arrive at the umbilicus U'', completing the geodesic line UPU''.

9. If instead of $C=+$ we have $C=-$, everything is similar, but the geodesic line proceeding from U in the direction UP will first cut the arc BA of the major mean section at a point P_1; then the arc BC' of the minor mean section at a point P_2; and, finally, arrive as before at the umbilicus U'''.

10. The intermediate case is when $C=0$, viz., we have here $\Pi(h)-\Psi(k)=0$; the geodesic line here passes from U in the direction UP to B (extremity of the mean axis, $h=-a$, $k=-c$); Π and Ψ then each change their sign, so that, considering them as positive, the equation still is $\Pi(h)-\Psi(k)=0$, and the geodesic line at last arrives at the umbilicus U''. It will be easily understood how in the like manner $\Pi(h)+\Psi(k)=C$ refers to the line $U'PU'''$.

11. Reverting to the equation $\Pi(h)-\Psi(k)=C$, or as I will now write it

$$\Pi(h)-\Psi(k)=\Pi(h_0)-\Psi(k_0),$$

which belongs to the portion UP of the geodesic line UPU'', we require when h is $=-b-u$, and $k=-b+v$ (u and v indefinitely small) to know the ratio of the increments u, v; this in fact serves to determine the direction at U of the geodesic line through the given point (h_0, k_0).

12. For this purpose writing $h=-b-u$, we find

$$\Pi(h)=\int_u^{a-b}\frac{du}{u}\sqrt{\frac{b+u}{(a-b-u)(b-c+u)}},$$

which is

$$=\int_u^{a-b}\frac{du}{u}\left\{\sqrt{\frac{b+u}{(a-b-u)(b-c+u)}}-\sqrt{\frac{b}{(a-b)(b-c)}}\right\}$$

$$+\sqrt{\frac{b}{(a-b)(b-c)}}\log\frac{a-b}{u},$$

and, when u is indefinitely small, this is

$$\Pi(h)=\int_0^{a-b}\frac{du}{u}\left\{\sqrt{\frac{b+u}{(a-b-u)(b-c+u)}}-\sqrt{\frac{b}{(a-b)(b-c)}}\right\}+\sqrt{\frac{b}{(a-b)(b-c)}}\log\frac{a-b}{u}.$$

Similarly, when $k=-b+v$, where v is indefinitely small

$$\Psi(k)=\int_0^{b-c}\frac{dv}{v}\left\{\sqrt{\frac{b-v}{(a-b+v)(b-c-v)}}-\sqrt{\frac{b}{(a-b)(b-c)}}\right\}+\sqrt{\frac{b}{(a-b)(b-c)}}\log\frac{b-c}{v}.$$

13. Each of the integrals is of the dimension $-\frac{1}{2}$ in a, b, c, and the difference of the integrals may be represented by

$$M\sqrt{\frac{b}{(a-b)(b-c)}};$$

we have therefore

$$\Pi(h)-\Psi(k)=\sqrt{\frac{b}{(a-b)(b-c)}}\left\{M+\log\frac{a-b}{b-c}\frac{v}{u}\right\},$$

where

$$\sqrt{\frac{b}{(a-b)(b-c)}}M=\int_0^{a-b}\frac{du}{u}\left\{\sqrt{\frac{b+u}{(a-b-u)(b-c+u)}}-\sqrt{\frac{b}{(a-b)(b-c)}}\right\}$$

$$-\int_0^{b-c}\frac{dv}{v}\left\{\sqrt{\frac{b-v}{(a-b+v)(b-c-v)}}-\sqrt{\frac{b}{(a-b)(b-c)}}\right\}.$$

14. Suppose the inferior limits replaced by the indefinitely small positive quantities ϵ, ϵ' respectively; and for the variable in the second integral write $-u$; then

$$M=\int_{-(b-c)}^{a-b}\left\{\frac{du}{u}\sqrt{\frac{b+u}{(a-b-u)(b-c+u)}}-\sqrt{\frac{b}{(a-b)(b-c)}}\right\},$$

it being understood that the values $u=-\epsilon'$ to $u=+\epsilon$ are omitted from the integration: this is

$$=\int_{-(b-c)}^{a-b}\frac{du}{u}\sqrt{\frac{b+u}{(a-b-u)(b-c+u)}}-\sqrt{\frac{b}{(a-b)(b-c)}}\log\frac{a-b}{\epsilon}\frac{\epsilon'}{b-c}$$

with the same convention as to the integral; or if $\epsilon'=\epsilon$, then

$$M=M'-\sqrt{\frac{b}{(a-b)(b-c)}}\log\frac{a-b}{b-c},$$

where

$$\sqrt{\frac{b}{(a-b)(b-c)}}M'=\int_{-(b-c)}^{a-b}\frac{du}{u}\sqrt{\frac{b+u}{(a-b-u)(b-c+u)}}=\int_{-a}^{-c}\frac{dh}{b+h}\sqrt{\frac{h}{(a+h)(c+h)}},$$

the omitted elements being from $u=-\epsilon$ to $u=+\epsilon$; that is (in the language of Cauchy) we take for the integral its *principal* value. And hence

$$\Pi(h)-\Psi(k)=\sqrt{\frac{b}{(a-b)(b-c)}}\left\{M'+\log\frac{v}{u}\right\}.$$

15. By what precedes this is $=\Pi(h_0)-\Psi(k_0)$; or if we write simply (h, k) instead of (h_0, k_0), that is, consider the geodesic line UP, which is drawn from the

point P, coordinates (h, k), to the umbilicus U, the coordinates of a point consecutive to the umbilicus are $-b-u$, $-b+v$, where u, v are connected by the last-mentioned equation, in which M' is a transcendental function depending on (a, b, c) but independent of the particular geodesic line.

16. If for the geodesic line through the point B, or say for the B-geodesic $\frac{v}{u}=\frac{v_0}{u_0}$, then $M'=-\log\frac{v_0}{u_0}$, and we have in general

$$\Pi(h)-\Psi(k)=\sqrt{\frac{b}{(a-b)(b-c)}}\log\frac{v\,u_0}{u\,v_0},$$

a result which I proceed to further transform as follows:

If x_0, y_0, z_0 refer to the umbilicus U, then considering first the consecutive point P on the geodesic line (coordinates $-b-u$, $-b+v$) and next the consecutive

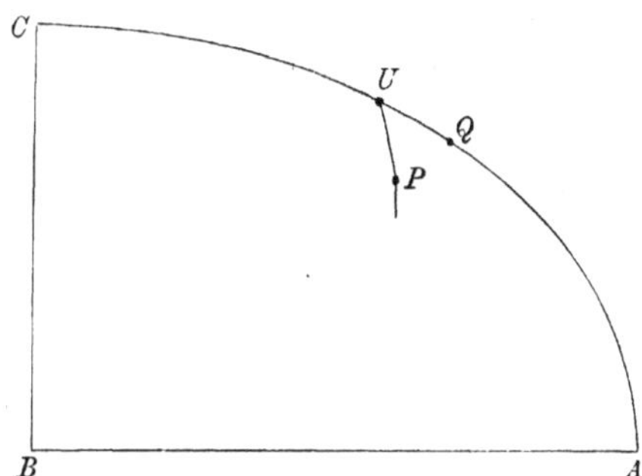

point Q on the umbilicar section, we have for these two points respectively,

$$dx_0=\frac{\frac{1}{2}\sqrt{a}\,(v-u)}{\sqrt{(a-b)(a-c)}},$$

$$dy_0=\frac{\sqrt{b}\,\sqrt{uv}}{\sqrt{(a-b)(b-c)}},$$

$$dz_0=\frac{\frac{1}{2}\sqrt{c}\,(u-v)}{\sqrt{(b-c)(a-c)}},$$

$$\delta x_0=\frac{\frac{1}{2}\sqrt{a}}{\sqrt{(a-b)(a-c)}},$$

$$\delta y_0=0,$$

$$\delta z_0=\frac{-\frac{1}{2}\sqrt{c}}{\sqrt{(b-c)(a-c)}},$$

say these are α, β, γ, and α', β', γ'; and then

$$\alpha\alpha' + \beta\beta' + \gamma\gamma' = \tfrac{1}{4}\left\{\frac{a}{(a-b)(a-c)} + \frac{c}{(b-c)(a-c)}\right\}(v-u),$$

$$= \frac{\frac{1}{4}(v-u)}{a-c}\left\{\frac{a}{a-b} + \frac{c}{b-c}\right\}, \quad = \frac{\frac{1}{4}b(v-u)}{(a-b)(b-c)},$$

$$\alpha^2 + \beta^2 + \gamma^2 = \left\{\frac{a}{(a-b)(a-c)} + \frac{c}{(b-c)(a-c)}\right\}.\tfrac{1}{4}(u-v)^2 + \frac{buv}{(a-b)(b-c)}$$

$$= \frac{b}{(a-b)(b-c)}.\tfrac{1}{4}(u+v)^2,$$

$$\alpha'^2 + \beta'^2 + \gamma'^2 = \frac{b}{(a-b)(b-c)}.\tfrac{1}{4},$$

whence

$$\sqrt{\alpha^2+\beta^2+\gamma^2}\sqrt{\alpha'^2+\beta'^2+\gamma'^2} = \frac{\frac{1}{4}(u+v)\,b}{(a-b)(b-c)},$$

and hence

$$\cos\phi = \frac{\alpha\alpha' + \beta\beta' + \gamma\gamma'}{\sqrt{\alpha^2+\beta^2+\gamma^2}\sqrt{\alpha'^2+\beta'^2+\gamma'^2}} = \frac{v-u}{v+u},$$

that is,

$$\cos(180^\circ - \phi) = \frac{u-v}{u+v}, \quad \text{or} \quad \tan^2\tfrac{1}{2}\phi = \frac{u}{v},$$

where if U is the umbilicus, P the consecutive point $-b-u$, $-b+v$, and UQ the element of the umbilicar principal section, $\phi = \angle PUQ$, $180^\circ - \phi = \angle PUQ'$. For the B-geodesic we have

$$2\log\tan\tfrac{1}{2}\phi_0 = \log\frac{u_0}{v_0} = M'.$$

17. The foregoing equation for $\Pi(h) - \Psi(k)$ now becomes

$$\Pi(h) - \Psi(k) = \sqrt{\frac{b}{(a-b)(b-c)}}\log\frac{\tan^2\frac{1}{2}\phi_0}{\tan^2\frac{1}{2}\phi};$$

viz., ϕ_0 is the south azimuth of the B-geodesic at the umbilicus, a mere function of (a, b, c) and ϕ is the south azimuth at the umbilicus, of the geodesic line under consideration, so that we may consider the geodesic line to be determined by the south azimuth ϕ as its parameter.

Formulæ for the case $ac - b^2 = 0$.

18. I annex the following investigation in regard to the case $ac - b^2 = 0$.

We have in general

$$\frac{1}{\sqrt{b(a-b)(b-c)}}\frac{d}{dx}\log\frac{\sqrt{-x(a-b)(b-c)}+\sqrt{-b(a+x)(c+x)}}{\sqrt{-x(a-b)(b-c)}-\sqrt{-b(a+x)(c+x)}}$$

$$= -\frac{1}{b}\frac{1}{\sqrt{x(a+x)(c+x)}}$$

$$+\frac{1}{b}\frac{1}{b+x}\sqrt{\frac{x}{(a+x)(c+x)}}$$

$$+\frac{1}{bx+ac}\sqrt{\frac{x}{(a+x)(c+x)}}.$$

In fact, denoting the logarithm by $\log\dfrac{P+Q}{P-Q}$, we have

$$\frac{d}{dx}\log\frac{P+Q}{P-Q}=\frac{2(PQ'-P'Q)}{P^2-Q^2}$$

where

$$2(PQ'-P'Q)=2PQ\left(\frac{Q'}{Q}-\frac{P'}{P}\right)=\sqrt{x(a+x)(c+x)b(a-b)(b-c)}\left\{\frac{1}{a+x}+\frac{1}{c+x}-\frac{1}{x}\right\}$$

$$=\frac{\sqrt{b(a-b)(b-c)}}{\sqrt{x(a+x)(c+x)}}(x^2-ac);$$

$$P^2-Q^2=-x(a-b)(b-c)+b(a+x)(c+x)$$

$$=(bx+ac)(b+x);$$

that is

$$\frac{2(PQ'-P'Q)}{P^2-Q^2}=-\frac{\sqrt{b(a-b)(b-c)}}{\sqrt{x(a+x)(c+x)}}\frac{x^2-ac}{(bx+ac)(b+x)},$$

$$=\frac{\sqrt{b(a-b)(b-c)}}{\sqrt{x(a+x)(c+x)}}\left\{-\frac{1}{b}+\frac{x}{b(b+x)}+\frac{x}{bx+ac}\right\},$$

which proves the theorem.

19. Hence in the particular case $ac = b^2$ we have

$$\frac{1}{\sqrt{b(a-b)(b-c)}}\log\frac{\sqrt{-h(a-b)(b-c)}+\sqrt{-b(a+h)(c+h)}}{\sqrt{-h(a-b)(b-c)}-\sqrt{-b(a+h)(c+h)}},$$

$$=-\frac{1}{b}\int_{-a}^{h}\frac{dh}{\sqrt{h(a+h)(c+h)}}$$

$$-\frac{2}{b}\int_{-a}^{h}\frac{dh}{b+h}\sqrt{\frac{h}{(a+h)(c+h)}}\left(=+\frac{2}{b}\Pi(h)\right),$$

that is

$$\Pi(h) = \tfrac{1}{2}\int_{-a}^{h} \frac{dh}{\sqrt{h(a+h)(c+h)}} + \tfrac{1}{2}\sqrt{\frac{b}{(a-b)(b-c)}} \log \frac{\sqrt{-h(a-b)(b-c)} + \sqrt{-b(a+h)(c+h)}}{\sqrt{-h(a-b)(b-c)} - \sqrt{-b(a+h)(c+h)}},$$

or say

$$= \tfrac{1}{2}\int_{-a}^{h} \frac{dh}{\sqrt{h(a+h)(c+h)}} + \tfrac{1}{2}\sqrt{\frac{b}{(a-b)(b-c)}} \log \frac{1+H}{1-H},$$

where

$$H^2 = \frac{b}{(a-b)(b-c)} \frac{(a+h)(c+h)}{h},$$

viz. we see that $\Pi(h)$ depends on the more simple integral

$$\int_{-a}^{h} \frac{dh}{\sqrt{h(a+h)(c+h)}}.$$

20. Similarly

$$\frac{1}{\sqrt{b(a-b)(b-c)}} \log \frac{\sqrt{-k(a-b)(b-c)} + \sqrt{-b(a+k)(c+k)}}{\sqrt{-k(a-b)(b-c)} - \sqrt{-b(a+k)(c+k)}}$$

$$= \frac{1}{b}\int_{k}^{-c} \frac{dk}{\sqrt{k(a+k)(c+k)}}$$

$$+ \frac{2}{b}\int_{k}^{-c} \frac{dk}{b+k}\sqrt{\frac{k}{(a+k)(c+k)}} \left(= + \frac{2}{b}\Psi(k)\right),$$

that is

$$\Psi(k) = -\tfrac{1}{2}\int_{k}^{-c} \frac{dk}{\sqrt{k(a+k)(c+k)}} + \tfrac{1}{2}\sqrt{\frac{b}{(a-b)(b-c)}} \log \frac{\sqrt{-k(a-b)(b-c)} + \sqrt{-b(a+k)(c+k)}}{\sqrt{-k(a-b)(b-c)} - \sqrt{-b(a+k)(c+k)}},$$

or say

$$\Psi(k) = -\tfrac{1}{2}\int_{k}^{-c} \frac{dk}{\sqrt{k(a+k)(c+k)}} + \tfrac{1}{2}\sqrt{\frac{b}{(a-b)(b-c)}} \log \frac{1+K}{1-K},$$

where

$$K^2 = \frac{b}{(a-b)(b-c)} \frac{(a+k)(c+k)}{k},$$

that is, $\Psi(k)$ depends on the more simple integral,

$$\int_{k}^{-c} \frac{dk}{\sqrt{k(a+k)(c+k)}}.$$

Write $h = -b-u$, $k = -b+v$, where u and v are indefinitely small, then

$$\Pi(h) - \Psi(k) = \tfrac{1}{2}\int_{-a}^{-c} \frac{dh}{\sqrt{h(a+h)(c+h)}} + \tfrac{1}{2}\sqrt{\frac{b}{(a-b)(b-c)}} \log \frac{1+U}{1-U}\,\frac{1-V}{1+V},$$

where

$$U^2 = \frac{\left(1 - \frac{u}{a-b}\right)\left(1 + \frac{u}{b-c}\right)}{1 + \frac{u}{b}} = 1 - \frac{bu^2}{(a-b)(b-c)(b+u)} \text{ (attending to } ac = b^2\text{)},$$

and

$$V^2 = \frac{\left(1 + \frac{v}{a-b}\right)\left(1 - \frac{v}{b-c}\right)}{1 - \frac{v}{b}} = 1 - \frac{bv^2}{(a-b)(b-c)(b-v)},$$

$$\Pi(h) - \Psi(k) = \tfrac{1}{2}\int_{-a}^{-c} \frac{dh}{\sqrt{h(a+h)(c+h)}} + \sqrt{\frac{b}{(a-b)(b-c)}} \log\frac{v}{u}.$$

21. Comparing with the result obtained for the general case the two agree, if only

$$\int_{-a}^{-c} \frac{dh}{b+h}\sqrt{\frac{h}{(a+h)(c+h)}} = \tfrac{1}{2}\int_{-a}^{-c} \frac{dh}{\sqrt{h(a+h)(c+h)}},$$

where on the left-hand side the integral has its principal value: a result which must therefore hold good when $ac = b^2$.

Calculation of the Umbilicar Geodesics for Ellipsoid $a : b : c = 4 : 2 : 1$.

22. As a specimen of the way in which we may, on a given ellipsoid, calculate the course of a geodesic line, I take the semiaxes to be as $2 : \sqrt{2} : 1$, or, for convenience, $a = 1000$, $b = 500$, $c = 250$; and, considering the geodesic lines through the umbilicus, I calculate by quadratures the functions

$$\Pi(h) = 100{,}000 \int_{-1000}^{h} \frac{-dh}{500+h}\sqrt{\frac{h}{(1000+h)(250+h)}},$$

$$\Psi(k) = 100{,}000 \int_{k}^{-250} \frac{dk}{500+k}\sqrt{\frac{k}{(1000+k)(250+k)}}.$$

The results do not pretend to minute accuracy: I have not attempted to estimate or correct for any error occasioned by the intervals (10 units) being too large; and there may possibly be accidental errors.

TABLE I.

$-h=$	Π′	Π (h)	$-h$	Π′	Π (h)	$-h$	Π′	Π (h)
1000	∞	0	840	27·6	6746	630	51·5	13972
999	231·4	462	830	27·8	7023	620	54·1	14499
998	164·0	659	820	27·9	7301	610	59·2	15066
997	134·2	809	810	28·1	7582	600	65·5	15689
996	116·5	934	800	28·4	7865	590	72·1	16377
995	104·4	1044	790	28·7	8151	580	80·9	17142
990	74·6	1492	780	29·2	8440	570	93·0	18011
980	54·0	2135	770	29·7	8735	560	106·8	19010
970	45·1	2630	760	30·3	9035	550	127·6	20183
960	40·1	3056	750	31·0	9341	540	159·1	21616
950	36·6	3439	740	31·8	9655	530	211·5	23469
940	34·2	3794	730	32·6	9977	520	316·7	26111
930	32·5	4127	720	33·6	10308	510	632·7	30858
920	31·2	4446	710	34·7	10650	505	1265·0	35602
910	30·2	4753	700	36·0	11004	504	1581·2	37014
900	29·4	5051	690	37·4	11371	503	2107·7	38834
890	28·8	5342	680	39·0	11754	502	3162·3	41398
880	28·4	5628	670	40·8	12153	501	6324·1	45792
870	28·1	5911	660	42·0	12567	500	∞	∞
860	27·9	6190	650	45·4	13005			
850	27·8	6469	640	48·2	13473			

TABLE II.

$-k$	Ψ'	$\Psi(k)$	$-k$	Ψ'	$\Psi(k)$	$-k$	Ψ'	$\Psi(k)$
250	∞	0	320	45·5	4655	440	107·1	12207
251	232·2	462	330	46·1	5114	450	127·9	13383
252	165·5	661	340	47·3	5581	460	159·2	14818
253	136·0	811	350	48·9	6062	470	211·6	16673
254	118·6	939	360	51·1	6562	480	316·7	19314
255	106·8	1051	370	53·8	7086	490	632·7	24062
260	78·1	1514	380	57·2	7641	495	1265·0	28806
270	60·5	2207	390	61·4	8235	496	1581·2	30218
280	51·7	2768	400	66·7	8875	497	2108·1	32037
290	48·2	3268	410	73·2	9575	498	3162·3	34602
300	46·3	3741	420	81·6	10349	499	6234·1	38995
310	45·5	4200	430	91·4	11214	500	∞	∞

23. But it is obviously convenient to revert these Tables so as to have for the common arguments a series of uniformly increasing values of Π or Ψ, viz., we obtain by interpolation the values of h and k belonging to the given values of Π or Ψ, and thus obtain the following Table. Here, in any line of the Table the values of h, k are such that $\Pi(h) - \Psi(k) = 0$, viz., the values in question belong to successive points of the B-geodesic. And to obtain the values for any other geodesic line $\Pi(h) - \Psi(k) = \pm 500\,m$, we have only to take each value of k from the line m lines above or below the line from which h is taken; and similarly the table gives at once the values belonging to a geodesic line $\Pi(h) + \Psi(k) = 500\,m$.

TABLE III.

Π = Ψ =	h	D.	k	D.	Π = Ψ =	h	D.	k	D.
0	1000		250		13000	650·1		446·7	
		1·2		1·4			20·6		7·8
500	998·8		251·4		14000	629·5		454·5	
		3·4		3·1			18·3		6·5
1000	995·4		254·5		15000	611·2		461·0	
		5·5		5·2			15·7		5·3
1500	989·9		259·7		16000	595·5		466·3	
		7·8		7·3			13·6		4·9
2000	982·1		267·0		17000	581·9		471·2	
		9·5		8·2			11·8		3·8
2500	972·6		275·2		18000	570·1		475·0	
		11·5		9·4			10·0		3·8
3000	961·1		284·6		19000	560·1		478 8	
		12·8		10·3			8·5		2·6
3500	948·3		294·9		20000	551·6		481·4	
		14·5		10·7			7·3		2·2
4000	933·8		305·6		21000	544·3		483·6	
		15·6		11·0			6·2		2·0
4500	918·2		316·6		22000	538·1		485·6	
		16·5		10·9			4·9		2·2
5000	901·7		327·5		23000	533·2		487·8	
		17·2		10·8			4·6		2·1
5500	884·5		338·3		24000	528·6		489·9	
		17·7		10·4			8·1		2·1
6000	866·8		348·7		26000	520·5		492·0	
		17·9		10·1			4·5		2·1
6500	848·9		358·8		28000	516·0		494·1	
		18·1		9·5			4·2		1·7
7000	830·8		368·3		30000	511·8		495·8	
		17·9		9·2			3·0		1·2
7500	812·9		377·5		32000	508·8		497·0	
		17·6		8·6			2·1		0·8
8000	795·3		386·1		34000	506·7		497·8	
		17·3		8·0			2·0		0·5
8500	778·0		394·1		36000	504·7		498·3	
		16·8		7·7			1·2		0·5
9000	761·2		401·8		38000	503·5		498·8	
		16·3		7·1			1·0		
9500	744·9		408·9		39000	. .		499·0	
		15·6		6·6					
10000	729·3		415·5		40000	502·5			
		14·9		6·2			0·6		
10500	714·4		421·7		42000	501·9			
		14·3		5·9			0·5		
11000	700·1		427·6		44000	501·4			
		13·5		5·3			0·4		
11500	686·6		432·9		45800	501·0			
		12·8		5·0					
12000	673·8		437·9		∞	500		500	
		23·7		8·8					

Graphical Construction: Projection on the Umbilicar Plane.

24. The most convenient mode of delineation of the geodesic lines is obtained by projecting them orthogonally on the umbilicar plane: the contour of the figure is here the umbilicar section, or ellipse $\frac{x^2}{a}+\frac{z^2}{c}=1$; and the curves of curvature of each series are projected into elliptic arcs lying within the ellipse in question, the one set cutting at right angles the axes AA', the other cutting at right angles the axes CC'; the equations of the complete ellipses being

$$x^2\frac{a-b}{a(a+h)}+z^2\frac{c-b}{c(c+h)}-1=0$$

and

$$x^2\frac{a-b}{a(a+k)}+z^2\frac{c-b}{c(c+k)}-1=0.$$

25. I constructed, by means of the table, a drawing of this kind for the ellipsoid a, b, $c=1000$, 500, 250, the lengths $\sqrt{a}$ and $\sqrt{c}$ being taken to be 12 inches and 6 inches respectively: the process consists in taking from the table for a series of values $\Pi=\Psi$ (say $\Pi=\Psi=1000$, $=2000$ &c.), the values of h and k, laying down for such values the elliptic arcs which represent the two curves of curvature respectively, thus dividing the bounding ellipse into a series of curvilinear rectangles, and then obtaining the geodesic lines by drawing the diagonals of these rectangles, and of course rounding off the corners so as to form continuous curves. The Plate shows on a reduced scale so much of the drawing as is comprised within a quadrant of the bounding ellipse (viz. it is a representation of an octant of the ellipsoid).

Elliptic-Function Formulæ.

26. I have in all that precedes abstained from the use of elliptic functions, since obviously the form $\sqrt{1-k^2\sin^2\phi}$ of the radical of an elliptic function is in nowise specially appropriate to the present question. But (more particularly in the above-mentioned case $ac-b^2=0$, where the radical is $\sqrt{h(a+h)(c+h)}$ without any exterior factor $b+h$ in the denominator) the formulæ are expressible easily and elegantly by elliptic functions, and it is desirable to make the transformation. Reverting to the formulæ which, in the case in question (viz. when $ac-b^2=0$), give the values of $\Pi(h)$ and $\Psi(k)$; and writing therein $h=-a+(a-c)\sin^2\phi$, $k=-a+(a-c)\sin^2\psi$, also

$$\kappa=\sqrt{1-\frac{c}{a}},\text{ or }\frac{c}{a}=1-\kappa^2,\ =\kappa'^2,$$

we have

$$\int_{-a}^{h}\frac{dh}{\sqrt{h(a+h)(c+h)}}=2\int_0^{\phi}\frac{d\phi}{\sqrt{a-(a-c)\sin^2\phi}}=\frac{2}{\sqrt{a}}F(\kappa,\phi),$$

$$\int_{k}^{-c}\frac{dk}{\sqrt{k(a+k)(c+k)}}=2\int_{\psi}^{\frac{1}{2}\pi}\frac{d\psi}{\sqrt{a-(a-c)\sin^2\psi}}=\frac{2}{\sqrt{a}}\left\{F_{\prime}(\kappa)-F(\kappa,\psi)\right\}.$$

27. Hence

$$\Pi(h)=\frac{1}{\sqrt{a}}F(\kappa,\ \phi)+\frac{1}{2\sqrt{a(1-\kappa')}}\log\frac{1+H}{1-H},$$

where

$$H=\frac{(1-\kappa')\sin\phi\cos\phi}{\sqrt{1-\kappa^2\sin^2\phi}};$$

(observe, as h passes from $-a$ to $-b$, ϕ passes from $\phi=0$ to $\sin^2\phi=\dfrac{1}{1+\kappa}$ and H from $H=0$ to $H=1$).

Similarly

$$\Psi(k)=\frac{1}{\sqrt{a}}\left\{F_{,}(\kappa)-F(\kappa,\ \psi)\right\}+\frac{1}{2\sqrt{a(1-\kappa')}}\log\frac{1+K}{1-K},$$

where

$$K=\frac{(1+\kappa')\sin\psi\cos\psi}{\sqrt{1-\kappa^2\sin^2\psi}},$$

and as k passes from $-c$ to $-b$, ψ passes from $\frac{1}{2}\pi$ to $\sin^2\psi=\dfrac{1}{1+\kappa}$, and K from 0 to 1.

28. The before-mentioned identical equation

$$\int_{-a}^{-c}\frac{dh}{b+h}\sqrt{\frac{h}{(a+h)(c+h)}}=\tfrac{1}{2}\int_{-a}^{-c}\frac{dh}{\sqrt{h(a+h)(c+h)}}$$

is by the same transformation converted into

$$\int_0^{\frac{1}{2}\pi}\frac{1-(1-\kappa')\sin^2\phi}{1-(1+\kappa')\sin^2\phi}\,\frac{d\phi}{\sqrt{1-\kappa^2\sin^2\phi}}=0.$$

To prove this, I remark that the equation is

$$0=\int_{-0}^{\frac{1}{2}\pi}d\phi\,\frac{\dfrac{1-\kappa'}{1+\kappa'}\{1-(1+\kappa')\sin^2\phi\}+\dfrac{2\kappa'}{1+\kappa'}}{1-(1+\kappa')\sin^2\phi}\,\frac{1}{\Delta\phi},$$

viz. this is

$$0=\frac{1-\kappa'}{1+\kappa'}F_{,}+\frac{2\kappa}{1+\kappa'}\Pi_{,}(-1-\kappa'),$$

or, what is the same thing,

$$\Pi_{,}(-1-\kappa')=-\frac{1-\kappa'}{2\kappa}F_{,},$$

where $\Pi_{,}(-1-\kappa')$ denotes the principal value of the integral

$$\int_0^{\frac{1}{2}\pi}d\phi\,\frac{1}{1-(1+\kappa')\sin^2\phi}\,\frac{1}{\Delta\phi}.$$

Now (Leg. *Fonct. Ellip.*, t. I., p. 71), we have

$$\Pi_{,}(-\kappa^2 \sin^2\theta) + \Pi_{,}\left(-\frac{1}{\sin^2\theta}\right) = F_{,},$$

where, upon examination, it will appear that $\Pi_{,}\left(-\frac{1}{\sin^2\theta}\right)$ in fact represents the principal value of the integral.

Writing herein $\sin^2\theta = \frac{1}{1-\kappa'}$, and therefore $\cos^2\theta = \frac{\kappa'}{1+\kappa'}$, or $\tan^2\theta = \kappa'$, this is

$$\Pi_{,}(-1+\kappa') + \Pi_{,}(-1-\kappa'), \ = F_{,},$$

and the formula (p'), p. 141, attributing therein to θ the foregoing value, becomes

$$\Pi_{,}(-1-\kappa') = E_{,} + \frac{1}{\kappa'}\left\{F_{,}E(\theta) - E_{,}F(\theta)\right\}.$$

But θ is the value for the bisection of the function $F_{,}$, viz., we have

$$2F(\theta) = F_{,},$$

$$2E(\theta) = E_{,} + 1 - \kappa',$$

whence

$$F_{,}E(\theta) - E_{,}F(\theta) = \tfrac{1}{2}(1-\kappa')F_{,},$$

or the formula in question gives

$$\Pi_{,}(-1+\kappa') = \frac{1+\kappa'}{2\kappa'}E_{,},$$

whence

$$\Pi_{,}(-1-\kappa') = -\frac{1-\kappa'}{2\kappa'}F_{,},$$

the result which was to be proved.

29. The value of M' $\left(\text{observing that } \frac{b}{(a-b)(b-c)} = \frac{1}{(\sqrt{a}-\sqrt{c})^2} = \frac{1}{a(1-\kappa')^2}\right)$ is

$$\frac{1}{\sqrt{a}(1-\kappa')}M' = \tfrac{1}{2}\int_{-a}^{-c}\frac{dh}{\sqrt{h(a+h)(c+h)}},$$

which is

$$= \tfrac{1}{2}\frac{2}{\sqrt{2}}F_{,}(\kappa),$$

that is we have

$$M' = (1-\kappa')F_{,}(\kappa),$$

or, what is the same thing,

$$\log\tan\tfrac{1}{2}\phi_0 = \frac{1-\kappa'}{2}F_{,}(\kappa),$$

that is

$$\tan\tfrac{1}{2}\phi_0 = \left(\frac{1-\kappa'}{2}F_{,}(\kappa)\right)$$

(ϕ_0 the South azimuth of the B-geodesic at the umbilicus).

30. I purposely calculated the Table by quadratures as being a method available where the equation $ac - b^2 = 0$ is not satisfied; but in the present case, where this

equation is satisfied, the table might have been calculated from Legendre's Tables of Elliptic Integrals. Observe that $a = 1000$, $b = 500$, $c = 250$, gives $\kappa = \frac{1}{2}\sqrt{3}$ or angle of modulus $= 60°$. As an instance of the comparison[1], suppose $h = -800$, then $\sin^2 \phi = \frac{200}{750} = \frac{4}{15}$, $\log \sin \phi = 9{\cdot}71298$, $\phi = 31° \, 5'$.

$$\sqrt{\frac{b}{(a-b)(b-c)}} = \sqrt{\frac{500}{500\,.\,250}} = \frac{\sqrt{10}}{50} = {\cdot}06326,$$

$$H^2 = \frac{500\,.\,200\,.\,550}{800\,.\,500\,.\,250} = \frac{110}{200}, \ \log = \bar{1}{\cdot}87018,$$

$$H = {\cdot}7416 \ \frac{1+H}{1-H} = \frac{1{\cdot}7416}{2{\cdot}584} = 6{\cdot}7582,$$

$$\Pi(h) = {\cdot}03163\, F(31° \, 5') + {\cdot}03163 \text{ h. l. } 6{\cdot}7582,$$

$F\,31° =$	·56166
	163
$F\,31° \, 5' =$	·56329
h. l. 6·7582 =	1·91075
	2·47404
× by	·03163
	·0782043

or multiplying by 100,000 (factor introduced into my Table) this is $= 7820{\cdot}43$. The value $\Pi(-800) = 7864$ given by my Table agrees sufficiently well with this, the correct value.

31. I calculate also the angle ϕ_0, viz. we have

$$\text{h. l. } \tan \tfrac{1}{2}\phi_0 = \frac{1-\kappa'}{2} F_{,}\kappa, \quad = \tfrac{1}{4} F_{,}(60°). \qquad \text{Leg. vol. III. Table VIII.}$$

$$= \tfrac{1}{4}\, 2{\cdot}15651 = {\cdot}53913,$$

whence by Leg. Table IV.

$$\tfrac{1}{2}\phi_0 = 45° + \tfrac{1}{2}\,.\,29° \, 29'{\cdot}64$$

$$= 59° \, 44'{\cdot}82$$

or

$$\phi_0 = 119° \, 29'{\cdot}64.$$

This exceeds 90°, and since at the umbilicus the tangent plane is at right angles to the plane of projection, the B-geodesic should in the drawing proceed (as it in fact does) from U in the sense UC, touching the bounding ellipse at the point U.

[1] In the present calculation, log denotes an ordinary logarithm, the hyperbolic logarithm being distinguished as h. l.

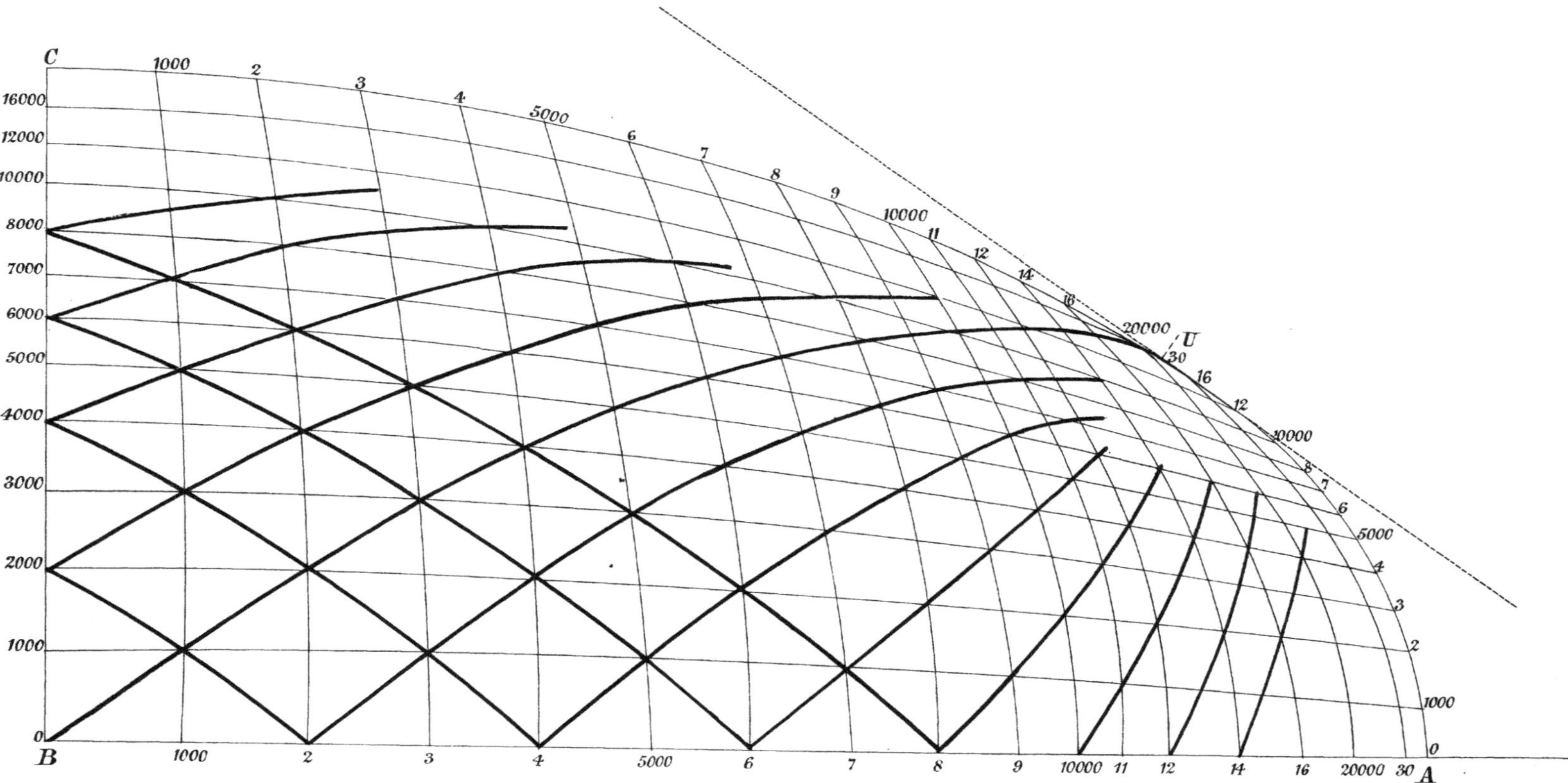

C
B
A
U

479.

THE SECOND PART OF A MEMOIR ON THE DEVELOPMENT OF THE DISTURBING FUNCTION IN THE LUNAR AND PLANETARY THEORIES.

[From the *Memoirs of the Royal Astronomical Society*, vol. XXXIX. (1872), pp. 55—74. Read January 12, 1872.]

THE present communication is a sequel to my paper, "The First Part of a Memoir on the Development of the Disturbing Function in the Lunar and Planetary Theories," *Memoirs R.A.S.*, vol. XXVIII. (1859), pp. 187—215, [214], and I have therefore entitled it as above, but it, in fact, relates only to the Planetary Theory. In the First Part, I gave in effect, but not explicitly, an expression for the general coefficient $D(j, j')$ in terms of the coefficients of the multiple cosines of θ in the expansions of the several powers $(r^2+r'^2-2rr'\cos\theta)^{-s-\frac{1}{2}}$, or say $(a^2+a'^2-2aa'\cos\theta)^{-s-\frac{1}{2}}$; viz., at the foot of page 208 I speak of the term involving $\cos(jU+j'U')$ as having a certain given value; the term in question is $D(j, j')\cos(jU+j'U')$; and consequently the expression for $D(j, j')$ is

$$D(j, j')=\Sigma\frac{\Pi_1(x-\frac{1}{2})}{\Pi x}\eta^{2x}\,\Sigma M_x{}^{\lambda}R_x{}^{\lambda};$$

the omission was, however, a material one, inasmuch as this expression for the general coefficient serves to connect my formulæ with Leverrier's development, *Annales de l'Observ. de Paris*, t. I. (1855), pp. 275—330 and 358—383, and I resume the question for the purpose of applying it.

Formula for the general Coefficient $D(j, j')$.

In the First Part, the reciprocal of the distance of the two planets, or function

$$\{r^2+r'^2-2rr'(\cos U\cos U'+\sin U\sin U'\cos\Phi)\}^{-\frac{1}{2}}$$

is taken to be developed in multiple cosines of U, U', the general term being

$$D(j, j') \cos(jU + j'U'),$$

where j, j' have each of them any integer value from $-\infty$ to $+\infty$ (zero not excluded), but so that j, j' are simultaneously even or simultaneously odd. We have $D(-j, -j') = D(j, j')$ and $D(j', j) = D(j, j')$; and it hence appears that the really distinct values of the coefficient may be taken to be those for which j is not negative, and as regards absolute magnitude is not less than j'; and for such values of j, j' we have the above-mentioned expression

$$D(j, j') = \Sigma \frac{\Pi_1(x-\frac{1}{2})}{\Pi x} \eta^{2x} \Sigma M_x^{\vartheta} R_x^{\vartheta},$$

which I proceed to explain and develope.

$\Pi_1(x-\frac{1}{2})$ and Πx (x being a positive integer) denote respectively $\frac{1}{2} \cdot \frac{3}{2} \ldots (x-\frac{1}{2})$, and $1 \cdot 2 \cdot 3 \ldots x$; in particular for $x = 0$, the value of each factorial is $= 1$.

η denotes $\sin \frac{1}{2} \Phi$.

The coefficients R_x^{ϑ} are those of the multiple cosines in certain developments, viz. we have

$$r^x r'^x \{r^2 + r'^2 - 2rr' \cos(U - U')\}^{-x-\frac{1}{2}} = \Sigma R_x^i \cos i(U - U'),$$

where, as usual, i extends from $-\infty$ to ∞ and $R_x^{-i} = R_x^i$. Writing with Leverrier

$$(a^2 + a'^2 - 2aa' \cos H)^{-\frac{1}{2}} = \tfrac{1}{2} \Sigma A^i \cos iH,$$

$$aa' (a^2 + a'^2 - 2aa' \cos H)^{-\frac{3}{2}} = \tfrac{1}{2} \Sigma B^i \cos iH,$$

$$a^2a'^2 (a^2 + a'^2 - 2aa' \cos H)^{-\frac{5}{2}} = \tfrac{1}{2} \Sigma C^i \cos iH,$$

$$a^3a'^3 (a^2 + a'^2 - 2aa' \cos H)^{-\frac{7}{2}} = \tfrac{1}{2} \Sigma D^i \cos iH,$$

then $2R_0^i$, $2R_1^i$, $2R_2^i$, $2R_3^i$ are the same functions of r, r' that A^i, B^i, C^i, D^i respectively are of a, a'.

The expression of M_x^{ϑ} is

$$M_x^{\vartheta} = (-)^{x-\frac{1}{2}(j+j')} \frac{\Pi x}{\Pi \frac{1}{2}(x-j-\vartheta)\, \Pi \frac{1}{2}(x+j'+\vartheta)} \; \frac{\Pi x}{\Pi \frac{1}{2}(x-j+\vartheta)\, \Pi \frac{1}{2}(x+j'-\vartheta)};$$

and, finally, in the expression for $D(j, j')$, x has every integer value from 0 to ∞, and, for any given value of x, ϑ extends by steps of two units from the inferior value $-(x-j')$ to the superior value $x-j$.

It is convenient to write $x = \frac{1}{2}(j+j') + s$; we have then ϑ extending from $-\frac{1}{2}(j-j') - s$ to $-\frac{1}{2}(j-j') + s$, or writing $\vartheta = -\frac{1}{2}(j-j') + \theta$, θ has the $s+1$ values s, $s-2$, $s-4, \ldots -s$, viz. for $s = 2p+1$ the values are ± 1, $\pm 3, \ldots \pm (2p+1)$, and for $s = 2p$ they are 0, ± 2, $\pm 4, \ldots \pm 2p$.

Making these changes we have

$$D(j,\ j') = \Sigma \frac{\Pi_1 \{\frac{1}{2}(j+j') + s - \frac{1}{2}\}}{\Pi \{\frac{1}{2}(j+j') + s\}} \eta^{j+j'+2s} \Sigma M^{-\frac{1}{2}(j-j')+\theta}_{\frac{1}{2}(j+j')+s} R^{-\frac{1}{2}(j-j')+\theta}_{\frac{1}{2}(j+j')+s},$$

where

$$M^{-\frac{1}{2}(j-j')+\theta}_{\frac{1}{2}(j+j')+s} = (-)^s \frac{\Pi \{\frac{1}{2}(j+j') + s\}}{\Pi \frac{1}{2}(s-\theta) \Pi \frac{1}{2}(j+j'+s+\theta)} \frac{\Pi \{\frac{1}{2}(j+j') + s\}}{\Pi \frac{1}{2}(s+\theta) \Pi \frac{1}{2}(j+j'+s-\theta)},$$

viz. this is $(-)^s$ into the product of two binomial coefficients, each belonging to the exponent $\frac{1}{2}(j+j')+s$.

Particular Cases, $j+j'=0$, 2, 4, 6, being those required in the Planetary Theory.

Considering successively the cases $j+j'=0$, 2, 4, 6, we have, first,

$$D(j,\ -j) = \Sigma \frac{\Pi_1 (s-\frac{1}{2})}{\Pi s} \eta^{2s} \Sigma (-)^s \left\{\frac{\Pi s}{\Pi \frac{1}{2}(s-\theta) \Pi \frac{1}{2}(s+\theta)}\right\}^2 R_s^{-j+\theta}$$

which, developed as far as η^6, is

$$(*) \qquad \begin{aligned} D(j,\ -j) = {} & \tfrac{1}{2} A^{-j} \\ & - \frac{1}{2} \eta^2 \tfrac{1}{2}(B^{-j+1} + B^{-j-1}) \\ & + \frac{1.3}{2.4} \eta^4 \tfrac{1}{2}(C^{-j+2} + 4C^{-j} + C^{-j-2}) \\ & - \frac{1.3.5}{2.4.6} \eta^6 \tfrac{1}{2}(D^{-j+3} + 9D^{-j+1} + 9D^{-j-1} + D^{-j-3}), \end{aligned}$$

where, and in what immediately follows, A, B, C, D are used to denote functions (not of $(a,\ a')$, but) of r, r'.

Secondly,

$$D(j,\ -j+2) = \Sigma \frac{\Pi_1 (s+\frac{1}{2})}{\Pi (s+1)} \eta^2 \Sigma \eta^{2s} \left\{(-)^s \frac{\Pi (s+1)}{\Pi \frac{1}{2}(s-\theta) \Pi \frac{1}{2}(s+\theta)+1} \times \frac{\Pi (s+1)}{\Pi \frac{1}{2}(s+\theta) \Pi \frac{1}{2}(s-\theta)+1} R_{s+1}^{-j+1+\theta}\right\},$$

which, developed to η^6, is

$$(*) \qquad D(j,\ -j+2) = \eta^2 \left\{ \begin{aligned} & \frac{1}{2} \cdot \tfrac{1}{2} B^{-j+1} \\ & - \frac{1.3}{2.4} \eta^2 \cdot \tfrac{1}{2}(2C^{-j+2} + 2C^{-j}), \\ & + \frac{1.3.5}{2.4.6} \eta^4 \cdot \tfrac{1}{2}(3D^{-j+3} + 9D^{-j+1} + 3D^{-j-1}) \end{aligned} \right\}.$$

Thirdly,

$$D(j,\ -j+4)=\Sigma\frac{\Pi_1(s+\frac{3}{2})}{\Pi\ (s+2)}\eta^4\ .\ \Sigma\eta^{2s}(-)^s\left\{\frac{\Pi\ (s+2)}{\Pi\frac{1}{2}(s-\theta)\ \Pi\frac{1}{2}(s+\theta)+2}\right.$$

$$\left.\times\frac{\Pi\ (s+2)}{\Pi\frac{1}{2}(s+\theta)\ \Pi\frac{1}{2}(s-\theta)+2}R_{s+2}^{-j+2\theta}\right\},$$

which, developed to η^6, is

(*) $$D(j,\ -j+4)=\eta^4\left\{\frac{1\ .\ 3}{2\ .\ 4}\ .\ \tfrac{1}{2}\,C^{-j+2}\right.$$

$$\left.-\frac{1\ .\ 3\ .\ 5}{2\ .\ 4\ .\ 6}\eta^2\ .\ \tfrac{1}{2}\,(3D^{-j+3}+3D^{-j+1})\right\}:$$

and, fourthly,

$$D(j,\ -j+6)=\Sigma\frac{\Pi_1(s+\frac{5}{2})}{\Pi\ (s+3)}\eta^6\ \Sigma\eta^{2s}(-)^s\left\{\frac{\Pi\ (s+3)}{\Pi\frac{1}{2}(s-\theta)\ \Pi\frac{1}{2}(s+\theta)+3}\right.$$

$$\left.\times\frac{\Pi\ (s+3)}{\Pi\frac{1}{2}(s+\theta)\ \Pi\frac{1}{2}(s-\theta)+3}R_{s+3}^{-j+3+\theta}\right\},$$

which, developed to η^6, is simply

(*) $$D(j,\ -j+6)=\eta^6\frac{1\ .\ 3\ .\ 5}{2\ .\ 4\ .\ 6}\ .\ \tfrac{1}{2}\,D^{-j+3}.$$

The foregoing formulæ, although obtained on the supposition $j=0$, or positive, apply without alteration to the case $j=$negative, and the entire series of terms of an order not exceeding 6 as regards η may be written,

$$\begin{array}{ll} D\,(j,\ -j) & \cos\,(jU-jU') \\ +\,2D\,(j,\ -j+2) & \cos\,(jU+(-j+2)\,U') \\ +\,2D\,(j,\ -j+4) & \cos\,(jU+(-j+4)\,U') \\ +\,2D\,(j,\ -j+6) & \cos\,(jU+(-j+6)\,U'), \end{array}$$

where j has every integer value from $-\infty$ to $+\infty$.

Comparison with Leverrier.

This is in fact what Leverrier's expression becomes on putting therein $e=e'=0$. To verify this, observe that Leverrier having defined his A^i, B^i, C^i, D^i, as above, writes further

$$\begin{aligned} E^i &= \tfrac{1}{2}\,(B^{i-1}+B^{i+1}), \\ G^i &= \tfrac{3}{8}\,(C^{i-2}+4C^i+C^{i+2}), \\ H^i &= \tfrac{5}{16}\,(D^{i-3}+9D^{i+1}+9D^{i+1}+D^{i+3}), \\ L^i &= \tfrac{3}{4}\,(C^{i-2}+C^i), \\ S^i &= \tfrac{15}{16}\,(D^{i-3}+3D^{i-1}+D^{i+1}), \\ T^i &= \tfrac{15}{16}\,(D^{i-3}+D^{i-1}), \end{aligned}$$

(consequently $E^{-i}=E^i$, $G^{-i}=G^i$, $H^{-i}=H^i$, $L^{-i+2}=L^i$, $S^{-i+2}=S^i$, $T^{-i+4}=T^i$), and that the terms in question, putting in the coefficients $e=e'=0$, are with him

$$\begin{aligned}
\{(1)^i+(11)^i\eta^2+(17\ )^i\eta^4+(20\ )^i\eta^6\}\quad &\cos(il'-i\lambda),\\
\{(212)^i\eta^2+(218)^i\eta^4+(221)^i\eta^6\}\quad &\cos[il'-(i-2)\lambda-2\tau'],\\
\{(372)^i\eta^4+(375)^i\eta^6\}\quad &\cos[il'-(i-4)\lambda-4\tau'],\\
\{(449)^i\eta^6\}\quad &\cos[il'-(i-6)\lambda-6\tau'],
\end{aligned}$$

where, substituting for $(1)^i$, $(11)^i$, &c., their values, the coefficients are

$$\tfrac{1}{2}A^i-\eta^2\tfrac{1}{2}E^i+\eta^4\,.\,\tfrac{1}{2}G^i-\eta^6\tfrac{1}{2}H^i,$$

$$=\tfrac{1}{2}A^i-\eta^2\,.\,\tfrac{1}{4}(B^{i-1}+B^{i+1})+\eta^4\,.\,\tfrac{3}{16}(C^{i-2}+4C^i+C^{i+2})-\eta^6\,.\,\tfrac{5}{32}(D^{i-3}+9D^{i-1}+9D^{i+1}+D^{i+3});$$

$$\eta^2\,.\,\tfrac{1}{2}B^{i-1}-\eta^4\,.\,L^i+\eta^6S^i,\ =\eta^2\,.\,\tfrac{1}{2}B^{i+1}-\eta^4(\tfrac{3}{4}C^{i-2}+C^i)+\eta^6\,.\,\tfrac{15}{16}(D^{i-3}+3D^{i-1}+D^{i+1});$$

$$\eta^4\,.\,\tfrac{3}{8}C^{i-2}-\eta^6T^i,\ =\eta^4\,.\,\tfrac{3}{8}C^{i-2}-\eta^6\,.\,\tfrac{15}{16}(D^{i-3}+D^{i-1});$$

and

$$\eta^6\,.\,\tfrac{5}{16}D^{i-3}.$$

Writing herein j in place of i, and for A^j, B^{j-1}, &c., the equal values A^{-j}, B^{-j+1}, &c., we have precisely the foregoing coefficients $D(j,\ -j),\ldots D(j,\ -j+6)$.

The Development in Powers of e, e′.

The complete expression of the reciprocal of the distance is obtained from

$$\begin{aligned}
&D(j,\ -j)\qquad \cos(jU-jU')\\
+\,&2D(j,\ -j+2)\ \cos\big(jU+(-j+2)\,U'\big)\\
+\,&2D(j,\ -j+4)\ \cos\big(jU+(-j+4)\,U'\big)\\
+\,&2D(j,\ -j+6)\ \cos\big(jU+(-j+6)\,U'\big),
\end{aligned}$$

by writing therein for r, r', U, U', instead of the circular, the elliptic values, that is the values

$$\begin{aligned}
r\ &=a\ \text{elqr}(e,\ L-\Pi)\qquad, &&=a\ (1+x),\\
r'\ &=a'\ \text{elqr}(e',\ L'-\Pi')\qquad, &&=a'\ (1+x'),\\
U\ &=\Pi-\Theta+\text{elta}(e,\ L-\Pi), &&=\Pi-\Theta+f,\\
U'&=\Pi'-\Theta'+\text{elta}(e',\ L'-\Pi'), &&=\Pi'-\Theta'+f';
\end{aligned}$$

L, Π, Θ the mean longitude in orbit, longitude of perihelion in orbit, and longitude of node; and the like for L', Π', Θ'; "elqr" = elliptic quotient radius, "elta" = elliptic true anomaly; or, what is the same thing, if we write $\text{elta}(e,\ L-\Pi)=L-\Pi+\text{eltt}(e,\ L-\Pi)$, and the like for $\text{elta}(e',\ L'-\Pi')$, then

$$\begin{aligned}
U\ &=L-\Theta+\text{eltt}(e,\ L-\Pi), &&=L-\Theta+y,\\
U'&=L'-\Theta'+\text{eltt}(e',\ L'-\Pi'), &&=L'-\Theta'+y'.
\end{aligned}$$

The process for doing this is explained, First Part, pp. 205—207, [214], viz., writing $r=a(1+x)$, $r'=a'(1+x')$, and restoring j' (instead of its value $-j, \ldots -j+6$, as the case may be), we have a general term

$$\frac{1}{\Pi\alpha\Pi\alpha'}a^{\alpha}a'^{\alpha'}\left(\frac{d}{da}\right)^{\alpha}\left(\frac{d}{da}\right)^{\alpha'}.D(j,j').x^{\alpha}x'^{\alpha'}\cos\left[j(\Pi-\Theta+f)+j'(\Pi'-\Theta'+f')\right],$$

where $D(j, j')$ now denotes the value obtained by writing a, a' in place of r, r' and f, f' are the true anomalies elta $(e, L-\Pi)$ and elta $(e', L'-\Pi')$. And the second factor, $x^{\alpha}x'^{\alpha'}$ into the cosine, is given as a series

$$\Sigma\Sigma\left([\cos]^{i}+[\sin]^{i}\right)\left([\cos]^{i'}+[\sin]^{i'}\right)\cos\left[i(L-\Pi)+i'(L'-\Pi')+j(\Pi-\Theta)-j'(\Pi'-\Theta')\right],$$

where $[\cos]^{i}$, $[\sin]^{i}$ are functions of e, $[\cos]^{i'}$, $[\sin]^{i'}$ functions of e'. Or, what is better, the term $x^{\alpha}x'^{\alpha'}$ into the cosine may be written $x^{\alpha}x'^{\alpha'}\cos\left[j(L-\Theta+y)+j'(L'-\Theta'+y')\right]$, and the expansion then is

$$\Sigma\Sigma\left([\cos]^{i}+[\sin]^{i}\right)\left([\cos]^{i'}+[\sin]^{i'}\right)\cos\left[i(L-\Pi)+i'(L'-\Pi')+j(L-\Theta)+j'(L'-\Theta')\right],$$

where as before $[\cos]^{i}$, $[\sin]^{i}$ are functions of e, $[\cos]^{i'}$, $[\sin]^{i'}$ are the same functions of e', viz. the e-functions are those given in the two "datum-tables" $(x^0 \ldots x^7)\cos jy$ and $(x^0 \ldots x^7)\sin jy$, taken from Leverrier, which I have given in my "Tables of the Developments of Functions in the Theory of Elliptic Motion," *Memoirs R.A.S.* vol. XXIX. (1861), pp. 191—306, [216]. In order to better show which are the symbols referred to, we may, instead of $[\cos]^{i}$, &c., write $[x^{\alpha}\cos jy]^{i}$, &c., the formula will then be

$$\begin{aligned}x^{\alpha}x'^{\alpha'}\cos\left[j(L-\Theta+y)+j'(L'-\Theta'+y')\right]&=\\ \Sigma\Sigma\left([x^{\alpha}\cos jy]^{i}+[x^{\alpha}\sin jy]^{i}\right)&\left([x'^{\alpha'}\cos j'y']^{i'}+[x'^{\alpha'}\sin j'y']^{i'}\right)\\ &\times\cos\left[i(L-\Pi)+i'(L'-\Pi')+j(L-\Theta)+j'(L'-\Theta')\right];\end{aligned}$$

and if we attribute to i, i' any given values, that is, attend to any particular multiple cosine,

$$\cos\left[i(L-\Pi)+i'(L'-\Pi')+j(L-\Theta)+j'(L'-\Theta')\right],$$

the coefficient hereof will be

$$\Sigma\frac{1}{\Pi\alpha\Pi\alpha'}a^{\alpha}\left(\frac{d}{da}\right)^{\alpha}a'^{\alpha'}\left(\frac{d}{da}\right)^{\alpha'}D(j,j').\left([x^{\alpha}\cos jy]^{i}+[x^{\alpha}\sin jy]^{i}\right)\left([x'^{\alpha'}\cos j'y']^{i'}+[x'^{\alpha'}\sin j'y']^{i'}\right),$$

where α, α' each extend from zero to infinity, but to obtain the expression up to a given order p in e, e', we take only the values up to $\alpha+\alpha'=p$.

Particular Case.

Thus, for instance, in $\cos\left[j(L-\Theta)-j'(L'-\Theta')\right]$ the terms independent of e' are

$$\begin{aligned}&D(j,-j)\left\{[x^0\cos jy]^0+[x^0\sin jy]^0\right\}\\ &+\frac{1}{1}a\left(\frac{d}{da}\right)D(j,-j')\left\{[x'\cos jy]^0+[x'\sin jy]^0\right\}\\ &+\frac{1}{1.2}a^2\left(\frac{d}{da}\right)^2D(j,-j)\left\{[x^2\cos jy]^0+[x^2\sin jy]^0\right\},\\ &+\&c.\end{aligned}$$

which, observing that in the present case the sine terms vanish, is

1	$\frac{e^2}{8}$	$\frac{e^4}{384}$	$\frac{e^6}{46080}$		
1	$-8j^2$	$+96j^4$ $-54j^2$	$-1280j^6$ $+3920j^4$ $-3440j^2$	1	. $D(j, -j)$
	$+4$	$-48j^2$	$-360j^2$	$\frac{1}{1}\,a\frac{d}{da}$	. ,,
	$+4$	$-96j^2$	$+1920j^4$ $-1320j^2$	$\frac{1}{1.2}\,a^2\left(\frac{d}{da}\right)^2$	. ,,
		$+144$	$-2880j^2$	$\frac{1}{1.2.3}\,a^3\left(\frac{d}{da}\right)^3$	. ,,
		$+144$	$-5760j^2$	$\frac{1}{1.2.3.4}\,a^4\left(\frac{d}{da}\right)^4$	. ,,
			$+14400$	$\frac{1}{1.2..5}\,a^5\left(\frac{d}{da}\right)^5$	. ,,
			$+14400$	$\frac{1}{1.2..6}\,a^6\left(\frac{d}{da}\right)^6$	. ,,
			0	$\frac{1}{1.2..7}\,a^7\left(\frac{d}{da}\right)^7$	. ,,

viz. the term in e^2 is

$$e^2\left\{-j^2+\tfrac{1}{2}\,a\frac{d}{da}+\tfrac{1}{4}a^2\left(\frac{d}{da}\right)^2\right\}D(j,\,-j):$$

viz. writing $\eta=0$, and therefore $D(j,\,-j)=\frac{1}{2}A^{-j}$, the term in e^2 is

$$e^2\left\{-j^2+\tfrac{1}{2}\,a\frac{d}{da}+\tfrac{1}{4}\,a^2\left(\frac{d}{da}\right)^2\right\}\tfrac{1}{2}A^{-j},$$

which, conformably with Leverrier's subscript notation

$$A_1{}^i=\frac{1}{1}\,a\frac{d}{da}A^i,\quad A_2{}^i=\frac{1}{1.2}\,a^2\left(\frac{d}{da}\right)^2A^i,\ \text{\&c.},$$

I write

$$e^2\{-j^2+\tfrac{1}{2}(\quad)_1+\tfrac{1}{4}2(\quad)_2\}\tfrac{1}{2}A^{-j}=e^2\{-\tfrac{1}{2}j^2A^{-j}+\tfrac{1}{4}A_1^{-j}+\tfrac{1}{4}A_2^{-j}\}.$$

The term in question is given by Leverrier as $(\frac{1}{2}e)^2(2)^i$, $=e^2.\frac{1}{4}(2)$, $h=i$ and $K^i=A^i$, $=e^2.\frac{1}{4}(-2i^2A^i+A_1^i+A_2^i)$, which agrees.

Similarly the term in e^4 is

$$\frac{e^4}{384}\{96j^4-54j^2-48j^2(\quad)_1-96j^2(\quad)_2+144(\quad)_3+144(\quad)_4\}\tfrac{1}{2}A^{-j},$$

$$=\frac{e^4}{768}\{(96j^4-54j^2)A^{-j}-48j^2A_1^{-j}-96j^2A_2^{-j}+144A_3^{-j}+144A_4^{-j}\},$$

and the term in question is given by Leverrier as $(\frac{1}{2}e)^4(4)^i=e^4.\frac{1}{16}(4)$, $h=i$ and $K^i=A^i$,

$$=e^4\tfrac{1}{16}\{\tfrac{1}{8}(-9i^2+16i^4)A^i-i^2A_1^i-2i^2A_2^i+3A_3^i+3A_4^i\},$$

which agrees. I have not made the comparison of any more terms.

LEVERRIER'S *Results expressed in terms of the Arguments*, $L'-\Theta'$, $L'-\Pi'$, $L-\Theta$, $L-\Pi$.

The angles which Leverrier uses in his arguments are l', λ, ω, ϖ', and τ', viz. we have,

$$\begin{aligned} l' &=\Theta'+(L'-\Theta'),\\ \lambda &=\Theta'+(L-\Theta),\\ \varpi' &=\Theta'+(\Pi'-\Theta'),\\ \omega &=\Theta'+(\Pi-\Theta),\\ \tau' &=\Theta', \end{aligned}$$

where L, Π, Θ are the mean longitude of the planet m, its perihelion and the mutual node, all in the orbit of m; and similarly L', Π', Θ' are the mean longitude of the planet m', of its perihelion and of the mutual node, all in the orbit of m'. On substituting the foregoing values of l', λ, &c., Θ', as it should do, disappears, and the arguments are all of them linear functions of $L'-\Theta'$, $\Pi'-\Theta'$, $L-\Theta$, $\Pi-\Theta$; or, if we please, of $L'-\Theta'$, $L'-\Pi'$, $L-\Theta$, $L-\Pi$, that is of the distances of each planet from its own perihelion and from the mutual node. It is, I think, convenient to use these last angular distances, and accordingly in Leverrier's arguments, I write,

$$\begin{aligned} l' &=\Theta'+(L'-\Theta'),\\ \lambda &=\Theta' \quad \ldots\ldots \quad +(L-\Theta),\\ \varpi' &=\Theta'+(L'-\Theta')-(L'-\Pi'),\\ \omega &=\Theta' \quad \ldots\ldots \quad +(L-\Theta)-(L-\Pi),\\ \tau' &=\Theta', \end{aligned}$$

and for the purpose of reference form as it were an Index to his result as follows:

Reciprocal of Distance = as follows:

Terms of order zero: terms of orders 2, 4, 6, *having the same arguments.*

				$L'-\Theta'$	$L'-\Pi'$	$L-\Theta$	$L-\Pi$
$(1)^i$	(1 .. 20)	..	cos	i	0	$-i$	0
$(21)^i(\frac{1}{2}e)(\frac{1}{2}e')$	(21 .. 30)	..	,,	i	+1	$-i$	−1
$(31)^i(\frac{1}{2}e)^2(\frac{1}{2}e')^2$	(31 .. 34)	..	,,	i	+2	$-i$	−2
$(35)^i(\frac{1}{2}e)^3(\frac{1}{2}e')^3$	(35 .. 35)	..	,,	i	+3	$-i$	−3
$(36)^i(\frac{1}{2}e)^2\eta^2$	(36 .. 39)	..	,,	i	0	$-i+2$	−2
$(40)^i(\frac{1}{2}e)(\frac{1}{2}e')\eta^2$	(40 .. 43)	..	,,	i	−1	$-i+2$	−1
$(44)^i(\frac{1}{2}e')^2\eta^2$	(44 .. 47)	..	,,	i	−2	$-i+2$	0
$(48)^i(\frac{1}{2}e)^3(\frac{1}{2}e')\eta^2$	(48 .. 48)	..	,,	i	+1	$-i+2$	−3
$(49)^i(\frac{1}{2}e)(\frac{1}{2}e')^3\eta^2$	(49 .. 49)	..	,,	i	−3	$-i+2$	+1

Terms of the first order: terms of orders 3, 5, 7, *having the same arguments.*

				$L'-\Theta'$	$L'-\Pi'$	$L-\Theta$	$L-\Pi$
$(50)^i\ \frac{1}{2}e$	(50 .. 69)	..	cos	i	0	$-i$	+1
$(70)^i\ \frac{1}{2}e'$	(70 .. 89)	..	,,	i	+1	$-i$	0
$(90)^i(\frac{1}{2}e)^2(\frac{1}{2}e')$	(90 .. 99)	..	,,	i	+1	$-i$	−2
$(100)^i(\frac{1}{2}e)(\frac{1}{2}e')^2$	(100 .. 109)	..	,,	i	+2	$-i$	−1
$(110)^i(\frac{1}{2}e)^3(\frac{1}{2}e')^2$	(110 .. 113)	..	,,	i	+2	$-i$	−3
$(114)^i(\frac{1}{2}e)^2(\frac{1}{2}e')^3$	(114 .. 117)	..	,,	i	+3	$-i$	−2
$(118)^i(\frac{1}{2}e)^4(\frac{1}{2}e')^3$	(118 .. 118)	..	,,	i	+3	$-i$	−4
$(119)^i(\frac{1}{2}e)^3(\frac{1}{2}e')^4$	(119 .. 119)	..	,,	i	+4	$-i$	−3
$(120)^i(\frac{1}{2}e)\eta^2$	(120 .. 129)	..	,,	i	0	$-i+2$	−1
$(130)^i(\frac{1}{2}e')\eta^2$	(130 .. 139)	..	,,	i	−1	$-i+2$	0
$(140)^i(\frac{1}{2}e)^3\eta^2$	(140 .. 143)	..	,,	i	0	$-i+2$	−3
$(144)^i(\frac{1}{2}e)^2(\frac{1}{2}e')\eta^2$	(144 .. 147)	..	,,	i	+1	$-i+2$	−2
$(148)^i(\frac{1}{2}e)^2(\frac{1}{2}e')\eta^2$	(148 .. 151)	..	,,	i	−1	$-i+2$	−2
$(152)^i(\frac{1}{2}e)(\frac{1}{2}e')^2\eta^2$	(152 .. 155)	..	,,	i	−2	$-i+2$	−1
$(156)^i\ \frac{1}{2}e\ (\frac{1}{2}e')^2\eta^2$	(156 .. 159)	..	,,	i	−2	$-i+2$	+1
$(160)^i(\frac{1}{2}e')^3\eta^2$	(160 .. 163)	..	,,	i	−3	$-i+2$	0
$(164)^i(\frac{1}{2}e)^4(\frac{1}{2}e')\eta^2$	(164 .. 164)	..	,,	i	+1	$-i+2$	−4
$(165)^i(\frac{1}{2}e)^3(\frac{1}{2}e')^2\eta^2$	(165 .. 165)	..	,,	i	+2	$-i+2$	−3
$(166)^i(\frac{1}{2}e)^2(\frac{1}{2}e')^3\eta^2$	(166 .. 166)	..	,,	i	−3	$-i+2$	+2
$(167)^i(\frac{1}{2}e)(\frac{1}{2}e')^4\eta^2$	(167 .. 167)	..	,,	i	−4	$-i+2$	+1
$(168)^i(\frac{1}{2}e)^3\eta^4$	(168 .. 168)	..	,,	i	0	$-i+2$	−3
$(169)^i(\frac{1}{2}e)^2(\frac{1}{2}e')\eta^4$	(169 .. 169)	..	,,	i	−1	$-i+4$	−2
$(170)^i(\frac{1}{2}e)(\frac{1}{2}e')^2\eta^4$	(170 .. 170)	..	,,	i	−2	$-i+4$	−1
$(171)^i(\frac{1}{2}e')^3\eta^4$	(171 .. 171)	..	,,	i	−3	$-i+4$	0

Terms of second order: terms of orders 4, 6, *having the same arguments.*

			$L'-\Theta'$	$L'-\Pi'$	$L-\Theta$	$L-\Pi$
$(172)^i (\frac{1}{2}e)^2$	(172 .. 181)	..	cos	i	0	$-i$	$+2$
$(182)^i (\frac{1}{2}e)(\frac{1}{2}e')$	(182 .. 191)	..	,,	i	$+1$	$-i$	$+1$
$(192)^i (\frac{1}{2}e')^2$	(192 .. 201)	..	,,	i	$+2$	$-i$	0
$(202)^i (\frac{1}{2}e)^3 (\frac{1}{2}e')$	(202 .. 205)	..	,,	i	$+1$	$-i$	-3
$(206)^i (\frac{1}{2}e) (\frac{1}{2}e')^3$	(206 .. 209)	..	,,	i	$+3$	$-i$	-1
$(210)^i (\frac{1}{2}e)^4 (\frac{1}{2}e')^2$	(210 .. 210)	..	,,	i	$+2$	$-i$	-4
$(211)^i (\frac{1}{2}e)^2 (\frac{1}{2}e')^4$	(211 .. 211)	..	,,	i	$+4$	$-i$	-2
$(212)^i \eta^2$	(212 .. 221)	..	,,	i	0	$-i+2$	0
$(222)^i (\frac{1}{2}e) (\frac{1}{2}e') \eta^2$	(222 .. 225)	..	,,	i	$+1$	$-i+2$	-1
$(226)^i (\frac{1}{2}e) (\frac{1}{2}e') \eta^2$	(226 .. 229)	..	,,	i	-1	$-i+2$	$+1$
$(230)^i (\frac{1}{2}e)^4 \eta^2$	(230 .. 230)	..	,,	i	0	$-i+2$	-4
$(231)^i (\frac{1}{2}e)^3 (\frac{1}{2}e') \eta^2$	(231 .. 231)	..	,,	i	-1	$-i+2$	-3
$(232)^i (\frac{1}{2}e)^2 (\frac{1}{2}e')^2 \eta^2$	(232 .. 232)	..	,,	i	-2	$-i+2$	-2
$(233)^i (\frac{1}{2}e) (\frac{1}{2}e')^3 \eta^2$	(233 .. 233)	..	,,	i	-3	$-i+2$	-1
$(234)^i (\frac{1}{2}e')^4 \eta^2$	(234 .. 234)	..	,,	i	-4	$-i+2$	0
$(235)^i (\frac{1}{2}e')^2 (\frac{1}{2}e')^2 \eta^2$	(235 .. 235)	..	,,	i	$+2$	$-i+2$	-2
$(236)^i (\frac{1}{2}e)^2 (\frac{1}{2}e')^2 \eta^2$	(236 .. 236)	..	,,	i	-2	$-i+2$	$+2$
$(237)^i (\frac{1}{2}e)^2 \eta^4$	(237 .. 237)	..	,,	i	0	$-i+4$	-2
$(238)^i (\frac{1}{2}e) (\frac{1}{2}e') \eta^4$	(238 .. 238)	..	,,	i	-1	$-i+4$	-1
$(239)^i (\frac{1}{2}e')^2 \eta^4$	(239 .. 239)	..	,,	i	-2	$-i+4$	0

Terms of third order: terms of orders 5, 7, *having the same arguments.*

			$L'-\Theta'$	$L'-\Pi'$	$L-\Theta$	$L-\Pi$
$(240)^i (\frac{1}{2}e)^3$	(240 .. 249)	..	cos	i	0	$-i$	$+3$
$(250)^i (\frac{1}{2}e)^2 (\frac{1}{2}e')$	(250 .. 259)	..	,,	i	$+1$	$-i$	$+2$
$(260)^i (\frac{1}{2}e) (\frac{1}{2}e')^2$	(260 .. 269)	..	,,	i	$+2$	$-i$	$+1$
$(270)^i (\frac{1}{2}e')^3$	(270 .. 279)	..	,,	i	$+3$	$-i$	0

Terms of third order (concluded):

			$L'-\Theta'$	$L'-\Pi'$	$L-\Theta$	$L-\Pi$
$(280)^i\,(\tfrac{1}{2}e)^4\,(\tfrac{1}{2}e')$	(280 .. 283)	..	cos	i	$+1$	$-i$	-4
$(284)^i\,(\tfrac{1}{2}e)\,(\tfrac{1}{2}e')^4$	(284 .. 287)	..	,,	i	$+4$	$-i$	-1
$(288)^i\,(\tfrac{1}{2}e)^3\,(\tfrac{1}{2}e')^2$	(288 .. 289)	..	,,	i	$+2$	$-i$	-5
$(290)^i\,(\tfrac{1}{2}e)^2\,(\tfrac{1}{2}e')^3$	(290 .. 299)	..	,,	i	0	$-i+2$	$+1$
$(300)^i\,(\tfrac{1}{2}e')\,\eta^2$	(300 .. 309)	..	,,	i	$+1$	$-i+2$	0
$(310)^i\,(\tfrac{1}{2}e)^2\,(\tfrac{1}{2}e')\,\eta^2$	(310 .. 313)	..	,,	i	-1	$-i+2$	$+2$
$(314)^i\,(\tfrac{1}{2}e)\,(\tfrac{1}{2}e')^2\,\eta^2$	(314 .. 317)	..	,,	i	$+2$	$-i+2$	-1
$(318)^i\,(\tfrac{1}{2}e)^3\,\eta^2$	(318 .. 318)	..	,,	i	0	$-i+2$	-5
$(319)^i\,(\tfrac{1}{2}e)^4\,(\tfrac{1}{2}e')\,\eta^2$	(319 .. 319)	..	,,	i	-1	$-i+2$	-4
$(320)^i\,(\tfrac{1}{2}e)^3\,(\tfrac{1}{2}e')^2\,\eta^2$	(320 .. 320)	..	,,	i	-2	$-i+2$	-3
$(321)^i\,(\tfrac{1}{2}e)^2\,(\tfrac{1}{2}e')^3\,\eta^2$	(321 .. 321)	..	,,	i	-3	$-i+2$	-2
$(322)^i\,(\tfrac{1}{2}e)\,(\tfrac{1}{2}e')^4\,\eta^2$	(322 .. 322)	..	,,	i	-4	$-i+2$	-1
$(323)^i\,(\tfrac{1}{2}e')^5\,\eta^2$	(323 .. 323)	..	,,	i	-5	$-i+2$	0
$(324)^i\,(\tfrac{1}{2}e)^3\,(\tfrac{1}{2}e')^2\,\eta^2$	(324 .. 324)	..	,,	i	-2	$-i+2$	$+3$
$(325)^i\,(\tfrac{1}{2}e)^2\,(\tfrac{1}{2}e')^3\,\eta^2$	(325 .. 325)	..	,,	i	$+3$	$-i+2$	-2
$(326)^i\,(\tfrac{1}{2}e)\,\eta^4$	(326 .. 329)	..	,,	i	0	$-i+4$	-1
$(330)^i\,(\tfrac{1}{2}e')\,\eta^4$	(330 .. 333)	..	,,	i	-1	$-i+4$	0
$(334)^i\,(\tfrac{1}{2}e)^2\,(\tfrac{1}{2}e')\,\eta^4$	(334 .. 334)	..	,,	i	$+1$	$-i+4$	-2
$(335)^i\,(\tfrac{1}{2}e)\,(\tfrac{1}{2}e')^2\,\eta^4$	(335 .. 335)	..	,,	i	-2	$-i+4$	$+1$

Terms of fourth order: terms of order 6, *and of same argument.*

			$L'-\Theta'$	$L'-\Pi'$	$L-\Theta$	$L-\Pi$
$(336)^i\,(\tfrac{1}{2}e)^4$	(336 .. 339)	..	cos	i	0	$-i$	$+4$
$(340)^i\,(\tfrac{1}{2}e)^3\,(\tfrac{1}{2}e')$	(340 .. 343)	..	,,	i	$+1$	$-i$	$+3$
$(344)^i\,(\tfrac{1}{2}e)^2\,(\tfrac{1}{2}e')^2$	(344 .. 347)	..	,,	i	$+2$	$-i$	$+2$
$(348)^i\,(\tfrac{1}{2}e)\,(\tfrac{1}{2}e')^3$	(348 .. 351)	..	,,	i	$+3$	$-i$	$+1$
$(352)^i\,(\tfrac{1}{2}e')^4$	(352 .. 355)	..	,,	i	$+4$	$-i$	0
$(356)^i\,(\tfrac{1}{2}e)^5\,(\tfrac{1}{2}e')$	(356 .. 356)	..	,,	i	$+1$	$-i$	-5
$(357)^i\,(\tfrac{1}{2}e)\,(\tfrac{1}{2}e')^5$	(357 .. 357)	..	,,	i	$+5$	$-i-1$	0

Terms of fourth order (concluded):

			$L'-\Theta'$	$L'-\Pi'$	$L-\Theta$	$L-\Pi$
$(358)^i\,(\tfrac{1}{2}e)^2\,\eta^2$	(358 .. 358) ..	cos	i	0	$-i+2$	$+2$
$(362)^i\,(\tfrac{1}{2}e)\,(\tfrac{1}{2}e')\,\eta^2$	(362 .. 364) ..	,,	i	$+1$	$-i+2$	$+1$
$(366)^i\,(\tfrac{1}{2}e')^2\,\eta^2$	(366 .. 369) ..	,,	i	$+2$	$-i+2$	0
$(370)^i\,(\tfrac{1}{2}e)^3\,(\tfrac{1}{2}e')\,\eta^2$	(370 .. 370) ..	,,	i	-1	$-i+2$	$+3$
$(371)^i\,(\tfrac{1}{2}e)\,(\tfrac{1}{2}e')^3\,\eta^2$	(371 .. 371) ..	,,	i	$+3$	$-i+2$	-1
$(372)^i\,\eta^4$	(372 .. 375) ..	,,	i	0	$-i+4$	0
$(376)^i\,(\tfrac{1}{2}e)\,(\tfrac{1}{2}e')\,\eta^4$	(376 .. 376) ..	,,	i	$+1$	$-i+4$	-1
$(377)^i\,(\tfrac{1}{2}e)\,(\tfrac{1}{2}e')\,\eta^4$	(377 .. 377) ..	,,	i	-1	$-i+4$	$+1$

Terms of fifth order: terms of order 7 having the same arguments.

			$L'-\Theta'$	$L'-\Pi'$	$L-\Theta$	$L-\Pi$
$(378)^i\,(\tfrac{1}{2}e)^5$	(378 .. 381) ..	cos	i	0	$-i$	$+5$
$(382)^i\,(\tfrac{1}{2}e)^2\,(\tfrac{1}{2}e')$	(382 .. 385) ..	,,	i	$+1$	$-i$	$+4$
$(386)^i\,(\tfrac{1}{2}e)^3\,(\tfrac{1}{2}e')^2$	(386 .. 389) ..	,,	i	$+2$	$-i$	$+3$
$(390)^i\,(\tfrac{1}{2}e)^2\,(\tfrac{1}{2}e')^3$	(390 .. 393) ..	,,	i	$+3$	$-i$	$+2$
$(394)^i\,(\tfrac{1}{2}e)\,(\tfrac{1}{2}e')^4$	(394 .. 397) ..	,,	i	$+4$	$-i$	$+1$
$(398)^i\,(\tfrac{1}{2}e')^5$	(398 .. 401) ..	,,	i	$+5$	$-i$	0
$(402)^i\,(\tfrac{1}{2}e)^6\,(\tfrac{1}{2}e')$	(402 .. 402) ..	,,	i	$+1$	$-i$	-6
$(403)^i\,(\tfrac{1}{2}e)\,(\tfrac{1}{2}e')^6$	(403 .. 403) ..	,,	i	$+6$	$-i$	-1
$(404)^i\,(\tfrac{1}{2}e)^3\,\eta^2$	(404 .. 407) ..	,,	i	0	$-i+2$	$+3$
$(408)^i\,(\tfrac{1}{2}e)^2\,(\tfrac{1}{2}e')\,\eta^2$	(408 .. 411) ..	,,	i	$+1$	$-i+2$	$+2$
$(412)^i\,(\tfrac{1}{2}e)\,(\tfrac{1}{2}e')^2\,\eta^2$	(412 .. 415) ..	,,	i	$+2$	$-i+2$	$+1$
$(416)^i\,(\tfrac{1}{2}e')^3\,\eta^2$	(416 .. 419) ..	,,	i	$+3$	$-i+2$	0
$(420)^i\,(\tfrac{1}{2}e)^4\,(\tfrac{1}{2}e')\,\eta^2$	(420 .. 420) ..	,,	i	-1	$-i+2$	$+4$
$(421)^i\,(\tfrac{1}{2}e)\,(\tfrac{1}{2}e')^4\,\eta^2$	(421 .. 421) ..	,,	i	$+4$	$-i+2$	-1
$(422)^i\,(\tfrac{1}{2}e)\,\eta^4$	(422 .. 425) ..	,,	i	0	$-i+4$	$+1$
$(426)^i\,(\tfrac{1}{2}e')\,\eta^4$	(426 .. 429) ..	,,	i	$+1$	$-i+4$	0
$(430)^i\,(\tfrac{1}{2}e)^2\,(\tfrac{1}{2}e')\,\eta^2$	(430 .. 430) ..	,,	i	-1	$-i+4$	$+2$
$(431)^i\,(\tfrac{1}{2}e)\,(\tfrac{1}{2}e')^2\,\eta^2$	(431 .. 431) ..	,,	i	$+2$	$-i+4$	-1
$(432)^i\,(\tfrac{1}{2}e)\,\eta^6$	(432 .. 432) ..	,,	i	0	$-i+6$	-1
$(433)^i\,(\tfrac{1}{2}e')\,\eta^4$	(433 .. 433) ..	,,	i	-1	$-i+6$	0

Terms of sixth order.

			$L'-\Theta'$	$L'-\Pi'$	$L-\Theta$	$L-\Pi$
$(434)^i\,(\tfrac{1}{2}e)^6$	(434 .. 434) ..	cos	i	0	$-i$	$+6$
$(435)^i\,(\tfrac{1}{2}e)^5\,(\tfrac{1}{2}e')$	(435 .. 435) ..	,,	i	$+1$	$-i$	$+5$
$(436)^i\,(\tfrac{1}{2}e)^4\,(\tfrac{1}{2}e')^2$	(436 .. 436) ..	,,	i	$+2$	$-i$	$+4$
$(437)^i\,(\tfrac{1}{2}e)^3\,(\tfrac{1}{2}e')^3$	(437 .. 437) ..	,,	i	$+3$	$-i$	$+3$
$(438)^i\,(\tfrac{1}{2}e)^2\,(\tfrac{1}{2}e')^4$	(438 .. 438) ..	,,	i	$+4$	$-i$	$+2$
$(439)^i\,(\tfrac{1}{2}e)\,(\tfrac{1}{2}e')^5$	(439 .. 439) ..	,,	i	$+5$	$-i$	$+1$
$(440)^i\,(\tfrac{1}{2}e')^6$	(440 .. 440) ..	,,	i	$+6$	$-i$	0
$(441)^i\,(\tfrac{1}{2}e)^4\,\eta^2$	(441 .. 441) ..	,,	i	0	$-i+2$	$+4$
$(442)^i\,(\tfrac{1}{2}e)^3\,(\tfrac{1}{2}e')\,\eta^2$	(442 .. 442) ..	,,	i	$+1$	$-i+2$	$+3$
$(443)^i\,(\tfrac{1}{2}e)^2\,(\tfrac{1}{2}e')^2\,\eta^2$	(443 .. 443) ..	,,	i	$+2$	$-i+2$	$+2$
$(444)^i\,(\tfrac{1}{2}e)\,(\tfrac{1}{2}e')^3\,\eta^2$	(444 .. 444) ..	,,	i	$+3$	$-i+2$	$+1$
$(445)^i\,(\tfrac{1}{2}e')^4\,\eta^2$	(445 .. 445) ..	,,	i	$+4$	$-i+2$	0
$(446)^i\,(\tfrac{1}{2}e)^2\,\eta^4$	(446 .. 446) ..	,,	i	0	$-i+4$	$+2$
$(447)^i\,(\tfrac{1}{2}e)\,(\tfrac{1}{2}e')\,\eta^4$	(447 .. 447) ..	,,	i	$+1$	$-i+4$	$+1$
$(448)^i\,(\tfrac{1}{2}e')^2\,\eta^4$	(448 .. 448) ..	,,	i	$+2$	$-i+4$	0
$(449)^i\,\eta^6$	(449 .. 449) ..	,,	i	0	$-i+6$	0

Terms of seventh order.

			$L'-\Theta'$	$L'-\Pi'$	$L-\Theta$	$L-\Pi$
$(450)^i\,(\tfrac{1}{2}e)^7$	(450 .. 450) ..	cos	i	0	$-i$	$+7$
$(451)^i\,(\tfrac{1}{2}e)^6\,(\tfrac{1}{2}e')^2$	(451 .. 451) ..	,,	i	1	$-i$	$+6$
$(452)^i\,(\tfrac{1}{2}e)^5\,(\tfrac{1}{2}e')^2$	(452 .. 452) ..	,,	i	2	$-i$	$+5$
$(453)^i\,(\tfrac{1}{2}e)^4\,(\tfrac{1}{2}e')^3$	(453 .. 453) ..	,,	i	3	$-i$	$+4$
$(454)^i\,(\tfrac{1}{2}e)^3\,(\tfrac{1}{2}e')^4$	(454 .. 454) ..	,,	i	4	$-i$	$+3$
$(455)^i\,(\tfrac{1}{2}e)^2\,(\tfrac{1}{2}e')^5$	(455 .. 455) ..	,,	i	5	$-i$	$+2$
$(456)^i\,(\tfrac{1}{2}e)\,(\tfrac{1}{2}e')^6$	(456 .. 456) ..	,,	i	6	$-i$	$+1$
$(457)^i\,(\tfrac{1}{2}e')^7$	(457 .. 457) ..	,,	i	7	$-i$	0
$(458)^i\,(\tfrac{1}{2}e)^5\,\eta^2$	(458 .. 458) ..	,,	i	0	$-i+2$	$+5$

Terms of seventh order (concluded):

		$L'-\Theta'$	$L'-\Pi'$	$L-\Theta$	$L-\Pi$
$(459)^i\,(\tfrac{1}{2}e)^4\,(\tfrac{1}{2}e')\,\eta^2\,(459\ ..\ 459)$ ..	cos	i	$+1$	$-i+2$	$+4$
$(460)^i\,(\tfrac{1}{2}e)^3\,(\tfrac{1}{2}e')^2\,\eta^2\,(460\ ..\ 460)$ ..	"	i	$+2$	$-i+2$	$+3$
$(461)^i\,(\tfrac{1}{2}e)^2\,(\tfrac{1}{2}e')^3\,\eta^2\,(461\ ..\ 461)$ ..	"	i	$+3$	$-i+2$	$+2$
$(462)^i\,(\tfrac{1}{2}e)\,(\tfrac{1}{2}e')^4\,\eta^2\,(462\ ..\ 462)$ ..	"	i	$+4$	$-i+2$	$+1$
$(463)^i\,(\tfrac{1}{2}e')^5\,\eta^2\,(463\ ..\ 463)$ ..	"	i	$+5$	$-i+2$	0
$(464)^i\,(\tfrac{1}{2}e)^3\,\eta^4\,(464\ ..\ 464)$ ..	"	i	0	$-i+4$	$+3$
$(465)^i\,(\tfrac{1}{2}e)^2\,(\tfrac{1}{2}e')\,\eta^4\,(465\ ..\ 465)$ ..	"	i	$+1$	$-i+4$	$+2$
$(466)^i\,(\tfrac{1}{2}e)\,(\tfrac{1}{2}e')^2\,\eta^4\,(466\ ..\ 466)$ ..	"	i	$+2$	$-i+4$	$+1$
$(467)^i\,(\tfrac{1}{2}e')^3\,\eta^4\,(467\ ..\ 467)$ ..	"	i	$+3$	$-i+4$	0
$(468)^i\,(\tfrac{1}{2}e)\,\eta^6\,(468\ ..\ 468)$ ..	"	i	0	$-i+6$	$+1$
$(469)^i\,(\tfrac{1}{2}e')\,\eta^6\,(469\ ..\ 469)$ ..	"	i	$+1$	$-i+6$	0

Here the several coefficients are ultimately given in terms of the before-mentioned quantities A^i, B^i, C^i, D^i, E^i, G^i, H^i, L^i, S^i, T^i (functions of a, a'), and their differential coefficients in regard to a

$$\left(A_1{}^i = \frac{1}{1}\,a\,\frac{d}{da}\,A^i,\quad A_2{}^i = \frac{1}{1\,.\,2}\,a\,\frac{d^2}{da^2}\,A^i,\ \text{\&c.}\right),$$

as follows:—we have Leverrier, pp. 299—330, a list of functions (1), (2), ... (154) of the form $(1) = \tfrac{1}{2}K^i$, $(2) = -2h^2K^i + K_1{}^i + K_2{}^i$, $(3) = -2i^2K^i + K_1{}^i + K_2{}^i$, &c., involving i, h, and K^i and its derived functions $K_1{}^i$, $K_2{}^i$, &c. The coefficients of the several cosines are given by means of the functions in question, thus, first coefficient, above denoted as $(1)^i\,(1\ ...\ 20)$, is

$$= (1)^i + (2)^i\,(\tfrac{1}{2}e) + (3)^i\,(\tfrac{1}{2}e')\ ... + (20)^i\,\eta^6$$

where $(1)^i = (1)$, $(2)^i = (2)$ ... writing in the functions (1), (2) ... (10), $h = i$, and $K^i = A^i$;

$$(11)^i = (1),\ (12)^i = (2),\ \text{\&c., writing } h = i \text{ and } K^i = -E^i,$$

$$(20)^i = (1),\qquad \text{writing } h = i \text{ and } K^i = -H^i,$$

and so on for the various component coefficients $(1)^i$, $(2)^i$ ... $(469)^i$.

But the resulting expressions, for the several integer values $i = -10$ to $+10$, are worked out in the Addition II. (*Numerical Tables for the Calculation of the Coefficients of the Development of the Disturbing Function*), pp. 358—383. And this Addition contains also, indicated by the letters δ and Δ respectively, the expressions of the terms which experience an alteration in passing from the development of the reciprocal of the distance to those of the disturbing functions m' upon m, and m upon m' respectively.

We have

Disturbing Function m' upon m

$$= m' \left\{ -\frac{r \cos H}{r'^2} + \frac{1}{\rho} \right\}.$$

Disturbing Function m upon m'

$$= m \left\{ -\frac{r' \cos H}{r^2} + \frac{1}{\rho} \right\}.$$

The expressions of $-\frac{r \cos H}{r'}$ and $-\frac{r' \cos H}{r^2}$, developed to the third order in the eccentricities and inclination, are given, Leverrier, pp. 272 and 274. Expressed in the terms of the foregoing arguments $L' - \Theta'$, &c., and in terms of a, a' in place of a and α, these are as follows:

$-\frac{r \cos H}{r'^2} = \frac{a}{a'^2}$ into		$L' - \Theta'$	$L' - \Pi'$	$L - \Theta$	$L - \Pi$
$-1 + \frac{1}{2}(e^2 + e'^2) + \eta^2$	cos	1	0	-1	0
$-ee'$	,,	$+1$	$+1$	-1	-1
$+\frac{3}{2}e - \frac{3}{4}ee'^2 - \frac{3}{2}e\eta^2$	,,	$+1$	0	-1	$+1$
$-\frac{1}{2}e + \frac{1}{4}ee'^2 + \frac{3}{8}e^3 + \frac{1}{2}e\eta^2$	,,	$+1$	0	-1	-1
$-2e' + e^2e' + \frac{3}{2}e'^3 + 2e'\eta^2$	,,	$+1$	$+1$	-1	0
$-\frac{3}{4}e^2e'$	,,	$+1$	$+1$	-1	-2
$+\frac{3}{16}ee'^2$	,,	-1	$+2$	$+1$	-1
$-\frac{27}{16}ee'^2$	,,	$+1$	$+2$	-1	-1
$+\frac{3}{2}e\eta^2$	,,	$+1$	0	$+1$	-1

$-\frac{r\cos H}{r'^2}=\frac{a}{a'^2}$ into		$L'-\Theta'$	$L'-\Pi'$	$L-\Theta$	$L-\Pi$
$-\frac{1}{8}e^2$	cos	+1	0	−1	+2
$-\frac{3}{8}e^2$	,,	+1	0	−1	−2
$+3\,ee'$	,,	+1	+1	−1	+1
$-\frac{1}{8}e'^2$	,,	−1	+2	+1	0
$-\frac{27}{8}e'^2$	,,	+1	−2	−1	0
$-\eta^2$	,,	+1	0	+1	0
$-\frac{1}{24}e^3$	,,	+1	0	−1	+3
$-\frac{1}{3}e^3$	,,	+1	0	−1	−3
$-\frac{1}{4}e^2e'$	,,	+1	+1	−1	+2
$-\frac{1}{16}ee'^2$	,,	−1	+2	+1	+1
$+\frac{81}{16}ee'^2$	,,	+1	+2	−1	+1
$-\frac{1}{6}e'^3$	,,	−1	+3	+1	0
$-\frac{16}{3}e'^3$	,,	+1	+3	−1	0
$-\frac{1}{2}e\eta^2$	,,	+1	0	+1	+1
$-2\,e'\eta^2$	,,	+1	+1	0	+1

$-\frac{r'\cos H}{r^2}=\frac{a'}{a^2}$ into		$L'-\Theta'$	$L'-\Pi'$	$L-\Theta$	$L-\Pi$
$-1+\frac{1}{2}(e^2+e'^2)+\eta^2$	cos	1	0	−1	0
$-ee'$	,,	+1	+1	−1	−1
$-2\,e+e\,e'^2+\frac{3}{2}e^3+2\,e\eta^2$	,,	+1	0	−1	−1
$+\frac{3}{2}e'-\frac{3}{4}e^2e'-\frac{3}{2}e'\eta^2$	,,	−1	+1	+1	0
$-\frac{1}{2}e'+\frac{1}{4}e^2e'+\frac{3}{8}e'^3+\frac{1}{2}e'\eta^2$	,,	+1	+1	−1	0
$+\frac{3}{16}e^2e'$	,,	+2	−1	−2	+2
$-\frac{27}{16}e^2e'$	,,	+1	+1	−1	−2
$-\frac{3}{4}ee'^2$	,,	+1	+2	−1	−1
$+\frac{3}{2}e'\eta^2$	,,	+1	−1	+1	0
$-\frac{1}{8}e^2$	,,	+1	0	−1	+2
$-\frac{27}{8}e^2$	,,	+1	0	−1	−2
$+3\,ee'$	,,	−1	+1	+1	+1
$-\frac{1}{8}e'^2$	,,	−1	+2	+1	0

$-\frac{r'\cos H}{r^2} = \frac{a}{a^2}$ into		$L'-\Theta'$	$L'-\Pi'$	$L-\Theta$	$L-\Pi$
$-\frac{3}{8}e'^2$	cos	+1	+2	−1	0
$-\eta^2$	,,	+1	0	+1	0
$-\frac{1}{6}e^3$	,,	+1	0	−1	+3
$-\frac{16}{3}e^3$	,,	+1	0	−1	−3
$+\frac{81}{16}e^2e'$	,,	−1	+1	+1	+2
$-\frac{1}{16}e^2e'$	,,	+1	+1	−1	+2
$-\frac{1}{4}ee'^2$	,,	−1	+2	+1	+1
$-\frac{1}{24}e'^3$	,,	−1	+3	+1	0
$-\frac{1}{3}e'^3$	,,	+1	+3	−1	0
$-2\,e\eta^2$	,,	+1	0	+1	+1
$-\frac{1}{2}e'\eta^2$	,,	+1	+1	+1	0

It is hardly necessary to observe that, to obtain the expressions of the Disturbing Functions, these additional terms are to be combined with the corresponding terms in the expression of the reciprocal of the distance: thus, in the Disturbing Function Ω (m' upon m), the entire term depending on $\cos[L'-\Theta'-(L-\Theta)]$ is

$$=m'\left\{2(1,\ldots 20)_{i=1}+\frac{a}{a'^2}\left(-1+\tfrac{1}{2}(e^2+e'^2)+\eta^2\right)\right\}\cos[(L'-\Theta')-(L-\Theta)],$$

where, however, the supplemental term is taken to the third order only.

480.

ON THE EXPRESSION OF DELAUNAY'S l, g, h, IN TERMS OF HIS FINALLY ADOPTED CONSTANTS.

[From the *Monthly Notices of the Royal Astronomical Society*, vol. XXXII. (1871—72), pp. 8—16.]

WE have in Delaunay's lunar theory,

l, the mean anomaly of the Moon,

g, the mean distance of perigee from ascending node,

h, the mean longitude of ascending node,

quantities which vary directly as the time, the coefficients of t, or values of $\frac{dl}{dt}$, $\frac{dg}{dt}$, $\frac{dh}{dt}$, being given in his *Théorie du Mouvement de la Lune*, vol. II. pp. 237, 238. But these values are not expressed in terms of his constants a (or n), e, γ, finally adopted as explained p. 800, and it seems very desirable to obtain the expressions of l, g, h, in terms of these finally adopted constants: I have accordingly effected this transformation (which I found less laborious than I had anticipated). It will be convenient to imagine the a, n, e, γ of pp. 237, 238 replaced by A, N, E, Γ respectively. This being so, and writing m for the $\frac{n}{n'}$ of p. 800 we have, p. 800,

$$\begin{aligned}
A = a \Big\{ 1 &+ [-\tfrac{3}{4} m^2 - \tfrac{825}{256} m^3] \frac{a^2}{a'^2} \\
&+ (-\tfrac{2}{3} + 3\gamma^2 - \tfrac{3}{4} e^2 - e'^2 - 2\gamma^2 + \tfrac{5}{2}\gamma^2 e^2 + \tfrac{9}{2}\gamma^2 e'^2 - \tfrac{1}{16} e^4 - \tfrac{9}{8} e^2 e'^2 - \tfrac{5}{4} e'^4)\, m^2 \\
&+ (-\tfrac{9}{4}\gamma^2 - \tfrac{225}{16} e^2 + \tfrac{45}{8}\gamma^4 + \tfrac{81}{2}\gamma^2 e^2 - \tfrac{23}{4}\gamma^2 e'^2 + \tfrac{675}{128} e^4 - \tfrac{825}{16} e^2 e'^2)\, m^3 \\
&+ (\tfrac{1705}{288} - \tfrac{1529}{64}\gamma^2 - \tfrac{14639}{256} e^2 + \tfrac{7469}{192} e'^2)\, m^4 \\
&+ (\tfrac{787}{48} - \tfrac{9323}{256}\gamma^2 - \tfrac{227555}{1024} e^2 + \tfrac{7083}{32} e'^2)\, m^5 \\
&+ \tfrac{5887}{162} m^6 \\
&+ \tfrac{29809}{432} m^7,
\end{aligned}$$

and hence calculating N from the formula $N^2A^3 = n^2a^3$, we find

$$
\begin{aligned}
N = n \Big\{ 1 &+ \left[\tfrac{9}{8} m^2 + \tfrac{2475}{512} m^3\right] \frac{a^2}{a'^2} \\
&+ (1 - \tfrac{9}{2}\gamma^2 + \tfrac{9}{8} e^2 + \tfrac{3}{2} e'^2 + 3\gamma^4 - \tfrac{15}{4}\gamma^2e^2 - \tfrac{27}{4}\gamma^2e'^2 + \tfrac{3}{32} e^4 + \tfrac{27}{16} e^2e'^2 + \tfrac{15}{8} e'^4)\, m \\
&+ (\tfrac{27}{8}\gamma^2 + \tfrac{675}{32} e^2 - \tfrac{135}{16}\gamma^4 - \tfrac{243}{4}\gamma^2e^2 + \tfrac{69}{8}\gamma^2e'^2 - \tfrac{2025}{256} e^4 + \tfrac{2475}{32} e^2e'^2)\, m^3 \\
&+ (-\tfrac{515}{64} + \tfrac{3627}{128}\gamma^2 + \tfrac{44877}{512} e^2 - \tfrac{7149}{128} e'^2)\, m^4 \\
&+ (-\tfrac{787}{32} + \tfrac{30849}{512}\gamma^2 + \tfrac{754665}{2048} e^2 - \tfrac{21249}{64} e'^2)\, m^5 \\
&+ (-\tfrac{13183}{192})\, m^6 \\
&+ (-\tfrac{20807}{144})\, m^7,
\end{aligned}
$$

$= n(1 + Q)$ suppose.

The values of E, Γ are given p. 800, but for the present purpose we only require E^2, and Γ^2 to the fifth order, viz. the values of these are at once found to be

$$
\begin{aligned}
E^2 &= e^2\,(1 + \tfrac{81}{64} m^2 - \tfrac{2595}{128} m^3), \\
\Gamma^2 &= \gamma^2\,(1 + \tfrac{57}{64} m^2 - \tfrac{129}{128} m^3),
\end{aligned}
$$

whence also $E^4 = e^4$ and $\Gamma^4 = \gamma^4$.

The formulæ of pp. 237—238 now give

$$
\begin{aligned}
l = nt \Big\{ 1 &+ \left[-\tfrac{81}{32} m^2 - \tfrac{2475}{512} m^3\right] \frac{a^2}{a'^2} + Q \\
+ &\left\{\begin{aligned} &(-\tfrac{7}{4} + \tfrac{21}{2}\gamma^2 - \tfrac{3}{4} e^2 - \tfrac{21}{8} e'^2 + \tfrac{33}{4}\gamma^4 - \tfrac{39}{8}\gamma^2e^2 + \tfrac{63}{4}\gamma^2e'^2 - \tfrac{9}{8} e^2e'^2 - \tfrac{105}{32} e'^4)\, m^2 \\ &+ (\tfrac{1197}{128}\gamma^2 - \tfrac{243}{256} e^2)\, m^4 \\ &+ (-\tfrac{2709}{256}\gamma^2 + \tfrac{7785}{512} e^2)\, m^5 \end{aligned}\right\} (1 + Q)^{-1} \\
+ &\left\{\begin{aligned} &(-\tfrac{225}{32} + \tfrac{81}{4}\gamma^2 - \tfrac{675}{64} e^2 - \tfrac{825}{32} e'^2 - \tfrac{243}{4}\gamma^4 + \tfrac{1863}{32}\gamma^2e^2 + \tfrac{629}{8}\gamma^2e'^2 + \tfrac{2025}{256} e^4 - \tfrac{2475}{64} e^2e'^2)\, m^3 \\ &\qquad + (\tfrac{4617}{256}\gamma^2 - \tfrac{54675}{4096} e^2)\, m^5 \end{aligned}\right\} (1 + Q)^{-2} \\
+ &(-\tfrac{3265}{128} + \tfrac{3345}{32}\gamma^2 - \tfrac{7089}{256} e^2 - \tfrac{48225}{256} e'^2)\, m^4 (1 + Q)^{-3} \\
+ &(-\tfrac{248925}{2048} + \tfrac{175425}{256}\gamma^2 - \tfrac{167835}{2048} e^2 - \tfrac{1502265}{1024} e'^2)\, m^5 (1 + Q)^{-4} \\
+ &(-\tfrac{12626759}{24576})\, m^6 (1 + Q)^{-5} \\
+ &(-\tfrac{1365131021}{589824})\, m^7 (1 + Q)^{-6}
\end{aligned}
$$

(Observe that writing herein $Q = 0$, and omitting the terms in m^4 and m^5 in the coefficient of $(1 + Q)^{-1}$, and the term in m^5 in the coefficient of $(1 + Q)^{-2}$, we have the original formula of p. 237)

$$
\begin{aligned}
g = nt \Big\{ &\left[\tfrac{45}{16} m^2 + \tfrac{585}{32} m^3\right] \frac{a^2}{a'^2} \\
+ &\left\{\begin{aligned} &(\tfrac{3}{2} - \tfrac{15}{2}\gamma^2 + \tfrac{9}{8} e^2 + \tfrac{9}{4} e'^2 - \tfrac{45}{4}\gamma^4 + 15\,\gamma^2e^2 - \tfrac{45}{4}\gamma^2e'^2 - \tfrac{27}{64} e^4 + \tfrac{27}{16} e^2e'^2 + \tfrac{45}{16} e'^4)\, m^2 \\ &\qquad + (-\tfrac{855}{128}\gamma^2 + \tfrac{729}{512} e^2)\, m^4 \\ &\qquad + (\tfrac{1935}{256}\gamma^2 - \tfrac{23355}{1024} e^2)\, m^5 \end{aligned}\right\} (1 + Q)^{-1}
\end{aligned}
$$

$$+\left\{\begin{array}{l}(\frac{27}{4}-\frac{351}{16}\gamma^2-\frac{297}{64}e^2+\frac{401}{16}e'^2+\frac{135}{2}\gamma^4-\frac{1053}{32}\gamma^2e^2-\frac{1297}{16}\gamma^2e'^2+\frac{675}{256}e^4-\frac{1079}{64}e^2e'^2)\,m^3\\ \qquad +(-\frac{20007}{1024}\gamma^2-\frac{24057}{4096}e^2)\,m^5\end{array}\right\}(1+Q)^{-2}$$

$$+(\frac{1995}{64}-\frac{7989}{64}\gamma^2-\frac{9969}{256}e^2+\frac{29535}{128}e'^2)\,m^4\,(1+Q)^{-3}$$

$$+(\frac{17709}{128}-\frac{376653}{512}\gamma^2-\frac{440787}{2048}e^2+\frac{883245}{512}e'^2)\,m^5\,(1+Q)^{-4}$$

$$+\ \frac{2431349}{4096}\,m^6\,(1+Q)^{-5}$$

$$+\ \frac{62329307}{24576}\,m^7\,(1+Q)^{-6}$$

(where writing $Q=0$, and omitting the terms in m^4 and m^5 in the coefficient of $(1+Q)^{-1}$, and the term in m^5 in the coefficient of $(1+Q)^{-2}$, we have the original formula of p. 237). And

$$h=nt\left\{[-\frac{45}{32}m^2-\frac{1935}{512}m^3]\frac{a^2}{a'^2}\right.$$

$$+\left\{\begin{array}{l}(-\frac{3}{4}+\frac{3}{2}\gamma^2-\frac{3}{2}e^2-\frac{9}{8}e'^2-\frac{51}{8}\gamma^2e^2+\frac{9}{4}\gamma^2e'^2+\frac{21}{64}e^2-\frac{9}{4}e^2e'^2-\frac{45}{32}e'^4)\,m^2\\ \qquad +(\frac{171}{128}\gamma^2-\frac{243}{128}e^2)\,m^4\\ \qquad +(-\frac{387}{256}\gamma^2+\frac{7785}{256}e^2)\,m^5\end{array}\right\}(1+Q)^{-1}$$

$$+\left\{\begin{array}{l}(\frac{9}{32}-\frac{27}{16}e^2-\frac{189}{32}\gamma^2+\frac{23}{32}e'^2+\frac{27}{16}\gamma^4+\frac{567}{16}\gamma^2e^2-\frac{99}{16}\gamma^2e'^2-\frac{675}{256}e^4-\frac{349}{16}e^2e'^2)\,m^3\\ \qquad +(-\frac{1539}{1024}\gamma^2-\frac{15309}{2048}e^2)\,m^5\end{array}\right\}(1+Q)^{-2}$$

$$+(\frac{177}{128}-\frac{195}{64}\gamma^2-\frac{699}{32}e^2+\frac{2685}{256}e'^2)\,m^4\,(1+Q)^{-3}$$

$$+(\frac{10949}{2048}-\frac{6369}{512}\gamma^2-\frac{133839}{1024}e^2+\frac{75759}{1024}e'^2)\,m^5\,(1+Q)^{-4}$$

$$+\ \frac{467977}{24576}\,m^6\,(1+Q)^{-5}$$

$$+\ \frac{26983045}{589824}\,m^7\,(1+Q)^{-6}$$

(where writing $Q=0$, and omitting the terms in m^4 and m^5 in the coefficient of $(1+Q)^{-1}$, and the term in m^5 in the coefficient of $(1+Q)^{-2}$, we have the original formula of p. 238). We hence have

$$\begin{aligned} l &= nt\,\{A+(1+B)\,Q+CQ^2\},\\ &= nt\,\{A+Q+BQ+CQ^2\},\\ g &= nt\,\{A'+B'Q+C'Q^2\},\\ h &= nt\,\{A''+B''Q+C''Q^2\},\end{aligned}$$

where $\left(\text{omitting the terms in } \frac{a^2}{a'^2}\right)$

$$\begin{aligned} A=1 &+(-\frac{7}{4}+\frac{21}{2}\gamma^2-\frac{3}{4}e^2-\frac{21}{8}e'^2+\frac{33}{4}\gamma^4-\frac{39}{8}\gamma^2e^2+\frac{63}{4}\gamma^2e'^2-\frac{9}{8}e^2e'^2-\frac{109}{32}e'^4)\,m^2\\ &+(-\frac{225}{32}+\frac{81}{4}\gamma^2-\frac{675}{64}e^2-\frac{825}{32}e'^2-\frac{243}{4}\gamma^4+\frac{1863}{32}\gamma^2e^2+\frac{629}{8}\gamma^2e'^2+\frac{2025}{256}e^4-\frac{105}{32}e'^4)\,m^3\\ &+(-\frac{3265}{128}+\frac{14577}{128}\gamma^2-\frac{1833}{64}e^2-\frac{48225}{256})\,m^4\\ &+(-\frac{243925}{2048}+\frac{177333}{256}\gamma^2-\frac{328065}{4096}e^2-\frac{1502265}{1024}e'^2)\,m^5\\ &+(-\frac{12626759}{24576})\,m^6\\ &+(-\frac{1365131021}{589824})\,m^7.\end{aligned}$$

$$\begin{aligned}
B = &\ \left(\tfrac{7}{4} - \tfrac{21}{2}\gamma^2 + \tfrac{3}{4}e^2 + \tfrac{21}{8}e'^2\right)m^2 \\
&+ \left(\tfrac{225}{16} - \tfrac{81}{2}\gamma^2 + \tfrac{675}{32}e^2 + \tfrac{825}{16}e'^2\right)m^3 \\
&+ \tfrac{9795}{128}m^4 \\
&+ \tfrac{243925}{512}m^5.
\end{aligned}$$

$$\begin{aligned}
C = &- \tfrac{7}{4}m^2 \\
&- \tfrac{675}{32}m^3.
\end{aligned}$$

$$\begin{aligned}
A' = &\ \left(\tfrac{3}{2} - \tfrac{15}{2}\gamma^2 + \tfrac{9}{8}e^2 + \tfrac{9}{4}e'^2 - \tfrac{45}{4}\gamma^4 + 15\gamma^2e^2 - \tfrac{45}{4}\gamma^2e'^2 - \tfrac{27}{16}e^4 + \tfrac{27}{16}e^2e'^2 + \tfrac{45}{16}e'^4\right)m^2 \\
&+ \left(\tfrac{27}{4} - \tfrac{351}{16}\gamma^2 - \tfrac{297}{64}e^2 + \tfrac{401}{16}e'^2 + \tfrac{135}{2}\gamma^4 - \tfrac{1053}{32}\gamma^2e^2 - \tfrac{1297}{16}\gamma^2e'^2 + \tfrac{675}{256}e^4 - \tfrac{1079}{64}e^2e'^2\right)m^3 \\
&+ \left(\tfrac{1995}{64} - \tfrac{16833}{128}\gamma^2 - \tfrac{19209}{512}e^2 + \tfrac{29535}{128}e'^2\right)m^4 \\
&+ \left(\tfrac{17709}{128} - \tfrac{765573}{1024}\gamma^2 - \tfrac{999051}{4096}e^2 + \tfrac{883245}{512}e'^2\right)m^5 \\
&+ \tfrac{2431349}{4096}m^6 \\
&+ \tfrac{62329307}{24576}m^7.
\end{aligned}$$

$$\begin{aligned}
B' = &\ \left(-\tfrac{3}{2} + \tfrac{15}{2}\gamma^2 - \tfrac{9}{8}e^2 - \tfrac{9}{4}e'^2\right)m^2 \\
&+ \left(-\tfrac{27}{2} + \tfrac{351}{8}\gamma^2 + \tfrac{297}{32}e^2 - \tfrac{401}{8}e'^2\right)m^3 \\
&- \tfrac{5985}{64}m^4 \\
&- \tfrac{17709}{32}m^5.
\end{aligned}$$

$$\begin{aligned}
C' = &\ \tfrac{3}{2}m^2 \\
&+ \tfrac{81}{4}m^3.
\end{aligned}$$

$$\begin{aligned}
A'' = &\ \left(-\tfrac{3}{4} + \tfrac{3}{2}\gamma^2 - \tfrac{3}{2}e^2 - \tfrac{9}{8}e'^2 - \tfrac{51}{8}\gamma^2e^2 + \tfrac{9}{4}\gamma^2e'^2 + \tfrac{21}{64}e^4 - \tfrac{9}{4}e^2e'^2 - \tfrac{45}{32}e'^4\right)m^2 \\
&+ \left(\tfrac{9}{32} - \tfrac{27}{16}\gamma^2 - \tfrac{189}{32}e^2 + \tfrac{23}{32}e'^2 + \tfrac{27}{16}\gamma^4 + \tfrac{567}{16}\gamma^2e^2 - \tfrac{99}{16}\gamma^2e'^2 - \tfrac{675}{256}e^4 - \tfrac{349}{16}e^2e'^2\right)m^3 \\
&+ \left(\tfrac{177}{128} - \tfrac{219}{128}\gamma^2 - \tfrac{3039}{128}e^2 + \tfrac{2685}{256}e'^2\right)m^4 \\
&+ \left(\tfrac{10949}{2048} - \tfrac{15825}{1024}\gamma^2 - \tfrac{220707}{2048}e^8 + \tfrac{75759}{1024}e'^2\right)m^5 \\
&+ \tfrac{467977}{24576}m^6 \\
&+ \tfrac{26983045}{589824}m^7.
\end{aligned}$$

$$\begin{aligned}
B'' = &+ \left(\tfrac{3}{4} - \tfrac{3}{2}\gamma^2 + \tfrac{3}{2}e^2 + \tfrac{9}{8}e'^2\right)m^2 \\
&+ \left(-\tfrac{9}{16} + \tfrac{27}{8}\gamma^2 + \tfrac{189}{16}e^2 - \tfrac{23}{16}e'^2\right)m^3 \\
&- \tfrac{531}{128}m^4 \\
&- \tfrac{10949}{512}m^5.
\end{aligned}$$

$$\begin{aligned}
C'' = &- \tfrac{3}{4}m^2 \\
&+ \tfrac{27}{32}m^3.
\end{aligned}$$

And in terms BQ, $B'Q$, $B''Q$, we have

$$\begin{aligned} Q = \ & (1 - \tfrac{9}{2}\gamma^2 + \tfrac{9}{8}e^2 + \tfrac{3}{2}e'^2)\,m^2 \\ & + (\tfrac{27}{8}\gamma^2 + \tfrac{675}{32}e^2)\,m^3 \\ & - \tfrac{515}{64}m^4 \\ & - \tfrac{787}{32}m^5, \end{aligned}$$

and in the terms CQ^2, $C'Q^2$, $C''Q^2$, simply $Q^2 = m^4$. Hence finally the required values of l, g, h, are

$$\begin{aligned} l = nt\Big\{ & 1 + [-\tfrac{45}{32}m^2 - \tfrac{7425}{512}m^3]\,\frac{a^2}{a'^2} \\ & + (-\tfrac{3}{4} + 6\gamma^2 + \tfrac{3}{8}e^2 - \tfrac{9}{8}e'^2 + \tfrac{45}{4}\gamma^4 - \tfrac{69}{8}\gamma^2e^2 + 9\gamma^2e'^2 + \tfrac{3}{32}e^4 + \tfrac{9}{16}e^2e'^2 - \tfrac{45}{32}e'^4)\,m^2 \\ & + (-\tfrac{225}{32} + \tfrac{189}{8}\gamma^2 + \tfrac{675}{64}e^2 - \tfrac{825}{32}e'^2 - \tfrac{1107}{16}\gamma^4 - \tfrac{81}{32}\gamma e^2 + \tfrac{349}{4}\gamma^2e'^2 + \tfrac{2475}{64}e^2e'^2)\,m^3 \\ & + (-\tfrac{4071}{128} + \tfrac{15852}{128}\gamma^2 + \tfrac{31605}{512}e^2 - \tfrac{61179}{256}e'^2)\,m^4 \\ & + (-\tfrac{265493}{2048} + \tfrac{335403}{512}\gamma^2 + \tfrac{1483665}{4096}e^2 - \tfrac{1767840}{1024}e'^2)\,m^5 \\ & + (-\tfrac{12822631}{24576})\,m^6 \\ & + (-\tfrac{1273925965}{589824})\,m^7, \end{aligned}$$

$$\begin{aligned} g = nt\Big\{ & \quad [\tfrac{45}{16}m^2 + \tfrac{585}{32}m^3]\,\frac{a^2}{a'^2} \\ & + (\tfrac{3}{2} - \tfrac{15}{2}\gamma^2 + \tfrac{9}{8}e^2 + \tfrac{9}{4}e'^2 - \tfrac{45}{4}\gamma^4 + 15\gamma^2e^2 - \tfrac{45}{4}\gamma^2e'^2 - \tfrac{27}{64}e^4 + \tfrac{27}{16}e^2e'^2 + \tfrac{45}{16}e'^4)\,m^2 \\ & + (\tfrac{27}{4} - \tfrac{351}{16}\gamma^2 - \tfrac{297}{64}e^2 + \tfrac{401}{16}e'^2 + \tfrac{135}{2}\gamma^4 - \tfrac{1053}{32}\gamma^2e^2 - \tfrac{1297}{16}\gamma^2e'^2 + \tfrac{675}{256}e^4 - \tfrac{1079}{64}e^2e'^2)\,m^3 \\ & + (\tfrac{1899}{64} - \tfrac{15009}{128}\gamma^2 - \tfrac{20649}{512}e^2 - \tfrac{28959}{128}e'^2)\,m^4 \\ & + (\tfrac{15981}{128} - \tfrac{663621}{1024}\gamma^2 - \tfrac{1152843}{4096}e^2 + \tfrac{847213}{512}e'^2)\,m^5 \\ & + \tfrac{2103893}{4096}m^6 \\ & + \tfrac{52802843}{24576}m^7, \end{aligned}$$

$$\begin{aligned} h = nt\Big\{ & \quad [-\tfrac{45}{32}m^2 - \tfrac{1935}{512}m^3)\,\frac{a^2}{a'^2} \\ & + (-\tfrac{3}{4} + \tfrac{3}{2}\gamma^2 - \tfrac{3}{2}e^2 - \tfrac{9}{8}e'^2 - \tfrac{51}{8}\gamma^2e^2 + \tfrac{9}{4}\gamma^2e'^2 + \tfrac{21}{64}e^4 - \tfrac{9}{4}e^2e'^2 - \tfrac{45}{32}e'^4)\,m^2 \\ & + (\tfrac{9}{32} - \tfrac{27}{16}\gamma^2 - \tfrac{189}{32}e^2 + \tfrac{23}{32}e'^2 + \tfrac{27}{16}\gamma^4 + \tfrac{567}{16}\gamma^2e^2 - \tfrac{99}{16}\gamma^2e'^2 - \tfrac{675}{256}e^4 - \tfrac{349}{16}e^2e'^2)\,m^3 \\ & + (\tfrac{273}{128} - \tfrac{843}{128}\gamma^2 - \tfrac{2739}{128}e^2 + \tfrac{3261}{256}e'^2)\,m^4 \\ & + (\tfrac{9797}{2048} - \tfrac{7185}{1024}\gamma^2 - \tfrac{165411}{2048}e^2 + \tfrac{73423}{1024}e'^2)\,m^5 \\ & + \tfrac{199273}{24576}m^6 \\ & + \tfrac{6657733}{589824}m^7, \end{aligned}$$

which values satisfy, as they should do, the equation $l+g+h=nt$. I recall that the precise signification of the constants is as follows: n is the coefficient of t in the expression of the Moon's longitude in terms of the time, a the corresponding elliptic value of the mean distance ($n^2a^3=$ sum of masses), e the eccentricity, such that in the expression of the longitude the coefficient of the leading term of the equation of the centre has its elliptic value

$$=2e-\tfrac{1}{4}e^3+\tfrac{5}{96}e^5$$

and γ the sine of the half-inclination, such that in the expression of the latitude the coefficient of the leading term has its elliptic value

$$=2\gamma-2\gamma e^2-\tfrac{1}{4}\gamma^3+\tfrac{7}{32}\gamma e^4+\tfrac{1}{4}\gamma^3e^2-\tfrac{5}{144}\gamma e^6$$

n', a' are the mean motion and mean distance of the Sun, $m=\dfrac{n'}{n}$, and e' is the eccentricity of the Sun's orbit, considered as constant.

481.

ON THE EXPRESSION OF M. DELAUNAY'S $h+g$ IN TERMS OF HIS FINALLY ADOPTED CONSTANTS.

[From the *Monthly Notices of the Royal Astronomical Society*, vol. XXXII. (1871—72), p. 74.]

I HAD the pleasure of receiving from M. Delaunay a letter dated Paris, 17th Dec. 1871, in which he informs me that, on referring to his papers, he had found there expressions for l, g, h, identical with those given by me in the November Number of the *Monthly Notices*,—with only a single typographical error, $\frac{23}{33}e'^2m^3$ instead of $\frac{23}{32}e'^2m^3$ [*ante* p. 532, corrected] in my expression of h.

M. Delaunay mentions also that he had obtained four additional terms in the expression for $h+g$ (longitude of the Moon's perigee), and that the complete expression in terms of the finally adopted constants is

$$
\begin{aligned}
h+g = \\
nt \Big\{ & \left(\tfrac{3}{4} - 6\gamma^2 - \tfrac{3}{8}e^2 + \tfrac{9}{8}e'^2 - \tfrac{45}{4}\gamma^4 + \tfrac{69}{8}\gamma^2e^2 - 9\gamma^2e'^2 - \tfrac{3}{32}e^4 - \tfrac{9}{16}e^2e'^2 + \tfrac{45}{32}e'^4\right)m^2 \\
& + \left(\tfrac{225}{32} - \tfrac{189}{8}\gamma^2 - \tfrac{675}{64}e^2 + \tfrac{825}{32}e'^2 + \tfrac{1107}{16}\gamma^4 + \tfrac{81}{32}\gamma^2e^2 - \tfrac{349}{4}\gamma^2e'^2 - \tfrac{2475}{64}e^2e'^2\right)m^3 \\
& + \left(\tfrac{4071}{128} - \tfrac{3963}{32}\gamma^2 - \tfrac{31605}{512}e^2 + \tfrac{61179}{256}e'^2\right)m^4 \\
& + \left(\tfrac{265493}{2048} - \tfrac{335403}{512}\gamma^2 - \tfrac{1483665}{4096}e^2 + \tfrac{1767849}{1024}e'^2\right)m^5 \\
& + \left(\tfrac{12822631}{24576} - \tfrac{25291729}{16384}e^2\right)m^6 \\
& + \left(\tfrac{1273925965}{589824} + \tfrac{3520388855}{1179648}e^2\right)m^7 \\
& + \tfrac{71028685589}{7077888}m^8 \\
& + \tfrac{32145914707741}{679477248}m^9 \\
& + \left[\tfrac{45}{32}m^2 + \tfrac{7425}{512}m^3\right]\frac{a^2}{a'^2}\Big\}.
\end{aligned}
$$

[Observe that $h+g$ is $=nt-l$, and compare with the expression for l, *ante* p. 532.]

482.

NOTE ON A PAIR OF DIFFERENTIAL EQUATIONS IN THE LUNAR THEORY.

[From the *Monthly Notices of the Royal Astronomical Society*, vol. XXXII. (1871—72), pp. 31—32.]

THE equations

$$\frac{d}{dt}\frac{d\rho}{dt} - \rho\left(\frac{dv}{dt}\right)^2 + \frac{1}{\rho^2} = km^2\rho\,\{\tfrac{1}{2} + \tfrac{3}{2}\cos(2v - 2mt)\},$$

$$\frac{d}{dt}\rho^2\frac{dv}{dt} \qquad = jm^2\rho^2\,\{\;-\tfrac{3}{2}\sin(2v - 2mt)\},$$

taking therein $j = k = 1$ in effect present themselves in the Lunar Theory, and particular integrals in series have been obtained, the development being carried to a great extent; but I give the results only as far as m^4, viz., writing

$$t - mt = D,$$

we have

$$\begin{aligned} v = t &+ (\tfrac{11}{8}m^2 + \tfrac{59}{12}m^3 + \tfrac{893}{72}m^4)\sin 2D \\ &\qquad\qquad\qquad + \tfrac{201}{256}m^4 \sin 4D, \\ \frac{1}{\rho} = 1 &+ \tfrac{1}{6}m^2 \qquad\qquad - \tfrac{179}{288}m^4 \\ &+ (\;m^2 + \tfrac{19}{6}m^3 + \tfrac{131}{18}m^4)\cos 2D \\ &\qquad\qquad\qquad + \tfrac{7}{8}m^4 \cos 4D. \end{aligned}$$

In the Lunar Theory j and k are properly each $= \dfrac{1}{1 + \dfrac{E}{m'}}$ (E the mass of the Earth, m' that of the Sun), but they are taken to be $= 1$; the numerical difference is inappreciable; but there would be a considerable theoretical advantage in retaining in the equations the coefficients j, k [regarded as each of them $=k$]: in fact, the developments could then be arranged according to the powers of k, that is according to the powers of the disturbing force; whereas, when k is taken $=1$, we have only a development in powers of m, and since m also presents itself through the coefficient $2-2m$ of t in $2v-2mt$, terms which are really of different orders in regard to the disturbing force, are united together into a single term: so that, instead of a term of the form $(Ak+Bk^2+\&c.)\,m^p$, where A, B, are numerical, we have the term $(A+B+..)\,m^p$, where of course $A+B..$ is given as a single numerical coefficient. There is no equal advantage in retaining the two coefficients k, j, as this only serves to show how a term arises from the central and tangential forces respectively; thus retaining these coefficients, the integrals as far as m^2 are

$$v = \quad t \quad + (\tfrac{1}{2}k+\tfrac{7}{8}j)\,m^2 \sin 2D,$$

$$\frac{1}{\rho} = 1 + \tfrac{1}{6}\,m^2k + (\tfrac{1}{2}k+\tfrac{1}{2}j)\,m^2 \cos 2D,$$

agreeing with the former result when $k=j=1$; but there is, nevertheless, some interest in retaining the two coefficients. I hope to develope the results somewhat further, and to communicate them to the Society.

483.

ON A PAIR OF DIFFERENTIAL EQUATIONS IN THE LUNAR THEORY.

[From the *Monthly Notices of the Royal Astronomical Society*, vol. XXXII. (1871—72), pp. 201—206.]

I CONSIDER the differential equations

$$\frac{d}{dt}\frac{d\rho}{dt} - \rho\left(\frac{dv}{dt}\right)^2 + \frac{1}{\rho^2} = km^2\rho\,\{\tfrac{1}{2} + \tfrac{3}{2}\cos(2v - 2mt)\},$$

$$\frac{d}{dt}\left(\rho^2\frac{dv}{dt}\right) \qquad = jm^2\rho^2\,\{\quad -\tfrac{3}{2}\sin(2v - 2mt)\},$$

which when $j = k = 1$ give the following equations in the lunar theory ($D = t - mt$):

$$\begin{aligned}\frac{1}{\rho} = 1 &+ \tfrac{1}{6}m^2 - \tfrac{179}{288}m^4 - \tfrac{97}{48}m^5 - \tfrac{757}{162}m^6 - \tfrac{4039}{432}m^7 - \tfrac{34751189}{1990656}m^8 - \tfrac{155067635}{4976640}m^9 \\ &+ \cos 2D\,[m^2 + \tfrac{19}{6}m^3 + \tfrac{131}{18}m^4 + \tfrac{383}{27}m^5 + \tfrac{510565}{20736}m^6 + \tfrac{23140781}{622080}m^7 \\ &\qquad + \tfrac{355021217}{9331200}m^8 + \tfrac{27888590059}{34992000}m^9] \\ &+ \cos 4D\,[\tfrac{7}{8}m^4 + \tfrac{2737}{480}m^5 + \tfrac{162869}{7200}m^6 + \tfrac{7554833}{108000}m^7 + \tfrac{2389416723}{12960000}m^8 + \tfrac{2335230125283}{5443200000}m^9] \\ &+ \cos 6D\,[\tfrac{219}{256}m^6 + \tfrac{151339}{17920}m^7 + \tfrac{29887443}{627200}m^8 + \tfrac{98978623957}{444528000}m^9] \\ &+ \cos 8D\,[\tfrac{2701}{3072}m^8 + \tfrac{70033633}{6021120}m^9],\end{aligned}$$

or as far as m^7,

$$\begin{aligned}\rho = 1 &- \tfrac{1}{6}m^2 + \tfrac{331}{288}m^4 + \tfrac{83}{16}m^5 + \tfrac{42775}{2592}m^6 + \tfrac{4787}{108}m^7 \\ &+ \cos 2D\,[-m^2 - \tfrac{19}{6}m^3 - \tfrac{125}{18}m^4 - \tfrac{709}{54}m^5 - \tfrac{485173}{20736}m^6 - \tfrac{24487949}{622080}m^7] \\ &+ \cos 4D\,[-\tfrac{3}{8}m^4 - \tfrac{1217}{480}m^5 - \tfrac{74069}{7200}m^6 - \tfrac{1749779}{54000}m^7] \\ &+ \cos 6D\,[-\tfrac{59}{256}m^6 - \tfrac{126193}{53760}m^7],\end{aligned}$$

($\frac{1}{\rho}$ is given by M. Delaunay only as far as m^5, the additional terms of $\frac{1}{\rho}$ and expression for ρ were kindly communicated to me by Prof. Adams); and

$$
\begin{aligned}
v = t & \\
& + \sin 2D\left(\tfrac{11}{8} m^2 + \tfrac{59}{12} m^3 + \tfrac{893}{72} m^4 + \tfrac{2855}{108} m^5 + \tfrac{8304449}{165888} m^6\right. \\
& \qquad \left. + \tfrac{102859909}{1244160} m^7 + \tfrac{7596606727}{746496000} m^8 - \tfrac{8051418161}{111974400} m^9\right) \\
& + \sin 4D\left(\tfrac{201}{256} m^4 + \tfrac{649}{120} m^5 + \tfrac{647623}{28800} m^6 + \tfrac{31363361}{432000} m^7 + \tfrac{123030377303}{414720000} m^8\right) \\
& + \sin 6D\left(\tfrac{3715}{6144} m^6 + \tfrac{664571}{107520} m^7\right)
\end{aligned}
$$

(Delaunay, t. II. pp. 815, 836, 845).

To integrate the original equations write

$$
\begin{aligned}
\rho &= 1 + \rho_1 + \rho_2 + \ldots, \\
v &= t + v_1 + v_2 + \ldots,
\end{aligned}
$$

where the suffixes indicate the degrees in the coefficients k, j conjointly: the equations for ρ_n, v_n take the form

$$
\begin{aligned}
\frac{d}{dt}\frac{d\rho_n}{dt} - 3\rho_n - 2\frac{dv_n}{dt} + V_n &= Q_n, \\
\frac{d}{dt}\left(\frac{dv_n}{dt} + 2\rho_n + U_n\right) \qquad &= P_n,
\end{aligned}
$$

where V_n, U_n, P_n, Q_n do not contain ρ_n or v_n. From the second equation we have

$$\frac{dv_n}{dt} + 2\rho_n + U_n = \Omega_n + \int P_n\,dt,$$

where Ω_n is a constant of integration, the integral $\int P_n\,dt$ containing only periodic terms; and then adding twice this to the first equation we have

$$\frac{d}{dt}\frac{d\rho_n}{dt} + \rho_n + V_n + 2U_n = 2\Omega_n + Q_n + 2\int P_n\,dt$$

which determines ρ_n; and substituting its value in the other equation we have $\frac{dv_n}{dt}$, and thence v_n; the constant Ω_n is determined so that $\frac{dv_n}{dt}$ may contain no constant term. We have

$$
\begin{aligned}
V_1 &= 0, \\
V_2 &= -\left(\frac{dv_1}{dt}\right)^2 - 2\rho_1\frac{dv_1}{dt} + 3\rho_1^2, \\
V_3 &= -2\frac{dv_1}{dt}\frac{dv_2}{dt} - 2\rho_1\frac{dv_2}{dt} - \rho_1\left(\frac{dv_1}{dt}\right)^2 \\
&\quad - 2\rho_2\frac{dv_2}{dt} + 6\rho_1\rho_2 - 4\rho_1^3,
\end{aligned}
$$

&c.

$$
\begin{aligned}
U_1 &= 0, \\
U_2 &= 2\rho_1\frac{dv_1}{dt} + \rho_1^2, \\
U_3 &= 2\rho_1\frac{dv_2}{dt} + (2\rho_2 + \rho_1^2)\frac{dv_1}{dt} + 2\rho_1\rho_2,
\end{aligned}
$$

&c.

$$Q_1 = km^2\left(\tfrac{1}{2} + \tfrac{3}{2}\cos 2D\right),$$
$$Q_2 = km^2\left\{3v_1 \sin 2D + \rho_1\left(\tfrac{1}{2} + \tfrac{3}{2}\cos 2D\right)\right\},$$
$$Q_3 = km^2\left\{- 3v_2 \sin 2D - 3v_1^2 \cos 2D + \rho_1 v_1 . 3 \sin 2D + \rho_2\left(\tfrac{1}{2} + \tfrac{3}{2}\cos 2D\right)\right\},$$
&c.

$$P_1 = jm^2\left(-\tfrac{3}{2}\sin 2D\right),$$
$$P_2 = jm^2\left(-3v_1\cos D - 3\rho_1 \sin 2D\right),$$
$$P_3 = jm^2\left\{-3v_2 \cos 2D + 3v_1^2 \sin 2D - 6\rho_1 v_1 \cos 2D + (2\rho_2 + \rho_1^2) . -\tfrac{3}{2}\sin 2D\right\},$$
&c.

In particular attending to the values of P_1, Q_1 the equations for ρ_1, v_1 are in their original form

$$\frac{d}{dt}\frac{d\rho_1}{dt} - 3\rho_1 + 2\frac{dv_1}{dt} = km^2\left(\tfrac{1}{2} + \tfrac{3}{2}\cos 2D\right),$$

$$\frac{d}{dt}\left(\frac{dv_1}{dt} + 2\rho_1\right) \qquad = jm^2\left(\quad -\tfrac{3}{2}\sin 2D\right),$$

whence in the transformed form they are

$$\frac{dv_1}{dt} + 2\rho_1 = \Omega_1 + \frac{3jm^2}{4(1-m)}\cos 2D,$$

and

$$\frac{d^2\rho_1}{dt^2} + \rho_1 = 2\Omega_1 + km^2\left(\tfrac{1}{2} + \tfrac{3}{2}\cos 2D\right) + \frac{\tfrac{3}{2}jm^2}{1-m}\cos 2D.$$

Thus the constant term of ρ_1 is $2\Omega_1 + \tfrac{1}{2}km^2$, giving in $\dfrac{dv_1}{dt}$ a constant term $-3\Omega_1 - km^2$ this must vanish, or we have $\Omega_1 = -\tfrac{1}{3}km^2$; and the equations thus become

$$\frac{dv_1}{dt} + 2\rho_1 = -\tfrac{1}{3}km^2 + \frac{3jm^2}{4(1-m)}\cos 2D,$$

$$\frac{d^2\rho_1}{dt^2} + \quad \rho_1 = -\tfrac{1}{6}km^2 + \left(\tfrac{3}{2}km^2 + \frac{\tfrac{3}{2}jm^2}{1-m}\right)\cos 2D,$$

and then completing the integration

$$\rho_1 = -\tfrac{1}{6}km^2 + \left\{\frac{-\tfrac{3}{2}km^2}{3-8m+4m^2} + \frac{-\tfrac{3}{2}jm^2}{(1-m)(3-8m+4m^2)}\right\}\cos 2D,$$

$$v_1 = \left\{\frac{\tfrac{3}{2}km^2}{(1-m)(3-8m+4m^2)} + \frac{\tfrac{3}{8}jm^2(7-8m+4m^2)}{(1-m)^2(3-8m+4m^2)}\right\}\sin 2D,$$

which are the accurate values of ρ_1 and v_1.

Expanding as far as m^6 we have

$$\rho_1 = k\left(-\tfrac{1}{6}m^2\right) + \cos 2D\left\{ k\left(-\tfrac{1}{2}m^2 - \tfrac{4}{3}m^3 - \tfrac{26}{9}m^4 - \tfrac{160}{27}m^5 - \tfrac{968}{81}m^6\right) + j\left(-\tfrac{1}{2}m^2 - \tfrac{11}{6}m^3 - \tfrac{85}{18}m^4 - \tfrac{575}{54}m^5 - \tfrac{3661}{162}m^6\right)\right\},$$

which for $j = k$ is

$$= k\left(- \quad m^2 - \tfrac{19}{6}m^3 - \tfrac{137}{18}m^4 - \tfrac{895}{54}m^5 - \tfrac{5597}{162}m^6\right),$$

and

$$v_1 = \sin 2D\{\ k(\ \tfrac{1}{2}m^2 + \tfrac{11}{6}m^3 + \tfrac{85}{18}m^4 + \tfrac{575}{54}m^5 + \tfrac{3661}{162}m^6)$$
$$+ j(\ \tfrac{7}{8}m^2 + \tfrac{37}{12}m^3 + \tfrac{589}{72}m^4 + \tfrac{1037}{54}m^5 + \tfrac{27331}{648}m^6)\},$$

which for $j = k$ is
$$= k(\ \tfrac{11}{8}m^2 + \tfrac{59}{12}m^3 + \tfrac{929}{72}m^4 + \tfrac{896}{27}m^5 + \tfrac{41975}{648}m^6)\ .$$

I have, not in general, but for the value $j = k$, calculated ρ_2 and v_2 as far as m^6: I have not made the calculation for ρ_3 and v_3, but their values may be deduced from the foregoing values of ρ, v; the final expressions (when $j = k$) of ρ, $= 1 + \rho_1 + \rho_2 + \rho_3 + \ldots$ and v, $= t + v_1 + v_2 + v_3 \ldots$ are

$$\begin{aligned}
\rho = 1 & \\
& + k\ (-\tfrac{1}{6}m^2 \qquad\qquad\qquad\qquad\qquad\qquad\qquad\qquad\qquad\qquad\qquad) \\
& + k^2(\qquad\qquad\qquad\qquad\quad \tfrac{331}{288}m^4 + \tfrac{83}{16}m^5 + \tfrac{5113}{288}m^6) \\
& + k^3(\qquad\qquad\qquad\qquad\qquad\qquad\qquad\qquad\quad - \tfrac{1621}{1296}m^6) \\
+ \cos 2D\{\ & k\ (-\ m^2 - \tfrac{19}{6}m^3 - \tfrac{137}{18}m^4 - \tfrac{895}{54}m^5 - \tfrac{5597}{162}m^6) \\
& + k^2(\qquad\qquad\qquad\qquad\quad \tfrac{2}{3}m^4 + \tfrac{31}{9}m^5 + \tfrac{329}{27}m^6) \\
& + k^3(\qquad\qquad\qquad\qquad\qquad\qquad\qquad\qquad\quad - \tfrac{2381}{2304}m^6)\} \\
+ \cos 4D\{\ & k^2(\qquad\qquad\qquad\qquad - \tfrac{3}{8}m^4 - \tfrac{1217}{480}m^5 - \tfrac{76589}{7200}m^6) \\
& + k^3(\qquad\qquad\qquad\qquad\qquad\qquad\qquad\qquad\quad + \tfrac{7}{20}m^6)\} \\
+ \cos 6D\{\ & k^3(\qquad\qquad\qquad\qquad\qquad\qquad\qquad\qquad\quad - \tfrac{59}{256}m^6)\},
\end{aligned}$$

and

$$\begin{aligned}
v = t & \\
+ \sin 2D\{\ & k\ (\tfrac{11}{8}m^2 + \tfrac{59}{12}m^3 + \tfrac{929}{72}m^4 + \tfrac{896}{27}m^5 + \tfrac{41975}{648}m^6) \\
& + k^2(\qquad\qquad\qquad\qquad - \tfrac{1}{2}m^4 - \tfrac{41}{12}m^5 - \tfrac{43}{3}m^6) \\
& + k^3(\qquad\qquad\qquad\qquad\qquad\qquad\qquad\qquad\quad - \tfrac{783}{2048}m^6)\} \\
+ \sin 4D\{\ & k^2(\qquad\qquad\qquad\qquad\quad \tfrac{201}{256}m^4 + \tfrac{649}{120}m^5 + \tfrac{665263}{28800}m^6) \\
& + k^3(\qquad\qquad\qquad\qquad\qquad\qquad\qquad\qquad\quad - \tfrac{49}{80}m^6)\} \\
+ \sin 6D\{\ & k^3(\qquad\qquad\qquad\qquad\qquad\qquad\qquad\qquad\quad + \tfrac{3715}{6144}m^6)\};
\end{aligned}$$

which for $k = 1$ agree with the foregoing formulæ (verifying them as far as m^5); the present formulæ exhibit the manner in which the expressions depend on the several powers of the disturbing force.

484.

ON THE VARIATIONS OF THE POSITION OF THE ORBIT IN THE PLANETARY THEORY.

[From the *Monthly Notices of the Royal Astronomical Society*, vol. XXXII. (1871—72), pp. 206—211.]

IT has always appeared to me that in the Planetary Theory, more especially when the method of the variation of the elements is made use of, there is a difficulty as to the proper mode of dealing with the inclinations and longitudes of the nodes, hindering the ulterior development of the theory. Considering the case of two planets m, m', and referring their orbits to any fixed plane and fixed origin of longitudes therein, let θ, θ' be the longitudes of the nodes, ϕ, ϕ' the inclinations ($p = \tan\phi \sin\theta$, $q = \tan\phi\cos\theta$, &c., as usual); then the disturbing functions for m, m' respectively are developed, not explicitly in terms of ϕ, ϕ', θ, θ', but in terms of Φ, the mutual inclination of the two orbits, and of Θ, Θ' the longitudes in the two orbits respectively of the mutual node of the two orbits; Φ and Θ, Θ' being functions (and complicated ones) of ϕ, ϕ', θ, θ'. Moreover, although in the general theory of the secular variations of the orbits of the planetary system, θ, ϕ, &c., are, as above, referred to one fixed plane (the ecliptic of a certain date), yet in the theory of each particular planet it is the practice, and obviously the convenient one, to refer for such planet the θ, ϕ to its own fixed plane (the orbit of the planet at a certain date), the effect of course being that ϕ, and consequently p, q, instead of being of the order of the inclinations to the ecliptic, are only of the order of the disturbing forces. It has occurred to me that the last-mentioned plan should be adhered to *throughout;* viz., that for each planet m, the position of its variable orbit should be determined by θ, the longitude of its node, and ϕ, the inclination in reference to the appropriate fixed plane (orbit of the planet at a certain date) and origin of longitude therein. The disturbing functions for the planets m and m' will of course depend not only on θ, θ', ϕ, ϕ', but on the quantities Φ, Θ, Θ' which determine the mutual positions of the two fixed planes of reference and origins of longitude therein, *these last being however absolute constants not affected by any variation of the elements;* so that as regards the variation of the elements the disturbing functions are in fact given as *explicit* functions of the variable elements θ, θ', ϕ, ϕ'; and where ϕ, ϕ' and therefore also p, q, p', q' are only of the order of the disturbing forces.

I proceed to work out this idea, for the present considering the development of the Disturbing Function only as far as the first powers of p, q, &c. For comparison with the ordinary theory, observe that in this theory the disturbing function contains only the *second* powers of the p, q, &c., made use of therein; these are in fact of a form such as $P+p$, $Q+q$, ... where P, Q are absolute constants and p, q, ... are the p, q, ... of the present theory; the ordinary theory gives therefore in the disturbing function a series of terms involving $(P+p)^2$, $(P+p)(Q+q)$, ... which I now take account of only as far as the first powers of p, q, ... viz., they are in effect reduced to P^2+2Pp, $PQ+Pq+Qp$, &c. ... The present theory is thus not now developed to the extent of giving the p, q, ... of the ordinary theory in the more complete form as the solutions of a system of simultaneous linear differential equations, but only to the extent of obtaining for these p, q, ... respectively the terms which are proportional to the time.

I commence with the following subsidiary problem. Consider a spherical triangle ABC (sides a, b, c, angles A, B, C, as usual), and taking the side c as constant, but the angles A and B as variable, let it be required to find the variations of C, a, b in terms of variations dA, dB and the variable elements C, a, b themselves. Although the geometrical proof would be more simple, I give the analytical one, as it may be useful.

We have

$$\cos C = -\cos A \cos B + \sin A \sin B \cos c,$$

and thence

$$\begin{aligned} -\sin C dC &= (\sin A \cos B + \cos A \sin B \cos c)\, dA \\ &\quad + (\sin B \cos A + \sin A \cos B \cos c)\, dB \\ &= \frac{\sin B \sin c}{\tan b}\, dA + \frac{\sin A \sin c}{\tan a}\, dB, \end{aligned}$$

that is

$$-\frac{\sin C}{\sin c}\, dC = \frac{\sin B \cos b}{\sin b}\, dA + \frac{\sin A \cos a}{\sin a}\, dB,$$

or finally

$$-dC = \cos b dA + \cos a dB.$$

Next

$$\sin a = \sin c \frac{\sin A}{\sin C},$$

or, differentiating,

$$\cos a\, da = \frac{\sin c}{\sin^2 C}(\sin C \cos A dA - \cos C \sin A dC)$$

or, substituting for dC its value,

$$= \frac{\sin c}{\sin^2 C}\left\{dA\,(\sin C\cos A + \cos C\sin A\cos b) + dB\cos C\sin A\cos a\right\},$$

$$= \frac{\sin c}{\sin^2 C}\left\{dA\,\frac{\sin A\sin b\cos a}{\sin a} + dB\cos C\sin A\cos a\right\},$$

that is

$$da = \frac{1}{\sin C}\left\{dA\,\frac{\sin A}{\sin a}\sin b + dB\cos C\sin A\right\} \div \frac{\sin C}{\sin c},$$

or, on the right-hand writing $\frac{\sin A}{\sin a}$ instead of $\frac{\sin C}{\sin c}$, this is

$$da = \frac{1}{\sin C}(dA\sin b + dB\cos C\sin a)\,;$$

and similarly

$$db = \frac{1}{\sin C}(dB\sin a + dA\cos C\sin b).$$

Now let the continuous lines represent the orbits of m, m' at certain dates, O, Q the origins of longitude therein; and the dotted lines the variable orbits of the planets respectively.

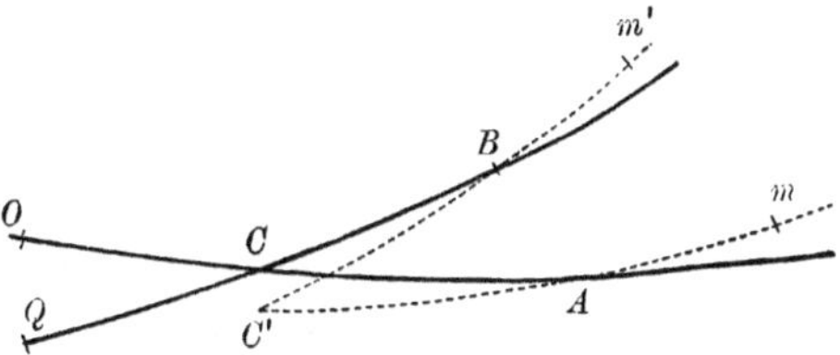

Write

$$OC = \Theta,\quad CA = \theta,\quad \angle CAC' = \phi,$$
$$QC = \Theta',\quad CB = \theta',\quad \angle CBC = \phi',$$
$$\angle C = \Phi.$$

Then, answering to the notation of the lemma, we have

$$a = \theta',\quad b = \theta,\quad C = \Phi,\qquad dA = \phi,\qquad dB = -\phi',$$
$$\text{or say}\qquad = \tan\phi,\qquad = -\tan\phi',$$

whence

$$C'B = a + da,$$

$$= \theta' + \frac{1}{\sin\Phi}(\tan\phi\sin\theta - \tan\phi'\cos\Phi\sin\theta'),$$

$$= \theta' + \frac{1}{\sin\Phi}(p - p'\cos\Phi),$$

$$C'A = b + db,$$

$$= \theta + \frac{1}{\sin \Phi}(-\tan \theta' \sin \phi' + \tan \phi \cos \Phi \sin \theta),$$

$$= \theta - \frac{1}{\sin \Phi}(p' - p \cos \Phi),$$

$$\angle C' = C + dC = \Phi - \cos \theta \tan \phi + \cos \theta' \tan \phi', \; = \Phi - q + q'.$$

Suppose v, v' are the longitudes of the planets in their two orbits respectively; that is

$$v = OA + Am = \Theta + \theta + Am,$$
$$v' = QB + Bm' = \Theta + \theta' + Bm',$$

whence

$$C'm = C'A + Am, \; = v - \Theta - \frac{1}{\sin \Phi}(p' - p \cos \Phi),$$

$$C'm' = C'B + Bm, \; = v' - \Theta' + \frac{1}{\sin \Phi}(p - p' \cos \Phi),$$

$$\angle C' = \Phi - q + q';$$

say these values are $v - \Theta + x$, $v' - \Theta' + x'$, $\Phi + y$. Then if $\overline{H}$ is the angular distance mm' of the two planets,

$$\cos \overline{H} = \cos (v - \Theta + x) \cos (v' - \Theta' + x') + \sin (v - \Theta + x) (\sin v' - \Theta' + x') \cos (\Phi + y),$$
$$= \cos (v - \Theta) \cos (v' - \Theta') + \sin (v - \Theta) \sin (v' - \Theta') \cos \Phi$$
$$+ x \, [- \sin (v - \Theta) \cos (v' - \Theta') + \cos (v - \Theta) \sin (v' - \Theta') \cos \Phi]$$
$$+ x' \, [- \cos (v - \Theta) \sin (v' - \Theta') + \sin (v - \Theta) \cos (v' - \Theta') \cos \Phi]$$
$$+ y \, [- \sin (v - \Theta) \sin (v' - \Theta') \sin \Phi],$$
$$= \cos H + \nabla \text{ suppose.}$$

The disturbing function for the planet m disturbed by m' is

$$\Omega = m' \left\{ \frac{1}{\sqrt{r^2 + r'^2 - 2rr' \cos \overline{H}}} - \frac{r \cos \overline{H}}{r'^2} \right\},$$

($\Omega = -R$, if R is the disturbing function of the *Mécanique Céleste*); and the term hereof which involves ∇ is

$$= \nabla \frac{d\Omega}{d \,.\, \cos \overline{H}}$$

where after the differentiation $\cos \overline{H}$ is replaced by $\cos H$,

$$= m' \left\{ \frac{rr'}{(r^2 + r'^2 - 2rr' \cos H)^{\frac{3}{2}}} + \frac{r}{r'^2} \right\} \nabla,$$

viz., this is a linear function of x, x', y, that is of p, q, p', q', with coefficients which of course involve the other variable elements and the time; but it will be remembered that Θ, Θ', Φ are not variable elements, but are absolute constants. The variations of p depend upon $\frac{d\Omega}{dq}$ and those of q on $\frac{d\Omega}{dp}$, and the quantities p, q, p', q',... disappear from these differential coefficients $\frac{d\Omega}{dq}$, $\frac{d\Omega}{dp}$; that is, disregarding periodic terms, and the variations of the elements, we obtain $\frac{dp}{dt}$, $\frac{dq}{dt}$ as absolute constants, or reckoning the time from the epoch belonging to the fixed orbit of m, we have p, q as mere multiples of the time ($p = At$, $q = Bt$, where A and B are constants); agreeing with the statement preceding the investigation.

Observe that the p, q, as used above, have reference not only to the fixed orbit of m, but also to the node thereon of the fixed orbit of m': we may, if we please, write $\mathrm{p} = \tan\phi \sin(\Theta + \theta)$, $\mathrm{q} = \tan\phi \cos(\Theta + \theta)$, that is, $\mathrm{p} = q \sin\Theta + p \cos\Theta$, $Q = q\cos\Theta - p\sin\Theta$ (or $p = \mathrm{p}\cos\Theta - \mathrm{q}\sin\Theta$, $q = P\sin\Theta + \mathrm{q}\cos\Theta$), and in place of p, q introduce into the formulæ p and q, which have reference only to the fixed orbit of m, and similarly writing $\mathrm{p}' - \tan\phi' \sin(\Theta' + \theta')$, $\mathrm{q}' = \tan\phi' \cos(\Theta + \theta')$, instead of p', q' introduce p′, q′ which have reference only to the fixed orbit of m'.

I remark that a table for the relative positions of the orbits of the eight Planets for the Epoch 1st January, 1850, is given in Leverrier's *Annales de l'Observ. de Paris*, t. II. (1856), pp. 64—66.

485.

PROBLEMS AND SOLUTIONS.

[From the *Mathematical Questions with their Solutions from the Educational Times*, vols. v. to xii. (1866—1869).]

[Vol. v., January to July, 1866, p. 17.]

1791. (Proposed by Professor Cayley.)—Given a quartic curve $U = 0$, to find three cubic curves $P = 0$, $Q = 0$, $R = 0$, each meeting the quartic in the same six points 1, 2, 3, 4, 5, 6, and such that $P = 0$, $R = 0$ may besides meet the quartic in the same three points a, b, c, and that $Q = 0$, $R = 0$ may besides meet the quartic in the same three points α, β, γ.

[Vol. v. pp. 25, 26.]

Note on the Problems in regard to a Conic defined by five Conditions of Intersection.

I use the word "intersection" rather than "contact" because it extends to the case of a 1-pointic intersection, which cannot be termed a contact. The conditions referred to are that the conic shall have with a given curve, at a point given or not given, a 1-pointic intersection, a 2-pointic intersection (= ordinary contact), a 3-pointic intersection, &c., as the case may be. It may be noticed that when the point on the curve is a given point, the condition of a k-pointic intersection is really only the condition that the conic shall pass through k given points; though from the circumstance that these are consecutive points on a conic, the formulæ for a conic passing through k discrete points require material alteration; for instance, in the two questions to find the equation of a conic passing through five given points, and to find the equation of a conic having at a given point of a given curve 5-pointic intersection with the curve, the forms of the solutions are very different from each other.

The foregoing remark shows, however, that it is proper to detach the conditions which relate to intersections at given points; and consequently attending only to the conditions which relate to intersection at an unascertained point (of course the intersections referred to must be at least 2-pointic, for otherwise there is no condition at all) we may consider the conics which pass through four points and satisfy one condition; or which pass through three points and satisfy two conditions; or which pass through two points and satisfy three conditions; or which pass through one point and satisfy four conditions; or which satisfy five conditions. Considering in particular the last case, let 1 denote that the conic has 2-pointic intersection, 2 that it has 3-pointic intersection, ... 5 that it has 6-pointic intersection with a given curve at an unascertained point.

Then the problems are in the first instance

5; 4, 1; 3, 2; 3, 1, 1; 2, 2, 1; 2, 1, 1, 1; 1, 1, 1, 1, 1.

But the intersections may be intersections with the same given curve or with different given curves; and we have thus in all 27 problems, viz. these are as given in the following table, where the colons (:) separate those conditions which refer to different curves:

No. of Prob.	Conditions.	No. of Prob.	Conditions.	No. of Prob.	Conditions.
1	5	10	3, 1 : 1	19	3 : 1 : 1
2	4, 1	11	3 : 1, 1	20	2 : 2 : 1
3	3, 2	12	2, 2 : 1	21	2, 1 : 1 : 1
4	3, 1, 1	13	2, 1 : 2	22	2 : 1, 1 : 1
5	2, 2, 1	14	2, 1, 1 : 1	23	1, 1, 1 : 1 : 1
6	2, 1, 1, 1	15	2, 1 : 1, 1	24	1, 1 : 1, 1 : 1
7	1, 1, 1, 1, 1	16	2 : 1, 1, 1	25	2 : 1 : 1 : 1
8	4 : 1	17	1, 1, 1, 1 : 1	26	1, 1 : 1 : 1 : 1
9	3 : 2	18	1, 1, 1 : 1, 1	27	1 : 1 : 1 : 1 : 1

Thus Problem 1 is to find a conic having 6-pointic intersection with a given curve: Problem 2 a conic having 5-pointic intersection and also 2-pointic intersection with a given curve... Problem 7 is to find a conic having five 2-pointic intersections with (touching at five distinct points) a given curve....Problem 27 is to find a conic having 2-pointic intersection with (touching) each of five given curves. Or we may in each case take the problem to be merely to find the number of the conics which satisfy the required conditions. This number is known in Prob. 1, for the case of a curve of the order m without singularities, viz. the number is $= m(12m-27)$. It is also known in Problems 25 and 26 in the case where the first curve (that to which the symbol 2, or 1, 1 relates) is a curve without singularities; and it is known in Prob. 27, viz. if m, n, p, q, r be the orders and M, N, P, Q, R the classes of the five curves respectively, then the number is $=(M, m)(N, n)(P, p)(Q, q)(R, r)\{1, 2, 4, 4, 2, 1\}$, that is, $1\,MNPQR + 2\,\Sigma MNPQr +$ &c. The number is not, I believe, known in any other of the problems. In particular, (Prob. 7) we do not as yet know the number of the conics which touch a given curve at five points. It would be interesting to obtain this number; but (judging from the analogous question of finding the double tangents of a curve) the problem is probably a very difficult one.

[Vol. v. p. 37.]

1857. (Proposed by Professor CAYLEY.)—If for shortness we put

$$P = x^3 + y^3 + z^3,\quad Q = yz^2 + y^2z + zx^2 + z^2x + xy^2 + x^2y,\quad R = xyz,$$
$$P_0 = a^3 + b^3 + c^3,\quad Q_0 = bc^2 + b^2c + ca^2 + c^2a + ab^2 + a^2b,\quad R_0 = abc;$$

then (α, β, γ) being arbitrary, show that the cubic curves $\begin{vmatrix} \alpha, & \beta, & \gamma \\ P, & Q, & R \\ P_0, & Q_0, & R_0 \end{vmatrix} = 0$ pass all of them through the same nine points, lying six of them upon a conic and three of them upon a line; and find the equations of the conic and line, and the coordinates of the nine points of intersection; find also the values of $(\alpha : \beta : \gamma)$ in order that the cubic curve may break up into the conic and line.

[Vol. v. p. 37.]

1730. (Proposed by Professor CAYLEY.)—Show that (I) the condition in order that the roots k_1, k_2, k_3 of the equation

$$\gamma k^3 + (-g - \tfrac{1}{2}\alpha + \tfrac{1}{2}\beta + \tfrac{3}{2}\gamma)\,k^2 + (-g - \tfrac{3}{2}\alpha - \tfrac{1}{2}\beta + \tfrac{1}{2}\gamma)\,k - \alpha = 0 \tag{A}$$

may be connected by a relation of the form

$$k_3(k_1 - k_2) - (k_2 - k_3) = 0, \tag{1}$$

and (II) the result of the elimination of a, b, c from the equations

$$a^2(b + c) = -2\alpha, \tag{2}$$
$$b^2(c + a) = 2\beta, \tag{3}$$
$$c^2(a + b) = -2\gamma, \tag{4}$$
$$(b - c)(c - a)(a - b) = -4g, \tag{5}$$

are each

$$4(\beta - \gamma)(\gamma - \alpha)(\alpha - \beta)\,g^3 + 4(-\Sigma\alpha^3\beta + 4\Sigma\alpha^2\beta^2 - 2\Sigma\alpha^2\beta\gamma)\,g^2$$
$$+ (\beta - \gamma)(\gamma - \alpha)(\alpha - \beta)\,g + 2(\beta - \gamma)^2(\gamma - \alpha)^2(\alpha - \beta)^2 = 0. \tag{B}$$

[Vol. v. pp. 38, 39.]

1834. (Proposed by Professor CAYLEY.)—1. It is required to find on a given cubic curve three points A, B, C, such that, writing $x=0$, $y=0$, $z=0$ for the equations of the lines BC, CA, AB respectively, the cubic curve may be transformable into itself by the inverse substitution $(\alpha x^{-1}, \beta y^{-1}, \gamma z^{-1})$ in place of x, y, z respectively, α, β, γ being disposable constants.

2. In the cubic curve $ax(y^2+z^2)+by(z^2+x^2)+cz(x^2+y^2)+2lxyz=0$ the inverse points (x, y, z) and (x^{-1}, y^{-1}, z^{-1}) are corresponding points (that is, the tangents at these two points meet on the curve).

Solution by the PROPOSER, S. ROBERTS, M.A., *and others.*

Since the points A, B, C are on the curve, the equation is of the form

$$fy^2z+ \quad gz^2x+ \quad hx^2y+ \quad iyz^2+ \quad jzx^2+ \quad kxy^2+2lxyz=0;$$

hence this equation must be equivalent to

$$\frac{f\beta^2\gamma}{y^2z}+\frac{g\gamma^2\alpha}{z^2x}+\frac{h\alpha^2\beta}{x^2y}+\frac{i\beta\gamma^2}{yz^2}+\frac{j\gamma\alpha^2}{zx^2}+\frac{k\alpha\beta^2}{xy^2}+\frac{2l\alpha\beta\gamma}{xyz}=0,$$

or,

$$j\frac{\alpha}{\beta}y^2z+k\frac{\beta}{\gamma}z^2x+i\frac{\gamma}{\alpha}x^2y+h\frac{\alpha}{\gamma}yz^2+f\frac{\beta}{\alpha}zx^2+g\frac{\gamma}{\beta}xy^2+2lxyz=0,$$

which will be the case if

$$f=j\frac{\alpha}{\beta},\quad g=k\frac{\beta}{\gamma},\quad h=i\frac{\gamma}{\alpha},\quad i=h\frac{\alpha}{\gamma},\quad j=f\frac{\beta}{\alpha},\quad k=g\frac{\gamma}{\beta}.$$

This implies $fgh=ijk$; and if this condition be satisfied, then $\alpha:\beta:\gamma$ can be determined, viz. we have $\alpha:\beta:\gamma=if:ij:hf$, which satisfy the remaining equations, so that the only condition is $fgh=ijk$.

Writing in the equation of the curve $x=0$, we find $fy^2z+iyz^2=0$, that is, the line $x=0$ meets the curve in the points $(x=0,\ y=0)$, $(x=0,\ z=0)$, and $(x=0,\ fy+iz=0)$. We have thus on the curve the three points

$$(x=0,\ fy+iz=0),\qquad (y=0,\ gz+jx=0),\qquad (z=0,\ hx+ky=0),$$

and in virtue of the assumed relation $fgh=ijk$, these three points lie in a line. Hence the points A, B, C must be such that BC, CA, AB respectively meet the curve in points A', B', C', which three points lie in a line; that is, we have a quadrilateral whereof the six angles A, B, C, A', B', C' all lie on the curve. It is well known that the opposite angles A and A', B and B', C and C' must be *corresponding points*, that is, points the tangents at which meet on the curve. And conversely taking A, C any two points on the curve, A' a corresponding point to A (any one of the four corresponding points), then AC, $A'C$ will meet the curve in the corresponding points B', B; and AB, $A'B'$ will meet on the curve in a point C' corresponding to C, giving the inscribed quadrilateral (A, B, C, A', B', C'); the triangle ABC is therefore constructed.

It is to be remarked that the equation $fgh = ijk$ being satisfied, we may without any real loss of generality write $f=j$, $g=k$, $h=i$, and therefore $\alpha=\beta=\gamma$; hence changing the constants we have the theorem: the inverse points (x, y, z), (x^{-1}, y^{-1}, z^{-1}) are corresponding points on the curve

$$ax(y^2+z^2)+by(z^2+x^2)+cx(x^2+y^2)+2lxyz=0.$$

[Vol. v. pp. 57, 58.]

Addition to the Note on the Problems in regard to a Conic defined by five Conditions of Intersection.

Since writing the Note in question, I have found that a solution of Problem 7 has been given by M. De Jonquières in the paper "Du Contact des Courbes Planes, &c.," *Nouvelles Annales de Mathématiques*, vol. III. (1864), pp. 218—222: viz. the number of conics which touch a curve of the order n in five distinct points is stated to be

$$\frac{n(n-1)(n-2)(n-3)(n-4)}{1.2.3.4.5}(n^5+15n^4-55n^3-495n^2+1584n+15).$$

There are given also the following results; the number of conics which pass through two given points and touch a curve of the order n in three distinct points is

$$\frac{n(n-1)(n-2)}{2}(n^3+6n^2-19n-12),$$

and the number of conics which pass through a given point and touch a curve of the order n in four distinct points is

$$\frac{n(n-1)(n-2)(n-3)}{1.2.3.4}(n^4+10n^3-37n^2-118n+282).$$

These formulæ are given without demonstration, and with an expression of doubt as regards their exactness—("elles sont exactes, je crois"); they apply, of course, to a curve of the order n without singularities; but assuming them to be accurate, the means exist for adapting them to the case of a curve with singularities.

[There is also a paper on the same subject in the *Annales* for January, 1866 (pp. 17—20), from the Editor's *Note* to which we have introduced a correction (+ 15 instead of − 35) in the formula given above.]

[Vol. v. pp. 58, 59.]

1876. (Proposed by R. BALL, M.A.)—If three of the roots of the equation $(a, b, c, d, e \between x, 1)^4 = 0$ be in arithmetical progression, show that

$$55296\, H^3J - 2304\, aH^2I^2 - 16632\, a^2HIJ + 625\, a^3I^3 - 9261\, a^3J^2 = 0,$$

where

$$H = ac - b^2, \quad I = ae - 4bd + 3c^2, \quad J = ace + 2bcd - ad^2 - b^2e - c^3.$$

Solution by PROFESSOR CAYLEY.

Write $(a, b, c, d, e \between x, 1)^4 = a(x-\alpha)(x-\beta)(x-\gamma)(x-\delta)$; then putting for a moment $\beta+\gamma+\delta = p$, $\beta\gamma+\beta\delta+\gamma\delta = q$, $\beta\gamma\delta = r$, and forming the equation

$$(\beta+\gamma-2\delta)(\beta+\delta-2\gamma)(\gamma+\delta-2\beta) = 0,$$

this is easily reduced to

$$-2p^3 + 9pq - 27r = 0.$$

But we have

$$a(x^3 - px^2 + qx - r)(x - \alpha) = (a, b, c, d, e \between x, 1)^4,$$

and hence

$$p = -\frac{4b}{a} - \alpha, \quad q = \frac{6c}{a} + \frac{4b}{a}\alpha + \alpha^2, \quad r = -\frac{4d}{a} - \frac{6c}{a}\alpha - \frac{4b}{a}\alpha^2 - \alpha^3.$$

Substituting these values of p, q, r, the foregoing equation becomes, after all reductions,

$$(20\, a^3,\ 20\, a^2b,\ -16\, ab^2 + 36\, a^2c,\ 128\, b^3 - 216\, abc + 108\, a^2d \between \alpha, 1)^3 = 0,$$

and from this and the equation $(a, b, c, d, e \between \alpha, 1)^4 = 0$, eliminating α, we should find the condition for three roots in arithmetical progression. But it appears from the theory of invariants that the result of the elimination may be obtained by writing $b = 0$, and expressing the result so obtained in terms of a, H, I, J. Hence, writing in the two equations $b = 0$, the first equation contains the factor $4a^2$, and throwing this out, the equations become

$$5a\alpha^4 + 27c\alpha + 27d = 0, \quad a\alpha^4 + 6c\alpha^2 + 4d\alpha + e = 0;$$

or multiplying the first by α and reducing by means of the second, the two equations become

$$5a\alpha^3 + 27c\alpha + 27d = 0, \quad 3c\alpha^2 - 7d\alpha + 5e = 0.$$

The result is of the degree 5 in the coefficients, but in order to avoid fractions in the final result it is proper to multiply it by a^4; it then becomes

$$625\, a^6e^3 - 4050\, a^5c^2e^2 + 6561\, a^4c^4e - 1890\, a^5ced^2 + 13122\, a^4c^3d^2 + 9261\, a^5d^4 = 0.$$

But writing as above $b = 0$, we have

$$a = a, \quad c = \frac{H}{a}, \quad e = \frac{I}{a} - \frac{3H^2}{a^3}, \quad d^2 = -\frac{J}{a} + \frac{HI}{a^2} - \frac{4H^3}{a^4};$$

and substituting these values, the result is found to contain the terms $\dfrac{IH^4}{a}$, $\dfrac{H^6}{a^3}$ with coefficients which vanish; viz. the coefficient of the first of these terms is

$$+16875+24300+6561+7560+18792-74088, =0;$$

and the coefficient of the second of the two terms is

$$-16875-36450-19683-75168+148176, =0.$$

The remaining terms give

$$\left.\begin{array}{lll} + \ 625 & = + \ 625\, a^3 I^3 \\ - \ 5625 - 4050 - 1890 + 9261 & = - \ 2304\, aH^2 I^2 \\ + \ 1890 - 18522 & = - \ 16632\, a^2 HIJ \\ - \ 18792 + 74088 & = + \ 55296\, H^3 J \\ + \ 9261 & = + \ 9261\, a^3 J^2 \end{array}\right\} = 0,$$

which is the required result; a more convenient form of writing it is

$$(55296\, J, \ -768\, I^2, \ -5544\, IJ, \ 625\, I^3 + 9261\, J^2 \between H, \ a)^3 = 0.$$

REMARK. If I and J denote as above the two invariants of the form $U=(a, b, c, d, e \between x, 1)^4$, and if we now use H to denote the Hessian of the form, viz.

$$H = (ac-b^2, \ \tfrac{1}{2}(ad-bc), \ \tfrac{1}{6}(ae+2bd-3c^2), \ \tfrac{1}{2}(be-cd), \ ce-d^2 \between x, \ 1)^4,$$

then it appears by the theory of invariants that the equation of the twelfth order

$$(55296\, J, \ -768\, I^2, \ -5544\, IJ, \ 625\, I^3 + 9261\, J^2 \between H, \ U)^3 = 0,$$

is such that each of its roots forms with some three of the roots of the equation $U=0$ a harmonic progression; viz. if the three roots are β, γ, δ, then we have

$$\frac{2}{x-\gamma} = \frac{1}{x-\beta} + \frac{1}{x-\delta}, \quad \text{or} \quad x = \frac{2\beta\delta - (\beta+\delta)\gamma}{\beta+\delta-2\gamma};$$

so that the roots of the equation of the twelfth order are the twelve values of the last-mentioned function of three roots.

[Vol. v. pp. 65, 66.]

On the Problems in regard to a Conic defined by five Conditions of Intersection.

There has been recently published in the *Comptes Rendus* (t. LXII. pp. 177—183, January, 1866) an extract of a memoir "Additions to the Theory of Conics," by M. H. G. Zeuthen (of Copenhagen). The extract gives the solutions of fourteen problems, with a brief indication of the method employed for obtaining them. Of these problems, four relate to intersections at given points, the remaining ten are included among the twenty-seven problems enumerated in my *Note* on this subject in the January Number of the *Educational Times* (*Reprint*, vol. V., p. 25); but two of these ten are the problems 25 and 26 which are in my *Note* stated to have been solved; there are, consequently, of the twenty-seven problems, in all twelve which are solved: viz. these are where it is to be observed that Zeuthen's solutions apply to the case

No. of Prob.	1,	8,	10,	12,	14,	17,	19,	21,	23,	25,	26,	27
Zeuthen's No.	—,	14,	13,	11,	8,	3,	12,	7,	2,	6,	1,	—

of a curve of a given order with given numbers of double points and cusps. The problems 25 and 26 had been previously solved only in the case of a curve without singularities. As to Prob. 27, the solution mentioned in my former *Note* is in fact applicable to the general case. The solution for Prob. 1 may also be extended to this general case, viz. for a curve of the order m with δ double points and κ cusps the required number is $= m(12m-27) - 24\delta - 27\kappa$; or, if n be the class, then this number is $= 12n - 15m + 9\kappa$; so that all the twelve problems are solved in the general case.

The results obtained by M. de Jonquieres, as stated in my *Note* in the March Number (*Reprint*, vol. V., p. 57), seem to be all of them erroneous. In fact, for the number of conics passing through two given points and touching a curve of the order m in three distinct points (which is a particular case of Prob. 23), Zeuthen's formula applied to a curve without singularities gives this

$$= \tfrac{1}{6} m(m-2)(m^4 + 5m^3 - 17m^2 - 49m + 108)$$

instead of the value

$$\tfrac{1}{2} m(m-1)(m-2)(m^3 + 6m^2 - 19m - 12)$$

which is

$$= \tfrac{1}{6} m(m-2)(m^4 + 5m^3 - 25m^2 + 7m + 12);$$

and I have by my own investigation verified Zeuthen's Number. So for the number of conics through a given point and touching a curve of the order m in four distinct points (which is a particular case of Prob. 17), Zeuthen's formula applied to a curve without singularities gives this

$$= \tfrac{1}{24} m(m-2)(m-3)(m^5 + 9m^4 - 15m^3 - 225m^2 + 140m + 1050)$$

instead of the value

$$\tfrac{1}{24} m(m-1)(m-2)(m-3)(m^4 + 10m^3 - 37m^2 - 118m + 282)$$

which is

$$= \tfrac{1}{24} m(m-2)(m-3)(m^5 + 9m^4 - 47m^3 - 81m^2 + 400m - 282),$$

and it may I think be inferred that the expression obtained for the number of conics which touch a given curve in five distinct points (Prob. 7), containing as it does the factor $(m-1)$, is also erroneous.

I have obtained for Prob. 2 a solution which I believe to be accurate; viz. the number of the conics (4, 1), (that is, the conics which have with a given curve a 5-pointic intersection and also a 2-pointic intersection, or ordinary contact), is

$$=10n^2+10nm-20m^2-130n+140m+10\kappa(m+n-9)-4[(n-3)\kappa+(m-3)\iota]$$

where ι (the number of inflexions) is $=3n-3m+\kappa$, but I prefer to retain the foregoing form, without effecting the substitution.

[Vol. v. pp. 88, 89.]

1890. (Proposed by Professor CAYLEY.)—Find the equation of a conic passing through three given points and having double contact with a given conic.

Solution by the PROPOSER.

Let the given points be the angles of the triangle ($x=0$, $y=0$, $z=0$), and let the equation of the given conic be $U=(a, b, c, f, g, h \between x, y, z)^2=0$; then the equation of the required conic is

$$U-(x\sqrt{a}+y\sqrt{b}+z\sqrt{c})^2=0,$$

for this is a conic having double contact with the conic $U=0$, and, since the terms in (x^2, y^2, z^2) each vanish, it is also a conic passing through the given points.

It is clear that there are four conics satisfying the conditions of the Problem, viz. putting for shortness

$$P=x\sqrt{a}+y\sqrt{b}+z\sqrt{c}, \qquad P_1=-x\sqrt{a}+y\sqrt{b}+z\sqrt{c},$$
$$P_2=x\sqrt{a}-y\sqrt{b}+z\sqrt{c}, \qquad P_3=x\sqrt{a}+y\sqrt{b}-z\sqrt{c},$$

the four conics are

$$U-P^2=0, \quad U-P_1^2=0, \quad U-P_2^2=0, \quad U-P_3^2=0.$$

It may be remarked that the conics P, P_1 have a fourth intersection lying on the line $y\sqrt{b}+z\sqrt{c}=0$, and the conics P_2, P_3 a fourth intersection lying on the line $y\sqrt{b}-z\sqrt{c}$; which two lines are harmonics in regard to the lines $y=0$, $z=0$.

Similarly the conics P_1, P_2 have a fourth intersection on the line $x\sqrt{a}+z\sqrt{c}=0$, and the conics P, P_3 a fourth intersection on the line $x\sqrt{a}-z\sqrt{c}=0$; which two lines are harmonics in regard to the lines $z=0$, $x=0$. And the conics P_1, P_3 have a fourth intersection on the line $x\sqrt{a}+y\sqrt{b}=0$, and the conics P, P_2 a fourth intersection on the line $x\sqrt{a}-y\sqrt{b}=0$; which two lines are harmonics in regard to the lines $x=0$, $y=0$. It may further be remarked that the equations of any two of the four conics may be taken to be

$$\alpha yz+\beta zx+\gamma xy=0, \quad \alpha' yz+\beta' zx+\gamma' xy=0.$$

The general equation of a conic having double contact with each of these conics then is

$$n^2z^2 - 2n(\gamma\alpha' + \gamma'\alpha)yz - 2n(\gamma\beta' + \gamma'\beta)zx - 4n\gamma\gamma'xy + [(\beta\gamma' - \beta'\gamma)x - (\gamma\alpha' - \gamma'\alpha)y]^2 = 0,$$

where n is arbitrary: and, having double contact with this conic, we have (besides the above-mentioned two conics) two new conics each passing through the angles of the triangle; viz. writing for greater convenience

$$n = \frac{(\beta\gamma' - \beta'\gamma)(\gamma\alpha' - \gamma'\alpha)}{K - \gamma\gamma'}, \quad \text{or } K = \gamma\gamma' + \frac{(\beta\gamma' - \beta'\gamma)(\gamma\alpha' - \gamma'\alpha)}{n},$$

then the equations of the two new conics are

$$\gamma'\alpha\, yz + \gamma\beta'\, zx + Kxy = 0, \quad \gamma\alpha'\, yz + \gamma'\beta\, zx + Kxy = 0.$$

In fact, writing the equation under the form

$$\begin{aligned} &[nz + (\beta\gamma' - \beta'\gamma)x + (\gamma\alpha' - \gamma'\alpha)y]^2 \\ &- 4\ (\beta\gamma' - \beta'\gamma)(\gamma\alpha' - \gamma'\alpha)xy - 4n\gamma\gamma'xy \\ &- 2n(\beta\gamma' - \beta'\gamma)xz - 2n(\beta\gamma' + \beta'\gamma)xz \\ &- 2n(\gamma\alpha' - \gamma'\alpha)yz - 2n(\gamma\alpha' + \gamma'\alpha)yz = 0, \end{aligned}$$

we at once see that this is a conic having double contact with the conic $\gamma'\alpha yz + \gamma\beta'zx + Kxy = 0$, the equation of the chord of contact being $nz + (\beta\gamma' - \beta'\gamma)x + (\gamma\alpha' - \gamma'\alpha)y = 0$: and similarly it has double contact with the conic $\gamma\alpha'\, yz + \gamma'\beta\, zx + Kxy = 0$, the equation of the chord of contact being $nz - (\beta\gamma' - \beta'\gamma)x - (\gamma\alpha' - \gamma'\alpha)y = 0$.

[Vol. v. pp. 99, 100.]

1554. (Proposed by Professor CAYLEY.)—Show that, in the ellipse and its circles of maximum and minimum curvature respectively, the semi-ordinates through the focus of the ellipse are

For the circle of maximum curvature $y_1 = a(1-e)(1+2e)^{\frac{1}{2}}$,

for the ellipse $y_2 = a(1-e^2)$,

for the circle of minimum curvature $y_3 = \dfrac{a\{(1-e^2+e^4)^{\frac{1}{2}} - e^2\}}{(1-e^2)^{\frac{1}{2}}}$,

and that these values are in the order of increasing magnitude.

[Vol. VI., July to December, 1866, pp. 18, 19.]

1931. (Proposed by Professor CAYLEY.)—Find the stationary tangents (or tangents at the inflexions) of the nodal cubic

$$x(y-z)^2 + y(z-x)^2 + z(x-y)^2 = 0.$$

Solution by the PROPOSER.

The equation may be transformed into the form

$$(-8x+y+z)^{\frac{1}{3}}+\ (x-8y+z)^{\frac{1}{3}}+\ (x+y-8z)^{\frac{1}{3}}=0,$$

and it thence follows immediately that the stationary tangents are the lines

$$-8x+y+z=0,\quad x-8y+z=0,\quad x+y-8z\ =0,$$

respectively, and that the three points of contact, or inflexions, are the intersections of these lines with the line $x+y+z=0$.

In fact, writing

$$X=kx+y+z,\quad Y=x+ky+z,\quad Z=x+y+kz,$$

we have identically

$$\begin{aligned}
&(X+Y+Z)^3-27XYZ\\
&=(k+2)^3(x+y+z)^3-27(kx+y+z)(x+ky+z)(x+y+kz),\\
&=\quad (x^3+y^3+z^3)\{(k+2)^3-27k\}\\
&\quad +3(yz^2+y^2z+zx^2+z^2x+xy^2+x^2y)\{(k+2)^3-9(k^2+k+1)\}\\
&\quad +3\,xyz\,\{2(k+2)^3-9(k^3+3k+2)\}\\
&=(k-1)^2(k+8)(x^3+y^3+z^3)+3(k-1)^3(yz^2+y^2z+zx^2+z^2x+xy^2+x^2y)-3(k-1)^2(7k+2)\,xyz.
\end{aligned}$$

Hence, writing $k=-8$, we have

$$\begin{aligned}
(X+Y+Z)^3-27XYZ&=-2187\,\{yz^2+y^2z+zx^2+z^2x+xy^2+x^2y-6xyz\},\\
&=-2187\,\{x(y-z)^2+y(z-x)^2+z(x-y)^2\}.
\end{aligned}$$

The equation of the given curve is therefore

$$(X+Y+Z)^3-27XYZ=0,\quad \text{or}\quad X^{\frac{1}{3}}+Y^{\frac{1}{3}}+Z^{\frac{1}{3}}=0,$$

where of course X, Y, Z have the values

$$X=-8x+y+z,\quad Y=x-8y+z,\quad Z=x+y-8z.$$

[Vol. VI. pp. 35—39.]

1990. (Proposed by Professor SYLVESTER.)—Prove that the three points in which a circular cubic is cut by any transversal are the foci of a Cartesian oval passing through the four foci of the cubic.

Solution by PROFESSOR CAYLEY.

Some preliminary explanations are required in regard to this remarkable theorem.

1. I call to mind that a circular cubic (or cubic through the two circular points at infinity) has 16 foci, which lie 4 together on 4 different circles; and that the

property of 4 concyclic foci is that taking any three of them A, B, C, the distances of a point P of the curve from these three foci are connected by a linear relation $\lambda . AP + \mu . BP + \nu . CP = 0$, where $\lambda \pm \mu \pm \nu = 0$, or if as is more convenient the distances are considered as $\pm$, then where $\lambda + \mu + \nu = 0$. A circular cubic may be determined so as to satisfy 7 conditions; having a focus at a given point is 2 conditions; hence a circular cubic may be determined so as to pass through three given points, and to have as foci two given points.

2. A Cartesian, or bicircular cuspidal quartic (that is a quartic having a cusp at each of the circular points at infinity) has nine foci, but of these there are three which lie in a line with the centre of the Cartesian (or intersection of the cuspidal tangents), and which are preeminently the foci of the Cartesian. We may, therefore, say that the Cartesian has three foci, which foci lie in a line, the axis of the Cartesian. A Cartesian may be determined to satisfy 6 conditions; having a focus at a given point is 2 conditions; but having for foci three given points on a line is 5 conditions; and hence a Cartesian may be found having for foci three given points on a line, and passing through a given point; there are in fact two such Cartesians, intersecting at right angles at the given point.

3. The theorem at first sight appears impossible; for take any three points F, G, H in a line and any other point A; then, as just remarked, there are, having F, G, H for foci and passing through A, two Cartesians. And we may draw through F, G, H, and with A for focus, a circular cubic depending upon two arbitrary parameters; the position of a second focus of the circular cubic is (on account of the two arbitrary parameters) *primâ facie* indeterminate; and this is confirmed by the remark that the circular cubic can actually be so determined as to have for focus an arbitrary point B; and yet the theorem in effect asserts that the foci concyclic with A, of the circular cubic, lie on one or other of the two Cartesians.

4. To explain this, it is to be remarked that the arbitrary point B is a focus which is either concyclic with A or else not concyclic with A. In the latter case, although B is arbitrary, yet the foci concyclic with A may and in fact do lie on one of the Cartesians; the difficulty is in the former case if it arises; viz., if we can describe a cubic through the points F, G, H in a line, and with A and B as *concyclic* foci; that is, if we can find a third focus C, such that the distances from A, B, C of a point P on the curve are connected by a relation $\lambda . AP + \mu . BP + \nu . CP = 0$, where $\lambda + \mu + \nu = 0$. It may be shown that this is in a sense possible, but that the resulting cubic is not a proper circular cubic, but is the cubic made up of the line FGH taken twice, and of the line infinity. To show this, since the required cubic passes through the points F, G, H we have

$$\begin{aligned} \lambda . AF + \mu . BF + \nu . CF &= 0 \\ \lambda . AG + \mu . BG + \nu . CG &= 0 \\ \lambda . AH + \mu . BH + \nu . CH &= 0 \\ \lambda \quad + \mu \quad + \nu \quad &= 0 \end{aligned} \quad \text{and thence} \quad \begin{vmatrix} AF, & AG, & AH, & 1 \\ BF, & BG, & BH, & 1 \\ CF, & CG, & CH, & 1 \end{vmatrix} = 0,$$

being two conditions for the determination of the position of the point C; these give CG, CH as linear functions of CF; the distances CF, CG, CH of the point C from the points F, G, H in the line FGH are connected by a quadratic equation, and hence substituting for CG, CH their values in terms of CF, we have a quadratic equation for CF; as the given conditions are satisfied when C coincides with A or with B, the roots of this equation are $CF = AF$ and $CF = BF$. But if $CF = AF$, then the linear relations give $CG = AG$ and $CH = AH$, that is, C is a point opposite to A in regard to the line FGH. And similarly if $CF = BF$, then C is a point opposite to B in regard to the line FGH. But C being opposite to A or B, the fourth concyclic focus D will be opposite to B or A; that is, the pairs A, B and C, D of concyclic foci lie symmetrically on opposite sides of the line FGH; this of course implies that the four points lie on a circle.

5. Taking $Y = 0$ as the equation of the line FGH, $x^2 + y^2 - 1 = 0$ as the equation of the circle through the four points A, B, C, D, then these lie on a proper cubic

$$(x^2 + y^2 + 1)\,x + lx^2 + ny^2 = 0$$

(not passing through the points F, G, H) and the four foci are given as the intersections with the circle $x^2 + y^2 - 1 = 0$ of the pair of lines

$$x^2 - 2nx - nl = 0.$$

But if we attempt to describe with the same four foci a cubic

$$(x^2 + y^2 + 1)\,y + l'x^2 + 2m'xy + n'y^2 = 0,$$

then the foci are given as the intersections with the circle $x^2 + y^2 - 1 = 0$ of the conic

$$y^2 + 2m'x - 2l'y + m'^2 - n'l' = 0.$$

In order that these may coincide with the points (A, B, C, D) we must have

$$(x^2 - 2nx - nl) + (y^2 + 2m'x - 2l'y + m'^2 - n'l') = x^2 + y^2 - 1\,;$$

that is

$$m' = n, \quad l' = 0, \quad -nl + n^2 - n'l' = -1.$$

The last equation is $n'l' = n^2 + 1 - nl$, which, assuming that nl is not equal to $n^2 + 1$, {in this case the cubic $(x^2 + y^2 + 1)\,x + lx^2 + my^2 = 0$ would reduce itself to the line and conic $(x + n)\left(x^2 + y^2 + \frac{x}{n}\right) = 0$}, since $l' = 0$, gives $n' = \infty$, and therefore the cubic

$$(x^2 + y^2 + 1)\,y + l'x^2 + 2m'xy + n'y^2 = 0,$$

reduces itself to $y^2 = 0$, that is, the cubic in question reduces itself to the line FGH twice repeated, and the line infinity.

6. The conclusion is that F, G, H being given points on a line, and A and B being any other given points, there is not any proper cubic passing through F, G, H and having A, B for concyclic foci: and the *primâ facie* objection to the truth of the theorem is thus removed.

7. Considering the points F, G, H on a line and the point A as given, it has been seen that there are *two* Cartesians through A with the foci F, G, H; and the theorem asserts that in the circular cubics through F, G, H with the focus A, the foci concyclic with A lie on one or other of the two Cartesians: there are consequently through F, G, H with the focus A two systems of circular cubics corresponding to the two Cartesians respectively, each system depending upon two arbitrary parameters. But if we attend only to one of the two Cartesians and to the corresponding system of cubics, then the Cartesian passes through the four foci of each cubic, and if (instead of taking as given the points F, G, H and the focus A) we take as given the four concyclic foci A, B, C, D of a cubic, the theorem asserts that we have through A, B, C, D a Cartesian depending on two arbitrary parameters (or having for its axis an arbitrary line), and such that the foci of the Cartesian are the points of intersection F, G, H of its axis with the cubic. And I proceed to the proof of the theorem in this form.

8. The equation of a circular cubic having four foci on the circle $x^2+y^2-1=0$ is

$$(x^2+y^2+1)(Px+Qy)+lx^2+2mxy+ny^2=0;$$

and this being so, the four foci are the intersections of the circle with the conic

$$(Qx-Py)^2+2(-nP+mQ)x+2(mP-lQ)y+m^2-nl=0.$$

9. The general equation of a Cartesian is

$$(x^2+y^2+2Ax+2By+C)^2+2Dx+2Ey+F=0,$$

and by assuming for A, B, C, D, E, F, the following values which contain the two arbitrary parameters α and θ, viz. by writing

$$2A=\theta Q,\ 2B=-\theta P,\ C=\alpha-1,\ D=-n\theta^2P+(m\theta^2-\alpha\theta)Q,$$
$$E=(m\theta^2+\alpha\theta)P-l\theta^2Q,\ F=-\alpha^2+\theta^2(m^2-nl),$$

we have the equation of a system (the selected one out of two systems) of Cartesians through the four foci; in fact, substituting the foregoing values, the equation of the Cartesian is

$$\{x^2+y^2+\theta(Qx-Py)+\alpha-1\}^2-2\alpha\theta(Qx-Py)$$
$$+2\theta^2(-nP+mQ)x+2\theta^2(mP-lQ)y-\alpha^2+\theta^2(m^2-nl)=0,$$

and writing herein $x^2+y^2-1=0$, the equation reduces itself to

$$\theta^2\{(Qx-Py)^2+2(-nP+mQ)x+2(mP-lQ)y+m^2-nl\}=0,$$

verifying that the Cartesian passes through the four foci.

The coordinates of the centre of the Cartesian are $x=-A$, $y=-B$, and the equation of its axis is $E(x+A)-D(y+B)=0$; we have therefore to show that the points of intersection of this line with the cubic are the foci of the Cartesian.

10. To find where the line in question meets the cubic

$$(x^2+y^2+1)(Px+Qy)+lx^2+2mxy+ny^2=0,$$

writing in this equation

$$x=-A+D\Omega,\quad y=-B+E\Omega,$$

we have for the determination of Ω the equation

$$\{A^2+B^2+1-2(AD+BE)\,\Omega+(D^2+E^2)\,\Omega^2\}\times$$
$$\{-AP-BQ+(DP+EQ)\,\Omega\}+(l,\ m,\ n\between -A+D\Omega,\ -B+E\Omega)^2=0,$$

or observing that we have $AP+BQ=0$, this equation becomes

$$\begin{aligned}&(D^2+E^2)(DP+EQ)\,\Omega^3\\ &+\{-2(AD+BE)(DP+EQ)\qquad +\ lD^2+2mDE+nE^2\}\,\Omega^2\\ &+\{\ \ (A^2+B^2+1)(DP+EQ)\qquad -2lAD-2m(AE+BD)-2nBE\}\,\Omega\\ &+\{\qquad\qquad\qquad\qquad\qquad lA^2+2mAB+nB^2\}=0.\end{aligned}$$

11. Substituting for A, B, D, E their values in terms of $(P,\ Q,\ \alpha,\ \theta)$, we find

$$\begin{aligned}DP+\ EQ&=-\theta^2(nP^2-2mPQ+lQ^2),\\ lA^2+2mAB+nB^2&=\tfrac{1}{4}\theta^2(nP^2-2mPQ+lQ^2),\\ lAD+m(AE+BD)+nBE&=-\tfrac{1}{2}\alpha\theta^2(nP^2-2mPQ+lQ^2),\\ lD^2+2mDE+nE^2&=\left((nl-m^2)\,\theta^4+\alpha^2\theta^2\right)(nP^2-2mPQ+lQ^2),\end{aligned}$$

and substituting these values in the equation for Ω, the whole equation divides by $\theta^2(nP^2-2mPQ+lQ^2)$, and it then becomes

$$4(D^2+E^2)\,\Omega^3+4\{-2(AD+BE)-(nl-m^2)\,\theta^2-\alpha^2\}\,\Omega^2+4\{A^2+B^2+1-\alpha\}\,\Omega-1=0,$$

or, putting for shortness

$$\begin{aligned}C'&=C-A^2-B^2, &&=\alpha-1-A^2-B^2,\\ F'&=F-2(AD+BE), &&=-\alpha^2-\theta^2(nl-m^2)-2(AD+BE),\end{aligned}$$

the equation in Ω is

$$4(D^2+E^2)\,\Omega^3+4F'\Omega^2-4C'\Omega-1=0,$$

so that, Ω satisfying this equation, the intersections of the axis with the cubic are given by $x=-A+D\Omega$, $y=-B+E\Omega$.

12. The equation of the Cartesian, writing therein $x+A=u$ and $y+B=v$, and attending to the values of C' and F', is

$$(u^2+v^2+C')^2+2Du+2Ev+F'=0.$$

And to find the foci, writing in this equation $u+\rho$, $v+i\rho$ in place of u, v, we find

$$\{u^2+v^2+C'+2(u+vi)\,\rho\}^2+2(D+Ei)\,\rho+2Du+2Ev+F'=0,$$

that is

$$(u^2+v^2+C')^2+2Du+2Ev+F'+\{2(u+vi)(u^2+v^2+C')+D+Ei\}\,2\rho+4(u+vi)^2\rho^2=0.$$

Expressing that this equation in ρ has two equal roots, we find

$$4(u+vi)^2\{(u^2+v^2+C')^2+2Du+2Ev+F'\}-\{2(u+vi)(u^2+v^2+C')+D+Ei\}^2=0,$$

that is

$$4(2Du+2Ev+F')(u+vi)^2-4(u^2+v^2+C')(u+vi)(D+Ei)-(D-Ei)^2=0,$$

which equation is in fact the equation of the three tangents from one of the circular points at infinity. Writing it under the form $U+Vi=0$, the nine foci of the Cartesian are given as the intersections of the two cubics $U=0$, $V=0$. But of these nine points, three, the foci that we are concerned with, lie on the axis, or line $Eu-Dv=0$; in fact, we have

$$\begin{aligned} U=&\,4(u^2-v^2)(2Du+2Ev+F')\\ &-4(uD-vE)(u^2+v^2+C')\\ &-(D^2-E^2), \end{aligned}\qquad\qquad \begin{aligned} V=&\,8uv(2Du+2Ev+F')\\ &-4(uE+vD)(u^2+v^2+C')\\ &-2DE; \end{aligned}$$

and hence

$$2DEU-(D^2-E^2)V=(Eu-Dv)\{8(Du+Ev)(2Du+2Ev+F')-4(D^2+E^2)(u^2+v^2+C')\}=0,$$

which shows that the nine points lie three of them on the line $Eu-Dv=0$, and the remaining six on the conic

$$2(Du+Ev)(2Du+2Ev+F')-(D^2+E^2)(u^2+v^2+C')=0.$$

13. We have thus the three foci given as the intersections of the axis $Eu-Dv=0$, with the cubic

$$U=4(u^2-v^2)(2Du+2Ev+F')-4(uD-vE)(u^2+v^2+C')-(D^2-E^2)=0;$$

or, writing in this last equation $u=D\Omega$, $v=E\Omega$, that is $x=-A+D\Omega$, $y=-B+E\Omega$, we have

$$u^2-v^2=(D^2-E^2)\Omega^2,\quad uD-vE=(D^2-E^2)\Omega.$$

The whole equation divides by (D^2-E^2), and omitting this factor, it is

$$4\Omega^2\{2(D^2+E^2)\Omega+F'\}-4\Omega\{(D^2+E^2)\Omega^2+C'\}-1=0,$$

that is

$$4(D^2+E^2)\Omega^3+4F'\Omega^2-4C'\Omega-1=0,$$

the same equation as the equation in Ω before obtained; that is the intersections of the cubic with the axis are the three foci of the Cartesian.

[Vol. VI. pp. 57—59.]

1949. (Proposed by Professor CAYLEY.)—Find the conic of five-pointic intersection at any point of the cuspidal cubic $y^3=x^2z$.

Solution by the PROPOSER.

The equation $y^3 = x^2z$, is satisfied by the values $x : y : z = 1 : \theta : \theta^3$; and conversely, to any given value of the parameter θ there corresponds a point on the cubic $y^3 = x^2z$. Consider the five points corresponding to the values $(\theta_1, \theta_2, \theta_3, \theta_4, \theta_5)$ respectively; the equation of the conic through these five points is

$$\begin{vmatrix} x^2, & y^2, & z^2, & yz, & zx, & xy \\ 1, & \theta_1^2, & \theta_1^6, & \theta_1^4, & \theta_1^3, & \theta_1 \\ \vdots & & & & & \end{vmatrix} = 0,$$

where the remaining four lines of the determinant are obtained from the second line by writing therein $\theta_2, \theta_3, \theta_4, \theta_5$ successively in place of θ_1. Writing for shortness $\zeta^{\frac{1}{2}}(\theta_1, \theta_2, \theta_3, \theta_4, \theta_5)$ to denote the product of the differences of the quantities $(\theta_1, \theta_2, \theta_3, \theta_4, \theta_5)$, the equation contains the factor $\zeta^{\frac{1}{2}}(\theta_1, \theta_2, \theta_3, \theta_4, \theta_5)$, and we may therefore write it in the simplified form

$$\frac{1}{\zeta^{\frac{1}{2}}(\theta_1, \theta_2, \theta_3, \theta_4, \theta_5)} \begin{vmatrix} x^2, & y^2, & z^2, & yz, & zx, & xy \\ 1, & \theta_1^2, & \theta_1^6, & \theta_1^4, & \theta_1^3, & \theta_1 \\ \vdots & & & & & \end{vmatrix} = 0.$$

Hence putting in this equation $\theta_1 = \theta_2 = \theta_3 = \theta_4 = \theta_5 = \phi$, we have the equation of the conic of five-pointic intersection at the point (ϕ). The result in its reduced form may be obtained directly without much difficulty, but it is obtained most easily as follows: let the function on the left hand of the foregoing equation be represented by

$$(a, b, c, f, g, h \between x, y, z)^2,$$

then writing $x : y : z = 1 : \theta : \theta^3$, we have

$$(a, b, c, f, g, h \between 1, \theta, \theta^3)^2$$

$$= \frac{1}{\zeta^{\frac{1}{2}}(\theta_1, \theta_2, \theta_3, \theta_4, \theta_5)} \begin{vmatrix} 1, & \theta^2, & \theta^6, & \theta^4, & \theta^3, & \theta \\ 1, & \theta_1^2, & \theta_1^6, & \theta_1^4, & \theta_1^3, & \theta_1 \\ \vdots & & & & & \end{vmatrix},$$

$$= \frac{(\theta-\theta_1)(\theta-\theta_2)(\theta-\theta_3)(\theta-\theta_4)(\theta-\theta_5)}{\zeta^{\frac{1}{2}}(\theta, \theta_1, \theta_2, \theta_3, \theta_4, \theta_5)} \begin{vmatrix} 1, & \theta^2, & \theta^6, & \theta^4, & \theta^3, & \theta \\ 1, & \theta_1^2, & \theta_1^6, & \theta_1^4, & \theta_1^3, & \theta_1 \\ \vdots & & & & & \end{vmatrix},$$

$$= (\theta-\theta_1)(\theta-\theta_2)(\theta-\theta_3)(\theta-\theta_4)(\theta-\theta_5)(\theta+\theta_1+\theta_2+\theta_3+\theta_4+\theta_5);$$

for the determinant, which is a function of the order 16 in the quantities $(\theta, \theta_1, \theta_2, \theta_3, \theta_4, \theta_5)$ conjointly, divides by $\zeta^{\frac{1}{2}}(\theta, \theta_1, \theta_2, \theta_3, \theta_4, \theta_5)$, which is a function of the order 15; and as the quotient is a symmetrical function of $\theta, \theta_1, \theta_2, \theta_3, \theta_4, \theta_5$, it must be equal, save to a numerical factor which may be disregarded, to $\theta+\theta_1+\theta_2+\theta_3+\theta_4+\theta_5$.

Hence if ϕ be the parameter of the given point, writing $\theta_1 = \theta_2 = \theta_3 = \theta_4 = \theta_5 = \phi$, we have

$$(a, b, c, f, g, h \between 1, \theta, \theta^3)^2 = (\theta - \phi)^5 (\theta + 5\phi)$$
$$= (1, 0, -15, +40, -45, +24, -5 \between \theta, \phi)^6,$$

where the left-hand side is

$$a + b\theta^2 + c\theta^6 + f\theta^4 + g\theta^3 + h\theta, = \{c, 0, f, g, b, h, a\} (\theta, 1)^6,$$

that is we have

$$c = 1, \quad f = -15\phi^2, \quad g = 40\phi^3, \quad b = -45\phi^4, \quad h = 24\phi^5, \quad a = -5\phi^6,$$

and the equation of the conic of five-pointic intersection therefore is

$$(-5\phi^6, -45\phi^4, 1, -15\phi^2, 40\phi^3, 24\phi^5 \between x, y, z)^2 = 0,$$

or, what is the same thing,

$$-5\phi^6 x^2 - 45\phi^4 y^2 + z^2 - 15\phi^2 yz + 40\phi^3 zx + 24\phi^5 xy = 0,$$

which is the required result.

NOTE. The condition in order that any six points $(\theta_1, \theta_2, \theta_3, \theta_4, \theta_5, \theta_6)$ of the cubic $y^3 = x^2 z$ may lie on a conic, is

$$\theta_1 + \theta_2 + \theta_3 + \theta_4 + \theta_5 + \theta_6 = 0.$$

[Vol. VI. p. 65.]

1872. (Proposed by Professor CAYLEY.)—Show that the surfaces

$$xyz = 1, \quad yz + zx + xy + x + y + z + 3 = 0,$$

intersect in two distinct cubic curves; and find the equations of the cubic cones which have their vertices at the origin and pass through these curves respectively.

[Vol. VI. pp. 67—69.]

1969. (Proposed by Professor SYLVESTER.)—In two given great circles of a sphere intersecting at O are taken respectively two points P and Q, the arc joining which is of given length: prove that S, H two fixed points, and M a fixed line, in a plane may be found such that, for all positions of the arc PQ, a point M in the fixed line may be found satisfying the equations

$$SM \pm HM = \sin OP, \quad SM \mp HM = \sin OQ.$$

Solution by PROFESSOR CAYLEY.

1. In the spherical triangle OPQ, whereof the sides OP, OQ, PQ are θ, ϕ, β and the angle O is $=\alpha$, the relation between these quantities is $\cos\alpha = \dfrac{\cos\beta - \cos\theta\cos\phi}{\sin\theta\sin\phi}$; hence treating α, β as constants, and θ, ϕ as variable angles connected by the foregoing equation, it is required to show that we can find two fixed points S, H and a fixed line, such that taking M a variable point in this line and writing $SM=r$, $HM=s$, the relation between r and s (or equation of the fixed line in terms of r, s as coordinates of a point thereof) is obtained by substituting in the foregoing equation for θ and ϕ the values given by the two equations

$$\sin\theta = (r+s),\quad \sin\phi = (r-s),$$

or as, for the sake of homogeneity, it will be more convenient to write these equations,

$$m\sin\theta = (r+s),\quad m\sin\phi = (r-s).$$

2. Suppose that the perpendicular distances of S, H from the fixed line are a and b, and that the distance between the feet of the two perpendiculars is $2c$, then if x denote the distance of the point M from the midway point between the feet of the two perpendiculars, we have

$$r = \sqrt{\{(c+x)^2+a^2\}},\quad s=\sqrt{\{(c-x)^2+b^2\}},$$

and (a, b, c) being properly determined, the elimination of x from these equations should give between (r, s) a relation equivalent to that obtained by the elimination of (θ, ϕ) from the before-mentioned equations. Or, what is the same thing, the elimination of (r, s, x) from the equations

$$m\sin\theta = r+s,\quad m\sin\phi = r-s,\quad r=\sqrt{\{(c+x)^2+a^2\}},\quad s=\sqrt{\{(c-x)^2+b^2\}}$$

should give between (θ, ϕ) the relation

$$\cos\alpha = \frac{\cos\beta - \cos\theta\cos\phi}{\sin\theta\sin\phi};$$

that is, the last-mentioned equation should be obtained by the elimination of x from the equations

$$m(\sin\theta+\sin\phi) = 2\sqrt{\{(c+x)^2+a^2\}},\quad m(\sin\theta-\sin\phi)=2\sqrt{\{(c-x)^2+b^2\}}.$$

3. The equation in (θ, ϕ) may be written

$$\cos\beta - \cos\alpha\sin\theta\sin\phi = \cos\theta\cos\phi,$$

or, squaring and reducing,

$$\sin^2\theta+\sin^2\phi = \sin^2\beta + 2\cos\alpha\cos\beta\sin\theta\sin\phi + \sin^2\alpha\sin^2\theta\sin^2\phi,$$

that is

$$\sin^2\theta+\sin^2\phi = \frac{1-\cos^2\alpha-\cos^2\beta}{\sin^2\alpha} + \left(\sin\alpha\sin\theta\sin\phi + \frac{\cos\alpha\cos\beta}{\sin\alpha}\right)^2.$$

But from the two equations in x, we have

$$m^2(\sin^2\theta+\sin^2\phi)=4c^2+2a^2+2b^2+4x^2,\quad m^2\sin\theta\sin\phi=4cx+a^2-b^2,$$

whence

$$2x=\frac{b^2-a^2+m^2\sin\theta\sin\phi}{2c},$$

therefore

$$\sin^2\theta+\sin^2\phi=\frac{4c^2+2b^2+2a^2}{m^2}+\left(\frac{b^2-a^2+m^2\sin\theta\sin\phi}{2cm}\right)^2.$$

Hence, comparing the two results, we have

$$\frac{1-\cos^2\alpha-\cos^2\beta}{\sin^2\alpha}=\frac{4c^2+2b^2+2a^2}{m^2},\quad \frac{\cos\alpha\cos\beta}{\sin\alpha}=\frac{b^2-a^2}{2cm},\quad \sin\alpha=\frac{m}{2c};$$

or, as these may also be written,

$$\sin\alpha=\frac{m}{2c},\quad \cos^2\alpha+\cos^2\beta=\frac{-b^2-a^2}{2c^2},\quad 2\cos\alpha\cos\beta=\frac{b^2-a^2}{2c^2};$$

whence

$$(\cos\alpha+\cos\beta)^2=\frac{-a^2}{c^2},\quad (\cos\alpha-\cos\beta)^2=\frac{-b^2}{c^2},\quad \sin\alpha=\frac{m}{2c};$$

so that m being put equal to unity, or otherwise assumed at pleasure, a, b, c are given functions of α, β. Or conversely, if a, b, c are assumed at pleasure, then α, β, m are given functions of these quantities.

5. It is to be remarked that (α, β) being real, a and b will be imaginary, and consequently the points S, H of Professor Sylvester's theorem are imaginary[1]; if, however, we write $-a^2$, $-b^2$ in place of a^2, b^2 respectively, then the radicals $\sqrt{\{(c+x)^2-a^2\}}$, $\sqrt{\{(c-x)^2-b^2\}}$ have a real geometrical interpretation. The system of relations between $(\alpha, \beta, a, b, c, m)$ becomes

$$(\cos\alpha+\cos\beta)^2=\frac{a^2}{c^2},\quad (\cos\alpha-\cos\beta)^2=\frac{b^2}{c^2},\quad \sin\alpha=\frac{m}{2c};$$

and considering (a, b, c) as given, we may write

$$\cos\alpha=\frac{a+b}{2c},\quad \cos\beta=\frac{a-b}{2c},\quad m=\sqrt{\{4c^2-(a+b)^2\}},$$

viz. we have either this system or the similar system obtained by writing $-b$ in place of b.

6. Consider two circles with the radii a, b and having the distance of their centres $=2c$, and to fix the ideas assume that $2c>a+b$, that is, that the circles are

[1] Prof. Sylvester remarks that according as β is less or greater than α, we may find real values of θ, ϕ equal to one another in the one case and supplementary in the other. Hence we must in any case be able to make $r=0$ and $s=0$ indifferently, which shows *à priori* that the line being supposed real, each point S, H must be imaginary, but so that the squared distance of either from the line must be a *real negative quantity*, conformably to Prof. Cayley's important observation in the text. W. J. M.

exterior to each other. The foregoing equations signify that $90^\circ - \alpha$, $90^\circ - \beta$ are the inclinations to the line of centres of the inverse and the direct common tangents respectively, and that m is the length of the inverse common tangent. And the theorem is, that considering two circles as above, and taking M a variable point in

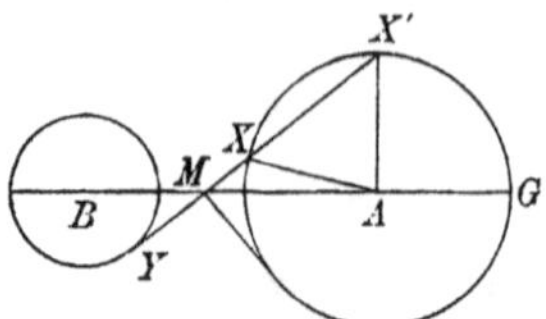

the line of centres, if r, s denote the tangential distances of M from the two circles respectively, and if m be the length of the inverse common tangent of the two circles, then the angles θ, ϕ determined by the equations

$$m \sin\theta = r + s, \quad m \sin\phi = r - s,$$

are connected by the relation

$$\cos\beta = \cos\theta\cos\phi + \sin\theta\sin\phi\cos\alpha,$$

(α, β) being constant angles, determined as above.

7. It is to be remarked that, assuming

$$k = \frac{\sin\alpha}{\sin\beta} = \frac{\sqrt{\{4c^2 - (a+b)^2\}}}{\sqrt{\{4c^2 - (a-b)^2\}}},$$

that is, k = inverse common tangent ÷ direct common tangent, then we have

$$\cos\alpha = \sqrt{(1 - k^2 \sin^2\beta)} = \Delta\beta,$$

or the equation in θ, ϕ becomes

$$\cos\beta = \cos\theta\cos\phi + \sin\theta\sin\phi\,\Delta\beta,$$

which is the algebraical equation connecting the amplitudes of the elliptic functions in the relation $F(\theta) + F(\phi) = F(\beta)$.

8. It is very noticeable that the above figure leads to another relation in elliptic functions, viz. it is the very figure employed for that purpose by Jacobi; in fact, considering therein YM as a variable tangent meeting the circle A in the two points X and X', then if 2ψ, $2\psi'$ denote the angles GAX, GAX' respectively, it is easy to see geometrically that we have $d\psi : d\psi' = YX : YX'$; where

$$(YX)^2 = (BX)^2 - b^2, \ = 4c^2 + a^2 + 4ac\cos 2\psi - b^2, \ = (2c+a)^2 - b^2 - 8ac\sin^2\psi,$$

and similarly $(YX')^2 = (2c+a)^2 - b^2 - 8ac\sin^2\psi'$, that is, writing $l^2 = \dfrac{8ac}{(2c+a)^2 - b^2}$, the differential equation is

$$\frac{d\psi}{\sqrt{(1 - l^2\sin^2\psi)}} - \frac{d\psi'}{\sqrt{(1 - l^2\sin^2\psi')}} = 0,$$

corresponding to an integral equation

$$F(\psi)-F(\psi')=F(\mu),$$

the modulus of the elliptic functions being

$$l, =\frac{\sqrt{8ac}}{\sqrt{\{(2c+a)^2-b^2\}}}.$$

In the problem above considered the modulus is

$$k, =\frac{\sqrt{\{4c^2-(a+b)^2\}}}{\sqrt{\{4c^2-(a-b)^2\}}},$$

and it is not very easy to see the connexion between the two results.

[Vol. VI. p. 81.]

Theorem: by PROFESSOR CAYLEY.

If (A, A'), (B, B') are four points (two real and the other two imaginary) related to each other as foci and antifoci (that is, if the lines AA', BB' intersect at right angles in a point O in such wise that $OA=OA'=i\,.\,OB=i\,.\,OB'$), then the product of the distances of any point P from the points A, A' is equal to the product of the distances of the same point P from the points B, B'.

In fact, the coordinates of A, A' may be taken to be $(\alpha, 0)$, $(-\alpha, 0)$, and those of B, B' to be $(0, \alpha i)$, $(0, -\alpha i)$; whence, if (x, y) are the coordinates of P, we have

$$\begin{aligned}(AP)^2&=(x-\alpha)^2+y^2=(x-\alpha+iy)(x-\alpha-iy),\\(A'P)^2&=(x+\alpha)^2+y^2=(x+\alpha+iy)(x+\alpha-iy),\\(BP)^2&=x^2+(y-i\alpha)^2=(x+iy+\alpha)(x-iy-\alpha),\\(B'P)^2&=x^2+(y+i\alpha)^2=(x+iy-\alpha)(x-iy+\alpha),\end{aligned}$$

from which the theorem is at once seen to be true.

An important application of the theorem consists in the means which it affords of passing from the foci (A, B, C, D) of a bicircular quartic, to the antifoci (A, B) and (C, D); viz. if these are (A', B', C', D'), then the equation $l\sqrt{(A)}+m\sqrt{(B)}+n\sqrt{(C)}=0$ must be transformable into $l'\sqrt{(A')}+m'\sqrt{(B')}+n'\sqrt{(C')}=0$. Writing these respectively under the forms

$$l^2A+m^2B-n^2C+2lm\sqrt{(AB)}=0,\quad l'^2A'+m'^2B'-n'^2C'+2l'm'\sqrt{(A'B')}=0,$$

the two radicals $\sqrt{(AB)}$, $\sqrt{(A'B')}$ are identical; and the remaining terms in the two equations respectively are rational functions, which when the ratios $l' : m' : n'$ are properly determined will be to each other in the ratio $lm : l'm'$; the two equations being thus identical.

[Vol. VI. p. 99.]

1970. (Proposed by Professor CAYLEY.)—Find the conditions in order that the conics

$$U = (a, b, c, f, g, h \between x, y, z)^2 = 0, \quad U' = (a', b', c', f', g', h' \between x, y, z)^2 = 0,$$

may have double contact.

Solution by the PROPOSER.

The coefficients of the two conics must be so related that for a properly determined value of θ we shall have identically $U - \theta U' = (\lambda x + \mu y + \nu z)^2$; but when this is so, the inverse coefficients of the quadric function $U - \theta U'$ are each $= 0$; that is, writing

$$\begin{aligned}
(A, B, C, F, G, H) &= (bc - f^2, ca - g^2, ab - h^2, gh - af, hf - bg, fg - ch)\\
(A', B', C', F', G', H') &= (b'c' - f'^2, \ldots \qquad g'h' - a'f', \ldots)\\
(\mathfrak{A}, \mathfrak{B}, \mathfrak{C}, \mathfrak{F}, \mathfrak{G}, \mathfrak{H}) &= (bc' + b'c - 2ff', \ldots \qquad gh' + g'h - af' - a'f, \ldots),
\end{aligned}$$

then we have the six equations $A - \theta\mathfrak{A} + \theta^2 A' = 0$, &c.

Or, eliminating θ, the required conditions are

$$\left\| \begin{matrix} A, & B, & C, & F, & G, & H \\ A', & B', & C', & F', & G', & H' \\ \mathfrak{A}, & \mathfrak{B}, & \mathfrak{C}, & \mathfrak{F}, & \mathfrak{G}, & \mathfrak{H} \end{matrix} \right\| = 0,$$

equivalent to three relations between the two sets of coefficients.

[Vol. VII., January to July, 1867, pp. 17—19.]

2110. (Proposed by Professor CAYLEY.)—Prove that the locus of the foci of the parabolas which pass through three given points is a unicursal quintic curve passing through the two circular points at infinity.

Solution by the PROPOSER.

More generally it may be shown that for the conics which pass through three given points and touch a given line, the locus of the intersection of the tangents drawn from two fixed points Q, Q' on this line to each conic of the series is a unicursal quintic passing through the two points Q and Q'.

Taking the three given points to be the angles of the triangle ($x = 0$, $y = 0$, $z = 0$), and the points Q, Q' to be the points (α, β, γ) and $(\alpha', \beta', \gamma')$ respectively, the equation of a conic through the three points is

$$fyz + gzx + hxy = 0,$$

which conic will touch the line through the points $(\alpha, \beta, \gamma)(\alpha', \beta', \gamma')$, if

$$\sqrt{\{f(\beta\gamma' - \beta'\gamma)\}} + \sqrt{\{g(\gamma\alpha' - \gamma'\alpha)\}} + \sqrt{\{h(\alpha\beta' - \alpha'\beta)\}} = 0.$$

The equation of the pair of tangents from (α, β, γ) to the conic is

$$(f^2, g^2, h^2, -gh, -hf, -bg \between \gamma y - \beta z, \alpha z - \gamma x, \beta x - \alpha y)^2 = 0,$$

that is

$$\begin{aligned} &x^2(g\gamma + h\beta)^2 + y^2(h\alpha + f\gamma)^2 + z^2(f\beta + g\alpha)^2 \\ &+ 2yz\{2gh\alpha^2 - (h\alpha + f\gamma)(f\beta + g\alpha)\} \\ &+ 2zx\{2hg\beta^2 - (f\beta + g\alpha)(g\gamma + h\beta)\} \\ &+ 2xy\{2fg\gamma^2 - (g\gamma + h\beta)(h\alpha + f\gamma)\} = 0, \end{aligned}$$

but one of the tangents through (α, β, γ) being

$$x(\beta\gamma' - \beta'\gamma) + y(\gamma\alpha' - \gamma'\alpha) + z(\alpha\beta' - \alpha'\beta) = 0,$$

it follows that the other tangent is

$$x\frac{(g\gamma + h\beta)^2}{\beta\gamma' - \beta'\gamma} + y\frac{(h\alpha + f\gamma)^2}{\gamma\alpha' - \gamma'\alpha} + z\frac{(f\beta + g\alpha)^2}{\alpha\beta' - \alpha'\beta} = 0.$$

Hence, writing for shortness

$$\begin{aligned} A &= g\gamma + h\beta, \quad B = h\alpha + f\gamma, \quad C = f\beta + g\alpha, \\ A' &= g\gamma' + h\beta', \quad B' = h\alpha' + f\gamma', \quad C' = f\beta' + g\alpha', \end{aligned}$$

the equations of the tangents from Q, Q' respectively are

$$A^2\frac{x}{\beta\gamma' - \beta'\gamma} + B^2\frac{y}{\gamma\alpha' - \gamma'\alpha} + C^2\frac{z}{\alpha\beta' - \alpha'\beta} = 0,$$

$$A'^2\frac{x}{\beta\gamma' - \beta'\gamma} + B'^2\frac{y}{\gamma\alpha' - \gamma'\alpha} + C'^2\frac{z}{\alpha\beta' - \alpha'\beta} = 0,$$

and for the coordinates of the intersection of these tangents, we have

$$\frac{x}{\beta\gamma' - \beta'\gamma} : \frac{y}{\gamma\alpha' - \gamma'\alpha} : \frac{z}{\alpha\beta' - \alpha'\beta} = B^2C'^2 - B'^2C^2 : C^2A'^2 - C'^2A^2 : A^2B'^2 - A'^2B^2.$$

$$\begin{aligned} BC' - B'C &= \qquad\quad f\{-f(\beta\gamma' - \beta'\gamma) + g(\gamma\alpha' - \gamma'\alpha) + h(\alpha\beta' - \alpha'\beta)\} \\ BC' + B'C &= 2gh\alpha\alpha' + f\{\ f(\beta\gamma' + \beta'\gamma) + g(\gamma\alpha' + \gamma'\alpha) + h(\alpha\beta' + \alpha'\beta)\}. \end{aligned}$$

To satisfy the equation

$$\sqrt{\{f(\beta\gamma' - \beta'\gamma)\}} + \sqrt{\{g(\gamma\alpha' - \gamma'\alpha)\}} + \sqrt{\{h(\alpha\beta' - \alpha'\beta)\}},$$

write

$$f = \frac{a^2}{\beta\gamma' - \beta'\gamma}, \qquad g = \frac{b^2}{\gamma\alpha' - \gamma'\alpha}, \qquad h = \frac{c^2}{\alpha\beta' - \alpha'\beta},$$

and therefore $a + b + c = 0$; we then have

$$-f(\beta\gamma' - \beta'\gamma) + g(\gamma\alpha' - \gamma'\alpha) + h(\alpha\beta' - \alpha'\beta), = -a^2 + b^2 + c^2, = -2bc;$$

and thence

$$f\{-f(\beta\gamma'-\beta'\gamma)+g(\gamma\alpha'-\gamma'\alpha)+h(\alpha\beta'-\alpha'\beta)\}=\frac{a^2}{\beta\gamma'-\beta'\gamma}(-2bc),$$

and the equations become

$$x:y:z=a(BC'+B'C):b(CA'+C'A):c(AB'+A'B),$$

where $BC'+B'C$, $CA'+C'A$, $AB'+A'B$, substituting therein for f, g, h the values $\frac{a^2}{\beta\gamma'-\beta'\gamma}$, $\frac{b^2}{\gamma\alpha'-\gamma'\alpha}$, $\frac{c^2}{\alpha\beta'-\alpha'\beta}$, are respectively functions of the fourth degree in a, b, c; hence (a, b, c) being connected by the relation $a+b+c=0$, x, y, z are proportional to quintic functions of (a, b, c), or what is the same thing, writing $a, b, c=1, \theta, -1-\theta$, then x, y, z are proportional to quintic functions of θ, that is, the locus is a unicursal quintic curve.

That the curve passes through the points $(\alpha', \beta', \gamma')$ and (α, β, γ) appears by considering the conics $fyz+gzx+hxy=0$, which pass through these points respectively.

For the first of these conics we have $f:g:h=\alpha(\beta\gamma'-\beta'\gamma):\beta(\gamma\alpha'-\gamma'\alpha):\alpha(\beta\gamma'-\beta'\gamma)$; the equation

$$A^2\frac{x}{\beta\gamma'-\beta'\gamma}+B^2\frac{y}{\gamma\alpha'-\gamma'\alpha}+C^2\frac{z}{\alpha\beta'-\alpha'\beta}=0,$$

reduces itself to $x(\beta\gamma'-\beta'\gamma)+y(\gamma\alpha'-\gamma'\alpha)+z(\alpha\beta'-\alpha'\beta)=0$, and as the other equation

$$A'^2\frac{x}{\beta\gamma'-\beta'\gamma}+B'^2\frac{y}{\gamma\alpha'-\gamma'\alpha}+C'^2\frac{z}{\gamma\alpha'-\gamma'\alpha}=0,$$

is that of a line through $(\alpha', \beta', \gamma')$ the two lines meet of course in the point $(\alpha', \beta', \gamma')$. And the like for the conic

$$f:g:h=\alpha'(\beta\gamma'-\beta'\gamma):\beta'(\gamma\alpha'-\gamma'\alpha):\gamma'(\alpha\beta'-\alpha'\beta).$$

If the triangle is equilateral, and (x, y, z) are respectively proportional to the perpendicular distances from the three sides, then we have for the circular points at infinity

$$(\alpha, \beta, \gamma)=(1, \omega, \omega^2),\quad (\alpha', \beta', \gamma')=(1, \omega^2, \omega),$$

where ω is an imaginary cube root of unity. These values give

$$\begin{aligned}&\beta\gamma'-\beta'\gamma=\gamma\alpha'-\gamma'\alpha=\alpha\beta'-\alpha'\beta=\omega^2-\omega\\&\alpha\alpha'=\beta\beta'=\gamma\gamma'=1,\quad \beta\gamma'+\beta'\gamma=\gamma\alpha'+\gamma'\alpha=\alpha\beta'+\alpha'\beta=-1;\end{aligned}$$

and the expressions for (x, y, z) take the form

$$\begin{aligned}x:y:z=&\,a\{2b^2c^2-a^2(a^2+b^2+c^2)\}\\&:b\{2c^2a^2-b^2(a^2+b^2+c^2)\}\\&:c\{2a^2b^2-c^2(a^2+b^2+c^2)\},\end{aligned}$$

or, what is the same thing, reducing by means of the relation $a+b+c=0$,

$$x:y:z=a(a^4-2a^2bc-2b^2c^2):b(b^4-2b^2ca-2c^2a^2):c(c^4-2c^2ab-2a^2b^2),$$

and the equation of the curve is obtained by eliminating (a, b, c) from these equations and the before mentioned equation $a+b+c=0$.

N.B. The above is a particular case of the following general theorem of M. Chasles: If the conics of a system (μ, ν) all of them touch the line QQ', the locus of the intersection of the tangents through Q, Q' to each conic of the series is a curve of the order $\frac{1}{2}\mu+\nu$, having a $(\frac{1}{2}\mu)$-tuple point at the points Q, Q' respectively.

[Vol. VII. pp. 26, 27.]

2250. (Proposed by Professor CAYLEY.)—From the focal equation $x^2+y^2=(lx+n)^2$ of a conic, deduce the remaining three focal equations.

Solution by the PROPOSER.

We are to find α, β, L, M, N such that the equation

$$(x-\alpha)^2+(y-\beta)^2=(Lx+My+N)^2$$

may be identical with the given equation. It is at once seen that we must have $M=0$ or else $L=0$; the first supposition gives two solutions, one of which is the given equation itself, the other is

$$\left(x-\frac{2ln}{1-l^2}\right)^2+y^2=\left\{lx-n\frac{1+l^2}{1-l^2}\right\}^2.$$

The second supposition, $L=0$, gives two solutions, which only differ by the sign of $i\ (=\sqrt{-1})$, viz. these are

$$\left(x-\frac{ln}{1-l^2}\right)^2+\left(y\mp\frac{lni}{1-l^2}\right)^2=\frac{-(ly\pm ni)^2}{1-l^2}.$$

There is, of course, no difficulty in *verifying* the identity of each of the three forms with the given form $x^2+y^2=(lx+n)^2$.

[Vol. VII. pp. 33, 34.]

1991. (Proposed by Professor CAYLEY.)—Given a point and three lines; it is required to draw through the point a plane meeting the three lines in three points equidistant from the given point.

Solution by the PROPOSER.

Let O be the given point, $OA'=a$, $OB'=b$, $OC'=c$ the perpendiculars let fall from O on the given lines respectively. Take θ an arbitrary line, and from the points

A', B', C' measure off on the three lines respectively the distances $A'A = \pm\sqrt{(\theta^2 - a^2)}$, $B'B = \pm\sqrt{(\theta^2 - b^2)}$, $C'C = \pm\sqrt{(\theta^2 - c^2)}$, or, considering each radical as containing implicitly the sign $\pm$, what is the same thing, the distances $A'A = \sqrt{(\theta^2 - a^2)}$, $B'B = \sqrt{(\theta^2 - b^2)}$, $C'C = \sqrt{(\theta^2 - c^2)}$, then we have $OA = OB = OC (= \theta)$; and consequently the problem is to determine θ in such wise that the plane ABC may pass through the given point O: for we shall then have through O a plane meeting the three given lines in the points A, B, C equidistant from O.

The coordinates of A, B, C are linear functions of the radicals $\sqrt{(\theta^2 - a^2)}$, $\sqrt{(\theta^2 - b^2)}$, $\sqrt{(\theta^2 - c^2)}$ respectively. Taking O as origin, the condition in order that the plane ABC may pass through O is

$$\begin{vmatrix} x_1, & y_1, & 1 \\ x_2, & y_2, & 1 \\ x_3, & y_3, & 1 \end{vmatrix} = 0,$$

and substituting for the coordinates their values in terms of θ, this is an equation linear in each of the three radicals, or say, an equation of the form

$$\left(\sqrt{(\theta^2 - a^2)},\ 1\right)\left(\sqrt{(\theta^2 - b^2)},\ 1\right)\left(\sqrt{(\theta^2 - c^2)},\ 1\right) = 0.$$

But we may represent any one of the three radicals, say $\sqrt{(\theta^2 - c^2)}$ by a single letter s; and this being so, we have $\sqrt{(\theta^2 - a^2)} = \sqrt{(s^2 + c^2 - a^2)} = \sqrt{P}$ suppose, and $\sqrt{(\theta^2 - b^2)} = \sqrt{(s^2 + c^2 - b^2)} = \sqrt{Q}$ suppose; and it is to be observed that there is no loss of generality in assuming that the distance $C'c = s$ is measured off from C' in a determinate sense, for as s passes from $-\infty$ to $+\infty$, we thus obtain for c every position whatever on the line in question; whereas the other two distances $A'A$, $B'B$, represented by the radicals $\sqrt{P}$ and $\sqrt{Q}$ respectively, remain each of them with the double sense $\pm$. The equation in s is of the form

$$(s,\ 1)(\sqrt{P},\ 1)(\sqrt{Q},\ 1) = 0,$$

or, what is the same thing, it is of the form

$$\alpha\sqrt{(PQ)} + \beta\sqrt{P} + \gamma\sqrt{Q} + \delta = 0,$$

where $(\alpha, \beta, \gamma, \delta)$ are respectively linear functions of s.

Proceeding to rationalise the equation, we have first

$$\alpha^2 PQ + 2\alpha\delta\sqrt{(PQ)} + \delta^2 = \beta^2 P + \gamma^2 Q + 2\beta\gamma\sqrt{(PQ)},$$

and then finally

$$(\alpha^2 PQ - \beta^2 P - \gamma^2 Q + \delta^2)^2 = 4(\beta\gamma - \alpha\delta)^2 PQ,$$

which, observing that P, Q are each of them of the second order in s, is an equation of the twelfth order in s; that is, the number of solutions is $= 12$.

The solution of the problem is greatly simplified when $a = b = c$, that is, when the three given lines are tangents to a sphere having its centre at the given point. We have in this case $\sqrt{P} = \pm s$, $\sqrt{Q} = \pm s$, or the equation in s is

$$(s,\ 1)(\pm s,\ 1)(\pm s,\ 1) = 0;$$

that is, the equation of the twelfth order breaks up into four equations each of the third order. The geometrical theory may also be further developed. In fact, assuming on each of the three lines respectively a certain sense as positive (and thus isolating a set of three solutions) the construction is, on the three lines, from the points A', B', C' respectively, measure off the distances $A'A = B'B = C'C = s$. Then the points A, B, C form on the three lines respectively three homographic series; that is, the lines BC, CA, AB are respectively generating lines of three hyperboloids, viz. hyperboloids which pass respectively through the second and third lines, the third and first lines, and the first and second lines. Taking the given point O as the centre of projection, and projecting the whole figure on any plane whatever, the projections of the lines BC are the tangents of a conic which is the projection of the visible contour of the hyperboloid generated by the lines BC; and the like for the lines CA and AB. Hence in the projection, or plane figure, we have a triangle whereof the sides A', B', C' are the projections of the three given lines respectively; inscribed in this triangle we have a variable triangle ABC, such that the side

BC envelopes a conic, say (A), which touches B' and C',
CA envelopes a conic, say (B), which touches C' and A',
AB envelopes a conic, say (C), which touches A' and B'.

The conics $(A)(B)(C)$ have three common tangents, say L, M, N; the conics

(B) and (C) having besides the common tangent A',
(C) and (A) having besides the common tangent B',
(A) and (B) having besides the common tangent C',

so that the common tangents of the conics (B) and (C), (C) and (A), (A) and (B) are the lines A', B', C' each once, and the lines L, M, N each three times. In the entire series of triangles ABC there are three triangles which degenerate into the lines L, M, N respectively, these being in fact the projections of the triangles ABC of the solid figure which lie in a plane with O. Or, what is the same thing, the planes of the required triangles ABC of the solid figure are the planes through O and the three lines L, M, and N, respectively.

[Vol. VII. pp. 34—36.]

1993. (Proposed by T. COTTERILL, M.A.)—If P is a point on a circle, in which A and B are fixed points on a diameter at equal distances from its centre, the curve envelope of lines cutting harmonically the two circles whose centres are A and B and radii AP, BP respectively, is independent of the position of P on the circle.

Solution by PROFESSOR CAYLEY.

1. More generally, the problem may be thus stated: If two conics touch at I, J the lines OI, OJ respectively; if P be a variable point on the first conic, and OAB

a fixed line through O meeting the second conic in the points A and B; then considering the conic which passes through P and touches at I, J the lines AI, AJ respectively, and also the conic which passes through P and touches at I, J the lines BI, BJ respectively; the envelope of the lines which cut harmonically the last-mentioned two conics is a conic independent of the position of P.

2. Taking $x=0$, $y=0$, $z=0$ for the equations of the lines OI, JI, and OJ respectively, the equations of the two given conics are

$$xz-y^2=0, \quad kxz-y^2=0;$$

hence the coordinates of P may be taken to be

$$x : y : z = 1 : \theta : \theta^2,$$

and the coordinates of the points A and B may be taken to be

$$x : y : z = 1 : k\alpha : k\alpha^2, \text{ and } x : y : z = 1 : -k\alpha : k\alpha^2.$$

The equations of the lines AI, AJ are

$$k\alpha x - y = 0, \quad z - \alpha y = 0;$$

hence the equation of the conic touching these lines at the points I, J respectively, and also passing through the point P, is

$$\frac{(k\alpha x-y)(z-\alpha y)}{(k\alpha-\theta)(\theta-\alpha)}=\frac{y^2}{\theta},$$

and similarly the equations of the lines BI, BJ being

$$k\alpha x + y = 0, \quad z + \alpha y = 0,$$

the equation of the conic touching these lines at the points I, J respectively, and also passing through the point P, is

$$\frac{(k\alpha x+y)(z+\alpha y)}{(k\alpha+\theta)(\theta+\alpha)}=\frac{y^2}{\theta},$$

or multiplying out and reducing, if the equations of the two conics are represented by

$$(a, b, c, f, g, h \between x, y, z)^2=0, \quad (a', b', c', f', g', h' \between x, y, z)^2=0,$$

respectively, then the values of the coefficients are

$$\begin{array}{ll} a=0, & a'=0, \\ b=2(k\alpha+\theta^2-k\alpha\theta), & b'=2(-k\alpha^2-\theta^2-k\alpha\theta), \\ c=0, & c'=0, \\ f=-\theta, & f'=\theta, \\ g=\theta k\alpha, & g'=\theta k\alpha, \\ h=-\theta k\alpha^2, & h'=\theta k\alpha^2. \end{array}$$

Now the tangential equation of the envelope of the line which cuts harmonically the last-mentioned two conics, is

$$(bc' + b'c - 2ff', \ldots gh' + g'h - af' - a'f, \ldots \between \xi, \eta, \zeta)^2;$$

or substituting for a &c. a' &c., their values, it is found that the coefficients of this equation have all of them the common factor $2\theta^2$, and that omitting this factor the equation is independent of θ, viz. the tangential equation of the envelope in question is

$$(1, -k^2\alpha^2, k^2\alpha^4, 0, k(2k-1)\alpha^2, 0 \between \xi, \eta, \zeta)^2 = 0,$$

which proves the theorem.

3. *In particular*, if $k = 1$, that is if the points A, B lie on the conic $xz - y^2 = 0$, then the tangential equation of the envelope is

$$(1, -\alpha^2, \alpha^4, 0, \alpha^2, 0 \between \xi, \eta, \zeta)^2 = 0,$$

that is

$$\xi^2 - \alpha^2\eta^2 + \alpha^4\zeta^2 + 2\alpha^2\xi\zeta = 0;$$

or, what is the same thing, the equation is

$$(\xi - \alpha\eta + \alpha^2\zeta)(\xi + \alpha\eta + \alpha^2\zeta) = 0,$$

and thus the envelope breaks up into the two points

$$\xi - \alpha\eta + \alpha^2\zeta = 0, \quad \xi + \alpha\eta + \alpha^2\zeta = 0;$$

that is, the points $(1, -\alpha, \alpha^2)$ and $(1, \alpha, \alpha^2)$, which are the points A and B respectively. That is, in the problem in its original form, if the points A and B are the extremities of a diameter of a given circle, then the two constructed circles are a pair of orthotomic circles with the centres A and B respectively; and the theorem is the very obvious one, that any line through the centre of either circle cuts the two circles harmonically.

[Vol. VII. pp. 52, 53.]

2270. (Proposed by Professor CAYLEY.)—To reduce the equation of a bicircular quartic into the form $SS' - k^3L = 0$, where $S = 0$, $S' = 0$ are the equations of two circles, $L = 0$ the equation of a line. (See Salmon's *Higher Plane Curves*, p. 128.)

Solution by the PROPOSER.

The equation of a bicircular quartic may be taken to be

$$(x^2 + y^2)^2 + (u_1 + u_0)(x^2 + y^2) + v_2 + v_1 + v_0 = 0,$$

where, and in what follows, the subscript numbers denote the degrees in the coordinates (x, y) of the several functions to which they are attached.

Introducing an arbitrary constant θ_0, and putting the equation under the form

$$(x^2+y^2)^2+(u_1+u_0-\theta_0)(x^2+y^2)+\theta_0(x^2+y^2)+v_2+v_1+v_0=0,$$

this may be identified with

$$(x^2+y^2+p_1+p_0)(x^2+y^2+q_1+q_0)+r_1+r_0=0\ ;$$

viz. the conditions in order to this identity are

$$p_1+p_0+q_1+q_0=u_1+u_0-\theta_0,$$

$$(p_1+p_0)(q_1+q_0)+r_1+r_0=\theta_0(x^2+y^2)+v_2+v_1+v_0,$$

that is

$$p_1+q_1=u_1,\quad p_0+q_0=u_0-\theta_0,$$

$$p_1q_1=\theta_0(x^2+y^2)+v_2,\quad p_1q_0+p_0q_1+r_1=v_1,\quad p_0q_0+r_0=v_0.$$

Hence

$$(p_1-q_1)^2=u_1^2-4v_2-4\theta_0(x^2+y^2),$$

where the right-hand side is a quadric function $(x, y)^2$, which, when the discriminant thereof is put $=0$, (that is, when θ_0 is determined as the root of a quadric equation,) is a perfect square, p_1-q_1 is then a known linear function, and p_1+q_1 being equal to the linear function u_1, we have p_1 and q_1 as linear functions of (x, y). We may take for the constants p_0 and q_0 any values satisfying the equation $p_0+q_0=u_0-\theta_0$; and we then have

$$r_1=v_1-p_1q_0-p_0q_1,\quad r_0=v_0-p_0q_0,$$

which completes the determination; the form

$$(x^2+y^2+p_1+p_0)(x^2+y^2+q_1+q_0)+r_1+r_0=0$$

is of course the same as the proposed form $SS'-k^3L=0$.

Cor. A somewhat more convenient form is $UU'-k^2V=0$, where $U=0$, $U'=0$ are the equations of two evanescent circles (pairs of imaginary lines), $V=0$ the equation of a circle; in fact the original form $SS'-k^3L=0$ may be written $(S-\alpha)(S'-\alpha')+(\alpha S'+\alpha' S-\alpha\alpha'-k^3L)=0$, which, when α, α' are so determined that $S-\alpha=0$, $S'-\alpha'=0$ may be evanescent circles, is of the required form $UU'-k^2V=0$. The equation $UU'=0$ is that of the two pairs of tangents to the curve at the circular points at infinity respectively; in fact, writing $U=pq$, $U'=p'q'$, each of the lines $p=0$, $q=0$, $p'=0$, $q'=0$ meets the circle $V=0$ in one or other of the circular points at infinity, and therefore only in a single point not at infinity; hence each of these lines meets the curve $UU'-k^2V=0$ three times in one of the circular points at infinity, that is, the line in question is a tangent to one of the two branches through the circular point at infinity.

[Vol. VII. pp. 87, 88.]

2309. (Proposed by Professor Cayley.)—Show that for n things

$1-$ (no. of partitions into 2 parts) $+\,1\,.\,2$ (no. of partitions into 3 parts)....

$\pm\,1\,.\,2\,.\,3\,..\,(n-1)$ (no. of partitions into n parts) $=0$.

For instance, $n=4$; partitions of (a, b, c, d) into two parts are (a, bcd), (b, cda), (c, dab), (d, abc), (ab, cd), (ac, db), (ad, bc); no. is $=7$. Partitions into three parts are (ab, c, d), (ac, b, d), (ad, b, c), (bc, a, d), (b, d, ac), (cd, a, b); no. is $=6$. Partition into 4 parts is (a, b, c, d); no. is $=1$. And we have

$$1-1.7+2.6-6.1=13-13=0.$$

Solution by the PROPOSER.

Write $n=a\alpha+b\beta+c\gamma+\dots$, where $\alpha, \beta, \gamma\dots$ are positive integers all of them different, and $a, \beta, \gamma\dots$ are positive integers; and consider the partitions wherein we have a parts each of α things, b parts each of β things, &c. Writing as usual $\Pi(n)=1.2.3\dots n$, the number of partitions of the form in question is

$$=\frac{\Pi n}{\Pi a\,.\,\Pi b\dots(\Pi\alpha)^a(\Pi\beta)^b\dots};$$

whence, putting for shortness $\alpha+\beta+\dots=p$, the theorem may be written

$$\Sigma(-)^{p-1}\frac{\Pi(p-1)\,\Pi n}{\Pi a\,.\,\Pi b\dots(\Pi\alpha)^a(\Pi\beta)^b\dots}=0,$$

the summation extending to all the partitions $n=a\alpha+b\beta+\dots$, as explained above.

Now if the n quantities $x, y, z,\dots$ are the nth roots of unity, we have $x+y+z\dots=0$, and therefore also $(x+y+z\dots)^n=0$, and the general term of the left-hand is

$$\frac{\Pi n}{(\Pi\alpha)^a(\Pi\beta)^b\dots}[\alpha^a\beta^b\dots],$$

where $[\alpha^a\beta^b\dots]$ denotes the symmetrical function $\Sigma x^\alpha y^\alpha\dots(a \text{ factors})\, u^\beta v^\beta\dots(b \text{ factors})\dots$ of the roots $x, y, z, u, v\dots$ of the equation $\theta^n-1=0$; where, as above, $n=a\alpha+b\beta+\dots$. Now by a formula not, I believe, generally known, but which is given on p. 175 of the translation of Hirsch's Algebra (Hirsch's *Collection of Examples &c. on the Literal Calculus and Algebra*, translated by the Rev. J. A. Ross, London, 1827), the value of the sum in question is $=(-)^{p-1}\dfrac{\Pi(p-1)}{\Pi a.\Pi b\dots}n$, where $p=a+b+\dots$, (the sign $\pm$, given in the formula as quoted, is at once seen to be $(-)^{p-1}$); whence, substituting and omitting the factor n, we have

$$\Sigma(-)^{p-1}\frac{\Pi(p-1)\,\Pi n}{\Pi a\,.\,\Pi b\dots(\Pi\alpha)^a(\Pi\beta)^b\dots}=0,$$

which is the required theorem.

OBSERVATION. In Cauchy's *Exercices d'Analyse &c.*, t. III., p. 173, is given a formula relating to the same mode of partition of the number n, viz. this is

$$\Sigma\frac{\Pi n}{\Pi a\,.\,\Pi b\dots\alpha^a\beta^b\dots}=\Pi n.$$

I have somewhere made the remark that, on the left-hand side, the terms which belong to the odd and the even values of $a+b+\ldots(=p)$ are equal, and that we have therefore

$$\Sigma(-)^{p-1}\frac{\Pi n}{\Pi a\,.\,\Pi b\ldots\alpha^a\beta^b\ldots}=0,$$

which is a theorem having a curious analogy with that demonstrated above.

[Vol. VII. pp. 99—102.]

2286. (Proposed by W. H. LAVERTY.)—If we have $(n-2)$ sets of n quantities each, $(\alpha_1, \alpha_2 \ldots \alpha_n)$, $(\beta_1, \beta_2 \ldots \beta_n)$, ... $(\lambda_1, \lambda_2 \ldots \lambda_n)$, connected with the n quantities $(r_1, r_2 \ldots r_n)$ by $\frac{1}{2}n(n-1)$ equations of which the type form is

$$(\alpha_k-\alpha_l)^2+(\beta_k-\beta_l)^2+\ldots(\lambda_k-\lambda_l)^2=r_k^2+r_l^2;$$

then show that

$$\frac{1}{r_1^2}+\frac{1}{r_2^2}+\ldots+\frac{1}{r_n^2}=0 \text{ and } \frac{P_1}{r_1^2}+\frac{P_2}{r_2^2}+\ldots\frac{P_n}{r_n^2}=0,$$

where P is any one of the quantities α, β, $\gamma \ldots \lambda$.

Solution by PROFESSOR CAYLEY.

Consider the case $n=4$; we have between $(\alpha_1, \alpha_2, \alpha_3, \alpha_4)$, $(\beta_1, \beta_2, \beta_3, \beta_4)$, ... (r_1, r_2, r_3, r_4) six equations, such as the equation

$$(\alpha_1-\alpha_2)^2+(\beta_1-\beta_2)^2=r_1^2+r_2^2; \tag{12}$$

and it is in effect required to show that these equations give

$$\frac{1}{r_1^2}:\frac{1}{r_2^2}:\frac{1}{r_3^2}:\frac{1}{r_4^2}=(234):-(341):(412):-(123),$$

where

$$(123)=\begin{vmatrix}\alpha_1, & \beta_1, & 1\\ \alpha_2, & \beta_2, & 1\\ \alpha_3, & \beta_3, & 1\end{vmatrix}, \text{ \&c.,}$$

viz. considering (α_1, β_1), (α_2, β_2), (α_3, β_3), (α_4, β_4) as the rectangular coordinates of four points in a plane, then (123) is the area (taken with a proper sign) of the triangle formed by the points 1, 2, 3; and the like for (234) &c.

Combining the equations as follows,

$$(12)+(34)-(13)-(24),$$

the r's disappear, and we have an equation

$$(\alpha_1-\alpha_4)(\alpha_2-\alpha_3)+(\beta_1-\beta_4)(\beta_2-\beta_3)=0,$$

which shows that the lines 14 and 23 intersect at right angles; similarly the lines 12 and 34, and also the lines 13 and 24, intersect at right angles; or starting from the given points 1, 2, 3, the point 4 is the intersection of the perpendiculars let fall from the angles 1, 2, 3 of the triangle 123 on the opposite sides respectively.

Again combining the equations as follows,

$$(12)+(13)-(23),$$

we obtain

$$r_1^2=(\alpha_1-\alpha_2)(\alpha_1-\alpha_3)+(\beta_1-\beta_2)(\beta_1-\beta_3).$$

The entire system of equations will remain unaltered if we pass from the original axes to any other system of rectangular axes; hence taking the axes of x in the sense from 1 to 2 along the line 12, $\beta_1-\beta_2$ becomes $=0$, and we have

$$\alpha_2-\alpha_1=12,\quad \alpha_3-\alpha_1=1\,(12,\ 34);$$

viz. $\alpha_2-\alpha_1$ is the distance 12 of the points 1 and 2, $\alpha_3-\alpha_1$ is the distance 1 (12, 34) of the point 1 from the point (12, 34) which is the intersection of the lines 12 and 34; we have therefore

$$r_1^2=12\,.\,1\,(12,\ 34).$$

But similarly

$$r_2^2=21\,.\,2\,(12,\ 34),\ =12\,.\,(12,\ 34)\,2,$$

(since $21=-12$ and $2\,(12,\ 34)=-(12,\ 34)\,2$). And we have therefore

$$r_1^2 : r_2^2=1\,(12,\ 34) : (12,\ 34)\,2,\ \text{or}\ \frac{1}{r_1^2} : \frac{1}{r_2^2}=(12,\ 34)\,2 : 1\,(12,\ 34).$$

Write

$$\lambda=\frac{(12,\ 34)\,2}{12},\qquad \mu=\frac{1\,(12,\ 34)}{12}$$

where 1 (12, 34) and (12, 34) 2 are as above the distances from 1 to (12, 34) and from (12, 34) to 2; and, in the denominators, 12 is the distance from 1 to 2; we have $\lambda+\mu=1$; the coordinates of (12, 34) are $\lambda\alpha_1+\mu\alpha_2$, $\lambda\beta_1+\mu\beta_2$, and the values of λ, μ are obtained by writing $\lambda\alpha_1+\mu\alpha_2$, $\lambda\beta_1+\mu\beta_2$, $\lambda+\mu$ for x, y, 1 in the equations

$$\begin{vmatrix} x, & y, & 1 \\ \alpha_3, & \beta_3, & 1 \\ \alpha_4, & \beta_4, & 1 \end{vmatrix}=0$$

of the line 34. Making this substitution, we find

$$\lambda\,(134)+\mu\,(234)=0,$$

where as above

$$(134)=\begin{vmatrix} \alpha_1, & \beta_1, & 1 \\ \alpha_3, & \beta_3, & 1 \\ \alpha_4, & \beta_4, & 1 \end{vmatrix},\ \&\text{c.},$$

we have therefore

$$\lambda : \mu = (234) : -(134) = (234) : -(341),$$

or, what is the same thing,

$$(12,\ 34)\,2 : 1\,(12,\ 34) = (234) : -(341);$$

and consequently

$$\frac{1}{r_1^2} : \frac{1}{r_2^2} \qquad = (234) : -(341);$$

or completing the system by symmetry

$$\frac{1}{r_1^2} : \frac{1}{r_2^2} : \frac{1}{r_3^2} : \frac{1}{r_4^2} = (234) : -(341) : (412) : -(123),$$

which is the required result.

In the case $n = 5$, we have between

$$(\alpha_1,\ \alpha_2,\ \alpha_3,\ \alpha_4,\ \alpha_5),\quad (\beta_1,\ \beta_2,\ \beta_3,\ \beta_4,\ \beta_5),\quad (\gamma_1,\ \gamma_2,\ \gamma_3,\ \gamma_4,\ \gamma_5),\quad (r_1,\ r_2,\ r_3,\ r_4,\ r_5)$$

ten equations such as the equation

$$(\alpha_1 - \alpha_2)^2 + (\beta_1 - \beta_2)^2 + (\gamma_1 - \gamma_2)^2 = r_1^2 + r_2^2. \qquad (12)$$

We obtain as before the equation

$$(\alpha_1 - \alpha_4)(\alpha_2 - \alpha_3) + (\beta_1 - \beta_4)(\beta_2 - \beta_3) + (\gamma_1 - \gamma_4)(\gamma_2 - \gamma_3) = 0,$$

which, considering $(\alpha_1,\ \beta_1,\ \gamma_1)$ &c. as the rectangular coordinates of five points 1, 2, 3, 4, 5 in space, signifies that the line 14 is at right angles to the line 23; the five points are therefore such that the line joining any two of them is at right angles to the line joining any other two of them, whence also the line joining any two is at right angles to the plane through the remaining three points. (The points 1, 2, 3, 4 form a tetrahedron such that that 12 and 34, also 13 and 42, also 14 and 23 are at right angles to each other, two of these conditions imply the third; and this being so, if a further condition be satisfied, the perpendiculars from 1, 2, 3, and 4 on the opposite faces respectively, will meet in a point 5, and we shall have the system of points 1, 2, 3, 4, 5 related as above.)

We further obtain as before

$$r_1^2 = (\alpha_1 - \alpha_2)(\alpha_1 - \alpha_3) + (\beta_1 - \beta_2)(\beta_1 - \beta_3) + (\gamma_1 - \gamma_2)(\gamma_1 - \gamma_3),$$

or taking the axis of x in the sense from 1 to 2 along the line 12, we have $\beta_1 - \beta_2 = 0$, $\gamma_1 - \gamma_2 = 0$, and the equation becomes

$$r_1^2 = 12\,.\,1\,(12,\ 345),$$

and similarly

$$r_2^2 = 12\,.\,(12,\ 345)\,2\,;$$

whence

$$\frac{1}{r_1^2} : \frac{1}{r_2^2} = (12,\ 345)\,2 : (12,\ 345)\,1.$$

Writing them $\lambda = \frac{(12.345)2}{12}$, $\mu = \frac{1(12,\ 345)}{12}$ (and therefore $\lambda + \mu = 1$) we find (λ, μ) by substituting $\lambda\alpha_1 + \mu\alpha_2$, $\lambda\beta_1 + \mu\beta_2$, $\lambda\gamma_1 + \mu\gamma_2$, $\lambda + \mu$ for x, y, z, 1 in the equation

$$\begin{vmatrix} x, & y, & z, & 1 \\ \alpha_3, & \beta_3, & \gamma_3, & 1 \\ \alpha_4, & \beta_4, & \gamma_4, & 1 \\ \alpha_5, & \beta_5, & \gamma_5, & 1 \end{vmatrix} = 0$$

of the plane 345; we have thus

$$\lambda(1345) + \mu(2345) = 0,$$

that is

$$\lambda : \mu = (2345) : -(1345) = (2345) : (3451),$$

whence

$$(12,\ 345)2 : 1(12,\ 345) = (2345) : (3451),$$

that is

$$\frac{1}{r_1^2} : \frac{1}{r_2^2} \qquad = (2345) : (3451),$$

or completing by symmetry

$$\frac{1}{r_1^2} : \frac{1}{r_2^2} : \frac{1}{r_3^2} : \frac{1}{r_4^2} : \frac{1}{r_5^2} = (2345) : (3451) : (4512) : (5123) : (1234),$$

which is the theorem for the case $n = 5$. The general case depends, it is clear, upon similar reasoning in a $(n-2)$dimensional geometry; leading to the conception in this geometry of a figure of $(n-1)$ points such that the line joining any two of them is at right angles to the line joining any other two of them.

[Vol. VII. p. 106.]

2331. (Proposed by Professor CAYLEY.)—Show that it is possible to find (X, Y, Z) linear functions of the trilinear coordinates (x, y, z) such that the equations $xX = yY = zZ$ may determine four given points.

[Vol. VIII., July to December, 1867, p. 26.]

2321. (Proposed by Professor CAYLEY.)—Given a conic, to find four points such that all the conics through the four points may have their centres in the given conic.

[Vol. VIII. p. 36.]

2371. (Proposed by Professor CAYLEY.)—(4). If P, Q be two points taken at random within the triangle ABC, what is the chance that the points A, B, P, Q may form a convex quadrangle?

[Vol. VIII. pp. 51, 52.]

Note on Question 1990. *By* PROFESSOR CAYLEY.

The theorem of paragraph 4 (*Reprint*, vol. VI. p. 88), (ascribed by Professor Sylvester to Mr Crofton), that "if a circle and a straight line be cut by any transversal in three points, these will be the foci of a system of Cartesian ovals having double contact with one another at two fixed points," may be enunciated under a more complete form, as follows:

If in a given circle the chords PP_1, BC meet in A, then each of the two Cartesians, foci A, B, C, which pass through P, will also pass through P_1; and moreover, if α, α' be the diametrals of the chord PP_1 (that is, the extremities of the diameter at right angles to PP_1) then the tangents at P, P_1 to one of the Cartesians will be αP, αP_1 respectively, and to the other of them $\alpha' P$, $\alpha' P_1$ respectively, these tangents being thus independent of the position of the chord BC; and thence also thus;

Given the points A, B, C *in lineâ*, and the point P;

through P, B, C	draw a circle	(A)	and let	PA	meet this in	P_1,
„ P, C, A	„	(B)	„	PB	„	P_2,
„ P, A, B	„	(C)	„	PC	„	P_3,

then each of the Cartesians, foci A, B, C, which pass through P will also pass through P_1, P_2, P_3; and if

α, α'	are the diametrals of	PP_1	in circle	(A),
β, β'	„	PP_2	„	(B),
γ, γ'	„	PP_3	„	(C),

then (the points of the several pairs being properly selected) the points (α, β, γ) and the points $(\alpha', \beta', \gamma')$ will each lie in a line through P, viz. the lines $P\alpha\beta\gamma$ and $P\alpha'\beta'\gamma'$ will be the tangents at P to the two Cartesians respectively.

The two Cartesians meet in the points P, P_1, P_2, P_3, and in the symmetrically situated points in regard to the axis ABC; the theorem contains as part of itself the well-known property that the two Cartesians cut at right angles at each of their points of intersection; it gives moreover the construction of the following problem:—given the foci A, B, C, and one intersection P of a pair of triconfocal Cartesians, to find the remaining intersections, and the tangents at each of the intersections.

[Vol. VIII. pp. 70—72.]

1911. (Proposed by Professor CAYLEY.)—Given four points, and also the "conic of centres"—viz. the conic which is the locus of the centres of the several conics which pass through the four given points; then if a conic through the four given points has for its centre a given point on the conic of centres, it is required to find a construction for the asymptotes of this conic.

Solution by the PROPOSER.

1. Consider four given points, and in connection therewith a given line IJ; the locus of the poles of IJ, in regard to the several conics which pass through the four points, is a conic, the "conic of poles." Consider a particular conic Θ, through the four points; the pole of IJ in regard to the conic Θ is a point C on the conic of poles, and the tangents from C to the conic Θ meet the conic of poles in two points H, K; the chord of intersection HK passes through the point Π which is the pole of IJ in regard to the conic of poles. Moreover, the polars of a point C', in regard to the several conics through the four points, meet in a point Ω', the "common pole" of C', and in particular if C' be the point C on the conic of poles, then the common pole is a point Ω on the line IJ; this being so, the line HK passes (as already mentioned) through Π, and the lines HK and $\Pi\Omega$ are harmonics in regard to the conic of poles.

2. Assuming the foregoing properties, then, given the four points, the line IJ, the conic of poles, and the point C on this conic; we may construct Π the pole of IJ in regard to the conic of poles; and also Ω the common pole of C; the line HK is then given as a line passing through Π, and harmonic to $\Pi\Omega$ in regard to the conic of poles; this line meets the conic of poles in the points H, K; and then CH, CK are the tangents from C to a conic Θ which passes through the four points.

3. In particular if IJ be the line infinity, then the conic of poles is the conic of centres; Π is the centre of this conic; Ω is as before the common pole of C; HK is given as the diameter of the conic of centres, conjugate to $\Pi\Omega$; H, K are the extremities of this diameter; and then CH, CK are the asymptotes of the conic through the four points, which has the point C for its centre; and the asymptotes are therefore constructed as required. If the points H, K are imaginary, the asymptotes will be also imaginary; the conic Θ is in this case an ellipse.

4. It is hardly necessary to remark, in regard to the construction of the point Ω, that we have among the conics through the four points, three pairs of lines meeting in points P, Q, R respectively (it is clear that the conic of poles passes through these three points); the harmonics of CP, CQ, CR in regard to the three pairs of lines respectively meet in a point, which is the required point Ω. In the particular case where the point C is on the conic of centres, the three harmonics are parallel; it is therefore sufficient to construct *one* of them; and the line HK is then the diameter of the conic of poles, conjugate to the harmonic so constructed.

5. It remains to prove the properties assumed in (1). We may take $z=0$ for the equation of the line IJ, $x=0$, $y=0$ for the equations of the tangents to the conic Θ at its intersections with the line IJ, so that we have $(x=0,\ y=0)$ for the coordinates of the point C; the equation of the conic Θ will be of the form $z^2-xy=0$, and the four points may then be taken to be the intersections of the conic $z^2-xy=0$, and the arbitrary conic

$$(a,\ b,\ c,\ f,\ g,\ h\between x,\ y,\ z)^2=0.$$

The equation of the conic of centres is found to be

$$x(ax+hy+gz)-y(hx+by+fz)=0,\ \text{ or }\ ax^2-by^2+gzx-hxy=0\,;$$

or, as it may also be written,

$$(2a, -2b, 0, -f, g, 0 \between x, y, z)^2 = 0;$$

and it is convenient to remark that the equation in line coordinates (or condition that this conic may be touched by the line $\xi x + \eta y + \zeta z = 0$) is

$$(-f^2, -g^2, -4ab, 2af, 2bg, -fg \between \xi, \eta, \zeta)^2 = 0.$$

The line $x = 0$ meets the conic of poles in the point $x = 0$, $by + fz = 0$, and the line $y = 0$ meets the same conic in the point $y = 0$, $ax + gz = 0$; hence the line HK, which is the line joining these two points, has for its equation

$$af\,x + bg\,y + fg\,z = 0;$$

and it only remains to be shown that this line passes through the point Π, and is the harmonic of the line $\Pi\Omega$ in regard to the conic of centres. The point Π is the pole of the line $z = 0$ in regard to the conic of centres, its coordinates are at once found to be

$$x : y : z = bg : af : -2ab;$$

and we thence see that Π is a point on the line HK. The point Ω is given as the intersection of the polars of C in regard to the conics $z^2 - xy = 0$, and $(a, b, c, f, g, h \between x, y, z)^2 = 0$ respectively; that is, as the intersection of the lines $z = 0$, and $gx + fy + cz = 0$; its coordinates therefore are

$$x : y : z = -f : g : 0.$$

Hence the equation of the line $\Pi\Omega$ is

$$2abg\,x + 2abf\,y + (af^2 + bg^2)\,z = 0.$$

Now, in general, if we have a conic the line-equation whereof is $(A, B, C, F, G, H \between \xi, \eta, \zeta)^2 = 0$, then the condition in order that, in regard thereto, the lines $\lambda x + \mu y + \nu z = 0$ and $\lambda' x + \mu' y + \nu' z = 0$ may be harmonics, is

$$(A, B, C, F, G, H \between \lambda, \mu, \nu \between \lambda', \mu', \nu') = 0;$$

that is

$$A\lambda\lambda' + B\mu\mu' + C\nu\nu' + F(\mu\nu' + \mu'\nu) + G(\nu\lambda' + \nu'\lambda) + H(\lambda\mu' + \lambda'\mu) = 0.$$

Hence, in order that the two lines HK and $\Pi\Omega$ may be harmonics in regard to the conic of centres, we should have

$$(-f^2, -g^2, -4ab, 2af, 2bg, -fg \between af, bg, fg \between 2abg, 2abf, af^2 + bg^2) = 0.$$

But developing, and omitting the common factor $abfg$, which enters into all the terms, this equation is

$$-(2af^2) - (2bg^2) - 4(af^2 + bg^2) + \{4af^2 + 2(af^2 + bg^2)\} + \{4bg^2 + 2(af^2 + bg^2)\} - 2(af^2 + bg^2) = 0,$$

which is identically true; and the lines HK and $\Pi\Omega$ are therefore harmonics in regard to the conic of centres.

[Vol VIII. p. 74.]

2371. (Proposed by Professor CAYLEY.)—If P, Q be two points taken at random within the triangle ABC, what is the chance that the points A, B, P, Q may form a convex quadrangle?

[Vol. VIII. pp. 86, 87.]

2466. (Proposed by H. MURPHY.)—If four points A, B, C, D be either in the same plane or not, and if the three rectangles $AB.CD$, $AC.DB$, $AD.BC$ be taken; the sum of any two of them is greater than the third, except when the points lie on the circumference of a circle.

Solution by PROFESSOR CAYLEY.

Write for shortness $BC=f$, $CA=g$, $AB=h$; $AD=a$, $BD=b$, $CD=c$; then, Lemma, if r be the radius of the sphere circumscribed about the tetrahedron $ABCD$, we have

$$4r^2\left\{\begin{array}{l} -a^4f^2-b^4g^2-c^4h^2-f^2g^2h^2 \\ +(a^2f^2+b^2c^2)(g^2+h^2-f^2) \\ +(b^2g^2+c^2a^2)(h^2+f^2-g^2) \\ +(c^2h^2+a^2b^2)(f^2+g^2-h^2) \end{array}\right\} = 2b^2c^2g^2h^2+2c^2a^2h^2f^2+2a^2b^2f^2g^2-a^4f^4-b^4g^4-c^4h^4,$$

where the left-hand side is $=576V^2r^2$, if V be the volume of the tetrahedron.

Suppose first that the points are not in the same plane, then the left-hand side $(=576\,V^2r^2)$ is positive; therefore the right-hand side is also positive, or putting for shortness $af=\alpha$, $bg=\beta$, $ch=\gamma$, we have

$$2\beta^2\gamma^2+2\gamma^2\alpha^2+2\alpha^2\beta^2-\alpha^4-\beta^4-\gamma^4=+,\ \text{that is, } 4\beta^2\gamma^2-(\alpha^2-\beta^2-\gamma^2)^2=+,$$

and thence $\alpha<\beta+\gamma$; for if α were equal to or greater than $\beta+\gamma$, say $\alpha=\beta+\gamma+x$, the left-hand side would be $4\beta^2\gamma^2-\{2\beta\gamma+2(\beta+\gamma)x+x^2\}^2$, which vanishes if $x=0$, and is negative for x positive. Similarly $\beta<\gamma+\alpha$, $\gamma<\alpha+\beta$; and the theorem is thus proved for the case where the four points are not in a plane.

Starting from this general case, if we imagine the point D continually to approach and ultimately to coincide with the plane ABC, but so as not to be in the circle ABC, then the expression $2\beta^2\gamma^2+2\gamma^2\alpha^2+2\alpha^2\beta^2-\alpha^4-\beta^4-\gamma^4$, which does not vanish in the limit, is throughout equal to the positive quantity $576\,V^2r^2$ (in the limit V is $=0$ and $r=\infty$, but Vr is finite, and of course V^2r^2 is positive), that is, the expression in question is $=+$, and the theorem follows as before. Of course when the four points are in a circle, then the expression is $=0$, and consequently one of the quantities α, β, γ is equal to the sum of the other two.

The lemma is at once proved by means of my theorem for the relation between the distances of five points in space, {*Cambridge Mathematical Journal*, vol. II. (1841), p. 269, [1],} viz. if the point 1 is the centre of the circumscribed sphere, and the points 2, 3, 4, 5 are the points A, B, C, D respectively, then the relation in question, viz.

$$\begin{vmatrix} 0\ , & (12)^2, & (13)^2, & (14)^2, & (15)^2, & 1 \\ (21)^2, & 0\ , & (23)^2, & (24)^2, & (25)^2, & 1 \\ (31)^2, & (32)^2, & 0\ , & (34)^2, & (35)^2, & 1 \\ (41)^2, & (42)^2, & (43)^2, & 0\ , & (45)^2, & 1 \\ (51)^2, & (52)^2, & (53)^2, & (54)^2, & 0\ , & 1 \\ 1\ , & 1\ , & 1\ , & 1\ , & 1\ , & 0 \end{vmatrix} = 0$$

becomes

$$\begin{vmatrix} 0, & r^2, & r^2, & r^2, & r^2, & 1 \\ r^2, & 0, & h^2, & g^2, & a^2, & 1 \\ r^2, & h^2, & 0, & f^2, & b^2, & 1 \\ r^2, & g^2, & f^2, & 0, & c^2, & 1 \\ r^2, & a^2, & b^2, & c^2, & 0, & 1 \\ 1, & 1, & 1, & 1, & 1, & 0 \end{vmatrix} = 0.$$

Multiplying the last line by $-r^2$ and adding it to the first line, this is

$$\begin{vmatrix} -r^2, & 0, & 0, & 0, & 0, & 1 \\ r^2, & 0, & h^2, & g^2, & a^2, & 1 \\ r^2, & h^2, & 0, & f^2, & b^2, & 1 \\ r^2, & g^2, & f^2, & 0, & c^2, & 1 \\ r^2, & a^2, & b^2, & c^2, & 0, & 1 \\ 1, & 1, & 1, & 1, & 1, & 0 \end{vmatrix} = 0,$$

and then proceeding in the same way with the first and last columns the equation is

$$\begin{vmatrix} -2r^2, & 0, & 0, & 0, & 0, & 1 \\ 0\ , & 0, & h^2, & g^2, & a^2, & 1 \\ 0\ , & h^2, & 0, & f^2, & b^2, & 1 \\ 0\ , & g^2, & f^2, & 0, & c^2, & 1 \\ 0\ , & a^2, & b^2, & c^2, & 0, & 1 \\ 1\ , & 1, & 1, & 1, & 1, & 0 \end{vmatrix} = 0,$$

which is in fact the equation of the Lemma. See my papers in the *Quarterly Journal of Mathematics*, vol. III. (1859), pp. 275—277, [286], and vol. V. (1861), pp. 381—384, [297].

COR.—It appears by the demonstration that for any four points not in the same plane, the expression

$$-a^4f^2-b^4g^2-c^4h^2-f^2g^2h^2$$
$$+(a^2f^2+b^2c^2)(g^2+h^2-f^2)+(b^2g^2+c^2a^2)(h^2+f^2-g^2)+(c^2h^2+a^2b^2)(f^2+g^2-h^2)$$

is always positive.

[Vol. VIII. pp. 105, 106.]

2472. (Proposed by Professor CAYLEY.)—Through four points on a circle to draw a conic such that an axis may pass through the centre of the circle.

Solution by the PROPOSER.

Let the equation of the conic be $(a, b, c, f, g, h \between x, y, 1)^2=0$, then if as usual the inverse coefficients are represented by (A, B, C, F, G, H), the equation of the two axes is

$$(a-b)(Cx-G)(Cy-F)+h[(Cx-G)^2-(Cy-F)^2]=0,$$

whence if an axis pass through the origin

$$(a-b)FG+h(G^2-F^2)=0.$$

Consider now the circle $x^2+y^2-1=0$ and on it the four points in which it is intersected by the conic $(a, b, c, f, g, h \between x, y, 1)^2=0$; then for any conics through the four points we have

$$(a, b, c, f, g, h \between x, y, 1)^2+\lambda(x^2+y^2-1)=0;$$

so that, taking this for the equation of the required conic, and representing it by

$$(a', b', c', f', g', h' \between x, y, 1)^2=0,$$

the values of the coefficients are

$$a'=a+\lambda,\quad b'=b+\lambda,\quad c'=c+\lambda,\quad f'=f,\quad g'=g,\quad h'=h,$$

and we thence have

$$F'=F-\lambda f,\quad G'=G-\lambda g,\quad a'-b'=a-b,\quad h'=h.$$

The required relation is

$$(a'-b')F'G'+h'(G'^2-F'^2)=0,$$

that is

$$(a-b)(F-\lambda f)(G-\lambda g)+h\{(G-\lambda g)^2-(F-\lambda f)^2\}=0,$$

a quadric equation in λ; and substituting for λ each of its two values, we have the two required conics

$$(a, b, c, f, g, h \between x, y, 1)^2+\lambda(x^2+y^2-1)=0,$$

for each of which an axis passes through the centre of the circle.

[Vol. IX., January to June, 1868, pp. 20, 21.]

Note on Question 2471. *By* PROFESSOR CAYLEY.

In the singularly beautiful solution which Mr Woolhouse has given of this question (see *Reprint*, vol. VIII. p. 100), it is important to note what is the analytical problem solved, and how the solution is obtained. Considering a plane area bounded by any closed convex curve, and in it three points P, P', P'', Mr Woolhouse investigates the average area of the triangle $PP'P''$, viz. this depends on the sextuple integral

$$\int \pm \{x'y'' - x''y' + x''y - xy'' + xy' - x'y\}\, dx\, dy\, dx'\, dy'\, dx''\, dy'',$$

where the sign $\pm$ has to be taken so that $\pm\{\ \}$ shall be positive, and where the integration in respect to each set of coordinates extends over the entire closed area; the difficulty is as to the mode of dealing with the discontinuous sign. It is remarked that the integral is

$$= 6 \int \pm \{x'y'' - x''y' + x''y - xy'' + xy' - x'y\}\, dx\, dy\, dx'\, dy'\, dx''\, dy'';$$

the variables in this last expression being restricted in such wise that x, x'', x' are in the order of increasing magnitude; the term $\pm\{\ \}$ is of the form $\pm(x'-x)(y''-\beta)$, where β is independent of y, and where (as is easily seen) if v'', u'' be the upper and lower ordinate corresponding to the abscissa x'', then β lies between the values u'' and v''. But $x'-x$ is positive, hence the sign $\pm$ must be so taken that $\pm(y''-\beta)$ shall be positive, that is, from $y''=u''$ to $y''=\beta$ the sign is $-$, and from $y''=\beta$ to $y''=v''$ the sign is $+$.

Hence for the integration in regard of y'' we have

$$\int \pm (y''-\beta)\, dy'' = \int_\beta^{v''} + (y''-\beta)\, dy'' + \int_{u'}^{\beta} - (y''-\beta)\, dy'', = \tfrac{1}{2}(v''-\beta)^2 + \tfrac{1}{2}(\beta - u'')^2;$$

and the discontinuous sign $\pm$ is thus got rid of. The remaining integrations are then effected in the order x'', y', y, x', x, the limits being for x'' from x to x', for y' from u' to v', and for y from u to v (if the upper and lower ordinates corresponding to the abscissa x and x' are v, u and v', u' respectively) and finally for x' from x to the maximum abscissa, and for x from the minimum to the maximum abscissa. The final result involves only single definite integrals between the extreme values of x, the functions under the integral sign containing indefinite integrations from the same arbitrary inferior limit, say $x=0$; the form of the result (previous to its simplification by taking the axes to be principal axes through the centre of gravity of the area) is however somewhat complicated; and it would not be easy to show *a posteriori*, that the value is invariantive, that is, independent of the position of the axes: that this is so is of course apparent from the original form of the integral.

[Vol. IX. pp. 38, 39.]

2530. (Proposed by Professor CAYLEY.)—Trace the curve

$$\frac{1}{\sqrt{z}}+\frac{1}{\sqrt{(x+iy)}}+\frac{1}{\sqrt{(x-iy)}}=0,$$

where the coordinates x, y, z are the perpendicular distances of the current point P from the sides of an equilateral triangle, the coordinates being positive for a point within the triangle.

Solution by the PROPOSER.

The form of the equation shows that the curve is a tricuspidal quartic, having a real cusp at the point $(x=0,\ y=0)$, and two imaginary cusps at the points $(z=0,\ x+iy=0)$ and $(z=0,\ x-iy=0)$. The rationalised form of the equation is

$$(x^2+y^2)^2-4zx(x^2+y^2)-4z^2y^2=0.$$

$x=0$ gives $y^2(y^2-4z^2)=0$, the point C twice, and two other real points α, α' on the line BC.

$y=0$ gives $x^3(x-4z)=0$, the point C three times, and a real point β on the line CA.

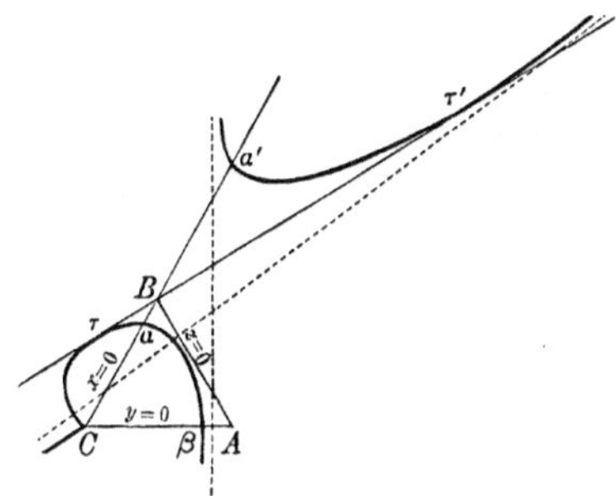

It is easy to find that there is a double tangent $z+2x=0$, viz. $z+2x=0$ gives $(3x^2-y^2)^2=0$, two points τ, τ' (each twice) on the line in question.

Laying down these points, it appears that the curve must have two real asymptotes, and that the form is as shown in the figure.

[Vol. IX. pp. 55, 56.]

2553. (Proposed by Professor CAYLEY.)—Show that the surface $y^2z^2+z^2x^2+x^2y^2-2xyz=0$ meets the sphere $x^2+y^2+z^2=1$ in four circles; and explain in a general manner the form of the curve of intersection of the surface by any other sphere having the same centre, and thence the form of the surface itself (being a particular case of Steiner's surface, and which by the homographic transformations $w^{-1}x$, $w^{-1}y$, $w^{-1}z$ for x, y, z gives $y^2z^2+z^2x^2+x^2y^2-2wxyz=0$, the general equation of Steiner's surface).

Solution by the PROPOSER.

Take X, X', Y, Y', Z, Z' the intersections of the sphere $x^2+y^2+z^2=1$ by the three axes respectively; then we have $x^2+y^2+z^2=1$, $x+y+z=-1$, the equations of the circle through the points X', Y', Z'; and from these two equations we deduce $yz+zx+xy=0$, and thence

$$y^2z^2+z^2x^2+x^2y^2+2xyz\,(x+y+z)=0,$$

that is

$$y^2z^2+z^2x^2+x^2y^2-2xyz=0\,;$$

so that the circle lies on the quartic surface; and by changing successively the signs of each two of the three coordinates, we have three other circles lying on the sphere and also on the quartic surface; viz. we have in all four circles, the above-mentioned circle through (X', Y', Z'), and three other circles through (X', Y, Z), (X, Y', Z), (X, Y, Z') respectively, making together a curve of the order 8, the complete intersection of the quartic surface by the sphere.

The quartic surface lies entirely in the four octants of space xyz, $xy'z'$, $x'yz'$, $x'y'z$; and as to the portion of the surface which lies in the octant xyz, this meets the sphere $x^2+y^2+z^2=1$ in portions of the three circles (X', Y, Z) (X, Y', Z) (X, Y, Z') constituting a tricuspidal form lying within the octant XYZ as shown in the figure. The intersection by a sphere, radius <1, projected on the octant XYZ, is a trinodal form, lying outside the tricuspidal one, as shown by a dotted line in the figure; the intersection by a sphere radius >1, projected in the same way, is a trigonoid form lying inside the tricuspidal one, as also shown by a dotted line in the figure; as the radius approaches to and ultimately becomes $=\dfrac{2}{\sqrt{3}}$, this diminishes, and becomes ultimately a mere point, and when the radius is greater than this value the intersection is imaginary.

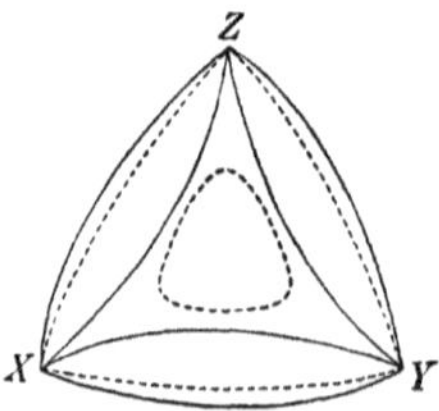

Imagine on the solid sphere, radius $=1$, the four tricuspidal forms lying in alternate octants as above; cut away down to the centre the portions lying without these tricuspidal forms; and build up on the tricuspidal forms, until the greatest distance from the centre becomes $=\frac{2}{\sqrt{3}}$; we have a solid figure with four prominences situate as the summits of a tetrahedron, the bounding surface whereof is the surface in question: it is to be added that the axes are nodal lines on the surface, viz. the portions which lie within the solid figure are the intersections of two real sheets of the surface, the portions which lie without the solid figure are isolated, or acnodal, lines on the surface.

[Vol. IX. pp. 73, 74.]

2573. (Proposed by Professor CAYLEY.)—The envelope of a variable circle having for its diameter the double ordinate of a rectangular cubic is a Cartesian.

{DEFINITION. The expression "a rectangular cubic" is used to express a cubic with three real asymptotes, having a diameter at right angles to one of the asymptotes and at an angle of 45° to each of the other two asymptotes, viz. the equation of such a cubic is $xy^2 = x^3 + bx^2 + cx + d$.}

Solution by the PROPOSER.

The equation of the variable circle may be taken to be

$$(x-\theta)^2 + y^2 = \theta^2 - 2m\theta + \alpha + \frac{2A}{\theta},$$

viz. θ being the abscissa of the rectangular cubic, the squared ordinate is taken to be $=\frac{1}{\theta}(\theta^3 - 2m\theta^2 + \alpha\theta + 2A)$, or, what is the same thing, the equation of the variable circle is

$$x^2 + y^2 - \alpha - 2(x-m)\theta - \frac{2A}{\theta} = 0.$$

Hence, taking the derived equation in regard to θ, we have

$$x - m - \frac{A}{\theta^2} = 0,$$

and thence

$$x^2 + y^2 - \alpha = \frac{4A}{\theta};$$

therefore

$$(x^2 + y^2 - \alpha)^2 = \frac{16A^2}{\theta^2} = 16A(x-m);$$

that is, the equation of the envelope is

$$(x^2 + y^2 - \alpha)^2 - 16A(x-m) = 0,$$

which is a known form of the equation of a Cartesian.

[Vol. IX. pp. 82, 83.]

2493. (Proposed by Professor CAYLEY.)—1. Given the conic $U=0$ (but observe that the *function* U contains implicitly an arbitrary constant factor which is *not* given) and also the conic $U+1=0$, to construct the conic $U+l=0$, where l is a given constant.

2. Given the conics $U=0$, $U+1=0$, $V=0$, $V+1=0$, and the constants θ, k, to construct the conic $\theta U+\theta^{-1}V+2k=0$.

Solution by the PROPOSER.

1. The conics $U=0$, $U+1=0$, $U+l=0$ are obviously concentric similar and similarly situated conics, and if drawing a line in any direction from the centre, the radius-vectors for the three conics respectively are r, r', R, then it is easy to see that we have

$$R^2=lr'^2+(1-l)r^2.$$

There is no difficulty in constructing geometrically the radius R, and then the conic $U+l=0$ is given as the concentric similar and similarly situated conic passing through the extremity of this radius.

2. To construct the conic $\theta U+\theta^{-1}V+2k=0$. By what precedes, we may construct the two conics $\theta U+k=0$, $\theta^{-1}V+k=0$; the four points of intersection of these lie on the required conic $\theta U+\theta^{-1}V+2k=0$, and also on the conic $\theta U-\theta^{-1}V=0$; which last conic is consequently given as a conic passing through the four points in question, and also through the four points of intersection of the given conics $U=0$, $V=0$. But the conic $\theta U-\theta^{-1}V=0$ being constructed, the conic $\theta U+\theta^{-1}V=0$ can also be constructed; viz. the tangents of these two conics and of the conics $U=0$, $V=0$, at each of the four intersections $U=0$, $V=0$, form a harmonic pencil; and we have thus the conic $\theta U+\theta^{-1}V=0$ a conic passing through four given points, and having at each of these a given tangent. And then finally the required conic $\theta U+\theta^{-1}V+2k=0$ is given as a conic concentric similar and similarly situated with the conic $\theta U+\theta^{-1}V=0$, and passing through the four given points

$$\theta U+k=0,\quad \theta^{-1}V+k=0.$$

3. Treating k as an absolute constant but θ as a variable parameter, the envelope of the conic $\theta U+\theta^{-1}V+2k=0$ is the quartic curve $UV-k^2=0$. This is a curve used by Plücker (in the *Theorie der algebraischen Curven*) for the purpose of showing that the 28 double tangents of a quartic curve may be all of them real. In fact, if $U=0$, $V=0$ be ellipses intersecting in four real points; and if, moreover, the implicit constants be such that U is positive for points *without* the first ellipse, V positive for points *within* the second ellipse, then since $UV, =k^2$, is positive for all points of the curve in question, the curve must be wholly situate in the four closed spaces which lie outside the one and inside the other of the two ellipses; consisting therefore of four detached portions. And when k is sufficiently small, then the figure of each portion is that of a concavo-convex lens with its angles rounded off: viz. each such portion has a real double tangent of its own. Any two portions have obviously four real double tangents, and hence the total number of real double tangents is $4+6\times 4, =28$.

4. A construction has been given by Aronhold (*Berl. Monatsber.*, July, 1864) by which, taking any 7 given lines as double tangents of a quartic curve, the remaining 21 double tangents can be constructed, and which, when the seven given lines are real, leads to a system of 28 real double tangents; but wishing to construct the figure of the 28 real double tangents, it occurred to me that the easier manner might be to construct Plücker's curve $UV-k^2=0$, as the envelope of the conic $\theta U+\theta^{-1}V+2k=0$, and then to draw the tangents of this curve: the construction is, however, practically one of considerable difficulty, and I have not yet accomplished it.

[Vol. IX. p. 87.]

2451. (Proposed by Professor CAYLEY.)—If A, B, C, D are the intersections of a conic by a circle, then the antipoints of A, B, and the antipoints of C, D, lie on a confocal conic.

N.B. If AB, $A'B'$ intersect at right angles in a point O in such wise that $OA'=OB'=i\,.\,OA=i\,.\,OB$ {where $i=\sqrt{(-1)}$ as usual}, then A', B' are the antipoints of A, B, and conversely.

[Vol. IX. pp. 101—103.]

2590. (Proposed by Professor CAYLEY.)—It is required to verify Professor Kummer's theorem that "if a quartic surface is such that every section by a plane through a certain fixed point is a pair of conics, the surface is a pair of quadric surfaces (except only in the case where it is a quartic cone having its vertex at the fixed point)."

Solution by the PROPOSER.

The theorem may be more generally stated as follows; if a surface is such that every section through a certain fixed point (is or) includes a proper conic, then the surface (is or) includes a proper quadric surface. In order to the demonstration, I premise the following *Lemma:* If a surface is such that every section through a certain fixed line includes a conic, then the line meets each of these conics in the same two points.

In fact, if the line meet the surface in any n points, then it is clear that each of the conics will meet the line in some two of these n points; and as the plane of the section passes continuously from any one to any other position, the two points of intersection with the conic cannot pass abruptly from being some two to being some other two of the n points, that is, they are always the same two points.

Consider now a surface which is such that every section through a fixed point O includes a conic; and consider three lines xx', yy', zz' meeting in the point O. Let the conics in the planes yz, zx, xy be A, B, C respectively; then since the conics through the line xx' all pass through the same two points, and since B, C are two of these conics, B and C meet xx' in the same two points X, X'; similarly C and A meet yy' in the same two points Y, Y'; and A, B meet zz' in the same two points Z, Z'; that is, we have the conics A, B, C intersecting

B, C in the two points X, X',
C, A „ „ Y, Y',
A, B „ „ Z, Z';

hence taking on the conics A, B, C the points α, β, γ respectively, and drawing a quadric surface Σ through the nine points X, X', Y, Y', Z, Z', α, β, γ, this meets the conic A in the five points Y, Y', Z, Z', α; that is, it passes through the conic A, and similarly it passes through the conic B, and through the conic C.

Consider how any plane whatever through O intersecting the conics A, B, C in the points L and L', M and M', N and N' respectively; the section of the quadric surface Σ by the plane in question is a conic through the six points L, L', M, M', N, N'. But the section of the surface includes a conic through these same six points, and which is consequently the same conic; in fact, the section of the surface by the plane in question includes a conic, and since every section through the line LL' includes a conic through the same two points, and one of these conics is the conic A which passes through the points L and L', the conic in question passes through the points L and L'; and similarly it passes through the points M and M', and through the points N and N'. That is, for any plane whatever through O, the section of the surface includes the conic which is the section of the quadric surface Σ, and the surface thus includes as part of itself the quartic surface Σ.

The foregoing demonstration ceases, however, to be applicable if O is a point on the surface, and the conic included in the section through O is always a conic passing through the point O. In the case where O is a non-singular point of the surface (that is, where there is at O a unique tangent plane) a like demonstration applies. Take through O a section, and let this include the conic A; on A take any point O' and through OO' a section including the conic B; we have thus the conics A, B intersecting in the points O, O'. Take through O any plane meeting the conics A, B in the points X, Y respectively—the section by this plane includes a conic C passing through the points O, X, Y; and each of the conics A, B, C touches at O the same plane, viz. the tangent plane of the surface. Hence, taking on the conic A the point α, consecutive to O, and any other point α'; on the conic B the point β, consecutive to O, and any other point β'; and on the conic C a point γ'; we may, through the nine points O, α, β, O', α', β', X, Y, γ' describe a quadric surface Σ; this will touch at O the tangent plane of the surface, that is, it will touch the conic C, or (what is the same thing) pass through a point γ of this conic consecutive to O. Hence the quadric surface meets the conic A in the five points O, O', α, α', X, that is, it entirely contains the conic A; similarly it meets the conic B in five points O, O', B, B', Y, that is, it entirely contains the conic B; and it meets the conic C in the five points O, γ, X, Y, γ', that is, it entirely contains this conic. And it may then be shown as before that the surface will include the quadric surface Σ. But there still remains for consideration the case where O is a conical point on the surface, and I do not at present see how this is to be treated.

I remark that, taking three lines xx', yy', zz' which meet in a point O, then if a surface be such that every section through xx' includes a conic, every section through yy' includes a conic, and every section through zz' includes a conic; and if besides the two points, say X, X', on the conics through the line xx' are ordinary points on the surface, then the surface will include a quadric surface. In fact, if the surface has at each of the points X, X' an ordinary tangent plane, then the conics through xx', and (as conics of the series) the two conics B, C all of them touch the two tangent planes; hence, constructing as before the quadric surface Σ, this also touches the two tangent planes: and taking through xx' a plane meeting the conic A in the points L, L', the section of the surface includes a conic which touches the section of the quadric surface Σ at the points X, X', and besides meets it in the points L, L'; such conic coincides therefore with the section of the quadric surface Σ; that is, every section of the surface through the line xx' includes the conic which is the section of the quadric surface Σ; and the surface thus includes as part of itself the quadric surface Σ.

[Vol. x., July to December, 1868, pp. 17—19.]

2609. (Proposed by Professor CAYLEY.)—Given three conics passing through the same four points; and on the first a point A, on the second a point B, and on the third a point C. It is required to find, on the first a point A', on the second a point B', and on the third a point C', such that the intersections of the lines

$A'B'$ and AC, $A'C'$ and AB, lie on the first conic;

$B'C'$ and BA, $B'A'$ and BC, lie on the second conic;

$C'A'$ and CB, $C'B'$ and CA, lie on the third conic.

Solution by the PROPOSER.

Let the six intersections in question be called α, α'; β, β'; γ, γ', respectively; then BC intersects second conic in β', third conic in γ; CA intersects third conic in γ', first conic in α; AB intersects first conic in α', second conic in β; and we have

A' the intersection of $\alpha\beta'$, $\gamma\alpha'$,

B' the intersection of $\beta\gamma'$, $\alpha\beta'$,

C' the intersection of $\gamma\alpha'$, $\beta\gamma'$;

and it has to be shown that the points A', B', C' so determined lie—A' on the first conic, B' on the second conic, C' on the third conic.

Taking $x=0$, $y=0$, $z=0$ for the equations of the sides of the triangle ABC, the equations of the three conics may be taken to be $U=0$, $V=0$, $W=0$, where the functions U, V, W are such that identically $U+V+W=0$; and then observing that

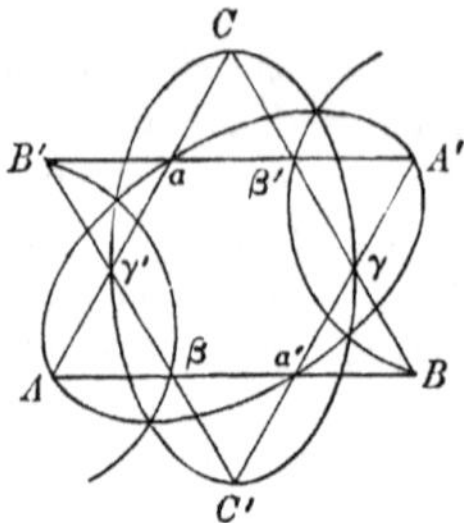

the conics pass through the points $(y=0,\ z=0)$, $(z=0,\ x=0)$, $(x=0,\ y=0)$, respectively, we see that the equations may be taken to be

$$\begin{aligned}
&(\ 0,\ -b,\ \ c,\ f_1,\ g_1,\ h_1 \between x,\ y,\ z)^2 = 0,\\
&(\ a,\ \ 0,\ -c,\ f_2,\ g_2,\ h_2 \between x,\ y,\ z)^2 = 0,\\
&(-a,\ \ b,\ \ 0,\ f_3,\ g_3,\ h_3 \between x,\ y,\ z)^2 = 0,
\end{aligned}$$

where

$$f_1+f_2+f_3=0,\quad g_1+g_2+g_3=0,\quad h_1+h_2+h_3=0.$$

The coordinates of the points α, β, γ, α', β', γ' are at once found to be

$$\begin{array}{llll}
\alpha, & (\ c,\ \ 0,\ -2g_1); & \alpha', & (\ b,\ 2h_1,\ \ 0\ )\\
\beta, & (-2h_2,\ \ a,\ \ 0\ ); & \beta', & (\ 0,\ \ c,\ 2f_2)\\
\gamma, & (\ 0,\ -2f_3,\ \ b\ ); & \gamma', & (2g_3,\ \ 0,\ \ a\ );
\end{array}$$

and hence the equations of $\beta\gamma'$, $\gamma\alpha'$, $\alpha\beta'$ are

$$\begin{array}{ll}
\beta\gamma'; & ax + 2h_2 y - 2g_3 z = 0,\\
\gamma\alpha'; & -2h_1 x + by + 2f_3 z = 0,\\
\alpha\beta'; & 2g_1 x - 2f_2 y + cz = 0.
\end{array}$$

Hence for the point A', which is the intersection of $\gamma\alpha'$, $\alpha\beta'$, coordinates are

$$bc + 4f_2 f_3,\quad 4f_3 g_1 + 2ch_1,\quad 4h_1 f_2 - 2bg_1;$$

and A' will be on the first conic if only

$$(0,\ -b,\ c,\ f_1,\ g_1,\ h_1 \between bc + 4f_2 f_3,\ \ 4f_3 g_1 + 2ch_1,\ \ 4h_1 f_2 - 2bg_1)^2 = 0,$$

viz. this equation is

$$\begin{aligned}
&-\ b\,(\ 16f_3^2 g_1^2 + 16f_3 g_1 h_1 c + 4h_1^2 c^2)\\
&+\ c\,(\ 16h_1^2 f_2^2 - 16f_2 g_1 h_1 b + 4g_1^2 b^2)\\
&+2f_1(\ 16g_1 h_1 f_2 f_3 - 8g_1^2 f_3 b + 8h_1^2 f_2 c - 4g_1 h_1 bc)\\
&+2g_1(+16h_1 f_2^2 f_3 - 8g_1 f_2 f_3 b + 4h_1 f_2 bc - 2g_1 b^2 c\ )\\
&+2h_1(+16g_1 f_2 f_3^2 + 8h_1 f_2 f_3 c + 4g_1 f_3 bc + 2h_1 bc^2\ ) = 0,
\end{aligned}$$

viz. this is easily found to be

$$8(2g_1f_3 + ch_1)(2h_1f_2 - bg_1)(f_1+f_2+f_3) = 0,$$

which is satisfied in virtue of $f_1+f_2+f_3=0$; that is, A' is on the first conic; and similarly, in virtue of $g_1+g_2+g_3=0$, B' is on the second conic; and in virtue of $h_1+h_2+h_3=0$, C' is on the third conic. But the same thing appears at once by the remark that the equations of the three conics are

$$\begin{aligned}
-y(-2h_1x + by + 2f_3z) + z(2g_1x - 2f_2y + cz) &= 0,\\
-z(2g_1x - 2f_2y + cz) + x(ax + 2h_2y - 2g_3z) &= 0,\\
-x(ax + 2h_2y - 2g_3z) + y(-2h_1x + by + 2f_3z) &= 0.
\end{aligned}$$

It may be added that, taking (x_1, y_1, z_1), (x_2, y_2, z_2), (x_3, y_3, z_3), (x_4, y_4, z_4) as the coordinates of the four points of intersection of the three conics, the first conic is given by means of these four points and the fifth point $(y=0, z=0)$; and similarly for the other two conics; whence, denoting the determinants formed with any four columns out of the matrix

$$\left\|\begin{matrix}
x_1^2, & y_1^2, & z_1^2, & y_1z_1, & z_1x_1, & x_1y_1\\
x_2^2, & y_2^2, & z_2^2, & y_2z_2, & z_2x_2, & x_2y_2\\
x_3^2, & y_3^2, & z_3^2, & y_3z_3, & z_3x_3, & x_3y_3\\
x_4^2, & y_4^2, & z_4^2, & y_4z_4, & z_4x_4, & x_4y_4
\end{matrix}\right\|$$

by 1234, 1235, &c., we easily find the equations of the three conics, viz. these may be written

$$\begin{array}{lcccccc}
 & x^2, & y^2, & z^2, & yz, & zx, & xy\\
1456\,(& 0, & +3456, & -2456, & +2356, & +2364, & +2345) = 0,\\
2356\,(& -3456, & 0, & +1456, & +3156, & +3164, & +3145) = 0,\\
3456\,(& 2456, & -1456, & 0, & +1256, & +1264, & +1245) = 0,
\end{array}$$

the exterior factors 1456, 2356, 3456 being introduced in order to bring the equations into the above-mentioned form, wherein the sum of the quadric functions is $=0$.

[Vol. x. pp. 88, 89.]

2743. (Proposed by M. JENKINS, M.A.)—Show that if p be a prime number and m and n any positive integers, the highest power of p contained in $\dfrac{\Pi(m+n)}{\Pi(m)\Pi(n)}$ may be obtained by expressing $m+n$ and either m or n in the scale of p; the number of times that it would be necessary to borrow in subtracting the latter number from the former being the index of the power of p required.

Solution by PROFESSOR CAYLEY.

1. In adding any two numbers, we carry a certain number of times; and it is easy to see that the sum of the digits of the two components, less the sum of the digits of the sum, is equal to nine times the number of carryings; moreover, that the number of carryings is equal to the number of borrowings, if either of the components be subtracted from the sum.

2. The same thing is true in any scale of notation, only, instead of nine, we have the radix of the scale, less unity: say the theorem is

$$S(m)+S(n)-S(m+n)=(p-1)x.$$

3. If p be a prime number, the number of times that the factor p occurs in $\Pi(m)$ is

$$E\left(\frac{m}{p}\right)+E\left(\frac{m}{p^2}\right)+E\left(\frac{m}{p^3}\right)+\&\text{c.},$$

where $E\left(\frac{m}{p}\right)$ denotes the integer part of $\frac{m}{p}$, and similarly $E\left(\frac{m}{p^2}\right)$ &c. the integer part of $\frac{m}{p^2}$, &c.; the series is, of course, finite.

Hence the number of times that the factor p occurs in $\frac{\Pi(m+n)}{\Pi(m)\,\Pi(n)}$ is

$$N=E\left(\frac{m+n}{p}\right)+E\left(\frac{m+n}{p^2}\right)+\&\text{c.}-E\left(\frac{m}{p}\right)-E\left(\frac{m}{p^2}\right)-\&\text{c.}-E\left(\frac{n}{p}\right)-E\left(\frac{n}{p^2}\right)-\&\text{c.}$$

4. Hence, expressing m, n, $m+n$ in the scale to the radix p, suppose

$$m=a+bp+cp^2+dp^3,\quad n=a'+b'p+c'p^2+d'p^3,\quad m+n=\alpha+\beta p+\gamma p^2+\delta p^3,$$

we have

$$E\left(\frac{m}{p}\right)+E\left(\frac{m}{p^2}\right)+\&\text{c.}=b+cp+dp^2+c+dp+d=d(p^2+p+1)+c(p+1)+b;$$

and similarly for $$E\left(\frac{n}{p}\right)+\&\text{c.},\ E\left(\frac{m+n}{p}\right)+\&\text{c.}\ldots;$$

whence

$$\begin{aligned}(p-1)N=&\ \ \delta\,(p^3-1)+\gamma\,(p^2-1)+\beta\,(p-1)\\ &-d\,(p^3-1)-c\,(p^2-1)-b\,(p-1)\\ &-d'\,(p^3-1)-c'\,(p^2-1)-b'\,(p-1)\\ =&\ \{m+n-S(m+n)\}-\{m-S(m)\}-\{n-S(n)\}\\ =&\ S(m)+S(n)-S(m+n),\ =(p-1)x,\end{aligned}$$

if x be the number of times of carrying for the sum $m+n$, or of borrowing for the difference $(m+n)-m$ or $(m+n)-n$; that is, $N=x$, the required theorem. I remark that although the foregoing expression of the number N is a very elegant and ingenious one, yet the original form of N, as given at the end of (3), is the natural and proper expression of the number of times that the factor p occurs in the binomial coefficient $\frac{\Pi(m+n)}{\Pi(m)\Pi(n)}$.

[Vol. x. p. 98.]

2756. (Proposed by J. GRIFFITHS, M.A.)—Show that an infinite number of triangles can be described such that each has the same circumscribing, nine-point, and self-conjugate circles as a given triangle.

Solution by PROFESSOR CAYLEY.

It is a known theorem that if two triads of points, say A, B, C and A', B', C', are self-conjugate in regard to a conic S, they lie in a conic Σ. Take the conic S and the points A, B, C as given; then Σ will be a conic passing through A, B, C; and if on this conic we take any point A', and then take B' to be either of the intersections of the conic Σ by the polar of A in regard to S, and finally construct C' as the pole of $A'B'$ in regard to S, then, by what precedes, C' will be on a conic through A, B, C, A', B', that is, on the conic Σ. Or given the conic S, the triangle ABC, and the conic Σ through A, B, C, we obtain an infinity of triangles $A'B'C'$, self-conjugate in regard to S and inscribed in Σ. That is, if S, Σ are circles, we have an infinity of triangles self-conjugate in regard to the circle S and inscribed in the circle Σ; and inasmuch as the nine-points circle can be constructed by means of the two circles S, Σ alone, the triangles have all of them the same nine-points circle.

[Vol. x. p. 108.]

2737. (Proposed by Professor CAYLEY.)—Find *in solido* the locus of a point P, such that from it two given points A, C, and two given points B, D, subtend equal angles.

[Vol. XI., January to June, 1869, pp. 33—38.]

2718. (Proposed by Professor CAYLEY.)—Find *in plano* the locus of a point P, such that from it two given points A, C, and two given points B, D, subtend equal angles.

2757. (Proposed by Professor CAYLEY.)—If

$$x_0^2 + y_0^2 = 1, \qquad \text{and} \qquad \begin{vmatrix} x, & y, & 1 \\ x_0, & y_0, & 1 \\ x_1, & y_1, & 1 \end{vmatrix} = L;$$
$$x_1^2 + y_1^2 = 1,$$

show that each of the equations

$$\frac{a^2(x-x_0)^2 + b^2(y-y_0)^2}{(xx_0 + yy_0 - 1)^2} = \frac{a^2(x-x_1)^2 + b^2(y-y_1)^2}{(xx_1 + yy_1 - 1)^2}, \tag{1}$$

$$\frac{a^2(x-x_0)^2 + b^2(y-y_0)^2}{(xy_0 - x_0y)^2 - (x-x_0)^2 - (y-y_0)^2} = \frac{a^2(x-x_1)^2 + b^2(y-y_1)^2}{(xy_1 - x_1y)^2 - (x-x_1)^2 - (y-y_1)^2}, \tag{2}$$

represents the right line $L = 0$ and a cubic curve.

1819. (Proposed by C. TAYLOR, M.A.)—From two fixed points on a given conic pairs of tangents are drawn to a variable confocal conic, and with the fixed points as foci a conic is described passing through any one of the four points of intersection. Show that its tangent or normal at that point passes through a fixed point.

Solution of the above Problems by PROFESSOR CAYLEY.

1. It is easy to see that drawing through the points A, C a circle, and through B, D a circle, such that the radii of the two circles are proportional to the lengths AC, BD, then that the required locus is that of the intersections of the two variable circles.

Take $AC = 2l$, MO perpendicular to it at its middle point M, and $= p$; a, b the coordinates of M, and λ the inclination of p to the axis of x; then

$$\text{coordinates of } O \text{ are} \quad a + p\cos\lambda, \quad b + p\sin\lambda,$$
$$\text{coordinates of } A, C \text{ are } a \pm l \sin\lambda, \quad b \mp l \cos\lambda,$$

and hence the equation of a circle, centre O and passing through A, C, is

$$(x - a - p\cos\lambda)^2 + (y - b - p\sin\lambda)^2 = l^2 + p^2;$$

or, what is the same thing,

$$(x-a)^2 + (y-b)^2 - l^2 = 2p\,[(x-a)\cos\lambda + (y-b)\sin\lambda].$$

If $2m$, q, c, d, μ refer in like manner to the points B, D, then the equation of a circle, centre say Q, and passing through B, D, is

$$(x-c)^2+(y-d)^2-m^2=2q\,[(x-c)\cos\mu+(y-d)\sin\mu];$$

and the condition as to the radii is $l^2+p^2 : m^2+q^2 = l^2 : m^2$, that is, $p^2 : q^2 = l^2 : m^2$, or $p : q = \pm l : m$. And we thus have for the equation of the required locus

$$\frac{(x-a)^2+(y-b)^2-l^2}{(x-a)\cos\lambda+(y-b)\sin\lambda}=\pm\frac{l}{m}\,\frac{(x-c)^2+(y-d)^2-m^2}{(x-c)\cos\mu+(y-d)\sin\mu},$$

viz. the locus is composed of two cubics, which are at once seen to be *circular* cubics. One of these will however belong (at least for some positions of the four points) to the case of the subtended angles being *equal*, the other to that of the subtended angles being *supplementary;* and we may say that the required locus is a circular cubic.

2. If two of the points coincide, suppose C, D at T; then, taking T as the origin, we may write

$$a = l\sin\lambda,\quad b=-l\cos\lambda,$$
$$c=-m\sin\mu,\quad d=m\cos\mu,$$

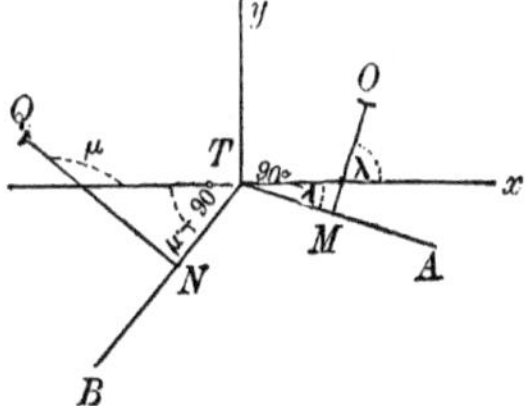

and the equation becomes

$$\frac{x^2+y^2+2l\,(x\sin\lambda-y\cos\lambda)}{x\cos\lambda+y\sin\lambda}=\pm\frac{l}{m}\,\frac{x^2+y^2+2m\,(x\sin\mu-y\cos\mu)}{x\cos\mu+y\sin\mu},$$

viz. this is

$$(x^2+y^2)\,[m\,(x\cos\mu+y\sin\mu)\mp l\,(x\cos\lambda+y\sin\lambda)]-2lm\,\{(x\sin\lambda-y\cos\lambda)(x\cos\mu+y\sin\mu)$$
$$\pm(x\sin\mu-y\cos\mu)(x\cos\lambda+y\sin\lambda)\}=0.$$

Taking the lower signs, the term in { } is $(x^2+y^2)\sin(\lambda-\mu)$, and the equation is

$$(x^2+y^2)\,\{m\,(x\cos\mu+y\sin\mu)+l\,(x\cos\lambda+y\sin\lambda)-2lm\sin(\lambda-\mu)\}=0,$$

viz. this is $x^2+y^2=0$, and a line which is readily seen to be the line AB; and in fact from any point whatever of this line the points A, T and the points B, T subtend *supplementary* angles.

Taking the upper signs, the equation is

$$(x^2+y^2)\,[m\,(x\cos\mu+y\sin\mu)-l\,(x\cos\lambda+y\sin\lambda)]$$
$$-\,2lm\,\{(x^2-y^2)\sin(\lambda+\mu)-xy\cos(\lambda+\mu)\}=0,$$

which is the locus for *equal* angles, a circular cubic as in the case of the four distinct points.

3. The question is connected with Question 1819, which is given above. In fact, taking A, B for the fixed points on the given conic, and P for the intersection of any two of the tangents, if in the conic (foci A, B) which passes through P, the tangent or normal at P passes through a fixed point T, then it is clear that at P the points A, T and B, T subtend equal angles, and consequently the locus of P should be a circular cubic as above. The theorem will therefore be proved if it be shown that the locus of P considered as the intersection of tangents from A, B to the variable confocal conic is in fact the foregoing circular cubic. I remark that the fixed point T is in fact the intersection of the tangents AT, BT to the given conic at the points A, B respectively.

4. Consider the points A, B, (which we may in the first instance take to be arbitrary points, but we shall afterwards suppose them to be situate on the conic $\frac{x^2}{a^2}+\frac{y^2}{b^2}=1$,) and from each of them draw a pair of tangents to the confocal conic $\frac{x^2}{a^2+h}+\frac{y^2}{b^2+h}=1$. Take $(x_0,\ y_0)$ for the coordinates of A, and $(x_1,\ y_1)$ for those of B; then the equation of the pair of tangents from A is

$$\left(\frac{x_0^2}{a^2+h}+\frac{y_0^2}{b^2+h}-1\right)\left(\frac{x^2}{a^2+h}+\frac{y^2}{b^2+h}-1\right)-\left(\frac{xx_0}{a^2+h}+\frac{yy_0}{b^2+h}-1\right)^2=0,$$

or, what is the same thing,

$$\frac{(xy_0-x_0y)^2}{(a^2+h)(b^2+h)}-\frac{(x-x_0)^2}{a^2+h}-\frac{(y-y_0)^2}{b^2+h}=0,$$

that is

$$(xy_0-x_0y)^2-(b^2+h)(x-x_0)^2-(a^2+h)(y-y_0)^2=0,$$

or as this may also be written

$$(xy_0-x_0y)^2-b^2(x-x_0)^2-a^2(y-y_0)^2=h\,[(x-x_0)^2+(y-y_0)^2];$$

and similarly for the tangents from B we have

$$(xy_1-x_1y)^2-b^2(x-x_1)^2-a^2(y-y_1)^2=h\,[(x-x_1)^2+(y-y_1)^2];$$

in which equations the points $(x_0,\ y_0)$, $(x_1,\ y_1)$ are in fact any two points whatever.

5. Eliminating h, we have as the locus of the intersection of the tangents

$$\frac{(xy_0-x_0y)^2-b^2(x-x_0)^2-a^2(y-y_0)^2}{(x-x_0)^2+(y-y_0)^2}=\frac{(xy_1-x_1y)^2-b^2(x-x_1)^2-a^2(y-y_1)^2}{(x-x_1)^2+(y-y_1)^2},$$

which is apparently a quartic curve; but it is obvious *à priori* that the locus includes as part of itself the line AB which joins the two given points. In fact, there is in the series of confocal conics one conic which touches the line in question, and since for this conic one of the tangents from A and also one of the tangents from B is the line AB, we see that every point of the line AB belongs to the required locus. The locus is thus made up of the line in question and of the cubic curve.

6. To effect the reduction it will be convenient to write ax, by in the place of x, y, (ax_0, by_0, ax_1, by_1 in place of x_0, y_0, x_1, y_1,) and thus consider the equation under the form

$$\frac{a^2(x-x_0)^2+b^2(y-y_0)^2}{(xy_0-x_0y)^2-(x-x_0)^2-(y-y_0)^2}=\frac{a^2(x-x_1)^2+b^2(y-y_1)^2}{(xy_1-x_1y)^2-(x-x_1)^2-(y-y_1)^2};$$

it is to be shown that this equation represents the line $L=0$, and a cubic curve.

Writing for a moment $x_0=x+\xi_0$, $y_0=y+\eta_0$, and $x_1=x+\xi_1$, $y_1=y+\eta_1$, the equation becomes

$$\frac{a^2\xi_0^2+b^2\eta_0^2}{(x\eta_0-y\xi_0)^2-\xi_0^2-\eta_0^2}=\frac{a^2\xi_1^2+b^2\eta_1^2}{(x\eta_1-y\xi_1)^2-\xi_1^2-\eta_1^2},$$

and hence, multiplying out, the equation is at once seen to contain the factor $\xi_0\eta_1-\xi_1\eta_0$ (which is in fact the determinant just mentioned), and when divested of this factor the equation is

$$a^2[(x^2-1)(\xi_0\eta_1+\xi_1\eta_0)-2xy\xi_0\xi_1]=b^2[(y^2-1)(\xi_0\eta_1+\xi_1\eta_0)-2xy\eta_0\eta_1].$$

Writing herein for ξ_0, η_0, ξ_1, η_1 their values, and consequently

$$\xi_0\xi_1=x^2-x(x_0+x_1)+x_0x_1,$$
$$\eta_0\eta_1=y^2-y(y_0+y_1)+y_0y_1,$$
$$\xi_0\eta_1+\xi_1\eta_0=2xy-x(y_0+y_1)-y(x_0+x_1)+x_0y_1+x_1y_0,$$

and arranging the terms, the equation is found to be

$$(a^2x^2+b^2y^2)[-x(y_1+y_0)-y(x_1+x_0)]+(a^2x^2+b^2y^2)(x_0y_1+x_1y_0)-2xy[a^2(1+x_0x_1)-b^2(1+y_0y_1)]$$
$$+(a^2-b^2)[x(y_1+y_0)+y(x_1+x_0)-(x_0y_1+x_1y_0)]=0,$$

which is the required cubic curve.

7. Restoring the original coordinates, or writing $\frac{x}{a}$, $\frac{y}{b}$, $\frac{x_0}{a}$, &c. in place of x, y, x_0, &c., we have

$$(x^2+y^2)[-x(y_1+y_0)+y(x_1+x_0)]+(x^2-y^2)(x_0y_1+x_1y_0)-2xy(a^2-b^2+x_0x_1-y_0y_1)$$
$$+(a^2-b^2)[x(y_1+y_0)+y(x_1+x_0)-(x_0y_1+x_1y_0)]=0,$$

which is a circular cubic the locus of the intersections of the tangents from the arbitrary points (x_0, y_0), (x_1, y_1) to the series of confocal conics $\frac{x^2}{a^2+h}+\frac{y^2}{b^2+h}=1$; the origin of the coordinates is at the centre of the conics.

8. Supposing that the points (x_0, y_0), (x_1, y_1) are on the conic $\frac{x^2}{a^2}+\frac{y^2}{b^2}=1$, and that we have consequently $\frac{x_0^2}{a^2}+\frac{y_0^2}{b^2}=1$, $\frac{x_1^2}{a^2}+\frac{y_1^2}{b^2}=1$, the equations of the tangents at these points respectively are $\frac{xx_0}{a^2}+\frac{yy_0}{b^2}=1$, $\frac{xx_1}{a^2}+\frac{yy_1}{b^2}=1$; and hence, writing for shortness $\alpha=y_0-y_1$, $\beta=x_1-x_0$, $\gamma=x_0y_1-x_1y_0$, we find $x=-\frac{a^2\alpha}{\gamma}$, $y=-\frac{b^2\beta}{\gamma}$ as the coordinates of the point of intersection T, of the two tangents; and in order to transform to this point as origin, we must in place of x, y write $x-\frac{a^2\alpha}{\gamma}$, $y-\frac{b^2\beta}{\gamma}$ respectively. Or what is more convenient, we may in the equation at the end of (6), in which it is to be now assumed that $x_0^2+y_0^2=1$, $x_1^2+y_1^2=1$, write $x-\frac{\alpha}{\gamma}$, $y-\frac{\beta}{\gamma}$ for x, y, and then restore the original coordinates by writing $\frac{x}{a}$, $\frac{y}{b}$, $\frac{x_0}{a}$, &c., for x, y, x_0, &c., and $\frac{\alpha}{b}$, $\frac{\beta}{a}$, $\frac{\gamma}{ab}$ for α, β, γ, these quantities throughout signifying $\alpha=y_0-y_1$, $\beta=x_1-x_0$, $\gamma=x_0y_1-x_1y_0$. I however obtained the equation referred to the point T as origin by a different process, as follows:

9. Starting from the equation at the commencement of (5), I found that the points (x_0, y_0), (x_1, y_1) being on the conic $\frac{x^2}{a^2}+\frac{y^2}{b^2}=1$, the equation could be transformed into the form

$$\frac{\left(\frac{xx_0}{a^2}+\frac{yy_0}{b^2}-1\right)^2}{(x-x_0)^2+(y-y_0)^2}=\frac{\left(\frac{xx_1}{a^2}+\frac{yy_1}{b^2}-1\right)^2}{(x-x_1)^2+(y-y_1)^2},$$

an equation which (not, as the original one, for all values of (x_0, y_0), (x_1, y_1), but) for values of (x_0, y_0), (x_1, y_1) such that $\frac{x_0^2}{a^2}+\frac{y_0^2}{b^2}=1$, $\frac{x_1^2}{a^2}+\frac{y_1^2}{b^2}=1$, breaks up into the line AB and a cubic curve.

10. To simplify the transformation, write as before ax, by, ax_0, &c., for x, y, x_0, &c. We have thus to consider the equation

$$\frac{a^2(x-x_0)^2+b^2(y-y_0)^2}{(xx_0+yy_0-1)^2}=\frac{a^2(x-x_1)^2+b^2(y-y_1)^2}{(xx_1+yy_1-1)^2},$$

where $x_0^2+y_0^2=1$, $x_1^2+y_1^2=1$, and which equation, I say, breaks up into the line $L=0$, and into a cubic.

Write for shortness $\alpha=y_0-y_1$, $\beta=x_1-x_0$, $\gamma=x_0y_1-x_1y_0$, so that the equation of the last-mentioned line is $\alpha x+\beta y+\gamma=0$. Then it may be verified that, in virtue of the relations between (x_0, y_0), (x_1, y_1), we have identically

$$(x-x_0)(xx_1+yy_1-1)+(x-x_1)(xx_0+yy_0-1)=(\alpha x+\beta y+\gamma)\frac{x_0+x_1}{\alpha\gamma}(\gamma x+\alpha),$$

$$(x-x_0)(xx_1+yy_1-1)-(x-x_1)(xx_0+yy_0-1)=\beta x^2-\alpha xy-\gamma y-\beta;$$

and, similarly,

$$(y-y_0)(xx_1+yy_1-1)+(y-y_1)(xx_0+yy_0-1)=(\alpha x+\beta y+\gamma)\frac{y_0+y_1}{\beta\gamma}(\gamma y+\beta),$$

$$(y-y_0)(xx_1+yy_1-1)-(y-y_1)(xx_0+yy_0-1)=\beta xy-\alpha y^2+\gamma x+\alpha.$$

11. The equation in question may be written $a^2P+b^2Q=0$, where

$$P=(x-x_0)^2(xx_1+yy_1-1)^2-(x-x_1)^2(xx_0+yy_0-1)^2,$$
$$Q=(y-y_0)^2(xx_1+yy_1-1)^2-(y-y_1)^2(xx_0+yy_0-1)^2,$$

values which are given by means of the formulæ just obtained; there is a common factor $\alpha x+\beta y+\gamma$ which is to be thrown out; and we have also, as is at once verified, $\frac{y_0+y_1}{\beta}=\frac{x_0+x_1}{\alpha}$, so that these equal factors may be thrown out. We thus obtain the cubic equation

$$a^2(\gamma x+\alpha)(\beta x^2-\alpha xy-\gamma y-\beta)+b^2(\gamma y+\beta)(\beta xy-\alpha y^2+\gamma x+\alpha)=0.$$

This is simplified by writing $x-\frac{\alpha}{\gamma}$ for x, $y-\frac{\beta}{\gamma}$ for y. It thus becomes

$$a^2x[(\gamma x-\alpha)(\beta x-\alpha y)-\gamma^2y]+b^2y[(\gamma y-\beta)(\beta x-\alpha y)+\gamma^2x]=0;$$

or, what is the same thing,

$$a^2x[\gamma x(\beta x-\alpha y)-\alpha\beta x+(\alpha^2-\gamma^2)y]+b^2y[\gamma y(\beta x-\alpha y)-(\beta^2-\gamma^2)x+\alpha\beta y]=0;$$

that is

$$\gamma(a^2x^2+b^2y^2)(\beta x-\alpha y)+a^2[-\alpha\beta x^2+(\alpha^2-\gamma^2)xy]+b^2[-(\beta^2-\gamma^2)xy+\alpha\beta y^2]=0.$$

12. Restoring $\frac{x}{a}, \frac{x_0}{a}, \frac{x_1}{a}$ for x, x_0, x_1, and $\frac{y}{a}, \frac{y_0}{a}, \frac{y_1}{a}$ for y, y_0, y_1; writing consequently $\frac{\alpha}{b}, \frac{\beta}{a}, \frac{\gamma}{ab}$ in place of α, β, γ, if α, β, γ are still used to denote $\alpha=y_0-y_1$, $\beta=x_1-x_0$, $\gamma=x_0y_1-x_1y_0$, the equation becomes

$$\gamma(x^2+y^2)[b^2\beta x-a^2\alpha y]+a^2[-b^2\alpha\beta x^2+(a^2\alpha^2-\gamma^2)xy]+b^2[-(b^2\beta^2-\gamma^2)xy+a^2\alpha\beta y^2]=0,$$

where now, as originally, $\frac{x_0^2}{a^2}+\frac{y_0^2}{b^2}=1$, $\frac{x_1^2}{a^2}+\frac{y_1^2}{b^2}=1$; viz. this is the equation, referred to the point T as origin, of the locus of the point P considered as the intersection of tangents from A, B to the variable confocal conic; and it is consequently the equation which would be obtained as indicated in (8). The locus is thus a circular cubic; the equation is identical in form with that obtained (2) for the locus of the point at which A, T and B, T subtend equal angles, and the complete identification of the two equations may be effected without difficulty.

13. I may remark that M. Chasles has given (*Comptes Rendus*, tom. 58, February, 1864) the theorem that the locus of the intersections of the tangents drawn from a fixed conic to the conics of a system (μ, ν) is a curve of the order 3ν. The confocal

conics, *quà* conics touching four fixed lines, are a system (0, 1); hence, taking for the fixed conic the two points A, B, we have, as a very particular case, the foregoing theorem, that the locus of the intersections of the tangents drawn from two fixed points to a variable confocal conic is a cubic curve.

[Vol. XI. p. 49.]

Note on Question 2740. *By* PROFESSOR CAYLEY.

The envelope of the curve

$$A\cos 2\theta + B\sin 2\theta + C\cos\theta + D\sin\theta + E = 0,$$

(where A, B, C, D, E are any functions of the coordinates, and θ is a variable parameter,) is obtained in the particular case $E=0$ (Salmon's *Higher Plane Curves*, p. 116), and the same process is applicable in the general case where E is not $=0$. From the great variety of the problems which depend upon the determination of such an envelope, the result is an important one, *and ought to be familiarly known* to students of analytical geometry. We have only to write $z = \cos\theta + i\sin\theta$, the trigonometrical functions are then given as rational functions of z, and the equation is converted into a quartic equation in z; the result is therefore obtained by equating to zero the discriminant of a quartic function. The equation, in fact, becomes

$$A\frac{1}{2}\left(z^2+\frac{1}{z^2}\right) + B\frac{1}{2i}\left(z^2-\frac{1}{z^2}\right) + C\frac{1}{2}\left(z+\frac{1}{z}\right) + D\frac{1}{2i}\left(z-\frac{1}{z}\right) + E = 0,$$

that is

$$A(z^4+1) - Bi(z^4-1) + C(z^3+z) - Di(z^3-z) + 2Ez^2 = 0;$$

or, multiplying by 12 to avoid fractions, this is

$$(a,\ b,\ c,\ d,\ e \between z,\ 1)^4 = 0,$$

where

$$a = 12(A-Bi),\quad b = 3(C-Di),\quad c = 4E,$$
$$e = 12(A+Bi),\quad d = 3(C+Di);$$

and substituting in

$$(ae - 4bd + 3c^2)^3 - 27(ace - ad^2 - b^2e + 2bcd - c^3)^2 = 0,$$

the equation divides by 1728, and the final result is

$$\{12(A^2+B^2) - 3(C^2+D^2) + 4E^2\}^3$$
$$- \{27A(C^2-D^2) + 54BCD - [72(A^2+B^2) + 9(C^2+D^2)]E + 8E^3\}^2 = 0.$$

It is to be noticed, that in developing the equation according to the powers of E, the terms in E^6, E^4 each disappear, so that the highest power is E^3; the degree in the coordinates, or order of the curve, is on this account sometimes lower than it would otherwise have been.

[Vol. XII., July to December, 1869, p. 69.]

2920. (Proposed by Professor CAYLEY.)—Imagine a tetrahedron $BB'CC'$ in which the opposite sides BB', CC' are at right angles to each other and to the line joining their middle points M, N; and in which moreover $\overline{CN}^2+\overline{NM}^2+\overline{MB}^2=0$, (or, what is the same thing, the sides CB, CB', $C'B$, $C'B'$ are each $=0$; the tetrahedron is of course imaginary; viz. the lines CC', BB' and points M, N may be real; but the distances $MB=MB'$ and $NC=NC'$ may be one real and the other imaginary, or both imaginary, but they cannot be both real) the points B, B' and C, C' are said to be "skew antipoints." Then it is required to prove that

1. A given system of skew antipoints may be taken to be the nodes (conical points) of a tetranodal cubic surface, passing through the circle at infinity, and which is in fact a Parabolic Cyclide.

2. The equation of the surface may be expressed in the form

$$x(x+\beta)(x+\gamma)+(x+\beta)y^2+(x+\gamma)z^2=0.$$

3. The section through either of the lines $(y=0,\ x+\gamma=0)$ and $(z=0,\ x+\beta=0)$ is made up of this line and a circle; the two systems of circles being the curves of curvature of the surface; it is required to verify this *à posteriori;* viz. by means of the equation of the surface to transform the differential equation of the curves of curvature in such manner that the transformed equation shall have the integrals

$$y=C(x+\gamma),\quad z=C'(x+\beta).$$

In the following Contents, the Problems are referred to each by its No. and the Proposer's name, and the Subject is briefly indicated. An asterisk shows that no solution was given. A line —— shows that there is no No.

NOTES AND REFERENCES.

445, 451, 454. We have the two papers by K. Rohn, "Die Flächen vierter Ordnung hinsichtlich ihrer Knotenpunkte und ihrer Gestalten," *Preisschr. der F. J. Gesell. zu Leipzig* (Leipzig, 1886, pp. 1—58), and same title *Math. Ann.* t. XXIX. (1887), pp. 81—97. I have not been able to examine the conclusions arrived at in these papers with as much care as would have been desirable.

I call to mind that for a k-nodal quartic surface the tangent cone from any node is a sextic cone with $(k-1)$ nodal lines, breaking up it may be into cones of lower orders—see table p. 265: and that we distinguish the quartic surfaces according to the forms of the sextic cones corresponding to the k nodes respectively. It will be recollected that (6) denotes a sextic cone, (6_1) a sextic cone with one nodal line, $(5_1, 1)$ a sextic cone breaking up into a quintic cone with one nodal line and a plane; and so in other cases.

There is a sort of break in the theory; in fact when the number of nodes is not greater than 7 these may be any given points whatever, and taking the 7 points at pleasure we have surfaces with 8 nodes, and 9 nodes, but not with any greater number of nodes, viz. for a surface with 10 or more nodes, it is not permissible to take 7 of these as points at pleasure, so that the theory of the surfaces with 10 or more nodes is so to speak separated off from that of the surfaces with a smaller number of nodes. For the case of 10 nodes we have the symmetroid $10\,(3, 3)$ and other forms, for 11 nodes Rohn finds 3 or ? 4 forms; for 12 nodes he has four forms, viz. my 3 forms and a fourth form 12_d; for 13 nodes he has two forms, 13_b agreeing with my 13_a, and 13_α which replaces my non-existent form 13_β; for 14 nodes, 15 nodes and 16 nodes he has in each case a single form, agreeing with my results. Without endeavouring to complete the theory, I write down a table as follows:

No. of Nodes		Form of Cones	Remarks
16		$16\,(1, 1, 1, 1, 1, 1)$	
15		$15\,(2, 1, 1, 1, 1)$	
14		$8\,(3_1, 1, 1, 1) + 6\,(2, 2, 1, 1)$	
13	$13_b = 13_a$	$3\,(4_3, 1, 1) + 1\,(3, 1, 1, 1) + 9\,(3_1, 2, 1)$	
,,	13_α	$1\,(2, 2, 2) + 12\,(4_3, 1, 1)$	13_α replaces my non-existent 13_β, $= 13\,(2, 2, 2)$
12	$12_b = 12_a$	$12\,(4_3, 2)$	
,,	$12_a = 12_\beta$	$2\,(5_6, 1) + 6\,(3_1, 3_1) + 4\,(3, 2, 1)$	

Table continued.

No. of Nodes		Form of Cones	Remarks
12	$12_c = 12_\gamma$	$12\,(4_2,\ 1,\ 1)$	$12_c = 12_\gamma$ is a peculiarly simple and elegant form; the equation is $A^2 - xyzw = 0$, where A is a quadric function of the coordinates.
"	12_d	$2\,(4_2,\ 1,\ 1) + 8\,(5_6,\ 1) + 2\,(4_3,\ 2)$	
11	$11_a = 11_\alpha$	$1\,(6_{10}) + 10\,(3_1,\ 3_1)$	
"	11_b	$8\,(6_{10}) +\ 3\,(4_2,\ 2)$	
"	11_c	$6\,(5_5,\ 1) + 5\,(6_{10})$	
"	11_d	?	
10		$10\,(3,\ 3)$	The quartic surface is here the symmetroid.
9			
8			
7			
6			
5		$5\,(6_4)$	
4		$4\,(6_3)$	
3		$3\,(6_2)$	
2		$2\,(6_1)$	
1		$1\,(6\ )$	

The suffixes a, b, c, d refer to Rohn's forms, the suffixes α, β, γ to my forms. The form 11_d is given in the first but not in the second of Rohn's two memoirs, and I am not sure as to the intended character of the sextic cones. I have not attempted to fill up the third column of the table for the Nos. of nodes 9, 8, 7, 6, as there may be particular cases which I have not considered. For the Nos. 5, 4, 3, 2, 1, the cone is a sextic cone with at most 4 nodal lines, and consequently in each case a proper sextic cone not breaking up into cones of inferior orders.

END OF VOL. VII.

CAMBRIDGE:

PRINTED BY C. J. CLAY, M.A. AND SONS,

AT THE UNIVERSITY PRESS.

www.ingramcontent.com/pod-product-compliance
Lightning Source LLC
LaVergne TN
LVHW011243110826
845149LV00001B/34

9781418185992